# Understanding the behaviour and improving the welfare of chickens

It is widely recognised that agriculture is a significant contributor to global warming and climate change. Agriculture needs to reduce its environmental impact and adapt to current climate change whilst still feeding a growing population, i.e. become more 'climate-smart'. Burleigh Dodds Science Publishing is playing its part in achieving this by bringing together key research on making the production of the world's most important crops and livestock products more sustainable.

Based on extensive research, our publications specifically target the challenge of climate-smart agriculture. In this way we are using 'smart publishing' to help achieve climate-smart agriculture.

Burleigh Dodds Science Publishing is an independent and innovative publisher delivering high quality customer-focused agricultural science content in both print and online formats for the academic and research communities. Our aim is to build a foundation of knowledge on which researchers can build to meet the challenge of climate-smart agriculture.

For more information about Burleigh Dodds Science Publishing simply call us on +44 (0) 1223 839365, email info@bdspublishing.com or alternatively please visit our website at www.bdspublishing.com.

**Related titles:**

Achieving sustainable production of poultry meat Volume 3: Health and welfare
Print (ISBN 978-1-78676-072-2); Online (ISBN 978-1-78676-075-3, 978-1-78676-074-6)

Improving gut health in poultry
Print (ISBN 978-1-78676-304-4); Online (ISBN 978-1-78676-306-8, 978-1-78676-307-5)

Advances in poultry genetics and genomics
Print (ISBN 978-1-78676-324-2); Online (ISBN 978-1-78676-326-6, 978-1-78676-327-3)

Chapters are available individually from our online bookshop:
https://shop.bdspublishing.com

climate•SMART•publishing
IN AGRICULTURAL SCIENCE

BURLEIGH DODDS SERIES IN AGRICULTURAL SCIENCE
NUMBER 91

# Understanding the behaviour and improving the welfare of chickens

Edited by Professor Christine Nicol, Royal Veterinary College – University of London, UK

burleigh dodds
SCIENCE PUBLISHING

Published by Burleigh Dodds Science Publishing Limited
82 High Street, Sawston, Cambridge CB22 3HJ, UK
www.bdspublishing.com

Burleigh Dodds Science Publishing, 1518 Walnut Street, Suite 900, Philadelphia, PA 19102-3406, USA

First published 2020 by Burleigh Dodds Science Publishing Limited


Library of Congress Control Number: 2020942303

British Library Cataloguing in Publication Data
A catalogue record for this book is available from the British Library

ISBN 978-1-78676-422-5 (Print)
ISBN 978-1-78676-425-6 (PDF)
ISBN 978-1-78676-424-9 (ePub)
ISSN 2059-6936 (print)
ISSN 2059-6944 (online)

DOI 10.19103/AS.2020.0078

Typeset by Deanta Global Publishing Services, Dublin, Ireland

# Contents

Series list xi
Introduction xvii

**Part 1 Behaviour**

1 Advances in understanding the genetics of poultry behaviour 3
*Dominic Wright and Rie Henriksen, IFM Biology - Linköping University, Sweden*

1 Introduction 3
2 The genetics of behaviour 5
3 Mapping genes for behaviour 6
4 Behavioural types and their genetic basis 11
5 Pleiotropy and the potential for selection 19
6 Epigenetics and behaviour 20
7 Commercial aspects and research 21
8 Conclusions 22
9 Where to look for further information 22
10 References 23

2 Understanding the sensory perception of chickens 37
*Birte L. Nielsen, INRAE, France*

1 Introduction 37
2 Vision 38
3 Olfaction 42
4 Taste 45
5 Hearing 47
6 Touch 49
7 Welfare implications 50
8 Where to look for further information 51
9 References 51

3 Understanding states of suffering with implications for improved management of poultry 59
*Ian J. H. Duncan, University of Guelph, Canada*

1 Introduction 59
2 Fear 61
3 Frustration 67
4 Pain 71
5 Other states of suffering 77
6 Conclusions and future trends 80
7 Where to look for further information 81
8 References 81

4 Understanding chicken learning and cognition and implications for improved management 91
*Rafael Freire, Charles Sturt University, Australia*

1 Introduction 91
2 Early learning 92
3 Habituation and associative learning 94
4 Cognition 95
5 Improving management through an understanding of learning and cognitive abilities 104
6 Conclusion 110
7 Where to look for further information 111
8 References 111

5 Understanding poultry social behaviour and its impact on animal welfare 117
*Inma Estevez, Neiker-Tecnalia Basque Institute for Agricultural Research and Development and IKERBASQUE, Basque Foundation for Science, Spain*

1 Introduction 117
2 Why do animals live in groups? 119
3 Imprinting and early-life experiences 121
4 Group size and social plasticity: implications for animal welfare 126
5 Managing social behaviour for better welfare 130
6 Understanding mating systems in poultry: implications for the welfare of breeding flocks 135
7 Strategies for better management of poultry flocks 139
8 References 140

6 Poultry welfare monitoring: wearable technologies 149
*Dana L. M. Campbell, CSIRO, Australia; and Marisa A. Erasmus, Purdue University, USA*

1 Introduction 149
2 Radio-frequency identification technology 151
3 Wearable sensors and accelerometers 157
4 Case study: outdoor stocking density in free-range laying hens 164
5 Conclusion 165
6 Future trends in research 166
7 Where to look for further information 168
8 References 169

7 Poultry welfare monitoring: group-level technologies 177
*Marian Stamp Dawkins and Elizabeth Rowe, University of Oxford, UK*

1 Introduction 177
2 Types of automated assessment 179
3 Automated measures of welfare as part of precision farming 183
4 Why isn't automated welfare assessment more widely used? 184
5 Conclusions 186
6 Future trends 186
7 References 189

8 Improving welfare assessment indicators and protocols for poultry 197
*Linda Keeling, Swedish University of Agricultural Sciences, Sweden*

1 Introduction 197
2 Welfare assessment protocols 200
3 Welfare assessment indicators 204
4 Examples of animal-based measures 206
5 Prioritising welfare indicators 213
6 Future trends 215
7 Conclusion 216
8 Where to look for further information 217
9 References 217

**Part 2 Welfare issues in breeding, management and housing**

9 Welfare issues affecting broiler breeders 227
*Anja Brinch Riber, Aarhus University, Denmark*

1 Introduction 227
2 Welfare issues in broiler breeders 232
3 Conclusion and future trends 251

4 Acknowledgements 252
5 Where to look for further information 252
6 References 253

10 Opportunities to improve the welfare of young chickens 261
*Elske N. de Haas, Utrecht University, The Netherlands*

1 Introduction 261
2 Welfare of parental stock and effects on offspring 262
3 Incubation practices to optimize chick welfare 264
4 Hatching practices to optimize chick welfare 274
5 Rearing practices to optimize pullet welfare 280
6 Conclusions 293
7 Where to look for further information 294
8 References 294

11 Welfare issues in poultry housing and management: broilers 313
*Ingrid C. de Jong, Wageningen Livestock Research, Wageningen University and Research, The Netherlands*

1 Introduction 313
2 Broiler behaviour and space use 317
3 Welfare issues in broiler chicken production: leg health, heat stress and behavioural restrictions 319
4 The relationship between growth rate and broiler welfare 323
5 Effects of environment and management on welfare 325
6 Conclusions and future trends 337
7 Where to look for further information 338
8 References 338

12 Welfare issues in poultry housing and management: laying hens 349
*Victoria Sandilands, Scotland's Rural College (SRUC), UK*

1 Introduction 349
2 Conventional cages 350
3 Enriched cages 354
4 Cage-free housing 357
5 Conclusion 365
6 Where to look for further information 366
7 References 366

13 The role of perches in chicken welfare 375
*Lars Schrader and Julia Malchow, Institute of Animal Welfare and Animal Husbandry - Friedrich-Loeffler-Institut, Germany*

1 Introduction 375
2 Why do chickens perch? 377

3 Ontogeny of perching 385
4 Anatomic prerequisites for perching 389
5 Properties of perches: the chicken's view 390
6 Elevated structures in pullet and layer housing 393
7 Elevated structures in broiler housing 400
8 Conclusions 404
9 Future trends in research 407
10 References 408

14 Improving welfare in catching and transport of chickens 417
*Leonie Jacobs, Virginia Tech, USA; and Frank A. M. Tuyttens, Institute for Agricultural and Fisheries Research (ILVO) and Ghent University, Belgium*

1 Introduction 417
2 Broiler chicken pre-slaughter phase and associated welfare concerns 417
3 Improving pre-slaughter welfare for broiler chickens 434
4 The pre-slaughter phase for laying hens 442
5 Conclusions and future trends 444
6 Where to look for further information 445
7 References 446

15 Improving welfare in poultry slaughter 459
*Dorothy McKeegan, Institute of Biodiversity, Animal Health and Comparative Medicine, University of Glasgow, UK; and Jessica Martin, The Royal (Dick) School of Veterinary Studies and The Roslin Institute, University of Edinburgh, UK*

1 Introduction 459
2 Lairage 463
3 Stunning methods 466
4 Conclusions 492
5 Where to look for further information 493
6 References 494

16 Cause and prevention of injurious pecking in chickens 509
*Nienke van Staaveren and Alexandra Harlander, University of Guelph, Canada*

1 Introduction 509
2 Identifying the underlying causes of injurious pecking 516
3 Practical attempts and management strategies to reduce injurious pecking 535
4 Conclusion 544
5 Future trends in research 544
6 Where to look for further information 546
7 References 546

17 Bone health and associated problems in layer hens 567
*Christina Rufener, University of California-Davis, USA; and Michael J. Toscano, University of Bern, Switzerland*

1 Introduction 567
2 Bone development, growth and remodelling 568
3 Identified bone health problems 570
4 Contributory factors to poor bone health 574
5 Influence of poor bone health on productivity and welfare 577
6 Strategies for improving bone health 579
7 Future trends in research 590
8 Where to look for further information 591
9 References 591

18 Poultry health monitoring and management: bone and skin health in broilers 603
*Gina Caplen, University of Bristol, UK*

1 Introduction 603
2 Leg disorders and lameness 604
3 Contact dermatitis 621
4 Conclusion and future trends 631
5 Where to look for further information 633
6 References 635

Index 653

# Series list

| Title | Series number |
|---|---|
| **Achieving sustainable cultivation of maize - Vol 1**<br>From improved varieties to local applications<br>Edited by: Dr Dave Watson, CGIAR Maize Research Program Manager, CIMMYT, Mexico | 001 |
| **Achieving sustainable cultivation of maize - Vol 2**<br>Cultivation techniques, pest and disease control<br>Edited by: Dr Dave Watson, CGIAR Maize Research Program Manager, CIMMYT, Mexico | 002 |
| **Achieving sustainable cultivation of rice - Vol 1**<br>Breeding for higher yield and quality<br>Edited by: Prof. Takuji Sasaki, Tokyo University of Agriculture, Japan | 003 |
| **Achieving sustainable cultivation of rice - Vol 2**<br>Cultivation, pest and disease management<br>Edited by: Prof. Takuji Sasaki, Tokyo University of Agriculture, Japan | 004 |
| **Achieving sustainable cultivation of wheat - Vol 1**<br>Breeding, quality traits, pests and diseases<br>Edited by: Prof. Peter Langridge, The University of Adelaide, Australia | 005 |
| **Achieving sustainable cultivation of wheat - Vol 2**<br>Cultivation techniques<br>Edited by: Prof. Peter Langridge, The University of Adelaide, Australia | 006 |
| **Achieving sustainable cultivation of tomatoes**<br>Edited by: Dr Autar Mattoo, USDA-ARS, USA & Prof. Avtar Handa, Purdue University, USA | 007 |
| **Achieving sustainable production of milk - Vol 1**<br>Milk composition, genetics and breeding<br>Edited by: Dr Nico van Belzen, International Dairy Federation (IDF), Belgium | 008 |
| **Achieving sustainable production of milk - Vol 2**<br>Safety, quality and sustainability<br>Edited by: Dr Nico van Belzen, International Dairy Federation (IDF), Belgium | 009 |
| **Achieving sustainable production of milk - Vol 3**<br>Dairy herd management and welfare<br>Edited by: Prof. John Webster, University of Bristol, UK | 010 |
| **Ensuring safety and quality in the production of beef - Vol 1**<br>Safety<br>Edited by: Prof. Gary Acuff, Texas A&M University, USA & Prof. James Dickson, Iowa State University, USA | 011 |
| **Ensuring safety and quality in the production of beef - Vol 2**<br>Quality<br>Edited by: Prof. Michael Dikeman, Kansas State University, USA | 012 |
| **Achieving sustainable production of poultry meat - Vol 1**<br>Safety, quality and sustainability<br>Edited by: Prof. Steven C. Ricke, University of Arkansas, USA | 013 |
| **Achieving sustainable production of poultry meat - Vol 2**<br>Breeding and nutrition<br>Edited by: Prof. Todd Applegate, University of Georgia, USA | 014 |

**Achieving sustainable production of poultry meat - Vol 3** 015
Health and welfare
Edited by: Prof. Todd Applegate, University of Georgia, USA

**Achieving sustainable production of eggs - Vol 1** 016
Safety and quality
Edited by: Prof. Julie Roberts, University of New England, Australia

**Achieving sustainable production of eggs - Vol 2** 017
Animal welfare and sustainability
Edited by: Prof. Julie Roberts, University of New England, Australia

**Achieving sustainable cultivation of apples** 018
Edited by: Dr Kate Evans, Washington State University, USA

**Integrated disease management of wheat and barley** 019
Edited by: Prof. Richard Oliver, Curtin University, Australia

**Achieving sustainable cultivation of cassava - Vol 1** 020
Cultivation techniques
Edited by: Dr Clair Hershey, formerly International Center for Tropical Agriculture (CIAT), Colombia

**Achieving sustainable cultivation of cassava - Vol 2** 021
Genetics, breeding, pests and diseases
Edited by: Dr Clair Hershey, formerly International Center for Tropical Agriculture (CIAT), Colombia

**Achieving sustainable production of sheep** 022
Edited by: Prof. Johan Greyling, University of the Free State, South Africa

**Achieving sustainable production of pig meat - Vol 1** 023
Safety, quality and sustainability
Edited by: Prof. Alan Mathew, Purdue University, USA

**Achieving sustainable production of pig meat - Vol 2** 024
Animal breeding and nutrition
Edited by: Prof. Julian Wiseman, University of Nottingham, UK

**Achieving sustainable production of pig meat - Vol 3** 025
Animal health and welfare
Edited by: Prof. Julian Wiseman, University of Nottingham, UK

**Achieving sustainable cultivation of potatoes - Vol 1** 026
Breeding improved varieties
Edited by: Prof. Gefu Wang-Pruski, Dalhousie University, Canada

**Achieving sustainable cultivation of oil palm - Vol 1** 027
Introduction, breeding and cultivation techniques
Edited by: Prof. Alain Rival, Center for International Cooperation in Agricultural Research for Development (CIRAD), France

**Achieving sustainable cultivation of oil palm - Vol 2** 028
Diseases, pests, quality and sustainability
Edited by: Prof. Alain Rival, Center for International Cooperation in Agricultural Research for Development (CIRAD), France

**Achieving sustainable cultivation of soybeans - Vol 1** 029
Breeding and cultivation techniques
Edited by: Prof. Henry T. Nguyen, University of Missouri, USA

**Achieving sustainable cultivation of soybeans - Vol 2** 030
Diseases, pests, food and non-food uses
Edited by: Prof. Henry T. Nguyen, University of Missouri, USA

**Achieving sustainable cultivation of sorghum - Vol 1** 031
Genetics, breeding and production techniques
Edited by: Prof. William Rooney, Texas A&M University, USA

**Achieving sustainable cultivation of sorghum - Vol 2** 032
Sorghum utilization around the world
Edited by: Prof. William Rooney, Texas A&M University, USA

**Achieving sustainable cultivation of potatoes - Vol 2** 033
Production, storage and crop protection
Edited by: Dr Stuart Wale, Potato Dynamics Ltd, UK

**Achieving sustainable cultivation of mangoes** 034
Edited by: Prof. Víctor Galán Saúco, Instituto Canario de Investigaciones Agrarias (ICIA), Spain & Dr Ping Lu, Charles Darwin University, Australia

**Achieving sustainable cultivation of grain legumes - Vol 1** 035
Advances in breeding and cultivation techniques
Edited by: Dr Shoba Sivasankar et al., formerly International Crops Research Institute for the Semi-Arid Tropics (ICRISAT), India

**Achieving sustainable cultivation of grain legumes - Vol 2** 036
Improving cultivation of particular grain legumes
Edited by: Dr Shoba Sivasankar et al., formerly International Crops Research Institute for the Semi-Arid Tropics (ICRISAT), India

**Achieving sustainable cultivation of sugarcane - Vol 1** 037
Cultivation techniques, quality and sustainability
Edited by: Prof. Philippe Rott, University of Florida, USA

**Achieving sustainable cultivation of sugarcane - Vol 2** 038
Breeding, pests and diseases
Edited by: Prof. Philippe Rott, University of Florida, USA

**Achieving sustainable cultivation of coffee** 039
Edited by: Dr Philippe Lashermes, Institut de Recherche pour le Développement (IRD), France

**Achieving sustainable cultivation of bananas - Vol 1** 040
Cultivation techniques
Edited by: Prof. Gert H. J. Kema, Wageningen University and Research, The Netherlands & Prof. André Drenth, University of Queensland, Australia

**Global Tea Science** 041
Current status and future needs
Edited by: Dr V. S. Sharma, formerly UPASI Tea Research Institute, India & Dr M. T. Kumudini Gunasekare, Coordinating Secretariat for Science Technology and Innovation (COSTI), Sri Lanka

**Integrated weed management** 042
Edited by: Emeritus Prof. Rob Zimdahl, Colorado State University, USA

**Achieving sustainable cultivation of cocoa** 043
Edited by: Prof. Pathmanathan Umaharan, Cocoa Research Centre - The University of the West Indies, Trinidad and Tobago

**Robotics and automation for improving agriculture** 044
Edited by: Prof. John Billingsley, University of Southern Queensland, Australia

**Water management for sustainable agriculture** 045
Edited by: Prof. Theib Oweis, ICARDA, Jordan

**Improving organic animal farming** 046
Edited by: Dr Mette Vaarst, Aarhus University, Denmark & Dr Stephen Roderick, Duchy College, UK

**Improving organic crop cultivation** 047
Edited by: Prof. Ulrich Köpke, University of Bonn, Germany

**Managing soil health for sustainable agriculture - Vol 1** 048
Fundamentals
Edited by: Dr Don Reicosky, Soil Scientist Emeritus USDA-ARS and University of Minnesota, USA

**Managing soil health for sustainable agriculture - Vol 2** 049
Monitoring and management
Edited by: Dr Don Reicosky, Soil Scientist Emeritus USDA-ARS and University of Minnesota, USA

**Rice insect pests and their management** 050
E. A. Heinrichs, Francis E. Nwilene, Michael J. Stout, Buyung A. R. Hadi & Thais Freitas

**Improving grassland and pasture management in temperate agriculture** 051
Edited by: Prof. Athole Marshall & Dr Rosemary Collins, IBERS, Aberystwyth University, UK

**Precision agriculture for sustainability** 052
Edited by: Dr John Stafford, Silsoe Solutions, UK

**Achieving sustainable cultivation of temperate zone tree fruit and berries - Vol 1** 053
Physiology, genetics and cultivation
Edited by: Prof. Gregory A. Lang, Michigan State University, USA

**Achieving sustainable cultivation of temperate zone tree fruit and berries - Vol 2** 054
Case studies
Edited by: Prof. Gregory A. Lang, Michigan State University, USA

**Agroforestry for sustainable agriculture** 055
Edited by: Prof. María Rosa Mosquera-Losada, Universidade de Santiago de Compostela, Spain & Dr Ravi Prabhu, World Agroforestry Centre (ICRAF), Kenya

**Achieving sustainable cultivation of tree nuts** 056
Edited by: Prof. Ümit Serdar, Ondokuz Mayis University, Turkey & Emeritus Prof. Dennis Fulbright, Michigan State University, USA

**Assessing the environmental impact of agriculture** 057
Edited by: Prof. Bo P. Weidema, Aalborg University, Denmark

**Critical issues in plant health: 50 years of research in African agriculture** 058
Edited by: Dr Peter Neuenschwander and Dr Manuele Tamò, IITA, Benin

**Achieving sustainable cultivation of vegetables** 059
Edited by: Emeritus Prof. George Hochmuth, University of Florida, USA

**Advances in breeding techniques for cereal crops** 060
Edited by: Prof. Frank Ordon, Julius Kuhn Institute (JKI), Germany & Prof. Wolfgang Friedt, Justus-Liebig University of Giessen, Germany

**Advances in Conservation Agriculture - Vol 1** 061
Systems and Science
Edited by: Prof. Amir Kassam, University of Reading, UK and Moderator, Global Conservation Agriculture Community of Practice (CA-CoP), FAO, Rome, Italy

**Advances in Conservation Agriculture - Vol 2** 062
Practice and Benefits
Edited by: Prof. Amir Kassam, University of Reading, UK and Moderator, Global Conservation Agriculture Community of Practice (CA-CoP), FAO, Rome, Italy

**Achieving sustainable greenhouse cultivation** 063
Edited by: Prof. Leo Marcelis & Dr Ep Heuvelink, Wageningen University, The Netherlands

**Achieving carbon-negative bioenergy systems from plant materials** 064
Edited by: Dr Chris Saffron, Michigan State University, USA

**Achieving sustainable cultivation of tropical fruits** 065
Edited by: Prof. Elhadi M. Yahia, Universidad Autónoma de Querétaro, Mexico

**Advances in postharvest management of horticultural produce** 066
Edited by: Prof. Chris Watkins, Cornell University, USA

**Pesticides and agriculture** 067
Profit, politics and policy
Dave Watson

**Integrated management of diseases and insect pests of tree fruit** 068
Edited by: Prof. Xiangming Xu and Dr Michelle Fountain, NIAB-EMR, UK

**Integrated management of insect pests: Current and future developments** 069
Edited by: Emeritus Prof. Marcos Kogan, Oregon State University, USA & Emeritus Prof. E. A. Heinrichs, University of Nebraska-Lincoln, USA

**Preventing food losses and waste to achieve food security and sustainability** 070
Edited by: Prof. Elhadi M. Yahia, Universidad Autónoma de Querétaro, Mexico

**Achieving sustainable management of boreal and temperate forests** 071
Edited by: Dr John Stanturf, Estonian University of Life Sciences , Estonia

**Advances in breeding of dairy cattle** 072
Edited by: Prof. Julius van der Werf, University of New England, Australia & Prof. Jennie Pryce, Agriculture Victoria and La Trobe University, Australia

**Improving gut health in poultry** 073
Edited by: Prof. Steven C. Ricke, University of Arkansas, USA

**Achieving sustainable cultivation of barley** 074
Edited by: Prof. Glen Fox, University of California-Davis, USA and The University of Queensland, Australia & Prof. Chengdao Li, Murdoch University, Australia

**Advances in crop modelling for a sustainable agriculture** 075
Edited by: Emeritus Prof. Kenneth Boote, University of Florida, USA

**Achieving sustainable crop nutrition** 076
Edited by: Prof. Zed Rengel, University of Western Australia, Australia

**Achieving sustainable urban agriculture** 077
Edited by: Prof. Johannes S. C. Wiskerke, Wageningen University, The Netherlands

**Climate change and agriculture** 078
Edited by Dr Delphine Deryng, NewClimate Institute/Integrative Research Institute on Transformations of Human-Environment Systems (IRI THESys), Humboldt-Universität zu Berlin, Germany

**Advances in poultry genetics and genomics** 079
Edited by: Prof. Samuel E. Aggrey, University of Georgia, USA; Prof. Huaijun Zhou, University of California-Davis, USA; Dr Michèle Tixier-Boichard, INRAE, France; and Prof. Douglas D. Rhoads, University of Arkansas, USA

**Achieving sustainable management of tropical forests** 080
Edited by: Prof. Jürgen Blaser, Bern University of Life Sciences, Switzerland; and Dr Patrick D. Hardcastle, Forestry Development Specialist, UK

**Improving the nutritional and nutraceutical properties of wheat and other cereals** 081
Edited by: Prof. Trust Beta, University of Manitoba, Canada

**Achieving sustainable cultivation of ornamental plants** 082
Edited by: Emeritus Prof. Michael Reid, University of California-Davis, USA

**Improving rumen function** 083
Edited by: Dr C. S. McSweeney, CSIRO, Australia; and Prof. R. I. Mackie, University of Illinois, USA

**Biostimulants for sustainable crop production** 084
Edited by Youssef Rouphael, Patrick du Jardin, Patrick Brown, Stefania De Pascale and Giuseppe Colla

**Improving data management and decision support systems in agriculture** 085
Edited by: Dr Leisa Armstrong, Edith Cowan University, Australia

**Achieving sustainable cultivation of bananas - Volume 2** 086
Germplasm and genetic improvement
Edited by: Prof. Gert Kema, Wageningen University, The Netherlands; and Prof. Andrè Drenth, The University of Queensland, Australia

**Reconciling agricultural production with biodiversity conservation** 087
Edited by: Prof. Paolo Bàrberi and Dr Anna-Camilla Moonen, Institute of Life Sciences - Scuola Superiore Sant'Anna, Pisa, Italy

**Advances in postharvest management of cereals and grains** 088
Edited by: Prof. Dirk E. Maier, Iowa State University, USA

**Biopesticides for sustainable agriculture** 089
Edited by: Prof. Nick Birch, formerly The James Hutton Institute, UK; and Prof. Travis Glare, Lincoln University, New Zealand

**Understanding and improving crop root function** 090
Edited by: Emeritus Prof. Peter Gregory, University of Reading, UK

**Understanding the behaviour and improving the welfare of chickens** 091
Edited by: Prof. Christine Nicol, Royal Veterinary College - University of London, UK

**Advances in measuring soil health for sustainable agriculture** 092
Edited by: Prof. Wilfred Otten, Cranfield University, UK

**Supporting smallholders in achieving sustainable agriculture** 093
Edited by: Dr Dominik Klauser and Dr Michael Robinson, Syngenta Foundation for Sustainable Agriculture (SFSA), Switzerland

**Advances in horticultural soilless culture** 094
Edited by: Prof. Nazim Gruda, University of Bonn, Germany

**Reducing greenhouse gas emissions from livestock production** 095
Edited by: Dr Richard Baines, Royal Agricultural University, UK

**Understanding the behaviour and improving the welfare of pigs** 096
Edited by: Emerita Prof. Sandra Edwards, University of Newcastle, UK

**Genome editing for precision crop breeding** 097
Edited by: Dr Matthew Willman, Cornell University, USA

# Introduction

This collection summarises the rich body of research on understanding the behaviour of chickens and using this knowledge to optimise welfare management of broilers and layers. Part 1 of this volume reviews advances in research on key aspects of poultry behaviour and welfare monitoring. Chapters include the genetics and epigenetics of poultry behaviour, sensory perception, pain and stress responses, learning and cognition as well as social behaviour. The section also includes coverage of wearable, video and acoustic technologies to monitor chicken behaviour and welfare as well as developments in welfare protocols. Part 2 discusses particular welfare issues affecting broilers and layers. Topics covered include welfare in hatcheries and during rearing, housing and management, the role of enrichment as well as optimising welfare during catching, transport and slaughter. The book also reviews advances in understanding specific welfare issues such as injurious pecking, bone and skin health.

## Part 1 Behaviour

The first chapter of the book focuses on advances in understanding the genetics of poultry behaviour. Chapter 1 begins by examining the genetics of behaviour and goes on to analyse the different approaches for the mapping of genes for behaviour, focusing on top-down and bottom-up approaches. The chapter provides a clear explanation and exposition of the terminology used in genetics, gene mapping and selection. It then reviews the genetic basis of different behavioural traits in chickens, such as anxiety, brooding, feather pecking and aggressive male mating behaviour. A section on pleiotropy and the potential for selection in chicken is also provided, followed by an analysis of epigenetics and behaviour. It also highlights how genetic and genomic techniques are used by commercial companies, then closes with an overall conclusion on the importance of refining the genetic and genomic tools and provides resources for further information on the subject.

The subject of Chapter 2 is understanding the sensory perception of chickens. Chickens perceive environmental stimuli via their senses. The affinity and capacity of the different sensory modalities are therefore of paramount importance for the behaviour and welfare of broilers and laying hens, and sensory perception needs to be taken into account when we house and handle domestic poultry. Emphasis is put on the importance of vision, olfaction, taste, hearing and touch for the perception of the environment by the birds, and how different ages and different contexts influence how a chicken responds to its surroundings. Finally, the influence of different sensory inputs is summarised, together with important aspects of the senses for the welfare of chickens.

Chapter 3 considers subjective and affective states in chickens, and the importance of these for understanding states of suffering. The chapter begins by discussing the causes of three potential states of suffering experienced by poultry species, namely fear, frustration and pain, and the behavioural symptoms associated with these states. Birds can be frightened by stimuli that are sudden and intense, novel and which signal special evolutionary dangers. The lack of a nesting site and lack of a roosting site in battery cages are the main causes of intense frustration in laying hens. Pain can be caused by feather pecking and cannibalism, the artificial environment, surgical procedures, and breeding practices.

The next chapter concentrates on understanding chicken learning and cognition and implications for improved management. Chapter 4 begins by discussing early learning in domestic chickens, focusing on maternal and brood companion effects, then addresses habituation and associative learning. The chapter moves on to examine cognition, specifically focusing on social learning, visual and spatial cognition, time perception, transitive inference, numerical abilities, affective states, communication and memory. The chapter then discusses various methods for improving management through understanding the learning and cognitive abilities at various stages of a chicken's life. It also considers how positive or negative affective states might result in changes (biases) in cognitive processing and decision making. The chapter concludes by emphasising how knowledge of chicken learning and cognition can help to re-frame the treatment of chickens and provides sources for further information on the subject.

Chapter 5 reviews our understanding of poultry social behaviour and its impact on animal welfare. Sociality was a prerequisite for domestication that allowed animals to remain in groups under human custody. The social group provides opportunities to find food, protection from predation or weather conditions. However, very dense or large groups normally found in many production environments may increase competition and be a potential source of social stress. Social relationships within small groups of domestic fowl are based on the establishment of social hierarchies, but social dynamics of large groups are less rigid. . Social plasticity allows animals to better adapt to the diversity of environmental and social conditions that may be encountered though life. The chapter discusses how different management and environmental factors may affect the social dynamics of the domestic fowl and analyses the potential impact for their welfare. Special attention is dedicated to imprinting processes that may determine how domestic fowl respond to social models, and to the relationships developed in breeding flocks.

The next chapter examines poultry welfare monitoring, specifically focusing on wearable technologies which provide the best opportunity for obtaining data on individual birds within large flocks. Chapter 6 begins by discussing the use of radio-frequency identification technology (RFID) systems in chickens and other housing systems. The chapter then goes on to address wearable sensors and accelerometers, focusing on the effects of these sensors on chicken behaviour and how different behavioural activities can be classified. It also highlights how these sensors can be used for monitoring disease and euthanasia in chickens, as well as monitoring perching, jumping, falls and collisions. A section on the importance of using wearable technologies for measuring physical activity levels is also included, followed by a case study that analyses outdoor stocking density in free-range laying hens. The chapter concludes by providing potential areas for future research, particularly those that might support the transition from research tool to commercial application. Examples of resources for further information on the subject are provided.

Expanding on topics previously covered in Chapter 6, Chapter 7 focuses on group level technologies used for poultry behaviour and welfare monitoring. Commercial poultry are frequently kept in groups of thousands of individuals where tagging or identifying every bird is logistically impossible. Group, rather than individual, level approaches to assessing their health and welfare are therefore currently the most feasible. This chapter covers developments in ways of automating welfare assessment for poultry with particular emphasis on broiler chickens and the use of visual images from CCTV and video, sound and temperature sensing. A specific example of camera technology to detect flocks with high levels of hockburn and other health issues is described. Despite considerable potential for using technology to assess poultry welfare it is not currently widely used in practice. Reasons for this and the potential costs and benefits of applying Precision Livestock Farming to poultry are discussed in relation to the importance of making sure that technology is used to improve rather than diminish animal welfare.

The final chapter of Part 1 concentrates on improving welfare assessment indicators and protocols for poultry. There have been considerable advances in welfare assessment in the past few decades. Chapter 8 explains some of the terminology related to welfare assessment and why the emphasis is moving towards including animal-based indicators of poultry welfare rather resource-based indicators. The chapter also reviews some of the more commonly used laying hen and broiler welfare assessment indicators, focusing on those that reflect the behaviour of birds. Among the clinical indicators discussed are assessments of pecking damage and bird cleanliness. Behaviour indicators include those that are recorded from undisturbed birds, such as vocalisation, and those that use a test situation. A welfare assessment protocol is a description of the procedures to collect the indicators. In the final sections of the chapter, methods to prioritise between different indicators are discussed. This remains an important question for animal welfare science in general, and understanding how different indicators relate to bird preferences and cognitive biases is an emerging area of research. Future trends to improve poultry welfare assessment are discussed.

## Part 2 Welfare issues in breeding, management and housing

Part 2 begins with a discussion of welfare issues affecting broiler breeders. The demand for broiler meat has been growing for decades, and broiler meat represents the major animal protein source in many countries around the world. To meet this demand, a consistent selection for fast growth is employed in the broiler industry. Chapter 9 first focuses on housing conditions of broiler breeders then reviews growth potential and feed restriction. Broiler breeders are often severely feed restricted and this chapter reviews the consequences in terms of physiological stress, hunger and repetitive pecking or drinking behaviours. The chapter moves on to address the current welfare issues in broiler breeders such as unfulfilled behavioural and physiological needs, aggression, mutilation and the associated welfare problems. It concludes by providing an analysis of the significant challenges that need to be addressed in the future and offers examples of resources that could be used for further information on the subject.

Chapter 10 provides an extensive review of an area where research effort is greatly needed. It considers the factors that affect the health, behaviour and welfare of young chicks and growing birds. It starts by assessing how the welfare of parental stock can influence offspring via direct or epigenetic effects. The chapter then reviews research on incubation and hatching practices to optimize chick welfare within commercial hatcheries. The potential of on-farm hatching to alleviate some of these problems is discussed. Finally, the chapter assesses rearing practices to optimize pullet welfare, including the importance of enrichment in, for example, in reducing the risk of developing injurious pecking behaviours, reducing risk of injury by improving spatial and navigation skills, and in producing birds that are more resilient and less fearful.

The next chapter focuses on welfare issues in poultry housing and management of broilers. Chapter 11 begins by providing an international perspective on broiler production systems and the differing legislative frameworks operating around the world. The chapter highlights how genetic selection influences growth potential and broiler welfare. The chapter moves on to discuss broiler behaviour and space use, then examines how differing production systems, breeds and stocking densities affect the prevalence of mortality, leg health, skin infections, heat stress, antibiotic usage and ability to perform behaviour. It also addresses the relationship between growth rate and broiler welfare, followed by a discussion on the effects of management practices and environmental conditions, including litter and air quality on broiler welfare. The chapter concludes by emphasising the importance of improving broiler welfare in different production systems and provides examples of resources for further information.

Chapter 12 examines welfare issues in poultry housing and management of laying hens. Laying hens produce a large number of eggs on an annual basis. In recent years, permitted housing methods in the EU have changed to enriched cages and cage-free systems (i.e. barn, free-range and organic production methods), but worldwide hens are still also housed in conventional cages. The chapter provides a clear description of the characteristics of each of these systems. Conventionally caged hens have weaker bones and are their behaviour is severely restricted by lack of resources and small space allowances. Hens from enriched cages benefit from some improvements such as nest boxes, perches, greater space, and litter for pecking and scratching. Cage-free hens have the most behavioural freedom and better bone strength, but they are also at risk of greater keel bone damage (particularly with multi-tier structures), exposure to pathogens (particularly with free-range and organic), and greater mortality. This chapter considers how current research is being directed to mitigate some of the welfare risks associated with cage-free systems. This is an important goal given the drivers favouring a move towards cage-free systems in many countries.

Chapter 13 focuses on the role of perches in chicken welfare. The chapter carefully distinguishes different aspects of perching that are often conflated. Night-time roosting and day-time perching are both adaptive strategies to avoid predators, but both may serve other functions. Perching motivation and perching ability are both influenced by genetics and age-related changes and there has been more research on layer strains than in broilers. Both are considered in this chapter. The chapter discusses the ontogeny of perching, the anatomic prerequisites for perching and the properties of perches from the view of chickens by focusing on perception and motivation. The primary motivation for chickens appears to be to gain access to an elevated position, with a lesser motivation to grasp a structure with the foot. The chapter concludes by emphasising the structural properties and arrangements of perches that can best meet the needs of laying hens (whilst reducing the risk of keel bone damage) and of broiler chickens.

The subject of Chapter 14 is improving welfare in catching and transport of chickens. It opens with a discussion of the broiler chicken pre-slaughter phase and the associated welfare concerns, focusing specifically on age of depopulation, thinning versus whole flock removal, feed and water withdrawal, catching and loading, transportation, lairage and the economic impact of the pre-slaughter phase. This is followed by an analysis of the various steps that can be taken to improve broiler welfare such as fitness assessments, the use of mechanical catching as opposed to manual, control of the thermal environment broilers are transported in and more efficient training of staff involved in the process. The chapter also briefly discusses the pre-slaughter phase for laying hens as a comparison, before concluding with a section on future research trends for broiler chicken welfare in terms of pre-slaughter and transportation.

Chapter 15 focuses on improving welfare in poultry slaughter. Poultry production involves the killing of very large numbers of birds so there is a compelling need to protect welfare at slaughter. In most countries, slaughter must be preceded by stunning to induce unconsciousness. The major stunning approaches used in chicken slaughter are electrical stunning and methods that modify the atmosphere (via introduction of gas or reduction in air pressure). The chapter opens with a discussion on welfare issues related to lairage and pre-slaughter handling, when directly relevant to the experience of birds. It briefly outlines some relevant regulatory frameworks, with a focus on the European Union which is widely recognised to have the most stringent legal protection for animals at the time of killing. It then discusses current and emerging methods, concluding with prospects for improvement of welfare based on available systems and identification of knowledge gaps for research.

The next chapter examines causes and prevention of injurious pecking in laying hens. The high prevalence of injurious pecking (IP) in laying birds is a major concern from animal welfare, societal, and economic points of view. IP is defined as bird-to-bird pecking that results in or has a high likelihood of causing integument injury and psychological harm to the victim. Chapter 16 describes three forms of IP – tissue pecking (TP), aggressive pecking (AP), and different forms of feather pecking (FP). Furthermore, it explores the two major views explaining the origin of severe FP, the most prevalent form of IP. The first, the traditional ethological view, emphasizes the role of the environment in creating motivational frustration, for example that inadequate foraging substrates will result in an increased tendency of birds to peck at feathers as a substitute. In contrast, the dysfunctional view identifies underlying neurobiological (and potentially gastrointestinal) dysfunctions induced by intense or sustained stress as the cause of severe FP. Finally, the chapter concludes by highlighting risk factors and management strategies that are used to reduce IP.

Chapter 17 explores bone health and associated problems in layer hens. This chapter provides a summary of the basic skeletal system and its development, specific problems of bone health, and efforts to reduce the problem. Adult hens must support normal biological functioning while producing a large number of eggs which require mobilization of resources including minerals and energy. Laying hens and associated commercial egg production is one of the most universal agricultural products across the globe. Likely as a consequence of this mobilization and related factors, laying hens have weakened skeletal systems leading to specific problems including fractured keels during the laying period, and a susceptibility to other bone injuries during removal from cages or barns at the end of lay. To combat these problems, multiple strategies are being investigated to try to resolve these problems. These strategies include revised breeding goals, and adaptations to nutrition, housing, and management. The importance of early rearing is also reviewed as some housing and management systems provide better opportunities for appropriate skeletal and spatial cognitive development than others.

The final chapter of the book addresses poultry health monitoring and management, specifically focusing on bone and skin health in broilers. Chapter 18 builds on the information provided in Chapter 11 and opens with a discussion on leg disorders and lameness. It differentiates developmental and infectious causes of lameness and shows how these are related to impaired walking ability. Traditional and novel, automated techniques for qualitative and quantitative assessment of leg health are reviewed. It also examines the prevalence of lameness and specific leg pathologies, followed by an analysis of key risk factors associated with lameness such as age, sex and body mass, genotype, stocking density and environmental conditions. The challenges of assessing the welfare impact of lameness in terms of bird movement, pain and behavioural restriction are considered and recent experimental results reviewed. This is followed by a section on the various strategies that can be used to prevent and control lameness. Contact dermatitis is also discussed in terms of the risk factors, welfare impact and methods of prevention and control. The chapter concludes by emphasising the importance of developing ways to monitor and manage these issues in broilers.

# Part 1

## Behaviour

# Chapter 1

# Advances in understanding the genetics of poultry behaviour

*Dominic Wright and Rie Henriksen, IFM Biology – Linköping University, Sweden*

1 Introduction
2 The genetics of behaviour
3 Mapping genes for behaviour
4 Behavioural types and their genetic basis
5 Pleiotropy and the potential for selection
6 Epigenetics and behaviour
7 Commercial aspects and research
8 Conclusions
9 Where to look for further information
10 References

## 1 Introduction

The ultimate aim when analysing the underlying genetics of behaviour, or indeed the genetics of any quantitative trait, is often to find the causal polymorphisms or mutations, and the genes that they modify, that give rise to the observed changes in the phenotypic response. However, it is often both highly complex and also very rare to identify these precise mutations (Flint, 2003). Indeed, it is hard enough to identify the causal genes with a high degree of certainty, even without identifying the precise mutations that affect them. More generally, we wish to identify the genetic architecture of a given trait (Falconer and Mackay, 1996, Lynch and Walsh, 1998). In this case, the genetic architecture refers to the number of genes (or rather the number of polymorphisms that affect the trait – the quantitative trait loci (QTL), where they are located, and what their effect size is (Falconer and Mackay, 1996, Lynch and Walsh, 1998). A QTL refers to a discrete genomic region that affects the trait. Note that the QTL can be a region containing a few or, often, many genes. A distinction is made between such loci, and the identification of the actual causal nucleotide or mutation (coined a Quantitative Trait Nucleotide (QTN) (Mackay et al., 2009)), or indeed the gene that the QTN controls – the Quantitative Trait Gene (QTG)

http://dx.doi.org/10.19103/AS.2020.0078.01

**Table 1** Glossary of terms used

| Term | Definition |
|---|---|
| Epistasis | An interaction by two (or more) separate genes that affect a phenotype. In the case of QTL mapping, this refers to a situation where the two QTLs interact with one another to affect the phenotype in a manner that differs from the standard effects of each QTL on its own. |
| Pleiotropy | Where one gene may have multiple different phenotypic effects. In the case of QTL mapping, this refers to a situation where a single QTL locus has effects on multiple different phenotypes. |
| Additive effects | Additive effects in QTL mapping refer to the cumulative effect of each allele that is present at a QTL. For example, if an individual is homozygous for a particular QTL genotype, the total effect on the phenotype is twice the additive effect of that particular locus. |
| Dominance effects | Similar to additive effects, dominance effects in QTL mapping refer to the interaction between two alleles at a particular QTL. If one allele is dominant over the other, the dominance effect describes the degree of dominance (partial or full) and the direction of effect (which allele is dominant). |
| Microsatellites | These are one of a number of different molecular markers that are used for gene mapping experiments. Microsatellites refer to small copy number variants (with a motif that is generally 2 or 3 base pairs long, but with this motif repeated many times over). They are generally present in non-conserved regions of the genome and are usually selectively neutral, allowing them to often be very variable in length between individuals, making them particularly amenable to genetic mapping experiments. |
| Resequencing | This refers to when a particular individual or group of individuals have their genome sequenced for the exact base pair composition that makes up their DNA molecules. Initially, sequencing used to be performed using methods such as Sanger sequencing, whereby 1 kilobase region would be sequenced. More recently massively parallel sequencing means small regions (generally 100–150 base pairs) are sequenced, with millions of reads being produced per individual and then aligned with the known genome of the species being sequenced. In this way, exact genomic differences (most commonly single nucleotide polymorphisms or SNPs) can be identified between individuals. |
| Haplotype blocks | This refers to a region of an individual's genome where there has been no or little recombination. Haplotypes are groups of alleles that are close to one another and inherited together as a block. When recombination occurs this will break up this haplotype. |
| Recombination Rate | Recombination rate refers to the probability of a recombination event occurring during meiosis in gamete formation. When a crossover or recombination event occurs, the genetic information on one chromosome is mixed with the genetic information from the other chromosome in the pair. Recombination rate is typically referred to as the probability of a recombination occurring in a given region of the genome. So for example the recombination rate in chickens is around 1 recombination every 350 kilobases, while in humans and mice the recombination rate is around 1 recombination every 1 megabase. |

(Mackay et al., 2009). The effect size refers to whether a few genes of large effect are controlling the trait, or many genes of small effect. For example, the most extreme possibility is the infinitesimal animal model (Fisher, 1958) that assumes that many thousands of loci each with a tiny effect size make up the trait. Similarly, are the loci that control the trait purely additive or dominant in their effect, or are epistatic effects apparent between loci (Lynch and Walsh, 1998)? See Table 1 for a glossary of these and other terms. In these instances, the exact location of the QTN is not known, but rather a region of the genome is identified. In the case of domestic birds such as the chicken, comparisons are often made between divergent populations, classically wild and domestic birds (Jensen and Wright, 2014, Wright, 2015), or commercial broiler and layer breeds (Nones et al., 2006). This is particularly important as when identifying the genetics of a trait, it is the variation present within that trait that is being mapped. Such comparisons maximise the degree of variation present and are a powerful tool for genetic mapping.

## 2 The genetics of behaviour

In general, the genes and mutations underlying behaviour are even harder to identify than more standard morphological characteristics, and as such very few QTN and QTG have been identified for behaviour. Successful examples of behavioural QTG identification are typically limited to *Drosophila* (Anholt and Mackay, 2004, Mackay, 2004, Fitzpatrick et al., 2005), mouse and rat models (Chiavegatto et al., 2008, Gyetvai et al., 2011, Kim et al., 2009, Tomida et al., 2009, Wang et al., 2012, Yalcin et al., 2004, Heyne et al., 2014), and the honeybee (Robinson et al., 2008, Robinson et al., 2005). The ramifications for the identification of such behavioural genes are many and varied. From a medical perspective, anxiety-based disorders are highly prevalent and damaging in modern society, being one of the top ten causes of disability world-wide (Murray and Lopez, 1996, Vos et al., 2015). However, the identification of susceptibility loci for such traits is highly limited (Kas et al., 2007), with usually only a handful of loci identified (though in some of the largest studies more loci are now being identified albeit with very low effect size (Ripke et al., 2014)). This is despite the often high heritability estimates for diseases such as schizophrenia (McGue and Bouchard Jr, 1998), bipolar disorder (Burmeister et al., 2008) and major depressive disorder (Burmeister et al., 2008). From a non-medical perspective, very little is known about what mutations or polymorphisms affect behaviour in a non-morbid fashion, that is, the alleles that are responsible for natural quantitative variation. These can be vital for understanding responses to stress and also natural resistance or susceptibility to particular conditions, which may be particularly relevant for domesticated chickens. With regard to evolutionary theory, behavioural personality studies have been performed on a wide range of species (Sih et al., 2004); however, the genes basis of such traits still remain largely unexplored.

## 3 Mapping genes for behaviour

The mapping of genes for behaviour is similar to mapping other quantitative traits, with the exception that the traits under examination are in general harder to define and measure reliably. The methods generally employ either a top-down or a bottom-up approach. Essentially this refers to whether one starts at the phenotype and attempts to work down to the gene level (top-down), or whether one starts at a gene level and attempts to work up to the phenotype (bottom-up) (Boake et al., 2002).

### 3.1 Top-down approaches

#### 3.1.1 Pedigree analysis and heritability studies

Initially, the use of statistical analysis of pedigrees can demonstrate an actual heritable component to the trait of interest, through breeding designs, artificial selection or the like. At this point the aim is simply to show that a genetic component exists for the trait to be studied. The requirements for this are only the behavioural phenotypes for all individuals and knowledge of their relatedness (the pedigree). Heritability studies can be used to further dissect traits down to distinguish between broad sense and narrow sense heritability, with the genetic component further divided into additive and dominance components (additive in this case refers to the effect from those loci that give a cumulative effect, whereas dominance represents the interactions between alleles at the same locus) (Lynch and Walsh, 1998, Falconer and Mackay, 1996).

#### 3.1.2 Quantitative trait loci (QTL) and association mapping

QTL mapping is now a very well-established technique, first becoming popular in the 1990s (Lander and Botstein, 1989). The first major step in increasing its use came with microsatellite markers. More recently, Single Nucleotide Polymorphism (SNP) markers and resequencing technology have increased the number of molecular markers available exponentially. These advances are enabling the step to gene discovery to be performed (previously a major failing with this approach). In essence, QTL mapping is the crossing of two distinct populations that differ for one or more traits of interest. This creates an inter-cross population containing a large degree of genetic variation (containing alleles that both increase and decrease the trait under examination). QTL mapping involves mapping the loci that differ between the two parental populations that are crossed. Thus, this technique is particularly amenable for domestic animals, where extremely large phenotypic differences frequently occur between wild and domestic populations (Andersson and Georges, 2004), with this extreme phenotypic diversity providing excellent power for gene mapping. After the intercross is continued to either an $F_2$ or backcross generation, the intercross individuals are then genotyped for multiple markers spread throughout the genome, with the genotype information correlated with the phenotypic data. This enables the identification of the number of loci affecting the traits that differ between the two populations, their effect size in terms of the variation present in the intercross that they explain, and their genomic location.

The QTL can be defined in terms of their additive and dominance components, while pleiotropy (one locus affects multiple different traits) and epistasis (when multiple genes are interacting with one another) can also be identified. As stated previously, the major issue in standard QTL mapping is one of resolution of the detected loci (with very large confidence intervals generally being the norm), while the sample sizes that are required for the detection of small effect loci may also be large (Beavis, 1998). The issue of the resolution of the detected loci is limited by the number of recombinations present in the test cross (Darvasi, 1998, Lynch and Walsh, 1998). Typically, as the number of recombinations in a backcross and F2 are rather restricted, this means in turn that the resolution of each QTL is rather low, covering several megabases (Mb), and relatively few QTL have been refined to the causative gene (QTG/ QTN).

Association mapping is a similar, but more recent, application. This typically uses a single outbred population in combination with high-density SNP genotyping per individual. The difference with this technique is that by using a single large, outbred population, many more recombinations are present in comparison to a standard QTL intercross population, and therefore the the resolution is far greater due to the smaller haplotype blocks. This does therefore require far more markers per individual, while population substructure can be an issue. Additionally, in such large populations, the genetic architecture can be very complicated. Far more polymorphisms and alleles are contained in the natural population as compared to a standard QTL cross which typically utilises inbred populations/individuals and most commonly only two different populations with a limited number of founder individuals. Results with association mapping in large populations in humans tend to show that very few QTLs with large effects are present, and relatively little of the variance present can be explained by the loci which are discovered (Carlson et al., 2004). However, the power of this approach is that rather than being constrained by a precise cross, natural populations can be used far more easily.

In the case of domestic animals, the possess many features that make them more amenable to this type of analysis (Goddard and Hayes, 2009). Firstly, the strong directional effects of domestication selection often mean that fewer markers are required as haplotype blocks are larger (for instance in the case of the dog (Karlsson et al., 2007)). The chicken itself has several highly advantageous genomic features when it comes to these types of linkage mapping and association studies. They have a rather compact genome (just over 1Gb in length, as compared to around 3Gb for humans – (Wallis et al., 2004)). Not only that, but they also have a higher recombination rate (one recombination every 350kb, as opposed to one recombination every 1Mb in humans and mice (Groenen et al., 2000)). This increased recombination rate translates into a higher resolution due to more recombinations accruing over a given area, as compared to mammals. Therefore, they are highly amenable to these forms of genetic mapping and can yield excellent results with a higher resolution, meaning far narrower confidence intervals are generated.

### 3.1.3 Selective sweep mapping

The identification of selective sweeps is potentially an extremely powerful tool for the dissection of genetic architecture, and can be particularly relevant for domestic populations (Andersson and Georges, 2004). The basis of selective sweep mapping occurs when a new mutation arises in a population and is then selected upon via strong directional selection. In this case, not only the mutation but also the neighbouring polymorphisms (i.e. the haplotype in which the mutation arose) will also go to fixation. Thus, the surrounding SNP markers will 'hitch-hike' along with the mutation (Smith and Haigh, 1974). Once the new mutation is in full fixation, relating to the strength of selection and the time that has elapsed since fixation, this signature of selection will then be slowly eroded. In the case of domestic populations, the strength of selection is often high, which in turn increases the likelihood of such sweeps occurring. However, these sweeps are also dependent on the genetic architecture of the trait (Pritchard and Di Rienzo, 2010). In particular if many genes of small effect are responsible for the bulk of trait variation, selective sweeps may be less prevalent due to polygenic adaptation (where small changes in allele frequency can have a large cumulative effect). Such small changes in allele frequency will not lead to the characteristic regions of fixation that are detected by selective sweep mapping. Selective sweeps have been observed in domestic chickens, where around 50 40kb regions were putatively identified as being under selection in one study (Rubin et al., 2010), while a more recent study also found a number of further broiler and layer-specific sweeps (New Leif ref). This approach is particularly powerful when multiple domestic populations are analysed, with shared regions of identity-by-descent identified over multiple different population types. It has been used to locate both discrete mutations (for example, the pea comb (Wright et al., 2009) and yellow skin (Eriksson et al., 2008) mutations in chickens) and, for mutations affecting quantitative traits (QTNs), see comb size in chickens (Johnsson et al., 2012).

### 3.2 Bottom-up approaches

With a bottom-up approach, research starts at the level of the gene and is built-up to the behavioural phenotype. In its most fundamental form, genes can be knocked-out (when gene expression is reduced but not removed entirely), knocked-down (when the gene expression is gene reduced but not removed entirely), and/or altered in some other way (with for example the insertion of a novel gene into a genome, or the up-regulation of an existing gene) (Flint and Mott, 2001) to directly test the effects of specific genes. However, even in this case the identification of the relevant gene is required and is often problematic. In many cases, such research starts with a mutagenesis screen. This is a process where large numbers of mutant individuals are generated and then bulk screened for alterations in the phenotype (in this case behaviour) of interest (Nadeau and Frankel, 2000). This approach produced many of the early successes with behavioural gene identification, though these were restricted to model organisms, with genes relating to circadian rhythm in *Drosophila* (Sawyer et al., 1997, Tully, 1996), social aggregation in *C. elegans* (Coates and De Bono, 2002, de Bono et al., 2002) and larval foraging in *Drosophila* (De Belle and Sokolowksi, 1989, Sokolowski, 1998) identified. In this case, the relevance to domesticated chickens may be lower, though their size and generation times make them more likely than many other domestic animals.

#### 3.2.1 Transgenic analysis

Transgenic animals are the gold standard of genetic analyses and the final proof of gene function. The ultimate aim is to target germ cells in order to stably transmit the desired modifications to the progeny of the initially targeted animal (Davey et al., 2018). In the chicken, this is performed via Primordial Germ Cells (PGCs). These cells will differentiate into the gametes of the adult chicken and provide a target for genetic modification. These PGCs can now be cultured over a long period of time (Van De Lavoir et al., 2006) and also be genetically modified in vitro (Whyte et al., 2015). Genome editors, such as TALENs and CRISPR/Cas9, are site-specific DNA nucleases and can target specific regions of the genome (Cong and Zhang, 2015). Small deletions/insertions or the introduction of exogenous DNA sequences from a few base pairs to a few kilobases in length can all be introduced using these editors. Gene knockout chickens have been performed (Oishi et al., 2016, Park et al., 2014). DDX4 chickens have also been produced that are sterile when they contain a specific insert, allowing modified PGCs to be inserted into these birds. In this case, by activating the gene switch the host chicken's own PGCs are removed from the testes, allowing the genetically modified PGCs to be inserted and implanted in their place. These modified PGCs are then permanently implanted into the testis tissue, continually producing genetically modified sperms. To date, transgenic birds have mostly been used to produce reporter chickens that express markers or dyes in specific tissues. For example, Green Fluorescent Proteins (eGFPs) (McGrew et al., 2008) or the MacReporter chicken (Balic et al., 2014) that expresses a reporter in a specific haeomatopoeitic cell lineage. In the future, targeted chicken editing *in ovo* opens up the possibility to use gene editing to generate tissue-specific alleles to assess gene function in a particular tissue without the need to generate an entire line of transgenic chickens. Although there are certain instances of this already occurring (Abu-Bonsrah et al., 2016), this is still not altogether straight forward.

### 3.2.2 Global gene expression analysis

Moving up from transgenic analysis, global gene expression analysis can be performed using microarrays and now direct RNA sequencing of alleles to give more precise measures of transcription in the genome (Wang et al., 2009). The issue with this approach is the amount of data generated and its interpretation. The choice of tissue and the time-point that it is sampled can be extremely important. The gene of interest may be very limited in its expression both in terms of the tissues in which it is expressed and the timing of its expression. Furthermore, if a threshold response is responsible, relatively small changes in expression could generate the different phenotypes. Furthermore, when analysing gene expression differences between populations or treatment groups, hundreds or even thousands of genes may be identified; however, determining the causal genes from those which are due to downstream changes of such key regulators can be extremely complicated (Verdugo et al., 2010).

## 3.3 Combining top-down and bottom-up approaches

It is also possible to merge both top-down and bottom-up approaches in the form of an expression QTL (eQTL) analysis. This combines transcriptomic microarrays or RNA sequencing of a particular tissue on each individual, with a standard QTL mapping (Gibson and Weir, 2005). With this analysis, in essence, the gene expression values derived from either microarrays or RNA sequencing are then used as phenotypes that are then mapped to the genotypic markers available for all individuals. This will identify the causal elements that control gene expression for each gene that is examined. The difference to standard QTL mapping is that in this instance the physical position of the gene is already identified, so linkage to the specific markers adjacent to each gene can be preferentially checked. This reduces the multiple testing corrections and identifies what is known as *cis*-acting eQTL (where the causal element controlling variation in that gene is located close to the gene itself). An eQTL that affects a gene that it is not adjacent to it is known as a *trans*-acting eQTL. With eQTL mapping, the same caveats that apply to transcriptomic analysis still apply here - with the choice of tissue and the time point of sampling crucial. However, you can also correlate the behavioural phenotypes directly with the gene expression phenotypes, which are much more persuasive evidences of causal genes having been identified (Mehrabian et al., 2005).

## 4 Behavioural types and their genetic basis

Most of the previously discussed techniques have been used at some point in the search for the genetic basis of a variety of different behaviours in chicken, some with more success than others. From an evolutionary and developmental perspective, some of the most mapped and dissected traits are related to anxiety or fearfulness. For example, anxiety-related behaviours related to open-field activity, duration of time spent in tonic immobility, and the like have been used successfully. Such anxiety behaviour is often used as a model for understanding the genetic basis of such traits in humans and other mammals (Sokolowski, 1998) with a surprising degree of conservation between birds and mammals in this regard (Johnsson et al., 2016). Similarly, from a more applied perspective, feather pecking has been investigated extensively due to the effects on animal welfare as well as economic costs (D'eath et al., 2010). Other behaviours have been subjected to some genetic analysis, though generally far less than the above-mentioned phenotypes.

### *4.1 Anxiety behaviour*

One of the principal differences induced by domestication in chickens is anxiety behaviour. Chickens have been selected for a decrease in fear behaviour towards humans and anxiety in general (Muir and Cheng, 2014, Jensen and Wright, 2014). Indeed, it has been conjectured that tameness was one of the first traits to be selected upon in domestic animals, in particular in wolves and cats (Driscoll et al., 2009). However, assessing anxiety can be potentially problematic, with various different tests used to test anxiety in animals. As an example of this, genetic chicken stocks that appeared less fearful using one set of behavioural metrics (escape and avoidance behaviour) were just as fearful as other lines using a separate metric (heart rate) (Duncan and Filshie, 1980). A lack of consistency between different behavioural measures in a particular strain is often seen, which can be problematic when defining one particular domesticated strain as being particularly flighty or anxious or the reverse (Murphy, 1978).

A further note should also be made on distinguishing anxiety and fearfulness behaviour. Depending on the field, anxiety behaviour (often considered a pathology, especially from a human perspective) can be distinguished from fearfulness/fear behaviour. In this case, anxiety is behavioural change without necessarily an explicit threat, whereas fearfulness is in response to an actual or perceived threat stimulus (Goldsmith and Lemery, 2000). This is a potentially interesting distinction, though it can be extremely hard to truly distinguish whether a test is actually more related to anxiety or fear. For example, tonic immobility may be considered to be more of a fearfulness-related behaviour, whereas open field could be considered more anxiety-related. However, even in this instance, the open field arena may well be perceived as threat-related, so could easily be able to be considered a fearfulness assay. Generally speaking, being able to actually distinguish these types of sub-behaviours is often an intractable problem. What we really know from a test is often very specific (i.e. how much an animal moves in this arena); however, we then go on to extrapolate based on these results. In the case of behavioural genetics, it is often recommended to identify significant loci as those that are identified by several different tests that all measure putatively the same behaviour (Flint, 2003).

### 4.1.1 Open field test

One of the most popular tests for measuring anxiety is the open field test (Belzung, 1999, Ramos and Mormède, 1997, Archer, 1973). This was initially devised by Hall in 1934 (Hall, 1934) as test of anxiety/emotionality in rats, using movement and position in a brightly lit novel arena. Typically, thigmotaxis (time spent close to the walls), locomotion (speed and distance), and types of grooming behaviours are recorded (see review in Prut and Belzung, (2003)). Anxiety is generally considered to be triggered by two conditions during the test - social isolation and agoraphobia. The test is now used in a wide variety of different animals ranging from chickens, pigs, lambs, rabbits, and primates (Forkman et al., 2007). Of the studies that have been performed using the original mouse model, probably the largest and most fine detailed (in terms of the size of the QTL identified) is the Hetergeneous Stock intercross (Valdar et al., 2006). This utilised over 2000 mice and identified regions less than 3Mb in size, using a classic open field assay and dense SNP genotyping. A selection study in the chicken over eight generations that was based on differential selection for open field behaviour in 2-3-day-old chicks found corresponding physiological differences at the same age of testing, though these physiological differences were not present in adult chickens (Faure, 1981). This study demonstrated that open field behaviour is heritable in chickens. Similarly, a QTL study in White Leghorn layers that tested open field activity at 5 weeks and 29 weeks found that the detected QTL for juvenile and adult open field behaviour were separate, indicating at least a partially distinct genetic architecture controls open field behaviour in young and old birds (Buitenhuis et al., 2004). To identify the actual genes that affect quantitative variation in open field behaviour, a study by Johnsson et al. (2016) utilised a large combined QTL and eQTL cross comprising of an eighth generation advanced intercross between wild and domestic chickens that measured gene expression in the hypothalamus. Using this approach, ten candidate genes for anxiety were identified that possessed an eQTL that overlapped an open field QTL and also directly correlated gene expression with the behavioural phenotype they overlapped. Of these identified genes, four were also found to have significant effects in a mouse-advanced intercross for open field activity, and three were found to correlate in human schizophrenia and bipolar Genome Wide Association Studies (GWAS), indicating that the genes for anxiety-related activity appear to be remarkably conserved over a number of different species.

### 4.1.2 Tonic immobility

Tonic immobility, originally considered a form of animal hypnosis, was first described in the mid-1600s. It has been attributed to be related to fear (Jöngren et al., 2010, Gallup Jr, 1979) and anti-predation behaviour relating to feigning death after initial capture (Humphreys and Ruxton, 2018). However, the behaviour is so widespread and present in such a wide variety of species that it has also been thought of as a reflex response, cerebral inhibition or even paralysis through fear (Ratner, 1967). Evidence for feigning death comes from potential increases in survival chances by up to 50% (Sargeant and Eberhardt, 1975). Tonic immobility has been studied in species ranging from blue crabs (O'Brien and Dunlap, 1975), chickens (Gilman et al., 1950), and even sharks (Henningsen, 1994) amongst many others. Although the longer the animal stays in an immobile state the more fearful it is considered (Gallup et al., 1972); the relationship between tonic immobility and anxiety is potentially less straight forward. It frequently does not correlate with other measures of anxiety within lines (Schutz et al., 2004, Wright et al., 2010, Craig and Adams, 1984, Craig et al., 1986, Fogelholm et al., 2019). Tonic immobility is prolonged when the environment it is conducted in is made to appear more threatening (Gagliardi et al., 1976, Gallup et al., 1972). The genetic basis of tonic immobility has been investigated using heritability studies and linkage assays. Heritability and selection experiments have been conducted in chickens (Gallup, 1974, Campo and Carnicer, 1993, Craig and Muir, 1989) and quails (Benoff and Siegel, 1976), giving heritability estimates between 30% and 90%, depending on the population. Quantitative Trait Loci (QTL) studies on tonic immobility have been based on a wild x domestic chicken intercross (Schutz et al., 2004), an intercross between selected quail lines for high- and low-tonic immobility times (Beaumont et al., 2005, Minvielle et al., 2005), and a quail intercross selected for differential social reinstatement that was also assayed for tonic immobility (Recoquillay et al., 2015). Amongst others, a QTL was identified on chromosome 1 in both chicken and quail studies. Using a combined eQTL and QTL approach in a further eighth generation of the wild x domestic chicken intercross, the genes *PRDX4* and *ACOT9* were identified as being the most likely causal candidates (Fogelholm et al., 2019). These genes were also implicated previously in other aspects of anxiety behaviour (Johnsson et al., 2018a), reinforcing the link between tonic immobility and anxiety-related behaviours despite the lack of phenotypic correlations.

### 4.1.3 Social reinstatement test

Sociality encompasses a range of highly diverse behaviours, ranging from communication behaviour to inter-individual interactions. Several genetic mapping studies have found associations with social behaviour in mammals (Donaldson and Young, 2008, McGraw and Young, 2010, Persson et al., 2016, Brodkin et al., 2002, vonHoldt et al., 2017, Takahashi et al., 2010), fish (Greenwood et al., 2016, Wright et al., 2006a,b, Burmeister et al., 2005), and fruit flies (Shorter et al., 2015, Wu et al., 2003). Social reinstatement, the tendency of an animal to seek out conspecifics, is one such sociality behaviour, and is considered to be both a sociality and anxiety-related behaviour (Mills and Faure, 1991). In birds, it has been used to measure the general strength of sociality, and is classically assayed using a 'runway' or treadmill test to assess an animal's social motivation to associate with conspecifics (Jones et al., 1991, Suarez and Gallup, 1983). In the case of the social reinstatement assay commonly used in poultry, some of the most in-depth work concerns two lines of Japanese quail selected for high and low social reinstatement. These selected bird lines have then been assessed for a wide variety of social assays to assess how selection for social reinstatement interacts and affects other aspects of sociality. Launay et al. (Launay et al., 1991) found birds selected for high social reinstatement birds spent longer associating with conspecifics when given a paired goal box (one box empty, one box containing conspecifics). Other studies found that pairs of high social reinstatement individuals in an open field arena had significantly shorter inter-individual distance separating them as compared to the low social reinstatement birds (Mills et al., 1992, François et al., 1999). High social reinstatement birds will also preferentially associate with conspecifics at the expense of food and water access; furthermore high social reinstatement birds will use greater social facilitation in being able to learn to eat a novel food source through copying a conspecific 'teacher' (Mills et al., 1997). More recently, high social reinstatement birds show a non-specific attraction for social conspecifics (Schweitzer et al., 2009) and have a consistently stable emotional reactivity, even in the face of high social instability (Schweitzer and Arnould, 2010).

The genetics assessing social reinstatement in the chicken and quail have been based on QTL studies using the selected quail lines (Recoquillay et al., 2015, Recoquillay et al., 2013, Mills and Faure, 1991) and a wild x domestic intercross of chickens (Johnsson et al., 2016, Schutz et al., 2004). Multiple QTLs in both crosses were detected, while in the case of the latter, five candidate genes for behaviour were also identified via expression assays, with the strongest being the gene *TTRAP* (Johnsson et al., 2018a). In both the chicken and quail studies, QTLs for different measures of anxiety (tonic immobility, open field behaviour, and social reinstatement) overlap with one another, suggesting that once again these related measures are controlled by the same loci (Johnsson et al., 2018a, 2016, Recoquillay et al., 2015).

### 4.2 Brooding behaviour

Brooding is used to refer to incubation behaviour when the female hen incubates her clutch. Typically, under natural conditions, a female will lay 5-8 eggs before incubation (Meijer, 1995). However, this behaviour has been strongly selected against domestic populations, in particular in layer populations where it would reduce egg production significantly (birds cease producing eggs while brooding). The genetics of brooding was analysed initially using breeding studies, with Roberts and Card (Roberts and Card, 1933) and Warren (Warren, 1930) suggesting that the trait is at least partly controlled by sex-linked genes using reciprocal crosses. However, Hays found no corroborating evidence for sex-linked broodiness in Rhode Island Red birds. Further work proposed the X-linked prolactin gene as a putative candidate for the partial control of brooding behaviour (Dunn, 1998, Jiang et al., 2005), especially given the role of prolactin in inducing brooding behaviour (Bates et al., 1935, Hutt, 1949). Heritability studies using the Japanese Nagoya fowl found a heritability of less than 20% (Saeki, 1957), though a selection experiment in Rhode Island Red chickens found a stronger response to selection (Goodale et al., 1920). Using a wild x domestic intercross and by comparing egg production when eggs were removed daily versus eggs remaining with the chickens over the corresponding two-week periods, Johnsson et al. (2015a,b) identified several QTL for potential brooding behaviour, and even identified a candidate gene expressed in bone as a putatively causal candidate gene. Tendency to brood was also found to negatively correlate with the proportional size of the cerebellum in these intercross chickens (Henriksen et al., 2016). The cerebellum is found to be proportionally larger in domestic birds, implying that some of the genetic control of brooding may also occur in the cerebellum, though this remains to be verified.

### *4.3 Feather pecking (FP) behaviour*

Feather pecking (FP) is a damaging behaviour where one bird uses its beak to peck and firmly pull at the feathers of a conspecific. This type of damaging behavior is most often directed at the body of a conspecific and differs from aggressive pecking (which is directed at the head of a conspecific) in that it does not play a role in conflict resolution (Bilcik and Keeling, 1999, De Haas and Van Der Eijk, 2018). The behavior is however problematic since continued FP can lead to increasing susceptibility to injury and impairing thermoregulation due to skin being denuded, and in extreme cases to cannibalistic pecking (Savory, 1995). FP is a worldwide problem in the poultry industry and is expected to increase in incidence in the future, due to increased use of free-range housing in large groups as a consequence of bans on conventional cages worldwide and an expected ban on beak trimming in many European countries. FP has been found to be related to motivation to forage (Blokhuis, 1986, Huber-Eicher and Wechsler, 1997, Dixon et al., 2008). Similarly, FP is also a social behaviour, in that the risk of being feather pecked not only depends on the individual's ability to avoid being pecked (the Direct Genetic Effect or victim effect (DGE)), but also the conspecific flock-mates desire to FP (the Indirect Genetic Effect (IGE)) (Bijma et al., 2007, Griffing, 1967, Muir, 2005). For plumage condition, DGEs account for 6–31% of the total heritable variation, while IGEs account for 70–94% (Brinker et al., 2014). Taken together, they can in combination explain 10–54% of the total phenotypic variation in plumage condition. These results highlight the importance of the role of the flock mates in controlling FP behaviour – these indirect components of selection actually contribute to more of the phenotypic variation than the direct effects (Ellen et al., 2014).

Despite extensive research on FP, the underlying causes that trigger/motivate this behavior are not yet fully understood (De Haas and Van Der Eijk, 2018, Ellen et al., 2019). The propensity to FP differs between individuals suggesting that differences at the individual level contribute to the development of this maladaptive behavior. Additionally, FP is heritable and studies have shown that various commercial and experimental egg laying breeds differ in the amount of feather pecking behavior they perform (Kjær and Sørensen, 2002), suggesting a genetic component to the behavior. Studies have shown that feather pecking is indeed under genetic control and identified genetic variants underlying complex behaviour. However, there are often very variable heritability estimates between different populations, with these heritability estimates ranging from as little as 10% up to 40% (Kjaer et al., 2001, Grams et al., 2015a, Rodenburg et al., 2003, Bennewitz et al., 2014). Genetic correlations have also been performed between FP, locomotion and feather eating. These correlations were found to significantly interact with the different environmental effects (Lutz et al., 2017) .

### 4.3.1 Feather pecking linkage analyses

Variation between strains or selected lines has been used to try and identify regions on the genome associated with propensity to FP. These populations have been used in both genetic and transcriptomic studies. By creating an experimental intercross, quantitative trait loci studies aimed at identifying polymorphisms in the DNA associated with the behavior and gene expression studies aimed at finding differences in the transcriptome have been applied. An across-line association study by Biscarini et al. (Biscarini et al., 2010) identified a total of 81 SNPs that were associated with the IGE score for FP traits, and 11 SNPs that were associated with direct effects. A serotonin gene was amongst those identified, which adds support for monoamine signaling (via the serotonin receptor 2C) being important in the regulation of this behaviour.

Two F2 intercross populations have also been produced to analyse the genetics of FP behaviour. Buitenhuis et al. (Buitenhuis et al., 2003) reported markers for feather pecking on chromosomes 1, 2, and 10. In contrast, Jensen et al. (Jensen et al., 2005) identified only a single QTL for FP on chromosome 3 in an $F_2$ cross of Red Junglefowl (the wild orogeny of chickens) and White Leghorn, which had previously been shown to differ in FP intensity. Mapping results based on selection signatures between the lines (Grams et al., 2015b) and association results from the $F_2$ cross with approximately 900 individuals were jointly analysed in a meta-study (Lutz et al., 2017), revealed that the behavioral disorder were controlled by many genes with small effects and no single SNP had effects large enough to justify its use in marker-assisted selection and pointing to a candidate gene that might also be related to monoamine signaling.

### 4.3.2 Feather pecking transcriptomics

Large-scale transcriptomic studies have also been performed in chicken lines divergently selected on FP propensity by compared whole-brain gene expression in high and moderate feather peckers from a high FP selection line. Within the high FP selection line, Labouriau et al. (Labouriau et al., 2009) found 456 genes to be differently expressed in birds showing high amounts of FP compared with birds performing moderate amounts of FP, but no obvious biological pathway was apparent. A further study in 2010 by Hughes and Buitenhuis (2010) found an overall reduction in gene expression in high FP birds, while there were distinct patterns that were seen in gentle and severe FP birds (Hughes and Buitenhuis, 2010). Serotinergic and neurotransmitters have been implicated both through association studies (Lutz et al., 2017) and also with gene expression assays (De Haas and Van Der Eijk, 2018). This link between neurotransmitters (including monoamine signaling) and FP was also found by Wysocki et al. (2013). As well as these neurotransmitters, a role has also been put forward for the immune system. For example, Brunberg et al. (2011) found that several immune-related genes were differentially expressed between lines in hypothalamus tissue, with a link between FP and behaviour also identified by other studies (Biscarini et al., 2010, Buitenhuis et al., 2004, Hughes and Buitenhuis, 2010, Parmentier et al., 2009).

### 4.4 Aggressive male mating behaviour

Mating in broilers in particular can be problematic, with males directing high levels of aggressive behaviour towards females during mating (Millman and Duncan, 2000). This has been found to be specific to broiler males, with, for example, layer males and even game cock males (selected for high male-male aggression) not displaying such behaviour (Millman and Duncan, 2000). Selection studies have also been used to look at the heritability of mating frequency (Dunnington and Siegel, 1983), with a high and low frequency mating strain being produced over a large number of generations. Results were somewhat confusing, however, with moderate heritability estimates of around 0.18 for the high line after 23 generations (but initially giving an estimate of 0 over the first few generations), while for the low line the heritability was initially high (0.32) but only when measured for the first 11 generations.

### 4.5 General activity in chickens

General activity is another behaviour that has been measured in chickens, related to both the welfare and production implications. Standard cages, which can impede or inhibit movement, can lead to a reduction in bone mass (Whitehead and Wilson, 1992), while free-range/non-cage systems allow increased movement, but can also lead to more injuries (Harlander-Matauschek et al., 2015). Therefore there are production costs associated with activity levels (energy and feed consumption) as well as welfare implications (bone quality as well as feather pecking and cannibalism – see earlier). In terms of the genetics underlying this behaviour, physical activity differences have been observed between layer strains (Kozak et al., 2016), indicating at least some genetic basis. More in-depth analysis has come from the production of selection lines to estimate the heritability of this trait (Kjaer, 2017). These selected lines indicated moderate-to-high heritability for this trait, with estimates of 0.38 and 0.33 for the low and high lines, respectively. Interestingly, the high activity line birds were significantly heavier than the low line birds. As well as correlating with bone health, activity levels are also correlated with feather pecking incidence. Using genetic lines differing in feather pecking behaviour, locomotor activity was found to be higher in the high feather pecking lines and lower in the low feather pecking lines, relative to the control animals (Kjaer, 2009).

## 5 Pleiotropy and the potential for selection

A particularly pertinent question with production and selection is how the genetic architecture of anxiety and other behaviours in chickens relate to production traits. In particular, is there pleiotropy between the genetic loci for behaviour and the genetic loci for production phenotypes (i.e., does the same locus affect both of the different traits), or are the loci for the different trait types in linkage (separate loci that are physically close to one another), or are they entirely distinct and separate from one another. In QTL studies conducted in both quails (Recoquillay et al., 2015) and chickens (Johnsson et al., 2012, Wright et al., 2010), both production and behavioural traits were mapped simultaneously in the same intercross, allowing the co-localisation of loci underlying these traits to be assessed. In both instances, behavioural and production traits overlap (Recoquillay et al., 2015, Schutz et al., 2002, Wirén et al., 2013). In the chicken intercross, a point analysis of the growth 1 locus on chromosome 1 demonstrated either pleiotropy or close linkage between growth and some aspects of fear behaviour (Wirén et al., 2013), while in the quail intercross negative correlations between growth and fear and positive correlations between sociability and egg traits were observed (Recoquillay et al., 2013). However, due to the large confidence intervals for QTL in an $F_2$ design, it is very easy to get an overlap of QTL. This can be due to genuine pleiotropy, but also due to close linkage (i.e. the loci controlling these traits are separate but are physically close to one another on the genome). Pleiotropy vs linkage statistical tests performed in the $F_2$ chicken intercross indicated that these hotspot loci in the chicken were more likely to be linked with QTL, but with a pleiotropic core (with the caveat that close linkage and pleiotropy are indistinguishable in a QTL intercross) (Wright et al., 2010). This type of modularity is commonly seen in domestication (Wright, 2015). In the case of the chicken intercross, further generations of crossing give an increase in the number of recombinations that in turn increases the resolution to allow a more fine-scale dissection. In this case, the QTL appears to disperse more, indicating a greater role for linkage. Similarly, candidate genes for behaviour in the chicken intercross (Johnsson et al., 2018a,b) are separate from candidate genes for bone and growth (Henriksen et al., 2016, Johnsson et al., 2015b).

Given the potential for selection, the ability to select against aberrant/welfare-related behaviour is one way to improve animal welfare (Wegner, 1990). This can not only decrease the prevalence of these behaviours, but can also have effects on productivity, especially in light of the potential for pleiotropy and linkage between behavioural and production traits. However, the relationship between selection on one trait and the response in another is often not clear cut. For example, non-aggressive traits may be negatively correlated with productivity (Webster and Hurnik, 1991). Selection for a particular trait may also be rather specific and of less general effect. For example, selection for an increased and decreased plasma corticosterone response in chickens in social situations (Gross and Siegel, 1985) did not yield any effect when the stressor was from a non-social cue. Furthermore, these selected lines were also more susceptible to bacterial infections (in the case of the low stress line) or viral infections (in the case of the high stress line) (Siegel, 1993). Selection for increased productivity on individual birds can also have potential problems when these birds are housed in groups rather than as individuals. For example, Craig et al. (Craig, 1982, Craig, 1994) found that selection on social dominance will reduce performance when group-housed, but increases productivity when housed singly. Selection for productivity can also have knock-on effects on production, with production potentially increasing or decreasing, depending on the trait selected and the particular production trait that also responds (see review in (Muir and Cheng, 2014)).

Some of the longest selection lines have been performed in quails, based on tonic immobility (TI) and social reinstatement (Faure and Mills, 2014, Launay et al., 1993), while a Red Junglefowl chicken population has also been selected for tameness/fear of humans over a number of generations (Agnvall and Jensen, 2016, Katajamaa et al., 2018). In these cases, other correlated behavioural changes were also seen, for example, the low TI quail line (reduced fear) had an even greater reduction of fear behaviour with handling and environmental enrichment cues (Candland et al., 1963, Jones and Faure, 1981). It was also easier to catch the low TI birds (Faure and Mills, 2014). However, even here there was a variable response to stress exposure effects. There was a lower corticosterone response in the long TI (high fear) line, while a restraint test induced an increase in plasma corticosterone in the low TI (low fear) line (Remignon et al., 1996). There is a link between stress and a reduction in meat quality (struggling induces a build-up of lactic acid that decreases pH and increases water loss). In this case long TI (high fear) birds showed a higher percentage of water loss and lower pH 24 hours post-mortem, indicating that there were some production effects from the selection process (Remignon et al., 1996).

## 6 Epigenetics and behaviour

Epigenetics is a term used to define changes that are separate from the actual genome, but which alter gene expression (Richards, 2006). This can include mechanisms such as methylation, as well as histone modification (though others exist (Richards, 2006)). In the case of methylation, a cytosine base may have a methyl group, with this most typically occurring when the cytosine base is adjacent to a guanine (often referred to as a CpG island). Depending on where these CpG regions are situated, these can modify gene expression. For example, methylation that occurs within a gene body typically up-regulates gene expression (or more specifically is positively correlated with gene expression) (Jjingo et al., 2012). In contrast, when a promoter (a small region just upstream the start of a gene, that drives expression) is methylated, this usually leads to a decrease in gene expression (a negative correlation between methylation and gene expression) (Gaston and Fried, 1995). Histones, which are the proteins that pack DNA, may also be modified, with the histone tails being modifiable via methylation or acetylation. These can then alter the packing of the DNA, with more densely packed regions having lower gene expression. This is due to these more packed regions being less accessible to ribosomes and the other cell machinery that is required to transcribe genes into RNA and proteins.

Despite being in some ways separate from the genome, epigenetic variation can still have a genetic basis in controlling it, with polymorphisms in particular regions either preferentially recruiting or removing methylation or acetylation (Kasowski et al., 2013, Kilpinen et al., 2013, Pértille et al., 2019). Indeed, this may be true for the majority of epigenetic variation in the genome. However, some epigenetic changes are more dynamic and are environmentally induced. In animals, one of the best described behavioural changes that has a methylation-based mechanism is pup-licking by rat mothers (Weaver et al., 2004), with licking inducing methylation changes in the promoter of the glucocorticoid receptor gene. This causes the licking behaviour to be mimicked by the pups when they reach adulthood and nurse their own young.

Epigenetic mechanisms have also been mooted to play a role in domestication. As an example, it has been suggested that domestication may have selected animals with an increased capacity to respond epigenetically to environmental stress (Carter et al., 2005), and the epigenome could also potentially play a role in providing additional sources of variation in cases of rapid evolution and adaptation to a new environment (Liebl et al., 2013). In the case of chickens, there is evidence of transgenerational effects on behaviour, with the offspring of chickens exposed to a stressor also exhibiting effects on behaviour (Lindqvist et al., 2007). As well as this, stable differences in methylation levels are also present between wild and domestic chickens (Pértille et al., 2019, Bélteky et al., 2018, Natt et al., 2012). However, no direct evidence has yet been shown of how epigenetic and genetic components interact to regulate gene expression in domestication.

## 7 Commercial aspects and research

The genetic and genomic techniques described above have been used extensively by commercial companies, in particular those relating to classical quantitative genetics and genomics selection. In contrast, causal gene identification and sweep mapping are more usually performed in an academic setting, as they have less economic ramifications. Genomic prediction is one technique that commercial companies use to determine which animals to use for breeding purposes (Momen et al., 2018, Wolc et al., 2016), for example this has been used for survival of influenza infection in layers (Wolc et al., 2018) as well as male fertility traits (Wolc et al., 2019). QTL mapping via Genome Wide Association Mapping has been used to identify loci affecting response to Newcastle Disease virus (Saelao et al., 2019) and body weight response in broilers (Tarsani et al., 2019), amongst others. An important caveat with all the industry genetic and genomic tools and applications, are that these are rarely applied to animal behaviour (D'eath et al., 2010).

## 8 Conclusions

The advent of more refined and powerful genetic and genomic tools will allow a greater potential for the identification of the genes and polymorphisms underlying variation in behaviour. These tools will also provide additional potential avenues for selection and even gene modification with which to reduce aberrant or anxiolytic behaviour. Some of the issues surrounding potential pleiotropy or close linkage that have surfaced with previous selection studies can, in this way, be circumnavigated. This will have ramifications from both welfare and production aspects.

## 9 Where to look for further information

### *9.1 General introduction to QTL mapping*

One of the best general introductions to QTL mapping is in the text book by Falconer and MacKay (Falconer and Mackay, 1996), although more recently Lynch and Walsh go into more detail (but has a higher level of complexity) - *Evolution and Selection of Quantitative Traits* (2018) Wlash, B., Lynch, M. OUP Oxford ISBN: 9780192566645.

### *9.2 Genome-wide association studies (GWAS)*

A good review of GWAS with a slant on its uses in domestic animals is presented in Sharma et al. (2015).

### *9.3 Genomic selection*

Although not really central to this book chapter, genomic selection is a large part of Industrial applications to domestic animals. An excellent review is presented in Wolc et al. (2016).

## 10 References

Abu-Bonsrah, K. D., Zhang, D. and Newgreen, D. F. 2016. CRISPR/Cas9 targets chicken embryonic somatic cells in vitro and in vivo and generates phenotypic abnormalities. *Scientific Reports* 6, 34524.

Agnvall, B. and Jensen, P. 2016. Effects of divergent selection for fear of humans on behaviour in Red Junglefowl. *PLoS ONE* 11, e0166075.

Andersson, L. and Georges, M. 2004. Domestic-animal genomics: deciphering the genetics of complex traits. *Nature Reviews Genetics* 5(3), 202–212.

Anholt, R. R. H. and Mackay, T. F. C. 2004. Quantitative genetic analyses of complex behaviours in Drosophila. *Nature Reviews Genetics* 5(11), 838–849.

Archer, J. 1973. Tests for emotionality in rats and mice: a review. *Animal Behaviour* 21(2), 205–235.

Balic, A., Garcia-Morales, C., Vervelde, L., Gilhooley, H., Sherman, A., Garceau, V., Gutowska, M. W., Burt, D. W., Kaiser, P., Hume, D. A. and Sang, H. M. 2014. Visualisation of chicken macrophages using transgenic reporter genes: insights into the development of the avian macrophage lineage. *Development* 141(16), 3255–3265.

Bates, R. W., Lahr, E. L. and Riddle, O. 1935. The gross action of prolactin and follicle-stimulating hormone on the mature ovary and sex accessories of fowl. *American Journal of Physiology-Legacy Content* 111(2), 361–368.

Beaumont, C., Roussot, O., Feve, K., Vignoles, F., Leroux, S., Pitel, F., Faure, J. M., Mills, A. D., Guémené, D., Sellier, N., Mignon-Grasteau, S., Le Roy, P. and Vignal, A. 2005. A genome scan with AFLPTM markers to detect fearfulness-related QTLs in Japanese quail. *Animal Genetics* 36(5), 401–407.

Beavis, W. D. 1998. QTL analyses: power, precision and accuracy. 49th Annual Corn and Sorghum Industry Research Conference. Washington, DC,: ASTA.

Bélteky, J., Agnvall, B., Bektic, L., Höglund, A., Jensen, P. and Guerrero-Bosagna, C. 2018. Epigenetics and early domestication: differences in hypothalamic DNA methylation between red junglefowl divergently selected for high or low fear of humans. *Genetics Selection Evolution* 50(1), 13.

Belzung, C. 1999. Measuring rodent exploratory behavior. *Techniques in the Behavioral and Neural Sciences* 13, 738–749.

Bennewitz, J., Bögelein, S., Stratz, P., Rodehutscord, M., Piepho, H. P., Kjaer, J. B. and Bessei, W. 2014. Genetic parameters for feather pecking and aggressive behavior in a large F2-cross of laying hens using generalized linear mixed models. *Poultry Science* 93(4), 810–817.

Benoff, F. H. and Siegel, P. B. 1976. Genetic analysis of tonic immobility in young Japanese quail (Coturnix cotunix japonica). *Animal Learning and Behavior* 4(2), 160–162.

Bijma, P., Muir, W. M. and Van Arendonk, J. A. 2007. Multilevel selection 1: quantitative genetics of inheritance and response to selection. *Genetics* 175(1), 277–288.

Bilcik, B. and Keeling, L. J. 1999. Changes in feather condition in relation to feather pecking and aggressive behaviour in laying hens. *British Poultry Science* 40(4), 444–451.

Biscarini, F., Bovenhuis, H., Van Der Poel, J., Rodenburg, T. B., Jungerius, A. P. and Van Arendonk, J. A. 2010. Across-line SNP association study for direct and associative effects on feather damage in laying hens. *Behavior Genetics* 40(5), 715–727.

Blokhuis, H. J. 1986. Feather-pecking in poultry: its relation with ground-pecking. *Applied Animal Behaviour Science* 16(1), 63–67.

Boake, C. R. B., Arnold, S. J., Breden, F., Meffert, L. M., Ritchie, M. G., Taylor, B. J., Wolf, J. B. and Moore, A. J. 2002. Genetic tools for studying adaptation and the evolution of behavior. *American Naturalist* 160 (Suppl. 6), S143–S159.

Brinker, T., Bijma, P., Visscher, J., Rodenburg, T. B. and Ellen, E. D. 2014. Plumage condition in laying hens: genetic parameters for direct and indirect effects in two purebred layer lines. *Genetics Selection Evolution* 46, 33.

Brodkin, E. S., Goforth, S. A., Keene, A. H., Fossella, J. A. and Silver, L. M. 2002. Identification of quantitative trait loci that affect aggressive behavior in mice. *The Journal of Neuroscience* 22(3), 1165–1170.

Brunberg, E., Jensen, P., Isaksson, A. and Keeling, L. 2011. Feather pecking behavior in laying hens: hypothalamic gene expression in birds performing and receiving pecks. *Poultry Science* 90(6), 1145–1152.

Buitenhuis, A. J., Rodenburg, T. B., Siwek, M., Cornelissen, S. J., Nieuwland, M. G., Crooijmans, R. P., Groenen, M. A., Koene, P., Bovenhuis, H. and Van Der Poel, J. J. 2003. Identification of quantitative trait loci for receiving pecks in young and adult laying hens. *Poultry Science* 82(11), 1661–1667.

Buitenhuis, A. J., Rodenburg, T. B., Siwek, M., Cornelissen, S. J., Nieuwland, M. G., Crooijmans, R. P., Groenen, M. A., Koene, P., Bovenhuis, H. and Van Der Poel, J. J. 2004. Identification of QTLs involved in open-field behavior in young and adult laying hens. *Behavior Genetics* 34(3), 325–333.

Burmeister, M., Mcinnis, M. G. and Zöllner, S. 2008. Psychiatric genetics: progress amid controversy. *Nature Reviews Genetics* 9(7), 527–540.

Burmeister, S. S., Jarvis, E. D. and Fernald, R. D. 2005. Rapid behavioral and genomic responses to social opportunity. *PLoS Biology* 3(11), e363.

Campo, J. L. and Carnicer, C. 1993. Realized heritability of tonic immobility in White Leghorn hens: a replicated single generation test. *Poultry Science* 72(12), 2193–2199.

Candland, D. K., Nagy, Z. M. and Conklyn, D. H. 1963. Emotional behavior in the domestic chicken (white leghorn) as a function of age and developmental environment. *Journal of Comparative and Physiological Psychology* 56, 1069–1073.

Carlson, C. S., Eberle, M. A., Kruglyak, L. and Nickerson, D. A. 2004. Mapping complex disease loci in whole-genome association studies. *Nature* 429(6990), 446–452.

Carter, A. J., Hermisson, J. and Hansen, T. F. 2005. The role of epistatic gene interactions in the response to selection and the evolution of evolvability. *Theoretical Population Biology* 68(3), 179–196.

Chiavegatto, S., Izidio, G. S., Mendes-Lana, A., Aneas, I., Freitas, T. A., Torrao, A. S., Conceicao, I. M., Britto, L. R. G. and Ramos, A. 2008. Expression of [alpha]-synuclein is increased in the hippocampus of rats with high levels of innate anxiety. *Molecular Psychiatry* 14, 894–905.

Coates, J. C. and De Bono, M. 2002. Antagonistic pathways in neurons exposed to body fluid regulate social feeding in *C.elegans*. *Nature* 419(6910), 925–929.

Cong, L. and Zhang, F. 2015. Genome engineering using CRISPR-Cas9 system. In: *Chromosomal Mutagenesis*. Humana Press, New York, NY, 197–217.

Craig, J. V. 1982. Behavioral and genetic adaptation of laying hens to high-density environments. *BioScience* 32(1), 33–37.

Craig, J. V. 1994. Genetic influences on behavior associated with well-being and productivity in livestock. Proceedings of the Fifth World Congress on Genetics Applied to Livestock Production 20, 150–157.

Craig, J. V. and Adams, A. 1984. Behaviour and well-being of hens (Gallus domesticus) in alternative housing environments. *World's Poultry Science Journal*, 40, 221–240.

Craig, J. V., Craig, J. A. and Vargas, J. V. 1986. Corticosteroids and other indicators of hens' well-being in four laying-house environments. *Poultry Science* 65(5), 856–863.

Craig, J. V. and Muir, W. M. 1989. Fearful and associated responses of caged White Leghorn hens: genetic parameter estimates. *Poultry Science* 68(8), 1040–1046.

D'eath, R., Conington, J., Lawrence, A., Olsson, I. A. S. and Sandøe, P. 2010. Breeding for behavioural change in farm animals: practical, economic and ethical considerations. *Animal Welfare* 19, 17–27.

Darvasi, A. 1998. Experimental strategies for the genetic dissection of complex traits in animal models. *Nature Genetics* 18(1), 19–24.

Davey, M. G., Balic, A., Rainger, J., Sang, H. M. and McGrew, M. J. 2018. Illuminating the chicken model through genetic modification. *International Journal of Developmental Biology* 62(1-2), 257–264.

De Belle, J. S. and Sokolowksi, M. B. 1989. Genetic Localization of *foraging* (*for*): a major gene for larval behavior in *Drosophila melanogaster*. *Genetics* 123(1), 157–163.

De Bono, M., Tobin, D. M., Davis, M. W., Avery, L. and Bargmann, C. I. 2002. Social feeding in *C.elegans* is induced by neurons that detect aversive stimuli. *Nature* 419(6910), 899–903.

De Haas, E. N. and Van Der Eijk, J. A. J. 2018. Where in the serotonergic system does it go wrong? Unravelling the route by which the serotonergic system affects feather pecking in chickens. *Neuroscience and Biobehavioral Reviews* 95, 170–188.

Dixon, L. M., Duncan, I. J. H. and Mason, G. 2008. What's in a peck? Using fixed action pattern morphology to identify the motivational basis of abnormal feather-pecking behaviour. *Animal Behaviour* 76(3), 1035–1042.

Donaldson, Z. R. and Young, L. J. 2008. Oxytocin, vasopressin, and the neurogenetics of sociality. *Science* 322(5903), 900–904.

Driscoll, C. A., Macdonald, D. W. and O'Brien, S. J. 2009. From wild animals to domestic pets, an evolutionary view of domestication. *Proceedings of the National Academy of Sciences of the United States of America* 106(Suppl. 1), 9971–9978.

Duncan, I. J. H. and Filshie, J. H. 1980. The use of radio telemetry devices to measure temperature and heart rate in domestic fowl. In: *A Handbook on Biotelemetry and Radio Tracking*. Pergamon, 579–588.

Dunn, I. 1998. Genetic mapping of the chicken prolactin receptor gene: a candidate gene for the control of broodiness. *British Poultry Science* 39, 23–24.

Dunnington, E. A. and Siegel, P. B. 1983. Mating frequency in male chickens: long-term selection. *Theoretical and Applied Genetics* 64(4), 317–323.

Ellen, E. D., Rodenburg, T. B., Albers, G. A., Bolhuis, J. E., Camerlink, I., Duijvesteijn, N., Knol, E. F., Muir, W. M., Peeters, K., Reimert, I., Sell-Kubiak, E., Van Arendonk, J. A., Visscher, J. and Bijma, P. 2014. The prospects of selection for social genetic effects to improve welfare and productivity in livestock. *Frontiers in Genetics* 5, 377.

Ellen, E. D., Van Der Sluis, M., Siegford, J., Guzhva, O., Toscano, M. J., Bennewitz, J., Van Der Zande, L. E., Van Der Eijk, J. A. J., De Haas, E. N., Norton, T., Piette, D., Tetens, J., De Klerk, B., Visser, B. and Rodenburg, T. B. 2019. Review of sensor technologies in animal breeding: phenotyping behaviors of laying hens to select Against feather pecking. *Animals* 9(3), 108.

Eriksson, J., Larson, G., Gunnarsson, U., Bed'hom, B., Tixier-Boichard, M., Stromstedt, L., Wright, D., Jungerius, A., Vereijken, A., Randi, E., Jensen, P. and Andersson, L. 2008. Identification of the yellow skin gene reveals a hybrid origin of the domestic chicken. *PLoS Genetics*, 4, e1000010.

Falconer, D. S. and Mackay, T. F. C. 1996. *Introduction to Quantitative Genetics*. Essex. UK: Longman Group.

Faure, J. M. 1981. Bidirectional selection for open-field activity in young chicks. *Behavior Genetics* 11(2), 135–144.

Faure, J. M. and Mills, A. D. 2014. Improving the adaptability of animals by selection. *Genetics and the Behavior of Domestic Animals*. Elsevier.

Fisher, R. A. 1958. *The Genetical Theory of Natural Selection*. New York: Dover.

Fitzpatrick, M. J., Ben-Shahar, Y., Smid, H. M., Vet, L. E. M., Robinson, G. E. and Sokolowski, M. B. 2005. Candidate genes for behavioural ecology. *Trends in Ecology and Evolution* 20(2), 96–104.

Flint, J. 2003. Analysis of quantitative trait loci that influence animal behaviour. *Journal of Neurobiology* 54(1), 46–77.

Flint, J. and Mott, R. 2001. Finding the molecular basis of quantitative traits: successes and pitfalls. *Nature Reviews Genetics* 2(6), 437–445.

Fogelholm, J., Inkabi, S., Höglund, A., Abbey-Lee, R., Johnsson, M., Jensen, P., Henriksen, R. and Wright, D. 2019. Genetical genomics of tonic immobility in the chicken. *Genes* 10(5), 341.

Forkman, B., Boissy, A., Meunier-Salaün, M. C., Canali, E. and Jones, R. B. 2007. A critical review of fear tests used on cattle, pigs, sheep, poultry and horses. *Physiology and Behavior* 92(3), 340–374.

François, N., Mills, A. D. and Faure, J. M. 1999. Inter-individual distances during open-field tests in Japanese quail (Coturnix japonica) selected for high or low levels of social reinstatement behaviour. *Behavioural Processes* 47(2), 73–80.

Gagliardi, G. J., Gallup, G. G. and Boren, J. L. 1976. Effect of different pupil to eye size ratios on tonic immobility in chickens. *Bulletin of the Psychonomic Society* 8(1), 58–60.

Gallup, G. G. 1974. Genetic influence on tonic immobility in chickens. *Animal Learning and Behavior* 2(2), 145–147.

Gallup, G. G., Rosen, T. S. and Brown, C. W. 1972. Effect of conditioned fear on tonic immobility in domestic chickens. *Journal of Comparative and Physiological Psychology* 78(1), 22–25.

Gallup Jr, G. G. 1979. Tonic immobility as a measure of fear in domestic fowl. *Animal Behaviour* 27, 316–317.

Gaston, K. and Fried, M. 1995. CpG methylation has differential effects on the binding of YY1 and ETS proteins to the bi-directional promoter of the Surf-1 and Surf-2 genes. *Nucleic Acids Research* 23(6), 901–909.

Gibson, G. and Weir, B. 2005. The quantitative genetics of transcription. *Trends in Genetics* 21(11), 616–623.

Gilman, T. T., Marcuse, F. L. and Moore, A. U. 1950. Animal hypnosis: a study in the induction of tonic immobility in chickens. *Journal of Comparative and Physiological Psychology* 43(2), 99–111.

Goddard, M. E. and Hayes, B. J. 2009. Mapping genes for complex traits in domestic animals and their use in breeding programmes. *Nature Reviews. Genetics* 10(6), 381-391.

Goldsmith, H. H. and Lemery, K. S. 2000. Linking temperamental fearfulness and anxiety symptoms: a behavior-genetic perspective. *Biological Psychiatry* 48(12), 1199-1209.

Goodale, H. D., White, D. and Sanborn, R. 1920. *Broodiness in Domestic Fowl: Data Concerning Its Inheritance in the Rhode Island Red Breed*. No. 199. Massachusetts Agricultural Experiment Station.

Grams, V., Bögelein, S., Grashorn, M. A., Bessei, W. and Bennewitz, J. 2015a. Quantitative genetic analysis of traits related to fear and feather pecking in laying hens. *Behavior Genetics* 45(2), 228-235.

Grams, V., Wellmann, R., Preus, S., Grashorn, M. A., Kjaer, J. B., Bessei, W. and Bennewitz, J. 2015b. Genetic parameters and signatures of selection in two divergent laying hen lines selected for feather pecking behaviour. *Genetics Selection Evolution* 47, 77.

Greenwood, A. K., Mills, M. G., Wark, A. R., Archambeault, S. L. and Peichel, C. L. 2016. Evolution of schooling behavior in threespine sticklebacks is shaped by the Eda gene. *Genetics* 203(2), 677-681.

Griffing, B. 1967. Selection in reference to biological groups I. Individual and group selection applied to populations of unordered groups. *Australian Journal of Biological Sciences* 20(1), 127-139.

Groenen, M. A., Cheng, H. H., Bumstead, N., Benkel, B. F., Briles, W. E., Burke, T., Burt, D. W., Crittenden, L. B., Dodgson, J., Hillel, J., Lamont, S., De Leon, A. P., Soller, M., Takahashi, H. and Vignal, A. 2000. A consensus linkage map of the chicken genome. *Genome Research* 10(1), 137-147.

Gross, W. B. and Siegel, P. B. 1985. Selective breeding of chickens for corticosterone response to social stress. *Poultry Science* 64(12), 2230-2233.

Gyetvai, B., Simonyi, A., Oros, M., Saito, M., Smiley, J. and Vadász, C. 2011. mGluR7 genetics and alcohol: intersection yields clues for addiction. *Neurochemical Research* 36(6), 1087-1100.

Hall, C. S. 1934. Emotional behavior in the rat. I. Defecation and urination as measures of individual differences in emotionality. *Journal of Comparative Psychology* 18(3), 385-403.

Harlander-Matauschek, A., Rodenburg, T. B., Sandilands, V., Tobalske, B. W. and Toscano, M. J. 2015. Causes of keel bone damage and their solutions in laying hens. *World's Poultry Science Journal* 71(3), 461-472.

Henningsen, A. D. 1994. Tonic immobility in 12 elasmobranchs: use as an aid in captive husbandry. *Zoo Biology* 13(4), 325-332.

Henriksen, R., Johnsson, M., Andersson, L., Jensen, P. and Wright, D. 2016. The domesticated brain: genetics of brain mass and brain structure in an avian species. *Scientific Reports* 6, 34031.

Heyne, H. O., Lautenschläger, S., Nelson, R., Besnier, F., Rotival, M., Cagan, A., Kozhemyakina, R., Plyusnina, I. Z., Trut, L., Carlborg, Ö., Petretto, E., Kruglyak, L., Pääbo, S., Schöneberg, T. and Albert, F. W. 2014. Genetic influences on brain gene expression in rats selected for tameness and aggression. *Genetics* 198(3), 1277-1290.

Huber-Eicher, B. and Wechsler, B. 1997. Feather pecking in domestic chicks: its relation to dustbathing and foraging. *Animal Behaviour* 54(4), 757-768.

Hughes, A. L. and Buitenhuis, A. J. 2010. Reduced variance of gene expression at numerous loci in a population of chickens selected for high feather pecking. *Poultry Science* 89(9), 1858–1869.

Humphreys, R. K. and Ruxton, G. D. 2018. A review of thanatosis (death feigning) as an anti-predator behaviour. *Behavioral Ecology and Sociobiology* 72(2), 22.

Hutt, F. B. 1949. *Genetics of the Fowl*. New York: McGraw-Hill.

Jensen, P., Keeling, L., Schütz, K., Andersson, L., Mormede, P., Brändström, H., Forkman, B., Kerje, S., Fredriksson, R., Ohlsson, C., Larsson, S., Mallmin, H. and Kindmark, A. 2005. Feather pecking in chickens is genetically related to behavioural and developmental traits. *Physiology and Behavior* 86(1–2), 52–60.

Jensen, P. and Wright, D. 2014. Behavioral genetics and animal domestication. In: Grandin, T. and Deesing, M. J. (Eds) *Genetics and Behavior of Domestic Animals*. London: Academic Press.

Jiang, R. S., Xu, G. Y., Zhang, X. Q. and Yang, N. 2005. Association of polymorphisms for prolactin and prolactin receptor genes with broody traits in chickens. *Poultry Science* 84(6), 839–845.

Jjingo, D., Conley, A. B., Yi, S. V., Lunyak, V. V. and Jordan, I. K. 2012. On the presence and role of human gene-body DNA methylation. *Oncotarget* 3(4), 462–474.

Johnsson, M., Gustafson, I., Rubin, C. J., Sahlqvist, A. S., Jonsson, K. B., Kerje, S., Ekwall, O., Kämpe, O., Andersson, L., Jensen, P. and Wright, D. 2012. A sexual ornament in chickens is affected by pleiotropic alleles at HAO1 and BMP2, selected during domestication. *PLOS Genetics* 8(8), e1002914.

Johnsson, M., Henriksen, R., Fogelholm, J., Höglund, A., Jensen, P. and Wright, D. 2018a. Genetics and genomics of social behavior in a chicken model. *Genetics* 209(1), 209–221.

Johnsson, M., Henriksen, R., Höglund, A., Fogelholm, J., Jensen, P. and Wright, D. 2018b. Genetical genomics of growth in a chicken model. *BMC Genomics* 19(1), 72.

Johnsson, M., Jonsson, K. B., Andersson, L., Jensen, P. and Wright, D. 2015a. Genetic regulation of bone metabolism in the chicken: similarities and differences to mammalian systems. *PLOS Genetics* 11(5), e1005250.

Johnsson, M., Jonsson, K. B., Andersson, L., Jensen, P. and Wright, D. 2015b. Quantitative trait locus and genetical genomics analysis identifies putatively causal genes for fecundity and brooding in the chicken. *G3: Genes|Genomes|Genetics* 6(2), 311–319.

Johnsson, M., Williams, M. J., Jensen, P. and Wright, D. 2016. Genetical genomics of behavior: a novel chicken genomic model for anxiety behavior. *Genetics* 202(1), 327–340.

Jones, R. B. and Faure, J. M. 1981. The effects of regular handling on fear responses in the domestic chick. *Behavioural Processes* 6(2), 135–143.

Jones, R. B., Mills, A. D. and Faure, J. M. 1991. Genetic and experiential manipulation of fear-related behavior in Japanese quail chicks (Coturnix coturnix japonica). *Journal of Comparative Psychology* 105(1), 15–24.

Jöngren, M., Westander, J., Nätt, D. and Jensen, P. 2010. Brain gene expression in relation to fearfulness in female red junglefowl (Gallus gallus). *Genes, Brain, and Behavior* 9(7), 751–758.

Karlsson, E. K., Baranowska, I., Wade, C. M., Salmon Hillbertz, N. H. C., Zody, M. C., Anderson, N., Biagi, T. M., Patterson, N., Pielberg, G. R., Kulbokas, E. J., Comstock, K. E., Keller, E. T., Mesirov, J. P., Von Euler, H., Kampe, O., Hedhammar, A., Lander, E. S., Andersson, G., Andersson, L. and Lindblad-Toh, K. 2007. Efficient mapping of

Mendelian traits in dogs through genome-wide association. *Nature Genetics* 39(11), 1321–1328.

Kas, M. J., Fernandes, C., Schalkwyk, L. C. and Collier, D. A. 2007. Genetics of behavioural domains across the neuropsychiatric spectrum; of mice and men. *Molecular Psychiatry* 12(4), 324–330.

Kasowski, M., Kyriazopoulou-Panagiotopoulou, S., Grubert, F., Zaugg, J. B., Kundaje, A., Liu, Y., Boyle, A. P., Zhang, Q. C., Zakharia, F., Spacek, D. V., Li, J., Xie, D., Olarerin-George, A., Steinmetz, L. M., Hogenesch, J. B., Kellis, M., Batzoglou, S. and Snyder, M. 2013. Extensive variation in chromatin states across humans. *Science* 342(6159), 750–752.

Katajamaa, R., Larsson, L. H., Lundberg, P., Sörensen, I. and Jensen, P. 2018. Activity, social and sexual behaviour in Red Junglefowl selected for divergent levels of fear of humans. *PLoS ONE* 13(9), e0204303.

Kilpinen, H., Waszak, S. M., Gschwind, A. R., Raghav, S. K., Witwicki, R. M., Orioli, A., Migliavacca, E., Wiederkehr, M., Gutierrez-Arcelus, M., Panousis, N. I., Yurovsky, A., Lappalainen, T., Romano-Palumbo, L., Planchon, A., Bielser, D., Bryois, J., Padioleau, I., Udin, G., Thurnheer, S., Hacker, D., Core, L. J., Lis, J. T., Hernandez, N., Reymond, A., Deplancke, B. and Dermitzakis, E. T. 2013. Coordinated effects of sequence variation on DNA binding, chromatin structure, and transcription. *Science* 342(6159), 744–747.

Kim, S., Zhang, S., Choi, K. H., Reister, R., Do, C., Baykiz, A. F. and Gershenfeld, H. K. 2009. An E3 ubiquitin ligase, really interesting new gene (RING) finger 41, is a candidate gene for anxiety-like behavior and β-carboline-induced seizures. *Biological Psychiatry* 65(5), 425–431.

Kjaer, J. B. 2009. Feather pecking in domestic fowl is genetically related to locomotor activity levels: implications for a hyperactivity disorder model of feather pecking. *Behavior Genetics* 39(5), 564–570.

Kjaer, J. B. 2017. Divergent selection on home pen locomotor activity in a chicken model: selection program, genetic parameters and direct response on activity and body weight. *PLoS ONE* 12(8), e0182103.

Kjaer, J. B. and Sørensen, P. 2002. Feather pecking and cannibalism in free-range laying hens as affected by genotype, dietary level of methionine+ cystine, light intensity during rearing and age at first access to the range area. *Applied Animal Behaviour Science* 76(1), 21–39.

Kjaer, J. B., Sørensen, P. and Su, G. 2001. Divergent selection on feather pecking behaviour in laying hens (Gallus gallus domesticus). *Applied Animal Behaviour Science* 71(3), 229–239.

Kozak, M., Tobalske, B., Springthorpe, D., Szkotnicki, B. and Harlander-Matauschek, A. 2016. Development of physical activity levels in laying hens in three-dimensional aviaries. *Applied Animal Behaviour Science* 185, 66–72.

Labouriau, R., Kjaer, J. B., Abreu, G. C., Hedegaard, J. and Buitenhuis, A. J. 2009. Analysis of severe feather pecking behavior in a high feather pecking selection line. *Poultry Science* 88(10), 2052–2062.

Lander, E. S. and Botstein, D. 1989. Mapping Mendelian factors underlying quantitative traits using RFLP linkage maps. *Genetics* 121(1), 185–199.

Launay, F., Mills, A. D. and Faure, J. M. 1991. Social motivation in Japanese quail Coturnix coturnix japonica chicks selected for high or low levels of treadmill behaviour. *Behavioural Processes* 24(2), 95–110.

Launay, F., Mills, A. D. and Faure, J. M. 1993. Effects of test age, line and sex on tonic immobility responses and social reinstatement behaviour in Japanese quail Coturnix japonica. *Behavioural Processes* 29(1-2), 1-16.

Liebl, A. L., Schrey, A. W., Richards, C. L. and Martin, L. B. 2013. Patterns of DNA methylation throughout a range expansion of an introduced songbird. *Integrative and Comparative Biology* 53(2), 351-358.

Lindqvist, C., Janczak, A. M., Nätt, D., Baranowska, I., Lindqvist, N., Wichman, A., Lundeberg, J., Lindberg, J., Torjesen, P. A. and Jensen, P. 2007. Transmission of stress-induced learning impairment and associated brain gene expression from parents to offspring in chickens. *PLoS ONE* 2(4), e364.

Lutz, V., Stratz, P., Preus, S., Tetens, J., Grashorn, M. A., Bessei, W. and Bennewitz, J. 2017. A genome-wide association study in a large F2-cross of laying hens reveals novel genomic regions associated with feather pecking and aggressive pecking behavior. *Genetics Selection Evolution* 49(1), 18.

Lynch, M. and Walsh, B. 1998. *Genetics and Analysis of Quantitative Traits*. Sunderland, MA: Sinauer Associates.

Mackay, T. F., Stone, E. A. and Ayroles, J. F. 2009. The genetics of quantitative traits: challenges and prospects. *Nature Reviews Genetics* 10(8), 565-577.

Mackay, T. F. C. 2004. The genetic architecture of quantitative traits: lessons from Drosophila. *Current Opinion in Genetics and Development* 14(3), 253-257.

McGraw, L. A. and Young, L. J. 2010. The prairie vole: an emerging model organism for understanding the social brain. *Trends in Neurosciences* 33(2), 103-109.

McGue, M., Bouchard Jr, T. J. 1998. Genetic and environmental influences on human behavioral differences. *Annual Review of Neuroscience* 21, 1-24.

McGrew, M. J., Sherman, A., Lillico, S. G., Ellard, F. M., Radcliffe, P. A., Gilhooley, H. J., Mitrophanous, K. A., Cambray, N., Wilson, V. and Sang, H. 2008. Localised axial progenitor cell populations in the avian tail bud are not committed to a posterior Hox identity. *Development* 135(13), 2289-2299.

Mehrabian, M., Allayee, H., Stockton, J., Lum, P. Y., Drake, T. A., Castellani, L. W., Suh, M., Armour, C., Edwards, S., Lamb, J., Lusis, A. J. and Schadt, E. E. 2005. Integrating genotypic and expression data in a segregating mouse population to identify 5-lipoxygenase as a susceptibility gene for obesity and bone traits. *Nature Genetics* 37(11), 1224-1233.

Meijer, T. 1995. Importance of tactile and visual stimuli of eggs and nest for termination of egg laying of red junglefowl. *The Auk* 112(2), 483-488.

Millman, S. T. and Duncan, I. J. H. 2000. Effect of male-to-male aggressiveness and feed-restriction during rearing on sexual behaviour and aggressiveness towards females by male domestic fowl. *Applied Animal Behaviour Science* 70(1), 63-82.

Mills, A. D., Crawford, L. L., Domjan, M. and Faure, J. M. 1997. The behavior of the Japanese or domestic quail Coturnix japonica. *Neuroscience and Biobehavioral Reviews* 21(3), 261-281.

Mills, A. D. and Faure, J. M. 1991. Divergent selection for duration of tonic immobility and social reinstatement behavior in Japanese quail (*Coturnix coturnix japonica*) chicks. *Journal of Comparative Psychology* 105(1), 25-38.

Mills, A. D., Launay, F., Turro, L., Faure, J. M. and Picard, M. 1992. Behavioural consequences of divergent selection for social reinstatement behaviour. *Journal of Animal Science* 70, 159.

Minvielle, F., Kayang, B. B., Inoue-Murayama, M., Miwa, M., Vignal, A., Gourichon, D., Neau, A., Monvoisin, J. L. and Ito, S. I. 2005. Microsatellite mapping of QTL affecting growth, feed consumption, egg production, tonic immobility and body temperature of Japanese quail. *BMC Genomics* 6, 87.

Momen, M., Mehrgardi, A. A., Sheikhi, A., Kranis, A., Tusell, L., Morota, G., Rosa, G. J. M. and Gianola, D. 2018. Predictive ability of genome-assisted statistical models under various forms of gene action. *Scientific Reports* 8(1), 12309.

Muir, W. M. 2005. Incorporation of competitive effects in forest tree or animal breeding programs. *Genetics* 170(3), 1247-1259.

Muir, W. M. and Cheng, H. W. 2014. Genetic Influences on the behavior of chickens associated with welfare and productivity. In: *Genetics and the Behavior of Domestic Animals*. Academic Press, 317-359.

Murphy, L. B. 1978. The practical problems of recognizing and measuring fear and exploration behaviour in the domestic fowl. *Animal Behaviour* 26, 422-431.

Murray, C. J. and Lopez, A. D. 1996. Evidence-based health policy—lessons from the global burden of disease study. *Science* 274(5288), 740-743.

Nadeau, J. H. and Frankel, W. N. 2000. The Roads from phenotypic variation to gene discovery: mutagenesis versus QTLs. *Nature Genetics* 25(4), 381-384.

Natt, D., Rubin, C. J., Wright, D., Johnsson, M., Belteky, J., Andersson, L. and Jensen, P. 2012. Heritable genome-wide variation of gene expression and promoter methylation between wild and domesticated chickens. *BMC Genomics* 13, 59.

Nones, K., Ledur, M. C., Ruy, D. C., Baron, E. E., Melo, C. M., Moura, A. S., Zanella, E. L., Burt, D. W. and Coutinho, L. L. 2006. Mapping QTLs on chicken chromosome 1 for performance and carcass traits in a broiler x layer cross. *Animal Genetics* 37(2), 95-100.

O'Brien, T. J. and Dunlap, W. P. 1975. Tonic immobility in the blue crab (Callinectes sapidus, Rathbun): its relation to threat of predation. *Journal of Comparative and Physiological Psychology* 89(1), 86-94.

Oishi, I., Yoshii, K., Miyahara, D., Kagami, H. and Tagami, T. 2016. Targeted mutagenesis in chicken using CRISPR/Cas9 system. *Scientific Reports* 6, 23980.

Park, T. S., Lee, H. J., Kim, K. H., Kim, J. S. and Han, J. Y. 2014. Targeted gene knockout in chickens mediated by TALENs. *Proceedings of the National Academy of Sciences of the United States of America* 111(35), 12716-12721.

Parmentier, H. K., Rodenburg, T. B., De Vries Reilingh, G., Beerda, B. and Kemp, B. 2009. Does enhancement of specific immune responses predispose laying hens for feather pecking? *Poultry Science* 88(3), 536-542.

Persson, M. E., Wright, D., Roth, L. S., Batakis, P. and Jensen, P. 2016. Genomic regions associated with interspecies communication in dogs contain genes related to human social disorders. *Scientific Reports* 6, 33439.

Pértille, F., Da Silva, V. H., Johansson, A. M., Lindström, T., Wright, D., Coutinho, L. L., Jensen, P. and Guerrero-Bosagna, C. 2019. Mutation dynamics of CpG dinucleotides during a recent event of vertebrate diversification. *Epigenetics* 14(7), 685-707.

Pritchard, J. K. and Di Rienzo, A. 2010. Adaptation - not by sweeps alone. *Nature Reviews. Genetics* 11(10), 665-667.

Prut, L. and Belzung, C. 2003. The open field as a paradigm to measure the effects of drugs on anxiety-like behaviors: a review. *European Journal of Pharmacology* 463(1-3), 3-33.

Ramos, A. and Mormède, P. 1997. Stress and emotionality: a multidimensional and genetic approach. *Neuroscience and Biobehavioral Reviews* 22(1), 33-57.
Ratner, S. 1967. Comparative aspects of hypnosis. In: Gordon, J. E. (Ed.) *Handbook of Clinical and Experimental Hypnosis*. New York: Macmillan, 550-587.
Recoquillay, J., Leterrier, C., Calandreau, L., Bertin, A., Pitel, F., Gourichon, D., Vignal, A., Beaumont, C., Le Bihan-Duval, E. and Arnould, C. 2013. Evidence of phenotypic and genetic relationships between sociality, emotional reactivity and production traits in Japanese quail. *PLoS ONE* 8(12), e82157.
Recoquillay, J., Pitel, F., Arnould, C., Leroux, S., Dehais, P., Moréno, C., Calandreau, L., Bertin, A., Gourichon, D., Bouchez, O., Vignal, A., Fariello, M. I., Minvielle, F., Beaumont, C., Leterrier, C. and Le Bihan-Duval, E. 2015. A medium density genetic map and QTL for behavioral and production traits in Japanese quail. *BMC Genomics* 16, 10.
Remignon, H., Desrosiers, V., Bruneau, A., Guémené, D. and Mills, A. 1996. Influence on an acute stress on muscle parameters in quail divergently selected for emotivity. 20th World Poultry Conference, New Delhi, 716-721.
Richards, E. J. 2006. Inherited epigenetic variation–revisiting soft inheritance. *Nature Reviews Genetics* 7(5), 395-401.
Ripke, S., Neale, B. M., Corvin, A., Walters, J. T., Farh, K.-H., Holmans, P. A., Lee, P., Bulik-Sullivan, B., Collier, D. A. and Huang, H. 2014. Biological insights from 108 schizophrenia-associated genetic loci. *Nature* 511(7510), 421-427.
Roberts, E. and Card, L. E. 1933. Inheritance of broodiness in the domestic fowl. In Proceedings of the 5th World's Poultry Congress, vol. 2, 353-358.
Robinson, G. E., Fernald, R. D. and Clayton, D. F. 2008. Genes and social behavior. *Science* 322(5903), 896-900.
Robinson, G. E., Grozinger, C. M. and Whitfield, C. W. 2005. Sociogenomics: social life in molecular terms. *Nature Reviews Genetics* 6(4), 257-270.
Rodenburg, T. B., Buitenhuis, A. J., Ask, B., Uitdehaag, K. A., Koene, P., Van Der Poel, J. J. and Bovenhuis, H. 2003. Heritability of feather pecking and open-field response of laying hens at two different ages. *Poultry Science* 82(6), 861-867.
Rubin, C.-J., Zody, M. C., Eriksson, J., Meadows, J. R. S., Sherwood, E., Webster, M. T., Jiang, L., Ingman, M., Sharpe, T., Ka, S., Hallbook, F., Besnier, F., Carlborg, O., Bed'hom, B., Tixier-Boichard, M., Jensen, P., Siegel, P., Lindblad-Toh, K. and Andersson, L. 2010. Whole-genome resequencing reveals loci under selection during chicken domestication. *Nature*, 464, 587-591.
Saeki, Y. 1957. Inheritance of broodiness in Japanese Nagoya fowl, with special reference to sex-linkage and notice in breeding practice. *Poultry Science* 36(2), 378-383.
Saelao, P. and Wang, Y. Chanthavixay, G. Gallardo, R. A., Wolc, A., Dekkers, J., Lamont, S. J., Kelly, T. and Zhou, H. 2019. Genetics and genomic regions affecting response to newcastle disease virus infection under heat stress in layer chickens. *Genes*, 10, 61.
Sargeant, A. B. and Eberhardt, L. E. 1975. Death feigning by ducks in response to predation by red foxes (Vulpes fulva). *American Midland Naturalist* 94(1), 108-119.
Savory, C. J. 1995. Feather pecking and cannibalism. *World's Poultry Science Journal* 51(2), 215-219.
Sawyer, L. A., Hennessy, J. M., Pexioto, A. A., Rosato, E., Parkinson, H., Costa, R. and Kyriacou, C. P. 1997. Natural variation in a Drosophila clock gene and temperature compensation. *Science* 278(5346), 2117-2120.

Schutz, K., Kerje, S., Carlborg, O., Jacobsson, L., Andersson, L. and Jensen, P. 2002. QTL analysis of a red junglefowl × white leghorn intercross reveals trade-off in resource allocation between behavior and production traits. *Behavior Genetics* 32(6), 423-433.

Schutz, K. E., Kerje, S., Jacobsson, L., Forkman, B., Carlborg, O., Andersson, L. and Jensen, P. 2004. Major growth QTLs in fowl are related to fearful behavior: possible genetic links between fear responses and production traits in a red junglefowl × white leghorn intercross. *Behavior Genetics* 34(1), 121-130.

Schweitzer, C. and Arnould, C. 2010. Emotional reactivity of Japanese quail chicks with high or low social motivation reared under unstable social conditions. *Applied Animal Behaviour Science* 125(3-4), 143-150.

Schweitzer, C., Poindron, P. and Arnould, C. 2009. Social motivation affects the display of individual discrimination in young and adult Japanese quail (Coturnix japonica). *Developmental Psychobiology* 51(4), 311-321.

Sharma, A., Lee, J. S., Dang, C. G., Sudrajad, P., Kim, H. C., Yeon, S. H., Kang, H. S. and Lee, S. H. 2015. Stories and challenges of genome wide Association Studies in livestock - a review. *Asian-Australasian Journal of Animal Sciences* 28(10), 1371-1379.

Shorter, J., Couch, C., Huang, W., Carbone, M. A., Peiffer, J., Anholt, R. R. H. and Mackay, T. F. C. 2015. Genetic architecture of natural variation in Drosophila melanogaster aggressive behavior. *Proceedings of the National Academy of Sciences of the United States of America* 112(27), E3555-E3563.

Siegel, P. B. 1993. Behavior-genetic analyses and poultry husbandry. *Poultry Science* 72(1), 1-6.

Sih, A., Bell, A. and Johnson, J. C. 2004. Behavioral syndromes: an ecological and evolutionary overview. *Trends in Ecology and Evolution* 19(7), 372-378.

Smith, J. M. and Haigh, J. 1974. The hitch-hiking effect of a favourable gene. *Genetical Research* 23(1), 23-35.

Sokolowski, M. B. 1998. Genes for normal behavioural variation: recent clues from flies and worms. *Neuron* 21(3), 463-466.

Suarez, S. D. and Gallup, G. G. 1983. Social reinstatement and open-field testing in chickens. *Animal Learning and Behavior* 11(1), 119-126.

Takahashi, A., Tomihara, K., Shiroishi, T. and Koide, T. 2010. Genetic mapping of social interaction behavior in B6/MSM consomic mouse strains. *Behavior Genetics* 40(3), 366-376.

Tarsani, E., Kranis, A., Maniatis, G., Avendano, S., Hager-Theodorides, A. L. and Kominakis, A. 2019. Discovery and characterization of functional modules associated with body weight in broilers. *Scientific Reports* 9(1), 9125.

Tomida, S., Mamiya, T., Sakamaki, H., Miura, M., Aosaki, T., Masuda, M., Niwa, M., Kameyama, T., Kobayashi, J., Iwaki, Y., Imai, S., Ishikawa, A., Abe, K., Yoshimura, T., Nabeshima, T. and Ebihara, S. 2009. Usp46 is a quantitative trait gene regulating mouse immobile behavior in the tail suspension and forced swimming tests. *Nature Genetics* 41(6), 688-695.

Tully, T. 1996. Discovery of genes involved with learning and memory: an experimental synthesis of Hirschian and Benzerian perspectives. *Proceedings of the National Academy of Sciences of the United States of America* 93(24), 13460-13467.

Valdar, W., Solberg, L. C., Gauguier, D., Burnett, S., Klenerman, P., Cookson, W. O., Taylor, M. S., Rawlins, J. N. P., Mott, R. and Flint, J. 2006. Genome-wide genetic association of complex traits in heterogeneous stock mice. *Nature Genetics* 38(8), 879-887.

Van De Lavoir, M. C., Diamond, J. H., Leighton, P. A., Mather-Love, C., Heyer, B. S., Bradshaw, R., Kerchner, A., Hooi, L. T., Gessaro, T. M., Swanberg, S. E., Delany, M. E. and Etches, R. J. 2006. Germline transmission of genetically modified primordial germ cells. *Nature* 441(7094), 766-769.

Verdugo, R. A., Farber, C. R., Warden, C. H. and Medrano, J. F. 2010. Serious limitations of the QTL/microarray approach for QTL gene discovery. *BMC Biology* 8, 96.

vonHoldt, B. M., Shuldiner, E., Koch, I. J., Kartzinel, R. Y., Hogan, A., Brubaker, L., Wanser, S., Stahler, D., Wynne, C. D. L., Ostrander, E. A., Sinsheimer, J. S. and Udell, M. A. R. 2017. Structural variants in genes associated with human Williams-Beuren syndrome underlie stereotypical hypersociability in domestic dogs. *Science Advances* 3(7), e1700398.

Vos, T., Barber, R. M., Bell, B., Bertozzi-Villa, A., Biryukov, S., Bolliger, et al. 2015. Global, regional, and national incidence, prevalence, and years lived with disability for 301 acute and chronic diseases and injuries in 188 countries, 1990-2013: a systematic analysis for the Global Burden of Disease Study 2013. *The Lancet* 386(9995):743-800.

Wallis, J. W., Aerts, J., Groenen, M. A., Crooijmans, R. P., Layman, D., Graves, T. A., Scheer, D. E., Kremitzki, C., Fedele, M. J., Mudd, N. K., Cardenas, M., Higginbotham, J., Carter, J., Mcgrane, R., Gaige, T., Mead, K., Walker, J., Albracht, D., Davito, J., Yang, S. P., Leong, S., Chinwalla, A., Sekhon, M., Wylie, K., Dodgson, J., Romanov, M. N., Cheng, H., De Jong, P. J., Osoegawa, K., Nefedov, M., Zhang, H., Mcpherson, J. D., Krzywinski, M., Schein, J., Hillier, L., Mardis, E. R., Wilson, R. K. and Warren, W. C. 2004. A physical map of the chicken genome. *Nature* 432(7018), 761-764.

Wang, X., Mozhui, K., Li, Z., Mulligan, M. K., Ingels, J. F., Zhou, X., Hori, R. T., Chen, H., Cook, M. N., Williams, R. W. and Lu, L. 2012. A promoter polymorphism in the Per3 gene is associated with alcohol and stress response. *Translational Psychiatry* 2, e73-e73.

Wang, Z., Gerstein, M. and Snyder, M. 2009. RNA-Seq: a revolutionary tool for transcriptomics. *Nature Reviews. Genetics* 10(1), 57-63.

Warren, D. C. 1930. *Crossbred Poultry*. Kansas State Agricultural College: Agricultural Experiment Station.

Weaver, I. C., Cervoni, N., Champagne, F. A., D'Alessio, A. C., Sharma, S., Seckl, J. R., Dymov, S., Szyf, M. and Meaney, M. J. 2004. Epigenetic programming by maternal behavior. *Nature Neuroscience* 7(8), 847-854.

Webster, A. B. and Hurnik, J. F. 1991. Behavior, production, and weil-being of the laying hen. 2. Individual variation and relationships of behavior to production and physical condition. *Poultry Science* 70(3), 421-428.

Wegner, R.-M. 1990. Poultry welfare-problems and research to solve them. *World's Poultry Science Journal* 46(1), 19-30.

Whitehead, C. and Wilson, S. 1992. Characteristics of osteopenia in hens. In Poultry Science Symposium, 265-280.

Whyte, J., Glover, J. D., Woodcock, M., Brzeszczynska, J., Taylor, L., Sherman, A., Kaiser, P. and McGrew, M. J. 2015. FGF, insulin, and SMAD signaling cooperate for avian primordial germ cell self-renewal. *Stem Cell Reports* 5(6), 1171-1182.

Wirén, A., Wright, D. and Jensen, P. 2013. Domestication-related variation in social preferences in chickens is affected by genotype on a growth QTL. *Genes, Brain, and Behavior* 12(3), 330-337.

Wolc, A., Arango, J., Settar, P., Fulton, J. E., O'Sullivan, N. P. and Dekkers, J. C. M. 2019. Genetics of male reproductive performance in White Leghorns. *Poultry Science* 98(7), 2729-2733.

Wolc, A., Drobik-Czwarno, W., Fulton, J. E., Arango, J., Jankowski, T. and Dekkers, J. C. M. 2018. Genomic prediction of avian influenza infection outcome in layer chickens. *Genetics Selection Evolution* 50(1), 21.

Wolc, A., Kranis, A., Arango, J., Settar, P., Fulton, J. E., O'Sullivan, N. P., Avendano, A., Watson, K. A., Hickey, J. M., De Los Campos, G., Fernando, R. L., Garrick, D. J. and Dekkers, J. C. M. 2016. Implementation of genomic selection in the poultry industry. *Animal Frontiers* 6(1), 23-31.

Wright, D. 2015. The genetic architecture of domestication in animals. *Bioinformatics and Biology Insights* 9(Suppl 4), 11-20.

Wright, D., Boije, H., Meadows, J. R. S., Bed'hom, B., Gourichon, D., Vieaud, A., Tixier-Boichard, M., Rubin, C. J., Imsland, F., Hallböök, F. and Andersson, L. 2009. Transient ectopic expression of SOX5 during embryonic development causes the Pea-comb phenotype in chickens. *PLoS Genetics* 5(6), e1000512.

Wright, D., Butlin, R. K. and Carlborg, Ö. 2006a. Epistatic regulation of behavioural and morphological traits in the zebrafish (*Danio rerio*). *Behavior Genetics* 36(6), 914-922.

Wright, D., Nakamichi, R., Krause, J. and Butlin, R. K. 2006b. QTL analysis of behavioural and morphological differentiation between wild and laboratory zebrafish (*Danio rerio*). *Behavior Genetics* 36(2), 271-284.

Wright, D., Rubin, C. J., Martinez Barrio, A., Schütz, K., Kerje, S., Brändström, H., Kindmark, A., Jensen, P. and Andersson, L. 2010. The genetic architecture of domestication in the chicken: effects of pleiotropy and linkage. *Molecular Ecology* 19(23), 5140-5156.

Wu, Q., Wen, T., Lee, G., Park, J. H., Cai, H. N. and Shen, P. 2003. Developmental control of foraging and social behavior by the Drosophila neuropeptide Y-like system. *Neuron* 39(1), 147-161.

Wysocki, M., Preuss, S., Stratz, P. and Bennewitz, J. 2013. Investigating gene expression differences in two chicken groups with variable propensity to feather pecking. *Animal Genetics* 44(6), 773-777.

Yalcin, B., Willis-Owen, S. A. G., Fullerton, J., Meesaq, A., Deacon, R. M., Rawlins, J. N. P., Copley, R. R., Morris, A. P., Flint, J. and Mott, R. 2004. Genetic dissection of a behavioral quantitative trait locus shows that Rgs2 modulates anxiety in mice. *Nature Genetics* 36(11), 1197-1202.

# Chapter 2

## Understanding the sensory perception of chickens

*Birte L. Nielsen, INRAE, France*

1 Introduction
2 Vision
3 Olfaction
4 Taste
5 Hearing
6 Touch
7 Welfare implications
8 Where to look for further information
9 References

### 1 Introduction

It is no surprise that sensory perception, as for all sentient animals, is important for the behaviour and welfare of chickens. Sensory modalities are, after all, the means with which all environmental stimuli are perceived. In order for us to understand - or at least estimate - how a chicken takes in its surroundings, we need to know more about how the senses used by chickens are similar to and different from our own. As humans, we have a tendency to focus on the sensory modalities that we find most important, also when studying non-human animals. We may also assume that animals see or hear the world in a similar manner to us.

It is, however, very important to try and put ourselves in the place of the animal and experience the immediate environment and current situation from the animal's perspective if we are to assess properly the welfare impact of different environmental factors (Nielsen, 2018). For this reason, this chapter does not go into detail with the physiological and neurological pathways involved in sensory perception of chickens, nor to any great extent of the morphological structures. Instead, the focus is directly on the behaviour and welfare of chickens, both broilers and layers, at different ages and in the environmental contexts usually surrounding these birds. The five main senses

http://dx.doi.org/10.19103/AS.2020.0078.02

of vision, olfaction, taste, hearing and touch are covered, each in their own section, and the final section deals with other senses as well as an assessment of the impact of different sensory modalities on behavioural studies and animal welfare science.

## 2 Vision

The ability to see is ranked as very important by most sighted humans, who often choose vision as the last sense they would want to lose. And when delving into the literature on vision in chickens, it becomes clear that this sensory modality has a significant impact on the welfare of the birds. Like humans, chickens are highly visual beings, and use their eyes for many of their interactions with the environment, including foraging and identification of conspecifics (Appleby et al., 2004; Nicol, 2015). The latter makes vision an important sensory modality for the maintenance of the social hierarchy in groups of chickens (D'Eath and Keeling, 2003).

Compared to humans, the eyes of chickens are placed further apart and more on the sides of the head, which give the birds a different range of vision to us (Fig. 1). This makes chickens able to see a greater fraction of their surroundings, although most of this vision is monocular. Unlike us, the chicken cannot move its eyeballs very much, and changes its point of view by moving its head (Prescott et al., 2004). When walking, the chicken moves its body forward while for a fraction of time keeping its head in the same position relative to the ground. This allows it to focus on its surroundings during this short period, when the head and, hence, the eyes remain fixed in space (Necker, 2007). The ability of chickens to hold their head still while the body moves can be found

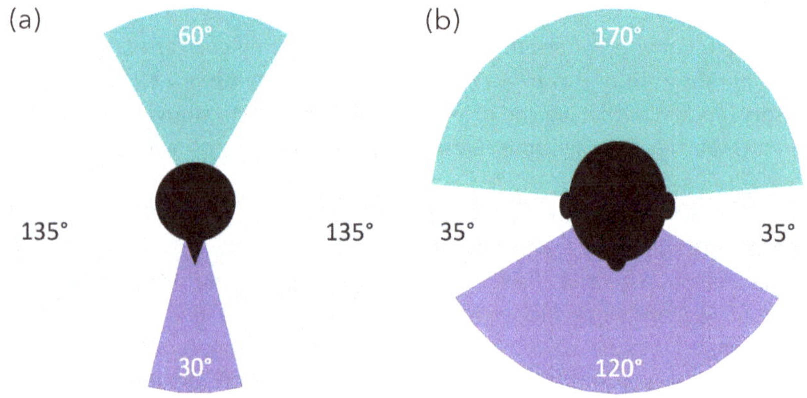


**Figure 1** Field of vision for (a) chickens and (b) humans, with the dark blue area showing binocular vision, light blue areas showing peripheral vision and green area depicting field that is not visible.

in numerous online videos, most famously in an awarded advertising feature for Mercedes Benz. Chickens also use their left and right eye for different tasks, so that the right eye is used preferentially for identifying food, whereas the left eye is involved in the spatial exploration and orientation (Rashid and Andrew, 1989). This differentiation is linked to the lateralization of the brain, as input to the left eye is processed in the right hemisphere of the brain and vice versa.

Using a so-called pebble-floor test (where food pellets are scattered onto a floor to which are glued pebbles of the same size and hue as the pellets), it is possible to assess the ability of chicks to distinguish between pebbles and pellets by counting how many times the birds peck at each type. When the left eye of chicks is covered, thus allowing them to use only their right eye, the chicks peck significantly more at the pellets than the pebbles than when using only their left eye (Mench and Andrew, 1986; Rogers et al., 2007). This laterality that gives the right eye an advantage in feed identification is dependent on exposure to light during the last days of incubation. Here, the posture of the embryo within the egg prevents light from reaching the left eye, whereas the right eye is facing the eggshell and hence exposed to a degree of brightness. We know this, because chicks that have had their head gently turned during the last days before hatching show a reversal of this lateralization (Rogers, 2008). In other words, whichever eye has been exposed to light before hatch is better at distinguishing feed from pebbles, and under natural conditions, this is usually the right eye.

One aspect of chicken vision associated with foraging is the so-called worm run, which is considered to be an innate response. A worm run is when 'the protrusion of a rod-like visual stimulus into the visual field away from the beak elicits running behaviour in the chick' (Rogers and Astiningsih, 1991). This stimulates other chicks to follow, while using their beaks to attempt to take hold of the protruding stimulus. Feeding boiled spaghetti to back-yard hens will yield the same response (personal observation).

Chickens have good colour vision, and a common practical exercise for students is to assess the colour preferences of newly hatched chicks. These usually consist of placing chicks on sheets of paper printed with different coloured dots, and noting down the colour of the first and subsequent dots pecked. Using this method, Ham and Osorio (2007) showed that 9-day-old chicks have a preference for orange over blue and for red over green, whereas only the former holds true for newly hatched chicks, as these did not show a clear preference when given a choice of red and green. The colour preferences of chickens in favour of hues in the orange/red over blue/green is thought to be associated with ripeness of fruit, although red is avoided if associated with an insect, thought to signal potential toxicity (Gamberale-Stille and Tullberg, 2001; Nicol, 2015). Chicks are also able to interpolate two colours: if they have previously been rewarded by pecking blue and green squares, they will peck even more on the squares of a turquoise colour, as if expecting increased award from pecking a mixture of the two colours (Jones et al., 2001).

In a trial investigating the importance of vision for body stability, laying hens with or without an opaque hood that blocked any visual input were placed on a moving perch (Leblanc et al., 2016). Although lack of vision did not result in more birds falling off the perch, none of the masked hens jumped off the perch and they kept a more crouched and stiff posture to keep balance compared to sighted birds. The authors speculate if the prior training of the hens to sit on the swaying perches had improved their balancing ability. In an intriguing experiment, Collins et al. (2011) investigated in more detail the importance of vision for the behaviour and welfare of chickens by comparing normal-sighted White Leghorn chickens with a genetically blind variety of the same strain. They found the blind chickens to sit and to preen themselves more than did the sighted birds, and the groups of blind chickens were less synchronized in their behaviour, aggregated less and were less stressed in a test of social isolation. The sighted birds had a faster growth than their blind counterparts, with the latter showing more abnormal behaviour, including air-pecking, circle-walking and stargazing, the latter denoting a condition where a young chick cannot hold its neck upright, and the head falls backwards onto the back of the bird. The authors conclude that the welfare of the blinded chickens is compromised, emphasising yet again the importance of vision for poultry.

Different light sources are used in commercial poultry production, including fluorescent and incandescent lighting as well as light-emitting diodes (LEDs) and lights mimicking natural lighting. Overall, no effects on production and welfare measures have been found when comparing fluorescent and incandescent lighting (Lewis and Morris, 1998). Broilers show a preference for cold-white over neutral-white LED lighting, with a slightly faster growth in the former, but with no differences in welfare parameters between the two lighting types (Riber, 2015). Others have found no differences between LED and fluorescent lighting in terms of production parameters and plumage condition, but with a higher activity levels in the LED-illuminated groups (Liu et al., 2018).

Chickens are able to see into the ultraviolet (UV) spectrum, and more than half a century ago studies found that egg production of laying hens improved when they were exposed to UV light (Barrott et al., 1951). When viewed under UV light, the plumage of some species of Galliformes shows markings that are not visible to humans under normal lighting (Sherwin and Devereux, 1999; Mullen and Pohland, 2008). These patches are also present on white strains of domestic chickens (personal observation), and may be involved in individual recognition or when birds peck the plumage of others, although this has never been demonstrated. More recently, the influence of UV light on the behaviour and welfare of domestic fowl has been investigated. Absence of UV wavelengths in the lighting used to raise chickens led to higher levels of basal corticosterone in the blood (Maddocks et al., 2001), indicating that UV-deficient rearing may have some welfare consequences for the birds. James et al. (2018) went the other way and provided extra UV lighting for broiler chickens. Exposure to supplementary UV light improved the feather condition, shortened the duration of tonic immobility and gave rise to better gait scores compared to control. Natural lighting, which contains UV light, has been found to increase activity and improve the gait score of broilers (Bailie et al., 2013). This may be mediated through improved vision leading to more activity in general, although the mechanisms are not fully known. Taken together, these results suggest that natural lighting, including light in the UV spectrum, has welfare benefits for chickens.

Another aspect of lighting is flicker rate, where a 60 kHz light bulb will increase and decrease its luminosity 60 times every second. Humans cannot see this, but as birds, including chickens, often have higher flicker sensitivity, it was thought to be a potential welfare problem in commercial poultry production, if the chickens were constantly exposed to a flickering light. Studies have shown that chickens are unlikely to detect the 100 Hz flicker of fluorescent light (Jarvis et al., 2002; Prescott et al., 2003, 2004), which is the most common light source in broiler and egg-laying facilities. However, more recently, Railton et al. (2009) trained hens to peck one key if two lights were flickering and another key if both lights were steady. The authors found that the hens on average perceived the lights as flickering at around 83 Hz (ranging from 69 Hz to 95 Hz).

The light schedule in itself also has significant importance for the welfare of broiler chickens and laying hens, and the different commercial goals of these two types of birds has led to very different lighting schedules being applied. Laying hens are bred and kept for egg production purposes, and the lighting schedule applied is aimed to maintain reproductive activity while at the same time preventing feather pecking and minimizing energy use. This has given rise to lighting schedules with intermittent or very low lighting (Boshouwers and Nicaise, 1987; Coenen et al., 1988). Broilers on the other hand are raised for fast and efficient meat production, and providing a lot of light can encourage feeding behaviour and promote activity, although periods of darkness are necessary to ensure the leg health of the birds. It should also be kept in mind that broilers and laying hens have similar visual ability, and the lighting schedules used aim to maximize production without compromising the health and welfare of the birds to the extent this is possible. Diurnal rhythms necessary for proper sleep cycles are dependent on periods of darkness, and ability to manoeuvre in three-dimensional space requires sufficient light (Taylor et al., 2003). It has long been known that absence of a dark period affects the development of the eye and that exposure to continuous light will cause severe corneal flattening and hyperopia, that is, difficulty in focussing on objects that are up close (Li et al., 1995; Lewis and Gous, 2009). As little as 4 h of continuous darkness per 24 h is sufficient to prevent this damage (Li et al., 2000). Broilers are able to adapt their behaviour to regular periods of darkness by increasing their feeding activity prior to the onset of the light period (Duve et al., 2011), and Prescott et al. (2003) propose that the minimum duration of the dark period should be 6 h. The presence of a dark period may increase the risk of foot-pad dermatitis in broilers, as the birds are less active giving rise to more wet litter, but the effects appear to be absent or minor within 8 h of darkness daily (Duve et al., 2011; Skrbic et al., 2015). The effects of lighting on the welfare of broilers and laying hens are covered in detail elsewhere in this book.

## 3 Olfaction

Originally, olfaction was thought to be absent or at least very poor in domestic chickens, but we have now known for some time that chickens have a well-developed sense of smell and that they perceive and react to olfactory stimuli (Jones and Roper, 1997; Steiger et al., 2008). The delay in acquiring this information on olfactory capacity of domestic fowl has contributed to odours being included to a lesser extent than other types of sensory stimuli in scientific studies of poultry (Nielsen et al., 2015; Krause et al., 2016). This scarcity is slowly being remedied, and as domestic chickens can be trained to detect specific volatiles in natural environments, they have recently been suggested as potential biological sensors in forensic applications (Prada and Furton, 2018), although the benefits over using dogs remain unclear.

Olfaction is closely associated with the sense of taste, and it is therefore not surprising that domestic chickens, like many other animals, not only forage based on visual stimuli, but also use odours to identify and distinguish feed items. Indeed, a significant reduction in weight gain occurs if chickens are deprived of their sense of smell by blocking their nares (Porter et al., 2002; Tallet et al., 2003). Chickens display neophobia and are thus reluctant to ingest feed that smells differently from what they are used to (Jones, 1987). Disguising novel feed by use of a masking odour can mitigate the problem when changes in diet are necessary (Dixon and Nicol, 2008). Indeed, food preferences can be influenced by different odorants during incubation, in particular the days before hatching (Bertin et al., 2012). Chicks exposed to low concentrations of orange and vanilla odours while still in the egg, eat more of a feed odorized by these odours post-hatch (Bertin et al., 2010). However, when strong odour concentrations are used during incubation, the chicks avoid the odorised feed. We know that chickens use their sense of smell to recognize aversive flavours, as has been demonstrated by conditioning domestic chicks to avoid ingesting bitter-tasting (quinine-flavoured) water when it has been odorized with the smell of orange or almonds but, surprisingly, not when vanilla is used as the odorant (Turro et al., 1994; Roper and Marples, 1997).

Chickens are attracted to the smell of litter from their home pen (Burne and Rogers, 1995), and will choose this over clean litter or litter from a pen of an unknown conspecific (Jones and Gentle, 1985). If exposed to an odour, such as strawberry, while in the egg, chickens will prefer environments in which this odour is present (Porter and Picard, 1998; Sneddon et al., 1998). Familiar odours also provide a degree of stress reduction when placed in a novel environment. There are some indications that broilers are less stressed if they are raised in the presence of an odorant resembling the smell of maternal uropygial gland secretions (Madec et al., 2008). Jones et al. (2002) found that pairs of chickens tested in an open field were more likely to move apart and to feed if they were tested in an arena odorized with the same odorant (vanillin) as used to odorize their home pen. There is also some indication that chickens exposed to stressful handling in the days following hatching have different volatile organic compounds in their droppings than do chicks that have been minimally handled, giving rise to differences in smell (Bombail et al., 2018). Whether chickens can detect these differences and respond to these stress odours have not yet been established.

Smells are also involved in individual recognition of conspecifics. Karlsson et al. (2010) found that individual birds of red junglefowl had their own characteristic body odour, as shown by gas chromatography and mass spectrometry analyses, as well as by discrimination tests performed by mice. Differences in fatty acid composition of secretions from the uropygial gland, also known as preen oil, have been found between hens that had been feather-pecked and those that had not (Sandilands et al., 2004), suggesting that odours might play a role in the predisposition of a bird to become the victim (or not) of feather pecking. However, in the junglefowl study, no consistent odorous characteristics were found to separate pecked and non-pecked birds (Karlsson et al., 2010). However, the odour emanating from preen oil is likely to play a role as a social cue and is important for mate choice. Cockerels are more likely to mount intact or sham-operated females than females that have had their uropygial gland removed, but this preference is not found in anosmic males (Hirao et al., 2009).

Whereas the attraction of chickens to familiar odours in the environment is learned, either pre- or post-hatch, some odours are innately aversive. Chickens have an innate aversion to the smell of blood from conspecifics. Jones and Black (1979) exposed 7-day-old chicks to different odours in the form of liquids presented in a petri-dish for 15 min (Fig. 2) with water used as the control. Compared to water, all of the liquids presented resulted in less time spent near the odour source, but with chicken blood being the most aversive. The inclusion of chicken blood in a sealed dish, and mouse blood in an open dish was the final proof that it was indeed the odour and the species of origin that caused the observed response. It is worth noting that

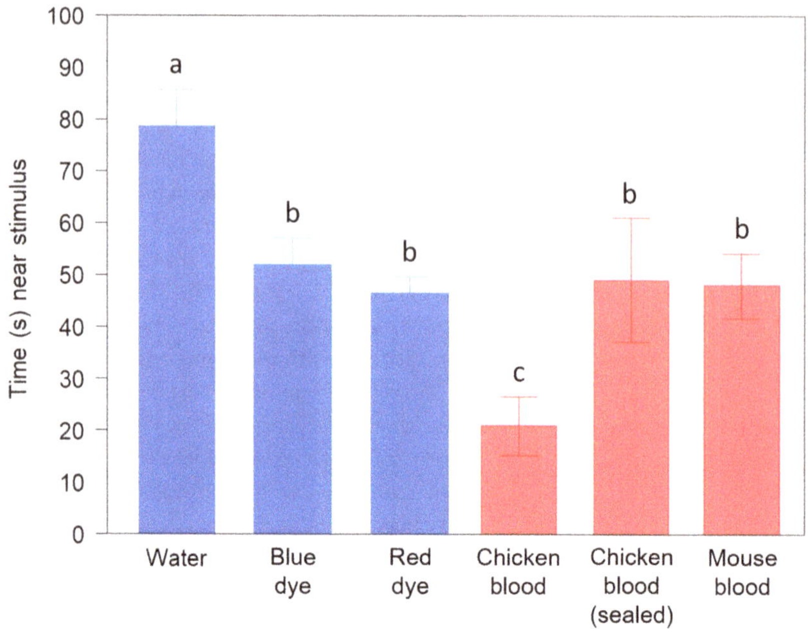


**Figure 2** Response of 7-day-old chicks to various odours presented in a petri dish for 15 min. The graph shows the average amount of time in seconds (± s.e.) that the birds spend within 75 mm of the stimulus dish. Bars with different superscripts differ significantly ($P$ <0.05). Adapted from Jones and Black (1979).

hens have been found to peck on blood-filled containers and eat the blood if they have seen other birds do this (Cloutier et al., 2002). Perhaps the aversion to the smell of blood diminishes with age? Alternatively, as shown earlier, the colour red is attractive to chickens in most situations, so consumption of blood associated with cannibalism may be initiated by an attraction to the colour red, with the smell being masked by other odours when outside an experimental setting.

Innate responses to odorants are most often seen in connection with the smell of predators (Nielsen, 2017). However, only few studies have been carried out to investigate the response of domestic chickens to odours from animals that prey on birds. Fluck et al. (1996) found that 7-day-old chicks responded with aversive behaviour when exposed to cat odour as compared to when a neutral odour was used. This effect was not seen at 4 days of age, because the chicks responded aversively in general to the test situation, nor was it found at 10 days of age. Others have tested domestic strains similar to junglefowl and found the birds being more vigilant and search for food less in the presence of faeces from tigers and wild dogs as compared to when faeces from elephants and antelopes were used (Zidar and Løvlie, 2012).

Noxious or aversive odours in the ambient environment are likely to affect the behaviour and welfare of domestic chickens. Day-old chicks respond with fewer pecks and more headshaking when exposed to increasing concentrations of various odorants, including isoamyl acetate (banana odour), allyl sulfide (garlic odour) and eugenol (clove-like odour; Burne and Rogers, 1996). They are thus able to detect these smells and find them aversive at high concentrations. Another compound, methyl anthranilate, which has a fruity grape odour, is used as a bird repellent, and domestic chicks are able to detect this smell (Marples and Roper, 1997). The most studied volatiles in this context are ammonia and carbon dioxide because they are, respectively, an aerial pollutant of the rearing environment and used for stunning poultry prior to slaughter. Exposure to ammonia in poultry may cause irritation of mucous membranes and give rise to respiratory diseases (Kristensen and Wathes, 2000). Kristensen et al. (2000) found that laying hens preferred to spend time in air containing 0 ppm compared to 25 and 45 ppm of ammonia. Jones et al. (2005) found that broilers would avoid environments with aerial ammonia concentrations above 10 ppm, independent of their prior experience with exposure to ammonia, and that the preference for lower ammonia levels (4 ppm) was stronger during the period when lights were dimmed and the birds rested. For carbon dioxide, domestic chickens can detect it at around 10% dilution, but show only mild-to-moderate aversion to this compound (McKeegan et al., 2006).

## 4 Taste

Gustation, or the sense of taste, is important for foraging and feeding, also in chickens. It is closely linked with the sense of olfaction, and what humans perceive as flavour when eating something is a combination of the smell and taste of the food. Like humans, chickens are able to taste acid, bitter, salt and sweet, as well as umami (Yoshida et al., 2018), but whereas we have roughly 10 000 taste buds, chickens have around 300, and these are primarily placed in the palate (Rajapaksha et al., 2016). Expression of taste receptor genes for bitter, sweet and umami in the gastrointestinal tract of chickens suggests that these are involved in the sensing of nutrient compositions, such as carbohydrates and amino acids (Cheled-Shoval et al., 2015).

Despite chickens having a well-developed sense of taste, they may rely on gustation much less than do mammals when foraging. It has been suggested that the primary function of taste in chickens is to enable the birds to reject foul-tasting (no pun intended) or noxious items (Appleby et al., 2004). Indeed, a chicken upon tasting a bitter food item will readily perform bill-wiping behaviour, where each side of the beak is vigorously wiped along the ground, indicating a certain degree of unpleasantness associated with the taste experience. This visual disgust reaction can lead to social learning of the aversion, at least in young chicks (Johnston et al., 1998), and such dislike of bitter taste can be exploited, as we shall see in the following example.

Chickens have three bitter taste receptors of which two (cT2R1 and cT2R7) are functional (Dey et al., 2017). Harlander-Matauschek and her colleagues (2009, 2010) wondered if aversiveness to a bitter taste could be used to reduce feather pecking in laying hens. The researchers housed hens in 12 groups of 10 hens each. In six of the groups, the hens came from a line selected for high feather-pecking behaviour, and the other six groups were composed of hens from a low feather-pecking line. Within each selection line, hens in three of the groups had their body plumage sprayed with a 4% quinine solution to give the feathers a bitter taste. The remaining groups were not sprayed, but the hens were handled in the same way as the quinine-treated birds.

The results showed a significant reduction in severe feather pecking in both feather-pecking lines for two weeks following the quinine treatment (Fig. 3a), indicating that the birds exposed to quinine-treated feathers learned that these were not attractive to peck, and that this learned behaviour persisted for at least two weeks (Harlander-Matauschek et al., 2009). In a subsequent experiment using a similar set-up, the birds were sprayed weekly from hatch until 25 weeks of age. Independent of the genetic line, severe feather pecking was reduced in the early lay period (18–25 weeks of age) while quinine spraying was still carried out (Fig. 3b). Unfortunately, the effect did not persist beyond the early laying period, as no differences in feather pecking were seen when the hens were no longer treated with quinine (Harlander-Matauschek et al., 2010).

One of the more intriguing aspects of the sense of taste in chickens is their inability to detect capsaicin, the component that makes chili peppers spicy hot (Marin-Burgin et al., 2000). This can be used to prevent other capsaicin-sensitive animals, such as rats, from eating feed intended for chickens. The compound has also been investigated for its capacity to reduce the effects of salmonella infections in chickens. Tellez et al. (1993) fed dietary capsaicin (18 ppm) to Leghorn chicks for the first 14 days after hatch, and challenged these and a control group with *Salmonella enteritidis* on day 13. The researchers found a significant reduction in the pH level inside the intestines of the chicks fed capsaicin, as well as a reduction in the number of *S. enteritidis* positive chicks compared to the control group.

Although little to do with the ability of chickens to taste, there has been attempts to change the flavour of eggs by feeding laying hens different forms of spices. When fed garlic, fennel, peppermint and marjoram, the eggs obtain a well-seasoned flavour according to a qualified taste panel (Richter et al., 2002). Other less-effectual spices tested were sage, caraway, thyme, paprika, cardamom and a spice referred to as sausage-mix. The authors question the economic feasibility and even the market for producing taste-manipulated eggs.

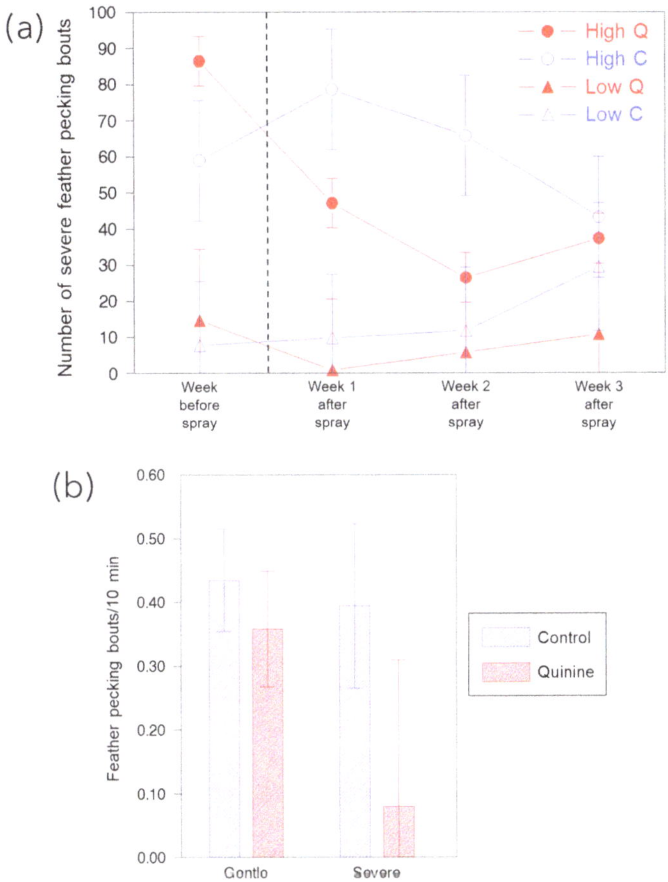


**Figure 3** (a) Number of severe feather-pecking bouts (± s.e.) registered during 8 periods of 10 min in each week before and after treatment with quinine (Q) compared to a non-treated control (C) for two lines of laying hens selected for high and low feather-pecking behaviour, respectively (adapted from Harlander-Matauschek et al., 2009); (b) Gentle and severe feather pecking bouts (per 10 min; ± s.e.) across selection lines during the period of early lay (18–25 weeks) in hens sprayed weekly with quinine or not (control) from hatch until 25 weeks of age (adapted from Harlander-Matauschek et al., 2010).

## 5 Hearing

Hearing in most animal species is closely associated with vocalization, and for chickens this is no exception. They are highly vocal animals, and depending on their age and sex they can produce around 30 different identifiable types of vocalization (Collias and Joos, 1953). These include chirps produced by small chicks, clucks emitted by broody hens to attract her offspring, calls made by hens at the point of lay and the crowing of the cockerels. Vocalizations by chickens serve as a means of social communication, and transfer information about danger, food or social status to conspecifics. If we can accurately measure and interpret (Collias, 1987) these vocalizations in a commercial poultry-production setting, monitoring of chicken sounds have the potential to be a way to assess the motivational state and welfare of the birds (McGrath et al., 2017). This could include detection of stress vocalisations or a more general monitoring of vocal changes caused by respiratory infections (Sufka et al., 2006; Mcloughlin et al., 2019) as discussed elsewhere in this book.

Chicken vocalizations have a large range of frequencies, with some being very low, equivalent to the second lowest octave on the piano (Collias and Joos, 1953). This also implies that chickens are able to hear these vocalizations. Unlike humans, chickens have no external ear but, like other birds, chickens are both able to regrow and move the vibration-sensitive hairs in the inner ear (Birkhead, 2012), which humans are unable to do. Human hearing is in the range from 20 Hz to 20 kHz, and some find that chickens can detect sounds below the human hearing range of 20 Hz (Warchol and Dallos, 1989). The range of hearing is likely to reflect the biological relevance of surrounding sounds. Naïve 3-day-old chicks show an innate attraction to sounds that are discrete (as opposed to continuous) and pulsating fast, thus resembling the characteristic clucking of the mother hen (De Tommaso et al., 2019). Figure 4 shows the hearing sensitivity of humans and chickens for frequencies from 200 Hz to 8000 Hz.

The importance of hearing is perhaps underestimated in chickens, and the noisy environment in which they are sometimes raised is likely to impair the social communication described above. Broiler chickens have been found to vocalize

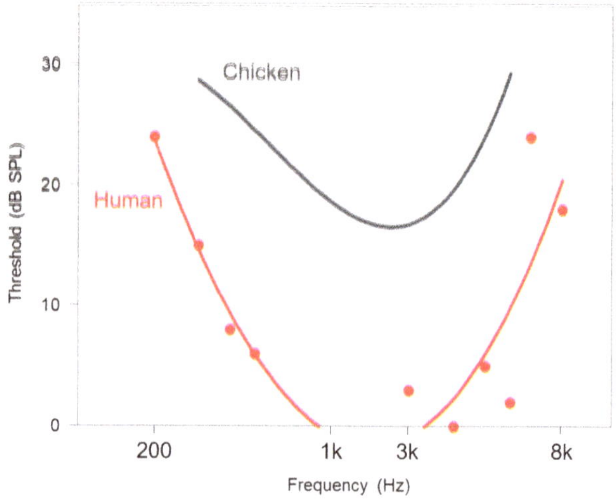


**Figure 4** Approximate hearing sensitivity threshold (dB SPL) as a function of frequency (Hz) for humans and chickens. Adapted from Hogan and Turner (1998) and Gleich and Langemann (2011).

louder when the background noise increases (Brumm et al., 2009). In commercial poultry production, noise intensity can at times be high as a result of sounds from feed trucks, ventilation fans and other equipment (MAFBC, 1999), and this may be aversive to the birds over and above impairing interspecific communication. A 10-min exposure to noise at 80 dB and 100 dB leads to a significant increase in plasma corticosterone level in broilers (Chloupek et al., 2009). One aspect of hearing includes the rapid reaction to certain auditory disturbances. Accidentally dropping a spanner on the concrete floor of a chicken shed will lead to the immediate cessation of all bird-produced noises as the flock listens intently for a few seconds before resuming their previous behavioural undertaking.

## 6 Touch

The beak forms an important part in the behaviour of chickens involving tactile sensory perception. The birds use their beaks to forage and feed as well as for preening. They also use their beaks when hatching, and pecking is among the first activities performed after hatch (Picard et al., 1997). The tip of the beak has a large number of mechanosensory neurons, used for touch and temperature perception. Chickens are able to detect the physical structure of feed and can distinguish between hard and soft pellets (Bouvarel et al., 2009). During preening, the beak is used to transfer secretions from the uropygial gland onto the feathers.

The importance of the beak for tactile perception in the chicken is thus well-known. When beak-trimming is performed, usually on day-old chicks, to inhibit feather pecking in laying hens, this severely impairs these sensory inputs from the physical environment. Indeed, in the days following beak trimming, the chicks perform less beak-related behaviour than do chicks with entire beaks (Gentle et al., 1997), and this is even more profound if the beak-trimming is done when the birds are adult. In the latter case, neuromas can form around the scar tissue (Gentle and Wilson, 2004), presumably associated with pain when the hen uses its beak (Cheng, 2006).

Another point of contact with the physical environment is via the legs and feet. Typical foraging behaviour of chickens involve scraping the surface of the ground, and stepping backwards to investigate if any potential feed items have been uncovered. The feet are also used when perching during night-time resting. If the litter surface is wet or sticky, this can result in foot-pad dermatitis, which in the more severe cases is likely to be associated with pain when using the feet during foraging or resting (Sinclair et al., 2015). Another example of nociceptive tactile input is during shackling when broilers are hung by the legs at the abattoir (Gentle and Wilson, 2004).

The plumage and skin of chickens also form part of the tactile sensory perception. Specialized feathers, such as bristles and filoplumes, have fewer and more spaced barbs than the commonly known type of feathers (Alibardi, 2017). Via the sense of touch these aid in the perception of the physical environment and the position of other feathers (Appleby et al., 2004). Like humans, the skin of the chicken can detect thermal and tactile stimuli. The pulling out of feathers that happens during severe feather pecking is painful.

## 7 Welfare implications

Before highlighting the welfare aspects of the different chicken senses covered in previous sections, it should be mentioned that the domestic chicken appears to have a sixth sense in the form of magnetoreception, that is, the ability to detect the direction of a magnetic field. Although chicks of a layer strain can be trained to use magnetic cues to find a hidden object, this is highly dependent on the strain of chicken tested (Freire et al., 2008; Denzau et al., 2013b) as well as age at training and the training method employed (Denzau et al., 2013a,b). For this reason, the magnetic sense of chickens is unlikely to play a role in their behaviour and welfare in a commercial setting, at least until we know more about their ability to detect magnetic fields.

From the research carried out on the five main chicken senses, it is clear that these are pertinent for most of the behavioural repertoire of the birds. We know that vision is important, with different eyes dedicated to different tasks. Chickens can see into the ultraviolet, a spectrum that may provide some welfare benefits, and they suffer eye damage if no dark periods are provided. Indeed, schedule, type and intensity of lighting are important for the welfare of broilers and layers in terms of rest and visibility. Chickens have a preference for red colours, but this preference is context specific, and they find the smell of chicken blood aversive, at least when young. Olfaction is likely to be much more important for domestic chickens than previously thought (McKeegan, 2004; Krause et al., 2016) and familiar odours may be used to reduce stress associated with novelty. Smells are also important for individual recognition of other flock members, and in a commercial broiler or layer house, the aerial ammonia concentrations may impair the olfactory ability of the birds (Anderson et al., 1964).

Taste, on the other hand, is thought to be used primarily to avoid noxious food items. This can be exploited to prevent or at least delay feather pecking in the same way that foul-tasting substances are used to reduce nail biting in humans. And just like important smells can be masked by odorant pollution, so can important sounds be drowned in noise: vocalisation is important for social communication in poultry and noisy environments may impair the ability to hear sounds made by conspecifics. The beak is an essential organ for tactile stimulation and sensing of the environment, in particular when foraging and preening. The beak of a hen has been compared to the human hand in terms of tactile importance, and the widespread use of beak trimming to prevent feather pecking in some layer strains and some egg-production systems can only be detrimental for the sensory input of the trimmed birds.

Overall, the sensory inputs of chickens are as varied, if not more, as those of humans, and are of immense importance for the way the birds perceive and hence respond to their environment. This should be kept in mind when we house and handle domestic poultry, as companion animals and for scientific and production purposes.

## 8 Where to look for further information

The following books, chapters and articles provide a good overview of the senses of birds, with particular emphasis on chickens:

- Birkhead, T. 2012. *Bird Sense - What It's Like to be a Bird*. Bloomsbury Publishing, London, 288 pp. ISBN: 9781408820131.
- Krause, E. T., Schrader, L. and Caspers, B. A. 2016, Olfaction in chicken (*Gallus gallus*): a neglected mode of social communication?. *Front. Ecol. Evol.* 4, 94. doi: 10.3389/fevo.2016.00094.
- McKeegan, D. E. F. 2004. Sensory perception: chemoreception, in G. Perry (Ed.), Welfare of the Laying Hen. *Poult. Sci. Symp. Series* 27, 139-53. doi: 10.1079/9780851998138.0139.
- Nicol, C. 2015. Sensory biology, Chapter 2 in *The Behavioural Biology of Chickens'*. CABI Publishing, Wallingford, UK, pp. 15-34.
- Prescott, N. B., Wathes, C. M. and Jarvis, J. R. 2003. Light, vision and the welfare of poultry. *Anim. Welf.* 12(2), 269-88.

## 9 References

Alibardi, L. 2017. Review: cornification, morphogenesis and evolution of feathers. *Protoplasma* 254(3), 1259-81. doi:10.1007/s00709-016-1019-2.

Anderson, D. P., Beard, C. W. and Hanson, R. P. 1964. The Adverse effects of ammonia on chickens including resistance to infection with Newcastle Disease virus. *Avian Dis.* 8(3), 369-79. doi:10.2307/1587967.

Appleby, M. C., Mench, J. A. and Hughes, B. O. 2004. *Poultry Behaviour and Welfare*. CABI Publishing, Wallingford, 276pp.

Bailie, C. L., Ball, M. E. E. and O'Connell, N. E. 2013. Influence of the provision of natural light and straw bales on activity levels and leg health in commercial broiler chickens. *Animal* 7(4), 618-26. doi:10.1017/S1751731112002108.

Barrott, H. G., Schoenleber, L. G. and Campbell, L. E. 1951. The effect of ultraviolet radiation on egg production of hens. *Poult. Sci.* 30(3), 409-16. doi:10.3382/ps.0300409.

Bertin, A., Calandreau, L., Arnould, C., Nowak, R., Levy, F., Noirot, V., Bouvarel, I. and Leterrier, C. 2010. In ovo olfactory experience influences post-hatch feeding behaviour in young chickens. *Ethology* 116(11), 1027–37. doi:10.1111/j.1439-0310.2010.01820.x.

Bertin, A., Calandreau, L., Arnould, C. and Lévy, F. 2012. The developmental stage of chicken embryos modulates the impact of *in ovo* olfactory stimulation on food preferences. *Chem. Senses* 37(3), 253–61. doi:10.1093/chemse/bjr098.

Birkhead, T. 2012. *Bird Sense – What It's Like to Be a Bird*. Bloomsbury Publishing, London, 288pp. ISBN: 9781408820131.

Bombail, V., Barret, B., Raynaud, A., Jerôme, N., Saint-Albin, A., Ridder, C., Collin, A., Leterrier, C., Guilloteau, L. A. and Nielsen, B. L. 2018. In search of stress odours across species: behavioural responses of rats to faeces from chickens and rats subjected to various types of stressful events. *Appl. Anim. Behav. Sci.* 205, 216–26. doi:10.1016/j.applanim.2017.10.013.

Boshouwers, F. M. G. and Nicaise, E. 1987. Physical activity and energy expenditure of laying hens as affected by light intensity. *Br. Poult. Sci.* 28(1), 155–63. doi:10.1080/00071668708416947.

Bouvarel, I., Chagneau, A. M., Lecuelle, S., Lescoat, P., Ferreira, G., Duvaux-Ponter, C. and Leterrier, C. 2009. Feed composition and hardness interact in preference and intake in chickens. *Appl. Anim. Behav. Sci.* 118(1–2), 62–8. doi:10.1016/j.applanim.2009.01.008.

Brumm, H., Schmidt, R. and Schrader, L. 2009. Noise-dependent vocal plasticity in domestic fowl. *Anim. Behav.* 78(3), 741–6. doi:10.1016/j.anbehav.2009.07.004.

Burne, T. H. J. and Rogers, L. J. 1996. Responses to odorants by the domestic chick. *Physiol. Behav.* 60(6), 1441–7. doi:10.1016/s0031-9384(96)00300-9.

Burne, T. H. J. and Rogers, L. J. 1995. Odors, volatiles and approach-avoidance behavior of the domestic chick (*Gallus gallus domesticus*). *Int. J. Comp. Psychol.* 8, 99–114.

Cheled-Shoval, S. L., Druyan, S. and Uni, Z. 2015. Bitter, sweet and umami taste receptors and downstream signaling effectors: expression in embryonic and growing chicken gastrointestinal tract. *Poult. Sci.* 94(8), 1928–41. doi:10.3382/ps/pev152.

Cheng, H. 2006. Morphopathological changes and pain in beak trimmed laying hens. *World Poult. Sci. J.* 62(1), 41–52. doi:10.1079/WPS200583.

Chloupek, P., Voslářová, E., Chloupek, J., Bedáňová, I., Pištěková, V. and Večerek, V. 2009. Stress in broiler chickens due to acute noise exposure. *Acta Vet. Brno* 78(1), 93–8. doi:10.2754/avb200978010093.

Cloutier, S., Newberry, R. C., Honda, K. and Alldredge, J. R. 2002. Cannibalistic behaviour spread by social learning. *Anim. Behav.* 63(6), 1153–62. doi:10.1006/anbe.2002.3017.

Coenen, A. M. L., Wolters, E. M. T. J., van Luijtelaar, E. L. J. M. and Blokhuis, H. J. 1988. Effects of intermittent lighting on sleep and activity in the domestic hen. *Appl. Anim. Behav. Sci.* 20(3–4), 309–18. doi:10.1016/0168-1591(88)90055-X.

Collias, N. and Joos, M. 1953. The spectrographic analysis of sound signals of the domestic fowl. *Behaviour* 5(1), 175–88. doi:10.1163/156853953X00104.

Collias, N. E. 1987. The vocal repertoire of the red junglefowl: a spectrographic classification and the code of communication. *Condor* 89(3), 510–24. doi:10.2307/1368641.

Collins, S., Forkman, B., Kristensen, H. H., Sandøe, P. and Hocking, P. M. 2011. Investigating the importance of vision in poultry: comparing the behaviour of blind and sighted chickens. *Appl. Anim. Behav. Sci.* 133(1–2), 60–9. doi:10.1016/j.applanim.2011.04.013.

D'Eath, R. B. and Keeling, L. J. 2003. Social discrimination and aggression by laying hens in large groups: from peck orders to social tolerance. *Appl. Anim. Behav. Sci.* 84(3), 197-212. doi:10.1016/j.applanim.2003.08.010.

Denzau, S., Niessner, C., Rogers, L. J. and Wiltschko, W. 2013a. Ontogenetic development of magnetic compass orientation in domestic chickens (*Gallus gallus*). *J. Exp. Biol.* 216(16), 3143-7. doi:10.1242/jeb.088815.

Denzau, S., Niessner, C., Wiltschko, R. and Wiltschko, W. 2013b. Different responses of two strains of chickens to different training procedures for magnetic directions. *Anim. Cogn.* 16(3), 395-403. doi:10.1007/s10071-012-0580-0.

De Tommaso, M., Kaplan, G., Chiandetti, C. and Vallortigara, G. 2019. Naive 3-day-old domestic chicks (*Gallus gallus*) are attracted to discrete acoustic patterns characterizing natural vocalizations. *J. Comp. Psychol.* 133(1), 118-31. doi:10.1037/com0000132.

Dey, B., Kawabata, F., Kawabata, Y., Yoshida, Y., Nishimura, S. and Tabata, S. 2017. Identification of functional bitter taste receptors and their antagonist in chickens. *Biochem. Biophys. Res. Commun.* 482(4), 693-9. doi:10.1016/j.bbrc.2016.11.096.

Dixon, G. and Nicol, C. J. 2008. The effect of diet change on the behaviour of layer pullets. *Anim. Welf.* 17(2), 101-9.

Duve, L. R., Steenfeldt, S., Thodberg, K. and Nielsen, B. L. 2011. Splitting the scotoperiod: effects on feeding behaviour, intestinal fill, and digestive transit time in broiler chickens. *Br. Poult. Sci.* 52(1), 1-10. doi:10.1080/00071668.2010.549671.

Fluck, E., Hogg, S., Mabbutt, P. S. and File, S. E. 1996. Behavioural and neurochemical responses of male and female chicks to cat odour. *Pharmacol. Biochem. Behav.* 54(1), 85-91. doi:10.1016/0091-3057(95)02170-1.

Freire, R., Munro, U., Rogers, L. J., Sagasser, S., Wiltschko, R. and Wiltschko, W. 2008. Different responses in two strains of chickens (*Gallus gallus*) in a magnetic orientation test. *Anim. Cogn.* 11(3), 547-52. doi:10.1007/s10071-008-0147-2.

Gamberale-Stille, G. and Tullberg, B. S. 2001. Fruit or aposematic insect? Context-dependent colour preferences in domestic chicks. *Proc. Biol. Sci.* 268(1485), 2525-9. doi:10.1098/rspb.2001.1814.

Gentle, M. and Wilson, S. 2004. Pain and the laying hen. In: Perry, G. (Ed.), *Welfare of the Laying Hen, Papers from the 27th Poultry Science Symposium of the World's Poultry Science Association, Bristol*, pp. 165-75 doi:10.1079/9780851998138.0165.

Gentle, M. J., Hughes, B. O., Fox, A. and Waddington, D. 1997. Behavioural and anatomical consequences of two beak trimming methods in 1- and 10-d-old domestic chicks. *Br. Poult. Sci.* 38(5), 453-63. doi:10.1080/00071669708418022.

Gleich, O. and Langemann, U. 2011. Auditory capabilities of birds in relation to the structural diversity of the basilar papilla. *Hear. Res.* 273(1-2), 80-8. doi:10.1016/j.heares.2010.01.009.

Ham, A. D. and Osorio, D. 2007. Colour preferences and colour vision in poultry chicks. *Proc. Biol. Sci.* 274(1621), 1941-8. doi:10.1098/rspb.2007.0538.

Harlander-Matauschek, A., Beck, P. and Piepho, H. P. 2009. Taste aversion learning to eliminate feather pecking in laying hens, *Gallus gallus domesticus*. *Anim. Behav.* 78(2), 485-90. doi:10.1016/j.anbehav.2009.05.020.

Harlander-Matauschek, A., Beck, P. and Rodenburg, T. B. 2010. Effect of an early bitter taste experience on subsequent feather-pecking behaviour in laying hens. *Appl. Anim. Behav. Sci.* 127(3-4), 108-14. doi:10.1016/j.applanim.2010.09.005.

Hirao, A., Aoyama, M. and Sugita, S. 2009. The role of uropygial gland on sexual behavior in domestic chicken *Gallus gallus domesticus*. *Behav. Processes* 80(2), 115–20. doi:10.1016/j.beproc.2008.10.006.

Hogan, C. A. and Turner, C. W. 1998. High-frequency audibility: benefits for hearing-impaired listeners. *J. Acoust. Soc. Am.* 104(1), 432–41. doi:10.1121/1.423247.

James, C., Asher, L., Herborn, K. and Wiseman, J. 2018. The effect of supplementary ultraviolet wavelengths on broiler chicken welfare indicators. *Appl. Anim. Behav. Sci.* 209, 55–64. doi:10.1016/j.applanim.2018.10.002.

Jarvis, J. R., Taylor, N. R., Prescott, N. B., Meeks, I. and Wathes, C. M. 2002. Measuring and modelling the photopic flicker sensitivity of the chicken (*Gallus G. domesticus*). *Vision Res.* 42(1), 99–106. doi:10.1016/s0042-6989(01)00268-1.

Johnston, A. N. B., Burne, T. H. J. and Rose, S. P. R. 1998. Observation learning in day-old chicks using a one-trial passive avoidance learning paradigm. *Anim. Behav.* 56(6), 1347–53. doi:10.1006/anbe.1998.0901.

Jones, C. D., Osorio, D. and Baddeley, R. J. 2001. Colour categorization by domestic chicks. *Proc. Biol. Sci.* 268(1481), 2077–84. doi:10.1098/rspb.2001.1734.

Jones, E. K. M., Wathes, C. M. and Webster, A. J. F. 2005. Avoidance of atmospheric ammonia by domestic fowl and the effect of early experience. *Appl. Anim. Behav. Sci.* 90(3–4), 293–308. doi:10.1016/j.applanim.2004.08.009.

Jones, R. B. 1987. Food neophobia and olfaction in domestic chicks. *Bird Behav.* 7(2), 78–81. doi:10.3727/015613887791918051.

Jones, R. B. and Black, A. J. 1979. Behavioral responses of the domestic chick to blood. *Behav. Neural Biol.* 27(3), 319–29. doi:10.1016/S0163-1047(79)92384-7.

Jones, R. B. and Gentle, M. J. 1985. Olfaction and behavioral modification in domestic chicks (*Gallus domesticus*). *Physiol. Behav.* 34(6), 917–24. doi:10.1016/0031-9384(85)90014-9.

Jones, R. B. and Roper, T. J. 1997. Olfaction in the domestic fowl: a critical review. *Physiol. Behav.* 62(5), 1009–18. doi:10.1016/s0031-9384(97)00207-2.

Jones, R. B., Facchin, L. and McCorquodale, C. 2002. Social dispersal by domestic chicks in a novel environment: reassuring properties of a familiar odourant. *Anim. Behav.* 63(4), 659–66. doi:10.1006/anbe.2001.1943.

Karlsson, A. C., Jensen, P., Elgland, M., Laur, K., Fyrner, T., Konradsson, P. and Laska, M. 2010. Red junglefowl have individual body odors. *J. Exp. Biol.* 213(10), 1619–24. doi:10.1242/jeb.040279.

Krause, E. T., Schrader, L. and Caspers, B. A. 2016. Olfaction in chicken (*Gallus gallus*): a neglected mode of social communication? *Front. Ecol. Evol.* 4, 94. doi:10.3389/fevo.2016.00094.

Kristensen, H. H. and Wathes, C. M. 2000. Ammonia and poultry welfare: a review. *World Poult. Sci. J.* 56(3), 235–45. doi:10.1079/WPS20000018.

Kristensen, H. H., Burgess, L. R., Demmers, T. G. H. and Wathes, C. M. 2000. The preferences of laying hens for different concentrations of atmospheric ammonia. *Appl. Anim. Behav. Sci.* 68(4), 307–18. doi:10.1016/S0168-1591(00)00110-6.

LeBlanc, S., Tobalske, B., Quinton, M., Springthorpe, D., Szkotnicki, B., Wuerbel, H. and Harlander-Matauschek, A. 2016. Physical health problems and environmental challenges influence balancing behaviour in laying hens. *PLoS ONE* 11(4), e0153477. doi:10.1371/journal.pone.0153477.

Lewis, P. D. and Morris, T. R. 1998. Responses of domestic poultry to various light sources. *Worlds Poult. Sci. J.* 54(1), 7–25. doi:10.1079/WPS19980002.

Lewis, P. D. and Gous, R. M. 2009. Photoperiodic responses of broilers. II. Ocular development. *Br. Poult. Sci.* 50(6), 667–72. doi:10.1080/00071660903342066.

Li, T., Troilo, D., Glasser, A. and Howland, H. C. 1995. Constant light produces severe corneal flattening and hyperopia in chickens. *Vision Res.* 35(9), 1203–9. doi:10.1016/0042-6989(94)00231-a.

Li, T., Howland, H. C. and Troilo, D. 2000. Diurnal illumination patterns affect the development of the chick eye. *Vision Res.* 40(18), 2387–93. doi:10.1016/s0042-6989(00)00098-5.

Liu, K., Xin, H. and Settar, P. 2018. Effects of light-emitting diode light v. fluorescent light on growing performance, activity levels and well-being of non-beak-trimmed W-36 pullets. *Animal* 12(1), 106–15. doi:10.1017/S1751731117001240.

Maddocks, S. A., Cuthill, I. C., Goldsmith, A. R. and Sherwin, C. M. 2001. Behavioural and physiological effects of absence of ultraviolet wavelengths for domestic chicks. *Anim. Behav.* 62(5), 1013–9. doi:10.1006/anbe.2001.1842.

Madec, I., Gabarrou, J. F., Saffray, D. and Pageat, P. 2008. Broilers (*Gallus gallus*) are less stressed if they can smell a mother odorant. *S. Afr. J. Anim. Sci.* 38(3), 201–6.

MAFBC 1999. *'Management of Noise on Poultry Farms'*, Poultry Factsheet from Ministry of Agriculture and Food, British Columbia, Canada. 4 pp.

Marin-Burgin, A., Reppenhagen, S., Klusch, A., Wendland, J. R. and Petersen, M. 2000. Low-threshold heat response antagonized by capsazepine in chick sensory neurons, which are capsaicin-insensitive. *Eur. J. Neurosci.* 12(10), 3560–6. doi:10.1046/j.1460-9568.2000.00275.x.

Marples, N. M. and Roper, T. J. 1997. Response of domestic chicks to methyl anthranilate odour. *Anim. Behav.* 53(6), 1263–70. doi:10.1006/anbe.1996.0433.

McGrath, N., Dunlop, R., Dwyer, C., Burman, O. and Phillips, C. J. C. 2017. Hens vary their vocal repertoire and structure when anticipating different types of reward. *Anim. Behav.* 130, 79–96. doi:10.1016/j.anbehav.2017.05.025.

McKeegan, D. E. F. 2004. Sensory perception: chemoreception. In: Perry, G. (Ed.), *Welfare of the Laying Hen, Papers from the 27th Poultry Science Symposium of the World's Poultry Science Association, Bristol*, pp. 139–53. doi:10.1079/9780851998138.0139.

McKeegan, D. E. F., McIntyre, J., Demmers, T. G. M., Wathes, C. M. and Jones, R. B. 2006. Behavioural responses of broiler chickens during acute exposure to gaseous stimulation. *Appl. Anim. Behav. Sci.* 99(3–4), 271–86. doi:10.1016/j.applanim.2005.11.002.

Mcloughlin, M. P., Stewart, R. and McElligott, A. G. 2019. Automated bioacoustics: methods in ecology and conservation and their potential for animal welfare monitoring. *J. R. Soc. Interface* 16(155), 20190225. doi:10.1098/rsif.2019.0225.

Mench, J. A. and Andrew, R. J. 1986. Lateralization of a food search task in the domestic chick. *Behav. Neural Biol.* 46(2), 107–14. doi:10.1016/s0163-1047(86)90570-4.

Mullen, P. and Pohland, G. 2008. Studies on UV reflection in feathers of some 1000 bird species: are UV peaks in feathers correlated with violet-sensitive and ultraviolet-sensitive cones? *Ibis* 150(1), 59–68. doi:10.1111/j.1474-919X.2007.00736.x.

Necker, R. 2007. Head-bobbing of walking birds. *J. Comp. Physiol. A Neuroethol. Sens. Neural. Behav. Physiol.* 193(12), 1177–83. doi:10.1007/s00359-007-0281-3.

Nicol, C. 2015, Chapter 2. Sensory biology. In: *The Behavioural Biology of Chickens*. CABI Publishing, Wallingford, pp. 15–34.

Nielsen, B. L., Jezierski, T., Bolhuis, J. E., Amo, L., Rosell, F., Oostindjer, M., Christensen, J. W., McKeegan, D., Wells, D. L. and Hepper, P. 2015. Olfaction: an overlooked sensory modality in applied ethology and animal welfare. *Front. Vet. Sci.* 2, 69. doi:10.3389/fvets.2015.00069.

Nielsen, B. L. 2017. Innateness and learning in olfactory behaviour and odour perception. In: Nielsen, B. L. (Ed.), *Olfaction in Animal Behaviour and Welfare*. CABI Publishing, Wallingford, pp. 16-25.

Nielsen, B. L. 2018. Making sense of it all: the importance of taking into account the sensory abilities of animals in their housing and management. *Appl. Anim. Behav. Sci.* 205, 175-80. doi:10.1016/j.applanim.2018.04.013.

Picard, M., Melcion, J. P., Bouchot, C. and Faure, J. M. 1997. Pecking and prehension of feed particles in domestic fowls. *Prod. Anim.* 10(5), 403-14.

Porter, R. H. and Picard, M. 1998. Effects of early odor exposure in domestic chicks. *Reprod. Nutr. Dev.* 38(4), 441-8. doi:10.1051/rnd:19980408.

Porter, R. H., Picard, M., Arnould, C. and Tallet, C. 2002. Chemosensory deficits are associated with weight gain in newly hatched chicks. *Anim. Res.* 51(4), 337-45. doi:10.1051/animres:2002024.

Prada, P. A. and Furton, K. G. 2018. Birds and dogs: toward a comparative perspective on odor use and detection. *Front. Vet. Sci.* 5, 188. doi:10.3389/fvets.2018.00188.

Prescott, N. B., Wathes, C. M. and Jarvis, J. R. 2003. Light, vision and the welfare of poultry. *Anim. Welf.* 12(2), 269-88.

Prescott, N. B., Jarvis, J. R. and Wathes, C. M. 2004. Vision in the laying hen. In: Perry, G. (Ed.), *Welfare of the Laying Hen, Papers from the 27th Poultry Science Symposium of the World's Poultry Science Association, Bristol*, pp. 155-64. doi:10.1079/9780851998138.0155.

Railton, R. C. R., Foster, T. M. and Temple, W. 2009. A comparison of two methods for assessing critical flicker fusion frequency in hens. *Behav. Processes* 80(2), 196-200. doi:10.1016/j.beproc.2008.11.016.

Rajapaksha, P., Wang, Z., Venkatesan, N., Tehrani, K. F., Payne, J., Swetenburg, R. L., Kawabata, F., Tabata, S., Mortensen, L. J., Stice, S. L., Beckstead, R. and Liu, H. X. 2016. Labeling and analysis of chicken taste buds using molecular markers in oral epithelial sheets. *Sci. Rep.* 6, 37247. doi:10.1038/srep37247.

Rashid, N. and Andrew, R. J. 1989. Right hemisphere advantage for topographical orientation in the domestic chicken. *Neuropsychologia* 27(7), 937-48. doi:10.1016/0028-3932(89)90069-9.

Riber, A. B. 2015. Effects of color of light on preferences, performance, and welfare in broilers. *Poult. Sci.* 94(8), 1767-75. doi:10.3382/ps/pev174.

Richter, T., Braun, P. and Fehlhaber, K. 2002. Influence of spiced feed additives on taste of hen's eggs. *Berl. Munch. Tierarztl. Wochenschr.* 115(5-6), 200-2.

Rogers, L. J. and Astiningsih, K. 1991. Social hierarchies in very young chicks. *Br. Poult. Sci.* 32(1), 47-56. doi:10.1080/00071669108417326.

Rogers, L. J., Andrew, R. J. and Johnston, A. N. B. 2007. Light experience and the development of behavioural lateralization in chicks III. Learning to distinguish pebbles from grains. *Behav. Brain Res.* 177(1), 61-9. doi:10.1016/j.bbr.2006.11.002.

Rogers, L. J. 2008. Development and function of lateralization in the avian brain. *Brain Res. Bull.* 76(3), 235-44. doi:10.1016/j.brainresbull.2008.02.001.

Roper, T. J. and Marples, N. M. 1997. Odour and colour as cues for taste-avoidance learning in domestic chicks. *Anim. Behav.* 53(6), 1241-50. doi:10.1006/anbe.1996.0384.

Sandilands, V., Powell, K., Keeling, L. and Savory, C. J. 2004. Preen gland function in layer fowls: factors affecting preen oil fatty acid composition. *Br. Poult. Sci.* 45(1), 109–15. doi:10.1080/00071660410001668932.

Sherwin, C. M. and Devereux, C. L. 1999. Preliminary investigations of ultraviolet-induced markings on domestic turkey chicks and a possible role in injurious pecking. *Br. Poult. Sci.* 40(4), 429–33. doi:10.1080/00071669987151.

Sinclair, A., Wyneken, C. W., Veldkamp, T., Vinco, L. J. and Hocking, P. M. 2015. Behavioural assessment of pain in commercial turkeys (*Meleagris gallopavo*) with foot pad dermatitis. *Br. Poult. Sci.* 56(5), 511–21. doi:10.1080/00071668.2015.1077204.

Skrbic, Z., Pavlovski, Z., Lukic, M. and Petricevic, V. 2015. Incidence of footpad dermatitis and hock burns in broilers as affected by genotype, lighting program and litter type. *Ann. Anim. Sci.* 15(2), 433–45. doi:10.1515/aoas-2015-0005.

Sneddon, H., Hadden, R. and Hepper, P. G. 1998. Chemosensory learning in the chicken embryo. *Physiol. Behav.* 64(2), 133–9. doi:10.1016/s0031-9384(98)00037-7.

Steiger, S. S., Fidler, A. E., Valcu,M. and Kempenaers, B. 2008. Avian olfactory receptor gene repertoires: evidence for a well-developed sense of smell in birds? *Proc. R. Soc. B* 275, 2309–17. doi:10.1098/rspb.200 8.0607.

Sufka, K. J., Feltenstein, M. W., Warnick, J. E., Acevedo, E. O., Webb, H. E. and Cartwright, C. M. 2006. Modeling the anxiety-depression continuum hypothesis in domestic fowl chicks. *Behav. Pharmacol.* 17(8), 681–9. doi:10.1097/FBP.0b013e3280115fac.

Tallet, C., Arnould, C., Picard, M. and Porter, R. 2003. Influence of olfaction on initial weight gain in two genotypes of chicks. *Br. Poult. Sci.* 44(5), 779–81. doi:10.1080/00071660410001666727.

Taylor, P. E., Scott, G. B. and Rose, P. 2003. The ability of domestic hens to jump between horizontal perches: effects of light intensity and perch colour. *Appl. Anim. Behav. Sci.* 83(2), 99–108. doi:10.1016/S0168-1591(03)00127-8.

Tellez, G. I., Jaeger, L., Dean, C. E., Corrier, D. E., Deloach, J. R., Williams, J. D. and Hargis, B. M. 1993. Effect of prolonged administration of dietary capsaicin on salmonella-enteritidis infection in leghorn chicks. *Avian Dis.* 37(1), 143–8. doi:10.2307/1591467.

Turro, I., Porter, R. H. and Picard, M. 1994. Olfactory cues mediate food selection by young chicks. *Physiol. Behav.* 55(4), 761–7. doi:10.1016/0031-9384(94)90057-4.

Warchol, M. E. and Dallos, P. 1989. Neural response to very low-frequency sound in the avian cochlear nucleus. *J. Comp. Physiol. A* 166(1), 83–95. doi:10.1007/BF00190213.

Yoshida, Y., Kawabata, F., Kawabata, Y., Nishimura, S. and Tabata, S. 2018. Short-term perception of and conditioned taste aversion to umami taste, and oral expression patterns of umami taste receptors in chickens. *Physiol. Behav.* 191, 29–36. doi:10.1016/j.physbeh.2018.03.020.

Zidar, J. and Løvlie, H. 2012. Scent of the enemy: behavioural responses to predator faecal odour in the fowl. *Anim. Behav.* 84(3), 547–54. doi:10.1016/j.anbehav.2012.06.006.

# Chapter 3

# Understanding states of suffering with implications for improved management of poultry

*Ian J. H. Duncan, University of Guelph, Canada*


1 Introduction
2 Fear
3 Frustration
4 Pain
5 Other states of suffering
6 Conclusions and future trends
7 Where to look for further information
8 References


## 1 Introduction

It is now generally accepted that an animal's welfare is all to do with its subjective experiences, and in particular, with the degree to which it suffers and feels pleasure (e.g. Duncan and Fraser, 1997). This view was originally proposed by Jeremy Bentham when he said about animals: 'The question is not, Can they reason? nor, Can they talk? but, Can they suffer?' (Bentham, 1789, p. 283). George John Romanes (1884) embraced this idea and included subjective experiences or feelings into Darwin's evolutionary framework. He said, 'Pleasures and pains must have been evolved as the subjective accompaniment of processes which are respectively beneficial or injurious to the organism, and so evolved for the purpose or to the end that the organism should seek the one and shun the other'. Unfortunately, this enlightened view was lost with the rise of Behaviorism during the first 70 years of the twentieth century. Proponents of Behaviorism argued strongly that subjective states could not be studied scientifically (e.g. James, 1904; Watson, 1928; Skinner, 1974) and behavioural scientists from different backgrounds fell in line with the Behaviorists.

When the welfare of animals used by human beings in intensive animal agriculture was beginning to be questioned by Ruth Harrison (1964) in her

http://dx.doi.org/10.19103/AS.2020.0078.03

book *Animal Machines*, the general public looked to behavioural scientists for answers. The committee formed by the British government to review this matter under the chairmanship of Professor Rogers Brambell (Command Paper 2836, 1965) suggested that, somehow, the feelings of the animals should be taken into account. However, when behavioural scientists started to investigate the welfare of farm animals, they eschewed any consideration of animal feelings (the influence of Behaviorism was pervasive) and concentrated on measuring the physiological stress response (e.g. Bareham, 1972; Bryant, 1972; Wood-Gush et al., 1975). Things started to change with the publication of Donald Griffin's (1976) book *The Question of Animal Awareness*. It suddenly became acceptable for behavioural scientists to consider animal consciousness and feelings when assessing welfare. Marian Dawkins (1980) pointed this out very clearly in her seminal book *Animal Suffering*.

A more technical term for feelings is 'subjective affective states'; the fact that they are *subjective* means that they are only available to the person or animal experiencing them. Nevertheless, if feelings are an important component of welfare, then, somehow, we should take account of them. By this line of reasoning, we seem to have reached an impasse: feelings are important for welfare but are not available for direct scientific investigation. However, it gradually became clear that, in the animal welfare debate, a direct investigation may not be necessary. We do not need to know (in fact, we cannot know) what an animal experiences when it is in pain, is frightened or frustrated. What would help enormously in the animal welfare debate would be information about how negatively an animal feels in particular situations in which we suspect that it is in pain, is frightened or frustrated. It would also be helpful to know how positively an animal feels when we think it might be experiencing pleasure. It was quickly realised that it might be possible to gain such information in two ways. First, animals may communicate their affective states to others of the same species, and it might be possible for us to interpret these signals. Second, it might be possible to find out how negatively or positively an animal feels about some particular aspect of its environment by seeing the extent to which it will avoid or approach that aspect.

'Suffering' implies a certain intensity and duration of a negative affective state. Thus a hen being kept in free-range conditions might experience some discomfort on a very hot day. If shade or other cooler conditions are available, the hen will move there, reducing or eliminating the negative feelings and will not suffer. If the hen cannot escape the heat, she will try to dissipate body heat by adopting certain postures and by panting. But these measures may not be sufficient and her body temperature may increase. The hen would then start to suffer. The same distinctions can be raised for other negative affective states. If the negative affective experience is mild and short-lasting, then it does not amount to suffering. Of course, there is no clear-cut line between mild negative affective state and suffering; one merges with the other. This was acknowledged by Marian Dawkins (1980) in her book *Animal Suffering*, when she stated (p. 76), 'Not all fear, frustration or conflict indicates suffering. But prolonged or intense occurrences of these same states may indicate great suffering'. This topic is covered in some detail by Weary (2014). He suggests that any assessment of animal suffering will not only have to consider the severity of the negative affective state, but also how concurrent negative states interact and the animal's ability to cope with adversity.

In this chapter, the major negative affective states of fear, frustration and pain are considered together with some other states that are more difficult to categorize. However, it should be pointed out that there are other ways of considering emotion. Mendl et al. (2010) suggest that dimensional theories of emotion have some advantages over theories that regard each emotion as being discrete. According to these dimensional theories, emotions can be represented in a two- or three-dimensional space with valence (positive or negative emotion) being represented in one dimension and arousal level in another (e.g. Carver, 2001; Russell, 2003). Since the main thrust of the current chapter is to aid the recognition of the states of suffering and to suggest methods of reducing the suffering, it was concluded that regarding the emotions as being discrete would be more helpful.

## 2 Fear

A definition of 'fear' is 'a state of motivation which is aroused by certain specific stimuli and normally gives rise to defensive behaviour or escape' (McFarland, 1981). Domestic fowl show a wide range of behavioural responses to fear, and in order to understand these, it is helpful to consider briefly the stimuli which lead to a fearful state. It has been suggested that all fear-inducing stimuli can be classified under the headings of (i) intensity, (ii) novelty, (iii) 'special evolutionary dangers' and (iv) stimuli arising from social interaction (Gray, 1971). Intense stimulation, particularly if it has a sudden onset, such as a loud noise, can clearly cause fear in chickens. It is also obvious that the adaptive significance of this is that a loud intensive stimulus may signal imminent danger. The advantages of avoiding novel stimuli or neophobia are also clear and well documented in chickens (Murphy, 1977). By 'special evolutionary dangers', we refer to sign stimuli that have particular significance to chickens in their evolutionary history, including those associated with predators. Finally, the stimuli arising from social interactions are those involved in agonistic encounters. In the case of domestic fowl, which form a social hierarchy, such stimuli are likely to be important (Banks et al., 1979; Lee et al., 1982).

## 2.1 Fear responses to intense and sudden stimuli

Working with very young Red Junglefowl chicks, Kruijt (1964) showed that they responded very little to sudden visual stimuli but did run away from auditory stimuli if these had few low frequencies and were of long duration. Avoidance responses to sudden visual stimuli developed only slowly over several days. Similar experiments with domestic chicks have not been well-controlled. Chicks have usually been placed in a novel environment, which is in itself frightening, and then exposed to the sudden stimulus. However, it seems that their responses are fairly similar to those of Junglefowl chicks. In general terms, the behavioural sequence is, first, a startled jerk followed by inhibition of peeping and any other ongoing behaviour and then frantic running followed by crouching or freezing in silence. This phase is usually ended by small head movements and then loud peeping and movements about the area, often accompanied by pushing or jumping at the walls and usually described as 'attempted escape' (Phillips and Siegel, 1966; Jones, 1980). This sequence of behaviour is usually interpreted as a progression from high to low fear levels.

Although the responses of chicks to fear-inducing stimuli have been studied fairly thoroughly, there have been no studies on the development of fear responses from hatching to adulthood. From the scant evidence that is available, it appears that fearfulness increases from hatching for a few days and then declines slightly to a plateau (Phillips and Siegel, 1966).

There have been many more studies on the effects of sudden and intense stimuli on adult birds; for example, a rapidly inflating balloon has been used in experiments to frighten adult hens. This has the property of 'looming' or appearing to rush towards the bird. It is a very potent fear-inducing stimulus. In response to the looming stimulus, hens showed a startled response with complete inhibition of all current behaviour. This was immediately followed by intense escape behaviour with wing-flapping and pushing against the sides of the cage (Duncan and Filshie, 1980). During this time, many of the birds gave high-intensity ground predator alarm calls (Collias and Joos, 1953). Hens of three different strains responded similarly to this treatment but with light hybrids showing more intense panic than broilers or medium hybrids. Heart rate, recorded using radiotelemetry, correlated well with the behavioural responses and increased to about 450 beats/min within 2–3 s of the balloon inflating. There were interesting strain differences in the recovery phase. Within 5 min, the light hybrid hens had calmed down, resumed normal maintenance activities and their heart rates had dropped to prestimulus levels. This contrasted with the broilers and medium hybrids, which also stopped performing escape behaviour within about 5 min. Thereafter, they tended to adopt a motionless standing alert posture and often appeared to be fixating visually on the deflated balloon. They often maintained this posture for several minutes during which time their heart rates remained elevated. This suggests that these birds were still in a fearful state. In another experiment, individual hens were kept in a two-compartment 'shuttle box', and the birds could be frightened in either compartment by a rapidly inflating balloon (Rutter and Duncan, 1991). When they were given a 20 s warning by a small signal light that the balloon was going to be inflated, medium hybrids learned quite quickly to move to the other compartment and avoid the inflating balloon. Light hybrid hens, on the other hand, were unable to learn this response, apparently because they were in such a panic. When the experimental set-up was changed from shuttle avoidance to one-way avoidance, the light hybrid hens were able to learn to avoid the inflating balloon (Rutter and Duncan, 1991).

In commercial chicken production practice, it is well recognized that sudden intense stimuli can have an adverse effect on welfare and on productivity. Fear causes a reduction in the welfare of all birds. In addition, the physiological alarm reaction burns extra energy which means that birds that are frightened, even occasionally, will be less efficient than birds that are calm. Moreover, the intense response shown by light hybrid hens can lead to physical injury and suffocation, as birds panic and rush away from the apparent source of the stimulus. For this reason, it is good husbandry practice to avoid sudden intense stimuli whenever possible. For example, it is good practice to (1) warn birds that a door into a barn is about to be opened by knocking on it, (2) have background music playing in a barn to cover up sudden noises and (3) use light dimming and brightening programs so that there is not a sudden change from light to dark and dark to light.

## 2.2 Fear responses to novelty

The responses of domestic fowl to novelty are generally similar to those elicited by sudden and intense stimuli but without a startle response and at a level suggesting a lower level of fear. Many of the studies claiming to investigate novelty have used an 'open field' test, and this test involves other variables in addition to novelty (e.g. Candland and Nagy, 1969; Jones, 1977, 1978; Faure et al., 1983). Chicks placed in an open field arena generally stayed immobile for a variable period of time and then walked or ran about the arena peeping and making what appeared to be attempts to escape. The more frightened the chicks, the longer the immobile period lasted. Another study carried out a novel environment test on two strains of growing birds at intervals from 4 days to 9 months of age (Murphy and Wood-Gush, 1978). The birds were tested alone and had not been isolated before. The chicks either stood or walked slowly and emitted peeps or lay down and remained immobile for long periods. The juveniles and adult birds spent longer periods remaining immobile. When they started to move, it was with slow wary steps. Some of the juveniles and adults gave ground predator alarm calls (Collias and Joos, 1953). In a second experiment, birds were less fearful when the novelty of the environment was reduced by making it more like the home environment (Murphy and Wood-Gush, 1978). A refinement of the open field test is the 'emergence test', which was developed to test fearfulness in rodents (e.g. Thiessen et al., 1962). Latency to emerge from an enclosed chamber with a doorway is measured. It has been shown that chicks which show high levels of fear in other tests also show long emergence latencies (Jones and Mills, 1983). However, when adult hens are considered, interpretation of 'emergence tests' becomes more problematic. When adult hens of a 'flighty' strain were tested, they tended to lie quite still in the chamber and then leap out into the pen. Hens of a 'docile' strain seemed much more relaxed in the chamber, moving about and showing exploratory pecking before eventually emerging slowly and cautiously after a much longer latency (Murphy and Wood-Gush, 1978). Using only this emergence test, hens of the 'docile' strain would be classified as more fearful but this does not correspond to the actual behaviour of the birds.

In order to avoid fear responses to novelty, care should be taken to make changes to the environment gradually. This is particularly true for feeding and drinking equipment, where avoidance of a feeder or a drinker for even 24 h could have an adverse effect on the welfare and production. A change to the feedstuff itself, for example, a change in texture or colour, could also lead to avoidance and a reduction in the welfare and production. Some thought should be given to the timing of husbandry routines. It might not always be possible to carry out husbandry tasks at exactly the same time each day, and so getting the birds used to some variation in timing might be useful. Finally, the same coloured clothing should be worn by all caretakers including relief workers.

## 2.3 Fear responses to special evolutionary dangers

The main danger to Junglefowl is predation by either aerial or ground predators. Domestic fowl have retained fearful responses to potential predators, although these have probably been greatly attenuated during domestication (Ratner and Boice, 1975). In response to an aerial predator, cockerels give a specific vocalization to which other domestic fowl respond by crouching and freezing (Collias and Joos, 1953). In response to a ground predator, adult birds of both sexes give the ground predator alarm call which triggers alertness, scanning of the environment and rapid movement away from the source of the alarm (Collias and Joos, 1953). Immature birds also vocalize in response to ground predators, but these calls have not been characterized.

It seems very likely that the original relationship between human beings and Junglefowl was a predator-prey one and that human approach induced high levels of fear in the birds (Duncan, 1992). The evidence suggests that although domestication has reduced this response a lot, it has not removed it completely (Duncan, 1985, 1990). Therefore the main (perhaps the only) predator-like stimulus that domestic fowl are exposed to in modern husbandry conditions are their caretakers. There is good evidence that domestic fowl that have had a 'normal' amount of contact with human beings are frightened by being handled (Hughes and Black, 1976; Murphy and Duncan, 1977).

White egg-laying strains of laying hens derived from the White Leghorn breed have the reputation of being 'flighty', whereas brown egg-laying strains derived from crosses of Light Sussex and Rhode Island Red breeds are classified as being more docile (Murphy, 1978). However, research has revealed that this difference in fearfulness is restricted to the responses of these strains to human beings. When birds of different ages of these two strains were tested with novel foods and novel objects, the 'flighty' birds showed less fear (Murphy, 1977). This difference between the two strains was confirmed in an experiment in which the behaviour and heart rate of individual hens were recorded as they were slowly approached by a human being (Jones et al., 1981). At the start of each observation, hens were presented with food in a novel way, and the docile hens showed more fear of this than did the flighty hens. During the human approach, the docile hens showed more of low-level fear than did the flighty birds when the human being was far from the birds (30–25 m). However, as the human being came close to the hens (<5 m), the flighty hens showed much higher levels of fear. The researchers speculated that when the human being was at some distance from the hens, it was perhaps seen as a strange object rather than a human being.

Attempts have been made to reduce the fearfulness of flighty birds towards human beings by linking human contact with a food reward (Murphy and Duncan, 1977). This training was not effective when carried out on adult birds. Hens of the flighty strain continued to withdraw from human contact whereas hens of the docile strain withdrew very little. However, when this experiment was repeated using chicks of a few days of age, chicks of the flighty strain continued to show fear and withdrawal but chicks of the docile strain actively approached the hand giving the food reward. It appears that the flightiness of the White Leghorn strain is very deep-seated and resistant to change. It was later shown that the difference between the strains is partly a genetically determined difference in withdrawal response present from hatching and partly a difference in learning with birds of the docile strain showing more habituation of withdrawal than those of the flighty strain (Murphy and Duncan, 1978).

What about fear in broilers, which are generally considered to be much less fearful than laying strains of domestic fowl? In a series of experiments with broilers, an attempt was made to simulate in the laboratory the catching, crating and trucking procedures that normally precede slaughter. It was found that the most stressful aspect of this train of events was the initial approach, catching and restraint by a human being (Duncan, 1989). It was also shown that broilers were actually less frightened by being caught, picked up and crated by a specially designed machine than they were by being manually caught (Duncan et al., 1986).

Some years ago, there were reports in the scientific literature of an exaggerated form of fearfulness called 'hysteria' (e.g. Prip, 1975; Hansen, 1976). In flocks of susceptible laying hens, there would be a spontaneous build-up of fear responses with alarm calls and intense panic and fleeing behaviour. This would reach a crescendo and then gradually subside only to erupt again 40–80 min later, and it could occur in battery cage and free run husbandry systems. This behaviour had a devastating effect on productivity. There were also many cracked eggs caused by the behavioural panic and deformed eggs caused by shell gland contractions during the calcification phase (Hughes et al., 1986). Attempts to solve the problem by increasing dietary tryptophan were fairly successful (Laycock and Ball, 1990), but, in fact, strains of laying hen with a reputation for hysteria were soon avoided by producers. The problem of hysteria today is therefore quite rare.

In any husbandry system, human contact can never be completely eliminated. Therefore, frequent regular exposure of the birds from an early age to human beings should help to minimize this particular source of fear. With regard to light hybrids (White Leghorns) which may not respond to regular contact, strains should be sought which have a reputation for being less flighty.

There are two other 'predator-like' sources of fear. One is a moving hosepipe which is presumably snake-like. However, this should be easy to avoid. The other is long-haired brushes or feather dusters which are used to dust battery cages. These seem to be seen, respectively, as ground and aerial predators and cause great panic in a flock. A simple solution to this problem is not obvious. Some other solution to dusty cages must be found.

As was pointed out in the introduction, not all instances of fear can be classified as suffering. As a prey species, domestic fowl have to be constantly on the lookout for signs of danger, and there will be frequent false alarms. These may be as mild as an alerting response, during which there is a transient increase in the heart rate. A distant stimulus that was perceived as danger and triggered a short-term alarm reaction may turn out to be non-threatening when it approaches and is seen more closely. A sudden intense stimulus may trigger a severe response that could be correctly classified as indicating suffering, but with repetition, that same stimulus may come to be ignored. The approach of a dominant bird may elicit some fear in a subordinate. However, if the subordinate hen crouches and the dominant hen moves away, there will be no suffering. On the other hand, if the dominant hen attacks, and this is just one of many attacks that the subordinate bird has experienced that day, the attacked bird will probably be suffering. There is no simple rule for deciding when minor alarms become serious enough to constitute suffering. Birds in a flock should be spending most of their time performing maintenance activities and should not be showing signs of alarm. It is inevitable that most flocks will occasionally be subjected to fear-inducing stimuli such as a vehicle backfiring or a thunderstorm. On such occasions, there may be a short period of suffering. On the other hand, flocks that show hysteria, as previously described, will be suffering most of the time.

## 3 Frustration

A definition of 'frustration' is the state of organism placed in a situation in which there is interference with a behaviour sequence that normally leads to a goal response (Duncan, 1970). The Brambell Committee (Command Paper 2836, 1965) thought that battery cages for laying hens would inevitably lead to a state of frustration:

> Much of the ingrained behaviour is frustrated by caging. The normal reproductive pattern of mating, hatching and rearing young is prevented and the only reproductive urge permitted is laying. They cannot fly, scratch, perch or walk freely. Preening is difficult and dust-bathing impossible .... The caged bird which is permitted only to fulfil the instinctive urges to eat and drink, to sleep, to lay and to communicate with its fellows, would appear to be exposed to considerable frustration.

The possibility of hens being frustrated in cages was investigated in detail by seeing exactly how domestic fowl responded when they were frustrated experimentally in various ways during feeding, nesting, incubation, brooding and sexual activity (Duncan, 1970). Frustrated birds showed a fairly limited repertoire of behaviour. If the frustration was severe, birds showed locomotion towards and away from the source of the frustration, and with repetition, this developed into stereotyped pacing behaviour (Duncan and Wood-Gush, 1972a). If two birds were frustrated simultaneously, the dominant bird showed an increase in aggression over the subordinate (Duncan and Wood-Gush, 1971). If the frustration was mild, birds showed an increase in 'displacement preening' (Tinbergen, 1952), that is, short bouts of frantic preening that seemed to be irrelevant to the frustrating situation (Duncan and Wood-Gush, 1972b). These symptoms of frustration are not generally seen in commercial husbandry systems. However, there is one very important exception: many hens show symptoms of severe frustration in battery cages 1–1½ h before they lay an egg (Duncan, 1970; Wood-Gush, 1972). It should be noted that nesting behaviour is triggered, not by the presence of an egg in the shell gland but by hormonal events that take place at ovulation, 24 h earlier (Wood-Gush and Gilbert, 1975). When nesting behaviour is triggered, hens normally separate themselves from their flock mates, adopt a characteristic posture, give the pre-laying vocalization, start to examine possible nest sites, eventually choose and enter one, sit and start to perform nest-building behaviour (Duncan and Kite, 1989).

Hens in battery cages start to perform this sequence of behaviour, but for most hens, the cage is inadequate and does not stimulate sitting behaviour. The searching phase is prolonged and soon develops into a stereotyped pacing behaviour (Wood-Gush, 1972). When this was investigated almost 50 years ago, strain differences in pre-laying frustration were found, with about 80% of light hybrid strains showing the symptoms compared to about 65–70% of medium hybrid strains (Wood-Gush, 1972; Mills et al., 1985). It has also been shown that the incidence of frustrated nesting in cages has remained at this high level for at least 30 years (Yue and Duncan, 2003) and that this was one of the main reasons for the European Union to ban battery cages from January 2012 (European Commission, 1999). The frustration associated with the lack of a nesting place is very severe and occurs for between one and one and a half hours on 7 days out of 8. It therefore accounts for a huge negative effect on the welfare of laying hens.

The occurrence of stereotyped pacing behaviour almost certainly signifies that hens are suffering. However, what about displacement preening, which occurs in experimental situations when frustration or conflict is less intense (Duncan and Wood-Gush, 1972b)? The fact that displacement activities are common in a wide range of 'natural situations' in wild bird species (e.g. McFarland, 1966a,b) and that they may have evolved with the function of distracting or 'cutting-off' the animal from a frustrating or conflict situation (Chance, 1962) suggests that they probably signify a less intense negative state than suffering.

The physiological stress response has also been measured in frustrated hens. Light hybrid hens in furnished cages were excluded from nests towards the end of their laying year, and several physiological stress indicators were measured, including faecal corticosterone metabolites, plasma corticosterone and yolk corticosterone (Alm et al., 2016). In addition, heterophil to lymphocyte ratio, eggshell irregularities, duration of tonic immobility and feather cover were recorded. The same measurements were taken from control hens which were not excluded from nests in neighbouring furnished cages. Most of the indicators revealed an increased stress response associated with the nest exclusion, but the pattern of the stress response varied between indicators. An interesting finding was that the control hens also showed indications of increased stress; it was hypothesized that the stress spread in some way to the control birds in adjacent cages which had access to nests (Alm et al., 2016).

The effect of frustrated nesting on the egg has been studied in some depth (Hughes et al., 1986). When hens cannot find a suitable nest, laying behaviour is often delayed and the retained egg can have extra calcium carbonate laid down on top of the cuticle, which is normally the last layer to be secreted. In brown-shelled eggs, this results in eggs that appear violet rather than brown. In hens that cannot find a suitable nesting place, the egg can be retained until the next egg reaches the shell gland. Shell formation starts on this second egg while it is pressed up against the first egg, and this results in a 'slab-sided egg'. So examination of eggs can often provide evidence of inadequacies in nest provision.

The solution to this problem is obvious: hens must be given a suitable nesting place. The nesting place should be secluded but not necessarily dark. It should also allow the performance of nest-building motor patterns, and in particular, rotating and scratching out with the feet (Fig. 1). Unfortunately, not all nesting sites in furnished cages (and even in free run systems) meet these simple criteria, and this can lead to frustration and, sometimes, eggs not being laid in the nest (Appleby et al., 1989).

There are several other behaviour patterns that are frustrated in modern husbandry systems. Feral domestic fowl roost high off the ground at night and during the day when they are resting (McBride et al., 1969; Wood-Gush et al., 1978). Modern strains of domestic fowl will also roost on high perches at night if given the opportunity (Blokhuis, 1984). It has also been shown that hens that are used to roosting on high perches show symptoms of frustration when they are denied access to the perches (Olsson and Keeling, 2000). In addition, hens

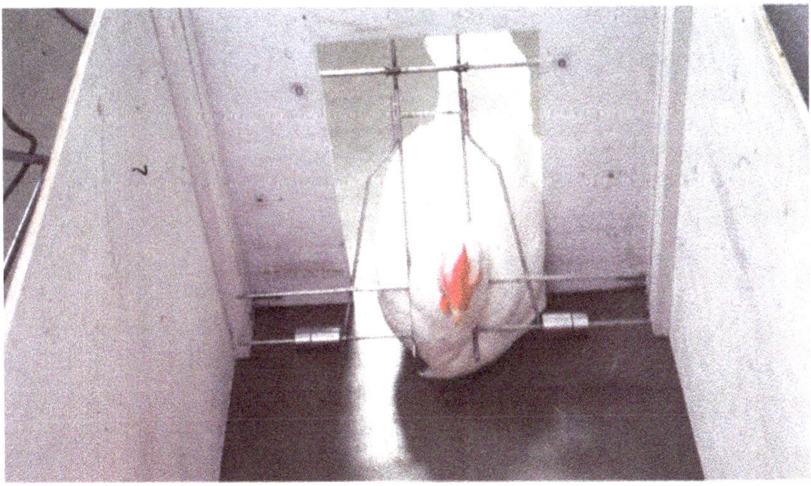

**Figure 1** A hen pushing open a weighted door in order to reach a nest box. Nesting behaviour is extremely important to hens and they will work hard to gain access to a suitable nesting place (Follensbee et al., 1992).

will work hard, by pushing open a weighted door, to gain access to perches for roosting (Olsson and Keeling, 2002).

This problem can be overcome by providing hens with roosting perches in all husbandry systems. The perches should preferably be raised well above the level where other activities such as feeding, drinking and foraging take place and where the roosting birds will not be disturbed (Appleby et al., 1998) (Fig. 2).

Jungle fowl (Dawkins, 1989) and feral fowl (Savory et al., 1978) spend a major part of their day foraging for food. Laying hens kept commercially in free run systems also spend from 7–25% of their time foraging in addition to the time they spend feeding from food troughs (Gibson et al., 1988; Appleby et al., 1989). Foraging is the appetitive component of feeding behaviour and comprises locomotion together with ground scratching, pecking and probing at items on the ground and tearing growing leafy material if this is available (Savory et al., 1978). Most of these motor patterns are denied to birds in cages or living in wire-floor or slatted-floor houses. It would therefore seem that frustration might be a distinct possibility in these systems. In fact, the previously described symptoms of frustration have not been reported in these circumstances. However, it has been suggested that outbreaks of feather pecking might be a response to the lack of a foraging substrate (Blokhuis, 1986; Aerni et al., 2000; Bilčík and Keeling, 2000; Rodenburg et al., 2004) and therefore could be regarded as frustration response. However, it has been shown that the motor patterns involved in feather pecking suggest that it is actually foraging behaviour simply re-directed at feathers (Dixon et al., 2008). Whether or not this is accompanied by a state of frustration is unclear at this time. Of course, birds that are feather

**Figure 2** Hens roosting at night in a house in a free-range system. It can be seen that height is important and the lower perches remain empty.

pecked may suffer pain, and this will be considered in the next section. In other experiments, frustrating laying hens experimentally by blocking their attempts to feed, did not increase feather pecking in birds with either a high or a low feather pecking genotype (Rodenburg et al., 2005).

Since foraging is a major component in domestic fowl time budgets, it would seem essential to provide an environment in which foraging can be freely engaged in. A deep litter husbandry system satisfies this requirement, but wire-floor and slatted floor systems do not. Some attempts have been made to encourage foraging behaviour in furnished cages but these have had limited success.

It has been suggested that frustration of behaviour in the laying hen leads to a particular vocalization, the gakel-call (Schenk et al., 1983; Koene and Wiepkema, 1991; Zimmerman and Koene, 1998a,b). However, this vocalization was not reported in many other investigations into frustration (Duncan, 1970; Duncan and Wood-Gush, 1971, 1972a,b; Olsson and Keeling, 2000). This difference in results is difficult to explain. The gakel-call or pre-laying call has been well characterized and, according to early publications, is very specific to the pre-laying period (Wood-Gush and Gilbert, 1969, 1973; Wood-Gush, 1975). These researchers showed that the call could be elicited in ovariectomized adult hens by injections of oestrogen and progesterone and that the call could be elicited in young immature females by injections of oestrogen alone. Other studies of laying behaviour have reported the occurrence of the pre-laying call, for example in feral fowl (McBride et al., 1969; Duncan et al., 1978) and in hens kept in quasi-commercial conditions (Meijsser and Hughes, 1989; Yue and Duncan, 2003). In these studies the call was given regularly but only in the pre-laying phase and, moreover, it was given whether or not there was frustration present. In a later paper Zimmerman et al. (2000) found that, in contrast to earlier experiments, frustration in a feeding test did not increase the incidence of gakel-calls casting some doubt on whether the call is a specific response to frustration. In the face of these discordant ideas, this topic desperately requires further research.

## 4 Pain

Pain should be distinguished from nociception. Nociception is the process through which information about noxious peripheral stimuli is transmitted by primary afferent nociceptors to the spinal cord, brainstem, thalamus and subcortical structures. Pain, on the other hand, is a subjective experience and is the perception of noxious stimulation in the cortex (Benson, 2004).

Definitions of pain that rely heavily on self-report are of limited use in animals. Because of this, Molony (1997) suggested this working definition: 'Animal pain is an aversive sensory and emotional experience representing an awareness by the animal of damage or threat to the integrity of its tissues; it changes the animal's physiology and behaviour to reduce or avoid the damage, to reduce the likelihood of reoccurrence and to promote recovery'.

Pain can occur in poultry following accidental injury resulting from housing or husbandry methods, following surgical procedures and as a result of disease processes, and these will be dealt with in order.

## 4.1 Pain caused by pecking

During the establishment of the peck-order, birds tend to peck each other aggressively. These pecks are directed at the head and neck region, are usually directed in a downward direction and are often vigorous enough to cause lesions on the head and comb of the recipient. There is little doubt that these pecks are painful. However, after the peck-order is formed, the incidence of aggressive pecking declines steeply and the peck-order is maintained by threats and avoidances (Gottier, 1968). Therefore the amount of pain attributable to agonistic encounters is probably minimal. Early research tried to establish how many individual flock-mates a hen could recognise thinking that this would limit how large a stable flock could be (Guhl, 1968). However, individual recognition is not necessary for the establishment of a stable hierarchy. During initial aggressive encounters, birds may learn their general status in the flock and thereafter adopt a subtle posture which other birds can compare to their own status. Whatever the mechanism, socially stable flocks of several thousand are common. It is also known that strange hens introduced to a stable flock will be attacked presumably because some aspect of their behaviour reveals them as strangers. Therefore, in order to minimise peck-order pecking, social groups of domestic fowl should be kept stable. This should not be a problem for commercial flocks of egg-laying hens or broilers; 'all in - all out' is the universal practice in order to maintain biosecurity. The exception to this rule is that broiler breeder flocks often use 'spiking' in which some of the older males are replaced with younger more vigorous males part-way through the breeding cycle. However, producers are aware that social friction may increase at this time and monitor their flocks closely in case remedial action is required.

Feather pecking is another form of pecking in which one bird pecks at the feathers of another bird sometimes pulling them out (Blokhuis and Arkes, 1984). Two forms of feather pecking have been recognized, gentle pecking which does not result in the removal of feathers and severe feather pecking which leads to feather loss (Savory, 1995). Severe feather pecking can develop into tissue pecking in denuded areas and this can lead to rank cannibalism (Savory, 1995). It is unlikely that gentle feather pecking leads to pain, but all other pecking in this category undoubtedly does (Gentle and Hunter, 1990). In contrast to aggressive pecking where the pecked bird will submit to or move away from, the aggressor and thus avoid further pecks, no such mechanism has evolved with feather pecking. This means that a pecked bird will often make no attempt to move away from the pecking bird until feather pulling or removal becomes painful. This results in greater damage and a quick spread of the problem through the flock. Even without active *avoidance*, it has been shown that inactive birds are more likely to become the target for feather pecks than active birds (Riber and Forkman, 2007). The tendency to feather peck and cannibalise is socially transmitted (Cloutier et al., 2002; McAdie and Keeling, 2002) and this results in more damage in larger groups (Wathes et al., 1985). It has been shown that severe feather pecking is exacerbated by high light intensity (e.g. 30 lux), whereas gentle feather pecking is more frequent at low light intensities (e.g. < 3 lux) (Kjaer and Vestergaard, 1999; Lambton et al., 2010).

Another form of damaging pecking is vent pecking in which pecks are directed at the skin and underlying tissue surrounding the vent (Savory, 1995). In a survey of 119 free-range and organic egg flocks in the UK, vent pecking was observed in 30% of flocks (Lambton et al., 2015). Vent pecking frequently occurs immediately after an egg has been laid when there is partial eversion of the cloaca for a few minutes after the egg is laid. The shiny red mucosal tissue that is exposed at this time seems to be a very potent stimulus for pecking. Vent pecking can quickly lead to cannibalism and even evisceration of the pecked bird and this aberrant behaviour clearly causes pain to the pecked bird (Pötzsch et al., 2001). There is some association between feather pecking and vent pecking, but the details are unclear (Lambton et al., 2015). The poultry industry's solution to feather pecking is beak trimming, but this is a painful procedure and another welfare problem is thus created.

The ultimate solution to feather pecking, cannibalism and vent pecking is genetic selection against these traits which are moderately heritable (Muir, 1996; Kjaer et al., 2001). Currently, there is not much incentive for primary breeding companies to incorporate selection against feather pecking into their programs. Perhaps if national governments placed a time limit on how long all forms of beak modification would be allowed, the primary breeding companies would be forced to take action.

### 4.2 Pain caused by the artificial environment

There are two common foot disorders in laying hens that are painful. They are bumblefoot (ulcerative pododermatitis) and toe pad hyperkeratosis. Bumblefoot tends to occur in hens kept in husbandry systems with perches. It appears that pressure on the sole of the foot in the same place can, over time, lead to small wounds allowing infection by opportunistic bacterial pathogens, the most common of which is *Staphylococcus aureus* (Tauson and Abrahamsson, 1994). This condition is undoubtedly painful.

There is continuing research into better design of perches (e.g. Struelens and Tuyttens, 2009; Pickel et al., 2011; Stratmann et al., 2015) but a non-damaging perch has not yet been perfected.

Toe pad hyperkeratosis is found only in caged hens and is caused by the hens slipping frequently on the sloping floor of the cage. The hyperkeratosis itself is not a welfare problem but in severe cases the thickened skin can split causing a painful wound and the wound can be a route for infection, inflammation and further pain (Tauson, 1980).

When conventional battery cages are eventually banned throughout the world, this problem will disappear. In the meantime, cage manufacturers should be encouraged to produce cages with as little slope on the floor as possible.

Broilers also suffer from foot and leg problems caused by the artificial environment in which they are raised. Broilers are normally kept on some form of litter material. If environmental conditions are ideal, then the faeces produced by the birds mix with the litter which remains dry and friable. However, if the litter becomes wet and sticky, there is a big risk that the birds will develop contact dermatitis in their foot pads and hocks (Berg, 2004).

The solution to this problem is to keep the litter dry and friable. Ventilation systems are often set to maximise feed efficiency which means keeping the temperature at or above 20–21°C. In cold weather the result may be that the fans do not operate frequently enough to remove excess moisture and this can lead to wet litter. This means that feed efficiency may have to be sacrificed and the fans run more frequently in order to ensure dry, friable litter. In some climates this may necessitate heating the poultry barn.

With the move away from battery cages in Europe, keel bone fractures have emerged as a major welfare problem. In one study, 36% of hens in furnished cages had keel bone fractures and in systems with multi-level perches the level rose to over 80% (Wilkins et al., 2011). Hens with keel bone fractures indubitably experience pain; in hens with keel fractures, an analgesic drug, butorphanol, increased mobility in the hens and decreased the latency to jump down from perches. Butorphanol had no effect on mobility indices in birds without keel fractures (Nasr et al., 2012). In a follow-up experiment it was shown that hens with healed keel bone fractures would select an environment which had been linked to the administration of an opioid analgesic drug, showing that even when healed, keel bone fractures are painful (Nasr et al., 2013). However, the claim made in the title of this paper (Nasr et al., 2013) that the drug induced a positive affective state was not proved; it was shown that the drug resulted in a less negative affective state.

It appears that most keel bone fractures result from hens crash-landing when attempting to change level in multi-level systems (Harlander-Matauschek et al., 2015). Hens are more accurate and have fewer crashes when jumping up than they do when jumping down (Moinard et al., 2004) but of course the impact of landing when jumping up is likely to be much less than when jumping down. In another study with hens in an aviary system, birds which were subsequently shown to have keel bone fractures, changed levels in the system by walking on ramps rather than flying or jumping. Treating the hens with paracetamol did not improve their mobility (Rentsch et al., 2019).

This problem can be tackled in two ways. First, the design of furnished cages and free run systems can probably be improved. The ideal distances, both horizontal and vertical, for hens to fly from one level to another need to be established. Also, the light levels in those husbandry systems probably need to be increased so that the birds can make full use of their ability to judge distance. Finally, the rearing environment needs to be such that chicks can start to learn the skills required for changing level from 2 to 3 weeks of age.

## 4.3 Pain following surgical procedures

It has already been pointed out that one of the 'solutions' to damaging pecking in flocks is to beak trim the birds. It has been shown that hot blade beak trimming causes both acute and chronic pain (Duncan et al., 1989; Gentle et al., 1991). However, hot blade beak trimming has been largely replaced with precision infra-red heat treatment (IRBT) to chicks' beaks at the hatchery. Although this causes some acute pain, there is no evidence of chronic pain (Gentle and McKeegan, 2007). It has also been shown that there is tissue healing after IRBT and that the long-term neurophysiological and anatomical changes are minimal (McKeegan and Philbey, 2012). In a review of the topic of beak trimming, McKeegan (2017) concludes that although it is a traumatic process for chicks, on balance it is the most reliable way to prevent severe feather pecking and cannibalism.

In the long term, the solution to this problem is in the hands of the primary breeding companies. As pointed out previously, a vigorous genetic selection program would probably pay dividends since severe feather pecking is moderately heritable (Muir, 1996; Kjaer et al., 2001).

Toe trimming is another surgery that is commonly carried out on some classes of poultry. Male broiler breeders often injure the females when they mount and tread them during copulation. There are probably two reasons why this can result in injury. First, there is a large size discrepancy between the sexes with the males being much bigger and heavier. Second, some strains of broiler breeder males are deficient in some aspects of courtship behaviour which means that the females may not crouch properly when courted and end up moving rather than squatting during copulation (Millman et al., 2000). In order to avoid these injuries to the female, it is common to amputate the last joint of the inside toes of the males at the hatchery. This used to be done by chopping and cauterizing but has been largely replaced by an infra-red treatment at the hatchery similar to that used in beak trimming. The evidence from turkeys suggests that there is probably some acute pain associated with this procedure (Fournier et al., 2015).

It is difficult to think of an easy solution to this problem. Mating behaviour generally does not cause injury in layer breeding stock in which the weight difference between male and female is much less than in broiler breeder stock. Male layer breeders do not have to be toe trimmed. So a possible solution is devising a breeding program in which there is much less weight discrepancy between males and females. However, this should probably be combined with ensuring that male breeding lines show full courtship behaviour so that females are fully crouched and still during mounting.

## 4.4 Pain as a result of breeding practices

There are several skeletal problems in broilers most of which are probably painful. These seem to have arisen as a result of intensive genetic selection for fast growth (Whitehead et al., 2003). There are several long-bone deformities of the leg including valgus-varus and bent or twisted tibiae and tarsometatarsi which are often accompanied by slippage of the gastrocnemius tendon. These deformities lead to lameness which is often severe and the birds usually become emaciated (Julian, 1998; Mench, 2004). Tibial dyschondroplasia (TD) is another common leg problem causing lameness. This is a heritable disorder involving abnormal cartilage development in the proximal end of the tibia (Farquharson and Jeffries, 2000). TD is surprisingly common. Surveys of broiler flocks in Denmark revealed incidences averaging 57% (Sanotra et al., 2001).

The question arises: are lame birds necessarily in pain? The evidence strongly suggests that they are. When given a choice between two feeds, one of which contained an anti-inflammatory drug, lame broilers ate more of the drugged feed than did broilers with no lameness. Moreover, the lame broilers consumed enough of this feed to reduce the lameness. Broilers in the same flock that were not lame consumed more of the standard diet (Danbury et al., 2000). In another series of experiments it was shown that lame broilers ran an obstacle course faster when treated with butorphanol than when given saline (Singh et al., 2017). The birds also took longer to lie down in a latency-to-lie test. These results strongly suggest that the lameness was caused by pain and not just mechanical inefficiency.

In all these cases, there has probably been too intensive genetic selection for fast growth with too little attention to the fitness of the final animal product. The solution would seem to be a more balanced approach to genetic selection.

## 5 Other states of suffering

### 5.1 Hunger

Extreme hunger can be considered as a state of suffering. Normally no class of poultry is exposed to extreme hunger, but there is one exception. Broiler breeder fowl (the parent stock that produce broilers) have the same huge appetites as their progeny. In order to reproduce efficiently they have to be very severely food restricted (Leeson and Summers, 2000; Hocking, 2009b). If broiler breeders are given free access to food, they overeat, become obese and suffer from all the problems of obesity which include low fertility and reduced life expectancy. However, in addition to those productivity problems, the birds experience metabolic states that are accompanied by pain or feelings of malaise and cause the birds to suffer (Renema and Robinson, 2000). Severe food restriction helps to avoid these problems.

At various times in their life, broiler breeders are food restricted to as much as 30–40% of what they would consume if given free access to food. They exhibit behavioural symptoms that strongly suggest greatly reduced welfare (Savory, 1989; Mench and Falcone, 2000). Attempts to alleviate this hunger problem by diluting the diet with non-nutritive substances reduce the symptoms of hunger very little (Savory et al., 1996; de Jong et al., 2005).

Broiler breeders on food-restricted programs tend to increase their water consumption and this can lead to wet litter (Leeson and Summers, 1979). However, as pointed out earlier, wet litter can lead to another welfare problem, contact dermatitis (Berg, 2004). For this reason, producers often limit the time that water is available to broiler breeders. This is probably not a welfare problem in itself in that the birds are not physiologically thirsty. However, it may be that filling the crop with water helps slightly to alleviate the feelings of hunger.

The long-term solution to this problem is to have breeding birds with a much smaller appetite. This would entail a genetic selection program moving in the opposite direction to the current one. It would seem very unlikely that primary breeding companies would be prepared to accept this suggestion. However, it might be possible to compensate for the slower growth of the resultant broilers by adding value in some other way, perhaps by producing a broiler chicken that had more flavour.

There is another procedure during which birds are subjected to extreme hunger. In many parts of the world, food deprivation is used to force laying hens into a moult at the end of a laying year. The deprivation can last as long as 14–21 days. Hens quickly stop laying, lose body fat and their reproductive tracts regress. After the deprivation period, the hens are put back on a layer diet and photo-stimulated and they fairly quickly start to lay for a second laying cycle. There have been attempts to justify this practice (Webster, 2000) but these have been countered by strong arguments that 14–21 days of starvation will cause great suffering (Duncan and Mench, 2000). Reviews of this topic have agreed that inducing hens to moult by starving them for an extended period will greatly reduce their welfare (e.g. McCowan et al., 2006) and effective, alternative, humane methods have quickly been proposed (e.g. Kuenzel et al., 2005). Being able to moult laying hens humanely is a big step forward. Because of the welfare problems associated with traditional moulting methods, many countries used to dispose of laying hens after one laying cycle. Humane moulting methods will allow hens to be kept through two or three cycles which means that many fewer hens need to be kept.

## 5.2 Suffering due to other behavioural restrictions

Keeping hens in battery cages prevents them from adopting certain postures. The most obvious one is the 'standing alert' posture which, in other environments, is very common (Dawkins, 1985). However there is no obvious attempt by hens in battery cages to adopt this posture and there are no accompanying symptoms of frustration. The posture is simply missing from the hens' repertoire. However, if given a choice, hens prefer an environment in which they can adopt this posture (Dawkins, 1985). These results are open to different interpretations, but we should probably assume that there is *some* suffering involved in keeping hens in an environment with such a restricted headroom.

Nicol (1987a) showed that other subtle behavioural changes took place when cage height and cage area were manipulated independently. When height was reduced to that of battery cages, in addition to the elimination of the standing alert posture, the incidence of body shaking, head stretching, head scratching and time spent sitting were reduced. When cage area was reduced to that of battery cages, body shaking, head scratching and feather raising were reduced whereas cage pecking was increased.

In another set of experiments, Nicol (1987b) showed that the behaviour patterns that had shown a reduced incidence when space was restricted, namely body shaking, head scratching and feather raising, showed a 'rebound' effect when the hens were given more space. This suggests that the spatial restriction had resulted in a build-up in motivation to perform these activities.

The results of all these investigations suggest that the dimensions of a normal battery cage may compromise welfare by restricting the performance of certain activities. The fact that if given a choice, hens prefer more space than that available in battery cages (Hughes, 1975; Dawkins, 1981) strongly suggests that battery cages provide insufficient space.

It is really only conventional battery cages that result in the behavioural restrictions described in this section. The solution, therefore, is to get rid of battery cages altogether – and the egg industry certainly seems to be moving in that direction.

It could be argued that the behavioural changes described in this section might have been included under 'Frustration'. However, the behavioural restrictions dealt with in this section resulted in fairly subtle changes in behaviour and were not accompanied by the symptoms of frustration previously described. Future research may suggest that the phenomena described in the 'Frustration' and 'Suffering due to other behavioural restrictions' sections should be combined into one category. In the meantime, there are sufficient differences to justify keeping them separate.

## 5.3 Suffering due to social stresses

Groups of domestic fowl do not arrange themselves at random in the space they occupy. For most activities, they space themselves out. But of course, they are a social species and do not spread out as much as possible. There appear to be other psychological forces that pull them together. Early experiments with small groups of hens kept in cages at commercial densities showed that decreasing the area per hen depressed egg production, reduced food consumption, lowered body weight and increased mortality, all of which indicate increased stress (Hill and Binns, 1973; Wells, 1973). It has also been shown, using overhead cameras, that when small groups of hens are kept at the usual cage space allowance they try to space themselves out as much as possible and orient themselves so that no two birds are standing directly facing each other (Keeling and Duncan, 1989). These results strongly suggest decreased welfare at normal cage densities.

When hens are kept in more complex environments, they tend to space themselves according to their activity. In a study of domestic fowl living in a semi-natural environment, it was found that birds were farthest apart when they were walking. They were closer when ground pecking during foraging, even closer when they were idly standing and closest when they were preening (Keeling and Duncan, 1991). It has also been observed that birds will come close enough for physical contact with each other when they are dust bathing and when they are roosting (Wood-Gush et al., 1978).

Of course in conventional battery cages, hens are extremely limited in the behaviour patterns that they can perform. The social spacing that hens prefer for walking, foraging and even just standing, is not available to them in battery cages at commercial bird densities. In addition, although hens in cages may be able to space themselves appropriately for roosting and dust-bathing, these activities cannot be performed in a natural way in cages. It would therefore seem that hens in battery cages are subject to suffering due to social stresses.

Once again it is conventional battery cages that result in the social stresses described in this section. If battery cages are eventually banned world-wide, the problem will largely disappear. However, there is one important proviso: Any of the alternative husbandry systems to battery cages or any of the husbandry systems used for meat poultry *could* be intensified to the extent that social stresses may start to impinge on the birds. For example, commercial broiler production is currently very close to that limit. If market forces pushed broiler production to delay slaughter to 7 weeks rather than 5 weeks, then social stress would become a problem for broilers. Therefore, intensity limits must be set for all classes of poultry.

## 6 Conclusions and future trends

In modern production systems there are several common practices that cause suffering in domestic fowl. The most important of these are as follows:

1. Keeping egg-laying hens in traditional battery cages leads to intense frustration in most birds for about 1½ h before each egg is laid. The solution to this problem is to move to husbandry systems in which a secluded nesting site is provided.
2. Painful keel bone fractures are common in all alternative husbandry systems to battery cages. These systems need to be better designed and better illuminated to allow hens to change levels safely by flying. In addition, chicks require to be reared from 2 to 3 weeks in an environment that allows them to learn how to change level safely.
3. Outbreaks of feather-pecking in laying strains cause severe suffering. Although infra-red beak trimming is minimally painful and currently fairly effective in preventing feather-pecking, that may not continue if the tendency to feather peck increases over generations. Breeding companies must be persuaded to select strains of bird that do not feather-peck and do not need to be beak trimmed.
4. Perch design and location in alternative husbandry systems for layers need to be improved to (a) prevent the occurrence of bumblefoot and (b) allow birds to perch as high off the ground as possible.
5. Food restriction of broiler breeders causes severe suffering. Breeding companies must be persuaded to take a step backwards and produce breeders with smaller appetites.
6. Toe trimming of male broiler breeders causes pain and problems with balance. Breeding companies must be persuaded to devise breeding programs in which there is (a) much less of a size discrepancy between males and females and (b) male lines that show full courtship behaviour that elicits crouching in the females. This should eliminate the need for toe trimming.

## 7 Where to look for further information

There are several textbooks available that give good information on various aspects of poultry welfare. *Welfare of the Laying Hen* edited by G. C. Perry is rather old (2004) but contains many excellent chapters. *Biology of Breeding Poultry* edited by Paul Hocking (2009a) has chapters that deal with various welfare problems of breeding birds. The most recent book on poultry welfare is *Advances in Poultry Welfare* edited by J. A. Mench (2018) and this gives in-depth coverage of all the topics covered in this chapter. Details of these three books are given in the References.

A good source of the most recent information on poultry welfare can be found in the journal *British Poultry Science* which has a section called Behaviour, Welfare, Husbandry & Environment. Papers on poultry welfare also commonly appear in the journals *Applied Animal Behaviour Science* and *Animal Welfare*.

The International Society for Applied Ethology (ISAE) meets annually. Papers on poultry welfare are often presented at these meetings and appear in the Proceedings of the ISAE.

## 8 References

Aerni, V., El-Lethey, H. and Wechsler, B. (2000). Effect of foraging material and food form on feather pecking in laying hens, *Br. Poult. Sci.* 41(1), 16-21.

Alm, M., Tauson, R., Holm, L., Wichman, A., Kalliokoski, O. and Wall, H. (2016). Welfare indicators in laying hens in relation to nest exclusion, *Poult. Sci.* 95(6), 1238-1247.

Appleby, M. C., Hughes, B. O. and Hogarth, G. S. (1989). Behaviour of laying hens in a deep litter house, *Br. Poult. Sci.* 30(3), 545-553.

Appleby, M. C., Hughes, B. O., McDonald, A. and Cordiner, L. S. (1998). Factors affecting use of perches in cages by laying hens, *Br. Poult. Sci.* 39, 186-190.

Banks, E. M., Wood-Gush, D. G. M., Hughes, B. O. and Mankovich, N. J. (1979). Social rank and priority of access to resources in domestic fowl, *Behav. Process.* 4(3), 197-209.

Bareham, J. R. (1972). Effects of cages and semi-intensive deep litter pens on the behaviour, adrenal response and production in two strains of laying hens, *Br. Vet. J.* 128(3), 153-163.

Benson, G. J. (2004). Pain in farm animals: nature, recognition and management. In: Benson, G. J. and Rollin, B. E. (Eds), *The Well-Being of Farm Animal: Challenges and Solutions*, Ames, IA: Blackwell Publishing, pp. 61-84.

Bentham, J. (1789). *Introduction to the Principles of Morals and Legislation*, (1996 Imprint), Oxford: Clarendon Press.

Berg, C. (2004). Pododermatitis and hock burn in broiler chickens. In: Weeks, C. and Butterworth, A. (Eds), *Measuring and Auditing Broiler Welfare*, Wallingford, UK: CABI Publishing, pp. 37-49.

Bilčík, B. and Keeling, L. J. (2000). Relationship between feather pecking and ground pecking in laying hens and the effect of group size, *Appl. Anim. Behav. Sci.* 68(1), 55-66.

Blokhuis, H. J. (1984). Rest in poultry, *Appl. Anim. Behav. Sci.* 12(3), 289-303.

Blokhuis, H. J. (1986). Feather-pecking in poultry: its relation with ground pecking, *Appl. Anim. Behav. Sci.* 16(1), 63–67.

Blokhuis, H. J. and Arkes, J. G. (1984). Some observations on the development of feather-pecking in poultry, *Appl. Anim. Behav. Sci.* 12(1–2), 145–157.

Bryant, J. (1972). The social environment: behaviour and stress in housed livestock, *Vet. Rec.* 90(13), 351–359.

Candland, D. K. and Nagy, Z. M. (1969). The open field; some comparative data, *Ann. N. Y. Acad. Sci.* 159(3), 831–851.

Carver, C. S. (2001). Affect and the functional bases of behavior: on the dimensional structure of affective experience, *Pers. Soc. Psychol. Rev.* 5(4), 345–356.

Chance, M. R. A. (1962). An interpretation of some agonistic postures, *Symp. Zool. Soc., Lond.* 8, 71–89.

Cloutier, S., Newberry, R. C., Honda, K. and Alldredge, J. R. (2002). Cannibalistic behaviour spread by social learning, *Anim. Behav.* 63(6), 1153–1162.

Collias, N. E. and Joos, M. (1953). The spectrographic analysis of signals of the domestic fowl, *Behaviour* 5, 176–188.

Command Paper 2836 (1965). *Report of the Technical Committee to Enquire into the Welfare of Animals kept under Intensive Livestock Husbandry Systems*, Command Report 2836, London: Her Majesty's Stationery Office.

Danbury, T. C., Weeks, C. A., Chambers, J. P., Waterman-Pearson, A. E. and Kestin, S. C. (2000). Self-selection of the analgesic drug Carprofen by lame broiler chickens, *Vet. Rec.* 146(11), 307–311.

Dawkins, M. S. (1980). *Animal Suffering: the Science of Animal Welfare*, London: Chapman and Hall.

Dawkins, M. S. (1981). Priorities in the cage size and flooring preferences of domestic hens, *Br. Poult. Sci.* 22(3), 255–263.

Dawkins, M. S. (1985). Cage height preference and use in battery-kept hens, *Vet. Rec.* 116(13), 345–347.

Dawkins, M. S. (1989). Time budgets in red jungle fowl as a basis for the assessment of welfare in domestic fowl, *Appl. Anim. Behav. Sci.* 24(1), 77–80.

de Jong, I. C., Enting, IH., van Voorst, A. and Blokhuis, H. J. (2005). Do low-density diets improve broiler breeder welfare during rearing and laying?, *Poult. Sci.* 84(2), 194–203.

Dixon, L. M., Duncan, I. J. H. and Mason, G. (2008). What's in a peck? Using fixed action pattern morphology to identify the motivational basis of abnormal feather-pecking behaviour, *Anim. Behav.* 76(3), 1035–1042.

Duncan, I. J. H. (1970). Frustration in the fowl. In: Freeman, B. M. and Gordon, R. F. (Eds), *Aspects of Poultry Behaviour*, Edinburgh, UK: British Poultry Science, pp. 15–31.

Duncan, I. J. H. (1985). How do fearful birds respond? In: Wegner, R.-M. (Ed.), Second European Symposium on Poultry Welfare, Celle, Germany: German Branch of the World's Poultry Science Association, pp. 96–106.

Duncan, I. J. H. (1989). The assessment of welfare during the handling and transport of broilers. In: Faure, J. M. and Mills, A. D. (Eds), Third European Symposium on Poultry Welfare, Tours, France: French Branch of the World's Poultry Science Association, pp. 93–107.

Duncan, I. J. H. (1990). Reactions of poultry to human beings. In: Zayan, R. and Dantzer, R. (Eds), *Social Stress in Domestic Animals*, Dordrecht, The Netherlands: Kluwer Academic Publishers, pp. 121–131.

Duncan, I. J. H. (1992). The effect of the researcher on the behavior of poultry. In: Davis, H. and Balfour, D. (Eds), *The Inevitable Bond: Examining Scientist-Animal Interactions*, Cambridge, UK: Cambridge University Press, pp. 285-294.

Duncan, I. J. H. and Filshie, J. H. (1980). The use of radiotelemetry devices to measure temperature and heartrate in the domestic fowl. In: Amlaner, C. J. and Macdonald, D. W. (Eds), *A Handbook on Biotelemetry and Radio Tracking*, London: Pergamon Press, pp. 579-588.

Duncan, I. J. H. and Fraser, D. (1997). Understanding animal welfare. In: Appleby, M. C. and Hughes, B. O. (Eds), *Animal Welfare*, Wallingford, UK: CABI Publishing, pp. 19-31.

Duncan, I. J. H. and Kite, V. G. (1989). Nest site selection and nest building behaviour in domestic fowl, *Anim. Behav.* 37, 215-231.

Duncan, I. J. H. and Mench, J. (2000). 'Does hunger hurt?', Letter to the Editor, *Poult. Sci.* 79(6), 934.

Duncan, I. J. H., Savory, C. J. and Wood-Gush, D. G. M. (1978). Observations on the reproductive behaviour of domestic fowl in the wild, *Appl. Anim. Ethol.* 4(1), 29-42.

Duncan, I. J. H., Slee, G. S., Kettlewell, P., Berry, P. and Carlisle, A. J. (1986). Comparison of the stressfulness of harvesting broilers by machine and by hand, *Br. Poult. Sci.* 27, 87-92.

Duncan, I. J. H., Slee, G. S., Seawright, E. and Breward, J. (1989). Behavioural consequences of partial beak amputation (beak trimming) in poultry, *Br. Poult. Sci.* 30(3), 479-488.

Duncan, I. J. H. and Wood-Gush, D. G. M. (1971). Frustration and aggression in the domestic fowl, *Anim. Behav.* 19(3), 500-504.

Duncan, I. J. H. and Wood-Gush, D. G. M. (1972a). Thwarting of feeding behaviour in the domestic fowl, *Anim. Behav.* 20(3), 444-451.

Duncan, I. J. H. and Wood-Gush, D. G. M. (1972b). An analysis of displacement preening in the domestic fowl, *Anim. Behav.* 20(1), 68-71.

European Commission (1999). Council Directive 1999/74/EC laying down minimum standards for the protection of laying hens, *Off. J. Eur. Commun.* L203, 53-57.

Farquharson, C. and Jeffries, D. (2000). Chondrocytes and longitudinal bone growth: the development of tibial dyschondroplasia, *Poult. Sci.* 79(7), 994-1004.

Faure, J. M., Jones, R. B. and Bessei, W. (1983). Fear and social motivation as factors in open-field behaviour of the domestic chick: a theoretical consideration, *Biol. Behav.* 8, 103-116.

Follensbee, M. E., Duncan, I. J. H. and Widowski, T. M. (1992). Quantifying nesting motivation of laying hens, *J. Anim. Sci.* 70, Suppl. 1, 50.

Fournier, J., Schwean-Lardner, K., Knezacek, T. D., Gomis, S. and Classen, H. L. (2015). The effect of toe trimming on behavior, mobility, toe length and other indicators of welfare in tom turkeys, *Poult. Sci.* 94(7), 1446-1453.

Gentle, M. J. and Hunter, L. N. (1990). Physiological and behavioural responses associated with feather removal in *Gallus gallus* var. *domesticus*, *Res. Vet. Sci.* 50(1), 95-101.

Gentle, M. J. and McKeegan, D. E. F. (2007). Evaluation of the effects of infrared beak trimming in broiler breeder chicks, *Vet. Rec.* 160(5), 145-148.

Gentle, M. J., Waddington, D., Hunter, L. N. and Jones, R. B. (1991). Behavioural evidence for persistent pain following partial beak amputation in chickens, *Appl. Anim. Behav. Sci.* 27(1-2), 149-157.

Gibson, S. W., Dun, P. and Hughes, B. O. (1988). The performance and behaviour of laying fowls in a covered strawyard system, *Res. Dev. Agric.* 5, 153-163.

Gottier, R. F. (1968). The dominance-submission hierarchy in the social behavior of the domestic chicken, *J. Gen. Psychol.* 112, 205-226.

Gray, J. (1971). *The Psychology of Fear and Stress*, London, UK: Weidenfeld and Nicolson.

Griffin, D. R. (1976). *The Question of Animal Awareness*, New York: The Rockefeller University Press.

Guhl, A. M. (1968). Social inertia and social stability in chickens, *Anim. Behav.* 16(2), 219-232.

Hansen, R. S. (1976). Nervousness and hysteria of mature female chickens, *Poult. Sci.* 55(2), 531-543.

Harlander-Matauschek, A., Rodenburg, T. B., Sandilands, V. and Tobalske, B. W. (2015). Causes of keel bone damage and their solutions in laying hens, *World's Poult. Sci. J.* 71, 461-472.

Harrison, R. (1964). *Animal Machines*, London, UK: Vincent Stuart.

Hill, A. T. and Binns, M. R. (1973). Effect of varying population density and size on laying performance, Proceedings of the 4th European Poultry Conference, London, pp. 605-609.

Hocking, P. (Ed.) (2009a). *Biology of Breeding Poultry*, Wallingford, UK: CABI Publishing.

Hocking, P. M. (2009b). Feed restriction. In: Hocking, P. M. (Ed.), *Biology of Breeding*, Poultry, Wallingford, UK: CABI Publishing, pp. 307-330.

Hughes, B. O. (1975). Spatial preference in the domestic hen, *Br. Vet. J.* 131(5), 560-564.

Hughes, B. O. and Black, A. J. (1976). The influence of handling on egg production, egg shell quality and avoidance behaviour of hens, *Br. Poult. Sci.* 17(2), 135-144.

Hughes, B. O., Gilbert, A. B. and Brown, M. F. (1986). Categorisation and causes of abnormal eggshells: relationship with stress, *Br. Poult. Sci.* 27(2), 325-337.

James, W. (1904). Does "consciousness" exist?, *J. Philos. Psychol. Sci. Method* 1(18), 477-491.

Jones, R. B. (1977). Repeated exposure of the domestic chick to a novel environment; effects on behavioural responses, *Behav. Process.* 2(2), 163-173.

Jones, R. B. (1978). Activities of chicks in their home cages and in an open field, *Br. Poult. Sci.* 19(6), 725-730.

Jones, R. B. (1980). Responses of male and female domestic chicks to a startling stimulus and the effects of a tranquilliser, *Behav. Process.* 5(2), 161-172.

Jones, R. B., Duncan, I. J. H. and Hughes, B. O. (1981). The assessment of fear in domestic hens exposed to a looming human stimulus, *Behav. Process.* 6(2), 121-133.

Jones, R. B. and Mills, A. D. (1983). Estimation of fear in two lines of the domestic chick: correlations between various measures, *Behav. Proc.* 8, 243-253.

Julian, R. J. (1998). Rapid growth problems: ascites and skeletal deformities in broilers, *Poult. Sci.* 77(12), 1773-1780.

Keeling, L. J. and Duncan, I. J. H. (1989). Inter-individual distances and orientation in laying hens housed in groups of three in two different-sized enclosures, *Appl. Anim. Behav. Sci.* 24(4), 325-342.

Keeling, L. J. and Duncan, I. J. H. (1991). Social spacing in domestic fowl under semi-natural conditions: the effect of behavioural activity and activity transitions, *Appl. Anim. Behav. Sci.* 32(2-3), 205-217.

Kjaer, J. B., Sørensen, P. and Su, G. (2001). Divergent selection of feather pecking behaviour in laying hens (*Gallus gallus domesticus*), *Appl. Anim. Behav. Sci.* 71(3), 229-239.

Kjaer, J. B. and Vestergaard, K. S. (1999). Development of feather pecking in relation to light intensity, *Appl. Anim. Behav. Sci.* 62(2-3), 243-254.

Koene, P. and Wiepkema, P. R. (1991). Pre-dustbath vocalizations as an indicator of need in domestic hens. In: Böhncke, E. and Molkenthin, V. (Eds), *Alternatives in Animal Husbandry*, Witzenhausen, Germany: Agrarkultur Verlag, pp. 95–103.

Kuenzel, W. J., Wideman, R. F., Chapman, M., Golden, C. and Hooge, D. M. (2005). A practical method for inducing moulting of caged layers that combines full access to feed and water, thyroactive protein, and short day-length, *World's Poult. Sci. J.* 61, 599–624.

Kruijt, J. P. (1964). Ontogeny of social behaviour in Burmese Red Junglefowl (*Gallus gallus spadiceous*), *Behaviour, Suppl.*, XII, 1-201.

Lambton, S. L., Knowles, T. G., Yorke, C. and Nicol, C. J. (2010). The risk factors affecting the development of gentle and severe feather pecking in loose housed laying hens, *Appl. Anim. Behav. Sci.* 123(1–2), 32–42.

Lambton, S. L., Knowles, T. G., Yorke, C. and Nicol, C. J. (2015). The risk factors affecting the development of vent pecking and cannibalism in free-range and organic laying hens, *Anim. Welf.* 24(1), 101–111.

Laycock, S. R. and Ball, R. O. (1990). Alleviation of hysteria in laying hens with dietary tryptophan, *Can. J. Vet. Res.* 54(2), 291–295.

Lee, Y. P., Craig, J. V. and Dayton, A. D. (1982). The social rank index as a measure of social status and its association with egg production in White Leghorn pullets, *Appl. Anim. Ethol.* 8(4), 377–390.

Leeson, S. and Summers, J. D. (1979). Wet litter with breeders? Restrict feed and water, *Broiler Ind.* 42, 44–46.

Leeson, S. and Summers, J. D. (2000). *Broiler Breeder Production*, Guelph, Canada: University Books.

McAdie, T. M. and Keeling, L. J. (2002). The social transmission of feather pecking in laying hens: effects of environment and age, *Appl. Anim. Behav. Sci.* 75(2), 147–159.

McBride, G., Parer, I. P. and Foenander, F. (1969). The social organization and behaviour of the feral domestic fowl, *Anim. Behav. Monogr.* 2, 125–181.

McCowan, B., Schrader, J., DiLorenzo, A. M., Cardona, C. and Klingborg, D. (2006). Effects of induced molting on the well-being of egg-laying hens, *J. Appl. Anim. Welf. Sci.* 9(1), 9–23.

McFarland, D. J. (1966a). On the causal and functional significance of displacement activities, *Z. Tierpsychol.* 23(2), 217–235.

McFarland, D. J. (1966b). The role of attention in the disinhibition of displacement activities, *Quart. J. Exp. Psychol.* 18(1), 19–30.

McFarland, D. (1981). *The Oxford Companion to Animal Behaviour*, Oxford: Oxford University Press, p. 180.

McKeegan, D. (2017). Beak trimming of laying hens: welfare costs and benefits. In: Roberts, J. (Ed.) *Achieving Sustainable Production of Eggs. Volume 2: Animal Welfare and Sustainability*, Cambridge: Burleigh Dodds Science Publishing, pp. 125–143.

McKeegan, D. E. F. and Philbey, A. W. (2012). Chronic neurophysiological and anatomical changes associated with infrared beak treatment and their implications for laying hen welfare, *Anim. Welf.* 21(2), 207–217.

Meijsser, F. M. and Hughes, B. O. (1989). Comparative analysis of pre-laying behaviour in battery cages and in three alternative systems, *Br. Poult. Sci.* 30(4), 747–760.

Mench, J. A. (2004). Lameness. In: Weeks, C. A. and Butterworth, A. (Eds), *Measuring and Auditing Broiler Welfare*, Wallingford, Oxon: CABI Publishing, pp. 3–17.

Mench, J. A. (Ed.) (2018). *Advances in Poultry Welfare*, Duxford, UK: Woodhead Publishing.

Mench, J. A. and Falcone, C. (2000). Welfare concerns in feed-restricted meat-type poultry parent stocks, Proceedings of the 21st World's Poultry Congress, (Paper S3.3.03), Montreal: World's Poultry Science Association.

Mendl, M., Burman, O. H. P. and Paul, E. S. (2010). An integrative and functional framework for the study of animal emotion and mood, *Proc. Biol. Sci.* 277(1696), 2895–2904.

Millman, S. T., Duncan, I. J. H. and Widowski, T. M. (2000). Male broiler breeder fowl display high levels of aggression towards females, *Poult. Sci.* 79(9), 1233–1241.

Mills, A. D., Duncan, I. J. H., Slee, G. S. and Clark, J. S. B. (1985). Heart rate and laying behaviour in two strains of domestic chicken, *Physiol. Behav.* 35(1), 145–147.

Moinard, C., Statham, P., Haskell, M. J., McCorquodale, C., Jones, R. B. and Green, P. R. (2004). Accuracy of laying hens in jumping upwards and downwards between perches in different light environments, *Appl. Anim. Behav. Sci.* 85(1–2), 77–92.

Molony, V. (1997). Comments on Anand and Craig, (Letters to the Editor), *Pain* 70(2–3), 293.

Muir, W. M. (1996). Group selection for adaptation to multi-hen cages: selection program and direct responses, *Poult. Sci.* 75(4), 447–458.

Murphy, L. B. (1977). Responses of domestic fowl to novel food and objects, *Appl. Anim. Ethol.* 3(4), 335–349.

Murphy, L. B. (1978). The practical problems of recognizing and measuring fear and exploration behaviour in the domestic fowl, *Anim. Behav.* 26, 422–431.

Murphy, L. B. and Duncan, I. J. H. (1977). Attempts to modify the responses of domestic fowl towards human beings. I. The association of human contact with a food reward, *Appl. Anim. Ethol.* 3(4), 321–334.

Murphy, L. B. and Duncan, I. J. H. (1978). Attempts to modify the responses of domestic fowl towards human beings. II. The effect of early experience, *Appl. Anim. Ethol.* 4(1), 5–12.

Murphy, L. B. and Wood-Gush, D. G. M. (1978). The interpretation of the behaviour of the domestic fowl in strange environments, *Biol. Behav.* 3, 39–61.

Nasr, M. A. F., Browne, W. J., Caplen, G., Hothersall, B., Murrell, J. C. and Nicol, C. J. (2013). Positive affective state induced by opioid analgesia in laying hens with bone fractures, *Appl. Anim. Behav. Sci.* 147(1–2), 127–131.

Nasr, M. A. F., Nicol, C. J. and Murrell, J. C. (2012). Do laying hens with keel bone fractures experience pain?, *PLoS ONE* 7(8), e42420. https://doi.org/10.1371/journal.pone.0042420.

Nicol, C. J. (1987a). Effect of cage height and area on the behaviour of hens housed in battery cages, *Br. Poult. Sci.* 28(2), 327–335.

Nicol, C. J. (1987b). Behavioural responses of laying hens following a period of spatial restriction, *Anim. Behav.* 35(6), 1709–1719.

Olsson, I. A. S. and Keeling, L. J. (2000). Night-time roosting in laying hens and the effect of thwarting access to perches, *Appl. Anim. Behav. Sci.* 68(3), 243–256.

Olsson, I. A. S. and Keeling, L. J. (2002). The push-door for measuring motivation in hens: laying hens are motivated to perch at night, *Anim. Welf.* 11, 11–19.

Perry, G. C. (Ed.) (2004). *Welfare of the Laying Hen*, Wallingford, UK: CABI Publishing.

Phillips, R. E. and Siegel, P. B. (1966). Development of fear in chicks of two closely related genetic lines, *Anim. Behav.* 14(1), 84–88.

Pickel, T., Schrader, L. and Scholz, B. (2011). Pressure load on keel bone and foot pads in perching laying hens in relation to perch design, *Poult. Sci.* 90(4), 715–724.

Pötzsch, C. J., Lewis, K., Nicol, C. J. and Green, L. E. (2001). A cross-sectional study of the prevalence of vent pecking in laying hens in alternative systems and its association with feather pecking, management and disease, *Appl. Anim. Behav. Sci.* 74(4), 259-272.

Prip, M. (1975). Hysteria in laying hens, *World's Poult. Sci. J.* 31, 306-308.

Ratner, S. C. and Boice, R. (1975). Effects of domestication on behaviour. In: Hafez, E. S. E. (Ed.), *The Behaviour of Domestic Animals*, London: Baillière Tindall, pp. 3-19.

Renema, R. and Robinson, F. E. (2000). Reproductive implications of full feeding female meat-type poultry parent stocks, Proceedings of the 21st World's Poultry Congress, (Paper S3.3.02), Montreal: World's Poultry Science Association.

Rentsch, A. K., Rufener, C. B., Spadavecchia, C., Stratman, A. and Toscano, M. J. (2019). Laying hen's mobility is impaired by keel bone fractures and does not improve with paracetamol treatment, *Appl. Anim. Behav. Sci.* 216, 19-25.

Riber, A. J. and Forkman, B. (2007). A note on the behaviour of the chicken that receives feather pecks, *Appl. Anim. Behav. Sci.* 12, 289-304.

Rodenburg, T. B., van Hierden, Y. M., Buitenhuis, A. J., Riestra, B., Koene, P., Korte, S. M., van der Poel, J. J., Groothuis, T. G. G. and Blokhuis, H. J. (2004). Feather pecking in laying hens: new insights and directions for research?, *Appl. Anim. Behav. Sci.* 86(3-4), 291-298.

Rodenburg, T. B., Koene, P., Bokkers, E. A. M., Bos, M. E. H., Uitdehaag, K. A. and Spruijt, B. M. (2005). Can short-term frustration facilitate feather pecking in laying hens?, *Appl. Anim. Behav. Sci.* 91(1-2), 85-101.

Romanes, G. J. (1884, reprinted 1969). *Mental Evolution in Animals*, New York: AMS Press.

Russell, J. A. (2003). Core affect and the psychological construction of emotion, *Psychol. Rev.* 110(1), 145-172.

Rutter, S. M. and Duncan, I. J. H. (1991). Shuttle and one-way avoidance as measures of aversion in the domestic fowl, *Appl. Anim. Behav. Sci.* 30(1-2), 117-124.

Sanotra, G. S., Lund, J. D., Erksboll, A. K., Petersen, J. S. and Vestergaard, K. S. (2001). Monitoring leg problems in broilers: a survey of commercial broiler production in Denmark, *World's Poult. Sci. J.* 57, 55-69.

Savory, C. J. (1989). Stereotyped behaviour as a coping strategy in restricted-fed broiler breeder stock. In: Faure, J. M. and Mills, A. D. (Eds). Proceedings of the Third European Symposium on Poultry Welfare, Tours, France: French Branch of the WPSA, pp. 261-264.

Savory, C. J. (1995). Feather pecking and cannibalism, *World's Poult. Sci. J.* 51, 215-219.

Savory, C. J., Hocking, P. M., Mann, J. S. and Maxwell, M. H. (1996). Is broiler breeder welfare improved by using qualitative rather than quantitative food restriction to limit growth, *Anim. Welf.* 5, 105-127.

Savory, C. J., Wood-Gush, D. G. M. and Duncan, I. J. H. (1978). Feeding behaviour in a population of domestic fowl in the wild, *Appl. Anim. Ethol.* 4(1), 13-27.

Schenk, P. M., Meysser, F. M. and Limpens, H. J. G. (1983). Gakeln als Indikator für Frustration bei Legehennen, *KTBL-Schrift* 299, 65-80.

Singh, P. M., Johnson, C. B., Gartrell, B., Mitchison, S., Jacob, A. and Chambers, P. (2017). Analgesic effects of morphine and butorphanol in broiler chickens, *Vet. Anaes. Analges* 44(3), 538-545.

Skinner, B. F. (1974). *About Behaviorism*, New York: Random House.

Stratmann, A., Fröhlich, E. K. F., Harlander-Matauschek, A., Schrader, L., Toscano, M. J., Würbel, H. and Gebhardt-Henrich, S. G. (2015). Soft perches in an aviary system reduce incidence of keel bone damage in laying hens, *PLoS ONE* 10(3), e0122568.

Struelens, E. and Tuyttens, F. A. M. (2009). Effects of perch design on behaviour and health of laying hens, *Anim. Welf.* 18, 533–538.

Tauson, R. (1980). Cages: how could they be improved? In: Moss, R. (Ed), *The Laying Hen and Its Environment*, The Hague: Martinus Nijhoff, pp. 269-304.

Tauson, R. and Abrahamsson, P. (1994). Foot and skeletal disorders in laying hens, *Acta Agri. Scand. A* 44, 110–119.

Thiessen, D. D., Zolman, J. F. and Rodgers, D. A. (1962). Relation between adrenal weight, brain cholinesterase activity, and hole-in-the-wall behaviour of mice under different living conditions, *J. Comp. Physiol. Psychol.* 55, 186–190.

Tinbergen, N. (1952). "Derived" activities: their causation, biological significance, origin and emancipation during evolution, *Quart. Rev. Biol.* 27(1), 1-32.

Wathes, C. M., Charles, D. R. and Clark, J. A. (1985). Group size and plumage damage in laying hens, *Br. Poult. Sci.* 26(4), 459–463.

Watson, J. B. (1928). *Behaviorism*, London: Routledge and Keegan Paul.

Weary, D. M. (2014). What is suffering in animals? In: Appleby, M. C., Weary, D. M. and Sandøe, P. (Eds), *Dilemmas in Animal Welfare*, Wallingford, UK: CAB International, pp. 188–202.

Webster, A. B. (2000). Behavior of White Leghorn laying hens after withdrawal of feed, *Poult. Sci.* 79(2), 192–200.

Wells, R. G. (1973). Stocking density and colony size for caged layers, Proceedings of the 4th European Poultry Conference, London, pp. 617–622.

Whitehead, C. C., Fleming, R. H., Julian, R. J. and Sørensen, P. (2003). Skeletal problems associated with selection for increased production. In: Muir, W. M. and Aggrey, S. E. (Eds), *Poultry Genetics, Breeding and Biotechnology*, Wallingford, UK: CAB International, pp. 29–52.

Wilkins, L. J., McKinstry, J. L., Avery, N. C., Knowles, T. G., Brown, S. N., Tarlton, J. and Nicol, C. J. (2011). Influence of housing system and design on bone strength and keel bone fractures in laying hens, *Vet. Rec.* 169(16), 414.

Wood-Gush, D. G. M. (1972). Strain differences in response to sub-optimal stimuli in the fowl, *Anim. Behav.* 20(1), 72-76.

Wood-Gush, D. G. M. (1975). Nest construction by the domestic hen: some comparative and physiological considerations. In: Wright, P., Caryl, P. G. and Vowles, D. M. (Eds), *Neural and Endocrine Aspects of Behaviour in Birds*, Amsterdam: Elsevier, pp. 35–49.

Wood-Gush, D. G. M., Duncan, I. J. H. and Fraser, D. (1975). Social stress and welfare problems in agricultural animals. In: Hafez, E. S. E. (Ed) *The Behaviour of Domestic Animals* (3rd edn), Baltimore,MD: Williams and Wilkins, pp. 182–200.

Wood-Gush, D. G. M., Duncan, I. J. H. and Savory, C. J. (1978). Observations on the social behaviour of domestic fowl in the wild, *Biol. Behav.* 3, 193–205.

Wood-Gush, D. G. M. and Gilbert, A. B. (1969). Oestrogen and pre-laying behaviour of the domestic hen, *Anim. Behav.* 17(3), 586–589.

Wood-Gush, D. G. M. and Gilbert, A. B. (1973). Some hormones involved in the nesting behaviour of hens, *Anim. Behav.* 21(1), 98–103.

Wood-Gush, D. G. M. and Gilbert, A. B. (1975). The physiological basis of a behaviour pattern in the domestic hen, *Symp. Zool. Soc. Lond.* 35, 261–276.

Yue, S. and Duncan, I. J. H. (2003). Frustrated nesting behaviour: relation to extra-cuticular shell calcium and bone strength in White Leghorn hens, *Br. Poult. Sci.* 44(2), 175–181.

Zimmerman, P. H. and Koene, P. (1998a). The effect of frustrative nonreward on vocalisations and behaviour in the laying hen, *Gallus gallus domesticus*, *Behav. Processes* 44(1), 73–79.

Zimmerman, P. H. and Koene, P. (1998b). The effect of loss of predictability and controllability of reward during frustration on behaviour in two strains of laying hens, *Gallus gallus domesticus*, *Neth. J. Zool.* 48, 255–265.

Zimmerman, P. H., Koene, P. and van Hooff, J. A. R. A. M. (2000). Thwarting of behaviour in different contexts and the gakel-call in the laying hen, *Appl. Anim. Behav. Sci.* 69(4), 255–264.

# Chapter 4

# Understanding chicken learning and cognition and implications for improved management

*Rafael Freire, Charles Sturt University, Australia*

1 Introduction
2 Early learning
3 Habituation and associative learning
4 Cognition
5 Improving management through an understanding of learning and cognitive abilities
6 Conclusion
7 Where to look for further information
8 References

## 1 Introduction

A great deal of research has revealed the complex and remarkable learning and cognitive abilities of chickens. Most of this research has been undertaken in the laboratory, often using commercial strains of chickens, and, in fact, surprisingly little is known about the cognition of wild Junglefowl. Many of the laboratory tests that reveal complex mental abilities in themselves rely on cognitive processes such as imprinting, associative learning and discrimination. Unusual in the comparative cognition research, many of these tests have been undertaken on newly hatched chicks, which have highly developed brains and are capable of learning from a young age. One advantage of testing these abilities in an artificial laboratory is that it is possible to rule out through careful experimental design and manipulations that such complex cognition has not arisen by simpler mechanisms. Although this may not necessarily reduce the impressiveness of the mental feats, it has important implications for those working on the mechanisms that are employed in complex cognition. Many of the studies cited in this chapter describe the key research that has made significant advances in providing evidence of the functioning of these cognitive mechanisms.

http://dx.doi.org/10.19103/AS.2020.0078.04

## 2 Early learning

### 2.1 Maternal effects

Domestic chicks are a precocial species that can respond to their environment and cope with new situations in ways that allow them to survive and thrive with little maternal input. Studies on precocial knowledge indicate that chicks are able to, without learning, distinguish and attend to stimuli and so guide innate approach and avoidance responses (Versace and Vallortigara 2015). For example, chicks have an innate preference for hollow as opposed to filled cylinders though the possible function of this preference, or what this finding reveals about the innate recognition processes, remains unclear (Versace et al. 2016). Domestic chicks can also show rapid and important learning during early life, which is briefly reviewed in this section, but also later in this chapter (learning what to peck (Section 4.1); commercial rearing (Section 5.1.2); learning about spatial cues (Section 4.6)). Soon after hatching, chicks learn to identify the hen, or the mother figure, from other prominent stimuli by a process termed 'imprinting'. Imprinting usually occurs within the first 24 h post-hatching and is facilitated by chicks' learning to discriminate calls by the mother from other noises 1 day before hatching (Fält 1981). Learning to discriminate the mother figure from other stimuli occurs rapidly and leaves a strong memory trace that is difficult to reverse once established. In a natural setting, or if the mother hen is present after hatching, then in most cases, the chick will imprint on the mother. However, if the mother is absent, chicks will approach other conspicuous visual stimuli as soon as they are physically able to do so and will follow such stimuli if they are moving (Bateson 1966).

The importance of the mother to the chick's behavioural and cognitive development has in recent times attracted researchers' considerable attention (Edgar et al. 2016). Hens produce a variety of calls which serve to direct appropriate chick behaviour. Cluck calls, for example, serve to attract the chick towards the hen, and chicks learn to discriminate between their own mother's cluck call and those of other hens by 4 days of age (Fält 1981). Of particular interest has been the influence of the mother's vocalisation and pecking behaviour which appear to shape the chick's foraging skills and ability to discriminate between edible and non-edible food items. Broody hens emit a characteristic vocalisation when pecking at new food items which attract and encourage chicks to direct their pecks at the food items (Wauters and Richard-Yris 2002). This latter effect is so pronounced that, in the first week of life, chicks reared with a foster hen rely on the hen to find food caches and appear to rarely discover hidden food caches by themselves (Gajdon et al. 2015). In one key study that revealed the possible complexity of the hen-chick interaction, Nicol and Pope (1996) examined the behaviour of hens and chicks feeding on two differently coloured foods. Chicks were pre-trained that food of one colour was palatable, and that food of the other colour was unpalatable, and crucially the mother had been pre-trained separately that the reverse colour combinations indicated palatability. When the hen watched chicks pecking from a food source that she had been trained to regard as unpalatable, but wasn't, the hen increased food pecking, ground pecking and scratching. The study indicated that hens are sensitive to foraging errors made by the chicks and adjust their behaviour to redirect chick pecks towards palatable foods. The mother, therefore, appears to play a key role in the chick learning what and what not to peck .

Additionally, considerable attention has been given to the possibility that maternal effects can have a buffering effect on chick fearfulness. Developing methods which reduce chick fearfulness can be expected to provide welfare benefits when placed in commercial systems since 'flightiness' can lead to injuries and smothering (Widowski and Torrey 2018). There is some evidence that behaviour between the mother and the chick in response to fear-eliciting events may be influenced by cognitive processes. Edgar et al. (2013) found that hens increased vocalisations and walking, and decreased preening, when they perceived their chicks threatened (an environmental cue signalling the delivery of a puff of air). Crucially, this response was shown when the chicks were naïve to the cue signal, indicating that the hen's response was based on her knowledge of the perceived threat to the chick and not by the chick's behaviour. The above-mentioned findings indicate that mother hens are sensitive to imminent threats for their chicks and appear to alter behaviour in order to elicit appropriate responses in their chicks. These findings indicate that the mother may contribute to the development of the behaviour of the chick for coping with adverse environmental challenges.

## 2.2 Brood companion effects

As a highly social animal, it is not surprising to find that chicks seek the company of other chicks, and contact with other chicks reduces fear responses in an open field test (Jones 1989). Domestic chicks learn to discriminate familiar chicks from unfamiliar chicks in the early life and will approach a brood mate rather than an unfamiliar chick in a runway (Vallortigara and Andrew 1991). Detour tests using different goal animal combinations also suggest that chicks may be able to interpret the social interactions between chicks. Regolin et al. (1994) found that chicks would solve a detour problem faster if the goal was two interacting chicks, rather than two chicks unable to interact due to a visual barrier between them. This finding indicates that chicks recognise the social interactions of other chicks and perhaps find 'interacting' chicks more attractive than non-interacting chicks. More recent work has indicated that chicks are able to discriminate between familiar and unfamiliar birds based on acoustic calls alone. Koshiba et al. (2013) showed that chicks elicited more intermediate contact calls and a distinct distress call to a familiar compared to an unfamiliar chick when the pair of chicks were separated by a visual barrier which allowed acoustic contact. This latter study also found that chicks approached and exhibited bonding calls to unfamiliar chicks just as much as to familiar chicks, indicating that isolated chicks are motivated to gain contact with other chicks, whether familiar or unfamiliar. Combined, all these findings indicate the importance of a social environment made up of familiar, interacting chicks to the development of appropriate behaviour in the chicks. The influence of the mother hen on the development of pecking preferences in the first week of life raises the possibility that chicks may also learn about what to peck and where to find food from other brood mates, and is discussed in Section 4.1.

## 3 Habituation and associative learning

Habituation is one of the simplest forms of learning and is the decreased response to the repeated presentation of the same stimulus. For example, repeated visual exposure to a human results in lower avoidance of humans in encounter tests, and this is seen in both docile and flighty strains of chickens (Jones 1995). It is likely that habituation learning occurs to many stimuli in commercially housed chickens and, as the previous example indicates, may attenuate responses to mildly stressful events. Along with habituation to the presence of stockpeople, chickens may habituate to machinery or other sounds, sight of vehicles or smells.

It is likely that birds are also continuously forming associations between events, termed 'classical or Pavlovian conditioning'. For example, the sound of automatic feeders starting can set off a range of anticipatory responses and vocalisations in chickens in commercial systems. Associations formed between events and locations can also influence approach and avoidance behaviour. For example, Jones et al. (2012) showed that chickens preferred a compartment in which they had previously received food but avoided a compartment in which they had previously received an aversive noise. In chickens, the response to the reinforcer is not fixed and unvarying, but can vary depending on the quality of the reinforcer. Poling et al. (1996) trained hens to peck one of two keys that matched the colour of a light presented just prior. Chickens were more accurate when the reward was greater (4 s access to food) than when the reward was less generous (1 s access to food), indicating that the outcome (reward) influenced their response accuracy.

Whereas in classical conditioning the animal learns to associate two events with each other, in operant or instrumental learning the behaviour is shaped by the previous outcome of a particular event. If the outcome is desirable, the animal will perform more of the behaviour, whereas if the outcome was negative, the animal will perform less of the behaviour. Early work, for example, showed that hens readily learnt to peck a plate in order to obtain access to food (Duncan and Hughes 1972). Operant conditioning has mostly been used by animal welfare scientists to investigate preferences for different resources. For example, Faure (1986) used an apparatus in which a hen had to peck a key to be provided with more feeder space and found that hens rarely pecked to obtain more than 10 cm of the feeder space per bird. Additionally, operant conditioning provides animals with some degree of anticipation and control over their environment that may have a positive effect on welfare. That is, since the event follows the behaviour, the animal not only has knowledge about when the event will occur but also may be able to control the timing of the event.

Additionally, it is not unusual to find that animals prefer to show a behaviour in order to receive an outcome rather than have that outcome freely available, a phenomenon called 'contrafreeloading'. Tests on many animals, including chickens (Duncan and Hughes 1972), have shown that they will readily learn some operant response, such as manipulating a lever or pressing a panel in order to receive food. Lindberg and Nicol (1994) found that chickens readily learnt to use operant feeders, and their use was associated with increased time spent on foraging behaviour. A positive effect on welfare appeared to be maximised when operant feeders were used in conjunction with ad lib feed, indicating that having to 'work' for food may be preferred by chickens compared to having food freely available. There is however some evidence that domestication has perhaps reduced the capacity of hens to continually work using operant feeders, as a White Leghorn commercial strain showed less feeding from an operant feeder than Red Junglefowl (Lindqvist and Jensen 2009). Lastly, although allowing chickens to form operant associations in order to achieve some control of their environment is generally considered to be good for welfare, if the behaviour does not achieve the desired outcome or if the outcome is negative, then the animal may stop trying to exert an influence over its environment, a condition called 'learned helplessness'.

## 4 Cognition

### 4.1 Social learning

The ability to learn about the environment, events or new situations from other individuals is well developed in group-living chickens (Nicol 2006). Even from a young age, domestic chicks will avoid pecking at an aversive-tasting bead after they have observed the disgust response of another chick (Johnston et al. 1998). As well as learning what and what not to peck by observing other chicks or their mother, chicks can use the foraging behaviour of other chicks to guide their own foraging behaviour in a more general way. In one of the earliest demonstrations of this influence, McQuoid and Galef (1992) showed that young Junglefowl preferred to eat from the same type of dish or in the same location after seeing a conspecific feeding from that dish 48 h earlier.

More recent work which has looked at the food preferences and location preferences of chicks in the company of a knowledgeable demonstrator suggests that chicks learn what to peck mostly from a foster hen rather than from other chicks, even when the chick demonstrator is slightly older and more knowledgeable. Gajdon et al. (2015) showed that chicks that had experienced finding food at distinct locations in the presence of an older, experienced chick were more likely to search at those locations than chicks that had been in the same arena but with a naïve companion. However, chicks that had experienced foraging with an older tutor chick were no better at avoiding pecking at a non-food particle (gravel) than control chicks that had the experience of a naïve chick. It should be noted that in natural situations a chick is unlikely to be exposed to 'older, experienced tutors', since all brood mates would be roughly the same age and have the same experience. Nonetheless, the finding that chicks learn where to find food from brood mates suggests that chicks may be regarding the tutor as a surrogate providing valuable information that would normally come from the mother hen. Adult laying hens are also similarly able to use the movement of another bird to guide their own food searching but appear to be able to evaluate the ability of the other hen. In a laboratory arena in which hens were required to approach and peck a bowl in order to obtain the food in it, Wichman (2018) showed that observer hens were more likely to follow hens that they previously saw being successful at obtaining food more than hens that had previously not been seen to obtain food. The latter findings support the hypothesis that hens are sensitive to variation in the foraging ability of different hens and are able to choose to follow and learn from successful rather than unsuccessful individuals.

Social learning is also observed in quite unnatural situations and in the learning of new responses, indicating that the cognitive process may be largely independent of the natural context. Nicol and Pope (1992) exposed naïve hens to various demonstrators that had either been trained, or not trained, to peck one of two keys to obtain access to a food hopper. They found that naïve hens that had observed a trained hen performed more correct key pecks than those that had observed an untrained hen. The above study also showed the importance of the social demonstrator, as hens that observed a rod 'pecking' the correct key did not peck the correct key when given access to it. As a further indication of the flexibility of social learning, naïve hens learn best from dominant hens (Nicol and Pope 1994). Although it has proved difficult to accurately determine why this may be so, it is possible that hens recognise that dominant hens may possess a greater variety of adaptive and beneficial responses and may be worth paying attention to and emulating more than subordinate hens.

## 4.2 Visual cognition

Extracting information from the visual system in order to guide behaviour is largely underestimated by humans who achieve this feat seemingly effortlessly yet has been studied in depth in chickens. Chickens adopt various strategies to achieve these feats, including altering how they move and view objects and features in the environment in order to recognise objects (Dawkins and Woodington 2000). In a 3D environment, a great deal of visual inputs are of fragments of objects, partly hidden by nearer objects. When viewing such 'partly occluded objects', chickens are able to recognise and respond to the entire object as a whole, a phenomenon termed 'amodal completion' (Regolin and Vallortigara 1995). In addition, chickens appear to recognise that the occluded object is further away than the occluding object. Forkman (1998) trained hens to peck at an object that was relatively further away than another object. In unrewarded probe tests in which one object partly occluded the other, hens correctly pecked at the occluded object indicating that they were deriving information about the relative distance of both objects purely from seeing the partial occlusion.

One interesting aspect of visual cognition in chickens is their ability to use information about the shape of an arena to orient themselves. Chicks that are trained to find food in the centre of an arena will, when placed in a differently shaped arena, also search in the centre, indicating that they are using geometric information to guide their behaviour (Tommasi et al. 1997). Interestingly, when similarly trained to find food in the 'centre' of an arena, but tested in a larger arena of the same shape (i.e. a bigger square-shaped arena), chicks searched not only in the centre but also in locations that were an equal distance from the walls as in the smaller training arena (Tommasi and Vallortigara 2001). These findings suggest that chicks are adopting two search strategies: one based on geometric information (i.e. centre) and the other based on the absolute value of distance from the walls to the potential food location.

## 4.3 Time perception

Research on the perception of time by chickens is scarce but has focussed on a range of issues, ranging from perhaps the simple determining of a sense of duration to being able to think about past and future events. Taylor et al. (2002) examined the hens' ability to estimate a 6-min time interval. This was achieved by training hens to peck a touch screen 6 min after a symbol was shown in order to receive a reward. In unrewarded probe tests, pecking responses peaked at around 6 min after the symbol was shown, showing that hens were able to estimate the time interval. Having found that chickens can estimate duration, the next question was to investigate if they can use the knowledge of the duration between events to alter their behaviour to their benefit. Abeyesinghe et al. (2005) trained hens to peck at either a red or green light and after a delay of varying intervals receive access to food for different durations. Following a number of trials, they found that hens preferred to peck the key that had an interval of 6 s but access to food for 22 s, over a key that had a 2-s delay but access to food for only 3 s. The latter findings showed that hens were able to refrain from pecking the key in a typically fast stimulus-response pattern but instead showed self-control and an appreciation that a delay in responding would result in a longer access to food.

## 4.4 Transitive inference

Transitive inference is a form of logical reasoning that has been studied in various animal species and involves deducing the relation between two items that have not been paired. For example, if item B is larger than item C, and item C is larger than item D, then item B must also be larger than item D. Initial work on this involved a hen first observing a fight between a known conspecific and an unknown hen, and then being introduced to the unknown hen (Hogue et al. 1996). Hens that observed a dominant conspecific losing to the unknown hen did not initiate any attacks when mixed with the unknown hen, indicating that they were possibly acknowledging that they would lose such encounters. In contrast, when the hen viewed a dominant conspecific defeating an unknown hen, the hen attacked in about half of the time when mixed with the unknown hen. Although Hogue et al.'s (1996) findings supported the ability of hens to infer status from observing pairwise interactions, they were not able to rule out the possibility that hens may have gauged the fighting potential of the stranger hen from watching its interaction. More recently, Daisley et al. (2010) trained chicks that a food reward could be received by pecking one of two items, and by presenting a series of pairs of items established the hierarchy of which item would lead to the reward, that is A > B > C > D > E. In the critical test, items B and D were presented, both of which had never been presented together and both had been rewarded and unrewarded an equal amount of time. They found that chicks in the first attempt pecked the correct item (i.e. B), showing that they had inferred the relationship between B and D from the previously established pairwise combination (i.e. B > C and C > D).

## 4.5 Numerical abilities

In most humans, the recognition of different quantities is closely linked to our numerical abilities. The simplest form of this is the ability to differentiate between different sets based purely on the number of objects in each set. Rugani et al. (2008) showed that chicks were able to discriminate between sets of one and two and sets of two and three stimuli. Furthermore, in these tests, the experimenters were able to rule out the possibility that chicks were using other information such as the overall surface or space occupied that may have provided information about quantity. Rather intriguingly, chickens appear to perform number–space mapping in a similar way to humans. Many humans represent numbers on a mental number line, with the smallest number to the left and larger numbers to the right. Rugani et al. (2015) trained chicks to walk behind a centrally placed screen of either a small number of elements (circles on a card) or a large number of elements. When given a choice of two screens both with the same number of circles, chicks chose the left side when there were a small number of circles but the right screen when there were a large number of circles. The findings suggest that chicks tend to map numerical quantities on a left-right axis, as many humans would.

## 4.6 Spatial cognition

Spatial cognition of chickens has usually been examined with respect to the visual system, and some of these abilities are discussed in Section 4.2. More recent research on chickens showing the involvement of the sun and magnetic field in spatial orientation has not yet examined how these sensory modalities impact spatial cognition.

One frequently used test of spatial cognition is the detour test, which requires the subject to leave sight of a goal object by moving around an obstacle in order to achieve closer proximity to that goal object. The detour test is usually considered to reveal both spatial cognition (the position of the chicken relative to the hidden goal object) and memory processes (that the object remains in a certain spatial location). Chickens are highly motivated to reinstate contact with social companions or an imprinting stimulus, and when such stimuli are used, even young chicks are able to solve detour problems to reach a goal (Regolin et al. 1994). In the detour test, chicks have repeatedly been shown to be faster at performing the detour in subsequent tests. Such ability indicates detour learning, rather than spontaneous detour behaviour. As an indication of the chicks' memory for spatial tests, they are better at a second trial (i.e learning) even after there has been a 24-h delay between the first and second trial (Regolin and Rose 1999).

Relative movement of the mother and the chick is important for cognitive development particularly around 11 days of age, when the chick actively moves out of sight of the mother (Workman and Andrew 1989). In the domestic chick, there is a shift towards the right hemisphere becoming dominant at this age, which would suggest a sensitive period for spatial learning since the right hemisphere is involved in spatial reasoning (Andrew 1991). The latter development stage has been examined in the laboratory by imprinting chicks onto artificial objects and housing in an environment that does, or does not, allow the chick to leave the sight of the stimulus. Chicks reared in environments with screens that allow them to move out of sight of an artificial imprinting stimuli showed that chicks that have the experience of moving out of sight at 11 days of age make fewer errors in a detour test and are better able to attend to distal (far away) spatial cues than chicks that are not able to leave sight of the imprinting stimulus (Freire et al. 2004; Freire and Rogers 2005). These findings indicate that rearing in a visually complex environment is important in shaping the development of visual cognition in the chick.

## 4.7 Affective states

In recent years, animal welfare researchers have enthusiastically explored the possibility of a wide range of animals having affective states. 'Affective state' refers to the subjective state that would allow for pleasurable or displeasure mental experiences, which may vary in intensity or arousal. In humans we would describe these states as emotions, and it has been known for some time that people in a state of depression or anxiety tend to make more negative cognitive judgements, such as interpreting ambiguous stimuli unfavourably. Harding et al. (2004) showed for the first time that rats too, after being housed in an unpredictable environment applied to induce a negative affective state, interpreted ambiguous stimuli more pessimistically than those housed in an enriched environment. Although there is still discussion as to the relationship between affect and emotions in other animals, most authors consider that affective states are at least required for the sensation of emotions, and perhaps there is little difference between them in animals.

Research examining affective states in chickens has mainly explored cognitive bias. Cognitive bias is a shift in a response or perception; typically in animal tests it is judgement, based on previous experiences predicted to alter the affective state. Typically, a judgement bias test involves training animals that approaching or responding to one stimulus provides a reward whereas responding to another stimulus provides a negative reinforcer (i.e. punishment), the latter of which animals usually quickly learn to avoid. Affective state is deduced by presenting animals with ambiguous stimuli which are usually intermediate between the two stimuli presented earlier. Animals in a positive affective state should approach the ambiguous stimuli, whereas those in a negative affective state should avoid it. Deakin et al. (2016) examined chicken affective states using such an approach, by training chickens to peck at orange circles of different saturation levels to receive either a reward (mealworm) or punishment (one second puff of air). Following housing hens at low temperature, a condition that they are known to find aversive, Deakin et al. found that hens were less likely to peck ambiguous stimuli (orange circles of intermediate saturation levels) than hens housed at a more highly preferred higher temperature. In similar tests in which location, left or right, is paired with a reward or punishment, meat chickens that had been fed corticosterone (Iyasere et al. 2017) and laying hens that had been denied access to litter for foraging (Seehuus et al. 2013) showed a more negative judgement than control birds to stimuli placed at intermediate locations.

Although these findings show biases towards a negative judgement following negative events such as receiving puffs of air or corticosterone administration, other results on judgement bias have been more mixed. In a test requiring less training, Salmeto et al. (2011) showed that chicks that had been isolated for only 5 min were slower at approaching an ambiguous stimuli (a morphed image of a chick and owl predator) in a runway test, suggesting that isolation induced a negative affect. In contrast, Hernandez et al. (2015) did not find that 5 min of isolation in a novel environment induced a negative affect in adult laying hens. Zidar et al. (2018) tested the effects of cold stress and environmental enrichment and found that neither influenced judgement bias. However, in the latter study in which the manipulations were applied successively, there was some evidence that housing in the enriched environment had a buffering effect on the events that may be considered to induce a negative affect. The study of affect in animals is a fast-growing area of research, and it is clear that methods and hypotheses will continue to be refined as this challenging area progresses. For example, it would be important to distinguish true bias effects (i.e. the proportion of times that an ambiguous stimuli is selected) from other measures such as the speed of responding, which may be related to motivation factors or personality rather than changes in affect (de Haas et al. 2017a).

One possible function of affective states in animals may be to combine complex inputs, some of which may be ambiguous and uncertain, in order to generate an adaptive and suitable response. If so, affective state would be the largely undetermined mechanism that regulates decision making and may be experienced by the animal as something akin to emotions in humans. One complex decision for many animals, including chickens, is how to balance the need to move and explore the environment in order to find food with the need to minimise encounters with potential predators. Nicol et al. (2011) examined the choice of hens for three environments that differed in foraging and predator threat levels. In general, hens chose environments with high foraging opportunities and little threats as expected. Of particular interest were the findings that the choice for the high foraging/low threat environment was associated with lower stress levels, and hens showed fewer behaviours indicative of a negative affect such as head shaking and standing alert in this environment. The result is consistent with the notion that hens are using a common 'currency', such as affect, for decision making across such different environments.

## 4.8 Communication

Communication in chickens involves a large number of vocal as well as visual signals such as changes in posture. Although the variety and use of these signals suggest impressive abilities in themselves, it is research that has examined exactly what information is transmitted and how it is received that has highlighted the extent of cognitive abilities. Referential communication refers to communication that conveys information from one individual to another that can be used by the recipient to generate an appropriate response and is generally accepted to be more advanced than signals that simply convey physical information about the emitter (i.e. its size, sex, etc.). Studies on the 'food' call of chickens, usually given following discovering food or a substrate that can direct foraging behaviour, reveal the complexity of chicken communication. When isolated hens were played food calls, they responded by fixating downwards with the frontal binocular field, a normal anticipatory feeding movement (Evans and Evans 1999). Although the response to the food call could be of a simple stimulus-response type, further studies indicate that this is not the case. When hens are presented with food calls in playback experiments, they respond only with food searching if they have not found food. If they have recently found food, the chickens do not respond with food searching, indicating that they recognise that the information provided by the call is redundant (Evans and Evans 2007).

More careful analysis of the food call itself made by chickens finding food indicates that they alter the rate and number of food calls based on the quality of the food found (Marler et al. 1986). Recent evidence suggests that these calls may be even more complex, possibly indicating the types of foraging substrate which appear linked to the emitter's motivation state. McGrath et al. (2017) conditioned hens that various sounds signalled the arrival of food or non-food. They found that in the interval between the sound cue and the arrival of the reward, hens varied the number of food calls and clucks which was consistent with the variation in arousal. However, again, the use of these calls does not appear to be a simple response to the imminent arrival of food. It has been known for some time that, when in pairs, males give food calls when they find food irrespective of how far the female is at the time of finding food. Males were, though, more likely to give food calls in the absence of food when the female was far away than when she was nearer, which is consistent with the males using the food call deceptively to attract the female back to his vicinity (Gyger and Marler 1988).

## 4.9 Memory

'Working memory' is a term given to short-term memory that encompasses both the storage of information and some processing. One way to study working memory is by using a delayed matching to a sample test, which involves the presentation of a sample for a certain amount of time before it disappears and a number of alternatives, one of which is identical to the sample. In order to obtain a reward, the animal has to peck or show some other response to the stimulus that matches the original sample. In the only such test published on poultry to date, chickens are able to match the stimulus to the sample even after a 16 s delay (Nakagawa et al. 2004). Although this may seem modest, this is a difficult test and at this early stage of study chickens' performance does not appear to be too dissimilar from the performance of many mammals (Lind et al. 2015). In other tests examining the duration of stored information, chicks are able to remember the trajectory of a hidden imprinting stimulus for up to 3 min (Vallortigara et al. 1998), which is comparable to mammals.

Earlier we revealed the chickens' ability to recognise partly occluded objects as entire objects, rather than a fragment. This raises the question as to what extent chickens are able to respond to an object that is completely hidden, so providing no direct sensory experience. This problem has been studied within the framework of object permanence, which refers to the ability to hold a mental representation of a vanished object. In one of the simplest tests of this phenomenon, domestic chicks will readily move towards an imprinting stimulus that has been displaced behind a screen (Freire et al. 2004). As was covered earlier, chickens will also readily move around obstacles in order to approach or re-join an imprinting stimulus or companions in detour tests (Regolin et al. 1994; Freire et al. 2004), suggesting that they are able to direct movement towards where the goal objects 'should be'. It appears that chickens are able to form a mental representation not only to direct their movements but also to use this representation to guide choices based on the physical properties of the hidden object. Chiandetti and Vallortigara (2011) examined the choice of screens which differed in height, width or slant made by chicks after the disappearance of an imprinting stimulus. Crucially, chicks consistently chose screens which had the right size and shape to hide the object and avoided screens which could not. These findings indicate that chicks are able to encode the physical properties of the object in their representations.

Investigations into the types of memory formed by chickens have focussed on the distinction between procedural or declarative representations. Procedural representations are when a stimulus triggers a particular set of memories which produce a set of responses and are considered relatively basic forms of memory. Declarative representations, however, contain more general information about events which can be accessed to guide the most appropriate response. Declarative representations, therefore, allow flexible decisions to be made to new situations. Forkman (2000) used a devaluing test technique to provide evidence that hens are able to form declarative representations. In this test, hens were trained to feed on two different foods from two containers placed in different parts of a test arena. One food item was then devalued, by feeding the hens that food prior to re-testing. Crucially in the test, after the devaluing of one food item, hens went to the food container that previously contained the non-devalued food. This study showed that hens are able to encode and store information about the type of food that is located at a particular location and use that information to guide their movement towards the non-devalued food.

## 5 Improving management through an understanding of learning and cognitive abilities

The scientific evidence presented in the previous section provides an insight into the complexity of chicken learning and cognitive processes. In many respects, the cognitive abilities of chickens are comparable to those of mammals and other animals regarded as 'highly' intelligent. Further studies will no doubt continue to reveal the similarities and differences between complex cognitive processes in chickens and other animals. In the meantime, a pressing question is whether we should re-frame our view and treatment of chickens both as a research animal and as a production animal based on the above-mentioned evidence (Marino 2017). The scientific studies presented earlier may challenge some people's ethical stance towards chickens and stimulate us to think about the possible level of sentience that chickens may possess. However, science cannot answer the ethical decision concerning how chickens should be treated, only inform it, and further research is necessary to continue to improve the quality, extent and validity of this information. One glimpse of such changes is being shown by experimental scientists themselves, with many researchers now opting to rehome their chicken subjects after behaviour experiments, particularly laying-strain domestic chicks which will grow into egg producers and are particularly sought after as backyard chickens in many people's homes.

In a more practical sense, an understanding of chicken learning and cognition provides considerable insight into how best to manage and house them. Knowledge of their cognitive abilities can assist us in three main ways. First, cognitive processes will have been designed and to some extent are constrained to operate in a certain way, and adapting the management of chickens to take this into account will facilitate the animals' ability to adapt to the environment and synchronise their behaviour to better interact with it. Second, many welfare problems such as feather pecking or uneven distributions may have solutions by taking into account the chicken's cognitive abilities. Third, one of the challenges for welfare scientists is to accurately assess the welfare of animals in a range of environments, and a greater understanding of chicken cognition can assist in considering what factors may compromise welfare and also in developing protocols to assess welfare. Specific examples of how cognition can assist with these three areas are given below.

## 5.1 Early life

### 5.1.1 Ensuring appropriate cognitive development

As discussed in Section 4.6, spatial cognition is important and well developed in the chickens, yet several studies have suggested that the early experience of the chicken is important in enabling the full development of spatial cognitive abilities. Chicks that have had the experience of walking behind barriers at 11 days of age, a development stage that appears important for the development of spatial memory, made fewer orientation errors in a detour test and were more likely to use distal (i.e. distant) rather than proximal spatial cues (Freire and Rogers 2007). Additionally, initial work examining the importance of access to perches on spatial cognition showed that chicks reared with early access to perches were better at finding food in high areas than chicks reared without access to perches, and the similar performance on moving between perches between these groups of birds indicated that this finding was due to an effect on spatial cognition rather than physical ability (Gunnarsson et al. 2000).

More recent work has confirmed that the early experience of a complex environment has an effect on spatial abilities, and although it is likely that a complex environment during rearing allows for the development of better locomotory abilities and reduces fearfulness (Brantsæter et al. 2016), there is evidence that a complex environment leads to improvements in spatial cognition. Norman et al. (2019) reared chicks with and without access to perches and elevated structures and found better performance of enriched-housed chicks in a detour test compared to control chicks reared without access to perches or elevated structures. In a test that is perhaps not affected by the locomotory ability, pullets that readily move onto top perches when moved into a layer pen were also found to have a better memory for previously visited locations, as revealed by a radial arm maze where chickens retrieved mealworms from eight arms of a maze (Wichman et al. 2007). Taharntani et al. (2018), also using a test that is unlikely to be influenced by differences in the locomotory ability, provided some indication that meat chickens reared with access to straw bales were better at learning to avoid to peck at inedible food presented on a coloured background than chickens reared in an barren pen. These findings on the development of movement suggest that the early experience of a complex environment may be important in providing hens with the cognitive abilities to move in complex, large-group systems.

### 5.1.2 Maternal influences on chicken welfare

As presented earlier, the development of the chicken is strongly influenced by the presence of a mother or broody hen during rearing. Although chicks have yolk reserves which sustain them for 2–3 days post-hatching, it is not uncommon to observe a spike in chick mortality in commercial rearing around 3-4 days of age. This spike in mortality is due to chicks failing to learn to eat, termed 'starve-out'. In many countries management changes have been implemented to reduce this problem, and the first-week mortality can be low in some countries (e.g. 1.5% in the Netherlands; Yassin et al. 2009). To date, manipulations to relieve starve-out have focussed on providing 'attractive' food which is round and small and placed with a contrasting background. Interestingly, beak trimming does not appear to impair pecking of potential food items in 2- to 3-day-old chicks, but it does significantly reduce the amount of feed ingested at this age (Workman and Rogers 1990). Other manipulations to try and reduce the incidence of starve-out have focussed on reducing the interval between hatching and access to food, and although this appears to have some welfare advantages, early findings do not suggest that this manipulation has a significant effect on the first-week mortality (de Jong et al. 2019). The extent to which starve-out may occur due to the absence of a mother hen has not been investigated. Comparison of the behaviour of brooded and non-brooded chicks indicates that chicks reared with a broody hen showed increased ground pecking and perches at a younger age than chicks reared without a broody hen (Riber et al. 2007). Further work could examine the application of a broody hen or some equivalent stimulus to ensure that all young chicks adopt the feeding behaviour in the first week of life.

The above-mentioned research showing the links between the maternal behaviour and the development of foraging and feeding preferences raises the important commercial question of whether feather pecking could be influenced by early maternal effects (Nicol 2004a). Riber et al. (2007) found that mortality due to feather pecking and cannibalism was higher in the non-brooded chickens than in brooded chickens. Admittedly it currently seems difficult to envisage how commercially reared chicks could be exposed to the beneficial influences of a broody hen. However, a better understanding of these maternal effects could lead to alternative solutions to challenges in rearing. For example, Jensen et al. (2006) compared the behaviour of chicks that were reared with heating sources that either emitted no light or provided some light. Chicks provided with a completely dark night phase showed less feather pecking, feather and skin damage and lower mortality than those that were reared with heaters that provided some light. Although the mechanism for the latter effect is undetermined, it may be related to the effects of a dark phase synchronising behaviour of the chicks.

## 5.2 Associations to signal events and shape correct responses

As indicated earlier, chickens have well-developed learning abilities. The ability of chickens to form associations, discriminate between events and predict a time interval raises the possibility of using signals to indicate forthcoming events in commercial conditions. Such an approach could be used in management in a number of ways. For example, providing animals with cues prior to aversive events may allow them to prepare for such events, with a possible reduction in stress levels. Cues could also be used to indicate changes that chickens may regard as positive, such as the opening of pop-holes on a range. In many cases, many of these cues and association are already common in commercial systems. For example, most chickens recognise the switching on of automatic feeders with the arrival for food. There is still scope, however, to consider learning and assist chickens in forming beneficial associations in more ambiguous situations.

One practical application of associative learning could be to devalue feather pecking behaviour, in an effort to reduce the chances of feather pecking outbreaks. Harlander-Matauschek et al. (2008) conditioned 26-week-old laying hens to two different associations. Half the birds were offered feathers soaked in sucrose and formed a positive association to feather pecking (i.e. the behaviour was rewarded with the favoured taste of sucrose). Other birds were offered feathers soaked in quinine, a substance that birds found distasteful and hence formed a negative association with pecking feathers. Feather pecking behaviour was recorded after these associations, and birds previously offered the quinine-laced feathers feather-pecked significantly less than those that had previously been offered the sucrose-laced feathers. Although testing lasted only 8 days, and it is unknown if the feather-avoiding behaviour would become extinguished after a longer period of time, it provides a promising approach for conditioning pullets to not perform feather pecking.

An alternative could be to condition chicks to avoid feather pecking at an earlier age, when food preferences are being developed. Chicks will avoid pecking the feathers of other chicks that are laced with quinine (Matauschek et al. 2010) though as soon as the quinine wears off, feather pecking rates are similar to chicks that have not had that experience. In the latter study, it could be that chicks have formed an association between the smell of quinine and avoidance of pecking, in which case future work in this area may benefit from establishing an association with cues that the bird cannot detect when they have disappeared.

In applying learning theory to improvements in management, it would be worthwhile to pay attention to conditions that enhance or inhibit learning. Many animals, including chickens, have constraints on learning that result in them learning some associations more easily than others. In general, associations which are similar to the natural behaviour of that particular goal are easier to attain than more dissociated responses. For example, chickens readily learn to search for food in a substrate by pecking and scratching, since this is the natural behaviour, but have more difficulty learning to walk to a specific location to obtain food. The physiological state of the animal can also enhance or inhibit learning, and the relationship between the physiological state and learning is complex and requires further study. However under stressful conditions, chickens, like many other animals, find it harder to discriminate between a cue that gives a large food reward and one that gives a smaller food reward compared to chickens not under stressful conditions (de Haas et al. 2017b).

## 5.3 Applying an understanding of chicken cognition to improve welfare

### 5.3.1 Social cognition

The ability of chickens to learn from others has the potential to greatly impact animal welfare, both positively and negatively, and is an active area of research. Behaviour such as feather pecking and cannibalism can spread rapidly through a flock. As presented earlier, chickens learn about what to peck from each other. Of particular concern has been the possibility that chickens may adopt feather pecking or cannibalistic behaviour after observing other birds performing these responses. Clearly, if social transmission of behaviour occurs in this way, then these behaviours could spread through a flock and may explain why these behaviours are indeed often seen to spread through a flock quickly. Providing unequivocal evidence that feather pecking can spread by social learning in this way is challenging, because it is difficult to discount the possibility that birds not doing this behaviour are responding to the same cues or to attractive feathers displaced by the feather pecking of other birds. In support of the idea that feather pecking is transmitted by social learning, Zeltner et al. (2000) introduced five chicks, which were either feather peckers or did not show prior feather pecking (control group) into groups of chicks. They found that groups that had received the addition of five feather pecking birds performed more feather pecking than the control group. These findings support the concern that birds showing feather pecking contribute to other birds adopting this behaviour, though at the moment the precise mechanism for this transmission is unclear.

Some evidence also exists that cannibalistic pecking may be transmitted by social learning. Using a model to proxy the behaviour involved in cannibalistic pecking, Cloutier et al. (2002) showed that pullets that had observed a demonstrator bird pecking through a membrane in order to access a dish of blood were better at performing this action when allowed to do so than chickens that had not had the opportunity to observe others doing this task. Although the above-mentioned task was quite unnatural, the findings suggest that it could have been passed to other birds by social learning. If feather pecking and cannibalism are transmitted by social learning, then it may be prudent to consider on-farm alternations and management strategies which restrict opportunities for social learning. Such changes could include the addition of visual barriers to limit the opportunities for social learning, removing birds performing these behaviours or adding demonstrator birds that show pecking at appropriate pecking substrates and objects.

### 5.3.2 Moving in and out of houses and up and down perches

In large-group systems, the distribution of birds is rarely uniform and birds often fail to utilise all available resources, due to not accessing them. Resources which are difficult to access could include nest boxes that require climbing up on perches, or the outdoor range, which would require navigating to the pop-holes to exit the building. There are many factors that limit the use of an outdoor range by free-range hens such as the pop-hole design, the quality of the range and the flock size (Pettersson et al. 2016). The possibility that inadequate spatial cognition development could be a factor in range use has been acknowledged for some time and was discussed earlier with respect to early rearing. Only recently have the links between spatial cognition and ranging behaviour been examined in a more practical sense. Campbell et al. (2018) tested pullets before they were given access to a range in a simple T maze test requiring the bird to move down a runway and then turn left or right to receive a food reward. In the first month after being given access to a range, birds that were faster at learning the T maze, on the whole, showed more visits to the range, indicating that range visits may be related to spatial learning abilities. The latter study also supported more fundamental research described earlier showing the importance of experience on spatial cognition and indicating that rearing in a spatially complex environment is likely to have beneficial effects on the ranging behaviour. Further research on the characteristics and the way in which chickens develop spatial cognitive skills would be useful to inform on-farm practices that facilitate a more uniform distribution and better use of ranging areas.

### 5.3.3 Implications of cognition to animal welfare

One question that has been explored previously (e.g. Nicol 2004b) is how the absence of important resources in barren environments may influence chickens. At perhaps a simple level, the absence of a key resource may not allow the chicken to perform a particular behaviour, and this in itself could lead to behavioural problems and impaired welfare. However, at a more complex level, it raises the question of whether chickens have the capacity to *miss*, in a cognitive sense, a resource that is lacking in their environment. Research on mental representations in chickens suggests that chickens may be able to hold mental representations of these missing resources, and the evidence that these representations can alter behaviour suggests that chickens can interpret these to guide decision making. Although this evidence suggests that chickens have the mechanisms to represent absent resources in a flexible way, whether these representations can also trigger processes that allow a chicken to experience the absence of a resource is still unknown.

Additionally, the ability of chickens to form long-term memories and manipulate these memories to guide behaviour raises the possibility that they could recall past unpleasant events, and not only experience decreased welfare from them but perhaps also respond in an unwanted way. On a more positive note, the opposite could also be true. That is, past memories of good associations or events may allow an animal to cope better with recurring events. Although understanding the mental experiences of animals is a complex and challenging area of research, it provides insight into the possible welfare states of chickens in commercial systems.

Lastly, preference and motivational tests are a useful way to determine what animals want and can help us determine the importance of resources such as nest boxes, perches and dustbaths for chickens. One limitation of preference tests is that they do not clarify whether the animal's choice is impulsive or the outcome of a careful appraisal. The implication of this is that it can be difficult to know just how important a preferred resource may be. Studies of transitive inference and self-control indicate that chickens are able to 'weigh up' the relationship between choices and can vary their responses depending on the expected outcome of those choices. Again, whether this indicates a careful appraisal is still to be determined, but chickens possess some of the mental abilities that would be necessary to achieve such appraisal.

## 6 Conclusion

As presented earlier, knowledge of chicken learning and cognition can help us re-frame our view and treatment of chickens in all our interactions with this species. This is no different to the development of our values with respect to all species we interact with. Knowledge of chicken learning and cognition can also be applied to improve welfare, and there has been significant progress in various areas of research for this to be able to be applied now. For example, knowledge of spatial cognition and its development could be applied to improve distribution and movement of chickens in large-group systems through changes in rearing practices. A pressing welfare and economic problem in laying flocks is feather pecking and cannibalism, and research suggests that facilitating the learning of what and what not to peck, possibly based on observation learning processes, could be useful in preventing or controlling outbreaks of these behaviours. No doubt new research will similarly have the potential to contribute to the improved management and care of chickens, and it would be pleasing to see a faster transition from the fundamental research on chicken learning and cognition to improvements in welfare in production systems.

After reading about the complexity of chicken learning and cognition, it may be sensible to ask why chickens should possess such well-developed cognitive abilities. There are a couple of lines of evidence as to why chickens should be so cognitively advanced. First is that the complexity and variability in the environment has been a driver for complex cognition. Ancestors of domestic chickens would have lived in a home range and would need to remember and navigate between areas that are good for foraging, roosting, nesting and so on. Ancestors of chickens would also have had a varied diet and benefit from being flexible in identifying, learning and exploiting new foods. Environmental challenges such as these have been postulated as possible drivers for higher cognition in many animals. Second, chickens live in a complex social world where social bonds are established, need to be remembered and adjusted as individuals join and leave the social group. These relations may require a good long-term memory, the ability to infer third-party relationships and good communication to achieve group cohesion and synchronise the use of resources, and again may have been a driver for the evolution of sophisticated cognitive abilities. Clearly these two theories are not mutually exclusive, and it may be that they have both exerted evolutionary pressure for complex cognition in the ancestors of chickens. The complex mental abilities described in this chapter are therefore in keeping with the life history of the ancestors of chickens and should not be a surprise to biologists.

## 7 Where to look for further information

- Marino, L. 2017. Thinking chickens: a review of cognition, emotion, and behavior in the domestic chicken. *Animal Cognition* 20(2), 127–47.
- Mench, J. 2017. *Advances in Poultry Welfare*. Woodhead Publishing.
- Nicol, C. J. 2015. *The Behavioural Biology of Chickens*. CABI.
- Nicol, C. J. and Davies, A. 2013. Poultry welfare in developing countries. Poultry Development Review. Food and Agriculture Organization of the United Nations, Rome, pp. 110–20.
- Nicol, C. J., Bestman, M., Gilani, A. M., De Haas, E. N., De Jong, I. C., Lambton, S., Wagenaar, J. P., Weeks, C. A. and Rodenburg, T. B. 2013. The prevention and control of feather pecking: application to commercial systems. *World's Poultry Science Journal* 69(4), pp.775–88.
- Sandilands, V. and Hocking, P. M. (Eds). 2012. *Alternative Systems for Poultry: Health, Welfare and Productivity* (vol. 30). CABI.

## 8 References

Abeyesinghe, S. M., Nicol, C. J., Hartnell, S. J. and Wathes, C. M. 2005. Can domestic fowl, *Gallus gallus domesticus*, show self-control? *Animal Behaviour* 70(1), 1–11. doi:10.1016/j.anbehav.2004.10.011.

Andrew, R. J. 1991. *Neural and Behavioural Plasticity: the Use of the Domestic Chick as a Model*. Oxford University Press, Oxford, UK.

Bateson, P. P. G. 1966. The characteristics and context of imprinting. *Biological Reviews of the Cambridge Philosophical Society* 41(2), 177-211. doi:10.1111/j.1469-185x.1966.tb01489.x.

Brantsæter, M., Nordgreen, J., Rodenburg, T. B., Tahamtani, F. M., Popova, A. and Janczak, A. M. 2016. Exposure to increased environmental complexity during rearing reduces fearfulness and increases use of three-dimensional space in laying hens (*Gallus gallus domesticus*). *Frontiers in Veterinary Science* 3, 14. doi:10.3389/fvets.2016.00014.

Campbell, D. L. M., Talk, A. C., Loh, Z. A., Dyall, T. R. and Lee, C. 2018. Spatial cognition and range use in free-range laying hens. *Animals: an Open Access Journal from MDPI* 8(2), 26. doi:10.3390/ani8020026.

Chiandetti, C. and Vallortigara, G. 2011. Intuitive physical reasoning about occluded objects by inexperienced chicks. *Proceedings of the Royal Society B: Biological Sciences* 278(1718), 2621-7. doi:10.1098/rspb.2010.2381.

Cloutier, S., Newberry, R. C., Honda, K. and Alldredge, J. R. 2002. Cannibalistic behaviour spread by social learning. *Animal Behaviour* 63(6), 1153-62. doi:10.1006/anbe.2002.3017.

Daisley, J. N., Vallortigara, G. and Regolin, L. 2010. Logic in an asymmetrical (social) brain: transitive inference in the young domestic chick. *Social Neuroscience* 5(3), 309-19. doi:10.1080/17470910903529795.

Dawkins, M. S. and Woodington, A. 2000. Pattern recognition and active vision in chickens. *Nature* 403(6770), 652-5. doi:10.1038/35001064.

Deakin, A., Browne, W. J., Hodge, J. J., Paul, E. S. and Mendl, M. 2016. A screen-peck task for investigating cognitive bias in laying hens. *PLoS ONE* 11(7), e0158222. doi:10.1371/journal.pone.0158222.

de Haas, E. N., Lee, C., Hernandez, C. E., Naguib, M. and Rodenburg, T. B. 2017a. Individual differences in personality in laying hens are related to learning a colour cue association. *Behavioural Processes* 134, 37-42. doi:10.1016/j.beproc.2016.11.001.

de Haas, E. N., Lee, C. and Rodenburg, T. B. 2017b. Learning and judgment can be affected by predisposed fearfulness in laying hens. *Frontiers in Veterinary Science* 4, 113. doi:10.3389/fvets.2017.00113.

de Jong, I. C., Gunnink, H., Van Hattum, T., Van Riel, J. W., Raaijmakers, M. M. P., Zoet, E. S. and Van Den Brand, H. 2019. Comparison of performance, health and welfare aspects between commercially housed hatchery-hatched and on-farm hatched broiler flocks. *Animal* 13(6), 1269-77.

Duncan, I. J. H. and Hughes, B. O. 1972. Free and operant feeding in domestic fowls. *Animal Behaviour* 20(4), 775-7. doi:10.1016/s0003-3472(72)80150-7.

Edgar, J. L., Paul, E. S. and Nicol, C. J. 2013. Protective mother hens: cognitive influences on the avian maternal response. *Animal Behaviour* 86(2), 223-9. doi:10.1016/j.anbehav.2013.05.004.

Edgar, J., Held, S., Jones, C. and Troisi, C. 2016. Influences of maternal care on chicken welfare. *Animals: an Open Access Journal from MDPI* 6(1), 2. doi:10.3390/ani6010002.

Evans, C. S. and Evans, L. 1999. Chicken food calls are functionally referential. *Animal Behaviour* 58(2), 307-19. doi:10.1006/anbe.1999.1143.

Evans, C. S. and Evans, L. 2007. Representational signalling in birds. *Biology Letters* 3(1), 8–11. doi:10.1098/rsbl.2006.0561.

Fält, B. 1981. Development of responsiveness to the individual maternal 'clucking' by domestic chicks (*Gallus gallus domesticus*). *Behavioural Processes* 6(4), 303–17. doi:10.1016/0376-6357(81)90048-6.

Faure, J. M. 1986. Operant determination of the cage and feeder size preferences of the laying hen. *Applied Animal Behaviour Science* 15(4), 325–36. doi:10.1016/0168-1591(86)90125-5.

Forkman, B. 1998. Hens use occlusion to judge depth in a two-dimensional picture. *Perception* 27(7), 861–7. doi:10.1068/p270861.

Forkman, B. 2000. Domestic hens have declarative representations. *Animal Cognition* 3(3), 135–7. doi:10.1007/s100710000074.

Freire, R. and Rogers, L. J. 2005. Experience-induced modulation of the use of spatial information in the domestic chick. *Animal Behaviour* 69(5), 1093–100. doi:10.1016/j.anbehav.2004.09.009.

Freire, R. and Rogers, L. J. 2007. Experience during a period of right hemispheric dominance alters attention to spatial information in the domestic chick. *Animal Behaviour* 74(3), 413–8. doi:10.1016/j.anbehav.2006.09.018.

Freire, R., Cheng, H. W. and Nicol, C. J. 2004. Development of spatial memory in occlusion-experienced domestic chicks. *Animal Behaviour* 67(1), 141–50. doi:10.1016/j.anbehav.2003.03.015.

Gajdon, G. K., Mundwiler, B. and Stauffacher, M. 2015. Development of early foraging behaviour of domestic chicks in varying social contexts. *Applied Animal Behaviour Science* 164, 29–40. doi:10.1016/j.applanim.2014.08.012.

Gunnarsson, S., Yngvesson, J., Keeling, L. J. and Forkman, B. 2000. Rearing without early access to perches impairs the spatial skills of laying hens. *Applied Animal Behaviour Science* 67(3), 217–28. doi:10.1016/s0168-1591(99)00125-2.

Gyger, M. and Marler, P. 1988. Food calling in the domestic fowl, *Gallus gallus*: the role of external referents and deception. *Animal Behaviour* 36(2), 358–65. doi:10.1016/S0003-3472(88)80006-X.

Harding, E. J., Paul, E. S. and Mendl, M. 2004. Animal behaviour: cognitive bias and affective state. *Nature* 427(6972), 312. doi:10.1038/427312a.

Harlander-Matauschek, A., Wassermann, F., Zentek, J. and Bessei, W. 2008. Laying hens learn to avoid feathers. *Poultry Science* 87(9), 1720–4. doi:10.3382/ps.2007-00510.

Hernandez, C. E., Hinch, G., Lea, J., Ferguson, D. and Lee, C. 2015. Acute stress enhances sensitivity to a highly attractive food reward without affecting judgement bias in laying hens. *Applied Animal Behaviour Science* 163, 135–43. doi:10.1016/j.applanim.2014.12.002.

Hogue, M. E., Beaugrand, J. P. and Laguë, P. C. 1996. Coherent use of information by hens observing their former dominant defeating or being defeated by a stranger. *Behavioural Processes* 38(3), 241–52. doi:10.1016/s0376-6357(96)00035-6.

Jensen, A. B., Palme, R. and Forkman, B. 2006. Effect of brooders on feather pecking and cannibalism in domestic fowl (*Gallus gallus domesticus*). *Applied Animal Behaviour Science* 99(3–4), 287–300. doi:10.1016/j.applanim.2005.10.017.

Johnston, A. N. B., Burne, T. H. J. and Rose, S. P. R. 1998. Observation learning in day-old chicks using a one-trial passive avoidance learning paradigm. *Animal Behaviour* 56(6), 1347–53. doi:10.1006/anbe.1998.0901.

Jones, R. B. 1989. Avian open-field research and related effects of environmental novelty: an annotated bibliography, 1960–1988. *The Psychological Record* 39(3), 397–420. doi:10.1007/BF03395891.

Jones, R. B. 1995. Habituation to human beings via visual contact in docile and flighty strains of domestic chicks. *International Journal of Comparative Psychology* 8(3), 88–98.

Jones, A. R., Bizo, L. A. and Foster, T. M. 2012. Domestic hen chicks' conditioned place preferences for sound. *Behavioural Processes* 89(1), 30–5. doi:10.1016/j.beproc.2011.10.007.

Koshiba, M., Shirakawa, Y., Mimura, K., Senoo, A., Karino, G. and Nakamura, S. 2013. Familiarity perception call elicited under restricted sensory cues in peer-social interactions of the domestic chick. *PLoS ONE* 8(3), e58847. doi:10.1371/journal.pone.0058847.

Lind, J., Enquist, M. and Ghirlanda, S. 2015. Animal memory: a review of delayed matching-to-sample data. *Behavioural Processes* 117, 52–8. doi:10.1016/j.beproc.2014.11.019.

Lindberg, A. C. and Nicol, C. J. 1994. An evaluation of the effect of operant feeders on welfare of hens maintained on litter. *Applied Animal Behaviour Science* 41(3–4), 211–27. doi:10.1016/0168-1591(94)90024-8.

Lindqvist, C. and Jensen, P. 2009. Domestication and stress effects on contrafreeloading and spatial learning performance in red jungle fowl (*Gallus gallus*) and White Leghorn layers. *Behavioural Processes* 81(1), 80–4. doi:10.1016/j.beproc.2009.02.005.

Iyasere, O. S., Beard, A. P., Guy, J. H. and Bateson, M. 2017. Elevated levels of the stress hormone, corticosterone, cause 'pessimistic' judgment bias in broiler chickens. *Scientific Reports* 7(1), 6860. doi:10.1038/s41598-017-07040-y.

Marino, L. 2017. Thinking chickens: a review of cognition, emotion, and behavior in the domestic chicken. *Animal Cognition* 20(2), 127–47. doi:10.1007/s10071-016-1064-4.

Marler, P., Dufty, A. and Pickert, R. 1986. Vocal communication in the domestic chicken: I. Does a sender communicate information about the quality of a food referent to a receiver? *Animal Behaviour* 34, 188–93. doi:10.1016/0003-3472(86)90022-9.

Matauschek, A. H., Beck, P. and Rodenburg, T. B. 2010. Effect of an early bitter taste experience on subsequent feather-pecking behaviour in laying hens. *Applied Animal Behaviour Science* 127(3–4), 108–14. doi:10.1016/j.applanim.2010.09.005.

McGrath, N., Dunlop, R., Dwyer, C., Burman, O. and Phillips, C. J. C. 2017. Hens vary their vocal repertoire and structure when anticipating different types of reward. *Animal Behaviour* 130, 79–96. doi:10.1016/j.anbehav.2017.05.025.

McQuoid, L. M. and Galef, B. G. 1992. Social influences on feeding site selection by Burmese fowl (*Gallus gallus*). *Journal of Comparative Psychology* 106(2), 137–41. doi:10.1037/0735-7036.106.2.137.

Nakagawa, S., Etheredge, R. J., Foster, T. M., Sumpter, C. E. and Temple, W. 2004. The effects of changes in consequences on hens' performance in delayed-matching-to-sample tasks. *Behavioural Processes* 67(3), 441–51. doi:10.1016/j.beproc.2004.07.005.

Nicol, C. J. 2004a. Development, direction, and damage limitation: social learning in domestic fowl. *Learning and Behavior* 32(1), 72–81. doi:10.3758/bf03196008.

Nicol, C. J. 2004b. Chicken cognition. In: Perry, G. C. (Ed.), *The Welfare of the Laying Hen*. Poultry Science Symposium Series (vol. 27). CABI Publishing, Wallingford, UK, pp. 177–89.

Nicol, C. 2006. How animals learn from each other. *Applied Animal Behaviour Science* 100(1-2), 58-63. doi:10.1016/j.applanim.2006.04.004.

Nicol, C. J. and Pope, S. J. 1992. Effects of social learning on the acquisition of discriminatory keypecking in hens. *Bulletin of the Psychonomic Society* 30(4), 293-6.

Nicol, C. J. and Pope, S. J. 1994. Social learning in small flocks of laying hens. *Animal Behaviour* 47(6), 1289-96. doi:10.1006/anbe.1994.1177.

Nicol, C. J. and Pope, S. J. 1996. The maternal feeding display of domestic hens is sensitive to perceived chick error. *Animal Behaviour* 52(4), 767-74. doi:10.1006/anbe.1996.0221.

Nicol, C. J., Caplen, G., Statham, P. and Browne, W. J. 2011. Decisions about foraging and risk trade-offs in chickens are associated with individual somatic response profiles. *Animal Behaviour* 82(2), 255-62. doi:10.1016/j.anbehav.2011.04.022.

Norman, K. I., Adriaense, J. E. C. and Nicol, C. J. 2019. The impact of early structural enrichment on spatial cognition in layer chicks. *Behavioural Processes* 164, 167-74. doi:10.1016/j.beproc.2019.05.008.

Pettersson, I. C., Freire, R. and Nicol, C. J. 2016. Factors affecting ranging behaviour in commercial free-range hens. *World's Poultry Science Journal* 72(1), 137-50. doi:10.1017/S0043933915002664.

Poling, A., Temple, W. and Foster, T. M. 1996. The differential outcomes effect: a demonstration in domestic chickens responding under a titrating-delayed-matching-to-sample procedure. *Behavioural Processes* 36(2), 109-15. doi:10.1016/0376-6357(95)00022-4.

Regolin, L. and Rose, S. P. 1999. Long-term memory for a spatial task in young chicks. *Animal Behaviour* 57(6), 1185-91. doi:10.1006/anbe.1999.1097.

Regolin, L. and Vallortigara, G. 1995. Perception of partly occluded objects by young chicks. *Perception and Psychophysics* 57(7), 971-6. doi:10.3758/bf03205456.

Regolin, L., Vallortigara, G. and Zanforlin, M. 1994. Perceptual and motivational aspects of detour behaviour in young chicks. *Animal Behaviour* 47(1), 123-31. doi:10.1006/anbe.1994.1014.

Riber, A. B., Wichman, A., Braastad, B. O. and Forkman, B. 2007. Effects of broody hens on perch use, ground pecking, feather pecking and cannibalism in domestic fowl (*Gallus gallus domesticus*). *Applied Animal Behaviour Science* 106(1-3), 39-51. doi:10.1016/j.applanim.2006.07.012.

Rugani, R., Regolin, L. and Vallortigara, G. 2008. Discrimination of small numerosities in young chicks. *Journal of Experimental Psychology. Animal Behavior Processes* 34(3), 388-99. doi:10.1037/0097-7403.34.3.388.

Rugani, R., Vallortigara, G., Priftis, K. and Regolin, L. 2015. Animal cognition. Number-space mapping in the newborn chick resembles humans' mental number line. *Science* 347(6221), 534-6. doi:10.1126/science.aaa1379.

Salmeto, A. L., Hymel, K. A., Carpenter, E. C., Brilot, B. O., Bateson, M. and Sufka, K. J. 2011. Cognitive bias in the chick anxiety-depression model. *Brain Research* 1373, 124-30. doi:10.1016/j.brainres.2010.12.007.

Seehuus, B., Mendl, M., Keeling, L. J. and Blokhuis, H. 2013. Disrupting motivational sequences in chicks: are there affective consequences? *Applied Animal Behaviour Science* 148(1-2), 85-92. doi:10.1016/j.applanim.2013.07.008.

Tahamtani, F. M., Pedersen, I. J., Toinon, C. and Riber, A. B. 2018. Effects of environmental complexity on fearfulness and learning ability in fast growing broiler chickens. *Applied Animal Behaviour Science* 207, 49-56. doi:10.1016/j.applanim.2018.04.005.

Taylor, P. E., Haskell, M., Appleby, M. C. and Waran, N. K. 2002. Perception of time duration by domestic hens. *Applied Animal Behaviour Science* 76(1), 41–51. doi:10.1016/S0168-1591(01)00210-6.

Tommasi, L. and Vallortigara, G. 2001. Encoding of geometric and landmark information in the left and right hemispheres of the Avian Brain. *Behavioral Neuroscience* 115(3), 602–13. doi:10.1037/0735-7044.115.3.602.

Tommasi, L., Vallortigara, G. and Zanforlin, M. 1997. Young chickens learn to localize the centre of a spatial environment. *Journal of Comparative Physiology. A, Sensory, Neural, and Behavioral Physiology* 180(5), 567–72. doi:10.1007/s003590050073.

Vallortigara, G. and Andrew, R. J. 1991. Lateralization of response by chicks to change in a model partner. *Animal Behaviour* 41(2), 187–94. doi:10.1016/S0003-3472(05)80470-1.

Vallortigara, G., Regolin, L., Rigoni, M. and Zanforlin, M. 1998. Delayed search for a concealed imprinted object in the domestic chick. *Animal Cognition* 1(1), 17–24. doi:10.1007/s100710050003.

Versace, E. and Vallortigara, G. 2015. Origins of knowledge: insights from precocial species. *Frontiers in Behavioral Neuroscience* 9, 338. doi:10.3389/fnbeh.2015.00338.

Versace, E., Schill, J., Nencini, A. M. and Vallortigara, G. 2016. Naïve chicks prefer hollow objects. *PLoS ONE* 11(11), e0166425. doi:10.1371/journal.pone.0166425.

Wauters, A. M. and Richard-Yris, M. A. 2002. Mutual influence of the maternal hen's food calling and feeding behavior on the behavior of her chicks. *Developmental Psychobiology: The Journal of the International Society for Developmental Psychobiology* 41(1), 25–36. doi:10.1002/dev.10042.

Wichman, A. 2018. Ability of laying hens to distinguish between companions according to their success in gaining access to food. *Frontiers in Veterinary Science* 5, 234. doi:10.3389/fvets.2018.00234.

Wichman, A., Heikkilä, M., Valros, A., Forkman, B. and Keeling, L. J. 2007. Perching behaviour in chickens and its relation to spatial ability. *Applied Animal Behaviour Science* 105(1-3), 165–79. doi:10.1016/j.applanim.2006.05.009.

Widowski, T. and Torrey, S. 2018. Rearing young birds for adaptability. In: Mench, J. (Ed.), *Advances in Poultry Welfare*. Woodhead Publishing, pp. 49–76.

Workman, L. and Andrew, R. J. 1989. Simultaneous changes in behaviour and in lateralization during the development of male and female domestic chicks. *Animal Behaviour* 38(4), 596–605. doi:10.1016/S0003-3472(89)80004-1.

Workman, L. and Rogers, L. J. 1990. Pecking preferences in young chicks: effects of nutritive reward and beak-trimming. *Applied Animal Behaviour Science* 26(1-2), 115–26. doi:10.1016/0168-1591(90)90092-R.

Yassin, H., Velthuis, A. G., Boerjan, M. and van Riel, J. 2009. Field study on broilers' first-week mortality. *Poultry Science* 88(4), 798–804. doi:10.3382/ps.2008-00292.

Zeltner, E., Klein, T. and Huber-Eicher, B. 2000. Is there social transmission of feather pecking in groups of laying hen chicks? *Animal Behaviour* 60(2), 211–6. doi:10.1006/anbe.2000.1453.

Zidar, J., Campderrich, I., Jansson, E., Wichman, A., Winberg, S., Keeling, L. and Løvlie, H. 2018. Environmental complexity buffers against stress-induced negative judgement bias in female chickens. *Scientific Reports* 8(1), 5404. doi:10.1038/s41598-018-23545-6.

# Chapter 5

# Understanding poultry social behaviour and its impact on animal welfare

*Inma Estevez, Neiker-Tecnalia Basque Institute for Agricultural Research and Development and IKERBASQUE, Basque Foundation for Science, Spain*

1 Introduction
2 Why do animals live in groups?
3 Imprinting and early-life experiences
4 Group size and social plasticity: implications for animal welfare
5 Managing social behaviour for better welfare
6 Understanding mating systems in poultry: implications for the welfare of breeding flocks
7 Strategies for better management of poultry flocks
8 References

## 1 Introduction

All domestic animals are social species. Sociality was a prerequisite for domestication that allowed animals to remain in groups under human custody (Weary and Fraser, 2017). In nature, the social organization of animals can be lax, with group members coming in or going out (Bertram, 1992), or highly structured in a stable dominance hierarchy (Schjelderup-Ebbe, 1921; cited in Rajecki, 1988). Social relationships can be affected by many factors such as group age composition, gender, family ties, resource availability, or season of the year, among others. In domestic species, confinements impose additional restrictions to social relationships, either by isolating highly social species or by saturating the social environment. Therefore, social aspects derived from confinement conditions can have a direct, sometimes severe, impact on the welfare of animals if not managed correctly.

Highly structured social hierarchy was first documented in the domestic fowl *Gallus gallus domesticus* (Schjelderup-Ebbe, 1921; cited in Rajecki, 1988), and has been studied heavily in chickens under varying rearing conditions. A more or less linear hierarchy is established through aggressive displays in early phases of development (Guhl, 1953; Rushen, 1982;

http://dx.doi.org/10.19103/AS.2020.0078.05

Buchholz, 1997). The purpose is to determine dominance in the group, so that once established, dominants will have priority of access to resources, and aggressive encounters will be drastically reduced. Subordinates, on the other hand, will benefit from not spending energy in futile aggressive interactions that they are unlikely to win. Early studies suggested that domestic fowl housed in large groups, conditions likely to occur in commercial poultry, are unable to form a stable social hierarchy. It was argued that such conditions will lead to social instability, higher levels of aggressive encounters and reduced performance (Guhl, 1953; Banks and Allee, 1957; Al-Rawi and Craig, 1975). Higher than expected levels of aggressive interactions and increased difficulties in accessing resources derived from social instability would impact on the welfare of the domestic fowl.

To minimize the perceived negative impact on welfare, research was conducted to determine the optimal group size for the domestic fowl (see review by Estevez et al., 2007). Some of the questions to be answered were: Is it really possible to determine an optimal group size for maximizing domestic fowl welfare? Is there a group size from which any departure implies a reduction in quality of life? The main role of behaviour is to permit animals' adaptation to changing environmental conditions which can, occasionally, be highly unpredictable. In fact, behavioural adaptation including social and reproductive aspects is the first line of response that permits animals to survive in changing environmental conditions (Tuomainen and Candolin, 2011). Adaptation to variable conditions intrinsically implies that behaviour (including social aspects) has to be plastic, otherwise animals would not be able to thrive unless maintained within a narrow range of pre-established conditions. It is well-documented that, in the wild, the structure or composition of a social group may change depending on how factors such as season, resource availability or predation pressure (among others) influence individual group members (Taylor et al., 2014). Even the mating strategies of individual group members may be altered in response to these changes. Perhaps, this inner plasticity of social behaviour that permits animals to adapt to particular environmental conditions is the reason why social behaviour traits have low heritability (Turner, 2011).

Early studies on the social behaviour of domestic fowl did not contemplate social plasticity as a possibility, but rather assumed that their social organization was based on the more or less linear hierarchy observed in small group sizes. It was thus, proposed that large group sizes implied a negative impact on the welfare of the domestic fowl due to increased aggression derived from social instability and competition for resources. A new hypothesis, social tolerance (Estevez et al., 1997, 2002, 2003), was developed in the late 1990s. However, for many, the idea still persists that groups of domestic fowl, even when maintained in large flocks with several thousand birds, must try to form hierarchical social structures.

Reproductive behaviour is a particular type of social behaviour that is established among females and males with the purpose of producing offspring. Evolutionary history has shaped each species' mating systems for thousands of years, at times giving place to male/female conflicts, arms race or cryptic competition in favour of maximizing reproductive success (McDonald et al., 2017). The mating system in the domestic fowl is the polygyny, in which a male (normally dominant) mates with several females. This mating system is characterized by an intense competition among males for female access, which may result in female coercion and subsequent welfare problems (Pizzari, 2016). Under commercial conditions, it is recommended to house 1 male per 8-10 females as the optimal ratio to maximize fertility. However, as it will be discussed later, this ratio does not always prevent negative consequences for the welfare of males and especially of females.

This chapter analyses the social behaviour of the domestic fowl from the perspective of animal welfare. It considers the management practices that can help to promote a better social environment to protect animal welfare, as well as presenting methods of addressing conflict in breeding flocks. Understanding the plasticity of social behaviour in production animals, their complex and sometimes shocking strategies to reproduce, is of primary relevance for an optimum management, and of essence to meet their biological requirements to assure welfare.

## 2 Why do animals live in groups?

Animals live in groups because it improves their chances for survival. The social group protects individuals from predators and adverse weather conditions and facilitates the rearing of the young, migration and the identification of new food sources. More individuals in the group can lower predation risk by dilution (Hamilton, 1971; Elgar, 1989; Roberts, 1996; Beauchamp, 2003) or simply because more eyes in the group on the lookout for danger will increase the probability of detecting predators sooner (Pulliam, 1973; Krause and Ruxton, 2002). For the domestic fowl, specific benefits include, for example, improved thermoregulation by huddling, feathering condition through social preening or reduced fear in novel situations (Newberry et al., 2001). Social learning and social facilitation enable the discovery of new feeding grounds or food sources, stimulate feeding, preening and dust bathing (Duncan et al., 1998; Hoppitt et al., 2007; Hoppitt and Laland, 2008). Living in groups also has a positive impact on the chances for mating and improves reproductive success (McDonald et al., 2019). These forms of social support have important benefits for the welfare of domestic fowl as the group makes them feel safer, helps them to remain healthier and better nourished (by social facilitation). Contrarily, isolation will create a situation of high alert for social species, which is evidenced by high-pitched distress calls emitted specially by young chicks when placed in isolation (Feltenstein et al., 2002). Even though isolated birds can be assured similar or better opportunities to feed, optimal temperature, no competition from conspecifics, and no risk of predation, their sense of danger and distress is high (Marx et al., 2001). This reaction is a fixed behaviour response because isolation probably had major negative impact on the chances of survival for the ancestors of the domestic fowl. The need to be in a social group is present and evident in current commercial lines even though they have been heavily selected for productive traits for generations.

Domestication studies have shown that most active behaviours may appear at a lower frequency in highly selected lines of chickens, but are unlikely to disappear (Eklund and Jensen, 2011; Agnvall and Jensen, 2016). Anti-predator behaviour in particular, has been highly relevant for survival, has not been selected against by artificial selection processes, and it is a product of the past evolution (Tuomainen and Candolin, 2011). In fact, domestic fowl have been (and continue to be) predated particularly in extensive production systems and small backyard flocks. Therefore, all mechanisms making the birds wary about their environment to avoid predation should remain highly relevant in today's commercial lines. Domestic fowl, in both free range and indoor housing systems, maintain in fact a robust anti-predator behaviour, as evidenced by their strong tendency to group together, especially near walls and other protective areas (Newberry and Hall, 1990; Newberry and Shackleton, 1997; Cornetto and Estevez, 2001; Rodriguez-Aurrekoetxea et al., 2014) by jumping to top-level perches to better scan their surroundings (Newberry et al., 2001) or to safely roost at night (EFSA, 2015). Another social strategy that reduces predation risk is the synchronization of risky behaviours such as feeding, foraging, dust bathing or preening, as these behaviours imply a reduction of the possibilities to scan for potential danger. In domestic fowl these behaviours continue to be highly synchronized in groups, especially when in small ones (Malleau et al., 2007; Keeling et al., 2003, 2017). This behaviour is interpreted as a form of protection against predation when birds might be more exposed to danger. Thus, consideration of ways to make the birds feel safe and protected should be considered as a relevant aspect for their welfare.

Of course, living in groups also has negative consequences (Pulliam and Caraco, 1984; Beauchamp, 2016). Under production conditions, food and water are normally provided *ad libitum* and are evenly distributed to minimize competition, permitting the co-existence of large groups (Estevez et al., 1997). Contrarily, if resources become limited or clumped in space, competition may become intense favouring the monopolization of resources by dominants (Hughes et al., 1997; Estevez et al., 2002), resulting in increased variability in body condition and health, and a welfare risk. In free living animals excessive competition or lack of resources will naturally regulate the size of social groups.

Animals under excessive competition pressure will opt to leave the group, option that it is not possible for captive animals. This mis-match between expected and viable response can be a serious risk for the welfare of captive animals, if they are unable to adopt alternative strategies to cope with the conditions of the production environment. However, adoption of alternative behaviours as a response to the new conditions (see review Tuomainen and Candolin, 2011; Stroeymeyt et al., 2018) permits animals to adapt to challenging environments, as long as conditions are not too severe. Stroeymeyt et al. (2018) for example, showed that eusocial insects, using an ant (*Lasius niger*) model, are able to detect the presence of pathogens and adjust their behaviour to reinforce disease - inhibitory spatial distribution, specially protecting high-value individuals. Perhaps, similar complex social mechanisms that permit reducing risk of disease, potential competition with other species or populations may also exist in the highly social domestic chickens. Acknowledging the role of social plasticity in the life history of the species may open new and interesting possibilities to explore.

Some studies have shown the capacity of the domestic fowl to adapt their level of aggressive encounters. Estevez et al. (2002) showed that the level of aggressive interactions was higher in small (15–30 birds) than in larger groups (60–120), where the cost of competition would have been higher because of the larger number of competitors (Fig. 1). In the latter, adopting alternative tactics like scramble competition would be more advantageous (Estevez et al., 1997, 2002). Thus, negative impact of living in large groups may be overcome by the adoption of alternative responses without negative consequences for their welfare.

On the other hand, the social relationships among breeding individuals are very complex and, depending on the group dynamics and sex ratio, the level of male-male competition and sexual coercion may increase causing failed matings, increased male interference and sexual conflict (Pizzari, 2016). Thus, understanding of the basic principles that brings animals together, and the forces that pull them apart, becomes particularly relevant for assuring the welfare of captive animals. This knowledge will facilitate to detect and reduce negative consequences if competition among group members becomes intense, or to better understand how to meet their social needs for better animal welfare.

## 3 Imprinting and early-life experiences

Coming to life with no knowledge of the surrounding world can be disastrous for the survival of any living creature. To increase survival possibilities all animals, from insects to mammals, have some sort of precocious knowledge, knowledge that is either transmitted between generations in the form of instincts, or have

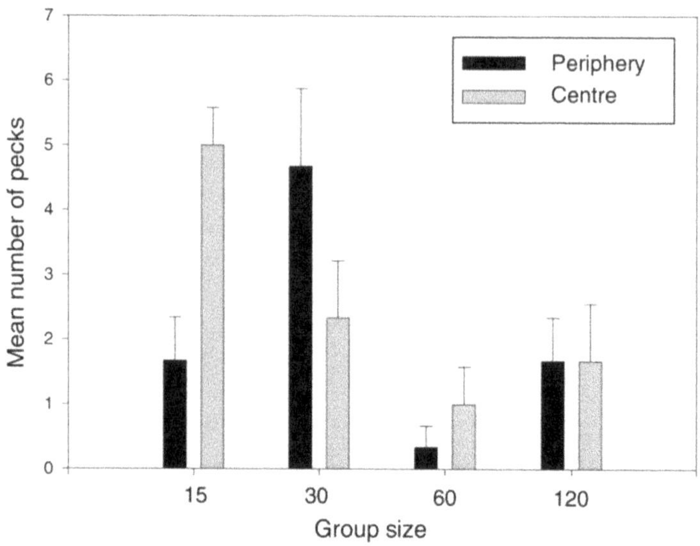


**Figure 1** An example of changes in the frequency of aggressive interactions as a result of an experimental situation of competition for resources (food). Higher frequency of interactions in smaller groups as compared to larger groups of pullets is observed. In larger groups with more competitors pullets adapt their behaviour to use less aggression in favour of scramble competition (Estevez et al., 2002).

some basic knowledge supported by simple learning mechanisms (Versace and Vallortigara, 2015). According to these authors, early knowledge can take place through early stimulation, including embryonic experience, or be transmitted as predispositions or imprinting mechanisms. Imprinting orients perception and learning during development towards biologically relevant models (Lorenz, 1935; Hess, 1958). This process takes place within a narrow window of opportunity (post-hatch days 1-4 in the domestic fowl), and it is considered the earliest expression of social behaviour (Di Giorgio et al., 2017).

Imprinting is the primary mechanism that facilitates early learning in precocial birds (i.e. that are able to move autonomously soon after hatching). For example, recognition of edible items is facilitated by following the mother hen, likely the nearest by individual that chicks encounter after hatching. Proximity to the mother also assures chicks can easily find a warm and safe place to rest, increasing their survival probabilities. Research has shown that imprint to moving entities, and specially to objects that show biological motion, occurs faster as compared to those with a rigid or random movement (Vallortigara et al., 2005). Preference for biological motion set in place the mechanisms to avoid imprinting with inanimate objects that may exist in the surrounding environment that would led to no biological benefits, while increasing chances of imprinting to progenitors (Di Giorgio et al., 2017). This mechanism, extended across distant taxonomical groups, favours establishing preferences for biological models, to which later develop a tighter social attachment (Miura and Matsushima, 2016). Interestingly, chickens also are predisposed to orient themselves towards face-like stimuli (Rosa-Salva et al., 2019) that appeared to be linked with the CONSPEC system by which animals learn the general features of care-givers and social patterns (Morton and Johnson, 1991). A subsequent learning mechanism, CONLERN, restricts the preferences towards more specific features, allowing the recognition of familiar individuals and social bonding (Rosa-Salva et al., 2015). According to Rosa-Salva et al., (2019), the predisposition towards head-like features, even of other animal species models, may be interpreted as evidence of the chicks' adaptive social capacities. The possibility of being attracted and imprinted with other animal models permits chicks reared in groups to maintain brood cohesion and thermal regulation capacity in the absence of the mother hen. Further, Rosa-Salva et al. (2019) suggest that social predisposition mechanisms that direct attention towards other chicks might be highly relevant to permit adaptation to social life conditions that they will encounter later on, and it is interpreted as a core knowledge system dedicated to social stimuli (Vallortigara, 2012). Of course, these mechanisms are supported by complex interplay of neuronal and hormonal changes that play key roles in the imprinting mechanisms (see Rosa-Salva et al., 2015, Batista et al., 2018). For example, androgens appear to change during the initial days and weeks after hatching (Di Giorgio et al., 2017) and thyroid hormones, when artificially provided, are capable of reopening the sensitivity window allowing chicks to imprint after the window of opportunity has theoretically elapsed (Yamaguchi et al., 2012). The neurophysiological mechanism involved in imprinting and the predisposition to focus on the surrounding social environment has long-term consequences for the adaptive capacity of the fowl as birds learn to recognize their co-specifics and sexual partners.

Perhaps the possibilities of accepting social diversity in phenotypic appearance of other birds might be closely related to imprinting mechanisms than forms, a more or less flexible, species patron depending on the diversity of phenotypes to which they are exposed during the imprinting window. The domestic fowl is highly plastic to phenotypic appearance during early rearing phases of development (Marin et al., 2014). On the contrary, they are reactive to changes in either the phenotypic appearance or to well-structured behavioural patterns if changes occur in adulthood (Campderrich et al., 2017a,b; Fig. 2). Probably as consequence of imprinting-like mechanisms during early rearing, chicks learn to identify the normal phenotypic appearance of co-specifics (Campderrich et al., 2017a,b) and the behaviour patterns that they will consider as 'normal phenotype' thereafter. Campderrich et al. (2017b) showed that little alterations in appearance of a small proportion of individuals (by placing a

(a)

(b)

**Figure 2** (a) Day-old pullets altered with a mark on the back of their head and unaltered for phenotypical appearance. (b) Adult layers with the two phenotypes. Altering phenotypes and individually altering birds in homogeneous groups in which all individuals were either marked or all unmarked and were in stable homogeneous groups until adulthood, produced a similar increment in the frequency of aggressive interactions (photos by I. Estevez).

small mark on the head or by removing the mark) in homogeneously reared groups of adult laying hens trigger aggressive interactions towards birds with the altered phenotype (Figs. 2 and 3). Of course, chicks and hens are able to react to more or less subtle differences in behaviour according, for example, to perceived danger (Edgar et al., 2013). It is the large frame of what it is within normality in appearance and behaviour that seems to be somewhat established during early development.

Diversity in the early rearing environment also contributes to build-up the robustness of young chickens, helping buffer against stress of unexpected future events (Campderrich et al., 2019). Perhaps it is possible to consider some

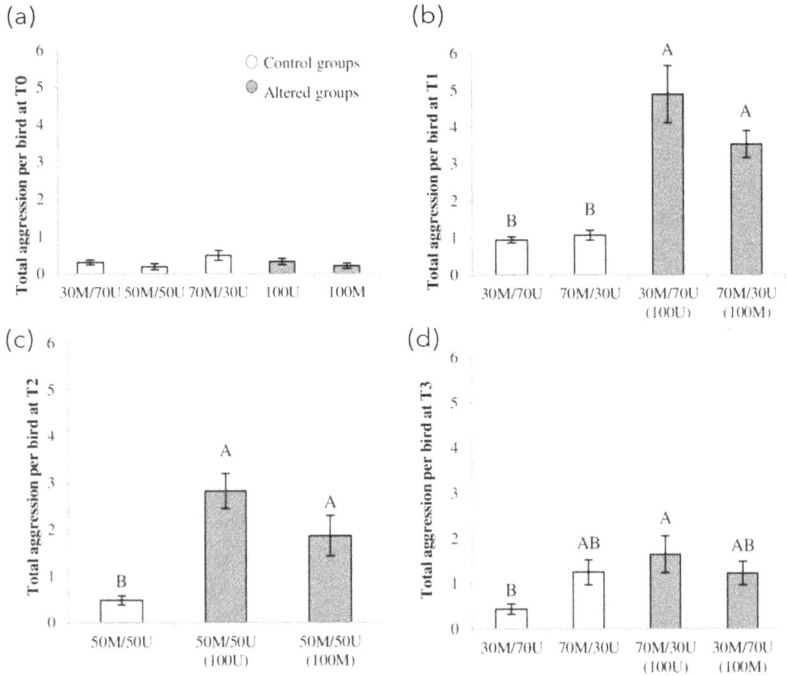


**Figure 3** Total aggression (interactions per bird/40 min). Frequency of total aggression per bird at T0 (27±28 weeks; 1A), T1 (35±36 weeks; 1B), T2 (39±40 weeks; 1C) and T3 (45±46 weeks; 1D). Bars represent means ± SE. M = marked; U = unmarked. Phenotypic appearance treatments: originally homogeneous (100U, 100M) and controls (30M/70U, 50M/50U, 70M/30U). Different letters denote significant differences among treatments at *P*<0.05. (In Campderrich et al., 2017b https://doi.org/10.1371/journal.pone.0188931.g001).

sort of 'imprinting' mechanisms towards the rearing environment, in the sense that during development birds construct the model of what will be considered a 'normal comfortable rearing environment'. As the mother hen and exposure to other chicks increases the opportunities to explore and learn about the features in their environment, early exposure to environmental complexity may widen opportunities to develop highly diverse behavioural repertoire, (for example perching, flying or jumping if perches or platforms are provided). More options also means better possibilities to control their responses (e.g. jumping on the perches if frightened), as opposed to being in a poor environment with no possibilities to execute behavioural changes. The results by Norman et al. (2019) indicate that early access to perches affects the birds' spatial navigation and physical capacity, providing experimental support on the relevance of the features of the early environment. Behavioural adaptation can also be increased if young birds are exposed to a dynamic environment, simulating changes that may take place in a natural environment. Preliminary results of ongoing research by Skånberg et al. (2019) in Prof. Keeling's lab have shown that both complexity and dynamic changes favour animal welfare, as chicks raised in complex and dynamic environments showed better ability to resolve a detour test and react in a new environment.

The acquired early abilities, with a wide spectrum of normalized phenotypic diversity, ample opportunities to explore a complex and perhaps changing environment, will be the best basis to produce adaptable chicks that are prepared for the unexpected, thereby improving their possibilities to react properly and assure their welfare. Early training will have major impact for the long-term welfare of the domestic fowl as they will be less reactive and better able to transition from the bi-dimensional rearing to the three-dimensional production environment as it occurs with broiler breeders (Estevez, 1999). For laying hens, current commercially available systems allow birds to experience perching, resting or feeding at different height levels, thus substantially reducing the guessing exercise that pullets had to do when transferred for the rearing to the production environment.

## 4 Group size and social plasticity: implications for animal welfare

The tendency of small groups to form stable social hierarchies has been proposed as one of the benefits of housing hens in small enriched cages (or in conventional cages, although these are prohibited in Europe because of the extreme spatial restriction). Despite the potential small-group-size advantage of enriched cages, practical experience led industry to drift towards larger colony cages with 60–100 birds. Most other forms of commercial flocks, including alternative production for meat or egg production are normally comprised by several thousand birds (Fig. 4). Studies conducted in larger flocks from the 1990s onwards evidenced that aggressive interactions are not particularly high or problematic (Estevez et al., 1997; Hughes et al., 1997; Nicol et al., 1999; Carmichael et al., 1999; Liste et al., 2015; Weimer et al., 2019). Some of the effects observed, such as reducing mobility, worsening plumage condition and bird performance are normally related to density (Estevez, 2008; Leone and Estevez, 2008a; Weimer et al., 2019) or to differential pen dimensions when controlled for density (Liste et al., 2015) rather than to group size per se. The new existent perspectives on the social dynamics of the domestic fowl portrayed the social behaviour of chickens as a more dynamic and plastic social organization than previously thought. The idea of social adaptation and social plasticity is actually more in line with the adaptive capacities that we know young chickens or pullets have when exposed to them at an early age.

**Figure 4** Flocks of several thousand laying hens and meat chickens in commercial conditions (photos by I. Estevez).

Large flocks might be at a higher risk of developing welfare problems, not because of the high number of birds but because of the tendency to maintain excessive density with increased risk to deteriorate environmental conditions faster (Dawkins, 2004; Estevez, 2008). Also if panic attacks were to occur, higher mortalities would be expected as compared to smaller flocks, because larger number of birds will pile in the end walls looking for protection. Nevertheless, this is not a problem that derives from the social skills of chickens, but is rather due to lack of consideration of their natural fear reaction (Richards, et al., 2012).

As we know, stable hierarchy requires individual recognition of group members (Guhl and Ortman, 1953) and remembering the outcome of all aggressive interactions (Pagel and Dawkins (1997). Meeting such requirements in flocks of several thousand birds would appear rather improbable. One possible suggested solution is the formation of subgroups with their own stable hierarchical system (McBride and Foenander, 1962). A prediction following this hypothesis is the limitation of movements to the close vicinity of their social subgroup to avoid conflicts with neighbouring strangers. Numerous studies confirm large individual variability in use of space patterns in flocks of domestic fowl, but no evidences support the idea of social restriction of movements that could be linked to the co-existence of social subgroups (Newberry and Hall, 1990; Estevez et al., 1997; Leone and Estevez, 2008b; Rodriguez-Aurrekoetxea et al., 2014; Fig. 5).

Domestic fowl do not engage in costly aggressive interactions in large groups, and competition at the feeder is more jostling for space rather than aggressive (Widowski et al., 2017). The tolerance hypothesis (Estevez et al., 1997) assumes intrinsic plasticity of the social behaviour (Tuomainen and Candolin, 2011), plasticity that it is also evidenced at the brain structural level (Batista et al., 2018), and permits fowl to adapt to ample ranges of physical and social environments. Social plasticity permits freedom of movement and minimal expenditure of time and energy in aggressive encounters in large groups. Actually, scientific evidences suggest that the amount of space used by chickens, independently of group size, is directly related to the size of the enclosure (Newberry and Hall, 1990, Leone and Estevez, 2008a; Liste et al., 2015). Even density has a much lower impact than pen size when considering the amount of space birds use, as demonstrated under experimental conditions in which group size, density and pen size were carefully controlled (Leone and Estevez, 2008a). The larger total space available to larger groups (at identical density) would favour overall bird mobility as the total available space for the birds to move will be proportionally larger (Liste et al., 2015). Although these studies were conducted in relatively small experimental pens, the consistency in results suggest that birds' mobility, and thus welfare, would be better in large enclosures independently for a constant density. In practical terms, fencing smaller groups in a large flock may facilitate inspection and the management of birds, however, would limit the potential for movement. Nevertheless, more data on spatial distribution of individual birds in commercial conditions will be necessary to confirm the impact of density, group size and enclosure size on bird movement.

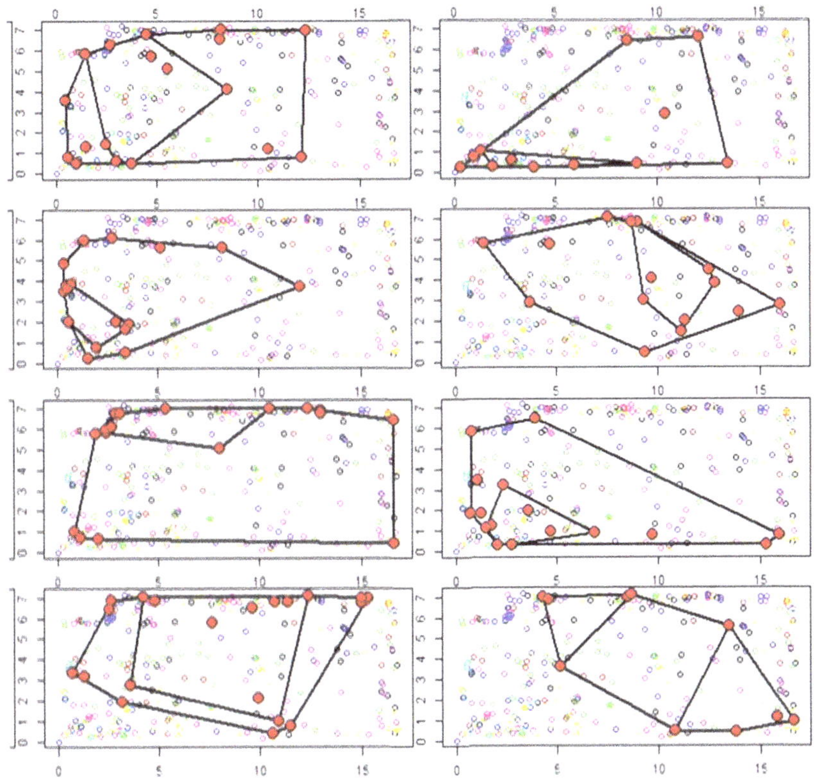

**Figure 5** Example of home ranges of slow-growing chickens in a commercial free-range farm in Northern Spain: approximate group size 1200. Different colour dots indicate the location of different individually tagged birds. Each graph shows the home range (90% of locations) and the 50% core area of individual birds. The higher dot density is evident in most figures showing preferences to stay close to walls. Data by A. Rodriguez-Aurrekoetxea and I. Estevez (unpublished figures).

Existing literature supports the idea that large group sizes do not have per se a negative impact on the social behaviour of the domestic fowl and on their welfare. This is, of course, as long as enough resources are evenly distributed in the environment. It is important to take caution with larger loose groups that are able to freely roam as they are more difficult to manage and supervise. In fact, the possibility of the birds to freely express their behaviour makes daily farm-care routines longer with caretakers needing to pay full attention to the expression of birds' behaviour in order to be able to detect early stages of feather pecking or health problems. It is also essential to carefully analyse potential flock problems to try to discern between problems that may be associated with density from those related to group size, as these are often confounded.

## 5 Managing social behaviour for better welfare

### 5.1 Managing conflict over feeding

Good nutrition is one of the main pillars of good welfare (Welfare Quality, 2009). Therefore, assuring adequate access to nutritious food and water is primordial to guarantee that all birds in the flock are well and homogeneously nourished. Poultry diets are optimized for macronutrients and micronutrients, and feeders are designed considering the birds' needs according to their age. Despite this, animals' behaviour, personalities and social dynamics may defeat all these efforts. Domestic fowl are normally feed *ad libitum* with some exceptions like broiler breeders , thus competition for resources is unlikely (Estevez et al., 1997). But certain situations may trigger competition. Optimal foraging theory literature evidenced that animals choose and remain in feeding areas depending on resources availability, nutrient quality and handling time (Lewis, 1980; Caraco, 1981; Alonso et al., 1995; Kacelnik and Brunner, 2002; Schaefer et al., 2003). Similar principles of optimal foraging apply to the domestic fowl. Leone and Estevez (2007) showed, in an experimental set-up in which access to feed was manipulated, that even highly selected fast-growing broiler chickens behave differently depending on the 'intrinsic value' of the resources available. Broilers behaved more aggressively at feeders in which feed was readily accessible as compared to those that required a longer handling time. Broilers also spent more time and ate more feed from the most efficient feeders. The differences in behavioural responses to feed accessibility found by Leone and Estevez (2007) suggest that, for example in field conditions, cluttered feeders (i.e. because of insufficient attention to maintain them clear of debris) may lead to higher levels of competition for resources, even though technically feeding space per bird might be according to standards. Domestic fowl will avoid eating in cluttered feeders, thus increasing feeding visits to cleaner ones, which may have an impact on the good nutrition of all birds.

It is also well known by egg producers that variability in performance and health may be due to subtle forms of competition. In laying hens, for example, heat stress can trigger increased competition as feeding patterns normally change to minimize heat stress (personal obs.). When high day temperature declines, hungry birds will compete jostling for access to feeder space. If the heat episode lasts for several days, this will result in increased flock heterogeneity and increased risk of behavioural problems such as feather pecking and cannibalism. In this situation it might be advisable to let birds to empty feeders so that bolder (dominant phenotype) birds get fully satiated and then run additional feeding rounds. In such a way shy (subordinate phenotype) birds may have a better chance to eat a complete diet, minimizing the risk for particle selection. Selection of particles is also a problem in aviaries in which feed is supplied in open chain feeders and layers have freedom to move around. Larger birds are known to patrol the beginning of the feeder so they get to choose the larger, nutritious particles. Quite frequently this can create large decompensations of the diet that would be unbalanced by the time the chain reaches the end of the aviary where less competitive individuals are located, impairing optimal bird nutrition.

The optimization of feeding strategies has been essential for survival of all animal species (Stephens and Krebs, 1986). Foraging decisions have been widely studied and modelled (see for example; Davidson and El Hady, 2019). It should not be surprising that animals, including domesticated species, became highly sensitive both to resource accessibility and distribution (Estevez et al., 2002) and to the feeding outcomes during their evolutionary history. A good understanding on how social dynamics affects birds' use of resources will facilitate a better management that will prevent subtle forms of competition. These are just a few examples of how different forms of competition for access to resources may be taking place affecting welfare of the birds, even though, in theory, enough feed and feeders appear to be available. This fact supports the idea that assessment of animal welfare is more realistically assessed on the animals themselves and not on the resources that are 'theoretically' at their disposal.

## 5.2 Management of factors associated with aggressive interactions

In commercial conditions, group size and density normally change simultaneously as dimensions of the poultry houses remain constant, but the number of birds within these houses may change. For example, densities may be lower in the summer months in commercial broilers to reduce heat stress impact. Although studies on social conflict often confound the effects of density with those of group size, or alternatively, with the effects of enclosure size, scientific evidence indicates that aggressive interactions in the domestic fowl declines or remains equally low with increased density and group size (Estevez et al., 1997, 2003; Hughes et al., 1997; Carmichael et al., 1999; Nicol et al., 1999; D'Eath & Keeling, 2003; Liste et al., 2015). As discussed earlier in this chapter, social conflict has been traditionally associated with competition for resources. But contrary to expectations, when the location of interactions is considered, data shows that aggressive encounters are more likely to take place in open areas, far from resources and where inter-individual distances are larger (Pettit-Riley et al., 2002; Ventura et al., 2012). Rodriguez-Aurrekoetxea and Estevez (2014) indicated that three components - proximity, orientation and activity - are required to trigger aggressive interactions in laying hen flocks in aviary systems. These characteristics were determined by analysing the behaviour of the givers and receivers of the aggressive interaction right prior to the encounter. A novelty of this study was the finding that the receiver of the aggression must have been in an active behavioural state like running or walking for the interaction to take place. The interaction, on the other hand, would not take place if birds come in contact while performing other behaviour like foraging. The need of all three components, orientation, distance and an active state to occur, simultaneously explains why most interactions take place in open areas and away from resources. The same reasoning explains the recurrent findings of reduced frequency of aggressive interactions at high population densities. The chance for two birds to confront each other while in an active state at high densities will be unlikely due to the presence of other birds in the path of movement (Newberry and Hall, 1990; Estevez et al., 1997), naturally limiting the chance that birds will confront each other while in an active state (Rodriguez-Aurrekoetxea and Estevez, 2014). The effectiveness of cover panels in reducing aggressive interactions in the domestic fowl (Cornetto et al., 2002) would be equally explained by the increased difficulties for two birds to confront each other directly in an active phase in a more complex environment. What remain unanswered still are the reasons that would trigger the aggressive encounter. In appearance, the giver of the interaction was not in a particular situation for defending resources, and there was not an apparent benefit from engaging in the aggressive interaction (Rodriguez-Aurrekoetxea and Estevez, 2014).

Tuomainen and Candolin (2011) indicated that maladaptive behavioural responses to human-induced environmental changes are relatively common, and animals may respond in an odd manner that makes no sense under the actual situation. An aggressive response when a bird feels threatened, even if there is no a clear benefit, may fall into this category. In the jungle, where the red junglefowl evolved, confrontation in a highly active state would appear to be unlikely as abundant vegetative structures would offer protection to the birds that, in addition, would spend most of their available time with their head down foraging. Thus, direct confrontation in an active state might have meant danger in their evolutionary history, behaviour that in the current situation seems not logical. It is noticeable that Agnvall et al. (2014) and Krause et al. (2019) found no effects on social behaviour when selecting birds for several generations according to fear and high activity levels, respectively. Despite finding major differences in fear or activity, no differences were detected in regard to the social component, even though active birds in the Krause et al. (2019) study may have had more chances of meeting other birds to interact with. Such results suggest that social behavioural responses are not likely to change by selection, probably because some aspects might have been highly structured by the evolutionary history of the domestic fowl.

Badges of status (Pagel and Dawkins, 1997), like comb colour and size (Cloutier and Newberry, 2000, 2002), have been proposed as social dominance signals when individual recognition is not possible. But these signals do not remain immutable; they may change naturally due to diseases, malnutrition, injuries or other causes. These changes have major consequences for the birds' welfare, as individuals with altered phenotype may be specifically targeted for aggression (Dennis et al., 2008; Campderrich et al., 2017a,b). Phenotypic diversity appeared to have a lower impact when phenotypic diversity is high from early rearing, but it may generate major social disruption if changes occur when the birds are adults (Campderrich et al., 2017a,b; Liste et al., 2015). Changes in phenotypic appearance are hypothesized to be associated with the relatively frequent bullying occurring towards sick chickens or layers. It can be directed towards young layers that delay egg laying, turning yellowish on the legs and around their eyes because of carotenoid accumulation (personal obs.). These healthy hens can be brought back to the flock once they start laying and the yellow coloration disappears. Thus, any situation that leads to changes in the phenotypical appearance of the birds may steer to welfare problems of specific individuals within the flock. Being aware of the changes taking place on the birds because of disease, accidents or any other factors that make birds to change appearance or normal behaviour is important as it is a high risk factor for their welfare.

## 5.3 Other considerations associated with social behaviour

Social facilitation and social learning processes are of fundamental relevance for social species to learn about their environment and increase the chances of survival. In chickens, the mother hen will encourage chicks to peck at edible particles by showing them to peck or will emit specific calls to attract their progeny to good feeding grounds. Maternal displays will also encourage chicks to get away from harmful or non-profitable items (see review by Nicol, 2004). It is possible to use social learning to improve flock management. A clear example of social learning that takes place when a new flock of chicks arrive to a poultry house is learning to drink from nipple drinkers, or the discovering of feeders. By training a few as they arrive to the facility, managers can be reassured that the entire flock soon will learn where to obtain food and water by learning from each other. The same principle can be used to accustom birds to use free-range areas, for example, by purposely training a few birds to range outdoors, or to use perches. Social facilitation refers to the higher probability for behaviour to take place when other birds perform the behaviour. For example, feeding bouts operate by social facilitation, as one bird goes to the feeder, soon others will follow. Social learning can also be a source for some management problems in free moving flocks of domestic fowl. Behaviours such as feather pecking and cannibalism with major impact on animal welfare (Lambton et al., 2010, 2015), although not directly caused by social factors, can easily spread in large groups through social learning (Cloutier & Newberry, 2000, Cloutier et al., 2002). Once dispersed in the population these problematic behaviours are difficult to control especially in large groups where birds have high mobility, like aviaries with or without outdoor access.

Occasional panic reactions when occurring in large groups can cause massive flock mortalities, with obvious implications for the welfare of poultry (Jones, 1996; Richards, et al., 2012). Panic reactions can occur during rearing or the production phase in laying hens and are caused by a sudden fear reaction. Triggering factors may include noise, human intervention (i.e. caretaker carrying a new element to the birds), or potential predators and tend to occur more frequently, early during production (Richards, et al., 2012). Panic reaction shows

(a)

**Figure 6** (a) Chickens unevenly distributed seeking the protection of the walls (b) experimental treatment with cover panels' help to get a more protective environment for the birds (photos by I. Estevez).

the strong need of social species to seek group or cover protection in case of perceived danger. Not all panic reactions involve the entire group, and some are limited to a few hundred or restricted to a certain location (Richards et al., 2012). Nonetheless, the larger the group the more severe the consequences of a panic reaction.

Bright and Johnson (2011) described three different types of smothering; the classic panic reactions that are one-off event that cause great mortality and it is caused by fear reactions due to unfamiliar noises, disturbances, or changes in light intensities. Nest box smothering, that is most common at the onset of lay, in which several hens will crowd into the same nest box, may cause mortalities of few to a large number of birds. Creeping/recurrent smothering is the most problematic for producers, as it may occur several times during lay, and may take place anywhere indoors and outdoors without a clear triggering factor. Bright and Johnson (2011) also indicated that all three are not mutually exclusive. When domestic fowl panic, they seek the protection of other birds and of barn structures. In open plan houses protection is only found at the end walls, where birds will crowd causing the death of potentially large number of birds. The use of discontinuous barriers or panels in poultry houses, or any other form of restraining structure may serve to gather birds around such partitions in smaller groups preventing the sudden migration of too many birds to the house end walls (Fig. 6) (Estevez, 1999), thus avoiding massive mortalities in the event of an unexpected frightening event.

## 6 Understanding mating systems in poultry: implications for the welfare of breeding flocks

Reproductive behaviour can be considered a specific aspect of the social relationships that are established among females and males with the purpose of producing progeny. The social dynamics of reproducing flocks is complex because of the differential evolutionary pressures among males and females (Pizzari, 2016). It is known that male fowl will compete for access to females by establishing territories to attract them and increase their reproductive success. By mate-guarding females the male fowl defend them from young male incursions and reduce risks of sperm competition (Pizzari, 2016; McDonald et al., 2017). The outcome of this polygynous mating system is a high variability in male fertility that depends greatly on males' social status (Jones and Mench, 1991) and their individual capacity to attract and retain a more or less important number of females. Contrarily, female fertility tends to be similar and normally high (McDonald et al., 2017).

It has been assumed that dominant domestic fowl males would establish mating territories, mate more often and have a higher reproductive success in an almost direct relationship (Guhl and Warren, 1946; Jones and Mench,

1991). Contrarily, Bilcik and Estevez (2005) and Bilcik et al. (2005) showed that male mating frequency in broiler breeders did not result in higher reproductive success in small experimental groups (with 10–12 females). In heavy bird strains, the lack of correlation between male mating success and progeny rate can be explained by the morphological changes resulting from the selection for large growth rate and breast size. A large breast may interfere with male mating success and, specifically, with sperm transference in heavy lines of domestic fowl (McGary et al., 2003a,b). There are also other reasons to explain the lack of apparent relationship between mating frequency and reproductive outcomes in the domestic fowl. Contrary to extended thoughts, females do not render fidelity to exclusive males. Despite the male efforts to monopolize females, they will normally mate with more than one male (polyandry) (Pizzari, 2016), even when in small groups, where access to females can be easily controlled. In this respect, Bilcik and Estevez (2005) showed, by DNA fingerprinting, that 72% of the females had progeny of at least two males of the three in the experimental group. It is likely that similar relationships exist in large commercial broiler breeder flocks, but evidences are so far inexistent because of the difficulties of tracking mating frequency and paternity in large flocks.

Abundant information is nonetheless available on mating strategies in small populations of red junglefowl. The outstanding work conducted by Pizzari's lab links social structure to sexual interactions and with fertility outcomes. McDonald et al. (2017), for example, showed that in red junglefowl young, aggressive individuals (towards other males) mated with more exclusive females, and were better able to safe-guard them from mating with competitors. On the other hand, older dominant males appeared to mate less than expected and followed the strategy of maintaining exclusivity on fewer females. Their low paternity share suggests that they produced less competitive ejaculates. Other component of the subtle competition levels include female behaviour that is often overlooked in reproductive flocks. Older dominant females, that normally have lower fertility, tend to isolate themselves from the group and mate, preferentially, with dominant males (McDonald et al., 2019). On the contrary, more fertile younger females tend to be highly promiscuous (McDonald et al., 2019). In these younger females the effects of post-copulatory competition through sperm competition will be higher (McDonald et al., 2017). The degree of sophistication of mating strategies in the red junglefowl is such that males can modulate the ejaculate expenditure in response to female novelty in order to optimize reproductive performance (Alvarez-Fernandez et al., 2018). These findings showcase the degree of complexity of the factors that ultimately will contribute to the end result of fertility in flocks in the red junglefowl.

It has been argued that genetic selection for growth in the domestic fowl has altered mating behaviour causing reduced fertility (Cheng et al., 1985) and increased aggressivity associated with sexual competition (Millman and

Duncan, 2000a,b) causing major problems to female welfare because of the harsh interactions. Other studies, however, have shown a low frequency of aggressive interactions even among lines of breeders with differential fertility (McGary et al., 2003b) and modulation of mating as result of the increased mating competition (Bilcik and Estevez, 2005). As in their wild counterparts, pressure to maximize reproductive success also determines when most matings take place from 19:00 to 21:00 (Bilcik and Estevez, 2005). This strategy maximizes the chances of egg fertilization (Pizzari and Birkhead, 2001). Similarities encountered, so far, suggest that it is unlikely that reproductive strategies in the domestic fowl would differ too much from the features described for the red junglefowl, as well as in regard to the complexity of factors, including social aspects, that ultimately affect reproductive success.

It can be argued that commercial reproductive populations are formed by birds of the same age, and therefore, fertility outcomes are expected to be more homogeneous across individuals. However, spiking, the introduction of younger males when flock fertility declines, is a common practice in broiler breeders. In addition, in large flocks with several thousand birds it may be more likely to find higher inter-individual variations among birds' mating strategies depending on their physical abilities or dominance potential, similar to what has been observed in the red jungle fowl. Results by Leone and Estevez (2008b) on movements of individual males in commercial broiler breeder flocks showed that, despite variation on use of space patterns, most tracked males moved throughout the facility. Interestingly, the range of movement was higher in enriched houses where fertility was also higher, possibly as chances to interact with new individuals increased (He et al., 2019). Although these results are far from being specific on mating systems, they might be interpreted as indirect evidence of the unlikely existence of male territoriality. Such spatial distribution would initially support the idea of multiple matings in a mix of polygynous-polyandrous system, where sperm competition may play a large role in males' reproductive success.

As has been presented, differential male fertility lays a strong pressure for access to females, especially of the most fecund, in order to maximize reproductive success (Pizzari, 2016). Sexual coercion and over mating have severe consequences for the welfare of females, causing, at times, major female mortalities (Estevez, 2009). Perhaps, female seclusion evolved as a biological adaptation to avoid male coercion. It is important to be aware that harsh sexual behaviour and gender conflict is not exclusive to the fowl. Reduction in female fitness, lifetime reproductive success and harm as consequence of male harassment has been often reported in nature (i.e. Amqvist and Rowe, 2005; Le Gallard et al., 2005; Eldakar et al., 2009). The strong evolutive roots in which this behaviour is engrained imply that it will be difficult to change, but it may be possible to control the negative impact on female welfare. Thus, research on commercial mating systems of domestic fowl should help to elucidate strategies to minimize the impact of male coercion on female welfare.

Severe female damage may occur as a result of wounds produced by repeated forced matings. Forced matings or coercive matings are more likely to occur in broiler breeders when genders are spatially segregated (personal obs.) and it is a warning sign to which attention should be paid (Estevez, 2009). Asynchrony in sexual maturity may be one of the causes for this situation, with males in commercial flocks usually maturing before females, because of the difficulties in managing male feeding. Normal commercial practice is to reduce the number of males in the flock. This practice, however, will have little to no effect, because male to female ratio in the litter area may still remain too high. As pointed by He et al. (2019) the structural complexity of the physical environment may have an impact on the rate of socio-sexual interactions. In structurally complex environments, females may have a better chance to avoid coercive matings by having more places to seclude themselves. Structural complexity may also reduce male-to-male mating disruptions, thereby reducing the rate of unsuccessful matings and helping to alleviate pressure on females. Estevez (1999), in a practical experience in broiler breeder flocks with severe female injuries and mortalities caused by male activity, placed a zigzag line of cover panels in the litter area. This intervention resulted in an effective attraction of a large number of females to the littered area (Fig. 7). The larger female availability resulted in a clear decline of coercive matings and female mortalities. Thus, we have major challenges ahead in order to assure the welfare of reproductive flock, females especially, because of the conflict of interest among males and females as the result of underlying differential evolutionary history.

**Figure 7** Broiler breeder males and females using protective cover in a commercial flock. Source: Photo by I. Estevez.

## 7 Strategies for better management of poultry flocks

The global flourishment of alternative production systems in which poultry can express a wide range of behaviours has increased the complexity of flock management. In cages, the limitations in space availability and the small group sizes precluded the expression of much social behaviours. If problems would occur, these were easy to identify and control in the restricted space of a small cage. The management of large flocks where birds can roam free, in aviaries with or without access to a veranda or in free-range, involves a much higher level of difficulty. A deep understanding on the complexity of the social interrelationships among birds and genders in breeding flocks of the domestic fowl is primordial to be able to overcome such challenges.

The domestic fowl are highly social species that determine largely how they behave. The group means protection and stimulates birds to conduct behaviours such as feeding, drinking or dust bathing. Isolation is highly stressful for social species, thus should be avoided. If it is absolutely necessary, perhaps there is an option to give the bird a companion or at least provide visual contact with other birds. Although in small groups they form a well-structured hierarchical system, in large groups they do not, and appear to function according to badges of status in a more lax and tolerant social system. This is why normally if resources are not restricted and are properly distributed in the poultry facility, aggressive interactions should not be an issue. If an unusually high level of interaction occurs, resource availability should be immediately checked. Nevertheless, domestic fowl are very receptive of changes on the phenotypic appearance of the flock members. Thus changes due to bird manipulation, wounds or changes due to sickness may be noticed and risk of bulling over those birds may increase. Note that feather pecking is not a form of aggressive interaction and it does not seem to be associated with social factors.

Early rearing is a critical time period for the domestic fowl as the conditions offered to the birds will determine their behaviour and plasticity. It will define what birds will consider as their normal environment and the conditions that make them feel safe. Thus, habituation to a rich social and physical environment will increase the range of tolerable conditions that will be considered within normality, reducing the risk of panic reactions. The negative, catastrophic consequences of such behaviour will be proportional to the size of the flock as there is a higher risk of pilling up. This is one of the reasons why it may be advisable to place physical divisions of large flocks into smaller groups, so that the potential damage can be somewhat limited if catastrophic effects were to occur. Alternatively, semipermeable barriers that do not preclude bird transit through the facility, and therefore facilitate greater bird mobility which is positive for their welfare; this may be a viable option depending on the facility design.

Caretakers should take advantage of the high plasticity that birds have during the early rearing period to accustom birds to conditions that may be later encountered in the production facility. In addition, increasing behavioural options early on is key to increasing birds' robustness and improves their capacity to overcome stressful events. In meat chickens, growers control this period and may adapt the birds to specific conditions. In pullets this is more of an issue because normally rearing and production take place at different facilities and most likely by different animal caretakers. Pullet growers should think on the range of conditions pullets may encounter in adulthood. Likewise, egg producers may want to inform pullet growers of any particular conditions that they will be experiencing.

Although the domestic fowl is a highly social species with high level of tolerance to a wide range of social and management conditions, they also have a rich behavioural background that was shaped through the evolutionary processes of the species. This background is particularly evident in their need to group as a response to avoid predation, and in breeding flocks where conflict of interest between females and males increased to a max the complexity of social relationships. Still this is an area where much more research is needed, especially under real commercial breeding conditions. Female coercion is a highly relevant welfare problem that should be minimized as much as possible. For now, increasing the environmental complexity by the provision of cover and perches may be an option to provide females with a way to safeguard from male harassment.

Our vision on the social behaviour of the domestic fowl has open to new ideas regarding its plasticity and possibilities for adaptation to different social environments. Nevertheless there are still research questions that need further advance, especially in regard to the relevance of the early environmental conditions, and its impact on social behaviour. Within a particular case of social behaviour is the reproductive behaviour of the domestic fowl, for which we have a lack of understanding of the social dynamics of large reproductive flocks.

## 8 References

Agnvall, B., Ali, A., Olby, S. and Jensen, P. 2014. Red Junglefowl (*Gallus gallus*) selected for low fear of humans are larger, more dominant and produce larger offspring. *Animal* 8(9), 1498–505. doi:10.1017/S1751731114001426.

Agnvall, B. and Jensen, P. 2016. Effects of divergent selection for fear of humans on behaviour in red junglefowl. *PLoS ONE* 11(11), e0166075. doi:10.1371/journal.pone.0166075.

Alonso, J. C., Alonso, J. A., Bautista, L. M. and Munoz-Pulido, R. 1995. Patch use in cranes: a field test of optimal foraging predictions. *Animal Behaviour* 49(5), 1367–79. doi:10.1006/anbe.1995.0167.

Al-Rawi, B. and Craig, J. V. 1975. Agonistic behaviour of caged chickens related to broiler size and area per bird. *Applied Animal Ethology* 2(1), 69–80. doi:10.1016/0304-3762(75)90066-8.

Alvarez-Fernandez, A., Borziak, K., McDonald, G. C., Dorus, S. and Pizzari, T. 2018. Female novelty and male status dynamically modulate ejaculate expenditure and seminal fluid proteome over successive matings in red jungle fowl. *Scientific Reports*, 9, 5852. doi:10.1038/s41598-019-41336-5.

Amqvist, G. and Rowe, L. 2005. *Sexual Conflict. Pincenton*. Princenton University Press, Princenton, NJ.

Banks, E. M. and Allee, W. C. 1957. Some relations between flock size and agonistic behavior in domestic hens. *Physiological Zoology* 30(3), 255–68. doi:10.1086/physzool.30.3.30160926.

Batista, G., Johnson, J. L., Dominguez, E., Costa-Mattioli, M. and Pena, J. L. 2018. Regulation of filial imprinting and structural plasticity by mTORC1 in newborn chickens. *Scientific Reports* 8(1), 8044. doi:10.1038/s41598-018-26479-1.

Beauchamp, G. 2003. Group-size effects on vigilance: a search for mechanisms. *Behavioural Processes* 63(3), 111–21. doi:10.1016/s0376-6357(03)00002-0.

Beauchamp, G. 2016. Function and structure of vigilance in a gregarious species exposed to threats from predators and conspecifics. *Animal Behaviour* 116, 195-201. doi:10.1016/j.anbehav.2016.04.008.

Bertram, B. C. R. 1992. *The Ostrich Communal Nesting System*. Princeton University Press, Princeton, NJ.

Bilcik, B. and Estevez, I. 2005. Impact of male-male competition and morphological traits on mating strategies and reproductive success in broiler breeders. *Applied Animal Behaviour Science* 92(4), 307-23. doi:10.1016/j.applanim.2004.11.007.

Bilcik, B., Estevez, I. and Russek-Cohen, E. 2005. Reproductive success of broiler breeders in natural mating systems: the effect of male-male competition, sperm quality and morphological characteristics. *Poultry Science* 84(9), 1453-62. doi:10.1093/ps/84.9.1453.

Bright, A. and Johnson, E. A. 2011.Smothering in commercial free-range laying hens: a preliminary investigation. *Veterinary Record* 168(19), 1-2 doi:10.1136/vr.c7462.

Buchholz, R. 1997. Male dominance and variation in fleshy head ornamentation in wild turkeys. *Journal of Avian Biology* 28(3), 223–30. doi:10.2307/3676973.

Campderrich, I., Liste, G. and Estevez, I. 2017a.Group size and phenotypic appearance: their role on the development of social dynamics in laying hens. *Applied Animal Behaviour Science* 189, 41–8. doi:10.1016/j.applanim.2017.01.014.

Campderrich, I., Liste, G. and Estevez, I. 2017b. The looks matter; Aggression escalation from changes in the phenotypic appearance in the domestic fowl. *PLoS ONE* December 20, 2017, pp 1–17. doi:10.1371/journal.pone.0188931.

Campderrich, I., Nazar, F. N., Wichman, A., Marín, R. H., Estevez, I. and Keeling, L. J. 2019. Environmental complexity; a buffer against stress in the domestic chick. *PLoS ONE* 14(1), e0210270. doi:10.1371/journal.pone.0210270.

Caraco, T. 1981. Risk-sensitivity and foraging groups. *Ecology* 62(3), 527–31. doi:10.2307/1937716.

Carmichael, N. L., Walker, A. W. and Hughes, B. O. 1999. Laying hens in large flocks in a perchery system: influence of stocking density on location, use of resources and behavior. *British Poultry Science* 40(2), 165–76. doi:10.1080/00071669987566.

Cheng, K. M., Burns, J. T. and Schoffner, R. N. 1985. Mating behaviour and fertility in domestic chickens. I. Inbreeding. *Applied Animal Behaviour Science* 13(4), 371–81. doi:10.1016/0168-1591(85)90016-4.

Cloutier S. and Newberry R. C. 2000. Recent social experience, body weight and initial patterns of attack predict the social status attained by unfamiliar hens in a new group. *Behaviour* 137(6), 705–26. doi:10.1163/156853900502303.

Cloutier, S. and Newberry, R. C. 2002. Differences in skeletal and ornamental traits between laying hen cannibals, victims and bystanders. *Applied Animal Behaviour Science* 77(2), 115–26. doi:10.1016/S0168-1591(02)00049-7.

Cloutier, S., Newberry, R. C., Honda, K. and Alldredge, J. R. 2002. Cannibalistic behaviour spread by social learning. *Animal Behaviour* 63(6), 1153–62. doi:10.1006/anbe.2002.3017.

Cornetto, T. L. and Estevez, I. 2001. Behavior of the domestic fowl in presence of vertical panels. *Poultry Science* 80(10), 1455–62. doi:10.1093/ps/80.10.1455.

Cornetto, T. L., Estevez, I. and Douglass, L. W. 2002. Using artificial cover to reduce aggression and disturbances in domestic fowl. *Applied Animal Behaviour Science* 75(4), 325–36. doi:10.1016/S0168-1591(01)00195-2.

Davidson, J. D. and El Hady, A. 2019. Foraging as an evidence accumulation process. *PLoS Computational Biology* 15(7), e1007060. doi:10.1371/journal.pcbi.1007060.

Dawkins, M. S., Donnelly, C. A. and Jones, T. A. 2004. Chicken welfare is influenced more by housing conditions than by stocking density. *Nature* 427(6972), 342–4. doi:10.1038/nature02226.

D'Eath, R. B. and Keeling, L. J. 2003. Social discrimination and aggression by laying hens in large groups: from peck orders to social tolerance. *Applied Animal Behaviour Science* 84(3), 197–212. doi:10.1016/j.applanim.2003.08.010.

Dennis, R. L., Newberry, R. C., Cheng, H. W. and Estevez, I. 2008. Appearance matters: artificial marking alters aggression and stress. *Poultry Science* 87(10), 1939–46. doi:10.3382/ps.2007-00311.

Di Giorgio, E., Loveland, J. L., Mayer, U., Rosa-Salva, O., Versace, E. and Vallortigara, G. 2017. Filial responses as predisposed and learned preferences: early attachment in chicks and babies. *Behavioural Brain Research* 325(B), 90–104. doi:10.1016/j.bbr.2016.09.018.

Duncan, I. J. H., Widowski, T. M., Malleau, A. E., Lindberg, A. C. and Petherick, J. C. 1998. External factors and causation of dustbathing in domestic hens. *Behavioural Processes* 43(2), 219–28. doi:10.1016/s0376-6357(98)00017-5.

Edgar, J. L., Paul, E. S. and Nicol, C. J. 2013. Protective mother hens: cognitive influences on the avian maternal response. *Animal Behaviour* 86(2), 223–9. doi:10.1016/j.anbehav.2013.05.004.

EFSA 2015. Scientific Opinion on welfare aspects of the use of perches for laying hens. (Panel on Animal Health and Welfare), *EFSA Journal* 13(6), 4131, 69 pp. doi:10.2903/j.efsa.

Eklund, B. and Jensen, P. 2011. Domestication effects on behavioural synchronization and individual distances in chickens (Gallus gallus). *Behavioural Processes* 86(2), 250–6. doi:10.1016/j.beproc.2010.12.010.

Eldakar, O. T., Dlugos, M. J., Wilcox, R. S. and Wilson, D. S. 2009. Aggressive mating as a tragedy of the commons in water strider *Aquarius remiges*. *Behaviour Ecology and Sociobiology* 64, 25–33. doi:101007/s00265-0090814-G.

Elgar, M. A. 1989. Predator vigilance and group size in mammals and birds: a critical review of the empirical evidence. *Biological Reviews of the Cambridge Philosophical Society* 64(1), 13–33. doi:10.1111/j.1469-185x.1989.tb00636.x.

Estevez, I. 1999. *Cover Panels for Chickens. Published in: Poultry Perspectives.* Maryland Cooperative Extension, UMD, College Park, MD, 1(2), 4–6.

Estevez, I. 2007. Density allowances for broilers; Where to set the limits? Poultry Science, 86, 1265–1272.

Estevez, I. 2007. *Cover Panels for Chickens. Published in: Poultry Perspectives.* Maryland Cooperative Extension, UMD, College Park, MD, 1(2), 4–6.

Estevez, I. 2009. Manejo de aves reproductoras para optimizar la fertilidad. Proceedings of the 46th Symposium Científico de Avicultura. Asociación Española de Ciencia Avícola, Zaragoza, pp. 45–55.

Estevez, I., Andersen, I. L., Nævdal, E. 2007. Group size, density and social dynamics in farm animals. *Applied Animal Behaviour Science* 103, 185–204.

Estevez, I., Keeling, L. J. and Newberry, R. C. 2003. Decreasing aggression with increasing group size in young domestic fowl. *Applied Animal Behaviour Science* 84(3), 213–8. doi:10.1016/j.applanim.2003.08.006.

Estevez, I., Newberry, R. C. and Arias de Reyna, L. 1997. Broiler chickens: a tolerant social system? *Etología* 5, 19–29.

Estevez, I., Newberry, R. C. and Keeling, L. J. 2002. Dynamics of aggression in the domestic fowl. *Applied Animal Behaviour Science* 76(4), 307–25. doi:10.1016/S0168-1591(02)00013-8.

Feltenstein, M. W., Ford, N. G., Freeman, K. B. and Sufka, K. J. 2002. Dissociation of stress behaviors in the chick social-separation-stress procedure. *Physiology and Behavior* 75(5), 675–9. doi:10.1016/s0031-9384(02)00660-1.

Guhl, A. M. 1953. Social behavior of the domestic fowl. *Kansas State Agricultural Experimental Station Technical Bulletin* 73, 3–48.

Guhl, A. M. and Warren, D. C. 1946. Number of offspring sired by cockerels related to social dominance in chickens. *Poultry Science* 25(5), 460–72. doi:10.3382/ps.0250460.

Hamilton, W. D. 1971. Geometry of the selfish herd. *Journal of Theoretical Biology* 31(2), 295–311. doi:10.1016/0022-5193(71)90189-5.

He, P., Maldonado-Chaparro, A. A. and Farine, D. R. 2019. The role of habitat configuration in shaping social structure: a gap in studies of animal social complexity. *Behavioral Ecology and Sociobiology* 73(1), 9. doi:10.1007/s00265-018-2602-7.

Hess, E. H. 1958. Imprinting in animals. *Scientific American* 198(3), 81–90. doi:10.1038/scientificamerican0358-81.

Hoppitt, W., Blackburn, L. and Laland, K. N. 2007. Response facilitation in the domestic fowl. *Animal Behaviour* 73(2), 229–38. doi:10.1016/j.anbehav.2006.05.013.

Hoppitt, W. and Laland, K. N. 2008. Social processes affecting feeding and drinking in the domestic fowl. *Animal Behaviour* 76(5), 1529–43. doi:10.1016/j.anbehav.2008.07.011.

Hughes, B. O., Carmichael, N. L., Walker, A. W. and Grigor, P. N. 1997. Low incidence of aggression in large flocks of laying hens. *Applied Animal Behaviour Science* 54(2–3), 215–34. doi:10.1016/S0168-1591(96)01177-X.

Jones, M. E. J. and Mench, J. A. 1991. Behavioral correlates of male mating success in multisire flocks as determined by DNA fingerprinting. *Poultry Science* 70(7), 1493–8. doi:10.3382/ps.0701493.

Jones, R. B. 1996. Fear and adaptability in poultry: insights, implications and imperatives. *World's Poultry Science Journal* 52(2), 131-74. doi:10.1079/WPS19960013.

Kacelnik, A. and Brunner, D. 2002. Timing and foraging: Gibbon's scalar expectancy theory and optimal patch exploitation. *Learning and Motivation* 33(1), 177-95. doi:10.1006/lmot.2001.1110.

Keeling, L. J., Estevez, I., Newberry, R. C. and Correia, M. G. 2003. Production-related traits of layers reared in different sized flocks: the concept of problematic intermediate group sizes. *Poultry Science* 82(9), 1393-6. doi:10.1093/ps/82.9.1393.

Keeling, L. J., Newberry, R. C. and Estevez, I. 2017. Flock size during rearing affects pullet behavioural synchrony and spatial clustering. *Applied Animal Behaviour Science* 194, 36-41. doi:10.1016/j.applanim.2017.04.002.

Krause, E. T., Phi-van, L., Dudde, A., Schrader, L. and Kjaer, J. B. 2019. Behavioural consequences of divergent selection on general locomotor activity in chickens. *Behavioural Processes* 169, art. no. 103980. doi:10.1016/j.beproc.2019.103980.

Krause, J. and Ruxton, G. D. 2002. *Living in Groups*. In: *Oxford Series in Ecology and Evolution*. Oxford University Press, Oxford, 210pp.

Lambton, S. L., Knowles, T. G., Yorke, C. and Nicol, C. J. 2010. The risk factors affecting the development of gentle and sever feather pecking in loose housed laying hens. *Applied Animal Behaviour Science* 123(1-2), 32-42. doi:10.1016/j.applanim.2009.12.010.

Lambton, S. L., Knowles, T. G., Yorke, C. and Nicol, C. J. 2015. The risk factors affecting the development of vent pecking and cannibalism in free-range and organic laying hens. *Animal Welfare* 24(1), 101-11. doi:10.7120/09627286.24.1.101.

Le Gallard, J. F., Fitze, P. S., Frerière, R. and Clobert, J. 2005. Sex ratio bias, male aggression, and population collapse in lizards. *Proceedings of the National Academy of Sciences of the United States of America* 102(18), 231-6. doi:10.1073/pnas.0505172102.

Leone, E. H. and Estevez, I. 2007. Patch Discrimination as a function of handling time. *Poultry Science* 86(3), 447-52. doi:10.1093/ps/86.3.447.

Leone, E. H. and Estevez, I. 2008a. Use of space in the domestic fowl: separating the effects of enclosure size, group size, and density. *Animal Behaviour* 76(5), 1673-82. doi:10.1016/j.anbehav.2008.08.004.

Leone, E. H. and Estevez, I. 2008b. Economic and welfare benefits of environmental enrichment for broiler breeders. *Poultry Science* 87(1), 14-21. doi:10.3382/ps.2007-00154.

Lewis, A. R. 1980. Patch use by gray squirrels and optimal foraging. *Ecology* 61(6), 1371-9. doi:10.2307/1939046.

Liste, G., Campderrich, I., Beltrán de Heredia, I. B. and Estevez, I. 2015. The relevance of variations in group size and phenotypic appearance on the behaviour and movement patterns of young domestic fowl. *Applied Animal Behaviour Science* 163, 144-57. doi:10.1016/j.applanim.2014.11.013.

Lorenz, K. 1935. Der Kumpan in der Umwelt des Vogels. Der Artgenosse als auslösendes Moment sozialer Verhaltensweisen. *Journal für Ornithologie* 83, 137-215, 289-413.

Malleau, A. E., Duncan, I. J. H., Widowski, T. M. and Atkinson, J. L. 2007. The importance of rest in young domestic fowl. *Applied Animal Behaviour Science* 106(1-3), 52-69. doi:10.1016/j.applanim.2006.06.017.

Marin, R. H., Liste, M. G., Campderrich, I. and Estevez, I. 2014. The impact of phenotypic appearance on body weight and egg production in laying hens: a group-size- and experience dependent phenomenon. *Poultry Science* 93(7), 1623–35. doi:10.3382/ps.2013-03705.

Marx, G., Leppelt, J. and Ellendorff, F. 2001. Vocalization in chicks (*Gallus gallus domesticus*) during stepwise social isolation. *Applied Animal Behaviour Science* 75(1), 61–74. doi:10.1016/S0168-1591(01)00180-0.

McBride, G. and Foenander, F. 1962. Territorial behaviour in flocks of domestic fowls. *Nature* 194(4823), 102. doi:10.1038/194102a0.

McDonald, G. C., Spurgin, L. G., Fairfield, E. A., Richardson, D. S. and Pizzari, T. 2017. Pre- and postcopulatory sexual selection favour aggressive, young males in polyandrous groups of red junglefowl. *Evolution*, 71(6), 1653–1669.

McDonald, G. C., Spurgin, L. G., Fairfiels, E. A., Richardson, D. S. and Pizzari, T. 2019. Differential female sociality is linked with the fine-scale structure of sexual interactions in replicate groups of red junglefowl, Gallus gallus. *Proceedings of the Royal Society B* 286(1913), 20191734. doi:10.1098/rspb.2019.1734.

McGary, S., Estevez, I. and Bakst, M. R. 2003a. Potential relationships between physical traits and male broiler breeder fertility. *Poultry Science* 82(2), 328–37. doi:10.1093/ps/82.2.328.

McGary, S., Estevez, I. and Russek-Cohen, E. 2003b. Reproductive and aggressive behaviour in male broiler breeders with varying fertility levels. *Applied Animal Behaviour Science* 82(1), 29–44. doi:10.1016/S0168-1591(03)00038-8.

Millman, S. T. and Duncan, I. J. H. 2000a. Effect of male-to-male aggressiveness and feed-restriction during rearing in sexual behavior and aggressiveness towards females by domestic fowl. *Applied Animal Behavioural Science* 70, 63–82.

Millman, S. T. and Duncan, I. J. H. 2000b. Strain differences in aggressiveness of male domestic fowl in response tomale model. *Applied Animal Behaviour Science* 66(3), 217–33. doi:10.1016/S0168-1591(99)00082-9.

Miura, M. and Matsushima, T. 2016. Biological motion facilitates filial impringting. *Animal Behaviour* 116, 171–80. doi:10.1016/j.anbehav.2016.03.025.

Morton, J. and Johnson, M. H. 1991. CONSPEC and CONLERN: a two-process theory of infants face recognition. *Psychological Review* 98(2), 164–81. doi:10.1037/0033-295x.98.2.164.

Newberry, R. C. and Hall, J. W. 1990. Use of pen space by broiler chickens: effects of age and pen size. *Applied Animal Behaviour Science* 25(1–2), 125–36. doi:10.1016/0168-1591(90)90075-O.

Newberry, R. C. and Shackleton, D. M. 1997. Use of visual cover by domestic fowl, a venetian blind effect. *Animal Behaviour* 54(2), 387–95. doi:10.1006/anbe.1996.0421.

Newberry, R. C., Estevez, I. and Keeling, L. J. 2001. Group size and perching behaviour in young domestic fowl. *Applied Animal Behaviour Science* 73(2), 117–29. doi:10.1016/s0168-1591(01)00135-6.

Nicol, C. J. 2004. Development, direction, and damage limitation: social learning in domestic fowl. *Learning and Behavior* 32(1), 72–81. doi:10.3758/bf03196008.

Nicol, C. J., Gregory, N. G., Knowles, T. G., Parkman, I. D. and Wilkins, L. J. 1999. Differential effects of increased stocking density, mediated by increased flock size, on feather pecking and aggression in laying hens. *Applied Animal Behaviour Science* 65(2), 137–52, doi:10.1016/S0168-1591(99)00057-X.

Norman, K. I., Adriaense, J. E. C. and Nicol, C. J. 2019. The impact of early structural enrichment on spatial cognition in layer chicks. *Behavioural Processes* 164, 167–74. doi:10.1016/j.beproc.2019.05.008.

Pagel, M. and Dawkins, M. S. 1997. Peck orders and group size in laying hens: 'futures contracts' for non-aggression. *Behavioural Processes* 40(1), 13–25. doi:10.1016/s0376-6357(96)00761-9.

Parker, G. A. 2000. Scramble in behaviour and ecology. *Philosophical Transactions of the Royal Society of London B*, 355(1403), 1637–45. doi:10.1098/rstb.2000.0726.

Pettit-Riley, R., Estevez, I. and Russek-Cohen, E. 2002. Effects of crowding and access to perches on aggressive behaviour in broilers. *Applied Animal Behaviour Science* 79(1), 11–25. doi:10.1016/S0168-1591(02)00116-8.

Pizzari, T. 2016. The Wood-Gush legacy: a sociobiology perspective to fertility and welfare in chickens. *Applied Animal Behaviour Science* 181, 12–8. doi:10.1016/j.applanim.2016.01.025.

Pizzari, T. and Birkhead, T. R. 2001. From whom does the hen cackle? The function of oviposition cackling. *Animal Behaviour* 61(3), 601–7. doi:10.1006/anbe.2000.1620.

Pulliam, H. R., 1973. *Journal of Theoretical Biology.* 38, 419–422.

Pulliam, H. R. and Caraco, T. 1984. Living in groups: is there an optimal group size? In: Krebs, J. R. and Davies, N. B. (Eds). *Behavioural Ecology: An Evolutionary Approach* (2nd edn.). Blackwell Scientific Publications Ltd., London, pp. 122–47.

Rajecki, D. W. 1988. Formation of leap orders in pairs of male domestic chickens. *Aggressive Behavior* 14, 425–36.

Richards, G. J., Brown, S. N., Booth, F., Toscano, M. J. and Wilkins, L. J. 2012. Panic in free-range laying hens. *Veterinary Record* 170(20), 519. doi:10.1136/vr.100685.

Roberts, G. 1996. Why individual vigilance declines as group size increases. *Animal Behaviour* 51(5), 1077–86. doi:10.1006/anbe.1996.0109.

Rodríguez-Aurrekoetxea, A. and Estevez, I. 2014. Aggressiveness in the domestic fowl: distance versus attitude. *Applied Animal Behaviour Science* 153, 68–74. doi:10.1016/j.applanim.2014.01.007.

Rodríguez-Aurrekoetxea, A., Leone, E. H. and Estevez, I. 2014. Environmental complexity and use of space in slow growing free range chickens. *Applied Animal Behaviour Science* 161, 86–94. doi:10.1016/j.applanim.2014.09.014.

Rosa-Salva, O., Mayer, U. and Vallortigara, G. 2015. Roots of a social brain: developmental models of emerging animacy-detection mechanisms. *Neuroscience and Biobehavioral Reviews* 50, 150–68. doi:10.1016/j.neubiorev.2014.12.015.

Rosa-Salva, O., Mayer, U. and Vallortigara, G. 2019. Unlearn visual preferences for the head region in domestic chicks. *PLoS ONE* 14(9), e0222079. doi:10.1371/journal.pone.0222079.

Rushen, J. 1982. The peck orders of chickens: how do they develop and why are they linear? *Animal Behaviour* 30(4), 1129–37. doi:10.1016/S0003-3472(82)80203-0.

Schaefer, H. M., Schmidt, V. and Bairlein, F. 2003. Discrimination abilities for nutrients: which difference matters for choosy birds and why? *Animal Behaviour* 65(3), 531–41. doi:10.1006/anbe.2003.2089.

Skånberg, L., Newberry, R. C., Estevez, I., Nazar, N. and Keeling, L. J. 2019. Environmental controllability and predictability affect the behavioural adaptability of laying hen chicks in novel situations. Newberry, R. C. and Braastad, B. O. (Eds). Proceedings of the 53th Congress of the International Society for Applied Ethology, Bergen,

5-9 August, Wageningen Academic Publishers, The Netherlands, p. 164. doi:10.3920/978-90-8686-889-6.

Stephens, D. W. and Krebs, J. R. 1986. *Foraging Theory*. Princeton University Press, Princeton, NJ.

Stroeymeyt, N., Grasse, A. V., Crespi, A., Mersch, D. P., Cremer, S. and Keller, L. 2018. Social network plasticity decreases disease transmission in a eusocial insect. *Science* 362(6417), 941-5. doi:10.1126/science.aat4793.

Taylor, M. L., Price, T. A. R. and Wedell, N. 2014. Polyandry in nature: a global analysis. *Trends in Ecology and Evolution* 29(7), 376-83. doi:10.1016/j.tree.2014.04.005.

Tuomainen, U. and Candolin, U. 2011. Behavioural responses to human-induced environmental change. *Biological Reviews of the Cambridge Philosophical Society* 86(3), 640-57. doi:10.1111/j.1469-185X.2010.00164.x.

Turner, S. P. 2011. Breeding against harmful social behaviours in pigs and chickens: state of the art and the way forward. *Applied Animal Behaviour Science* 134(1-2), 1-9. doi:10.1016/j.applanim.2011.06.001.

Vallortigara, G. 2012. Core knowledge of object, number and geometry: a comparative and neural approach. *Cognitive Neuropsychology* 29(1-2), 213-36. doi:10.1080/02643294.2012.654772.

Vallortigara, G., Regolin, L. and Marconato, F. 2005. Visually inexperienced chicks exhibit spontaneous preference for biological motion patterns. *PLoS Biology* 3(7), e208. doi:10.1371/journal.pbio.0030208.

Ventura, B. A., Siewerdt, F. and Estevez, I. 2012. Access to barrier perches improves behavior repertoire in broilers. *PLoS ONE* 7(1), e29826 doi:10.1371/journal.pone.0029826.

Versace, E. and Vallortigara, G. 2015. Origins of knowledge: insights from precocial species. *Frontiers in Behavioral Neuroscience* 9, 338. doi:10.3389/fnbeh.2015.00338.

Weary, D. M. and Fraser, D. 2017. Social and reproductive behaviour. In: Jensen, P. (Ed.). *The Ethology of Domesticated Animals: An Introductory Text* (3rd edn.). CAB International, pp. 79-89.

Weimer, S. L., Robison, C. I., Tempelman, R. J., Jones, D. R. and Karcher, D. M. 2019. Laying hen production and welfare in enriched colony cages at different stocking densities. *Poultry Science* 98(9), 3578-86 doi:10.3382/ps/pez107.

Welfare Quality 2009. *Welfare Quality Assessment Protocol for Poultry (Broilers, Laying Hens)*. Welfare Quality Consortium, Lelystad, the Netherlands.

Widowski, T. M., Caston, L. J., Casey-Trott, T. M. and Hunniford, M. E. 2017. The effect of space allowance and cage size on laying hens housed in furnished cages, Part II: behavior at the feeder. *Poultry Science* 96(11), 3816-23. doi:10.3382/ps/pex198.

Yamaguchi, S., Aoki, N., Kitajima, T., Likubo, E., Katagiri, S., Matsushima, T. and Homma, K. J. 2012. Thyroid hormone determines the start of the sensitive period of imprinting and primes later learning. *Nature Communications* 3, 1081. doi:10.1038/ncomms2088.

# Chapter 6

# Poultry welfare monitoring: wearable technologies

*Dana L. M. Campbell, CSIRO, Australia; and Marisa A. Erasmus, Purdue University, USA*

1 Introduction
2 Radio-frequency identification technology
3 Wearable sensors and accelerometers
4 Case study: outdoor stocking density in free-range laying hens
5 Conclusion
6 Future trends in research
7 Where to look for further information
8 References

## 1 Introduction

Globally, the trend for changes in modern poultry farming of broilers and layers is towards improved welfare. This is in response to both consumer pressures for welfare-friendly products originating from ethical farming systems, and to ensure sustainability of the industries in a changing climate with a growing population. With these goals in mind, industry trends or legislative changes have seen many countries moving towards alternative housing systems to the conventional layer cage (reviewed in Sandilands, 2019, Chapter 11) or indoor broiler shed (reviewed in de Jong, 2019, Chapter 10). These farming system changes have increased the environmental complexity in which the chickens are housed as well as amplified the number and intricacy of management decisions that must be made. For example, aviary systems for laying hens typically have multiple tiers and a litter area that require birds to be competent at moving around to avoid injury or death (Stratmann et al., 2015; Campbell et al., 2016a; Fulton, 2019). In broiler systems, for example, enrichments must be designed so that birds will be able to use them (Norring et al., 2016; Bergmann et al., 2017; Bailie et al., 2018) and provided in sufficient quantities for optimal impact (Rodriguez-Aurrekoetxea et al., 2014).

The variation within these alternative systems allows free choices by the birds for the use of distinct system areas (e.g. indoors, outdoors, litter, structural

http://dx.doi.org/10.19103/AS.2020.0078.06

tiers), which can result in sub-populations of birds with different behavioural patterns, consequently, leading to varying welfare states. For example, free-range housing systems for both broilers and layers have indoor and outdoor areas that each have their own environmental conditions that impact the birds' welfare. Not all birds will access the outdoor areas, and those that do, spend variable amounts of time outside (Rodriguez-Aurrekoetxea et al., 2014; Campbell et al., 2017a; Larsen et al., 2017; Taylor et al., 2017a), which changes depending on daily weather and/or season (Richards et al., 2011; Taylor et al., 2017b, 2018). This individual variation can sometimes be reflected in bird health (Rodriguez-Aurrekoetxea and Estevez, 2016; Campbell et al., 2017b; Larsen et al., 2018; Taylor et al., 2018), but, to date, the full management implications of these subpopulations, and best strategies to optimise the flock overall, are not well understood. Finally, loose-housing and large flock sizes result in high numbers of chickens in contact with each other, which can increase the risk of detrimental social behaviours such as feather pecking, cannibalism, or smothering (Sherwin et al., 2010; Barrett et al., 2014). Increased environmental complexity and poorer environmental control can also decrease hygiene and elevate disease risk (Jones et al., 2015; Courtice et al., 2018).

All of these systems are aimed at improving chicken welfare, but to measure and validate this for either customer assurance or continued improvement of the systems, assessing the state of the chickens at both the flock and the individual level is necessary. Animal behaviour is one of the most important animal-based measures of animal well-being. Animals use behaviour as a coping mechanism to survive and first respond to threats and stressors in their environment (Mench, 1998). Animal behaviour is an important tool for identifying animals that are unwell and is often the first sign of poor welfare that animal caretakers notice. However, the use of visual behavioural observations is subjective, error-prone and often unreliable (Weary et al., 2009). In contrast, using precision technology to quantify behavioural changes may provide a more reliable method for identifying sickness and other risks to animal welfare. Additionally, the sensitivity of precision technology may open up the possibility of detecting more subtle changes in behaviour that could be less resilient to poor health, such as play, exploration, or grooming, and be the first indicators of disease onset (e.g. mice: Litten et al., 2008; dairy cows: Mandel et al., 2017). The application of the combination of existing and new technologies to poultry farming makes it possible to identify changes in behaviour early, non-invasively and objectively, and identify specific types of animal welfare risks.

Given the vast numbers of birds that are typically housed within a farm, the application of modern precision technology to enable automated behavioural and welfare assessment is a rapidly growing sector. Precision technology is utilised across all major livestock industries with many sophisticated on-animal devices available in the commercial market to assist with monitoring individual ruminant animal welfare, such as collars, ankle straps, ear-tags, and implantable electronics. Poultry systems represent a more unique challenge because of the sheer number of birds to be monitored, their small size, and the economic value of the individual chicken. Technology for monitoring poultry welfare can be based at the flock level, such as the detection of disease or leg health through video-based monitoring (Colles et al., 2016; Dawkins, 2019, Chapter 7; Silvera et al., 2017; Fernández et al., 2018). Technology can also be based right at the level of the individual bird using wearable electronics. The current development status and availability of technology for monitoring poultry welfare was reviewed by Sassi and colleagues (2016), with the conclusion that there are multiple potential forms of monitoring, including acoustic analysis, image analysis, environmental sensors, thermal monitoring, and movement tracking. As yet, few technologies are widely commercially used in practice. Siegford and colleagues (2016) also reviewed the role of radio signal monitoring technology to obtain individual bird data within large groups of birds. More recently, Ellen and colleagues (2019) reviewed how the use of monitoring technologies such as radio-frequency identification (RFID), sensor technologies, and computer vision could identify birds at risk for feather pecking problems and the potential for genetic selection against this. Two other review papers focus on aspects of animal health. Neethirajan (2017) argues for the integration of available sensor technologies, which include sensors to monitor animal behaviour to detect animal health risks in real time. Similarly, Astill et al. (2018) reviewed the use of various types of sensors, including biosensors and wearable sensors, specifically to detect avian influenza and concluded that the implementation of technology in poultry farming is crucial as poultry production intensifies to meet increasing demand for food.

This chapter specifically reviews the use of wearable technologies applied to chickens and how these can be used to assess and monitor chicken welfare. Discussion is included for RFID technology, accelerometers, and wearable sensors with a case study presented on how RFID technology has been used to address an industry question of optimal outdoor stocking density for free-range hens within Australia. Finally, a summary of the use of research to improve agriculture, future research directions, and sources for further information are presented.

## 2 Radio-frequency identification technology

Radio-frequency identification (RFID) technology is used across multiple livestock systems for various purposes of automated animal identification and tracking (Brown-Brandl et al., 2019; Ruiz-Garcia and Lunadei, 2011). Within poultry systems, it is a key example of a wearable technology placed on individual chickens that results in extensive automated data collection to monitor behaviour and subsequently measure and improve welfare. RFID systems operate via radio waves that communicate between an RFID tag (worn by the chicken) and a tag reader or antenna that are all connected to a host system for data collection (Roberts, 2006). Tags are typically worn on the leg (Fig. 1) but can also be attached on the wings, on backpacks, or as injectable transponders (Zaninelli et al., 2016). The systems come in different forms, falling broadly into two classes based on the type of tags that the animals wear, passive or active. Passive tags, in comparison to active tags, operate without their own battery power source and thus are smaller, cheaper, and with an unlimited life-span, but have a shorter reading range (Ruiz-Garcia and Lunadei, 2011; Brown-Brandl et al., 2019; Ellen et al., 2019). There are also different frequencies tags can operate at (Ruiz-Garcia and Lunadei, 2011; Ellen et al., 2019), with low-frequency tags having a slower read rate compared to tags operating at high frequencies, but their signals are less impeded by water and metal. Ultra-high-frequency tags have an even faster read rate but a corresponding reduced performance around water and metals. Finally, ultra-wideband tags operate across multiple frequencies and are minimally affected by any surrounding metal. Improvements in the accuracy of tag reading and reading distance generally require trade-offs regarding tag size and power supply. RFID systems within poultry research are used to automatically track the locational movement of individual birds within a housing system and/or their use of specific resources to gain a better understanding of where chickens spend their time, what factors may lead to resource-use choices, and both individual and inter-individual behavioural patterns. One of the most frequent uses of this technology has been to examine the ranging patterns of free-range hens and broilers, but research has also looked at the automated tracking of movement patterns in aviaries, perch use, nest box use, and feeding and drinking behaviour. Examples of free-range applications of RFID technology, followed by RFID monitoring of resource use in other housing systems and what can be gained for understanding behaviour and improving welfare, are presented henceforth.

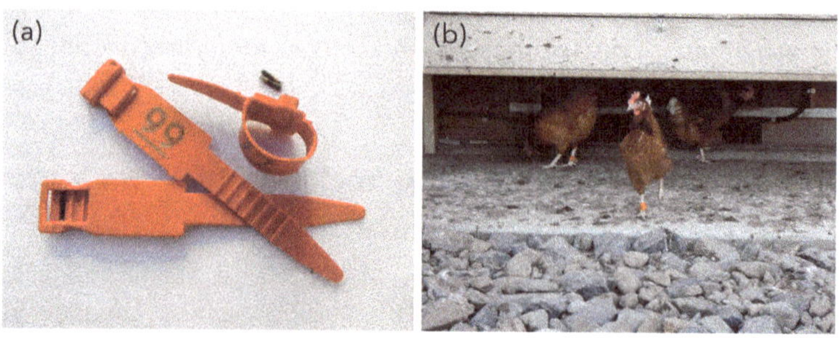


**Figure 1** (a) An adjustable leg band and a passive microchip that can be glued into the band. (b) The leg bands are worn by the hens in an experimental free-range system.

### 2.1 Radio-frequency identification technology systems and range use of chickens

To date, RFID systems using passive RFID tags (attached to wings or legs) have been applied globally to measure range usage in both laying hens and broilers, on commercial farms, and within experimental settings (Richards et al., 2011; Gebhardt-Henrich et al., 2014; Hartcher et al., 2016; Campbell et al., 2017a; Larsen et al., 2017; Taylor et al., 2017a), but only for research purposes. Free-range systems are unique from exclusively indoor-based systems as birds have a choice to go outdoors or not. Levels of range usage can vary extensively between farms, and precise range utilisation can be challenging to estimate visually because not all birds will range simultaneously (Dawkins et al., 2003; Pettersson et al., 2016). Thus, RFID technology has been increasingly used as it allows for accurate measures of when birds go outside, how long they go outside for, variables that may affect whether they range or not, and the impacts of ranging on bird health. Data gained from this chicken tracking are valuable for optimising how free-range systems are designed and managed.

The RFID systems applied to free-range farms vary, with antennas placed at the pop holes (e.g. Thurner et al., 2010; Richards et al., 2011; Campbell et al., 2017a; Taylor et al., 2017a) or at additional locations within the range area to track how far birds may move once they venture outside (e.g. Gebhardt-Henrich et al., 2014; Larsen et al., 2017). Tagged birds register on the system as they pass over an antenna. The number of antennas required is directly dependent on the size of the area to be monitored, which can sometimes restrict assessments to only a portion of a commercial facility, particularly for antennas placed on the range (Larsen et al., 2017). Antennas are typically less than 1 m in length, and read distances for wearable passive tags are generally less than 50 cm. To illustrate, an RFID system installed in an experimental free-range facility (Campbell et al., 2017a) contains decoding units which can each connect to four antennas; thus five decoders are needed to monitor nine pop holes (a total of 18 antennas, two per pop hole that are the width of a single hen only), which total approximately 50 kAUD in cost (Fig. 2). Generally, only a proportion of the total flock is tagged, allowing the estimation of flock-level patterns. Data are stored for the date/time/ID of every bird passage, often including the ability to determine the direction of travel through the use of two closely placed antennas or sensor pairs and algorithms that can filter out unpaired or 'false' readings (e.g. if a bird sits in a pop hole or if the bird does not complete an intended direction of travel).

RFID technology applied to free-range laying hen systems has been able to identify similar population-level patterns across multiple free-range settings, revealing what previous flock-level live observations could not. Not all birds will range, but the majority of hens do (67–81% ranged daily, Campbell et al.,

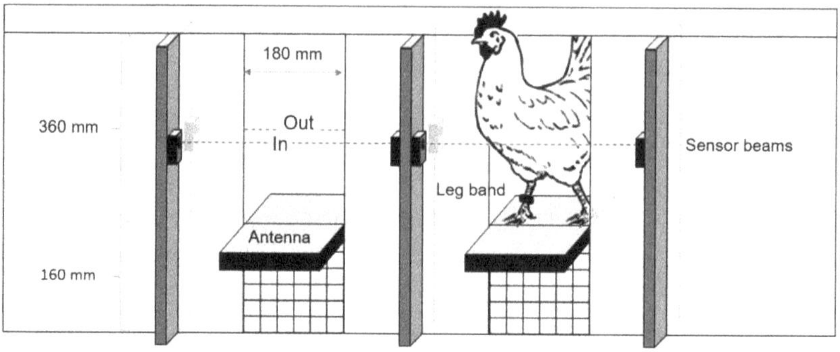


**Figure 2** The RFID system set up within the pop hole of a single pen. The antennas, sensor beams and micro-chipped leg band are indicated. Two pairs of sensor beams allow the determination of the direction of travel out of or into the pen. Reproduced from Campbell et al. (2018).

2017a; 69–82% ranged daily, Larsen et al., 2017; 80% ranged regularly, Richards et al., 2011), with individual variation in the length of ranging time and changes dependent on age, weather, and season (Pettersson et al., 2016). Of the time spent outside, less time may be spent in range areas farthest away from the shed (Larsen et al., 2017). There are fewer studies that have tracked the ranging of broilers, but work on commercial farms shows a dramatic effect of season on range usage across broiler flock replicates, where up to 33% of birds accessed the range in winter and up to 87% ranged in summer (Taylor et al., 2017a). Ranging by layers is also impacted by weather, but a single-layer flock cycle spans across multiple seasons (Richards et al., 2011).

The individual identification of chickens on free-range farms has substantially changed our understanding of bird behaviour in these systems and allowed further assessment of why this variation may exist, including the welfare implications of such variation. Behavioural tests on hens that vary in range use have shown that indoor-preferring hens are typically more fearful than hens that range daily (Campbell et al., 2016b; Hartcher et al., 2016; Larsen et al., 2018). There are also health parameters that correlate with range use, including darker combs (Larsen et al., 2018) and shorter toenails (Campbell et al., 2017b) in hens that range more or reduced range use in hens that suffer from keel bone damage (Richards et al., 2012). Broilers with greater daily range use and who range farther have reduced body weight, improved gait score, and reduced corticosterone responses to stress (Taylor et al., 2018, 2020).

Despite the wide application of RFID tracking, several questions remain. It is still unclear why some birds choose to remain indoors, if there are health, welfare, and performance implications of this across the flock or growth cycle, and if this is affected by the rearing conditions that the broilers or layers develop in. The causal relationships between ranging and welfare are also unclear; longer-term prospective studies may resolve this. The use of accelerometers may be valuable in examining the relative activity levels of birds that range or stay inside. Similarly, precisely where birds go outdoors and what they do when they are outside cannot be measured in current passive low-frequency RFID systems, but position tracking, at least, may be possible through the development of alternative tag types. Stadig and colleagues (2018a) demonstrated the application of active tags operating at ultra-wideband frequencies, worn on backpacks, which registered on seven fixed anchors located around the range site in a square formation to determine the position of individual slow-growing broilers. The system showed a mean of 68% successful registrations; future refinements would aim to improve that success rate, but, overall, the technology showed promise for measuring what passive RFID systems currently do not. However, Nakarmi et al. (2014) were able to track the locations of individuals in small group pens using passive RFID tags and a grid of antennas placed underneath the pen floor. The use of a backpack sensor versus a small leg band is potentially a greater physical strain on the chicken, but some assessments have shown minimal, short-lived differences in behaviour between tagged and untagged individuals (Stadig et al., 2018b; see Section 3).

## 2.2 Radio-frequency identification technology systems and resource use in other housing systems

RFID systems have also been applied to measure other movement patterns or resource use of both layers and broilers in other types of housing systems. Their ability to non-invasively, automatically, and continuously measure movements ensures their wide application to a range of behavioural and welfare research questions. Registration between different tag detection nodes can be compiled to determine relative activity levels of chickens within indoor housing systems, including changes across time as birds grow (Feiyang et al., 2016; Kjaer, 2017; Malchow et al., 2019). Similarly, a wireless system with backpack on-bird sensors, detection nodes, and a base station demonstrated that individual hens in a pen setting utilised the environment differently (Daigle et al., 2014). This was confirmed by Rufener and colleagues (2018), where birds equipped with active tags that were read via infrared beams in a custom-designed electronic system within an aviary housing environment exhibited patterns of movement that were consistent intra-individually, but distinct inter-individually. Classification of birds with different activity levels have consequently been used to breed high- or low-activity lines (Kjaer, 2017). Further to this, ultra-wideband tracking of where individuals go within indoor housing systems could be applied to detect behavioural differences in activity that may help identify birds susceptible to, or performing, damaging feather pecking behaviour (Rodenburg et al., 2017; Ellen et al., 2019). Correlations of movement activity with automated body weight scales and feeder sensors can also be modelled to potentially identify groups of sick, healthy, active, or inactive birds (Feiyang et al., 2016). Similarly, an ultra-high-frequency RFID system was developed by Li and colleagues (2019) to automatically detect feeding and drinking behaviour of broilers to quantify individual variation that could potentially inform management strategies. A similar system was also developed to measure feeding and nesting behaviour of laying hens in enriched cages to gain benchmark data of how hens spend their time (Li et al., 2017). This system was further used to test the impacts of varying amounts of feeder space on feeding behaviour, water intake, and egg production, which could inform both management strategies and system design refinements (Oliveira et al., 2019). A patented broiler breeder pullet precision feeding RFID system that allocated additional feed to pullets of lower body weight was demonstrated to reduce body weight CV (measure of flock uniformity) from 14% to less than 2% across the study period (Zuidhof et al., 2017). Nest boxes that automatically detect an individual hen entering can quantify nest box use, individual production rates (Oliveira et al., 2016; Zaninelli et al., 2016; Li et al., 2017; Chien and Chen, 2018), and relationships with other welfare variables such as keel fractures (Gebhardt-Henrich and Fröhlich, 2015). RFID sensors can also be placed to assess nest box type preferences (Ringgenberg et al., 2015) or preferences for different cage compartment areas (Sales et al., 2015). A system has also been applied to automatically quantify perch visits and perching duration, which could potentially be used to assess the relationships between individual perching variation and keel bone deviations or footpad dermatitis (Wang et al., 2019). The individual automated tracking of perching was able to demonstrate significant inter-pen variation despite the birds being from the same flocks, as well as significant individual variation in perch use, and showed some individuals did not perch at night despite sufficient available perching space (Wang et al., 2019). Finally, sensors on breeding cages can monitor entry of hens to facilitate the feasibility of natural mating over artificial insemination and improve animal husbandry techniques (Yu and Li, 2017).

These examples demonstrate that RFID systems are a valuable tool for research purposes where they permit the automatic quantification of movement behaviour and resource use that will provide insight into how the chickens are utilising and adapting to their housing environments. The technology could revolutionise behavioural research with precise data collected per animal relatively non-invasively. However, to date, RFID systems are not widely available for commercial implementation. The technology is not yet at a stage of being financially or logistically feasible to easily implement on farm as it is labour-intensive to install, maintain, trouble-shoot, tag birds, and monitor for lost/broken tags. Sensitive electronics in poultry barns can be challenging to maintain as the harsh environments with high dust volume and humidity degrade the equipment. Unprotected wires or system parts can also be chewed by rodents or pecked by chickens. Additionally, following system installation, there are thousands of data points being registered daily, which present challenges for compiling and interpreting that data, a point which will be discussed further in Section 6.

## 3 Wearable sensors and accelerometers

The majority of wearable motion sensors that have been investigated for poultry consist of accelerometers. Accelerometers measure the direction and magnitude of acceleration along one (single axis), two (biaxial; x and y axes), or three (triaxial; x, y and z axes) axes. Different types of accelerometers are available that can measure motion, shock, and vibration. When attached to an animal, accelerometers can measure surge, heave and sway, and provide a representation of the three-dimensional movement of the animal (reviewed in Brown et al., 2013). Advantages of accelerometers are that they are portable, wireless, and vary in price, so some relatively inexpensive accelerometers are available. However, the price is associated with the battery power and amount of memory that the device can store, which determines for how long data can be collected (Rushen et al., 2012). Therefore, unless accelerometers are used that can be paired with a receiver for continuous data downloading, which costs more and limits the size of the environment that animals can be monitored within, accelerometers have to be removed from animals to retrieve data (Rushen et al., 2012).

Accelerometers provide useful information about the animal's motion, but combining accelerometers with other types of sensors and data analysis methods can provide further context to understanding chicken behaviour. For example, the combination of temperature sensors with accelerometers enabled Okada et al. (2014) to conclude that changes in activity levels were a better early indicator of highly pathogenic avian influenza infection than a change in body temperature. Others have combined accelerometers with sensors that can track chicken location (e.g. Mica2Dot mote radio mobile sensors, Daigle et al., 2012; Tile Mate location devices, Buijs et al., 2018). Using accelerometers in conjunction with location tracking sensors may enable prediction or identification of chicken behaviour, without a need to observe chicken behaviour directly. However, the utility of combining location sensors with accelerometers has yet to be fully investigated. Instead of using combinations of sensors, another approach is to combine machine learning techniques with accelerometers to classify chicken behaviour (e.g. Abdoli et al., 2018), with the potential to track and monitor chicken behaviour automatically.

Studies that have been conducted to date indicate the potential application of wearable sensors to a wide range of behaviour and welfare scenarios. Specific applications that have been studied include the effects of wearable sensors on chicken behaviour, the ability to classify and compare behavioural activities, monitoring disease and euthanasia, and studying keel bone fractures and injuries. A discussion of each of these applications is presented in the following sections.

### 3.1 Effects of wearable sensors on chicken behaviour

Before wearable sensors can be used, they must be validated to ensure that they are not affecting the behaviour or physical ability of the animals that are under observation. Several studies have indicated that data from accelerometers can successfully be collected after a period of habituation. In addition, features of the sensor that can potentially influence bird behaviour and movement, such as sensor weight, need to be considered. As a general guide, based on similar studies in animal ecology (Murray and Fuller, 2000), Siegford et al. (2016) recommended that wearable technology for poultry should weigh less than 5% of the animal's body weight because the sensors can change bird behaviour.

The effects of wearable sensor technology on chicken behaviour and welfare have been examined in studies with laying hens (Daigle et al., 2012; Buijs et al., 2018) and one study with broiler chickens (Stadig et al., 2018b). All three studies utilised some version of a backpack to attach the sensors to the backs of the chickens. Reasons for attaching the device to the back of the chicken are that this arrangement maximises the stability of the sensor, maintains the quality of the signal, and does not result in tissue injuries to the birds (Quwaider et al., 2010). Results are generally in agreement that chickens habituate to wearing backpack sensors within a few days after sensor attachment; however, the genetic strains of chickens used, behavioural activities examined, and methods used differ across studies.

Using the sensors described in Quwaider et al. (2010), Daigle et al. (2012) examined the behaviour of commercial Hy-Line Brown laying hen pullets before and after attaching sensors at 11 weeks of age. Sensors and equipment weighed 1.07% of hens' body weight at the beginning of the study and 0.77% of hens' body weight at the end of the study. Several changes in behaviour were observed after sensors were attached. A greater percentage of hens wearing sensors used nest boxes and perches after sensor attachment than before sensors were attached. In addition, hens that wore sensors used the feeders less on 1 and 2 days after sensor attachment but used feeders more 16 days after sensors had been attached compared to hens without sensors. The percentage of agonistic interactions between hens wearing sensors and hens without sensors, where sensor-wearing hens were the aggressors, was higher than that expected at 8 and 16 days following sensor attachment. Agonistic interactions among pairs of sensor hens and pairs of non-sensor hens were no different from what would be expected. Based on their results, Daigle et al. (2012) concluded that hens habituated to the presence of the sensors after 2 weeks. In another study, Buijs et al. (2018) found no differences in aggressive behaviour or body weight when comparing sensor-wearing hens to hens that did not have sensors attached. The average body weight of birds in their study was not stated, so the weight of the sensor equipment relative to the birds' body weight is unknown. Their results revealed that there were some differences in behaviour (increased preening, sitting/lying and sidestepping/reversing, and less time standing and walking) when observed in a holding pen after being fitted with sensors than when observed in the holding pen but wearing no sensor. Drinking behaviour was also reduced after returning hens to their home pen, compared to when hens were not wearing sensors. Based on these results, Buijs et al. (2018) concluded that hens habituated to sensors relatively quickly, with few effects seen on behaviour after 2 days of wearing the sensors. Unlike the studies of Daigle et al. (2012) and Buijs et al. (2018), Stadig et al. (2018b) examined the effects of a body-mounted ultra-wideband (UWB) tag on the behaviour, leg health, and production of slow-growing male and female broiler chickens. At the time of tag attachment, the equipment weighed 1.8% of the birds' body weight. Some behavioural differences were noted, including age-related differences in how many pecks control birds directed toward birds wearing tags, and tag-wearing birds initially walked less than control birds. No differences were found between tag-wearing birds and control birds for gait, footpad dermatitis, hock dermatitis, body weight increase, or cleanliness, indicating that the tags did not adversely affect body condition and foot/leg health. Results from the aforementioned studies indicate that the attachment of wearable sensors using backpacks has some effects on chicken behaviour after sensor attachment, which illustrate the importance of providing chickens with a period of habituation before data are collected from wearable sensors. Sensor attachment does not seem to influence body weight in laying hens or broiler chickens, or footpad dermatitis and hock dermatitis of broiler chickens.

### 3.2 Classification of behavioural activities

Traditional methods of studying animal behaviour include live observations or video recordings that are then manually analysed by human observers. The use of technology to extract behavioural data and eliminate the need for labour-intensive and subjective manual analyses of video will greatly advance the study of chicken behaviour and welfare. As new technologies and data analysis methods are developed, it is becoming more feasible to extract information about animal behaviour from video and other sources, such as wearable sensors. Banerjee et al. (2012) applied machine learning to biaxial accelerometer data from laying hens to identify six distinct activities which included sitting/sleeping, standing, walking/running, feeding, drinking, and dustbathing. Accelerometers were attached to the backs of hens using a figure eight harness. In addition to the six behavioural categories, data were grouped together for higher-level classification where sitting/sleeping and standing were classified as static activities, walking/running was classified as dynamic, and feeding and drinking were classified as resource use. Data obtained from the accelerometers included time, x-axis acceleration, and y-axis acceleration; entropy and mean were extracted from the x- and y-axis acceleration. Their results indicated that machine learning techniques were able to classify behavioural activities with relatively high accuracy. More recently, Abdoli et al. (2018) developed an algorithm to detect behavioural activities from triaxial accelerometers fitted to the backs of laying hens infested with ectoparasites. Behavioural activities, including feeding/pecking, preening, and dustbathing were identified with a relatively high degree of accuracy. The authors concluded that their algorithm was able to 'learn from real, noisy and complex datasets'. These studies provide important advances in analysing chicken behaviour and have the potential to greatly improve chicken welfare once the methods become more widely available.

### 3.3 Monitoring disease and euthanasia

As poultry production increases and intensifies to meet global demands for food, risks of disease also increase. When chickens become infected with disease, especially with avian influenza, mass depopulation of flocks is conducted to prevent disease from spreading and to reduce the suffering of infected birds. The development of technologies to detect disease early and monitor disease outbreaks will be important in containing disease and safeguarding animal welfare. Furthermore, technologies that can assist with monitoring the depopulation process, especially when large numbers of chickens are killed, has the potential to ensure that depopulation is done humanely.

In order to detect avian influenza outbreaks early, Okada et al. (2009) developed a wireless, wearable sensor consisting of a triaxial accelerometer and thermistor probe (sensor to monitor chicken body temperature) to detect changes in chicken activity levels and body temperature associated with highly pathogenic avian influenza infection. Using the sensor, avian influenza infection was accurately detected several hours before death. However, not all avian influenza virus strains were associated with an increase in chicken body temperature. Consequently, Okada et al. (2014) conducted a follow-up study to investigate the feasibility of using only a triaxial accelerometer for the early detection of avian flu. Based on the differences in activity levels between pre-infection and post-infection, they were able to detect avian flu an average of 14.9 h before death occurred, which is twice as fast as what is possible using only the body temperature sensors.

Changes in broiler chicken activity levels have also been measured using accelerometers. Dawson et al. (2007, 2009) attached accelerometers to the legs of broiler chickens to (1) examine motion cessation in relation to heart relaxation recorded with an electrocardiogram (ECG) (Dawson et al., 2007) and (2) examine motion cessation in relation to brain death recorded using an electroencephalogram (EEG) (Dawson et al., 2009). Their results indicated that motion ceased before cardiac relaxation and cardiac arrest occurred following cervical dislocation (Dawson et al., 2007). Similarly, results from their other study revealed that accelerometer activity stopped before EEG activity stopped (Dawson et al., 2009). During mass depopulation with foam, it is not possible to observe chickens for signs of death. Therefore, accelerometers can provide a means of determining when the loss of brain activity occurs (Dawson et al., 2009). While it is not feasible to attach accelerometers to all chickens in a flock, attaching accelerometers to some chickens has the potential to improve disease detection and monitoring, as well as monitor mass depopulation, which can help reduce the need for people to enter the barn, further reducing the risk of spreading disease.

### 3.4 Perching, jumping, falls, and collisions

Keel bone damage and fractures are important concerns in all housing systems for laying hens, but systems that include perches are associated with greater levels of keel bone damage and fractures (reviewed in Harlander-Matauschek et al., 2015). Keel bone fractures are associated with pain (Nasr et al., 2012a) and reduced productivity (Nasr et al., 2012b). Therefore, several studies have used wearable sensors to determine factors associated with potential keel bone damage. Banerjee et al. (2014) used a backpack-mounted Mica2Dot mote radio node and triaxial accelerometer to examine when hens jumped from a perch and to calculate the hen's resulting landing force. Using their system, they were able to identify when hens had jumped from the perch, the force with which the hens landed and approximately from what height hens had jumped. Banerjee et al. (2014) suggested that the information gained from their study could be applied to other studies aimed at determining how much force results in keel bone damage and identifying particular situations that place hens at risk of keel bone fractures. More recently, LeBlanc et al. (2016) took a different approach to investigate risk factors for keel bone fractures in laying hens. In their study, triaxial accelerometers and triaxial gyroscopes were attached to hens using backpacks. Hens with various pre-existing conditions which included keel damage, footpad dermatitis, and damage of the wing feathers, as well as apparently healthy birds, were subjected to conditions that were postulated to affect their ability to maintain balance while perching. These conditions included crowding whereby 'dummy' birds were placed on either side of the hen with or without a hood placed over the head to obscure vision, and uncrowded conditions with or without the hood. Hens were tested in these conditions on a static perch and a swaying perch, operated by a motor. Hens' ability to maintain their balance while perching was affected by their physical condition; healthy hens displayed less complex movement to maintain balance while perching compared to hens with footpad dermatitis and hens with feather damage. In addition to health status, crowding negatively affected hens' balancing ability. Because hens had been habituated and trained prior to testing, LeBlanc et al. (2016) also concluded that training hens can be effective in improving their ability to maintain their balance while perching. Their study was the first to examine how both environmental factors and the health status of hens can influence their ability to balance on a perch, which provides insights into factors that may place hens at a higher risk of keel bone damage. In contrast to Banerjee et al. (2014) and LeBlanc et al. (2016) who examined factors associated with perching, Ali and Siegford (2018) did not examine perching behaviour, but used triaxial accelerometers to examine whether hens fell at night. In addition, they examined whether falling depended on whether birds were placed into aviary systems at 17 weeks compared to 25 weeks of age after being reared in floor pens, or whether being reared with access to perches compared to no perches in floor pens and placed into aviaries at 25 weeks affected falls. Birds reared in floor pens without perches and placed into aviaries at 17 weeks had more instances of falling, whereas birds reared without perches but placed into aviaries at 25 weeks had higher forces of collision and acceleration. Based on these results, Ali and Siegford (2018) concluded that accelerometers can be used to successfully detect hens' falls and collisions in aviaries at night. Through the use of wearable sensors, these studies (1) indicate that events associated with potential keel bone damage and injury can be detected non-invasively and (2) reveal important factors that can both help hens in non-cage systems, such as training (LeBlanc et al., 2016) and rearing environments (Ali and Siegford, 2018), and factors that can place hens at a greater risk of injury in non-cage systems, such as hens' physical condition (LeBlanc et al., 2016).

### 3.5 Physical activity levels

Physical activity level can be an indicator of health and welfare or can affect an animal's risk of sustaining injury. Studies have documented that chickens exhibit reduced activity levels in response to disease (e.g. Okada et al., 2014) and pain (e.g. Duncan et al., 1989), which has led to interest in using accelerometers to quantify and measure how activity levels change when chickens experience conditions expected to adversely affect their welfare. In particular, Casey-Trott and Widowski (2018) first validated that accelerometers (attached using a necklace) could detect inactivity in hens and then examined whether inactivity levels differed among hens with keel damage and hens without keel damage. Their results demonstrated that accelerometers can be used as a tool to assess chicken welfare, but contrary to what they expected, hens with keel bone damage spent less time inactive compared to healthy hens (Fig. 3).

Possible reasons for the decreased level of inactivity in hens with keel fractures are that the hens may be more restless, may spend less time standing, or may experience pain due to keel damage differently compared to how pain from other conditions would be experienced (Casey-Trott and Widowski, 2018). The accelerometers may also be useful in examining individual differences in behaviour and activity level (Casey-Trott and Widowski, 2018), which Kozak et al. (2016) suggest could be one factor affecting hens' risk of sustaining keel bone damage. Kozak et al. (2016) used accelerometers (attached using backpacks) to compare the intensity of activity level (low, moderate, and high) among laying hens of four commercial strains at three ages (10-16, 17-24, and 25-37 weeks). Their findings indicated that the accelerometers detected low-, moderate-, and high-intensity activities with a high degree of accuracy (97-98%). In addition, their results provide

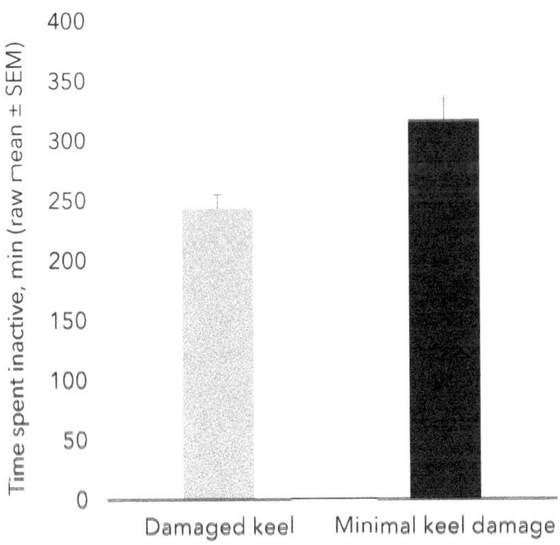


**Figure 3** The average time that hens with damaged keels and hens with little to no keel bone damage spent inactive over a 4-6 day period, determined using Actical® accelerometers. Figure produced using data from Casey-Trott and Widowski (2018). The time spent inactive was determined from activity counts obtained from the accelerometers, which differed significantly among hens with keel bone damage and hens with minimal keel bone damage.

insight into how activity levels of hens change depending on their stage of production and age, with hens entering the laying period (17–24 weeks) spending less time performing low-intensity activities and hens spending less time performing high-intensity activities as they age. Intensity level of activity differed among genetic strains as well, with white-feathered hens spending more time engaged in low-intensity activities compared to brown-feathered hens (Kozak et al., 2016). Results from Casey-Trott and Widowski (2018) and Kozak et al. (2016) indicate the potential value of using accelerometers to assess hen welfare but highlight that there may be important strain- and age-related effects that need to be considered when designing experiments and drawing conclusions about hens' activity levels.

## 4 Case study: outdoor stocking density in free-range laying hens

Within Australia, consumer preferences have driven increases in free-range laying hen systems as they are perceived to provide better hen welfare, but perceptions of how hens behave in these systems can be mismatched with reality. Free-range farms in Australia vary from mobile caravans with small flock sizes and extensive pasture access to larger-scale farms that house thousands of hens in fixed sheds with less outdoor range space per bird. Visual observations on large farms may show few birds using the range area simultaneously, which suggests not many birds go outdoors. Both these small and large-scale farms, and regardless of prevalence of hen ranging, may label their eggs as 'free-range'. Subsequently, Australian consumers felt misled by the 'free-range' label and precisely what product they were getting at a premium cost due to discrepancies between perception and system reality. As a resolution to this consumer dissatisfaction, the Australian Consumer Affairs Ministers developed an egg labelling information standard that requires egg cartons to have the hens' outdoor stocking density stated on the label and no hens are to be stocked above 10 000 hens/ha. However, at the time of this consumer debate, there was limited research on how much space hens need outdoors and how outdoor area may impact individual hen range usage. Outdoor stocking densities are based on simultaneous access by all hens (i.e. 10 000 hens/ha, gives 1 $m^2$ per hen when they are all outside together), but hens have a choice throughout the day of whether to range or not; thus, actual densities outdoors are fluid. Our experiment was designed to use RFID technology to measure individual range usage in small experimental free-range flocks with varying space outdoors as well as assess individual hen welfare and its relationship with ranging (Campbell et al., 2017a,b).

To answer the question of the impact of outdoor stocking density on range use and hen welfare, six identical indoor pens were set up to house 150

ISA Brown hens per pen. Each pen connected to a range area of a different size to test densities of 2000, 10000, or 20000 hens/ha. RFID systems were placed within the pop holes, and 50% of all hens were tagged with a leg band containing a microchip. The RFID system could detect each hen that moved through the pop holes recording exact time, the direction of travel, and microchip ID. The stored data enabled the calculation of the precise duration of time each bird spent outdoors, how often they visited the range, and the length of their range visits. These hens were tracked across a period of 15 weeks, following first range access at 21 weeks of age, with periodic weighing and external welfare scoring of each hen, following a modified version of the Welfare Quality® protocol (Welfare Quality®, 2009).

The RFID tracking data showed that hens with more space outside (lowest stocking density) had longer individual visits and spent more time outside daily (32–36 weeks of age: 2000 hens/ha: 4.78 h, 10000 hens/ha: 4.11 h, 20000 hens/ha: 3.82 h) during ranging hours, but hens within all densities still used the range area on average for 4 h around 35 weeks of age. Across all densities there was individual variation in how often each bird went outside, including 2% of hens that were never registered as visiting the range (Campbell et al., 2017a). The birds were assessed only up to peak lay, and thus most birds were in visibly good condition, but longer ranging hours did reduce toenail length (Campbell et al., 2017b). This research showed that more area outdoors increased time spent outside, which may assist consumers in making decisions about which eggs to purchase. The individual range tracking showed there were some welfare impacts of outdoor range usage but the health benefits of ranging still require further research on more flocks. The study is an example of how wearable precision technology can be used to gather information on an industry question and provide feedback to both producers and consumers.

## 5 Conclusion

Using RFID technology and wearable sensors to analyse animal behaviour and activity levels, respectively, has several advantages, including being non-invasive, collecting quantitative information and enabling longer-term, automated monitoring of animals, at a relatively lower cost than what has previously been possible. In addition to long-term monitoring, wearable sensors enable data to be collected in real time to detect changes in animal behaviour that may be early indicators of welfare problems, with the potential to automatically issue warnings (Berckmans et al., 2015) based on these behavioural changes. Scientists envision systems that can automatically monitor and detect threats rapidly and in real time (Astill et al., 2018) to diagnose and contain the threats faster than what is currently possible. Another advantage of using technology is that it may, in some circumstances, limit the need for humans to enter the barn, which is advantageous under conditions where disease risks need to be contained.

In addition to the early identification of animal health and welfare risks, studying individual animal behaviour and behavioural characteristics can guide genetic selection strategies to improve productivity and animal welfare. With wearable sensor technologies, it is possible to collect a range of information about an individual. Sensors also make it possible to collect information from individual animals that are housed in large groups or that are housed in environments where features of the environment may impede an animal's location from being determined.

To date, the application of wearable sensor technology to measure poultry behaviour has significantly advanced our understanding of individual chickens while housed in large groups, with multiple studies highlighting how each bird may adapt and utilise the same environment in different ways. The technology has also allowed insight into impacts that chicken behaviour may have on their health, allowing strategies to be sought for improving bird welfare. If current technology continues to improve, then it could be implemented on a wider scale to inform researchers, genetic companies, and housing system design companies and enable producers to make decisions to improve their own management strategies.

## 6 Future trends in research

Wearable technologies for poultry are gaining increasing interest. Their application to date has provided a wealth of previously unattainable information, yet the technology is still in its infancy of how it could be applied for research, and commercially. As the sophistication of technology advances, smaller devices with improved battery power and operating frequencies could potentially track precisely where specific chickens are located, as well as their behavioural repertoires across large sample numbers. Thus, future research with wearable technologies will continue to have a focus at the level of the individual bird.

At the individual bird level, wearable sensors such as accelerometers provide higher resolution about certain aspects of animal behaviour than what can be observed visually; for example, accelerometers can measure changes in the number of steps, activity levels, and changes in body posture that cannot be easily detected by the human eye. However, output from accelerometers is still mainly used in conjunction with manual observations of animal behaviour because it is not yet feasible to identify specific behaviours using accelerometers alone. Future research into particular aspects of animal behaviour that can be measured and quantified using accelerometers and wearable sensors, for example, force of steps, small changes in activity levels, and distance travelled, without the need to combine accelerometer output with manual visual observations, will provide valuable information about the utility of accelerometers and other wearable sensors.

Along with the improvements in technology and interpretation of technology, wearable electronics small enough and cost effective enough for chickens that can provide real-time data on heart rate and/or body temperature could be combined with changes in behaviour to monitor for disease onset or undetected stressors. For example, these types of sensors could be used to detect changes in physiological parameters such as heart rate variability, which has been linked to stress and animal welfare (see review by Von Borell et al., 2007). Sensors that can monitor chickens' physiological as well as behavioural status may also provide insights into individual differences in animal temperament (e.g. fearfulness or flightiness) and provide new ways to investigate associations among temperament, animal welfare, and productivity. Automated tracking and monitoring of chickens across their lifetime will further elucidate the relationship between their behaviour and resulting welfare, particularly establishment of any causal relationships. While this focus on the individual will remain, it is also necessary to start looking at how that individual behaves in response to its surrounding flock mates. As more chickens are managed in loose-housed systems, their social environments correspondingly increase in complexity. Understanding the individual within their flock will help identify impacts of social facilitation on individual behaviour, the impact of an individual on the group dynamic and vice versa, or impacts of other birds on health risks of disease and injury. The establishment of subpopulations of birds within loose-housed systems may require separate management strategies such as different feed diets, to optimise the performance and welfare of all birds. With laying hens in particular, separate flocks of birds in the same environments can be vastly different in their behaviour and welfare; precise understanding of individual interactions within the flock may reveal causes for these differences and solutions to optimise every flock.

In line with this need to understand individuals within flocks, one of the major trends of future research is the necessity for a merge between animal behaviour and data science. The data collected from precision monitoring can result in millions of data points which need to be interpreted in a meaningful manner. Big data techniques such as machine learning or novel analysis and data presentation methods will be critical in maximising the knowledge gained from the data and to model both individual and flock-level patterns (e.g. Campbell et al., 2018). Researchers with data modelling skills, programming skills, and expertise in working with large datasets will be necessary for deciphering long-term precision monitoring data. Additionally, individual electronic identification of animals could contribute to the development of

Internet of Animals where real-time data on individual animals are able to be viewed online across the world (Ellen et al., 2019; Pschera, 2016).

While the focus on the individual bird is a key trend in research, the investment in the individual bird to a producer is less clear. Multiple commercial products for individual animals exist for ruminant livestock species, such as collar devices that can, for example, monitor animal location, reproductive status, or behavioural signs of illness. The economic cost and benefit of a single chicken is far less than that of a single beef cow. Thus, commercial implementations of wearable sensor technology would need to be at a comparatively minimal cost to attract producer investment. Alternatively, devices could be placed on sub-samples of birds that may be representative of the flock overall, but the exact numbers that would be required for this are still to be determined. In these cases, the need for data to be presented in a user-friendly manner is critical and likely requires the development of automated algorithms to provide simple feedback on farm. An example of application of the technology commercially could be in free-range settings to provide objective evidence of range use to consumers, or for producers to modify their management strategies accordingly. Additionally, movement pattern deviations from flock averages or individual behavioural pattern deviations could be indicative of sick individuals, or onset of disease. With increasing development of meatless protein products, animal production may become a more niche market where the consumer price will warrant a technological investment that could, for example, provide lifetime quality assurance of an individual chicken.

## 7 Where to look for further information

In addition to the literature cited in this chapter, there are several good poultry technology review articles available (Ellen et al., 2019; Sassi et al., 2016; Siegford et al., 2016; Xin and Liu, 2017). Recommendations of key societies, organisations, and websites for the latest information include the following:

- World's Poultry Science Association (www.wpsa.com) organises regular national and international poultry conferences and publishes scientific reviews in the *World's Poultry Science Journal*.
- International Society of Applied Ethology (www.applied-ethology.org) hosts both international and regional conferences annually and has several hundred members that work on welfare research including the use of poultry monitoring technologies.
- Australian Poultry Science Symposium and Poultry Science Association (www.poultryscience.org) host annual conferences presenting a wide range of poultry research.

- Poultry Hub Australia (www.poultryhub.org) based at the University of New England, Armidale, NSW, Australia, provides up-to-date scientific data on multiple aspects of poultry systems.
- The American Society of Agricultural and Biological Engineers (https://www.asabe.org/) hosts annual conferences that feature research addressing agricultural systems.

Top research centres for future research and potential collaboration include the *Animal Behaviour and Welfare Team*, CSIRO, Australia; *Center for Animal Welfare Science*, Purdue University, USA; *Department of Animals in Science and Society*, Utrecht University, The Netherlands; *Center for Animal Welfare*, University of California-Davis, USA; *Campbell Center for the Study of Animal Welfare*, University of Guelph, Canada; *Animal Behavior and Welfare Group*, Michigan State University, USA; *Department of Agricultural and Biosystems Engineering*, Iowa State University, USA; and *ZTHZ, Division of Animal Welfare*, University of Bern, Switzerland.

## 8 References

Abdoli, A., Murillo, A. C., Yeh, C.-C. M., Gerry, A. C. and Keogh, E. J. 2018. Time series classification to improve poultry welfare. *7th IEEE International Conference on Machine Learning and Applications*, pp. 635–42. doi:10.1109/ICMLA.2018.00102.

Ali, A. and Siegford, J. 2018. An approach for tracking directional activity of individual laying hens within a multi-tier cage-free housing system (aviary) using accelerometers. *Meas. Behav*, 11, p. 176–80.

Astill, J., Dara, R. A., Fraser, E. D. G. and Sharif, S. 2018. Detecting and predicting emerging disease in poultry with the implementation of new technologies and big data: a focus on avian influenza virus. *Front. Vet. Sci.* 5, 263. doi:10.3389/fvets.2018.00263.

Baillie, C. L., Baxter, M. and O'Connell, N. E. 2018. Exploring perch provision options for commercial broiler chickens. *Appl. Anim. Behav. Sci.* 200, 114–22. doi:10.1016/j.applanim.2017.12.007.

Banerjee, D., Biswas, S., Daigle, C. and Siegford, J. M. 2012. Remote activity classification of hens using wireless body mounted sensors. *Proceedings of the Ninth International Conference on Wearable and Implantable Body Sensor Networks*, 9–12 May, London, UK, pp. 107–12.

Banerjee, D., Daigle, C. L., Dong, B., Wurtz, K., Newberry, R. C., Siegford, J. M. and Biswas, S. 2014. Detection of jumping and landing force in laying hens using wireless wearable sensors. *Poult. Sci.* 93(11), 2724–33. doi:10.3382/ps.2014-04006.

Barrett, J., Rayner, A. C., Gill, R., Willings, T. H. and Bright, A. 2014. Smothering in UK free-range flocks. Part 1: incidence, location, timing and management. *Vet. Rec.* 175(1), 19. doi:10.1136/vr.102327.

Berckmans, D., Hemeryck, M., Berckmans, D., Vranken, E. and Waterschoot, T. V. 2015. Animal sound talks! Real-time sound analysis for health monitoring in livestock. In: Ni, J.-Q., Lim, T.-T. and Wang, C. (Eds), *Animal Environment and Welfare – Proceedings of the International Symposium*. China Agriculture Press, Beijing, China, pp. 215–22.

Bergmann, S., Schwarzer, A., Wilutzky, K., Louton, H., Bachmeier, J., Schmidt, P., Erhard, M. and Rauch, E. 2017. Behavior as welfare indicator for the rearing of broilers in an enriched husbandry environment - a field study. *J. Vet. Behav.* 19, 90–101. doi:10.1016/j.jveb.2017.03.003.

Brown, D. D., Kays, R., Wikelski, M., Wilson, R. and Klimley, A. P. 2013. Observing the unwatchable through acceleration logging of animal behavior. *Anim. Biotelem.* 1(1), 20. doi:10.1186/2050-3385-1-20.

Brown-Brandl, T. M., Adrio, F., Maselyne, J., Kapun, A., Hessel, E. F., Saeys, W., Van Nuffel, A. and Gallmann, E. 2019. A review of passive radio frequency identification systems for animal monitoring in livestock facilities. *Appl. Eng. Agric.* 35(4), 579–91. doi:10.13031/aea.12928.

Buijs, S., Booth, F., Richards, G., McGaughey, L., Nicol, C. J., Edgar, J. and Tarlton, J. F. 2018. Behavioural and physiological responses of laying hens to automated monitoring equipment. *Appl. Anim. Behav. Sci.* 199, 17–23. doi:10.1016/j.applanim.2017.10.017.

Campbell, D. L. M., Goodwin, S. L., Makagon, M. M., Swanson, J. C. and Siegford, J. M. 2016a. Failed landings after laying hen flight in a commercial aviary over two flock cycles. *Poult. Sci.* 95(1), 188–97. doi:10.3382/ps/pev270.

Campbell, D. L. M., Hinch, G. N., Downing, J. A. and Lee, C. 2016b. Fear and coping styles of outdoor-preferring, moderate-outdoor and indoor-preferring free-range laying hens. *Appl. Anim. Behav. Sci.* 185, 73–7. doi:10.1016/j.applanim.2016.09.004.

Campbell, D. L. M., Hinch, G. N., Dyall, T. R., Warin, L., Little, B. A. and Lee, C. 2017a. Outdoor stocking density in free-range laying hens: radio-frequency identification of impacts on range use. *Animal* 11(1), 121–30. doi:10.1017/S1751731116001154.

Campbell, D. L. M., Hinch, G. N., Downing, J. A. and Lee, C. 2017b. Outdoor stocking density in free-range laying hens: effects on behaviour and welfare. *Animal* 11(6), 1036–45. doi:10.1017/S1751731116002342.

Campbell, D. L. M., Horton, B. J. and Hinch, G. N. 2018. Using radio-frequency identification technology to measure synchronised ranging in free-range laying hens. *Animals* 8(11), 210. doi:10.3390/ani8110210.

Casey-Trott, T. M. and Widowski, T. M. 2018. Validation of an accelerometer to quantify inactivity in laying hens with or without keel-bone fractures. *Anim. Welf.* 27(2), 103–14. doi:10.7120/09627286.27.2.103.

Chien, Y. R.. and Chen, Y. X. 2018. An RFID-based smart nest box: an experimental study of laying performance and behavior of individual hens. *Sensors* 18(3), 859. doi:10.3390/s18030859.

Colles, F. M., Cain, R. J., Nickson, T., Smith, A. L., Roberts, S. J., Maiden, M. C., Lunn, D. and Dawkins, M. S. 2016. Monitoring chicken flock behaviour provides early warning of infection by human pathogen Campylobacter. *Proc. Biol. Sci.* 283(1822). doi:10.1098/rspb.2015.2323.

Courtice, J. M., Mahdi, L. K., Groves, P. J. and Kotiw, M. 2018. Spotty liver disease: a review of an ongoing challenge in commercial free-range egg production. *Vet. Microbiol.* 227, 112–8. doi:10.1016/j.vetmic.2018.08.004.

Daigle, C. L., Banerjee, D., Biswas, S. and Siegford, J. M. 2012. Noncaged laying hens remain unflappable wearing body-mounted sensors: levels of agonistic behaviors remain unchanged and resource use is not reduced after habituation. *Poult. Sci.* 91(10), 2415–23. doi:10.3382/ps.2012-02300.

Daigle, C. L., Banerjee, D., Montgomery, R. A., Biswas, S. and Siegford, J. M. 2014. Moving GIS research indoors: spatiotemperoal analysis of agricultural animals. *PLoS ONE* 9(8), e104002. doi:10.1371/journal.pone.0104002.

Dawkins, M. S., Cook, P. A., Whittingham, M. J., Mansell, K. A. and Harper, A. E. 2003. What makes free-range broiler chickens range? *In situ* measurement of habitat preference. *Anim. Behav.* 66(1), 151–60. doi:10.1006/anbe.2003.2172.

Dawson, M. D., Lombardi, M. E., Benson, E. R., Alphin, R. L. and Malone, G. W. 2007. Using accelerometers to determine the cessation of activity in broilers. *J. Appl. Poult. Res.* 16(4), 583–91. doi:10.3382/japr.2007-00023.

Dawson, M. D., Johnson, K. J., Benson, E. R., Alphin, R. L., Seta, S. and Malone, G. W. 2009. Determining cessation of brain activity during depopulation or euthanasia of broilers using accelerometers. *J. Appl. Poult. Res.* 18(2), 135–42. doi:10.3382/japr.2008-00072.

Duncan, I. J. H., Slee, G. S., Seawright, E. and Breward, J. 1989. Behavioural consequences of partial beak amputation (beak-trimming) in poultry. *Br. Poult. Sci.* 30(3), 479–88. doi:10.1080/00071668908417172.

Ellen, E. D., van der Sluis, M., Siegford, J., Guzhva, O., Toscano, M. J., Bennewitz, J., van der Zande, L. E., van der Eijk, J. A. J., de Haas, E. N., Norton, T., Piette, D., Tetens, J., de Klerk, B., Visser, B. and Rodenburg, T. B. 2019. Review of sensor technologies in animal breeding: phenotyping behaviors of laying hens to select against feather pecking. *Animals* 9(3), 108. doi:10.3390/ani9030108.

Feiyang, Z., Yueming, H., Liancheng, C., Lihong, G., Wenjie, D. and Lu, W. 2016. Monitoring behavior of poultry based on RFID radio frequency network. *Int. J. Agr. Biol. Eng.* 9, 139–47. doi:10.3965/j.ijabe.20160906.1568.

Fernández, A. P., Norton, T., Tullo, E., van Hertem, T., Youssef, A., Exadaktylos, V., Vranken, E., Guarino, M. and Berckmans, D. 2018. Real-time monitoring of broiler flock's welfare status using camera-based technology. *Biosyst. Eng.* 173, 103–14. doi:10.1016/j.biosystemseng.2018.05.008.

Fulton, R. M. 2019. Health of commercial egg laying chickens in different housing systems. *Avian Dis.* 63(3), 420–6. doi:10.1637/11942-080618-Reg.1.

Gebhardt-Henrich, S. G. and Fröhlich, E. K. F. 2015. Early onset of laying and bumblefoot favour keel bone fractures. *Animals* 5(4), 1192–206. doi:10.3390/ani5040406.

Gebhardt-Henrich, S. G., Toscano, M. J. and Fröhlich, E. K. F. 2014. Use of outdoor ranges by laying hens in different sized flocks. *Appl. Anim. Behav. Sci.* 155, 74–81. doi:10.1016/j.applanim.2014.03.010.

Harlander-Matauschek, A., Rodenburg, T. B., Sandilands, V., Tobalske, B. W. and Toscano, M. J. 2015. Causes of keel bone damage and their solutions in laying hens. *World. Poult. Sci. J.* 71(3), 461–72. doi:10.1017/S0043933915002135.

Hartcher, K. M., Hickey, K. A., Hemsworth, P. H., Cronin, G. M., Wilkinson, S. J. and Singh, M. 2016. Relationships between range access as monitored by radio frequency identification technology, fearfulness, and plumage damage in free-range laying hens. *Animal* 10(5), 847–53. doi:10.1017/S1751731115002463.

Jones, D. R., Cox, N. A., Guard, J., Fedorka-Cray, P. J., Buhr, R. J., Gast, R. K., Abdo, Z., Rigsby, L. L., Plumblee, J. R., Karcher, D. M., Robison, C. I., Blatchford, R. A. and Makagon, M. M. 2015. Microbiological impact of three commercial laying hen housing systems. *Poult. Sci.* 94(3), 544–51. doi:10.3382/ps/peu010.

Kjaer, J. B. 2017. Divergent selection on home pen locomotor activity in a chicken model: selection program, genetic parameters and direct response on activity and body weight. *PLoS ONE* 12(8), e0182103. doi:10.1371/journal.pone.0182103.

Kozak, M., Tobalske, B., Springthorpe, D., Szkotnicki, B. and Harlander-Matauschek, A. 2016. Development of physical activity levels in laying hens in three-dimensional aviaries. *Appl. Anim. Behav. Sci.* 185, 66–72. doi:10.1016/j.applanim.2016.10.004.

Larsen, H., Cronin, G. M., Gebhardt-Henrich, S. G., Smith, C. L., Hemsworth, P. H. and Rault, J. L. 2017. Individual ranging behaviour patterns in commercial free-range layers as observed through RFID tracking. *Animals* 7(3), 21. doi:10.3390/ani7030021.

Larsen, H., Hemsworth, P. H., Cronin, G. M., Gebhardt-Henrich, S. G., Smith, C. L. and Rault, J. L. 2018. Relationship between welfare and individual ranging behaviour in commercial free-range laying hens. *Animal* 12(11), 2356–64. doi:10.1017/S1751731118000022.

LeBlanc, S., Tobalske, B., Quinton, M., Springthorpe, D., Szkotnicki, B., Wuerbel, H. and Harlander-Matauschek, A. 2016. Physical health problems and environmental challenges influence balancing behaviour in laying hens. *PLoS ONE* 11(4), e0153477. doi:10.1371/journal.pone.0153477.

Li, L., Zhao, Y., Oliveira, J., Verhoijsen, W., Liu, K. and Xin, H. 2017. A UHF RFID system for studying individual feeding and nesting behaviors of group-housed laying hens. *Trans. ASABE* 60(4), 1337–47. doi:10.13031/trans.12202.

Li, G., Zhao, Y., Hailey, R., Zhang, N., Liang, Y. and Purswell, J. L. 2019. An ultra-high frequency radio frequency identification system for studying individual feeding and drinking behaviors of group-housed broilers. *Animal* 13(9), 2060–9. doi:10.1017/S1751731118003440.

Litten, K., Acevedo, A., Browne, W., Edgar, J., Mendl, M., Owen, D., Sherwin, C., Würbel, H. and Nicol, C. 2008. Towards humane end points: behavioural changes precede clinical signs of disease in a Huntington's disease model. *Proc. Biol. Sci.* 275(1645), 1865–74. doi:10.1098/rspb.2008.0388.

Malchow, J., Puppe, B., Berk, J. and Schrader, L. 2019. Effects of elevated grids on growing male chickens differing in growth performance. *Front. Vet. Sci.* 6, 203. doi:10.3389/fvets.2019.00203.

Mandel, R., Nicol, C. J., Whay, H. R. and Klement, E. 2017. Short communication: detection and monitoring of metritis in dairy cows using an automated grooming device. *J. Dairy. Sci.* 100(7), 5724–8. doi:10.3168/jds.2016-12201.

Mench, J. 1998. Why it is important to understand animal behavior. ILAR J. 39, 20–6. doi:10.1093/ilar.39.1.20.

Murray, D. L. and Fuller, M. R. 2000. A critical review of the effects of marking on the biology of vertebrates. In: Boitani, L. and Fuller, T. (Eds), *Research Techniques in Animal Ecology Controversies and Consequences*. Columbia University Press, New York, NY, pp. 15–64.

Nakarmi, A. D., Tang, L. and Xin, H. 2014. Automated tracking and behavior quantification of laying hens using 3D computer vision and radio frequency identification technologies. *Trans. ASABE* 57, 1455–72. Available at: https://lib.dr.iastate.edu/abe_eng_pubs/612.

Nasr, M. A., Nicol, C. J. and Murrell, J. C. 2012a. Do laying hens with keel bone fractures experience pain? *PLoS ONE* 7(8), e42420. doi:10.1371/journal.pone.0042420.

Nasr, M., Murrell, J., Wilkins, L. and Nicol, C. 2012b. The effect of keel fractures on egg-production parameters, mobility and behaviour in individual laying hens. *Anim. Welf.* 21(1), 127–35.doi:10.7120/096272812799129376.

Neethirajan, S. 2017. Recent advances in wearable sensors for animal health management. *Sens. Biosensing. Res.* 12, 15–29. doi:10.1016/j.sbsr.2016.11.004.

Norring, M., Kaukonen, E. and Valros, A. 2016. The use of perches and platforms by broiler chickens. *Appl. Anim. Behav. Sci.* 184, 91-6. doi:10.1016/j.applanim.2016.07.012.

Okada, H., Suzuki, K., Tsukamoto, K. and Itoh, T. 2009. Wireless sensor system for detection of avian influence outbreak farms at an early stage. *Proceedings of the IEEE Sensation*, 25-28 October, pp. 1374-7.

Okada, H., Suzuki, K., Kenji, T. and Itoh, T. 2014. Applicability of wireless activity sensor network to avian influenza monitoring system in poultry farms. *J. Sens. Technol.* 4(1), 18-23. doi:10.4236/jst.2014.41003.

Oliveira, J., Xin, H., Zhao, Y., Li, L., Liu, K. and Glaess, K. 2016. Nesting behavior and egg production pattern of laying hens in enriched colony housing. *Trans. ASABE*. doi:10.13031/aim.20162456546, 162456546.

Oliveira, J. L., Xin, H. and Wu, H. 2019. Impact of feeder space on laying hen feeding behavior and production performance in enriched colony housing. *Animal* 13(2), 374-83. doi:10.1017/S1751731118001106.

Pettersson, I. C., Freire, R. and Nicol, C. J. 2016. Factors affecting ranging behaviour in commercial free-range hens. *Worlds Poult. Sci. J.* 72(1), 137-50. doi:10.1017/S0043933915002664.

Pschera, A. 2016. *Animal Internet: Nature and the Digital Revolution* (Trans. Lauffer, E.). New Vessel Press, New York, NY.

Quwaider, M., Daigle, C., Biswas, S., Siegford, J. and Swanson, J. 2010. Development of a wireless body-mounted sensor to monitor location and activity of laying hens in a non-cage housing system. *Trans. ASABE* 53(5), 1705-13. doi:10.13031/2013.34890.

Richards, G. J., Wilkins, L. J., Knowles, T. G., Booth, F., Toscano, M. J., Nicol, C. J. and Brown, S. N. 2011. Continuous monitoring of pop hole usage by commercially housed free-range hens throughout the production cycle. *Vet. Rec.* 169(13), 338. doi:10.1136/vr.d4603.

Richards, G. J., Wilkins, L. J., Knowles, T. G., Booth, F., Toscano, M. J., Nicol, C. J. and Brown, S. N. 2012. Pop hole use by hens with different keel fracture status monitored throughout the laying period. *Vet. Rec.* 170(19), 494. doi:10.1136/vr.100489.

Ringgenberg, N., Fröhlich, E. K. F., Harlander-Matauschek, A., Toscano, M. J., Würbel, H. and Roth, B. A. 2015. Effects of variation in nest curtain design on pre-laying behaviour of domestic hens. *Appl. Anim. Behav. Sci.* 170, 34-43. doi:10.1016/j.applanim.2015.06.008.

Roberts, C. M. 2006. Radio frequency identification (RFID). *Comput. Secur.* 25(1), 18-26. doi:10.1016/j.cose.2005.12.003.

Rodenburg, T. B., Bennewitz, J., de Haas, E. N., Košťál, L., Pichová, K., Piette, D., Tetens, J., Van Der Eijk, J., Visser, B. and Ellen, E. D. 2017. The use of sensor technology and genomics to breed for laying hens that show less damaging behaviour. *8th EC-PLF*, Nantes, France, 12-14 September.

Rodriguez-Aurrekoetxea, A. and Estevez, I. 2016. Use of space and its impact on the welfare of laying hens in a commercial free-range system. *Poult. Sci.* 95(11), 2503-13. doi:10.3382/ps/pew238.

Rodriguez-Aurrekoetxea, A., Leone, E. H. and Estevez, I. 2014. Environmental complexity and use of space in slow growing free range chickens. *Appl. Anim. Behav. Sci.* 161, 86-94. doi:10.1016/j.applanim.2014.09.014.

Rufener, C., Berezowski, J., Maximiano Sousa, F., Abreu, Y., Asher, L. and Toscano, M. J. 2018. Finding hens in a haystack: consistency of movement patterns within and across individual laying hens maintained in large groups. *Sci. Rep.* 8(1), 12303. doi:10.1038/s41598-018-29962-x.

Ruiz-Garcia, L. and Lunadei, L. 2011. The role of RFID in agriculture: applications, limitations and challenges. *Comput. Electron. Agric.* 79(1), 42–50. doi:10.1016/j.compag.2011.08.010.

Rushen, J., Chapinal, N. and de Passille, A. M. 2012. Automated monitoring of behavioural-based animal welfare indicators. *Anim. Welf.* 21(3), 339-50. doi:10.7120/09627286.21.3.339.

Sales, G. T., Green, A. R., Gates, R. S., Brown-Brandl, T. M. and Eigenberg, R. A. 2015. Quantifying detection performance of a passive low-frequency RFID system in an environmental preference chamber for laying hens. *Comput. Electron. Agric.* 114, 261–8. doi:10.1016/j.compag.2015.03.008.

Sassi, N. B., Averós, X. and Estevez, I. 2016. Technology and poultry welfare. *Animals* 6(10), 62. doi:10.3390/ani6100062.

Sherwin, C. M., Richards, G. J. and Nicol, C. J. 2010. Comparison of the welfare of layer hens in 4 housing systems in the UK. *Br. Poult. Sci.* 51(4), 488–99. doi:10.1080/00071668.2010.502518.

Siegford, J. M., Berezowski, J., Biswas, S. K., Daigle, C. L., Gebhardt-Henrich, S. G., Hernandez, C. E., Thurner, S. and Toscano, M. J. 2016. Assessing activity and location of individual laying hens in large groups using modern technology. *Animals* 6(2), 10. doi:10.3390/ani6020010.

Silvera, A. M., Knowles, T. G., Butterworth, A., Berckmans, D., Vranken, E. and Blokhuis, H. J. 2017. Lameness assessment with automatic monitoring of activity in commercial broiler flocks. *Poult. Sci.* 96(7), 2013–7. doi:10.3382/ps/pex023.

Stadig, L. M., Ampe, B., Rodenburg, T. B., Reubens, B., Maselyne, J., Zhuang, S., Criel, J. and Tuyttens, F. A. M. 2018a. An automated positioning system for monitoring chickens' location: accuracy and registration success in a free-range area. *Appl. Anim. Behav. Sci.* 201, 31–9. doi:10.1016/j.applanim.2017.12.010.

Stadig, L. M., Rodenburg, T. B., Ampe, B., Reubens, B. and Tuyttens, F. A. M. 2018b. An automated positioning system for monitoring chickens' location: effects of wearing a backpack on behaviour, leg health and production. *Appl. Anim. Behav. Sci.* 198, 83–8. doi:10.1016/j.applanim.2017.09.016.

Stratmann, A., Fröhlich, E. K. F., Gebhardt-Henrich, S. G., Harlander-Matauschek, A., Würbel, H. and Toscano, M. J. 2015. Modification of aviary design reduces incidence of falls, collisions and keel bone damage in laying hens. *Appl. Anim. Behav. Sci.* 165, 112–23. doi:10.1016/j.applanim.2015.01.012.

Taylor, P. S., Hemsworth, P. H., Groves, P. J., Gebhardt-Henrich, S. G. and Rault, J. L. 2017a. Ranging behaviour of commercial free-range broiler chickens 2: individual variation. *Animals* 7(7), 55. doi:10.3390/ani7070055.

Taylor, P. S., Hemsworth, P. H., Groves, P. J., Gebhardt-Henrich, S. G. and Rault, J. L. 2017b. Ranging behaviour of commercial free-range broiler chickens 1: factors related to flock variability. *Animals* 7(7), 54. doi:10.3390/ani7070054.

Taylor, P. S., Hemsworth, P. H., Groves, P. J., Gebhardt-Henrich, S. G. and Rault, J. L. 2018. Ranging behavior relates to welfare indicators pre- and post-range access in commercial free-range broilers. *Poult. Sci.* 97(6), 1861–71. doi:10.3382/ps/pey060.

Taylor, P. S., Hemsworth, P. H., Groves, P. J., Gebhart-Henrich, S. G. and Rault, J. L. 2020. Frequent range visits further from the shed relate positively to free-range broiler chicken welfare. *Animal* 14(1), 138–49. doi:10.1017/S1751731119001514.

Thurner, S., Maier, S., Icken, W., Wendle, G. and Preisinger, R. 2010. Identification reliability of laying hens at the wide electronic pop hole. *Landtechnik* 65, 139-41.

Von Borell, E., Langbein, J., Despres, G., Hansen, S., Leterrier, C., Marchant-Forde, J., Marchant-Forde, R., Minero, M., Mohr, E., Prunier, A., Valance, D. and Veissier, I. 2007. Heart rate variability as a measure of autonomic regulation of cardiac activity for assessing stress and welfare in farm animals: a review. *Physiol. Behav.* 92(3), 293-316. doi:10.1016/j.physbeh.2007.01.007.

Wang, K., Liu, K., Xin, H., Chai, L., Wang, Y., Fei, T., Oliveira, J., Pan, J. and Ying, Y. 2019. An RFID-based automated individual perching monitoring system for group-housed poultry. Agricultural and Biosystems Engineering Publications. 1025. Available at: https://lib.dr.iastate.edu/abe_eng_pubs/1025.

Weary, D. M., Huzzey, J. M. and von Keyserlingk, M. A. G. 2009. Board-invited review: using behavior to predict and identify ill health in animals. *J. Anim. Sci.* 87, 770-7. doi.org/10.2527/jas.2008-1297

Welfare Quality®. 2009. Welfare Quality® assessment protocol for poultry (broilers, laying hens). Welfare Quality® Consortium, Lelystad, the Netherlands.

Xin, H. and Liu, K. 2017. Precision livestock farming in egg production. *Anim. Front.* 7(1), 24-31. doi:10.2527/af.2017.0105.

Yu, Y. and Li, L. 2017. RFID-based laying hen individual behavior monitoring system. *Rev. Fac. Ing.* 32, 904-9.

Zaninelli, M., Rossi, L., Costa, A., Tangorra, F. M., Guarino, M. and Savoini, G. 2016. Performance of injected RFID transponders to collect data about laying performance and behaviour of hens. *Large Anim. Rev.* 22, 77-82.

Zuidhof, M. J., Fedorak, M. V., Ouellette, C. A. and Wenger, I. I. 2017. Precision feeding: innovative management of broiler breeder feed intake and flock uniformity. *Poult. Sci.* 96(7), 2254-63. doi:10.3382/ps/pex013.

# Chapter 7

## Poultry welfare monitoring: group-level technologies

*Marian Stamp Dawkins and Elizabeth Rowe, University of Oxford, UK*


1 Introduction
2 Types of automated assessment
3 Automated measures of welfare as part of precision farming
4 Why isn't automated welfare assessment more widely used?
5 Conclusions
6 Future trends
7 References


## 1 Introduction

In one sense, the health and welfare of commercial poultry is already highly automated. Modern broiler houses, for example, have sophisticated climate control systems that keep birds within the temperature and humidity guidelines recommended by the breeding companies (Yahav et al., 2005; Jones et al., 2005), delivery of food and water is automated and alarms are fitted to ensure a constant supply and light is regulated for both intensity and photoperiod (Corkery et al., 2013). There are environmental sensors to measure temperature (D'Alfonso et al., 1996; Carvalho et al., 2013; Coelho et al., 2016; Curi et al., 2017), ambient dust (Zhao et al., 2009), relative humidity (Carvalho et al., 2013; Coelho et al., 2016; Curi et al., 2017), vibration (Chen et al., 2010), ammonia concentration (Ji et al., 2016) and carbon dioxide concentration (Carvalho et al., 2013; Ji et al., 2016). In other words, the birds' basic needs for food, water and physical comfort can be catered for automatically (Fournel et al., 2017), thus fulfilling at least some of the requirements for good welfare as set out in the Five Freedoms (Brambell, 1965; Webster, 2001), Welfare Quality® (2009), World Organization for Animal Health General Principles (Fraser et al., 2013) and other welfare systems.

But, in another sense, welfare is very far from being automated for any poultry. Welfare is not just defined by the environment provided for an animal (inputs) but what the welfare outcomes actually are (Main et al., 2014).

http://dx.doi.org/10.19103/AS.2020.0078.07

Measuring health outcomes such as hockburn and pododermatitis in broilers at the slaughterhouse (Vanderhasselt et al., 2013) is a step in the right direction, but there are many aspects of welfare that simply cannot be measured on dead birds, such as whether they are able to walk without difficulty (Kestin et al., 1992). Currently, the welfare of living birds is generally assessed by human auditors visiting poultry houses and making observations (Ekstrand et al., 1998; de Jong et al., 2012) but this is very labour intensive and gives only a 'snapshot' of the state of the flock on the day of the visit (Wathes et al., 2008; Winckler, 2019). It can also risk spreading disease if the same team visits different sites. What would improve poultry welfare generally would be automated ways of measuring welfare outcomes that can be routinely applied to all flocks. Automated monitoring systems allow for non-intrusive assessment of welfare, where information can be collected without the stress of disturbing or handling animals (Wathes et al., 2008). Continuous automated monitoring can also provide a more complete picture of the overall welfare state of animals, and alerting farmers to problems as they arise in real time allows for fast and targeted interventions which will benefit the current flock, compared to traditional welfare assessments that occur at the end of the production cycle (Winckler, 2019; Buscher, 2019). This would benefit producers, consumers and the birds themselves.

In addition to these advantages, automated measures of welfare are often seen as more 'objective' than the 'subjective' assessment of a human (Banhazi et al., 2012). While this is potentially a benefit, it is important not to confuse the objectivity of the measurement with how these measurements are to be interpreted in welfare terms, which often contains a subjective or at least a contestable element of what is desirable. Certainly machines can more accurately and more continuously record what animals do than a person observing with the naked eye and writing down what they see. In the limited sense of being more accurate, then, automated measurements are indeed more objective. But the problem of interpreting the data that has been collected and deciding whether or not that is a good measure of welfare remains and is still a matter of controversy. Machine learning for welfare outcomes still needs to have specified what a 'good' or desirable welfare outcome is (Morota et al., 2018). In exactly the same way that the use of machine learning in human medicine needs to be related to desirable outcomes such as longer life, ability to walk or the satisfaction of patients, so too automated assessment of animal welfare still has to have an end-point - what is seen as good or desirable. Part of the reason that automated measures seem objective is that they force us humans to be clearer than we otherwise might be about what our welfare goals are, because we then have to instruct machines what to look for.

So what exactly do we mean by welfare, and, more importantly, in the context of this chapter, can machines measure it? Although there is no universal agreement about how to define animal welfare (Mason and Mendl,

1993; Fraser, 2008; Mellor and Beausoleil, 2015), it is possible to discern some common threads amongst the various definitions that are in use. An emphasis on good physical health, for example, is universally accepted as the foundation of good animal welfare. For example, the Five Freedoms (Brambell, 1965; Webster, 2001), a widely used system for assessing welfare around the world, lists freedom from disease and injury as one of the key indicators of welfare. This is emphasized just as strongly in more recent versions such as the Ten General Principles (OIE, 2016; Fraser et al., 2013), the Five Provisions or Domains (Mellor, 2016) and the Four Principles put forward by the European Welfare Quality assessment (Welfare Quality, 2009). Furthermore, many of the other criteria used by these and other welfare schemes are specifically aimed at maintaining good health - such as making sure that animals have adequate food and water and are kept in safe comfortable environments in which they are not injured. Many of the most pressing animal welfare issues such as feather-pecking in laying hens (Gunnarsson et al., 1999) are seen as serious precisely because of the injuries that result. Conversely, finding ways of preventing disease and stopping hens from injuring themselves by breaking their bones would be widely agreed as improving their welfare. Death, injury and disease are clear welfare outcomes with results that can be measured objectively and are therefore particularly suited to automation.

Of course, there is more to good welfare than simply the absence of disease and injury, as emphasized even more strongly in ideas such as Quality of Life (Bono and Mori, 2005; Broom, 2007) and lives worth living (Wathes, 2010; Mellor, 2016), but, as with humans, these more complex ideas are difficult to define and even more difficult to capture in algorithms (Scott et al., 2007; Taylor and Mills, 2007). This chapter will therefore be mainly based on measures that are the easiest to automate, such as lameness; mortality; disease; leg, foot and skin damage; and vocalizations. Then, as machine algorithms are improving all the time (Nasirahmadi et al., 2017; Liakos et al., 2018), this chapter will additionally look at the steps that have been taken to automate the measurement of other, more complex measures of welfare such as behaviour. The chapter will use examples largely, although not exclusively, from applications to broiler (meat) chickens because this is where the group-level technology has been most developed.

## 2 Types of automated assessment

### 2.1 Introduction: group versus individual

Unlike larger animals such as dairy cows or sows where each animal can be monitored and treated individually (Caja et al., 2016; Buscher, 2019), commercial poultry are kept in groups of many thousands and treated as a group (Winckler, 2019). Feed, vaccinations, medication, drinker height, lighting and other factors are not administered to a single individual but set or adjusted to the needs of the whole flock. Even when a particular bird is observed (e.g. during gait scoring; Kestin et al., 1992), this is only a part of a sample to give an estimate of a whole flock score. Individual chickens can be marked or tagged as part of research projects and yield valuable results about individual behaviour (Campbell et al., 2017; Stadig et al., 2018). For routine commercial farming, however, tagging all birds in a flock of either broilers or layers would present major logistical problems in applying and retrieving the tags, while computer identification and tracking of thousands of birds would be both computationally difficult and expensive. Even if such logistical problems could be overcome, it would currently be impossible to pick up and medicate a single bird among thousands. The large groups in which commercial broilers are kept have therefore led to group-level measures of welfare and to whole group approaches to automating welfare monitoring (Ben Sassi et al., 2016; Berckmans, 2017; Buscher, 2019). It is these group-level measures that are discussed here. In discussing the sensors used for such welfare monitoring (most commonly visual, sound or thermal imaging), we will ask two questions: what are the sensors detecting and how valid is the interpretation of this data in welfare terms? In general, finding sensors to collect data is the easy part. Turning the vast quantities of data that the sensors collect into output that is meaningful in welfare terms is much more difficult.

## 2.2 Visual information from CCTV or video

There are several different approaches to the processing of information coming from cameras. One is to use digital image processing technology to distinguish individual chickens from their background and then to analyse different postures of the birds to identify sick or healthy individuals (Zhuang et al., 2018); to identify particular behaviours (Leroy et al., 2006) such as wing spreading, scratching and preening (Pereira et al., 2013); to track their location in preference tests (Kashiha et al., 2015); and even to detect abnormalities of body oscillation, step frequency and step length in lame broilers (Aydin, 2017a). An alternative approach is not to recognize individual birds as such but to simply see a flock of birds as something that makes patterns of light against a darker background and then to analyse the patterns as the rate of change of brightness (or 'optical flow') in different parts of a moving visual image (Fleet and Weiss, 2005; Fig. 1). This gives a flock-level, rather than an individual bird, measure but has the great advantage that whole frames containing tens or hundreds of individual birds can be assessed together, and the statistical properties of these flow patterns can be derived automatically and continuously (Sonka et al., 1999) using algorithms that are simple enough to deliver results continuously in real time. Clustering of birds can be used to detect that they

| (a) Image frame at time $t$ | (b) Image frame at time $t + 1$ |
|---|---|

**Figure 1** 'Optical flow' is the rate of change of image brightness in different parts of a moving visual image, measured by comparing patterns of light and dark in successive video frames over time (Fleet and Weiss, 2005). These changes are combined to give an estimate of local velocity vectors. If all white chickens on the darker background remain stationary from one image to the next, there will no change in the pattern of brightness and no 'flow'. But if some of the chickens move between frames, some white areas will become dark and *vice versa*, and this is registered as a net 'flow' (Dawkins et al., 2009).

are too cold (Pereira et al., 2012). Yet other approaches involve a combination of individual and group measures so that 'events' such as feather pulling can be recognized but not which individual bird was responsible (Lee et al., 2011).

Using one or more of these techniques, it is now possible to automate the assessment of basic health measures and to validate the results against welfare outcomes assessed in more conventional ways. For example, the eYenamic system takes camera image data every minute, calculates activity and occupation indices for different areas of the screen and is able to link deviations in these indices to welfare assessment of hockburn, pododermaitis and gait score as carried by human observers (Aydin et al., 2015; De Montis et al., 2013; Norton et al., 2016; Aydin, 2017b; Fernandez et al., 2018).

Also using commercial flocks of broiler chickens, Dawkins et al. (2009, 2012) showed that flock movements, as measured by the statistical properties of optical flow patterns, were correlated with key physical health outcomes. Flocks that had lower mean levels of movement and higher kurtosis (a measure of variability) were scored with higher levels of hockburn and foot pad dermatitis at the end of life. Furthermore, these differences were already apparent in young chicks of less than 4 days old (Roberts et al., 2012; Dawkins et al., 2017) before there were any external signs of leg or foot damage. Optical flow patterns were also shown to differ between flocks subsequently testing positive or negative for *Campylobacter*, one of the major bacterial sources of food poisoning in humans, when chicks were less than a week old (Colles et al., 2016). *Campylobacter*-positive flocks showed lower mean and higher kurtosis of optical flow than flocks that remained free of infection, in line with the previous result that good welfare, including freedom from disease, is associated with more movement (higher mean) and more uniform movement (lower kurtosis). Since infection with *Campylobacter* is not detectable by normal culture methods until birds are around 21 days old, optical flow patterns were not only correlated with bird health but were also predicting future health status.

A similar ability of optical flow patterns to predict problems before they have detectable clinical effects is suggested by a study by Lee et al. (2011) on feather damage in laying hens. Optical flow patterns of the behaviour of 18-week-old birds were analysed from videos, this time using a hidden Markov model to quantify the number of anomalies or unusual events. The results correlated well with those recorded by a human observer and were predictive of the amount of feather damage when the same flocks were 38 weeks old. Feather damage is often a precursor for outbreaks of feather-pecking (Drake et al., 2010), so these preliminary results suggest that there is a potential for using automated monitoring for giving the warning of flocks at risk of welfare problems.

## 2.3 Sound

The technology to record and analyse sound is well developed, and sound can be used in welfare assessment in a variety of ways. The most obvious is to record sounds made by the birds themselves (Zimmerman et al., 2000). For example, very young chicks are unable to regulate their own body temperature and give high-pitched or distress calls when they are cold (Wood-Gush, 1971). By putting individual chicks into a climate-controlled chamber and altering the temperature, Moura et al. (2008) showed that both the amplitude and the frequency of these calls increased when the temperature dropped. In farm conditions, heat stress leads to distinctive calls by both broilers (Pereira et al., 2014) and layers (Lee et al., 2015). Du et al. (2018) monitored the number of vocalisations/bird both during the day and at night and argued that anomalies in this number of calls given by each bird, particularly at night, could be used for welfare assessment.

Another approach is to use non-communicatory sounds such as the noise made when birds peck food. In one study, a microphone positioned on the side of a feeder was used to estimate when and how much birds were eating (Aydin and Berckmans, 2016).

## 2.4 Temperature

Infra-red thermography is a non-invasive technique for measuring body surface temperature which has been related to stress, emotional arousal and leg pathologies in many bird species, including poultry (McCafferty, 2013). Changes in the surface temperature of the areas of bare skin can be symptomatic of short-term stress. For example, Edgar et al. (2013) showed that handling individual hens lead to an initial decrease of as much as 2°C in the surface head, eye and comb temperatures followed by an increase. Moe et al. (2017) found that during a 10-minute restraint, footpad temperatures dropped by nearly 0.5°C, while head region temperatures (e.g. nostrils, comb, wattle, eye) rose by 0.76°C. Both of these studies suggest that surface temperature changes can be used as measures of welfare but both were carried out in individual birds. Successfully translating these individual measurements into flock-level measures of 'stress' and distinguishing these from changes in the ambient temperature have yet to be achieved. Thermal imaging can also be used to assess feather cover as feathers insulate a bird and prevent heat loss. Cook et al. (2006) showed that the percentage of a thermal image that was within the temperature range of 17-24 C was negatively associated with feather cover, while the percentage within the range of 28-31 C was associated with a lack of feather cover or bare skin. Thermographic images have also been used to register multiple occupancies by laying hens, and the analysis of thermal images is now able count the number of birds in a given nest (Zaninelli et al., 2016, 2017, 2018).

As well as being used to assess the physical state of the birds themselves, thermal imaging cameras can also identify, with high precision, areas in a chicken house with different radiant temperatures and so alert the farmer to particular problems (Coelho et al., 2016).

### 2.5 Conclusions

It is clear that there are now, at least in the development stage, an increasing number of ways in which the welfare of poultry can be monitored automatically. Different sensors give different sorts of information, but for this to be used to maximum effect, it will be necessary to integrate the different sorts of information that are now available and to display them in a form that is accessible to farmers and easy for them to use (Van Hertem et al., 2017; Fournel et al., 2017). The full potential of many of these methods to give an automated picture of welfare has yet to be realized.

## 3 Automated measures of welfare as part of precision farming

The automated measures of welfare that are now being developed can potentially provide farmers with valuable information that they can then use to make decisions about how to manage their flocks for better health, for example, whether to put down extra litter or to apply medication. Using such technology, farmers are still in control but are given an extension of their own expertise in the form of extra eyes and ears that operate continuously when they cannot be present.

More ambitiously, however, the same information can also potentially be used to feed into more integrated automated control systems in which management decisions are taken largely or entirely by a computer. Such precision livestock systems (Werner et al., 2003; Banhazi and Black, 2009; Berckmans, 2017) have been inspired by the now-widespread use of precision agriculture in the growing of crops, where many different variables about both the environment and the growing plants are measured and then actions are automatically taken to match practice to crop needs with the goal of optimizing returns while preserving resources. Here, human decisions often defer to machine algorithms.

However, the idea that the concepts used to grow plants successfully can be transferred directly to poultry farming is controversial. On the one hand, Werkheiser (2018) argues that precision farming allows 'modern, large-scale farms to replicate and even improve on the benefits of caring farmers who know their animals, transferred to a much larger scale. This could be done via closer monitoring than farmers can provide to even a few animals, as well as integration of the data via decision algorithms that improve on the guesswork of traditional stockpersons'. On the other hand, precision farming may make the continued use of intensive systems more likely through making them more efficient and, therefore, more viable, which is of concern to some animal welfare advocates (Stevenson, 2017). For poultry, housed together in large groups and highly sensitive to the details of diet, temperature and air and litter quality, there is a vast potential for not only improving welfare through the use of technology but also reducing welfare through an emphasis on efficiency at the expense of welfare. While improving production and improving welfare are not mutually exclusive (Dawkins, 2016), they do not always coincide. When there is a conflict, what matters is the priority given to welfare.

There are encouraging signs that improving bird welfare is now a priority in much of the research to developing PLF for poultry. In a systematic review Rowe et al. (2019) concluded that 64% of PLF publications had animal health and welfare as one of their goals, the other being increased production. For the publications that only had one goal, more publications had animal health and welfare as the only goal of the study (40%) compared to production (27%). However, with PLF receiving increasing attention around the world and particularly in the USA, China and Belgium, there is an on-going need to ensure that animal welfare is seen as a key priority for sustainability in general and poultry production in particular (Rowe et al., 2019).

## 4 Why isn't automated welfare assessment more widely used?

The development of increasingly sophisticated monitoring systems for poultry farming with their potential to improve both welfare and productivity raises the question of why there are not more automated welfare monitoring systems in place on commercial poultry farms. This cannot be explained simply by farmers being unwilling to invest in technology because, as pointed out at the beginning of this chapter, many poultry houses now have complex computer-controlled heating and ventilation systems. Poultry farmers are clearly not averse to technology if it works and gives them something they could not obtain otherwise, It is the next step - the widespread use of automated ways of assessing behaviour and welfare - that is so far not been adopted by poultry producers.

Over 10 years ago, Wathes et al. (2008) suggested that this lack of take-up might be because the equipment was expensive or unreliable or because it was untested under commercial farm conditions, but there has been plenty of time since then for these problems to have been resolved. Yet poultry farming still appears to lag behind other sectors of agriculture in its use of basic health monitoring. In the dairy sector, for example, it is common to use technology to obtain diagnostic data about a range of health- and performance-related criteria (Caja et al., 2016). Poultry farming, with its need to monitor the health and welfare of large numbers of animals, its tight economic margins and the need for transparency in the face of public concern over bird welfare would seem to be even more in need of any help that technology can bring.

A possible answer for the fact that automated welfare is not more widely adopted is that the benefits - economic, production or even welfare - of available automated monitoring systems just simply have not been sufficiently demonstrated for farmers to see the point of investing in them. Caja et al. (2016) pointed out that 'dairy farmers will pay for and use technologies that provide what is, to them, a straightforward answer to a straightforward question (should I inseminate cow x?) when they believe it will have positive economic impact'. Although many of the poultry monitoring systems described previously have the potential to improve bird health and welfare, their actual benefits, over and above those that could be obtained by simpler methods, have not yet been demonstrated in practice. For example, keeping growing chicks within the correct temperature and humidity range is critical for their health and welfare (Dawkins et al., 2004), so it may be seen as much simpler to measure this directly with temperature and humidity sensors rather than indirectly by using the spatial distribution of the birds from cameras to determine whether they are too hot or too cold. In other words, good control of the environmental input may be more effective at maintaining the optimal temperature than measuring the results of temperature deviations using a complex and expensive camera or a sound system. If the aim is to monitor how much food chickens are eating, the most obvious way is to measure this directly rather than using the sound the birds make by pecking at food dishes. Water use, another variable that is easily monitored, has been recommended as an easy-to-use measure of welfare (RSPCA, 2013; OIE, 2016) since both increases and decreases in water consumption can indicate health problems and the total amount of water consumed by a flock over its lifetime is positively correlated with the prevalence of foot pad dermatitis at slaughter (Manning et al., 2007). Automated assessment of flock behaviour using optical flow also predicts hockburn and pododermatitis and is able to do so more effectively and when the birds are only a few days old (Dawkins et al., 2017) but for farmers to move from the relatively simple method of metering water to automated welfare monitoring with optical flow patterns, they would need to be convinced that the automated system gave them substantially better flock outcomes. Unless and until it can be demonstrated that there are additional benefits from automated welfare assessment, farmers will not invest in or use it. The current situation is that although there is a considerable potential for welfare improvement through automated monitoring, the evidence that welfare is actually improved by using such technology is still missing, even for the most basic health elements of welfare.

When it comes to more complex measures such as activity, or the recognition of specific behaviours such as dustbathing, there is even less evidence of benefit. Unless farmers can be convinced that knowing how much dustbathing or other behaviours their birds are actually doing helps them to keep their flocks healthier, or demonstrably improves the welfare of their birds in some tangible way, it is unlikely that they will invest in the automated ways of measuring it.

## 5 Conclusions

As this chapter shows, technology has a considerable potential for improving welfare through its ability to provide constant monitoring of the health and welfare of poultry throughout their lives and to provide early warning of problems before they become serious. The most promising existing systems are those that measure clearly defined health outcomes in broiler chickens such as lameness and leg and foot defects that are universally agreed to indicate poor welfare and also have economic implications for farmers, but even these have yet to be implemented on a commercial scale. Comparable systems for welfare monitoring in other poultry have yet to be developed.

## 6 Future trends

Looking into the future, prime targets for using automated monitoring include occasional events that can have serious welfare consequences such as smothering and feather-pecking in laying hens. 'Smothering' or 'piling'

(Campbell et al., 2016) is when hens sometimes pile on top of each other to the point that some of them may be killed, while feather-pecking is where hens damage the feathers of other birds, sometimes escalating to the point of injury or even death (Savory, 1995). The causes of these serious welfare issues are incompletely understood (Green et al., 2000; Gebhardt-Henrich and Stratmann, 2016) and although the risk of feather-pecking can be reduced by good management (Nicol et al., 2013), it still occurs, often with devastating effects. If automated monitoring of behaviour could give advance warning of when outbreaks are likely to occur or at least which flocks were most at risk (Lee et al., 2011), this would be a major contribution to improving poultry welfare.

Another use of automated systems could be to monitor the numbers of layers or broilers going outside to fulfil legal or auditing standard requirements for being 'free range' and to monitor a wider range of bird behaviours (Valletta et al., 2017) such as dustbathing and perching to satisfy a public demand for a fully comprehensive picture of bird welfare.

Properly used, automated welfare assessment could combine the wisdom of good stockmanship with far greater attention to individual birds than is currently possible. Smart computing and sensing technology offers the opportunity for shifting poultry production from an entirely group-based approach to a greater consideration of the needs of individual birds. While this shift has been relatively easy for sows and dairy cows where tagging can facilitate individual diets and treatments (Halachmi and Guarino, 2016), future developments may extend this to large groups of untagged poultry. Even crop agriculture does not consider a whole field as a single homogenous unit but treats different parts of the same field differently depending on soil type, moisture levels or the health of the plants growing there. Precision farming for poultry offers the possibility of being similarly 'smart' in not having to treat a flock as a single homogenous whole but rather to be able to apply heat, ventilation or extra litter at least down to a small area of the house.

Whatever the future holds for automated welfare assessment, however, its full potential has yet to be realised in practice. We have argued in this chapter that one of the main reasons for this is that its benefits have not been demonstrated sufficiently to convince poultry farmers to adopt the technology on a large scale. Currently, it is not seen as adding much, if anything, over and above what a stockperson could do on his or her own or with equipment that is already available, such as ventilation control and even water meters. Automated welfare assessment is not yet seen as a worth-while practice, either as a commercial proposition (Jukan et al., 2017), for increasing efficiency or even as a way of achieving its primary purpose of making real improvements to bird welfare. The following are suggested as ways of ensuring that automated monitoring fulfils its potential for achieving higher standards of poultry welfare within the context of sustainable food production:

1 The quality of information provided by automated monitoring needs to be improved above its present level so that it is clearly and demonstrably offering farmers something that is better than anything they have available at present. This means that the data obtained from sensors such as cameras and microphones should not be viewed in isolation but rather be part of a fully integrated system that includes all other kinds of information known to impact bird welfare, such as temperature, light and humidity (Van Hertem et al., 2018). To fully achieve this, new and better systems will be needed for analysing and sharing data sets (Jukan et al., 2017) so that best practice can be identified and recommended with the confidence that it is based on the largest possible evidence base. The collection and analysis of such 'Big Data' sets will not only mean better understanding of the factors leading to improved welfare but also the environmental, economic and food safety as well as other advantages of improving welfare (Gocsik et al., 2016; Henningsen et al., 2018). Linking welfare to these other benefits is an important part of making sure animal welfare is given high priority in the decision-making of farmers, governments and the public (Garnett et al., 2013; Dawkins, 2016; Jukan et al., 2017).
2 Automated welfare assessment needs to develop smarter algorithms so that it is both better at detecting welfare problems and able to include more in the definition of 'good welfare' than just the physical health outcomes it currently measures. There is a need for more powerful algorithms that take account of a wider range of welfare measure, but these will need to go hand in hand with more tightly defined and clearly demonstrated definitions of what good welfare is. An algorithm that correctly detects a specific behaviour will not carry much weight as a welfare indicator if that behaviour itself is not linked firmly to what is meant by good welfare, for example by being either an indicator of good health or something that the animals have been shown to want. In other words, the correct *interpretation* of what automated systems can tell us in welfare terms will be as important to their success as the development of new algorithms themselves.

Another key factor in the welfare value of automation will be the extent to which the technology is able to provide information about which interventions are most appropriate at any given time. Knowledgable systems that can tell farmers when there is a genuine problem with a flock and do so reliably without setting off too many false alarms will help the technology to be widely accepted. But even smarter systems that not only detect problems but can also indicate what actions are needed will be an important future development.

3 Public doubts about the welfare implications of automated assessment need to be taken seriously and directly addressed. This is best done by making sure that animal welfare achieves universally high priority with all stakeholders, which in turn means turning the potential welfare advantages of automated assessment into real improvements in the ways listed earlier. As has been emphasized throughout this chapter, automated welfare assessment offers a great promise of improving welfare through constant monitoring, greater care for the individual and early warning of problems and, through the use of large data sets, identifying best practice from around the world. Indeed, as automated measurements become more widespread and provide ever larger data sets, machine learning may provide us with new measures of 'welfare' itself based on a far wider range of measures than is currently possible and using information collected across animals' entire lifetimes.

In addition to improving bird welfare, automated systems have the potential to bring other benefits such as reduced susceptibility to diseases through improved immunity, earlier detection of infection, less waste through reduced mortality, higher standards of food quality and safety and greater job satisfaction for those in charge of poultry (Dawkins, 2016). But the promise has yet to be convincingly fulfilled. Changes in poultry production that make use of automated measurements are likely to continue in the face of rising demand for poultry products, increasing scarcity of human labour to look after birds and increasing public demands for higher welfare standards and transparency by producers.

It is important that the benefits that smart technology appears to offer us are ultimately translated into a reality of better lives for the birds themselves.

## 7 References


Aydin, A. 2017a. Development of an early detection system for lameness of broilers using computer vision. *Computers and Electronics in Agriculture* 136, 140-6. doi:10.1016/j.compag.2017.02.019.

Aydin, A. 2017b. Using 3D vision camera system to automatically assess the level of inactivity in broiler chickens. *Computers and Electronics in Agriculture* 135, 4-10. doi:10.1016/j.compag.2017.01.024.

Aydin, A. and Berckmans, D. 2016. Using sound technology to automatically detect the short-term feeding behaviours of broiler chickens. *Computers and Electronics in Agriculture* 121, 25-31. doi:10.1016/j.compag.2015.11.010.

Aydin, A., Bahr, C. and Berckmans, D. 2015. Automatic classification of measures of lying to assess the lameness of broilers. *Animal Welfare* 24(3), 335-43. doi:10.7120/09627286.24.3.335.

Banhazi, T. M. and Black, J. L. 2009. Precision livestock farming: a suite of electronic systems to ensure the application of best practice management on livestock farms. *Australian Journal of Multi-Disciplinary Engineering* 7(1), 1-14. doi:10.1080/14488388.2009.11464794.

Banhazi, T. M., Lehr, H., Black, J. L., Crabtree, H., Schofield, P., Tscharke, M. and Berckmans, D. 2012. Precision Livestock Farming: an international review of scientific and commercial aspects. *International Journal of Agricultural and Biological Engineering* 5, 1-9.

Ben Sassi, N., Averos, X. and Estevez, I. 2016. Technology and poultry welfare. *Animals* 6(10), 62. doi:10.3390/ani6100062.

Berckmans, D. 2017. General introduction to precision livestock farming. *Animal Frontiers* 7(1), 6-11. doi:10.2527/af.2017.0102.

Bono, G. and Mori, B. D. 2005. Animals and their quality of life: considerations beyond mere welfare. *Veterinary Research Communications* 29(Suppl. 2), 165-8.

Brambell, F. W. R. 1965. Report of the technical committee to enquire into the welfare of animals kept under intensive husbandry systems. Doc. 2836. Her Majesty's Stationery Office, London, UK. Reprinted 1970.

Broom, D. M. 2007. Quality of Life means welfare: how is it related to other concepts and assessed? *Animal Welfare* 16, 43-53.

Buscher, W. 2019. Digitization of the barn. *Zuchtungskunde* 91, 35-44.

Caja, G., Castro-Costa, A. and Knight, C. H. 2016. Engineering to support wellbeing of dairy animals: background and current scenario. *Journal of Dairy Research* 83(2), 136-47. doi:10.1017/S0022029916000261.

Campbell, D. L. M., Makagon, M. M., Swanson, J. C. and Siegford, J. M. 2016. Litter use by laying hens in a commercial aviary: dustbathing and piling. *Poultry Science* 95(1), 164-75. doi:10.3382/ps/pev183.

Campbell, D. L. M., Hinch, G. N., Dyall, T. R., Warin, L., Little, B. A. and Lee, C. 2017. Outdoor stocking density in free-range laying hens: radio-frequency identification of impacts on range use. *Animal* 11(1), 121-30. doi:10.1017/S1751731116001154.

Carvalho, T. M. R., Massari, J. M., Sabino, L. A. and Moura, D. J. 2013. Sensor placement to reach thermal comfort and air quality in broiler housing. *Precision Livestock Farming 2013 - Papers Presented at the 6th European Conference on Precision Livestock Farming, ECPLF 2013*, pp. 945-52.

Chen, Y., Ni, J. Q., Diehl, C. A., Heber, A. J., Bogan, B. W. and Chai, L. L. 2010. Large scale application of vibration sensors for fan monitoring at commercial layer hen houses. *Sensors (Basel, Switzerland)* 10(12), 11590-604. doi:10.3390/s101211590.

Coelho, D. J. de R., Tinôco, I. de F. F., Baptista, F. J., Souza, C. de F., Sousa, F. C. de, Cruz, V. F. da, Vieira, M. de F. A., Mendes, M. A. dos S. A., Oliveira, K. P. de and Barbari, M. 2016. Mapping the thermal comfort index in laying hens facilities. Organising Committee, CIGR 2016, Aarhus, Denmark.

Colles, F. M., Cain, R. J., Nickson, T., Smith, A. L., Roberts, S. J., Maiden, M. C. J., Lunn, D. and Dawkins, M. S. 2016. Monitoring chick flock behaviour provides early warning of infection by human pathogen Campylobacter. *Proceedings of the Royal Society of London, Series B* 283(1822).

Cook, N. J., Smykot, A. B., Holm, D. E., Fasenko, G. and Church, J. S. 2006. Assessing feather cover of laying hens by infrared thermography. *Journal of Applied Poultry Research* 15(2), 274-9. doi:10.1093/japr/15.2.274.

Corkery, G., Ward, S., Kenny, C. and Hemmingway, P. 2013. Incorporating smart sensing technology into the poultry industry. *Journal of World's Poultry Research* 3, 106-28.

Curi, T. M. RdC., Conti, D., Vercellino, RdA., Massari, J. M., Moura, D. Jd, Souza, Z. Md and Montanari, R. 2017. Positioning of sensors for control of ventilation systems in broiler houses: a case study. *Scientia Agricola* 74(2), 101-9. doi:10.1590/1678-992x-2015-0369.

D'Alfonso, T. H., Manbeck, H. B. and Roush, W. B. 1996. A case study of temperature uniformity in three laying hen production buildings. *Transactions of the American Society of Agricultural Engineers* 3, 669.

Dawkins, M. S. 2016. Animal welfare and efficient farming: is conflict inevitable? *Animal Production Science* 57(2), 201. doi:10.1071/AN15383.

Dawkins, M. S., Donnelly, C. A. and Jones, T. A. 2004. Chicken welfare is influenced more by housing conditions than by stocking density. *Nature* 427(6972), 342-4. doi:10.1038/nature02226.

Dawkins, M. S., Lee, H.-J.,Waitt, C. D. and Roberts, S. J. 2009. Optical flow patterns in broiler chicken flocks as automated measures of behaviour and gait. *Applied Animal Behaviour Science* 119(3-4), 203-9. doi:10.1016/j.applanim.2009.04.009.

Dawkins, M. S., Cain, R. and Roberts, S. J. 2012. Optical flow, flock behaviour and chicken welfare. *Animal Behaviour* 84(1), 219-23. doi:10.1016/j.anbehav.2012.04.036.

Dawkins, M. S., Roberts, S. J., Cain, R. J., Nickson, T. and Donnelly, C. A. 2017. Early warning of footpad dermatitis and hock burn in broiler chicken flocks using optical flow, body weight and water consumption. *Veterinary Record* 180(20), 499. doi:10.1136/vr.104066.

de Jong, I. C.,van Harn, J., Gunnink, H., Hindle, V. A. and Lourens, A. 2012. Footpad dermatitis in Dutch broiler flocks: prevalence and factors of influence. *Poultry Science* 91(7), 1569-74. doi:10.3382/ps.2012-02156.

De Montis, A., Pinna, A., Barra, M. and Vranken, E. 2013. Analysis of poultry eating and drinking behavior by software eYeNamic. *Journal of Agricultural Engineering* 44(2s), 166-72. doi:10.4081/jae.2013.275.

Drake, K. A.,Donnelly, C. A. and Dawkins, M. S. 2010. Influence of rearing and lay risk factors on propensity for feather damage in laying hens. *British Poultry Science* 51(6), 725-33. doi:10.1080/00071668.2010.528751.

Du, X., Lao, F. and Teng, G. 2018. A sound source localization analytical method for monitoring the abnormal night vocalisations of poultry. *Sensors* 18(9), Article No. 2906. doi:10.3390/s18092906.

Edgar, J. L., Nicol, C. J., Pugh, C. A. and Paul, E. S. 2013. Surface temperature changes in response to handling in domestic chickens. *Physiology and Behavior* 119, 195-200. doi:10.1016/j.physbeh.2013.06.020.

Ekstrand, C., Carpenter, T. E., Andersson, I. and Algers, B. 1998. Prevalence and control of foot-pad dermatitis in broilers in Sweden. *British Poultry Science* 39(3), 318-24. doi:10.1080/00071669888845.

Fernandez, A. P., Norton, T., Tullo, E., van Hertem, T., Youssef, A., Exadaktylos, V., Vranken, E., Guarino, M. and Berckmans, D. 2018. Real-time monitoring of broiler flock's welfare status using camera-based technology. *Biosystems Engineering* 173, 103-14. doi:10.1016/j.biosystemseng.2018.05.008.

Fleet, D. J. and Weiss, Y. 2005. Optimal flow estimation. In: Paragios, N., Chen, Y. and Faugeras, O. (Eds), *Mathematical Models for Computer Vision*. Springer, New York, NY, pp. 239-58.

Fournel, S., Rousseau, A. N. and Laberge, B. 2017. Rethinking environmental control strategy of confined animal housing systems through precision livestock farming. *Biosystems Engineering* 155, 96–123. doi:10.1016/j.biosystemseng.2016.12.005.

Fraser, D. 2008. *Understanding Animal Welfare: the Science in Its Cultural Context*. Wiley-Blackwell, Oxford, UK.

Fraser, D., Duncan, I. J. H., Edwards, S. A., Grandon, T., Gregory, N. G., Guyonnet, V., Hemsworth, P. H., Huertas, S. M., Huzzey, J. M., Mellor, D. J., Mench, J. A., Spinka, M. and Whay, H. R. 2013. General principles for the welfare of animals in production systems: the underlying science and its application. *Veterinary Journal* 198(1), 19-27. doi:10.1016/j.tvjl.2013.06.028.

Garnett, T., Appleby, M. C., Balmford, A., Bateman, I. J., Benton, T. G., Bloomer, P., Burlingame, B., Dawkins, M., Dolan, L., Fraser, D., Herrero, M., Hoffmann, I., Smith, P., Thornton, P. K., Toulmin, C., Vermeulen, S. J. and Godfray, H. C. 2013. Sustainable intensification in agriculture: premises and policies. *Science* 341(6141), 33-4. doi:10.1126/science.1234485.

Gebhardt-Henrich, S. G. and Stratmann, A. 2016. What is causing smothering in laying hens? *Veterinary Record* 179(10), 250-1. doi:10.1136/vr.i4618.

Gocsik, É., Brooshooft, S. D., de Jong, I. C. and Saatkamp, H. W. 2016. Cost-efficiency of animal welfare in broiler production: a pilot study is the Welfare Quality® assessment protocol. *Agricultural Systems* 146, 55-69. doi:10.1016/j.agsy.2016.04.001.

Green, L. E., Lewis, K., Kimpton, A. and Nicol, C. J. 2000. Cross-sectional study of the prevalence of feather pecking of laying hens in alternative systems and its associations with management and disease. *Veterinary Record* 147(9), 233-8. doi:10.1136/vr.147.9.233.

Gunnarsson, S., Keeling, L. J. and Svedberg, J. 1999. Effect of rearing factors on the prevalence of floor eggs, cloacal cannibalism and feather pecking in commercial flocks of loose housed laying hens. *British Poultry Science* 40(1), 12-8. doi:10.1080/00071669987773.

Halachmi, I. and Guarino, M. 2016. Editorial: precision livestock farming: a 'per animal' approach using advanced monitoring technologies. *Animal* 10(9), 1482-3. doi:10.1017/S1751731116001142.

Henningsen, A., Czekaj, T. G., Forkman, B., Lund, M. and Nielsen, A. S. 2018. The relationship between animal welfare and economic perfomance at farm level: a quantitative study of Danish pig producers. *Journal of Agricultural Economics* 69(1), 142-62. doi:10.1111/1477-9552.12228.

Ji, B., Zheng, W., Gates, R. S. and Green, A. R. 2016. Design and performance evaluation of the upgraded portable monitoring unit for air quality in animal housing. *Computers and Electronics in Agriculture* 124, 132–40. doi:10.1016/j.compag.2016.03.030.

Jones, T. A., Donnelly, C. A. and Dawkins, M. S. 2005. Environmental and management factors affecting the welfare of chickens on commercial farms in the United Kingdom and Denmark stocked at five densities. *Poultry Science* 84(8), 1155–65. doi:10.1093/ps/84.8.1155.

Jukan, A.,Masip-Bruin, X.and Amla, N. 2017. Smart computing and sensing technologies for animal welfare: a systematic review. *ACM Computing Surveys* 50(1), 1-27. doi:10.1145/3041960.

Kashiha, M. A., Green, A. R., Sales, T. G., Bahr, C., Berckmans, D. and Gates, R. S. 2015. Application of image processing on hen tracking in an environmental preference chamber. *Precision Livestock Farming 2015 - Papers Presented at the 7th European Conference on Precision Livestock Farming*, ECPLF 2015, pp. 185-94.

Kestin, S. C., Knowles, T. G., Tinch, A. E. and Gregory, N. G. 1992. Prevalence of leg weakness in broiler chickens and the relationship with genotype. *Veterinary Record* 133, 190-4.

Lee, H. J., Roberts, S. J., Drake, K . A. and Dawkins, M. S. 2011. Prediction of feather damage in laying hens using optical flows and Markov models. *Journal of the Royal Society, Interface* 8(57), 489-99. doi:10.1098/rsif.2010.0268.

Lee, J., Noh, B., Jang, S., Park, D., Chung, Y. and Chang, H. H. 2015. Stress detection and classification of laying hens by sound analysis. *Asian-Australasian Journal of Animal Sciences* 28(4), 592-8. doi:10.5713/ajas.14.0654.

Leroy, T., Vranken, E., Van Brecht, A., Struelens, E., Sonck, B. and Berckmans, D. 2006. A computer vision method for on-line behavioral quantification of individually caged poultry. *Transactions of the ASABE* 49(3), 795-802. doi:10.13031/2013.20462.

Liakos, K. G., Busato, P., Moshou, D., Pearson, S. and Bochtis, D. 2018. Machine learning in agriculture: a review. *Sensors* 18(8), Article No. 2674. doi:10.3390/s18082674.

Main, D. C. J., Mullan, S., Atkinson, C., Cooper, M., Wrathhall, J. H. M. and Blokhuis, H. J. 2014. Best practice framework for animal welfare certification schemes. *Trends in Food Science and Technology* 37(2), 127-36. doi:10.1016/j.tifs.2014.03.009.

Manning, L., Chadd, S. A. and Baines, R. N. 2007. Water consumption in broiler chickens: a welfare indicator. *World's Poultry Science Journal* 63(1), 63-71. doi:10.1017/S0043933907001274.

Mason, G. and Mendl, M. 1993. Why is there no simple way of measuring animal welfare? *Animal Welfare* 2, 301-20.

McCafferty, D. J. 2013. Applications of thermal imaging in avian science. *Ibis* 155(1), 4-15. doi:10.1111/ibi.12010.

Mellor, D. J. 2016. Updating animal welfare thinking: moving beyond the 'Five Freedoms' towards 'a life worth living'. *Animals* 6(3), 21. doi:10.3390/ani6030021.

Mellor, D. J. and Beausoleil, N. J. 2015. Extending the 'Five Domains' model for animal welfare assessment to incorporate positive welfare states. *Animal Welfare* 24(3), 241-53. doi:10.7120/09627286.24.3.241.

Moe, R. O., Bohlin, J., Flo, A., Vasdal, G. and Stubsjoen, S. M. 2017. Hot chicks, cold feet. *Physiology and Behavior* 179, 42-8. doi:10.1016/j.physbeh.2017.05.025.

Morota, G., Ventura, R. V., Silva, F. F., Koyama, M. and Fernando, S. C. 2018. Machine learning and data mining advance predictive big data in precision animal agriculture. *Journal of Animal Science* 96(4), 1540-50. doi:10.1093/jas/sky014.

Moura, C. J., Naas,I., Alves, E. C. S., Carvalho, T. M. R., Vale, M. M. and Lima, K. A. O. 2008. *Scientific Agriculture* 119, 178-83.

Nasirahmadi, A., Edwards, S. A. and Sturm, B. 2017. Implementation of machine vision for detecting behaviour of cattle and pigs. *Livestock Science* 202, 25-38. doi:10.1016/j.livsci.2017.05.014.

Nicol, C. J., Bestman, M.,Gilani, A. M., De Haas, E. N., De Jong, I. C., Lambton, S., Wagenaar, J. P., Weeks, C. A. and Rodenburg, T. B. 2013. The prevention and control of feather pecking: applications to commercial systems. *World's Poultry Science Journal* 69(4), 775-88. doi:10.1017/S0043933913000809.

Norton, T., Vranken, E., Exadaktylos, V., Berckmans, D., Lehr, H., Vessier, I., Blokhuis, H. and Berckmans, D. 2016. Implementation of Precision Livestock Farming (PLF) technology on EU farms: results from the EU-PLF project. In: *CIGR-AgEng Conference*, 26-29 June 2016, Aarhus, Denmark. Abstracts and full papers, 2016, pp. 1-7.

Pereira, D. F., Nääs, I. D. A., Gabriel Filho, L. R. A. and Neto, M. M. 2012. Cluster index for accessing thermal comfort for broiler breeders. *ASABE - 9th International Livestock Environment Symposium 2012*, ILES 2012, pp. 207-12.

Pereira, D. F., Miyamoto, B. C. B., Maia, G. D. N., Tatiana Sales, G., Magalhães, M. M. and Gates, R. S. 2013. Machine vision to identify broiler breeder behavior. *Computers and Electronics in Agriculture* 99, 194-9. doi:10.1016/j.compag.2013.09.012.

Pereira, E. M., Naas, IdA. and Garcia, R. G. 2014. Identification of acoustic parameters for broiler welfare estimates. *Engenharia Agricola* 34(3), 413-21. doi:10.1590/S0100-69162014000300004.

OIE. 2016. Chapter 7. Animal Welfare and broiler chicken production systems. Terrestrial Animal Health Code. OIE. World Organisation for Animal Health, pp. 1-7. Available at: http://www.oie.int/index.php?id=169&L=0&htmfile=chapitre_aw:broiler_chicken.ht (accessed on 08 July 2016).

Roberts, S. J., Cain, R. and Dawkins, M. S. 2012. Prediction of welfare outcomes for broiler chickens using Bayesian regression on continuous optical flow data. *Journal of the Royal Society, Interface* 9(77), 3436-43. doi:10.1098/rsif.2012.0594.

Rowe, E., Dawkins, M. S. and Gebhardt-Heinrich, S. 2019. A systematic review of precision livestock farming in the poultry sector: is technology focussed on improving animal welfare? *Animals* 9(9), Article 614.

RSPCA. 2013. *Welfare Standards for Chickens*. Royal Society for the Prevention of Cruelty to Animals, Horsham, UK.

Savory, C. J. 1995. Feather pecking and cannibalism. *World's Poultry Science Journal* 51(2), 215-19. doi:10.1079/WPS19950016.

Scott, E. M., Nolan, A., Reid, J. and Wiseman-Orr, M. L. 2007. Can we really measure animal QoL? Methodologies for measuring Qol in people and other animals. *Animal Welfare* 16, 17-24.

Sonka, M., Hlavac, V. and Boyle, R. 1999. *Image Processing Analysis and Machine Vision* (2nd edn.). PWS Publishing, London, UK.

Stadig, L. M., Ampe, B., Rodenburg, T. B., Reubens, B., Maselyne, J., Zhuang, S., Criel, J. and Tuyttens, F. A. M. 2018. An automated positioning system for monitoring chickens' location: accuracy and registration success in a free-range area. *Applied Animal Behaviour Science* 201, 31-9. doi:10.1016/j.applanim.2017.12.010.

Stevenson, P. 2017. Precision livestock farming: could it drive the livestock sector in the wrong direction? Available at: https://www.ciwf.org.uk/media/7431928/plf-could-it-drive-the-livestock-sector-in-the-wrong-direction.pdf (accessed on 9 May 2019).

Taylor, K. D. and Mills, D. S. 2007. Is quality of life a useful concept for companion animals? *Animal Welfare* 16, 55-65.

Valletta, J. J., Torney, C., Kings, M., Throbton, A. and Madden, J. 2017. Applications of Machine learning in animal behaviour studies. *Animal Behaviour* 124, 203-20. doi:10.1016/j.anbehav.2016.12.005.

Vanderhasselt, R. F., Sprenger, M., Duchateau, L. and Tuyttens, F. A. 2013. Automated assessment of footpad dermatitis in broiler chickens at the slaughterline: evaluation and correspondence with human expert scores. *Poultry Science* 92(1), 12-8. doi:10.3382/ps.2012-02153.

Van Hertem, T., Rooijahkers, L., Berckmans, D., Peña Fernández, A., Norton, T., Berckmans, D. and Vranken, E. 2017. Appropriate data visualisation is key to Precision Livestock

Farming acceptance. *Computers and Electronics in Agriculture* 138, 1–10. doi:10.1016/j.compag.2017.04.003.

Van Hertem, T., Norton, T., Berckmans, D. and Vranken, E. 2018. Predicting broiler gait scores from activity monitoring and flock data. *Biosystems Engineering* 173, 93–102. doi:10.1016/j.biosystemseng.2018.07.002.

Wathes, C. M. 2010. Lives worth living? *Veterinary Record* 166(15), 468-9. doi:10.1136/vr.c849.

Wathes, C. M., Kristensen, H. H., Aerts, J.-M. and Berckmans, D. 2008. Is precision livestock farming an engineer's daydream or nightmare, an animal's friend of foe and a farmer's panacea or pitfall? *Computers and Electronics in Agriculture* 64(1), 2-10. doi:10.1016/j.compag.2008.05.005.

Webster, A. J. F. 2001. Farm animal welfare: the five freedoms and the free market. *Veterinary Journal* 161(3), 229–37. doi:10.1053/tvjl.2000.0563.

Welfare Quality. 2009. *Assessment Protocol for Poultry, Broiler and Laying Hens.* Welfare Quality, Lelystad, The Netherlands. Available at: http://www.animalwelfareplatfo rm.eu/Twelve-farm-animal-welfare-criteria.php.

Werkheiser, I. 2018. Precision livestock farming and farmers' duties to livestock. *Journal of Agricultural and Environmental Ethics* 31(2), 181-95. doi:10.1007/s10806-018-9720-0.

Werner, A., Jarfe, A., Stafford, J. V. and Cox, S. W. R. 2003. *European Conference on Precision Livestock Farming (1st : 2003 : Berlin G) and European Conference on Precision Agriculture (4th : 2003 : Berlin G) 2003 Programme Book of the Joint Conference of ECPA-ECPLF.* Wageningen Academic, Wageningen, the Netherlands.

Winckler, C. 2019. Assessing animal welfare at the farm level: do we care sufficiently about the individual? *Animal Welfare* 28(1), 77-82. doi:10.7120/09627286.28.1.077.

Wood-Gush, D. G. M. 1971. *The Behaviour of the Domestic Fowl.* Heinemann, London, UK.

Yahav, S., Shinder, D., Tanny, J. and Cohen, S. 2005. Sensible heat loss: the broiler's paradox. *World's Poultry Science Journal* 61(3), 419-34. doi:10.1079/WPS200453.

Zaninelli, M., Redaelli,V., Tirloni, E., Bernardi, C., Dell'Orto, V. and Savoini, G. 2016. First results of a detection sensor for the monitoring of laying hens reared in a commercial organic egg production farm based on the use of infrared technology. *Sensors (Basel, Switzerland)* 16(10). doi:10.3390/s16101757.

Zaninelli, M., Redaelli, V., Luzi, F., Bontempo, V., Dell'Orto, V. and Savoini, G. 2017. A monitoring system for laying hens that uses a detection sensor based on infrared technology and image pattern recognition. *Sensors* 17(6). doi:10.3390/s17061195.

Zaninelli, M., Redaelli, V., Luzi, F., Mitchell, M., Bontempo, V., Cattaneo, D., Dell'Orto, V. and Savoini, G. 2018. Development of a machine vision method for the monitoring of laying hens and detection of multiple nest occupations. *Sensors* 18(1). doi:10.3390/s18010132.

Zhao, Y., Aarnink, A. J. A., Hofschreuder, P. and Groot Koerkamp, P. W. G. 2009. Evaluation of an impaction and a cyclone pre-separator for sampling high PM10 and PM2.5 concentrations in livestock houses. *Journal of Aerosol Science* 40, 868–78.

Zhuang, X. L., Bi, M. N., Guo,J. L.,Wu, S. Y. and Zhang, T. 2018. Development of an early warning algorithm to detect sick broilers. *Computers and Electronics in Agriculture* 144, 102-13. doi:10.1016/j.compag.2017.11.032.

Zimmerman, P. H., Koene, P. and Van Hooff, J. A. 2000. The vocal expression of feeding motivation and frustration in domestic laying hens. *Applied Animal Behaviour Science* 69, 265-73. doi:10.1016/s0168-1591(00)00136-2.

# Chapter 8

# Improving welfare assessment indicators and protocols for poultry

*Linda Keeling, Swedish University of Agricultural Sciences, Sweden*

1 Introduction
2 Welfare assessment protocols
3 Welfare assessment indicators
4 Examples of animal-based measures
5 Prioritising welfare indicators
6 Future trends
7 Conclusion
8 Where to look for further information
9 References

## 1 Introduction

There are many different welfare assessment protocols. Some refer to best practices and advice for farmers, while others check compliance with legislation, and an increasing number are related to quality assurance and to give farmers access to particular markets. There is no single accepted definition of a welfare assessment protocol, but it can generally be considered as a list of welfare indicators and a description of the procedures to collect the required data on them. The differing purposes of assessments mean that they also differ in the type and number of welfare indicators that the protocols contain. This chapter reviews some of the advances made in recent years and highlights areas where more research is needed to improve welfare assessment on poultry farms. There have also been studies to evaluate whether or not these assessments have led to any demonstrable improvements in welfare, for example, Mullan et al. (2016).

Generally speaking, one can split welfare assessment indicators into two types; input and outcome indicators (Table 1). The input indicators refer to those features of the environment of the animal that will affect its welfare. These are often referred to as resource-based measures (e.g. space allowance, type of floor) and management-based measures (e.g. biosecurity measures, feeding times). Outcome indicators, on the other hand, relate to the animal itself and

http://dx.doi.org/10.19103/AS.2020.0078.08

**Table 1** Some of the different types of indicators used in welfare assessment. Modified from EFSA (2012). The terms indicator and measures are sometimes used interchangably.

| Type of indicators | Explanation |
|---|---|
| Welfare outcome indicator | An observation, a record or a measurement used to obtain information on an individual animal's welfare that can be reliably used in practice by trained people. |
| Welfare input indicator | An observation, record or measurement about a factor (usually something related to the resources available to the animal or the way it is managed) that has the potential to affect the animal's welfare. Whether this factor has a detrimental or a beneficial effect (or no affect at all) on the animal's welfare will depend on which other factors are acting on the animal at the same time and the duration the animal is exposed to the factor. |
| Animal-based measure | A response of an animal or an effect on an animal used to assess its welfare. It can be taken directly on the animal or indirectly and includes the use of animal records. It can result from a specific event, for example, an injury, or be the cumulative outcome of many days, weeks or months, for example, body condition. |
| Management-based measure | An evaluation of what the animal unit manager or stockperson does and which management processes or tools are used. |
| Resource-based measure | An evaluation of a feature of the environment in which the animal is kept or to which it is exposed. |

are therefore often referred to as animal-based measures, for example, whether the animal has any injuries, its level of fearfulness. Most researchers agree that it is the animal-based measures that tell us about the welfare of the animal. Indeed almost all of the animal welfare research in the laboratory or under controlled conditions has been on animal-based measures, although this terminology was not used at the time. Since research on animal welfare started to flourish in the late 1960s, ethologists, physiologists and others have been comparing the responses of birds in different situations and quantifying the motivation to perform different behaviour patterns (Weeks and Nicol, 2006; EFSA 2005; De Jong et al., 2012). The step from valid and reliable indicators under controlled conditions to valid and reliable indicators that can be collected under farm conditions has not been easy. It has been difficult for practical reasons, but in part also because views about what is important to measure have differed. It was easiest initially to include welfare indicators that were also relevant for production and economy such as mortality, gait scoring (broilers) and plumage condition (layers). It has taken time for the benefits of assessing fearfulness or the quality of the human–animal relationship to become more widely accepted. Collaboration between researchers, advisers, veterinarians and farmers has been important in gaining acceptance for some measures, but researchers have also had to accept practical realities and to identify quicker, simpler indicators.

We have been keeping poultry for centuries and much is already known about how environment and management affect bird health and production. Although 'old' knowledge can be shown to be wrong in controlled experiments, it is rather obvious that systems or practices that result in high mortality are to be avoided. That said, it was only when researchers started to systematically collect data on the number of dead birds found trapped in different battery cage systems that details of design were improved (Tauson, 1985). The fact that the industry has been one of the targets in the early animal welfare debate, has further contributed to poultry welfare research, such as that towards optimising the length of the feeding trough, so that all birds can feed at the same time and aggression reduced, as well as keeping the litter dry and friable in broiler houses to reduce hock burns and breast blisters. Traditionally, protocols have therefore focussed on welfare input indicators, such as cage design and litter quality on the assumption that if birds are well-managed and given the appropriate resources, then their welfare will be good. Indeed many current protocols still focus on these types of indicators, since they are easy to measure and tend not to change much over time. Most legislation is also focussed on resource and management-based measures, for example, the EU laying hen directive (EC, 1999). Specifying the housing system is the absolute simplest protocol and is the basis for the labelling system of eggs (EC, 2008) (Fig. 1).

Nevertheless, not all birds respond to the environment in the same way. There are well-documented differences in behaviour, for example, Ali et al. (2019) and Campbell et al. (2019). Thus, even given the same housing system, same allocation of resources and the same management, there may be large variations in the welfare of the birds from one flock to another, and there will

**Figure 1** Cages are made largely of metal and insufficient attention was paid in early designs to places where birds can be trapped. Later designs increased the length of the feed trough and included perching, nesting and dust bating opportunities. Small details in the design can have large consequences for the welfare of the birds. Photo: Linda Keeling.

be variation in the welfare of individual birds within the flock. Illustrating this point, the Animal Needs Index (ANI-35L), which is a largely resource-based protocol, distinguished between housing systems, but correlated poorly with animal-based measures from farms with the same housing system (Mollenhorst et al., 2005; Zaludik et al., 2007). However, the situation is even more complex than this, since even if all birds did respond in the same way, most systems have some advantages and some disadvantages. Thus it can be difficult to decide which is the 'best system' even when they are compared using the same indicators (e.g. EFSA, 2005; Rodenburg et al., 2012; Sherwin et al., 2010).

There is a steady move towards using animal-based measures in welfare assessment protocols, for laying hens and broilers, something strongly supported by the large Welfare Quality and AssureWel research projects (Blokhuis et al., 2013; Anonymous, 2018). Looking at animal-based measures makes it possible to compare the welfare of birds in different housing systems and under different management practices. This is especially useful when considering welfare across regions with diverse traditions of poultry husbandry. For example, the broiler directive (EC, 2007) specifies that each EU member state records outcome indicators (e.g. mortality rates and slaughterhouse downgrades). The World Animal Health Organisation (OIE) is also developing science-based international standards, which are adopted by the OIE World Assembly of Delegates. Their proposed standards for broilers are mainly animal-based (OIE, 2019). Nevertheless, in the absence of standardisation of how animal-based measures are recorded, it is difficult to compare the data collected in different protocols so these data can be used efficiently. This was one of the conclusions from the LayWel project (Blokhuis et al., 2007) as well as an EFSA internal project on the use of animal-based measures to assess animal welfare (EFSA, 2015). Standardising welfare indicators is therefore one of the areas that will probably do most to improve poultry welfare assessment in the near future.

## 2 Welfare assessment protocols

As stated in the introduction, there are different types of welfare assessment protocols intended for different purposes, and so it is important they be 'fit for the purpose' for which they are intended. Animal-based measures are those that reflect the welfare of the bird and some of them can change rather quickly, for example, vocalisations. This makes these types of indicators especially useful for the farmer, who is probably in the building, checking birds several times per day. Other indicators change more slowly, for example body condition, and some, such as lameness in broilers or poor plumage condition in layers, may be more or less of a reoccurring problem for some farmers. In these cases regular monitoring can facilitate early detection, so interventions can be made to minimise the welfare consequences for the birds and the economic consequences for the farmer.

Inspectors from a private assurance scheme, checking that the farm satisfies the criteria to be certified, or from the competent authority, checking compliance with animal welfare legislation, visit farms considerably less often. They can observe animal-based measures to assess the current welfare status of birds in the flock, even if they may also record resource-based or management-based measures as a guide to what welfare status might be expected in the future, that is, assess the risk of poor welfare.

There are several issues to be considered when including more animal-based measures in animal welfare assessment protocols (Roe et al., 2011), although two important ones in practice are the issues of feasibility and reliability. Checking animal-based measures often takes more time than checking the environment and there can be a larger variation between assessors (Keeling, 2009). This may be a concern when checking compliance with legislation, although it is less of a concern if it is farmers doing their own checking for management purposes. The sample size should be large enough, and there is some evidence that farmers with laying hens prefer a large sample size of birds to be controlled, as they perceive that as more representative of their situation (Van Dijk et al., 2018). Observations of a smaller number of individuals are more feasible on a scheme level, but can be limiting for farm-level decisions (Main et al., 2012). For broilers, the feasibility of including animal-based measures is increased if the measures are taken at the slaughterhouse. It saves travelling time for the assessor, and fewer people have to be trained. Also, as discussed elsewhere, it opens up for automated recording of some indicators. To increase efficiency in audits still further, the assessor may be checking compliance within several areas and not only animal welfare, for example, requirements for an organic label or for marketing a particular region. There are several studies that propose combining welfare assessments at the slaughterhouse with the official food safety controls, for example, Iannetti et al. (2020).

Although assessing the welfare of birds is the focus, also important are the views of the people being audited. It is one thing to make an official control, or check compliance, but it is another thing to encourage a pro-active management by the farmer to promote improvement of poultry welfare. This is an expanding area of research and an increasing number of social science studies have investigated the role and attitudes of the farmer (Anneberg et al., 2013; Roe et al., 2011). It is proposed in a best practice framework for animal welfare certification that schemes should promote continuous improvement (Main et al., 2014). In a consultation exercise, Van Dijk et al. (2018) focussed on understanding farmers' perspective when introducing animal-based measures into welfare assessment protocols. While many farmers highlighted the positive management aspects, especially if the assessment was done jointly with the assessor, there was some criticism regarding perceived bureaucracy around self-assessments focussing on welfare outcomes, which they regarded as a duplication of something they felt they were already doing (Van Dijk et al., 2018). The large variation between the ability and skills of farmers should nevertheless be better considered and methods to promote sharing good practices between farmers have been explored (Van Dijk et al., 2019).

## 2.1 Covering all dimensions of welfare in a protocol

In experimental studies, the reason for an assessment protocol may be a single aspect of welfare, for example, an experiment to assess fear or identify early indicators of feather pecking. When giving advice to farmers, a broader but still rather targeted protocol may be used, depending on the problem or type of advice requested by farmers (Main et al., 2014). For the majority of welfare protocols and for almost all quality assurance schemes, however, the intention is an assessment of the overall welfare of birds on that farm.

Whichever precise definition of animal welfare is used as a basis for the welfare assessment protocol, there is considerable consensus on the main areas of welfare that need to be assessed. Sometimes these are covered by general phrases such as physical and mental well-being, for example, Hughes (1988), as a list of things to be achieved, for example, the UK Farm Animal Welfare Council's five freedoms (FAWC, 2012), split into different principles and criteria (Keeling et al., 2013) or domains (Mellor, 2017) or expressed in descriptive text, for example, the OIE definition (OIE, 2018). In one respect there are many similarities between these different frameworks, and that is reassuring, but they differ in how effectively they can be used as a basis for a welfare assessment protocol. This likely reflects their different intentions, but also changing opinions and experiences over time. The five freedoms are a list of welfare consequences to be avoided, such as hunger and pain, and they are still considered important for animals today. Each negative welfare consequence is followed by a list of ways in which it can be avoided, by specifying the resources that should be provided or the actions that should be taken. The Welfare Quality protocol reshuffles the five freedoms into four principles, related to the animal (its behaviour and health), its housing and its feeding with further subdivision into 12 mutually exclusive criteria. A main contribution of the Welfare Quality project was its focus on animal-based measures to assess each of these criteria and a science-based, consensus-building process towards protocols that could be implemented on a commercial farm or slaughterhouse (Forkman and Keeling, 2009). For the first time there was a clear inclusion of things to be promoted, for example, a positive emotional state, as well as things to be avoided, for example, pain. The Five Domains Model involves a further reshuffle, resulting in four physical and functional domains (covering the same areas as the four Welfare Quality principles) and a fifth mental domain. In this way, separating the objective (potentially measurable) domains from the subjective ones that cannot be studied directly. However, it steps back from specific measures or aggregation procedures, even though certain operational proposals were suggested by Mellor (2017). Both the Welfare Quality and the Five Domains Model have formed the basis of welfare assessment protocols.

A consequence of the multidimensional nature of welfare is the diverse range of welfare assessment indicators necessary in protocols to take account of the different dimensions. These indicators can come from different disciplines and may require different amounts of time, expertise or equipment to record them. There are some areas where there are many well-developed indicators, for example, assessing health and for clinical scoring, but there are also areas where there is a lack of welfare indicators, for example, emotional states. Thus there is still a lot of room for improvement in some areas of welfare assessment and protocols should be upgraded on a regular basis to take account of new welfare assessment indicators. One of the greatest challenges for the future of welfare assessment is how to integrate these different measures to give a single overall assessment. Even if methods are proposed there is still considerable discussion, for example, Botreau et al. (2007) and Sandoe et al. (2019). One approach is by getting experts to rank indicators or welfare dimensions (Rodenburg et al., 2008). An average can then be taken of the rankings, alternatively, more or less formalised procedures can be used to come to a consensus. Crucial in this approach is the choice of experts so that they reflect different disciplines, stakeholder categories and so on. Another approach is to use mathematical models to convert the different measures into a standardised score and then to integrate these. Although, even here, there are many ethical aspects to be considered (Veissier et al., 2011). None of these approaches yet, however, addresses integration over a longer period of time (see Section 5 for a discussion on overall quality of life).

Fortunately, integration of indicators gathered during a single visit is not necessary for a protocol to be implemented in practice and for it to be useful. Results from individual indicators, when gathered on many farms, can be used to highlight the advantages and disadvantages with different housing systems. This in turn can be used to improve a particular system or to design new ones. For the farmers, performance in each animal welfare indicator can be valuable information upon which to base management decisions. Such information is especially useful if it is possible for a farmer to benchmark the results for that indicator against that of other farmers, or against their own earlier results for the indicator, for example, to see if an intervention has led to the desired improvements.

## 3 Welfare assessment indicators

Critical thinking around welfare assessment indicators suggests that they should be developed according to the same criteria as for any other diagnostic test (EFSA, 2012). Definitions of some of these criteria, as they relate to welfare assessment, are given in Table 2.

Most important is the validity of the indicator. Some animal-based measures of welfare are intuitively linked with pain, for example, broken bones and large wounds, or with reduced welfare, for example, fear responses and abnormal behaviour. It is nevertheless important to confirm the validity of the indicator. This is not always straightforward as it is the emotional state associated with the indicator that is the main interest. For example, there has long been a discussion on which leg conditions are associated with pain in broilers (Skinner-Noble and Teeter, 2009; Hothersall et al., 2016) and see Riber et al. (2018) for a recent review on the welfare implications of keel bone fractures and keel bone deviations in laying hens. Also important when developing animal-based measures is that they are reliable. In most cases this refers to inter-observer reliability, meaning that a different person taking the same measure would get the same result. Reliability can also include an aspect that is sometimes referred to as robustness, meaning that the indicator should not be so sensitive that it is affected by the time of day or time of year. This is a particular challenge when

**Table 2** Some of the criteria considered to be important for welfare assessment indicators. (Modified from EFSA, 2012)

| Criteria | Explanation |
|---|---|
| Validity | The fitness for purpose of a measure that has been properly developed, optimised, and standardised for an intended purpose. Validation includes estimates of the analytical and diagnostic performance characteristics of the measure/indicator (i.e. sensitivity and specificity). |
| Reliability | A general term referring to the ability of the animal-based measure/welfare outcome indicator to be applied under various conditions, by different personnel while still providing similar (correct) results. |
| Repeatability | Level of agreement between repeated measurements of the animal-based measure/welfare outcome indicator on the same sample by the same assessor, on different occasions. |
| Robustness | The extent to which an animal-based measure/welfare outcome indicator is affected by changes in things such as environment or time of day. |
| Sensitivity | The probability that the consequence is detected by the animal-based measure/welfare outcome indicator when the adverse effect is present in the population (i.e. probability of a correct positive test). |
| Specificity | The degree with which the animal-based measure/welfare outcome indicator is related to a single welfare consequence or whether it relates (responds) to several different consequences (i.e. probability of a correct negative test). |

it comes to using behavioural indicators of welfare, where diurnal and season rhythms are common. On the other hand, the measure should be sensitive enough that it can be used to discriminate between different levels of welfare. One often refers to indicators according to their specificity and their sensitivity. The final criterion important to consider is the feasibility of the indicator. What was feasible to observe previously, when flock sizes were smaller, may no longer be feasible, or vice versa. With modern technology indicators can often be collected continuously (BenSassi et al., 2016). The issues of validity and reliability are especially important when the indicators are the basis for decisions that will affect the farmer economically, such as whether they are compliant or not with legislation or whether they satisfy the requirements to sell their eggs or meat birds within a specific labelling scheme.

Another important factor is that although animal welfare is a characteristic of the individual, many of the indicators used in practice are recorded at the group level, for example, vocalisation and stocking density. Others, even if they are recorded at the individual level, for example, gait score and plumage condition, are presented as a percentage or proportion within the flock. Some of the consequences of this are discussed by Winckler (2019). These include, for example, that the distribution within the flock may be skewed (a few very affected birds) and this could result in the same overall score as a flock with many mildly affected birds, plus that affected birds are not necessarily identified (to be treated if that is necessary). Some of these disadvantages can be minimised, for example, by giving an indication of the distribution variation or the proportion of birds in each category, but it seems inevitable given the large numbers of birds in poultry flocks, that welfare assessment indicators will usually be presented at the group level. Crucial though, if results are to be compared, is that data on a particular indicator are collected when birds are of the same age. The prevalence of many of the welfare problems seen in poultry, for example, lameness in broilers increases with age (and bodyweight), or they are associated with particular phases in the production period, for example, feather pecking is influenced among other things by hormonal changes around the onset of lay and up to peak egg production. Most welfare assurance schemes therefore recommend a particular age window when a visit should take place.

Finally, benchmarking of animal-based indicators on a large scale can be particularly important for early detection of any potential problems leading to poor welfare as a result of trends in the poultry sector, for example, changes in breeding goals, changes in raw ingredients in feed. These welfare changes might not otherwise be detected, or would not have been detected until much later. Benchmarking of important animal-based measures on a large scale would also give quicker feedback to policymakers on the effectiveness of legislation or other initiatives to improve animal welfare. Surveillance of outcome indicators is already established in other areas and there are similarities between what is discussed here and diagnoses in animal health based on clinical signs and meat inspection. However, as discussed previously, this benchmarking and surveillance require the collection of animal-based measures in a standardised manner.

## 4 Examples of animal-based measures

There are many different animal-based measures and as an increasing number of researchers become interested in welfare assessment, more measures are developed and older measures are refined to improve their reliability or are discarded in the light of new knowledge. In this section, some of the more commonly used animal-based measures in practical welfare assessment protocols are presented. They are divided into three categories; clinical measures, behavioural indicators in undisturbed situations and those involving monitoring the response of a bird in a test situation.

### *4.1 Clinical measures*

These can be indicators of health problems, for example, enteritis, parasites or a direct result of the environmental conditions, for example, contact dermatitis, which is mainly on the footpads or hocks of the bird. Although identifying diseases, inflammation and infection are an important part of welfare assessment, clinical indicators of poor health will not be discussed here. However, there are some clinical measures that directly reflect a behavioural welfare indicator, but which are quicker and easier to record than by observing the behaviour itself. They are usually recorded at the individual level and for that reason it is important to take a representative sample of birds from the flock. For layers in cages this is easily done by selecting cages in different rows and tiers. For birds in loose housing systems this can be done by catching a sample from different areas of the house, for example, Welfare Quality (2009) or by walking through the barn in a standardised route (transect method), for example, BenSassi et al. (2019).

#### *4.1.1 Mortality*

This is probably the most commonly recorded welfare indicator, and the ambition is of course to have as low mortality in a flock as possible. But it is a blunt tool at the individual bird level, since birds can die for many different reasons and it is how the bird dies that is most important from a welfare point of view. In some cases mortality is calculated as the difference in the number of birds entering the house and the number leaving it at the end of the production period. In this case, the number of birds that die is combined with those that are culled by the farmer. Thus a diligent farmer, who culls many birds, may have a seemingly poorer bird welfare than a farmer who does not. Also, it is not unusual for a hatchery to add a small extra percentage onto the actual number of birds ordered, which could result in a seemingly lower mortality by the end of production than is actually the case. Even if mortality is based on the actual number of birds found dead or culled, birds can go missing (predated, eaten or buried in the litter). Despite these issues, mortality can be a useful indicator if the method of recording it is standardised.

### *4.1.2 Plumage condition*

Feather pecking is the pecking and pulling out of feathers of another bird. Birds, especially young birds, also show gentle feather pecking, but this does not usually result in feathers being removed and does not seem to be a reliable early indicator of later severe feather pecking (Newberry et al., 2007). If the intention is to identify the number of birds that develop the abnormal behaviour of severe feather pecking, then it is necessary to observe the birds directly. However, the consequences of feather pecking can be recorded by scoring the plumage condition of birds. In laying hen flocks, recording the feather cover of birds is one of the most commonly used welfare indicators.

There are several different feather scoring schemes but, in principle, they differ in whether this is a whole body assessment or whether different areas of the body are scored separately (Campe et al., 2018). Schemes such as those proposed by LayWel (2006) record several parts of the body, whereas schemes such as AssureWel (Main et al., 2012) and Welfare Quality (2009) focus on just two or three key areas. The reason for scoring different regions separately is that not all feather damage is attributable to feather pecking. For example, damage to the underside of the neck is most likely attributable to abrasion against the food trough. Damage to the plumage on the head is more likely to be an indicator of the aggressive pecks received by the bird. Most feather pecks are directed towards the back, rump, tail and abdomen of a bird (Bilcik and Keeling, 1999) (Fig. 2) and these areas can be scored separately and the score summed, so a total for just these areas can be given. The schemes also differ in whether the birds are handled when assessing plumage condition, for example, Welfare Quality, or whether they are not handled, for example, AssureWel, (Fig. 3). A reduced (non-handled) version of the LayWel scoring system was shown to be a valid and reliable scoring method, involving less stress for the birds than the original procedure (Kjaer et al., 2011). The benefits of introducing plumage condition scoring in laying hens within a farm assurance scheme, in combination with feedback to farmers, and other industry initiatives is well-documented in the paper by Mullan et al. (2016). Feather pecking is less common in broilers although it can be observed.

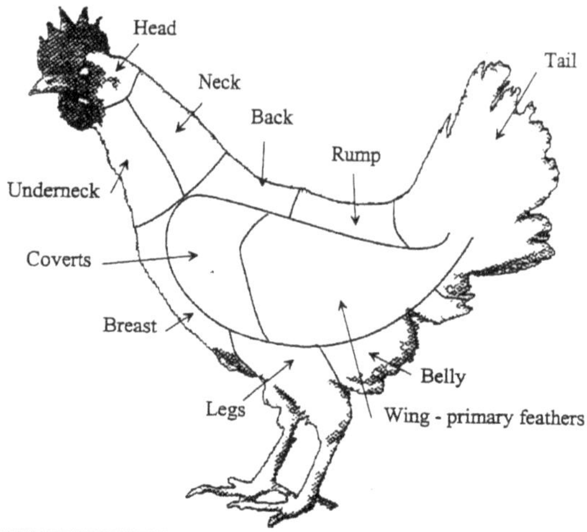


| Score | Body | Flight feathers | Skin injuries |
|---|---|---|---|
| 0 | Intact feathers | Intact feathers | No injuries or scratches |
| 1 | Some feathers scruffy, up to 3 missing feathers | Few feathers separated but non broken or missing | < 5 pecks or scratches |
| 2 | More damaged feathers, > 3 missing | A lot of feathers separated and/or a few broken or missing | 5 or more pecks/scratches or 1 wound < 1 cm diameter |
| 3 | Bald patch< 5 cm diameter or 50% of area | All feathers separated, a lot of broken or missing feathers | Wound > 1 cm in diameter, but < 2 cm |
| 4 | Bald patch > 5 cm diameter or > 50% of area | Most of the feathers missing or broken | Wound > 2 cm in diameter |
| 5 | Completely denuded area | Almost all feathers missing | ---------------------- |

Ex: < 1= less than 1 > 1= more than 1

**Figure 2** Example of a detailed scoring system of the body, flight feathers and the skin of laying hens, where each body part is scored separately. Bilcik and Keeling (1999).

There are other methods to assess plumage condition rather than visual inspection on the farm. These include using infrared thermography (Pichova et al., 2017) and scoring at slaughter, for example, Grafl et al. (2017).

### *4.1.3 Pecking damage on the comb*

Most aggressive pecks towards a bird are directed towards the head. Counting the number of peck wounds on the comb can therefore be an indicator of aggression. As for feather pecking, this is an indicator of the amount of aggression received, not given, but if taken on a representative sample of birds it can give an impression of the overall level of aggression in the flock much more quickly than behavioural observations of agonistic interactions.

**Figure 3** Person holding a laying hen to perform a clinical scoring. The naked area on the back of this bird is typical of feather pecking damage. Protocols can differ in whether it is necessary to pick up the bird or whether the assessment can be carried out while walking though the flock. Sometimes the welfare assessment is done at the slaughterhouse. Photo: Linda Keeling.

### 4.1.4 Pecking wounds and cloacal cannibalism

Cannibalism, which is the pecking and tearing of the skin of another bird, is a major welfare problem when it occurs (Nicol, 2018) (Bestman et al., 2017). Recording the number of peck wounds on the cloaca can give an early indication, although in many cases a cloacal cannibalism outbreak develops very quickly and it may be mortality and number of culled cannibalised birds that are the welfare indicators in such cases. Cannibalism wounds can also be seen sometimes on other parts of the body, for example, on the wings or base of the tail, and these are thought to be related to feather pecking, probably when one of the large feathers is removed. Cloacal cannibalism is not related to feather pecking even though the causal factors for both feather pecking and cannibalism are similar. Cloacal cannibalism is usually a problem only in laying hens, and not in broilers.

### 4.1.5 Bird cleanliness

Scoring cleanliness (or dirtiness) is a common animal-based measure for cattle and pigs and is part of the broiler Welfare Quality protocol (Welfare Quality, 2009). From an evolutionary point of view having clean feathers has been important for survival, so birds spend a considerable amount of time in preening and dust bathing. The reliability of cleanliness of the feathers as a welfare indicator has been further confirmed by Souza et al. (2018). In broilers a dirty plumage is often

**Figure 4** Broilers spend a large proportion of their time sitting and if the litter material is wet, then this can result in the birds' feathers on their breast becoming soiled. Photo: Linda Keeling.

correlated with contact dermatitis on the feet or hocks, reflecting poor litter quality (Fig. 4), whereas for layers in loose housing systems, it is more likely to be due to faeces from birds on perches dropping onto birds walking below.

#### 4.1.6 Body condition score

Body condition scoring has already been well developed in several species and poultry is no exception. It is usually determined by assessing the curvature of the breast muscle close to the keel bone (Sherwin et al., 2010). In poultry, very thin birds are the main welfare concern, since it implies either they are not getting sufficient access to the feeder or they have a health problem. Bodyweight, however, is not a reliable indicator of the welfare of an individual bird since it is strongly influenced by the size of the bird. Nevertheless, the variation in the distribution of bodyweight within a flock of birds of the same age can be a useful indicator.

### 4.2 Behavioural indicators

These can be undisturbed observations of the birds, which are presented here, or they can be observations in a specific test situation, which are presented in the next section.

### 4.2.1 Time budget

Although legislation often refers to the opportunity for birds to show natural behaviour or behaviour for which there is a behavioural need, observations of the general behaviour of birds, as, for example, in a time budget, are often difficult to interpret from a welfare point of view. The difficulty lies in whether the absence of behaviour is due to restriction in that environment or an adaptation to it. Also there are differences according to the strain of bird, especially between fast-growing and slow-growing broilers in their time budgets (Bokkers et al., 2007), but even between white and brown strains of laying hens (Ali et al., 2016). How the birds are reared, and whether they have early experience of perches or not, can also affect their time budgets. Determining a time budget is therefore not usually included in welfare assessment protocols. Nevertheless there are some behaviour patterns which have been repeatedly documented in experimental studies to be important, such as egg laying in an enclosed area (a nest box), dustbathing in an appropriate litter material and perching especially at night time (EFSA, 2005), and is the reason why provision of these resources is now required by legislation within the European Union. Such knowledge can guide assessors, for example, to observe whether birds are seen to be using perches and nest boxes, how they are moving through an aviary system or spacing themselves at the feeder. An example related to broilers is the need to rest, and so recording the extent to which they disturb one another while resting, for example Cornetto et al. (2002), could be used as a welfare indicator. Comfort behaviours in general and self-grooming in particular have, on several occasions, been identified in experimental studies as potential welfare indicators (e.g. Nicol et al., 2009), but these are rarely included in protocols. There is also increasing interest in the extent to which behavioural synchrony can be a welfare indicator, and play behaviour, but these are not yet validated reliable indicators of bird welfare (Keeling et al., 2017; Vasdal et al., 2019).

### 4.2.2 Vocalisation in poultry

Poultry have specific vocalisations that can be regarded as emotional expressions of distinct states and so can be used as welfare indicators (Kim et al., 2017; Fontana et al., 2016). Several of these vocalisations are easily recognisable as having distinct temporal and spectral patterns, even to the human ear, for example, the gakel-call (Zimmerman and Koene, 1998), which can indicate frustration, or general high-pitched (distress) vocalisations when birds experience fear or pain (Bright, 2008). One can listen for these in the barn or at the slaughterhouse, but qualitative assessment under commercial conditions is difficult because of other noises (e.g. fans and feeder chain). There is work to develop automated ways to record these vocalisations (Du et al., 2018) and this should make it more likely to include them in welfare assessment protocols in the future. There is evidence that hens vary their vocal repertoire also in positive situations (McGrath et al., 2017), which implies that this welfare assessment indicator may have potential in the future as an indicator of positive welfare.

### 4.2.3 Thermal regulation

Observing the spatial distribution of birds in a flock (e.g. huddling) to assess whether the environmental temperature is appropriate is a classic welfare indicator when rearing chicks, but it can also be used in adult birds. Panting and the body posture of lifting the wings are clear indicators of heat stress, although they can also be related to poor air quality. Fast-growing broilers can be particularly susceptible to warm temperatures. Spatial distribution may also indicate the existence of areas of wet litter or uneven provision of light, food or water.

## 4.3 Test situations in the building

### 4.3.1 Novel object test – fear

Tests for general fear have been reviewed by Forkman et al. (2007). The most commonly used fear tests for poultry are the novel object test, tonic immobility and open field tests. Of these, the novel object test is the easiest and most commonly used in assessment protocols. The novel object (e.g. a brightly coloured stick) is placed on the ground and the number of birds approaching the stick or within a certain distance in the following minutes is observed. This test can be used for broilers, layers and for loose-housed or caged birds. However, it should be highlighted that there is generally no evidence for an association between fear, for example, as measured by reluctance to approach a novel stimuli in a test situation and stress levels in birds (Nicol et al., 2011).

### 4.3.2 Human-poultry relationship

The responses of laying hens to an approaching human were first described for laying hens in cages (Barnett et al., 1992). Variations in response to human approach have also been developed for loose-housed birds, but a touch test is also sometimes used, for example, the number of birds within an arm's length (Graml et al., 2008). Even more recently, a test based on how birds move away as the person walks through the flock and how quickly they refill the area behind the person has been proposed as an indicator and this measure has also been linked to automated measures using cameras in the barn (Silvera et al., 2017). These tests have been validated by habituating birds to humans and then comparing the results of the tests between human-habituated and non-human-habituated birds.

### 4.3.3 Thirst in chickens

Although there are physiological tests to assess dehydration, there has not been a feasible animal-based measure and most welfare assessment protocols refer to the presence, cleanliness and accessibility of drinkers. However, Vanderhasselt et al. (2014) showed that a test of voluntary consumption from an unfamiliar open drinker was sensitive to water deprivation in broilers. It was not affected by location within the house suggesting that it could be a rather robust measure of thirst.

## 5 Prioritising welfare indicators

There are many welfare indicators. They continue to be improved to become more reliable and feasible, and new indicators continue to be identified and validated. It is clear that not all welfare indicators can be included in every welfare assessment protocol. Once the different dimensions of welfare are covered, it is usually necessary to prioritize between welfare indicators according to the purpose of the assessment. If the aim is to investigate a specific welfare issue, for example, reduced feather pecking or the number of broilers dead on arrival at the slaughterhouse, then the purpose will itself limit the number of indicators. However, if the aim is to obtain an overall welfare assessment of birds in a flock on that farm, the key question is how short can the list of welfare assessment indicators be and the protocol still be reliable i.e. result in the same overall welfare assessment of the farm or the same rank in relation to other farms?

One way, that does not necessarily shorten the protocol, but which makes welfare assessment less time-consuming is if the protocol does not always need to be completed in full. This two-step approach may involve an initial preliminary screening of the flock or farm, perhaps based on slaughterhouse data or mortality records, that is only followed up in the second step (using the full protocol) if there is evidence that bird welfare is poor. This approach may be appropriate if the purpose of the assessment is to identify and perhaps focus advisory attention on farms with poor bird welfare. It does not allow comparisons or ranking of farms, nor serve the purpose of checking if bird welfare is of the standard required for the farm to be included in a specific certification scheme.

A more structured approach to prioritise indicators is to try to reduce the number of measures that are indicators of the same or closely related welfare issues. Shortening the protocol can be done by classifying welfare into different domains or criteria and then having a small number of the 'best' indicators for each criterion. This was the approach taken in the Welfare Quality project (Blokhuis et al., 2013), see also Buijs et al. (2017) and Sandoe et al. (2019). Another way is to shorten the list of welfare indicators statistically by identifying which indicators co-vary or, for example, are linked in a cluster or principle component analysis. One example of this approach is by De Jong et al. (2016), who found a positive association between severe hock burns and gait scores on farms and then explored strategies to predict gait scores using hock burn scores. Agreement was found between the simplified protocols and the full protocol, although there was a loss of specificity. Associations between indicators are also important in the case of negative correlations. An example of this is that flocks with more lameness tend to score better on the human touch test (Vasdal et al., 2018). This implies that either assessment of the human–poultry relationship should focus on those tests that do not require broilers to move, or that the outcome needs to be corrected for the level of lameness in the flock, in which case both measures need to be recorded. A final example of how we can prioritise welfare indicators is to systematically examine associations between welfare indicators and subsequent choices for different environments, as a form of convergent validity. Nicol and colleagues found that responses associated with a positive choice included lower body temperature, blood glucose, heterophyl-lymphocyte ratio and response to novelty, and greater feed digestibility and self-grooming (Nicol et al., 2009). Further analysis of this dataset to investigate relationships between the many measures recorded identified two clusters, each containing relatively strong links (Nicol et al., 2011). One of these clusters implicated a relatively nervous behaviour and the other a relatively confident behaviour leading the authors to suggest that novel object responses may be a good summary indicator of welfare. They also noted that comfort behaviour, which is known to be important for welfare, appeared uncorrelated to other measures and so suggested that comfort behaviour should therefore be included in welfare assessment protocols (Nicol et al., 2011).

Another, although related approach, is that of identifying a few key welfare indicators, sometimes called iceberg indicators. The idea is that the indicator, by being multifactorial, reflects several dimensions of welfare at the same time. This is the reasoning behind the focus on footpad dermatitis in broilers, where the indicator reflects the condition of the litter in the house, which in turn is a risk factor for several other welfare problems. Better than reflecting several resource-based indicators, however, would be to identify an iceberg indicator that reflects several other animal-based measures. Feather pecking is an abnormal behaviour and is multifactorial in its causation, so is a potential candidate, but plumage scores are strongly influenced by age (Comin et al., 2019). Another potential candidate relates to fearfulness in birds (see the work of Nicol et al. (2011) above).

As the science of welfare continues to emphasise the animal's interpretation of its own situation and focus on affective states (emotion) in animals, there is increased interest in developing welfare indicators where the animal has done the integrating of its situation rather than we humans weighing the importance of different indicators against each other. The recent work to investigate the startle response (involuntary responses to sudden aversive stimuli) in birds as an indicator of their affective state (Ross et al., 2019) is one such example. The most studied area, however, is that of judgement bias. The cognitive judgement bias test is based on the fact that animals in a more negative affective state tend to interpret ambiguous cues more negatively than animals in a more positive state (Paul et al., 2005). Tests to investigate birds' responses to ambiguous cues can therefore be used as a proxy measure of their affective state (Deakin et al., 2016). It has even been proposed that the stress hormone corticosterone causes pessimistic judgement bias (Lyasere et al., 2017), which would support recording pessimism as a valid welfare indicator in chickens. A remaining issue with judgement bias tests is its poor feasibility under commercial conditions. Although there is work to develop a variation of the test that does not require training (Jones et al., 2018), it is not yet developed for poultry.

## 6 Future trends

Automated recording of welfare indicators has been mentioned several times earlier in this chapter and there is little doubt that advances in this area can be expected in the future (e.g. BenSassi et al., 2016). Poultry farms are already often highly technical, and so there is good potential for implementing automatic recording of animal-based measures, although cost and reliability is always an issue. There are clear advantages with this, as long as welfare assessment does not become restricted to only those indicators that are easy to assess automatically, to the disadvantage of continuing to record and further develop indicators related to other important aspects of bird welfare. Automatic recording of some animal-based measures, for example, footpad dermatitis is already taking place in some slaughterhouses.

More automation and better collation of data open up for the relatively new research area of animal welfare epidemiology and to improved animal welfare risk assessment methodologies. Epidemiological methods can help move towards more efficient protocols by helping to identify indicators that co-vary and by focussing our attention on specificity and sensitivity in overall welfare assessment. This was mentioned before as a way to reduce the number of animal-based measures in a protocol, but more data (potentially big data) will allow new associations between risk factors and outcome indicators to be identified.

So far in this chapter, the focus has been on indicators of poor welfare and reducing the risk of poor welfare in the future. Interest in indicators of good welfare has nevertheless increased and this trend is likely to continue, although the number of indicators of good welfare being investigated by researchers is currently considerably less than those of poor welfare. Implicit when developing and using indicators that already reflect overall welfare across several dimensions is that these reflect the combined effect of positive and negative welfare. Some examples of candidate integrated animal-based measures (e.g. judgement bias, startle response and mapping indicators to choices) were mentioned earlier, and work is likely to continue in these and other areas, especially linking them to epigenetic changes (Pertille et al., 2017).

There is also a great need for welfare indicators that can measure impact over the long term as well as the short term. So the final trend to be discussed in this chapter is that of quality of life. The term has been around for a long time, but if used for animals at all, it tended to be used for pet animals in a veterinary clinic setting. It became more common in the agricultural sector following the publication of the UK Farm Animal Welfare Council's report where terms such as 'a life worth living' were used (FAWC, 2009). Animal-based measures usually reflect the bird's welfare at a specific moment in time. Some may reflect longer periods of time (days and weeks), but there are few that can be considered to reflect the whole lifetime - which is what is implied by the term 'quality of life'. Thus some integration over time may be required, in addition to the already complex issue of integrating across the different dimensions of welfare mentioned earlier. One approach is the Extended Welfare Assessment Grid proposed by Wolfensohn et al. (2015). Scores for the four parameters, physical, psychological, environmental and procedural, are marked on the arms of the two-dimensional grid of these parameters and joined to form a polygon. The shape and area of this polygon can then be compared with assessments made at different points in time. This gives a good visual representation of welfare over time, but there is still the issue of how best to integrate the factors that contribute to the different parameters. An alternative approach is to consider the biological age of the animal (Bateson and Poirier, 2019). Biological age is a measure of the cumulative experiences of the animal and has the advantage that it does not require any subjective decisions about how different dimensions of welfare should be weighted in an overall assessment. However, determining biological age is not without problems and the two biomarkers of biological age the authors consider as most promising are the measures of telomere length (Bateson, 2016) and hippocampal volume (Poirier et al., 2019). While they are unlikely to be feasible welfare assessment indicators for poultry in commercial practice, that does not distract from their potential to contribute to our understanding of the cumulative experiences of an animal and so, perhaps, towards assessing the overall quality of its life.

## 7 Conclusion

There are several well-validated and reliable indicators of poultry welfare that can be used in welfare assessment protocols. These can be used to identify welfare problems, sometimes at early stages so that corrective action can be taken. The data collected could be used even more efficiently, for example, in epidemiological studies and for benchmark trends, if the indicators were collected in a standardised way. New welfare assessment indicators are being developed, for example, startle response and judgement bias tests, that seem to reflect the animal's interpretation of its situation. Established indicators, such as responses to novel objects and comfort behaviour, seem to become more appreciated, and some neurological measures, such as telomere length and hippocampal volume as indicators of biological age, may have potential in the future. These more general indicators may be used as a complement to specific welfare assessment indicators or they may contribute to a better understanding of how animals experience their lives. Other areas where improvements can be expected in the near future are in the automated collection of welfare indicators and in validating indicators of positive welfare.

## 8 Where to look for further information

- AssureWel project website http://www.assurewel.org.
- Blokhuis, H. J., Miele, M., Veissier, I. and Jones, B. (Eds). 2013. Improving farm animal welfare: science and society working together: the Welfare Quality approach. Wageningen Academic Publishers.
- Mench J. (Ed.). 2018. *Advances in Poultry Welfare*. Elsevier.
- Sandilands, V. and Hocking, P. 2012. Alternative systems for poultry. Poultry Science Symposium Series, Vol 30. (especially Chapter 12 (Laying hens) and Chapter 14 (Broilers)).
- Weeks, C. A. and Butterworth, A. 2004. *Measuring and Auditing Broiler Welfare*. CABI Publishing, Wallingford, UK.
- Welfare Quality Network site: http://www.welfarequality.net.

## 9 References

Ali, A. B. A., Campbell, D. L. M., Karcher, D. M. and Siegford, J. M. 2016. Influence of genetic strain and access to litter on spatial distribution of 4 strains of laying hens in an aviary system. *Poultry Science* 95(11), 2489-2502 [doi:10.3382/ps/pew236].

Ali, A. B. A., Campbell, D. L. M., Karcher, D. M. and Siegford, J. M. 2019. Nighttime roosting substrate type and height among 4 strains of laying hens in an aviary system. *Poultry Science* 98(5), 1935-1946 [doi:10.3382/ps/pey574].

Anneberg, I., Vaarst, M. and Sandoe, P. 2013. To inspect, to motivate - or to do both? A dilemma for on-farm inspection of animal welfare. *Animal Welfare* 22(2), 185-194 [doi:10.7120/09627286.22.2.185].

Anonymous 2018. *The AssureWel Manual: The AssureWel Approach to Improving Farm Animal Welfare: the Development and Use of Welfare Outcome Assessment in Farm Assurance*. Available at: http://www.assurewel.org/aboutassurewel/assurewelmanual.html; accessed April 2020.

Barnett, J. L., Hemsworth, P. H. and Newman, E. A. 1992. Fear of humans and its relationships WITH productivity in laying hens at commercial farms. *British Poultry Science* 33(4), 699–710 [doi:10.1080/00071669208417510].

Bateson, M. 2016. Cumulative stress in research animals: telomere attrition as a biomarker in a welfare context? *BioEssays* 38(2), 201-212 [doi:10.1002/bies.201500127].

Bateson, M. and Poirier, C. 2019. Can biomarkers of biological age be used to assess cumulative lifetime experience? *Animal Welfare* 28(1), 41-56 [doi:10.7120/09627286.28.1.041].

BenSassi, N., Averos, X. and Estevez, I. 2016. Technology and poultry welfare. *Animals* 6(10) [doi:10.3390/ani6100062].

BenSassi , N., Averos, X. and Estevez, I. 2019. Broiler chickens on-farm welfare assessment: estimating the robustness of the transect sampling method. *Frontiers in Veterinary Science* 6, 236 [doi:10.3389/fvets.2019.00236].

Bestman, M., Verwer, C., Brenninkmeyer, C., Willett, A., Hinrichsen, L. K., Smajlhodzic, F., Heerkens, J. L. T., Gunnarsson, S. and Ferrante, V. 2017. Feather-pecking and injurious pecking in organic laying hens in 107 flocks from eight European countries. *Animal Welfare* 26(3), 355-363 [doi:10.7120/09627286.26.3.355].

Bilcik, B. and Keeling, L. J. 1999. Changes in feather condition in relation to feather pecking and aggressive behaviour in laying hens. *British Poultry Science* 40(4), 444-451 [doi:10.1080/00071669987188].

Blokhuis, H., Jones, B., Veissier, I. and Miele, M. 2013. *Improving Farm Animal Welfare Science and Society Working Together: the Welfare Quality Approach Introduction*. Springer Academic Publishers, Wageningen.

Blokhuis, H. J., Van Niekerk, T. F., Bessei, W., Elson, A., Guémené, D., Kjaer, J. B., Maria Levrino, G. A., Nicol, C. J., Tauson, R., Weeks, C. A. and Van De Weerd, H. A. 2007. The LayWel project: welfare implications of changes in production systems for laying hens. *World's Poultry Science Journal* 63(1), 101–114 [doi:10.1017/S0043933907001328].

Bokkers, E. A. M., Zimmerman, P. H., Rodenburg, T. B. and Koene, P. 2007. Walking behaviour of heavy and light broilers in an operant runway test with varying durations of feed deprivation and feed access. *Applied Animal Behaviour Science* 108(1-2), 129–142 [doi:10.1016/j.applanim.2006.10.011].

Botreau, R., Veissier, I., Butterworth, A., Bracke, M. B. M. and Keeling, L. J. 2007. Definition of criteria for overall assessment of animal welfare. *Animal Welfare* 16, 225–228.

Bright, A. 2008. Vocalisations and acoustic parameters of flock noise from feather pecking and non-feather pecking laying flocks. *British Poultry Science* 49(3), 241–249 [doi:10.1080/00071660802094172].

Buijs, S., Ampe, B. and Tuyttens, F. A. M. 2017. Sensitivity of the Welfare Quality® broiler chicken protocol to differences between intensively reared indoor flocks: which factors explain overall classification? *Animal* 11(2), 244–253 [doi:10.1017/S1751731116001476].

Campbell, D. L. M., Dickson, E. J. and Lee, C. 2019. Application of open field, tonic immobility, and attention bias tests to hens with different ranging patterns. *PeerJ* 7, e8122 [doi:10.7717/peerj.8122].

Campe, A., Hoes, C., Koesters, S., Froemke, C., Bougeard, S., Staack, M., Bessei, W., Manton, A., Scholz, B., Schrader, L., Thobe, P. and Knierim, U. 2018. Analysis of the influences on plumage condition in laying hens: how suitable is a whole body plumage score as an outcome? *Poultry Science* 97(2), 358–367 [doi:10.3382/ps/pex321].

Comin, A., Jeremiasson, A., Kratzer, G. and Keeling, L. 2019. Revealing the structure of the associations between housing system, facilities, management and welfare of commercial laying hens using Additive Bayesian Networks. *Preventive Veterinary Medicine* 164, 23–32 [doi:10.1016/j.prevetmed.2019.01.004].

Cornetto, T., Estevez, I. and Douglass, L. W. 2002. Using artificial cover to reduce aggression and disturbances in domestic fowl. *Applied Animal Behaviour Science* 75(4), 325–336 [doi:10.1016/S0168-1591(01)00195-2].

de Jong, I., Berg, C., Butterworth, A. and Estevéz, I. 2012. Scientific report updating the EFSA opinions on the welfare of broilers and broiler breeders. *Supporting Publications* 2012:EN-295 [116 pp]. Available at:www.efsa.europa.eu/publications.

De Jong, I. C., Hindle, V. A., Butterworth, A., Engel, B., Ferrari, P., Gunnink, H., Moya, T. P., Tuyttens, F. A. M. and Van Reenen, C. G. 2016. Simplifying the Welfare Quality® assessment protocol for broiler chicken welfare. *Animal* 10, 117–127.

Deakin, A., Browne, W. J., Hodge, J. J. L., Paul, E. S. and Mendl, M. 2016. A screen-peck task for investigating cognitive bias in laying hens. *PLoS ONE* 11(7), e0158222 [doi:10.1371/journal.pone.0158222].

Du, X. D., Lao, F. D. and Teng, G. H. 2018. A sound source localisation analytical method for monitoring the abnormal night vocalisations of poultry *Sensors* 18, 2906.

EC 1999. Council Directive 1999/74/EC of 19th July 1999 laying down minimum standards for the protection of laying hens.

EC 2007. Council Directive 2007/43/EC of 28 June 2007 laying down minimum rules for the protection of chickens kept for meat production.

EC 2008. Commission Regulation (EC) No 589/2008 of 23 June 2008 laying down detailed rules for implementing Council Regulation (EC) No 1234/2007 as regards marketing standards for eggs.

EFSA 2005. The welfare aspects of various systems of keeping laying hens. *The EFSA Journal* 197, 1–23.

EFSA 2012. Statement on the use of animal-based measures to assess the welfare of animals. *EFSA Journal* 10, 2767.

EFSA 2015. EFSA's internal project on the use of animal-based measures to assess animal welfare in EU - state of art of the 10 years of activities and analysis of the gaps. *EFSA Supporting Publication* 2015:EN-884 [21 pp].

FAWC 2009. *Report on Farm Animal Welfare in Great Britain: Past, Present and Future*. Farm Animal Welfare Council, London.

FAWC 2012. Five Freedoms. Available at: https://webarchive.nationalarchives.gov.uk/20121010012427/http://www.fawc.org.uk/freedoms.htm; accessed August 2020.

Fontana, I., Tullo, E., Scrase, A. and Butterworth, A. 2016. Vocalisation sound pattern identification in young broiler chickens. *Animal* 10(9), 1567–1574 [doi:10.1017/S1751731115001408].

Forkman, B., Boissy, A., Meunier-Salauen, M. C., Canali, E. and Jones, R. B. 2007. A critical review of fear tests used on cattle, pigs, sheep, poultry and horses. *Physiology and Behavior* 92(3), 340–374 [doi:10.1016/j.physbeh.2007.03.016].

Forkman, B. and Keeling, L. J. 2009. Assessment of Animal Welfare measures for Layers and Broilers. *Welfare Quality Reports* (9), 176.

Grafl, B., Polster, S., Sulejmanovic, T., Purrer, B., Guggenberger, B. and Hess, M. 2017. Assessment of health and welfare of Austrian laying hens at slaughter demonstrates influence of husbandry system and season. *British Poultry Science* 58(3), 209–215 [doi:10.1080/00071668.2017.1280723].

Graml, C., Niebuhr, K. and Waiblinger, S. 2008. Reaction of laying hens to humans in the home or a novel environment. *Applied Animal Behaviour Science* 113(1–3), 98–109 [doi:10.1016/j.applanim.2007.10.004].

Hothersall, B., Caplen, G., Parker, R. M. A., Nicol, C. J., Waterman-Pearson, A. E., Weeks, C. A. and Murrell, J. C. 2016. Effects of carprofen, meloxicam and butorphanol on broiler chickens' performance in mobility tests. *Animal Welfare* 25(1), 55–67 [doi:10.7120/09627286.25.1.055].

Hughes, B. O. 1988. Current knowledge of bird welfare requirements. *World's Poultry Science Journal* 44(1), 62–63 [doi:10.1079/WPS19880006].

Iannetti, L., Neri, D., Santarelli, G. A., Cotturone, G., Vulpiani, M. P., Salini, R., Antoci, S., Di Serafino, G., Di Giannatale, E., Pomilio, F. and Messori, S. 2020. Animal welfare and microbiological safety of poultry meat: impact of different at-farm animal welfare levels on at-slaughterhouse Campylobacter and Salmonella contamination. *Food Control* 109, 106921 [doi:10.1016/j.foodcont.2019.106921].

Jones, S., Neville, V., Higgs, L., Paul, E. S., Dayan, P., Robinson, E. S. J. and Mendl, M. 2018. Assessing animal affect: an automated and self-initiated judgement bias task based on natural investigative behaviour. *Scientific Reports* 8(1), 12400 [doi:10.1038/s41598-018-30571-x].

Keeling, L., Evans, A., Forkman, B. and Kjaernes, U. 2013. 'Welfare Quality Principles and Criteria' in H. Blokhuis, M. Miele, I. Veissier and B. Jones (Eds), Improving Farm Animal Welfare: Science and Society Working Together: *The Welfare Quality Approach*, 91–114.

Keeling, L. J. 2009. An analysis of animal-based versus resource-based comments in official animal welfare inspection reports from organic and conventional farms in Sweden. *Animal Welfare* 18, 391–397.

Keeling, L. J., Newberry, R. C. and Estevez, I. 2017. Flock size during rearing affects pullet behavioural synchrony and spatial clustering. *Applied Animal Behaviour Science* 194, 36–41 [doi:10.1016/j.applanim.2017.04.002].

Kim, N. Y., Jang, S. Y., Kim, S. J., Jeon, B. T., Oh, M. R., Kim, E. K., Seong, H. J., Tang, Y. J., Yun, Y. S. and Moon, S. H. 2017. Behavioral and vocal characteristics of laying hens under different housing and feeding conditions. *Journal of Animal and Plant Sciences* 27, 65–74.

Kjaer, J. B., Glawatz, H., Scholz, B., Rettenbacher, S. and Tauson, R. 2011. Reducing stress during welfare inspection: validation of a non-intrusive version of the LayWel plumage scoring system for laying hens. *British Poultry Science* 52(2), 149–154 [doi:10.1080/00071668.2011.554799].

LayWel 2006. LayWel research project description and use of a new scoring system. Available at: https://www.laywel.eu/; accessed April 2020.

Lyasere, O. S., Beard, A. P., Guy, J. H. and Bateson, M. 2017. Elevated levels of the stress hormone, corticosterone, cause 'pessimistic' judgment bias in broiler chickens. *Scientific Reports* 7(1), 6860 [doi:10.1038/s41598-017-07040-y].

Main, D. C. J., Mullan, S., Atkinson, C., Bond, A., Cooper, M., Fraser, A. and Browne, W. J. 2012. Welfare outcomes assessment in laying hen farm assurance schemes. *Animal Welfare* 21(3), 389–396 [doi:10.7120/09627286.21.3.389].

Main, D. C. J., Mullan, S., Atkinson, C., Cooper, M., Wrathall, J. H. M. and Blokhuis, H. J. 2014. Best practice framework for animal welfare certification schemes. *Trends in Food Science and Technology* 37(2), 127–136 [doi:10.1016/j.tifs.2014.03.009].

McGrath, N., Dunlop, R., Dwyer, C., Burman, O. and Phillips, C. J. C. 2017. Hens vary their vocal repertoire and structure when anticipating different types of reward. *Animal Behaviour* 130, 79–96 [doi:10.1016/j.anbehav.2017.05.025].

Mellor, D. J. 2017. Operational details of the five domains model and its key applications to the assessment and management of animal welfare. *Animals* 7(8) [doi:10.3390/ani7080060].

Mollenhorst, H., Rodenburg, T. B., Bokkers, E. A. M., Koene, P. and De Boer, I. J. M. 2005. On-farm assessment of laying hen welfare: a comparison of one environment-based and two animal-based methods. *Applied Animal Behaviour Science* 90(3–4), 277–291 [doi:10.1016/j.applanim.2004.08.011].

Mullan, S., Szmaragd, C., Cooper, M. D., Wrathall, J. H. M., Jamieson, J., Bond, A., Atkinson, C. and Main, D. C. J. 2016. Animal welfare initiatives improve feather cover of cage-free laying hens in the UK. *Animal Welfare* 25(2), 243–253 [doi:10.7120/09627286.25.2.243].

Newberry, R. C., Keeling, L. J., Estevez, I. and Bilcik, B. 2007. Behaviour when young as a predictor of severe feather pecking in adult laying hens: the redirected foraging hypothesis revisited. *Applied Animal Behaviour Science* 107(3–4), 262–274 [doi:10.1016/j.applanim.2006.10.010].

Nicol, C. 2018. Feather pecking and cannibalism: can we really stop beak trimming? In Mench, J. A. (Ed.) *Advances in Poultry Welfare*. Woodhead Publishing, Duxford, UK, pp. 175–197.

Nicol, C. J., Caplen, G., Edgar, J. and Browne, W. J. 2009. Associations between welfare indicators and environmental choice in laying hens. *Animal Behaviour* 78(2), 413–424 [doi:10.1016/j.anbehav.2009.05.016].

Nicol, C. J., Caplen, G., Edgar, J., Richards, G. and Browne, W. J. 2011. Relationships between multiple welfare indicators measured in individual chickens across different time periods and environments. *Animal Welfare* 20, 133–143.

Paul, E. S., Harding, E. J. and Mendl, M. 2005. Measuring emotional processes in animals: the utility of a cognitive approach. *Neuroscience and Biobehavioral Reviews* 29(3), 469–491 [doi:10.1016/j.neubiorev.2005.01.002].

Pertille, F., Brantsaeter, M., Nordgreen, J., Coutinho, L. L., Janczak, A. M., Jensen, P. and Guerrero-Bosagna, C. 2017. DNA methylation profiles in red blood cells of adult hens correlate with their rearing conditions. *Journal of Experimental Biology* 220(19), 3579–3587 [doi:10.1242/jeb.157891].

Pichova, K., Bilcik, B. and Kost'al, L. 2017. Assessment of the effect of housing on feather damage in laying hens using IR thermography. *Animal* 11(4), 661–669 [doi:10.1017/S1751731116001981].

Poirier, C., Bateson, M., Gualtieri, F., Armstrong, E. A., Laws, G. C., Boswell, T. and Smulders, T. V. 2019. Validation of hippocampal biomarkers of cumulative affective experience. *Neuroscience and Biobehavioral Reviews* 101, 113–121 [doi:10.1016/j.neubiorev.2019.03.024].

Riber, A. B., Casey-Trott, T. M. and Herskin, M. S. 2018. The influence of keel bone damage on welfare of laying hens. *Frontiers in Veterinary Science* 5, 6 [doi:10.3389/fvets.2018.00006].

Rodenburg, T. B., Reu, K. D. and Tuyttens, F. A. M. 2012. Performance, welfare, health and hygiene of laying hens in non-cage systems in comparison with cage systems. In Sandilands, V. and Hocking, P. (Eds), *Alternative Systems for Poultry: Health, Welfare and Productivity*. CABI, Wallingford, UK, pp. 210–224.

Rodenburg, T. B., Tuyttens, F. A. M., De Reu, K., Herman, L., Zoons, J. and Sonck, B. 2008. Welfare assessment of laying hens in furnished cages and non-cage systems: assimilating expert opinion. *Animal Welfare* 17, 355–361.

Roe, E., Buller, H. and Bull, J. 2011. The performance of farm animal assessment. *Animal Welfare* 20, 69–78.

Ross, M., Garland, A., Harlander-Matauschek, A., Kitchenham, L. and Mason, G. 2019. Welfare-improving enrichments greatly reduce hens' startle responses, despite little change in judgment bias. *Scientific Reports* 9(1), 11881 [doi:10.1038/s41598-019-48351-6].

Sandoe, P., Corr, S., Lund, T. B. and Forkman, B. 2019. Aggregating animal welfare indicators: can it be done in a transparent and ethically robust way? *Animal Welfare* 28(1), 67–76 [doi:10.7120/09627286.28.1.067].

Sherwin, C. M., Richards, G. J. and Nicol, C. J. 2010. Comparison of the welfare of layer hens in 4 housing systems in the UK. *British Poultry Science* 51(4), 488–499 [doi:10.1080/00071668.2010.502518].

Silvera, A. M., Knowles, T. G., Butterworth, A., Berckmans, D., Vranken, E. and Blokhuis, H. J. 2017. Lameness assessment with automatic monitoring of activity in commercial broiler flocks. *Poultry Science* 96(7), 2013–2017 [doi:10.3382/ps/pex023].

Skinner-Noble, D. O. and Teeter, R. G. 2009. An examination of anatomic, physiologic, and metabolic factors associated with well-being of broilers differing in field gait score. *Poultry Science* 88(1), 2–9 [doi:10.3382/ps.2006-00450].

Souza, A. P. O., Soriano, V. S., Schnaider, M. A., Rucinque, D. S. and Molento, C. F. M. 2018. Development and refinement o f three animal-based broiler chicken welfare indicators. *Animal Welfare* 27(3), 263–274 [doi:10.7120/09627286.27.3.263].

Tauson, R. 1985. Mortality in laying hens caused by differences in cage design. *Acta Agriculturae Scandinavica* 35(2), 165–174 [doi:10.1080/00015128509435772].

Van Dijk, L., Buller, H. J., Blokhuis, H. J., Van Niekerk, T., Voslarova, E., Manteca, X., Weeks, C. A. and Main, D. C. J. 2019. Hennovation: learnings from promoting practice-led multi-actor innovation networks to address complex animal welfare challenges within the laying hen industry. *Animals* 9(1) [doi:10.3390/ani9010024].

Van Dijk, L., Elwes, S., Main, D. C. J., Mullan, S. M. and Jamieson, J. 2018. Farmer perspectives on welfare outcome assessment: learnings from four farm assurance scheme consultation exercises. *Animal Welfare* 27(1), 1–11 [doi:10.7120/09627286.27.1.001].

Vanderhasselt, R. F., Goethals, K., Buijs, S., Federici, J. F., Sans, E. C. O., Molento, C. F. M., Duchateau, L. and Tuyttens, F. A. M. 2014. Performance of an animal-based test of thirst in commercial broiler chicken farms. *Poultry Science* 93(6), 1327–1336 [doi:10.3382/ps.2013-03720].

Vasdal, G., Moe, R. O., De Jong, I. C. and Granquist, E. G. 2018. The relationship between measures of fear of humans and lameness in broiler chicken flocks. *Animal* 12(2), 334–339 [doi:10.1017/S1751731117001434].

Vasdal, G., Vas, J., Newberry, R. C. and Moe, R. O. 2019. Effects of environmental enrichment on activity and lameness in commercial broiler production. *Journal of Applied Animal Welfare Science* 22(2), 197-205 [doi:10.1080/10888705.2018.1456339].

Veissier, I., Jensen, K. K., Botreau, R. and Sandoe, P. 2011. Highlighting ethical decisions underlying the scoring of animal welfare in the Welfare Quality (R) scheme. *Animal Welfare* 20, 89-101.

Weeks, C. A. and Nicol, C. J. 2006. Behavioural needs, priorities and preferences of laying hens. *World's Poultry Science Journal* 62(2), 296-307 [doi:10.1079/WPS200598].

Welfare Quality® 2009. *Welfare Quality® Assessment Protocol for Poultry (Broilers, Laying Hens)*. Welfare Quality® consortium, Lelystad, Netherlands.

Winckler, C. 2019. Assessing animal welfare at the farm level: do we care sufficiently about the individual? *Animal Welfare* 28(1), 77-82 [doi:10.7120/09627286.28.1.077].

Wolfensohn, S., Sharpe, S., Hall, I., Lawrence, S., Kitchen, S. and Dennis, M. 2015. Refinement of welfare through development of a quantitative system for assessment of lifetime experience. *Animal Welfare* 24(2), 139-149 [doi:10.7120/09627286.24.2.139].

World Organisation for Animal Health (OIE) 2018. What is animal welfare? Available at: https://www.oie.int/en/animal-welfare/animal-welfare-at-a-glance/; accessed April 2020.

World Organisation for Animal Health (OIE) 2019. Terrestrial Animal Health Code, Chapter 7.10. Animal welfare and broiler chicken production systems. Available at: https://www.oie.int/standard-setting/terrestrial-code/access-online/; accessed April 2020.

Zaludik, K., Lugmair, A., Baumune, R., Troxer, J. and Niebuhir, K. 2007. Results of the Animal Needs Index (ANI-35L) compared to animal-based parameters in free-range and organic laying hen flocks in Austria. *Animal Welfare* 16, 217-219.

Zimmerman, P. H. and Koene, P. 1998. The effect of frustrative nonreward on vocalisations and behaviour in the laying hen, Gallus gallus domesticus. *Behavioural Processes* 44(1), 73-79 [doi:10.1016/s0376-6357(98)00035-7].

# Part 2

## Welfare issues in breeding, management and housing

# Chapter 9

# Welfare issues affecting broiler breeders

*Anja Brinch Riber, Aarhus University, Denmark*

1 Introduction
2 Welfare issues in broiler breeders
3 Conclusion and future trends
4 Acknowledgements
5 Where to look for further information
6 References

## 1 Introduction

The demand for broiler meat has been growing for decades, and broiler meat represents the major animal protein source in many countries around the world (OECD/FAO, 2018). One key factor for this popularity is the low cost of broiler meat compared to other meat types. To meet the demand for cheap and increasing amounts of broiler meat, the broiler industry has consistently selected for fast growth, a body composition with a high yield of breast meat and increased feed efficiency.

The structure of the broiler industry is divided into a primary breeding sector and a production sector (Fig. 1; Pollock, 1999). In the primary breeding sector, pedigree stock (i.e. pure lines, PL) produces progeny that goes on to the great grandparent (GGP) generation, followed by the grandparent (GP) generation. The grandparents produce the parent stock (PS) which passes to the production sector. Parent stock males and females are different breeds. The progeny of the parent stock is the final product, that is, the broiler chicken produced for meat. Worldwide, the parent stock consisted of around 410 million birds (Elfick, 2010) in 2009 with Europe being responsible for an estimated 60 million (Hiemstra and Napel, 2013). The global market for the parent stock in 2018 was estimated to have increased to 530–539 million birds (Elfick, pers. comm.). This chapter regards the parent stock, also referred to as the broiler breeders.

Around the world, the far majority of broiler breeders are of the fast growing genotypes such as Ross 308 and Cobb500. The growth rate of the offspring of these breeders is presently around 63 g per day, averaged over the lifespan of birds slaughtered at a target weight of 2.2 kg (Aviagen, 2019a; Cobb, 2018).

http://dx.doi.org/10.19103/AS.2020.0078.18

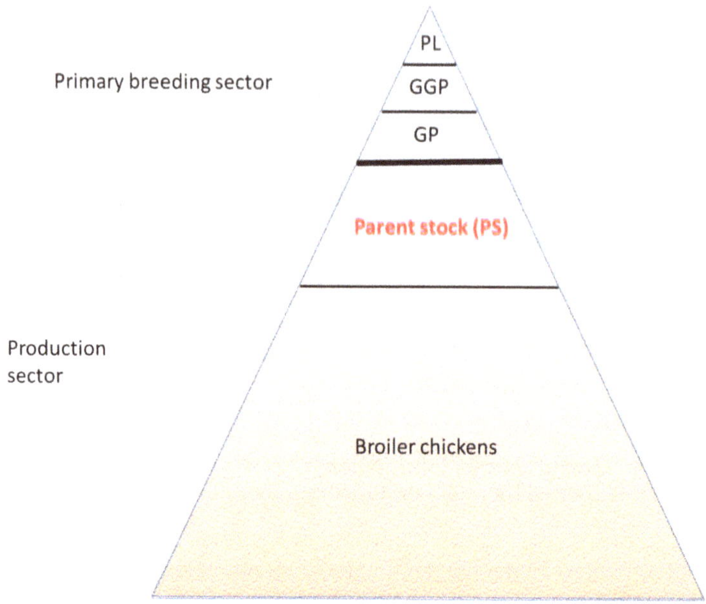


**Figure 1** The structure of the broiler industry, showing the pure lines (PL), great grandparents (GGP) and grandparents (GP) belonging to the primary breeding sector, and the parent stock (PS) and the end product, the broiler chickens, belonging to the production sector. ©Anja Brinch Riber, Aarhus University, Denmark.

Due to the continued selection for fast growth, the growth rate is expected to keep increasing. Different genotypes, usually with a slower growth rate, are normally used for the middle segment or organic production, but presently these genotypes constitute a small fraction of the worldwide production of broiler meat. Of the approximately 60 million broiler breeders in Europe, an estimated 4.5 million are alternative genotypes (Horne, 2018).

### 1.1 Housing conditions of broiler breeders

Conventional broiler breeders are reared in single-sex flocks. In Europe, broiler breeders are mainly housed in fully littered floor-house systems, but rearing in cages is practised in other parts of the world. Depending on the climate and part of the world, the barn may either be a closed-house system or an open-sided system. Most closed-house systems are lit with artificial light at a light intensity down to approximately 5 lux (EFSA, 2010). The environment is typically barren, for example, there are often no perches or raised slats available (Fig. 2a).

At around 18-23 weeks of age, the birds are moved to the production farm where the two sexes are mixed with a ratio of around one male per 10-14 hens in partly littered floor-house systems (Aviagen, 2019b). Exceptions exist

**Figure 2** Examples of (a) a rearing farm using a fully littered floor-house system and (b) a production farm using a partly littered floor-house system. In the latter, the feeders for the males are seen to the left in the photo. The three feed lines covered with grids are accessible by the females. A water line with yellow water nipples are visible to the right in the photo and behind those, the nest boxes can be seen. ©Rick van Emous, Wageningen Livestock Research, The Netherlands.

with the sexes being housed separately in cages and artificial insemination being practised. From day-old, the surplus males are typically reared for meat in broiler production systems. Like during the rearing period, the barn may be either a closed-house system or an open-sided system. A light intensity of 30–60 lux is recommended (Aviagen, 2019b). When housed in floor-house systems, both drinking lines and nest boxes are typically placed on raised slats (Fig. 2b). The production period usually lasts until 60–65 weeks of age. To increase the fertility, a procedure called spiking is often practised where older males (>40 weeks of age) are replaced by younger males (Ordas et al., 2015).

### 1.2 Growth potential and feed restriction of broiler breeders

Genetic selection for growth parameters in broiler breeders has caused an increased appetite due to modulation of mechanisms of hunger regulation (Denbow, 1989; Siegel and Wolford, 2003). As a result, *ad libitum* access to feed will result in obesity and consequently in serious health and reproduction problems during the breeding period (Heck et al., 2004; Renema and Robinson, 2004). This negative correlation between the fast growth and reproductive success has been identified as the broiler breeder paradox (Decuypere et al., 2010).

To prevent health and reproductive problems, broiler breeders are feed restricted, except for the first 7–10 days of life. The level of feed restriction depends on age and sex of the birds. In terms of the amount of feed ingested, feed restriction is at its most severe level around age 10–16 weeks (Arrazola, 2018) where restrictively fed female broiler breeders are allocated down to four times less than *ad libitum* fed individuals will eat (de Jong et al., 2002; Savory et al., 1996). Of importance is also the level of feed restriction in terms of nutrient intake. It has been suggested that the feed restriction in terms of nutrient intake is most severe around age 5–7 weeks where female broiler breeders are allocated around 20% of the *ad libitum* nutrient intake (van Emous, unpublished data).

During the production period, feed restriction continues, but at a less restricted level for the female broiler breeders, with an allowance of 40–80% of the *ad libitum* intake depending on the age of the birds (Arrazola, 2018; Bruggeman et al., 1999). Males are typically fed less restrictively during rearing but more during the production period compared to females (Renema et al., 2007; EFSA, 2010). When the sexes are housed together during the production period, this is solved by using two kinds of feed troughs – feed troughs for males are hanged at a height where only the males can reach them, while the feed troughs for females are constructed in a way that makes the males unable to reach the feed because of their wider heads (Laughlin, 2009).

The alternative broiler breeder production differs markedly from the conventional, as other genotypes with lower growth potentials are used (see example shown in Fig. 3). Therefore, the level of feed restriction is not as severe and can even be omitted for some genotypes. One alternative type of broiler breeders is the dwarf lines where the females are affected by dwarfism (Decuypere et al., 2010). Dwarf broiler breeder females have a reduced growth, resulting in limited need or, for some lines, no need for feed restriction (Decuypere et al., 2010). As the dwarf gene is sex linked, the males will still

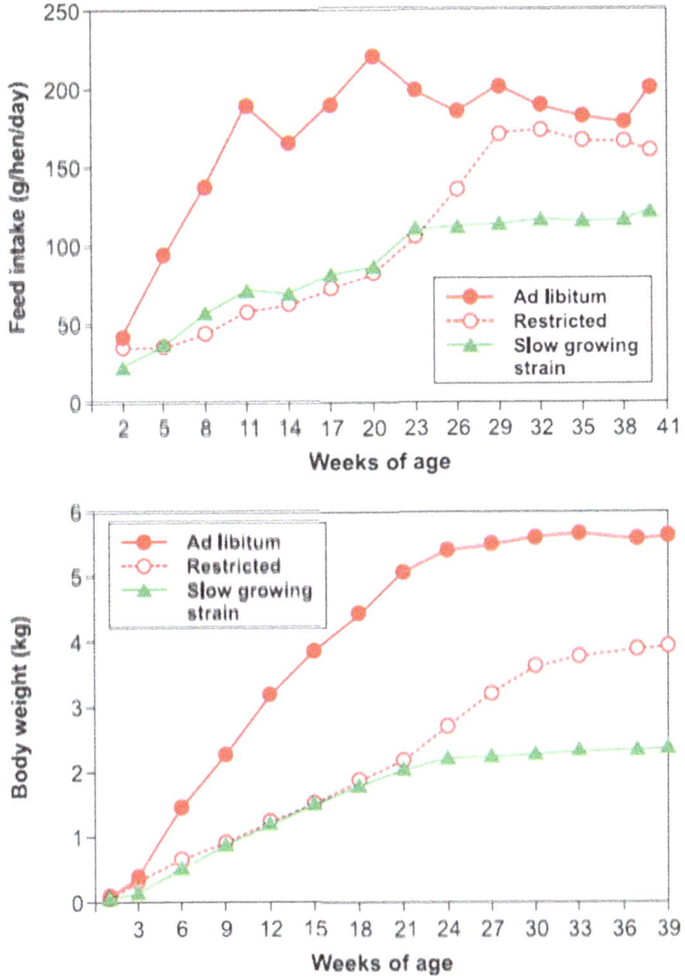


**Figure 3** An example of the mean daily feed intake per broiler breeder female (top) and body weight (bottom) in a fast-growing genotype fed *ad libitum* (continuous red line) or restrictively (dashed red line) and in a slow-growing genotype (green line). ©William (2012).

need to be fed restrictively, but as the number of breeder males compared to breeder females is considerably lower, feed restriction will only be applied to a minor fraction of the broiler breeders.

In terms of welfare, the parent stock of fast-growing genotypes is facing significantly more welfare challenges due to the necessity of keeping them on a restricted diet to avoid obesity and the associated health deterioration and infertility. In addition, knowledge on welfare issues affecting organic broiler breeders is very sparse. The remaining part of this chapter will therefore focus on broiler breeders of the fast-growing genotypes.

## 2 Welfare issues in broiler breeders

The severe feed restriction introduces another paradox in broiler breeder production: the negative correlation between keeping the birds healthy and reproductive by the use of feed restriction and the welfare problems introduced by feed restricting the birds. Basic behavioural and physiological needs are commonly not fulfilled, and, as a result, the broiler breeders develop abnormal behaviour indicative of frustration and hunger (see review in D'Eath et al., 2009). Physiological stress responses in feed-restricted broiler breeders are also frequently found (D'Eath et al., 2009). The practise of limiting access to water and the barren housing environment, where elevated resting places and foraging material are scarce or absent resources, further contributes to the lack of satisfying behavioural and physiological needs.

Another major welfare issue in broiler breeders is aggression, particularly sexual aggression. Studies have reported on the loss of courtship behaviour, resulting in males chasing females, forced matings and increased levels of fear in females (Millman et al., 2000; Jones et al., 2001; de Jong et al., 2009). Linked to this is the mutilation of broiler breeders as a preventive measure to reduce the skin damage inflicted to flock mates. The beak of both sexes is typically trimmed, and, for the males, also the spurs and/or the outermost part of the toes facing backwards are commonly trimmed (EFSA, 2010). The mutilation of the males is to the benefit of the hens, but the procedures have negative consequences for the welfare of the males, introducing yet another broiler breeder paradox.

In contrast to the bulk of knowledge on welfare issues affecting laying hens and broilers, research conducted on the welfare challenges in broiler breeders is rather limited. There are probably multiple reasons for this, including limited accessibility to broiler breeder farms, limited awareness among NGOs, authorities and consumers of the welfare issues affecting broiler breeders and low priority when it comes to funding broiler breeder studies due to the relatively low number of individuals (as compared to laying hens or broilers). Anyway, the above-mentioned welfare issues in the broiler breeder industry are described in further detail in this section. Furthermore, methods of reducing the extent of the problems are discussed.

### *2.1 Unfulfilled behavioural and physiological needs*

Physiological needs include the need for food, water and thermal comfort. The term 'behavioural needs' has been debated more, but, generally, a behavioural need is the need to perform a specific behaviour pattern regardless of the quality of the environment or whether the physiological needs, which the behaviour serves, are already fulfilled (Jensen and Toates, 1993). Thus, the animal's motivation to perform the behaviour plays a key role in whether the behavioural needs are perceived as satisfied. Animals that are highly motivated to perform a specific behaviour pattern may experience frustration and suffering if the performance of the behaviour is thwarted.

For conventional broiler breeders that are feed restricted and typically kept in barren environments, the behaviour pattern that is most commonly thwarted is feeding, including both the appetitive phase and the consummatory phase. Furthermore, in some parts of the world, resting in elevated positions is impeded during the rearing period due to the lack of perches or raised slats. With regard to physiological needs, broiler breeders experience metabolic hunger and may also suffer from the water restriction implemented. These welfare issues are discussed below.

### 2.1.1 Restrictive feeding and associated welfare problems

Being a motivated behaviour, feeding consists of two phases: the appetitive phase where the bird searches for food and gains food-related knowledge, that is, foraging, and the consummatory phase where the bird ingests food, that is, eating (Craig, 1918). Due to the feed restriction, the possibility of fulfilling the behavioural need for the consummatory phase of feeding is thwarted in broiler breeders. The daily ration of feed is quickly consumed, which may be as fast as in 4–16 min (Kostal et al., 1992). After ingestion of the daily ration, the broiler breeders are still highly motivated to continue feeding, partly because of hunger, and partly because of the unfulfilled behavioural need for carrying out feeding behaviour.

Thwarted possibility of performing the consummatory phase of feeding behaviour naturally causes an increase in time spent on the appetitive phase of feeding, that is, searching for food. Thus, as feed becomes restricted, more foraging behaviour in terms of pecking and scratching is performed, and also other oral activities may be increased. This has indeed been reported in several studies (Puterflam et al., 2006; Hocking et al., 1996, 2001, 2002; Savory and Maros, 1993; de Jong et al., 2003; Merlet et al., 2005). It may be questioned whether the increased foraging activities have any truly fulfilling effect on the behavioural need for feeding, particularly as the environment of broiler breeders is typically kept barren with limited access to foraging materials of high quantity and quality.

Certainly, feed-restricted broiler breeders show behaviour indicating frustration and hunger during both the rearing period and the production period. This involves increased general activity, in particular locomotion, indicating restlessness (Puterflam et al., 2006; de Jong et al., 2003). This comes at the expense of time spent on resting, eating and performing comfort behaviour (Puterflam et al., 2006; de Jong et al., 2003; Hocking et al., 2001, 2002) with the exception that at low levels of feed restriction, the broiler breeders may stand more, presumably to promote loss of heat resulting from the increased metabolism.

Furthermore, the feed-restricted broiler breeders often develop stereotypic pecking, that is, unvarying, repetitive pecking that has no obvious goal or function (Mason, 1991), directed towards objects in the environment (Puterflam et al., 2006; Nielsen et al., 2011). Stereotypic pecks may be directed towards the drinking nipples, resulting in both water spillage and excessive drinking (i.e. polydipsia, Hocking et al., 1993). In order to avoid water spillage and wet manure, which results in wet bedding and increased risk of contact dermatitis, the water consumption is controlled. This is partly done by reducing the water pressure, and partly by time limitations of access to the water. Few studies have investigated the effect on welfare of water restriction specifically. In general, water restriction is considered negative with respect to animal welfare, although one study showed that water restriction during rearing did not influence physiological welfare indicators in broiler breeders (Hocking et al., 1993).

The competition for feed results in an increased level of aggressive pecks, both among breeder males and among breeder females (Shea et al., 1990; Hocking and Jones, 2006; Hocking et al., 2005). Feather pecking, particularly of the tail feathers of conspecifics, is observed in broiler breeders (Morrissey et al., 2014; Girard et al., 2017a). It may have a stereotypic nature and take the form of 'tail sucking' (Morrissey et al., 2014; Nielsen et al., 2011; Girard et al., 2017a). Cannibalism is also reported, and studies show that alleviating hunger may reduce the risk of cannibalism, although not consistently (Hocking et al., 2004b; Nielsen et al., 2011). Although mating is thought to be the major cause of feather and skin damage in broiler breeders, feather pecking and cannibalistic pecks may also be important factors (de Jong and Guemene, 2011).

Physiologically, feed-restricted broiler breeders show signs of stress in terms of an increased relation between hetero granulocytes and lymphocytes (H/L) (Hocking et al., 1993, 1996; Bowling et al., 2018) and a higher level of plasma corticosterone (Hocking et al., 1993, 1996; de Jong et al., 2003). However, it remains debated whether this can be assigned to stress alone, as increased levels of plasma corticosterone may also occur due to metabolic reasons (de Jong et al., 2003; D'Eath et al., 2009).

### 2.1.2 The alleviation of hunger and fulfilment of behavioural needs in broiler breeders

Different feeding strategies have been used to alleviate the hunger felt by broiler breeders and to fulfil their behavioural need for feeding behaviour with varying success. Among these are qualitative feed restrictions, manipulating the number of daily meals, scattering of the feed by the use of spin feeders and appetite suppressants.

#### 2.1.2.1 Qualitative feed restriction

The idea of qualitative feed restriction is to reduce the quality of the feed in terms of nutrient content by adding non-nutritious or poor-nutritious diluents to the standard feed. This can be one or more types of fibres, and examples of fibres used are sugar beet, out hulls, sunflower flour and even sawdust. When applying qualitative feed restriction, a larger ration of feed is needed to obtain the same nutrient intake, and the content in different parts of the gastrointestinal system of the birds increases (Steenfeldt and Nielsen, 2012; Hocking et al., 2004a).

Further increase of gut fill may be obtained by adding soluble types of fibres to the feed. Soluble fibres can absorb more water, thus increasing the intestinal content more than insoluble fibres (Hocking et al., 2004b). However, the use of soluble fibres in the feed has to be done with care, as too high levels of soluble fibres leads to a swelling of the feed in the intestines in prolonged periods after feeding, which may be painful to the birds (Nielsen et al., 2011). The higher consumption of water in broiler breeders fed a diet high on soluble fibres may also result in a deterioration of the bedding, which increases the risk of contact dermatitis, compromises resting comfort and reduces foraging and dust bathing opportunities.

By increasing the intestinal contents, the broiler breeders are expected to obtain an improved feeling of satiety. In addition to increased satiety, applying qualitative feed restriction increases the chance of meeting the behavioural need of the broiler breeders for performing feeding behaviour, as the time spent on feeding increases when a larger amount of feed is to be ingested (Zuidhof et al., 1995; Savory et al., 1996; de Jong et al., 2005a; Moradi et al., 2013).

Indeed, studies comparing broiler breeders fed restrictively, either qualitatively or quantitatively (standard), have found positive effects of the former feeding strategy on a range of welfare indicators. For instance, qualitative feed restriction results in less water intake and reduced time spent on drinking behaviour (Zuidhof et al., 1995; Savory et al., 1996; Hocking, 2006). Other studies have found fewer stereotypic pecks on objects (Savory et al., 1996; Hocking et al., 2004b; de Jong et al., 2005a; Nielsen et al., 2011) and less stereotypic tail feather pecking (Nielsen et al., 2011) in the qualitatively feed-restricted birds.

To assess hunger in qualitatively feed-restricted broiler breeders, de Jong et al. (2005a) used a test of compensatory food intake where the amount of feed eaten besides the ordinary daily allowance was examined. The authors found that the compensatory food intake was reduced at 12 weeks of age in broiler breeder pullets allocated low-density feed compared to standard feed. Likewise, Nielsen et al. (2011) showed that regardless of the type of fibres, the compensatory feed intake was lower in a test of broiler breeders fed a fibre-rich feed compared to standard feed. The compensatory food intake test is a validated method for quantifying the level of hunger, as it has been shown that the more feed restricted the broiler breeders have been, the higher feed intake they will have for an extended period (at least 4 days) when allowed *ad libitum* access to feed (de Jong et al., 2003).

In regard to the effect of qualitative feed restriction on physiological stress, contradicting results have been found concerning both the relation between hetero granulocytes and lymphocytes (Zuidhof et al., 1995; Savory et al., 1996; Hocking et al., 2004b; de Jong et al., 2005a; Hocking, 2006; Jones et al., 2004) and the level of plasma corticosterone (Savory et al., 1996; Hocking, 2006; Moradi et al., 2013). van Emous et al. (2014) explained it as a reflection of a general increase in the content of plasma corticosterone of broiler breeders due to an increased feed restriction during the last 30 years. As a result, the broiler breeders respond less to differences in feed allowance.

To summarise, the qualitative feed restriction appears to have a positive effect on behaviour and welfare, although dependent on the type and amount of the fibres used. However, common for the studies conducted up until now is that none of the fibre-rich diets examined can be characterised as being very effective in terms of alleviating hunger without negatively affecting other welfare aspects. Even though there are positive effects of qualitative feed restriction on satiety, it is uncertain whether it fully satisfies the broiler breeders' motivation for feeding behaviour. Besides, there is a risk that qualitatively feed-restricted broiler breeders continue to experience metabolic hunger (de Jong et al., 2005a).

#### 2.1.2.2 Number of daily meals

Due to the feed restriction applied in the conventional broiler breeder industry, feeding programmes are always used. These programmes differ in the number of meals allocated daily (Fig. 4). Feeding once a day may be applied - either by choice or due to daily feeding being a requirement in the national legislation (e.g. UK, Sweden and Denmark). As an alternative, skip-a-day feeding programmes may be used (EFSA, 2010). It involves either one, two or three days (non-consecutive) without feed per week. On feeding days, the birds are fed a larger portion than if they were fed daily.

The effect on welfare of skip-a-day feeding programmes is, based on the few studies available, inconclusive. Lindholm et al. (2018) found signs of increased

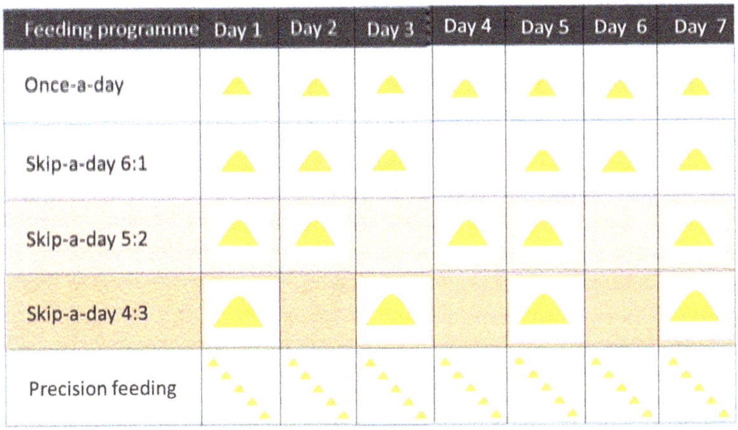


**Figure 4** Outline of the number and portions of meals during a week on the different feeding programmes. Portions not to scale. ©Anja Brinch Riber, Aarhus University, Denmark.

physiological stress in broiler breeder pullets experiencing 2 days a week without food compared to birds fed daily. This included elevated heterophil to lymphocyte ratios, increased adiposity and reduced muscle growth. On the other hand, the skip-a-day birds showed signs of lower anxiety before feeding times, which may be a result of the lower feed competition associated with larger portion sizes. Skip-a-day birds generally showed more interest in a novel object in the home pen, which indicates increased risk taking and reduced fear while fasting. No signs were found of the skip-a-day birds learning the feeding schedule, and this unpredictability may also increase stress.

In contrast, Skinner-Noble and Teeter (2009a cited in EFSA, 2010) found no signs of increased stress levels in skip-a-day birds compared to daily-fed birds, neither when using behavioural or physiological indicators. Furthermore, the plumage condition of broiler breeders has been found to be better during the early production period in skip-a-day birds compared to birds fed daily (Morrissey et al., 2014).

A new feeding system has recently been developed for broiler breeders: the precision feeding system (Zuidhof et al., 2017). In this system, small meals are provided to the individual bird multiple times each day. It works such that a bird can enter the system voluntarily at any time. The bird is automatically weighed, and if its body weight is below the target weight, access to feed is allowed for a limited period. When using the precision feeding system, the uniformity of the flock has been shown to increase (Zuidhof, 2018; Zuidhof et al., 2017).

Compared to skip-a-day feeding, broiler breeders with access to the precision feeding system show less restlessness behaviour, that is, more sitting and less standing and walking (Girard et al., 2017a). In addition, they perform less foraging and feather pecking behaviour (Girard et al., 2017a). However, increased aggression and more stereotypic pecking at the drinkers have been observed in precision-fed broiler breeders compared to those fed on the skip-a-day feeding programme (Girard et al., 2017a,b).

The effect of feeding twice daily, compared to once daily, on hunger and frustration in broiler breeder pullets has been examined by de Jong et al. (2005b). The result was unclear. The broiler breeders fed twice daily showed more locomotion, which may indicate increased foraging behaviour but also frustration because of an unfulfilled behavioural need of foraging. The other indicators of hunger examined (i.e. concentration of plasma-corticosterone, compensatory food intake and relation between plasma and glucose) were not influenced by whether the birds were fed once or twice daily.

Based on the existing knowledge, it is clear that the applied feeding programme in the broiler breeder industry is likely to affect the behaviour and welfare of the birds. However, deriving clear recommendations of which feeding programme to apply is hampered by the inconsistency of results from different studies of the same feeding programme and by the different directions in which welfare indicators are pointing when comparing different feeding programmes.

#### 2.1.2.3 Scattering of feed in the bedding

Scattering of feed in the litter during rearing of broiler breeder pullets is a feeding method practised in some parts of the world, particularly Europe and Northern America (EFSA, 2010). The scattering of feed is commonly done by using the so-called spin feeders. Using this method, the feed is quickly and evenly distributed to all the birds within the flock, promoting uniform feed intake (Hocking, 2004). In addition, the feeding method stimulates foraging behaviour. de Jong et al. (2005b) studied the effect of scattering all the feed in the litter during rearing of broiler breeder pullets. Stereotypic pecking on objects was reduced, but no other indicators of hunger, that is, the concentration of plasma corticosterone, compensatory food intake and the relation between plasma and glucose, were influenced by whether the feed was scattered in the bedding or allocated in feed troughs.

#### 2.1.2.4 Appetite suppressants

Appetite suppressants have been suggested as a method to alleviate the hunger felt by feed-restricted broiler breeders. The idea is that the use of these additives suppresses the appetite to a level where quantitative feed restriction is no longer necessary for the birds to remain on the target body weight. However, the method has only proved partly successful in improving the welfare of the birds (Sandilands et al., 2006; Savory et al., 1996). Concern has been expressed whether the reducing effect on appetite derives from discomfort felt by the birds (EFSA, 2010). Ethically this may not be acceptable to consumers or society as such.

### 2.1.3 Lack of elevated resting places and associated welfare problems

Elevated resting places, that is, raised slats or perches, are used for resting both during the day and at night. Indeed, broiler breeders will use elevated resting places when given access (Gebhardt-Henrich et al., 2017). Although slow-growing genotypes will use the elevated resting places more frequently, fast-growing genotypes also have a high use initially, yet this will reduce with age (Gebhardt-Henrich and Oester, 2014). Raised slats are preferred to perches by broiler breeders (Gebhardt-Henrich and Oester, 2014).

Thus, similar to other types of domestic fowl, broiler breeders are motivated to use elevated resting places. However, in some parts of the world, elevated resting places are not provided during the rearing period, that is, the first access to elevated structures occurs after transfer to the production farm. In addition to fulfilling a behavioural need, provision of elevated resting places also has positive effects on other welfare aspects. For instance, fear has been shown to be reduced in broiler breeders reared with access to perches (Brake et al., 1994). Furthermore, it is known from laying hens that elevated resting places also function as shelters during the day to avoid aggressive interactions with flock mates (Cordiner and Savory, 2001).

Providing elevated resting places during the rearing period prepares the broiler breeders for the more complex environment in the production houses. Usually, the nest boxes are elevated from the ground and often water nipples are placed above elevated slats, forcing both males and females to use the three-dimensional space. Rearing in houses with elevated resting places will promote the broiler breeders' use of the three-dimensional space in the production houses, as early access to three-dimensional structures improves the use later in life (Gunnarsson et al., 2000; Norman et al., 2018).

Negative effects in terms of keel bone damage have been associated with the use of perches in laying hens (Stratmann et al., 2015), but Gebhardt-Henrich et al. (2017) found no difference between adult broiler breeder females reared with or without perches. Provision of elevated resting places is considered to have an overall positive effect on behaviour and welfare of broiler breeders.

## 2.2 Aggression and associated welfare problems

Non-sexual aggression is common both among broiler breeder males and among broiler breeder females (Shea et al., 1990; Hocking and Jones, 2006;

Hocking et al., 2005). One explanation for this is the increased competition for feed. The level of non-sexual aggression has been reported to be highest at the time of the day where feed is allocated (Shea et al., 1990). Furthermore, feed-restricted broiler breeders display more non-sexual aggression at the feeder than do broiler breeders fed *ad libitum* (Mench, 1988). Frustration resulting from feed restriction could be another causal factor for the increased non-sexual aggression. With skip-a-day feeding, it has been shown that aggression is most pronounced on mornings where no feed is allocated (Mench, 1988), although this was not confirmed in a later study (Shea et al., 1990).

A high level of sexual aggression is also present in the broiler breeder production (Millman and Duncan, 2000; Millman et al., 2000). The sexual behaviour in broiler breeders lacks many of the elements of normal sexual behaviour, which both the red jungle fowl and laying hens exhibit, with courtship behaviour prior to mating being almost non-existing (Millman et al., 2000; Jones et al., 2001; de Jong et al., 2009). The mating behaviour of the males is described as being rough where they chase and peck the hens and force them to mate (Millman et al., 2000; Jones et al., 2001).

The rough mating behaviour induces fear and reduces welfare in broiler breeder females (de Jong et al., 2009; Millman et al., 2000; Leone and Estevez, 2008). The plumage of the hens is typically considerably deteriorated, especially in the second part of the production period, because of wear and tear caused by the males (de Jong and Guemene, 2011; Jones and Prescott, 2000). A poor plumage increases the risk of skin damage, as the protective layer of feathers is diminished. The skin damage inflicted is typically found along the torso and thigh beneath the wings and on the back of the head and neck on the females (Fig. 5a–c) (Millman et al., 2000).

According to Millman et al. (2000), the broiler breeder industry associates the problem with breeder male aggression with one particular parent strain that came into use when the high breast-yielding lines were introduced. Subsequently, sexual aggression was reported in all commercial broiler breeder strains, suggesting that it was a by-product resulting from selection for sexual vigour or a result of genetic drift from the inbreeding of foundation lines (Millman et al., 2000). No studies on the extent of sexual aggression in broiler breeders have been published within the last decade, that is, up-to-date information is unavailable.

### 2.3 Mutilation and associated welfare problems

In order to limit the damage inflicted by bird-to-bird interactions, including within same sex, mutilation of specific body parts is practised in broiler breeders. Mutilation is the removal or damage to a part or parts of the body,

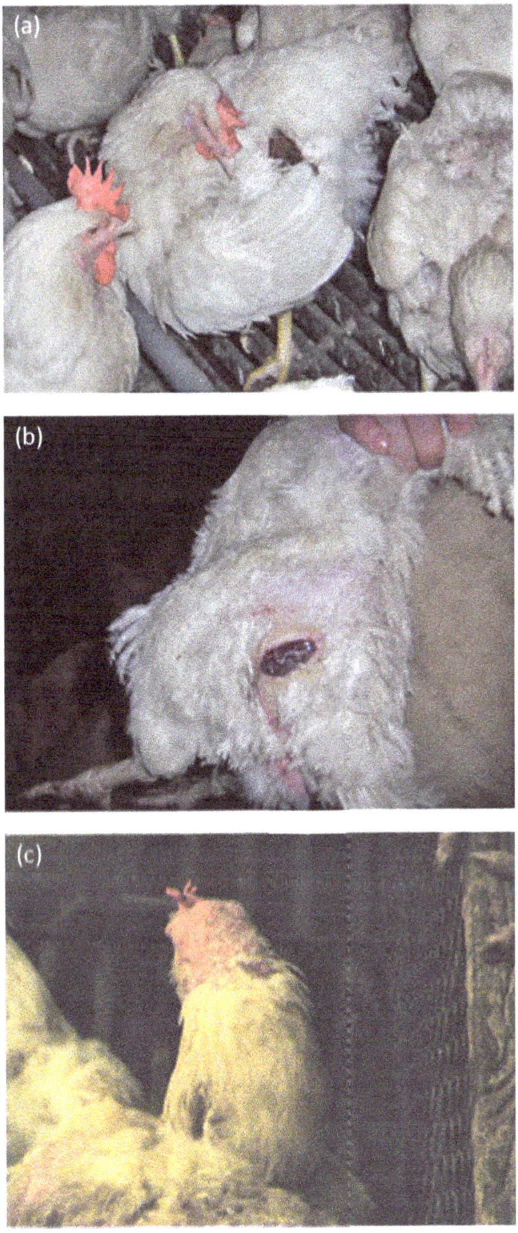

**Figure 5** Examples of broiler breeder females having the typical injuries inflicted during mating by the males: Wounds along the torso (a) and thigh (b) that are likely to have been caused by the spur or toes of the males. The denuded skin on the back of the head (c) is likely to have been inflicted when the males during mating use the beak to grab the female by the neck or back of the head. ©Ingrid de Jong, Wageningen Livestock Research, The Netherlands.

not being the horny dead body tissue and feathers, due to either an operation or a trauma (van Niekerk and de Jong, 2007).

In broiler breeders, the body parts that are mutilated differ between countries (EFSA, 2010). Trimming of the beak is routinely practised in both sexes. Likewise, the removal of spurs and toe trimming are routinely practised in males, with the exception of some genotypes where the growth of the spurs is reduced, eliminating the need for spur trimming. Toe trimming is performed on the outermost joint of the toe pointing backwards, digit I, and occasionally also on digit II (Fig. 6). The final type of mutilation is dubbing of the male comb, but it is no longer commonly practised.

In a full-scale study on a commercial broiler breeder production farm, where the males were beak-intact and toe-intact, mortality and culls due to wounds of breeder females were found to be increased (de Jong et al., 2018). In addition, more damage to the skin was observed in the breeder females housed with beak-intact and toe-intact breeder males. The damage was most frequently located on the flank, thighs and back of the females, suggesting that the intact toes of the breeder males caused the damage.

Interestingly, research from the Netherlands indicates that beak-intact breeder females have reduced mortality, and in some EU countries (e.g. Poland and UK) beak-intact broiler breeders are housed without increased damage to the skin (van Emous and de Jong, 2013). A recent study showed that omitting beak trimming of the breeder males had no effect on feather damage of the

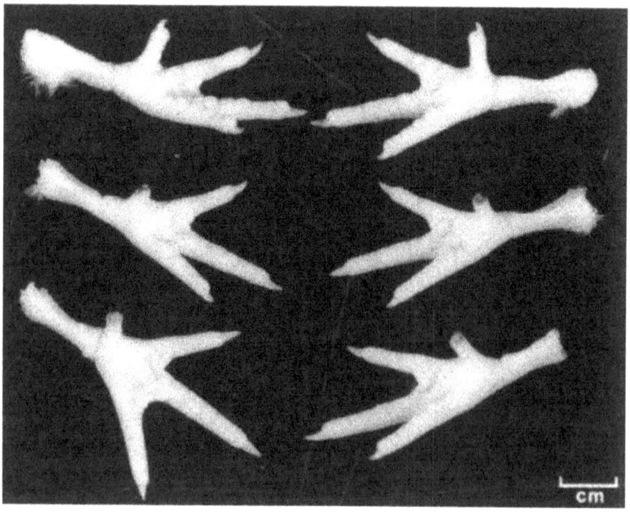


**Figure 6** Left and right feet 24 hours after hatching of toe-intact chicks (top) and toe-trimmed chicks (digit I; middle and bottom). Toe trimming is performed immediately after hatching. ©(Gentle and Hunter, 1988).

breeder females but increased the feather damage of the breeder males (van Niekerk and de Jong, 2017).

Most research on the effect of the procedure of mutilation on bird welfare concerns beak trimming, whereas the other types of mutilations have received little attention. However, all mutilations are performed on living and thus innervated tissue, for which reason acute pain is very likely to be caused by all types of mutilations. The pain involved and the impact of the various mutilations may, however, vary, as the degree of innervation of the tissue mutilated may differ.

The beak of a domestic fowl is highly innervated and contains nociceptors, thermoreceptors and mechanoreceptors for sensation of pain, temperature and pressure/texture, respectively (Gentle, 1989). Beak trimming therefore results in pain and sensory loss (Gentle, 2011; Gentle et al., 1997). In addition, as the beak is a sensitive tool used during natural behaviour, such as for grasping food, preening and nest building, beak trimming is considered problematic, as it causes a reduction in the bird's ability to manipulate items. For instance, beak-trimmed laying hens have been observed to have higher infestations of ectoparasites than beak-intact individuals (Mullens et al., 2010; Chen et al., 2011; Vezzoli et al., 2015).

Innervation of toes in domestic fowl and the consequences in terms of pain involved when trimming toes are only addressed in one study. Gentle and Hunter (1988) investigated the damage inflicted due to toe trimming in broiler breeder males. The toe trimming procedure was done at hatch by cutting one-third of the backward-facing toes and did not involve cautery. The toes were found to be well-innervated, and some neural regeneration took place after toe trimming. Some regenerating nerves became trapped in scar tissue and formed small discrete neuromas, which persisted over the 60-day observation period. Gentle and Hunter (1988) concluded that toe trimming is likely to cause acute pain at the time of amputation and the immediate period following, but, compared to beak trimming, it is less likely to be followed by chronic pain.

Evidence of pain involved in toe trimming also comes from toe trimming of turkey toms. Fournier et al. (2015) showed that on day 1 after toe trimming, the toms spent more time resting, less time walking, standing and at the feeder than intact birds, indicating pain due to the toe trimming procedure. Some of these behavioural consequences persisted through till day 5. In addition to the pain, toe trimming may also cause difficulties in the performance of scratching behaviour and problems with balance. At 19 weeks of age, toe-trimmed turkey toms have been found to walk less than toe-intact birds, which may reflect difficulty of maintaining balance when lacking parts of the toes (Fournier et al., 2015). Whether toe trimming has similar effects on broiler breeder males remains to be investigated.

To summarise, beak trimming and toe trimming can lead to acute and chronic pain. Therefore, reducing the risk of damage to the broiler breeder females by mutilation of the males brings about yet another broiler breeder paradox - mutilation of the breeder males results in improved welfare of the females but reduced welfare of the males, whereas omitting mutilation results in reduced welfare of the females. With regard to beak trimming of broiler breeder females, it can be questioned whether there is a true need for this procedure, as the research, although sparse, seems to show that it makes no difference to the condition of the plumage and skin of the females, but the procedure itself has negative welfare consequences.

### 2.3.1 Methods of reducing aggression to eliminate the need for mutilation

Limited research aiming directly at finding alternatives to mutilation as a strategy of preventing damage due to aggression in broiler breeders has been conducted. However, there are studies that indirectly address this topic. For instance, provision of elevated resting places may improve the possibility of escaping aggression as discussed previously in this chapter. In this section, the few studies that have examined the possibilities of reducing the extent of sexual aggression are covered. These include investigations of the effect on occurrence of sexual aggression when providing panels, reducing the stocking density, using UV lighting or housing the sexes separately either part of the day or fulltime. Furthermore, the potential of breeding towards less-aggressive broiler breeders is briefly discussed.

#### 2.3.1.1 The effect of panels

In the production system, broiler breeder males typically congregate on the littered area, while females appear fearful of the males and tend to avoid them by staying at the raised slats (Estevez, 1999; Millman et al., 2000). Leone and Estevez (2008) proposed that females in breeder flocks housed with increased possibilities of shelter in the littered area would have greater behavioural control over their environment, as the shelters would offer the opportunity to avoid aggressive interactions. As a result, male dispersal throughout the house would increase, and more females would therefore be attracted to the littered area, resulting in improved sexual interaction with fewer forced matings, less damage to females inflicted during mating and improved reproductive performance.

They investigated the hypothesis in an on-farm study where vertical cover panels were provided on the littered area centrally placed in the house during the production period (Leone and Estevez, 2008). The presence of cover panels was found to improve reproductive performance from 25 to 60 weeks of age. Egg production was increased by 2.1%, and hatchability and fertility improved, resulting in additional 4.5 chicks per female. The home range of the broiler breeder males housed with cover panels increased compared to males housed without panels. In addition, the broiler breeder males housed with cover panels were located more often on the slatted area than males housed without panels.

No observations of mating behaviour, damage inflicted on the broiler breeder females or the females' use of the littered area were recorded. Anecdotal results exist on these parameters from a demonstration (i.e. not an experiment in a controlled setting) where cover panels were introduced at 23 weeks of age in a commercial broiler breeder flock suffering from severe skin damage and high mortality of breeder females (Estevez, 1999). The introduction of cover panels was reported to have an immediate effect. Twenty-four hours after introducing the cover panels, the number of females in the scratching area increased by several hundred. The behaviour of the males was reported to be more relaxed, and mortality of the females was reduced.

Thus, providing cover panels seems to have the potential for reducing forced matings without compromising reproductive performance. Cover panels seem to improve the distribution within the house of both males and females whereby the mating opportunities of the males increases, and the damage and stress to the females caused by forced matings potentially may be decreased. As such, the provision of cover panels should be considered in the strategy for preventing damage inflicted on broiler breeder females during the production period.

### 2.3.1.2 The effect of stocking density

Kratzer and Craig (1980) found that layer breeder males showed more courtship behaviour at a reduced stocking density. They suggested that the lower stocking density facilitated an improvement of the females' recognition of male courtship behaviour. Following up on this hypothesis, de Jong et al. (2011) did an on-farm study where broiler breeders were housed at a standard stocking density of 8.8 birds/m$^2$ and a low stocking density of 5.2 birds/m$^2$. They found that a reduced stocking density resulted in more matings preceded by courtship behaviour (Fig. 7a), fewer forced matings (Fig. 7b) and more successful matings, that is, fewer hens struggled and escaped during mating (Fig. 7c). The effect was greatest if the stocking density was reduced during both the rearing period and the production period. The damage to the plumage and skin was less severe in both females and males housed at the reduced stocking density. Furthermore, more eggs were fertilised, more eggs hatched and a higher number of day-old chicks per hen were gained in the flocks housed at reduced stocking density.

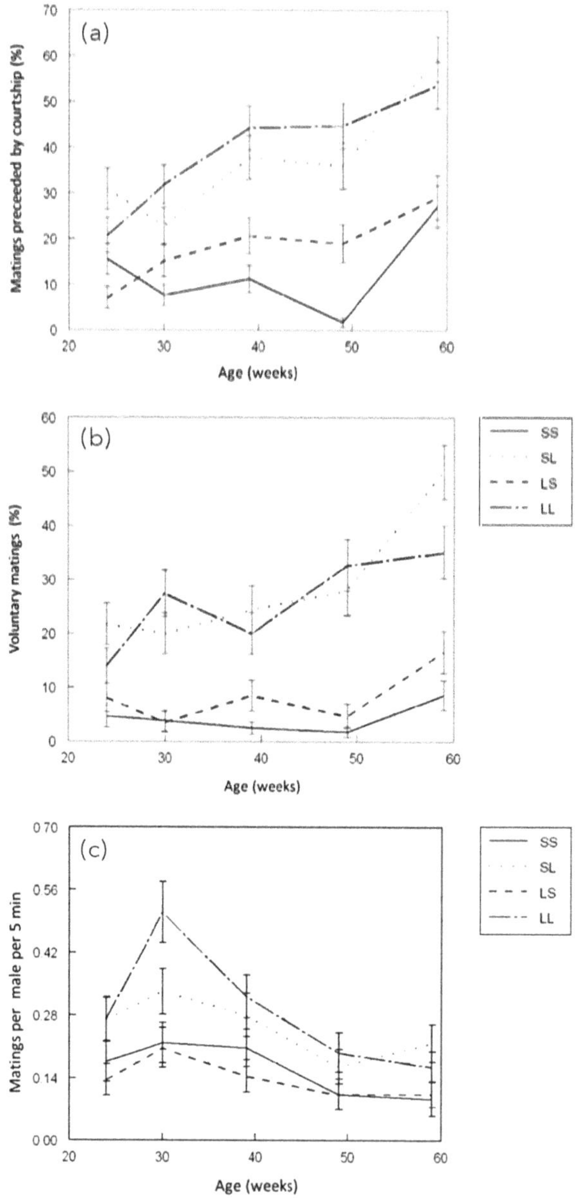


**Figure 7** The effect of stocking density (SD) during both the rearing period and the production period on (a) percentage of matings preceded by courtship behaviour, (b) percentage of voluntary matings and (c) mean number of successful matings per breeder male per 5 min. SS = standard SD both during rearing and production, LL = low SD both during rearing and production, SL = standard SD during rearing, low SD during production, LS = low SD during rearing and standard SD during production. ©de Jong et al. (2011).

Thus, reducing stocking density increases the proportion of voluntary matings and thus reduces the risk of damage inflicted on the broiler breeder females during mating. In addition, reduced stocking density improves the condition of the plumage and skin of the females, which may further protect against damage inflicted during mating. Importantly, reducing the stocking density can be done without compromising reproductive performance. Reduced stocking density may be used as a management tool to lower the risk of damage inflicted during mating. However, economically, this solution appears costly.

#### 2.3.1.3 The effect of ultraviolet light

Domestic fowls, including broiler breeders, are capable of seeing ultraviolet light with a wavelength between 320 and 400 mm, equalling $UV_A$ light. From other bird species it is known that $UV_A$ light plays a role for social communication and choice of partner through reflections of the plumage (see review in Jones et al., 2001). The light sources traditionally used in poultry houses do not include $UV_A$ light.

In broiler breeders, the behaviour in a $UV_A$-enriched environment has been examined (Jones et al., 2001). The number of sexual attempts and locomotion was shown to increase in the environment enriched with $UV_A$ light, and the females' choice of partner depended on the level of $UV_A$ light. The conclusion was that $UV_A$ light is involved in the communication of mating signals in broiler breeders and that $UV_A$ light, therefore, potentially may influence the level of sexual aggression in broiler breeders.

#### 2.3.1.4 Housing systems separating sexes

Using housing systems that separate the sexes either temporarily or permanently may be a solution to reduce/prevent the sexual aggression in broiler breeder production. In one system, the Quality Time Concept®, broiler breeder males and females are kept in large sex-mixed flocks, ensuring natural mating, but periodically the system separates the birds by taking advantage of the different types of feeders used for the two sexes. Another system is cages in which sexes are kept apart for lifetime, for which reason artificial insemination is necessary. Below, these two systems are described and discussed in terms of welfare implications to the broiler breeders.

The Quality Time Concept®: One alternative housing system for broiler breeders is the Quality Time Concept® (QTC; van Emous, 2010). The system was developed to adapt measures to improve fertility, based on variation in mating frequency over the day. In this housing system, broiler breeder males and females are separated for 5 hours a day during the light period, using separate feeding systems and a movable partition. When it is feeding time, the sexes move towards the feeders they are able to feed from, and while they are busy feeding, a partition is placed, separating the two sexes (Fig. 8).

Mating behaviour in the QTC compared to the standard housing system has been shown to be more voluntary, resulting in more successful matings (van Emous, 2010). In the QTC flocks, a better plumage of females aged 37 to 48 weeks was found, and no increase in aggressive behaviour among males occurred. In addition, increased mixing of males and females took place 2 hours before the dark period, indicating an improved sexual interaction. At this time period, mating activity normally peaks (Harris et al., 1980; Duncan et al., 1990; Bilcik and Estevez, 2005). In an initial trial, the fertility was reduced

(a)

(b)

**Figure 8** The Quality Time Concept® (QTC) separating broiler breeder males and females daily for some hours during the light period. In the first version of the system (a) males are separated from females in three male enclosures at the long side of the house, whereas in the second improved version (b) all males are separated in the rear end of the house. ©Rick van Emous, Wageningen Livestock Research, The Netherlands.

by 1-2% in the QTC, but after improving the concept (Fig. 8b) the mean fertility rate increased (van Emous and Gunnink, 2011). Separating males from females was rather effective, as only 2.7% females and 5.2% males were found in the wrong compartment.

The cage system: In some parts of the world, broiler breeders are kept in single-sex cages during the production period. Artificial insemination is practised, and the need for despurring and toe trimming of breeder males to avoid injuries on females during mating is thereby eliminated. However, housing broiler breeders in cages may affect other aspects of welfare negatively. Fulfilling behavioural needs of domestic fowls is more difficult in cage systems (Lay et al., 2011), and it may not be acceptable to the consumers.

Artificial insemination may also stimulate selection of heavy males, which has negative consequences for male welfare (Laughlin, 2009; Brillard, 2001). In addition, artificial insemination involves regular handling of the birds. Often, the handling is rough/perceived as rough by the birds, which may have negative implications for bird welfare (Rushen et al., 1999; Laurence et al., 2014). The production of chicks may also be negatively affected by housing in cages. A study from 1983 showed a reduced fertility and hatchability of eggs from artificially inseminated-caged breeder females compared to naturally mated females (Petitte et al., 1983).

To summarise, QTC reduces forced matings and is likely to decrease the frequency of overmating. Consequently, the risk of inflicting damage on the breeder females is reduced. QTC may therefore be part of a solution to reduce damage to the broiler breeder females caused by sexual aggression. Housing broiler breeders in single-sex cages solves the issue with damage inflicted by broiler breeder males on the females during mating and eradicates the need for toe trimming and despurring, but it generates other welfare problems and may affect production of chicks negatively. In addition, semen collection from breeder males and artificial insemination of breeder females are cost-intensive solutions because of the very time-consuming nature.

### 2.3.1.5 Breeding for normal sexual behaviour

Studies have shown that the aggression displayed by broiler breeder males has a genetic component. Millman et al. (2000) compared layer and broiler breeder males and found that the latter displayed higher levels of both non-sexual and sexual aggression. Compared to layer breeder males, broiler breeder males chased and pecked females more (Fig. 9a). They also forced matings more than three times as often as did layer breeder males (Fig. 9b). Unsuccessful attempts of mating, often interrupted due to struggling females, leading to escape, were almost twice as common in broiler breeder males compared

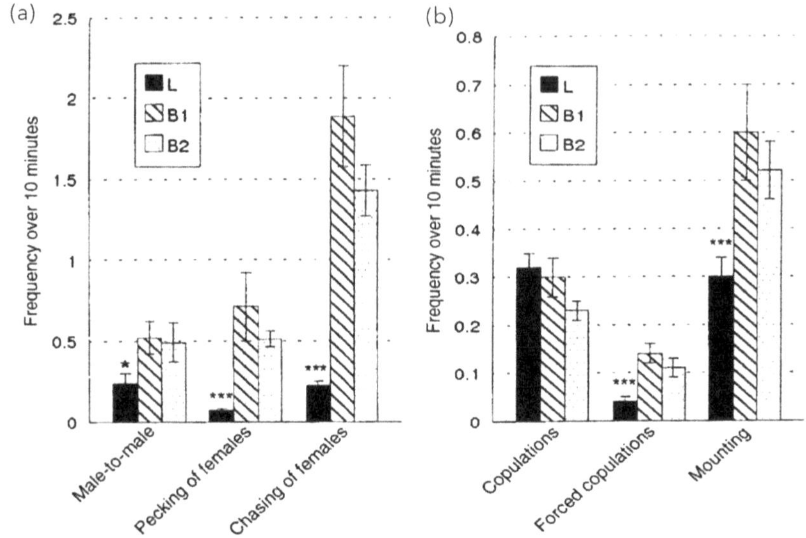


**Figure 9** Differences between males from a strain of layer breeders (L: Isa Brown) and from two strains of broiler breeders (B1: Peterson, B2: Ross) in (a) aggressive behaviour and (b) mating behaviour (copulation = voluntary and forced matings; mounting = unsuccessful matings where the female escaped). ©Millman et al. (2000).

to layer breeder males (Fig. 9b). Since sexual aggression was not affected by feeding regimes (*ad libitum* fed vs. feed restricted), it was concluded that the differences between the broiler breeder strains and the layer breeder strain were associated with genetic factors.

The higher level of sexual aggression may also be related to the underdeveloped courtship behaviour displayed by broiler breeders. For instance, in the study by Millman et al. (2000), courtship behaviour in the two strains of broiler breeders was virtually absent before mating (Fig. 10). In the early stage of development of sexual behaviour, sexual aggression may also occur in both jungle fowl and layer breeders, but as males become sexually experienced, the aggressive elements disappear (Wood-Gush, 1958; Kruijt, 1964). However, this development in sexual behaviour is not observed in broiler breeder males, as sexual aggression persists even with attainment of considerable sexual experience (Millman et al., 2000).

Millman et al. (2000) suggested that it is possible that broiler breeder males are developmentally retarded and have become halted at an early stage of sexual development. Adding the fact that de Jong et al. (2011) found fewer forced matings when recognition of mating behaviour was improved, this indicates that performance of normal courtship behaviour may potentially lead to less damage inflicted on the breeder females during mating. Thus, breeding for normal sexual behaviour, including display of courtship behaviour

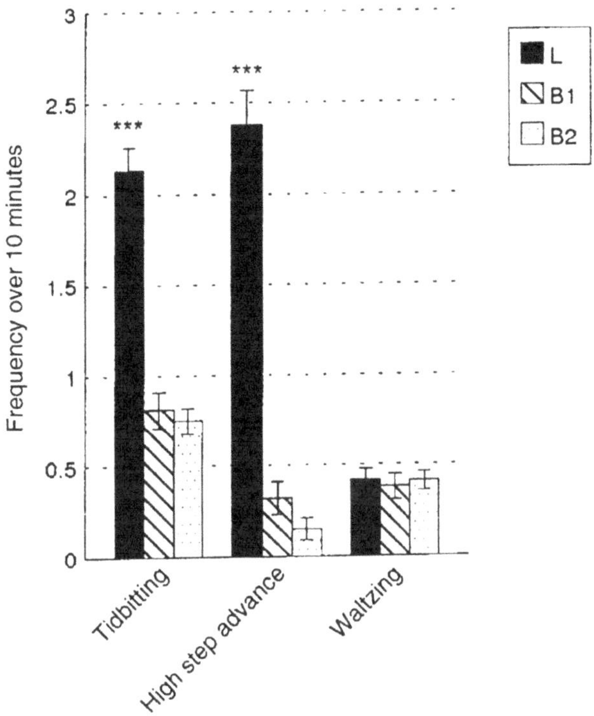


**Figure 10** Differences between males from a strain of layer breeders (L: Isa Brown) and from two strains of broiler breeders (B1: Peterson, B2: Ross) in display of courtship behaviour. ©Millman et al. (2000).

and reduced sexual aggression, seems overly important and essential in the achievement of improved welfare of broiler breeder females.

## 3 Conclusion and future trends

It appears from the exposition of knowledge on welfare issues affecting broiler breeders that there are a number of significant challenges which need to be addressed in the near future. Among these are the severe feed restrictions and the associated welfare problems in terms of water restriction, physiological stress and development of abnormal behaviour, indicative of frustration and hunger. In addition to the unfulfilled behavioural need for ingesting feed, other behavioural needs, for example, for foraging and elevated resting, are also compromised. Furthermore, welfare challenges associated with the high level of aggression, particularly sexual aggression, exist, causing increased levels of fear, pain and mortality. Linked to this is the mutilation of broiler breeders, which is used as a preventive measure to reduce damage inflicted to the plumage and skin of flock mates, but the mutilation is a welfare problem in itself.

As mentioned previously, research into welfare issues affecting broiler breeders is rather limited. However, this may change to some extent, as political attention has been drawn to the topic, which may open up for more research. Examples include the attention that has been drawn to the mutilation procedure in the domestic fowl during recent years, mainly in Northern Europe and Northern America. With a few exceptions, focus has been on beak trimming in laying hens, but Denmark and the Netherlands have taken it a step further by including mutilation in broiler breeders. In the Netherlands, it has been politically decided that mutilation in broiler breeders should be phased out by 2021. Moreover, the authorities in some countries, for example, Denmark and the Netherlands, have initiated work aiming at finding solutions to reduce the welfare problems associated with feed restriction.

Consumer-driven forces may also result in more research into how to improve welfare of broiler breeders. During recent years, consumer awareness of animal welfare has increased worldwide, though particularly in Europe and North America. Concurrently, an increasing number of companies within the food industry have included or improved the policy on animal welfare in their corporate social responsibility (CSR). The consumer demand for organic or middle-segment broiler meat has resulted in alternative genotypes of broiler breeders with reduced growth rates that have become available in some parts of the world, although still at low numbers.

## 4 Acknowledgements

Part of this chapter originates from a report, 'Alternatives to mutilation of the outermost joint of the backward-facing toe in broiler breeder males', that was commissioned by the Ministry of Environment and Food of Denmark as part of the 'Contract between Aarhus University and Ministry of Environment and Food for the provision of research-based policy advice at Aarhus University, 2017-2020'. The report is available online at: https://pure.au.dk/portal/da/persons/anja-brinch-riber(b746c493-a840-48f9-af06-0e6ac28ae8ba)/publications/alternatives-to-mutilation-of-the-outermost-joint-of-the-backwardfacing-toe-in-broiler-breeder-males(71baf919-7137-4084-a8dc-866cd1265eb0).html.

## 5 Where to look for further information

Further information on welfare issues affecting broiler breeders may be found in the following review papers:

- D'Eath et al. (2009): 'Freedom from hunger' and preventing obesity: the animal welfare implications of reducing food quantity or quality.
- de Jong and Guemene (2011): Major welfare issues in broiler breeders.

- Mench (2002): Broiler breeders: feed restriction and welfare.
- Riber et al. (2017): Environmental enrichment for broiler breeders: An undeveloped field.

Also recommended is the report containing the EFSA Scientific Opinion on welfare aspects of the management and housing of the grandparent and parent stocks raised and kept for breeding purposes (EFSA, 2010). The report can be found freely available on the web.

## 6 References

Arrazola, A. 2018. The effect of alternative feeding strategies for broiler breeder pullets throughout the production cycle. PhD Thesis. The University of Guelph.

Aviagen. 2019a. *Ross308-308FF: Performance Objectives*. Available at: http://en.aviagen.com/assets/Tech_Center/Ross_Broiler/Ross308-308FF-BroilerPO2019-EN.pdf.

Aviagen. 2019b. *Ross Parent Stock Management Handbook*. Available at: http://en.aviagen.com/assets/Tech_Center/Ross_PS/RossPSHandBook2018.pdf.

Bilcik, B. and Estevez, I. 2005. Impact of male-male competition and morphological traits on mating strategies and reproductive success in broiler breeders. *Applied Animal Behaviour Science* 92(4), 307-23. doi:10.1016/j.applanim.2004.11.007.

Bowling, M., Forder, R., Hughes, R. J., Weaver, S. and Hynd, P. I. 2018. Effect of restricted feed intake in broiler breeder hens on their stress levels and the growth and immunology of their offspring. *Translational Animal Science* 2(3), 263-71. doi:10.1093/tas/txy064.

Brake, J., Keeley, T. P. and Jones, R. B. 1994. Effect of age and presence of perches during rearing on tonic immobility fear reactions of broiler breeder pullets. *Poultry Science* 73(9), 1470-4. doi:10.3382/ps.0731470.

Brillard, J. P. 2001. Future strategies for broiler breeders: an international perspective. *World's Poultry Science Journal* 57(3), 243–50. doi:10.1079/WPS20010017.

Bruggeman, V., Onagbesan, O., D'hondt, E., Buys, N., Safi, M., Vanmontfort, D., Berghman, L., Vandesande, F. and Decuypere, E. 1999. Effects of timing and duration of feed restriction during rearing on reproductive characteristics in broiler breeder females. *Poultry Science* 78(10), 1424-34. doi:10.1093/ps/78.10.1424.

Chen, B. L., Haith, K. L. and Mullens, B. A. 2011. Beak condition drives abundance and grooming-mediated competitive asymmetry in a poultry ectoparasite community. *Parasitology* 138(6), 748-57. doi:10.1017/S0031182011000229.

Cobb. 2018. *Cobb500 Broiler Performance and Nutrition Supplement*. Available at: https://www.cobb-vantress.com/assets/Cobb-Files/product-guides/bdc20a5443/70dec630-0abf-11e9-9c88-c51e407c53ab.pdf.

Cordiner, L. S. and Savory, C. J. 2001. Use of perches and nestboxes by laying hens in relation to social status, based on examination of consistency of ranking orders and frequency of interaction. *Applied Animal Behaviour Science* 71(4), 305-17. doi:10.1016/s0168-1591(00)00186-6.

Craig, W. 1918. Appetites and aversions as constituents of instincts. *Biological Bulletin* 34(2), 91–107.

D'Eath, R. B., Tolkamp, B. J., Kyriazakis, I. and Lawrence, A. B. 2009. 'Freedom from hunger' and preventing obesity: the animal welfare implications of reducing food quantity or quality. *Animal Behaviour* 77(2), 275–88. doi:10.1016/j.anbehav.2008.10.028.

Decuypere, E., Bruggeman, V., Everaert, N., Li, Y., Boonen, R., de Tavernier, J., Janssens, S. and Buys, N. 2010. The broiler breeder paradox: ethical, genetic and physiological perspectives, and suggestions for solutions. *British Poultry Science* 51(5), 569–79. doi:10.1080/00071668.2010.519121.

de Jong, I. C. and Guemene, D. 2011. Major welfare issues in broiler breeders. *World's Poultry Science Journal* 67(1), 73–82. doi:10.1017/S0043933911000067.

de Jong, I. C., van Voorst, S., Ehlhardt, D. A. and Blokhuis, H. J. 2002. Effects of restricted feeding on physiological stress parameters in growing broiler breeders. *British Poultry Science* 43(2), 157–68. doi:10.1080/00071660120121355.

de Jong, I. C., van Voorst, A. S. and Blokhuis, H. J. 2003. Parameters for quantification of hunger in broiler breeders. *Physiology and Behavior* 78(4–5), 773–83. doi:10.1016/s0031-9384(03)00058-1.

de Jong, I. C., Enting, H., van Voorst, A. and Blokhuis, H. J. 2005a. Do low-density diets improve broiler breeder welfare during rearing and laying? *Poultry Science* 84(2), 194–203. doi:10.1093/ps/84.2.194.

de Jong, I. C., Fillerup, M. and Blokhuis, H. J. 2005b. Effect of scattered feeding and feeding twice a day during rearing on indicators of hunger and frustration in broiler breeders. *Applied Animal Behaviour Science* 92(1–2), 61–76. doi:10.1016/j.applanim.2004.10.022.

de Jong, I. C., Wolthuis-Fillerup, M. and van Emous, R. A. 2009. Development of sexual behaviour in commercially-housed broiler breeders after mixing. *British Poultry Science* 50(2), 151–60. doi:10.1080/00071660802710124.

de Jong, I. C., Lourens, A., Gunning, H., Wokel, L. and van Emous, R. 2011. Effect of stocking density on (the development of) sexual behaviour and technical performance in broiler breeds. (In Dutch: Effect van bezettingsdichtheid op (de ontwikkeling van) het paargedrag en de technische resultaten bij vleeskuikenouderdieren). Wageningen UR Livestock Research.

de Jong, I., van Hattum, T. and Gunnink, H. 2018. Meer schade en uitval bij volledig onbehandelde hanen (In English: More damage and loss with completely untreated roosters). *Pluimveehouderij,* 8 November 2018, pp. 26–8.

Denbow, D. M. 1989. Peripheral and central control of food intake. *Poultry Science* 68(7), 938–47. doi:10.3382/ps.0680938.

Duncan, I. J. H., Hocking, P. M. and Seawright, E. 1990. Sexual behavior and fertility in broiler breeder domestic fowl. *Applied Animal Behaviour Science* 26(3), 201–13. doi:10.1016/0168-1591(90)90137-3.

EFSA. 2010. Scientific Opinion on welfare aspects of the management and housing of the grand-parent and parent stocks raised and kept for breeding purposes. *EFSA Journal* 8(7), 1667.

Elfick, D. 2010. 50 years of selection in the broiler breeder industry and beyond. In: Ravindran, V. (Ed.), *Proceedings of the NZ Poultry Industry Conference*. Monogastric Research Centre, Massey University, Palmerston North, New Zealand.

Estevez, I. 1999. Cover panels for chickens: a cheap tool that can help you. *Poultry Perspectives* 1, 4–6.

Fournier, J., Schwean-Lardner, K., Knezacek, T. D., Gomis, S. and Classen, H. L. 2015. The effect of toe trimming on behavior, mobility, toe length and other indicators of welfare in tom turkeys. *Poultry Science* 94(7), 1446-53. doi:10.3382/ps/pev112.

Gebhardt-Henrich, S. G. and Oester, H. 2014. Do broiler breeders prefer elevated sleeping sites? In: Estevez, I., Manteca, X., Marin, R. H. and Averós, X. (Eds), *Proceedings of the 48th Congres of the International Society for Applied Ethology*, Vittoria-Gasteiz, Spain.

Gebhardt-Henrich, S. G., Toscano, M. J. and Wurbel, H. 2017. Perch use by broiler breeders and its implication on health and production. *Poultry Science* 96(10), 3539-49. doi:10.3382/ps/pex189.

Gentle, M. J. 1989. Cutaneous sensory afferents recorded from the nervus intramandibularis of *Gallus gallus* var *domesticus*. *Journal of Comparative Physiology. A, Sensory, Neural, and Behavioral Physiology* 164(6), 763-74. doi:10.1007/bf00616748.

Gentle, M. J. 2011. Pain issues in poultry. *Applied Animal Behaviour Science* 135(3), 252-8. doi:10.1016/j.applanim.2011.10.023.

Gentle, M. J. and Hunter, L. H. 1988. Neural consequences of partial toe amputation in chickens. *Research in Veterinary Science* 45(3), 374-6. doi:10.1016/S0034-5288(18)30968-8.

Gentle, M. J., Hughes, B. O., Fox, A. and Waddington, D. 1997. Behavioural and anatomical consequences of two beak trimming methods in 1- and 10-d-old domestic chicks. *British Poultry Science* 38(5), 453-63. doi:10.1080/00071669708418022.

Girard, M. T. E., Zuidhof, M. J. and Bench, C. J. 2017a. Feeding, foraging, and feather pecking behaviours in precision-fed and skip-a-day-fed broiler breeder pullets. *Applied Animal Behaviour Science* 188, 42-9. doi:10.1016/j.applanim.2016.12.011.

Girard, T. E., Zuidhof, M. J. and Bench, C. J. 2017b. Aggression and social rank fluctuations in precision-fed and skip-a-day-fed broiler breeder pullets. *Applied Animal Behaviour Science* 187, 38-44. doi:10.1016/j.applanim.2016.12.005.

Gunnarsson, S., Yngvesson, J., Keeling, L. J. and Forkman, B. 2000. Rearing without early access to perches impairs the spatial skills of laying hens. *Applied Animal Behaviour Science* 67(3), 217-28. doi:10.1016/s0168-1591(99)00125-2.

Harris, G. C., Benson, J., Sellers, R. and Cole, T. A. 1980. The mating activity of broiler breeder cockerels. *Arkansas Farm Research* 29, 15.

Heck, A., Onagbesan, O., Tona, K., Metayer, S., Putterflam, J., Jego, Y., Trevidy, J. J., Decuypere, E., Williams, J., Picard, M. and Bruggeman, V. 2004. Effects of ad libitum feeding on performance of different strains of broiler breeders. *British Poultry Science* 45(5), 695-703. doi:10.1080/00071660400006537.

Hiemstra, S. J. and Napel, J. T. 2013. Study of the impact of genetic selection on the welfare of chickens bred and kept for meat production. Available at: https://ec.europa.eu/food/sites/food/files/animals/docs/aw:practice_farm_broilers_653020_final-report_en.pdf.

Hocking, P. M. 2004. Measuring and auditing the welfare of broiler breeders. In: Weeks, C. A. and Butterworth, A. (Eds), *Measuring and Auditing Broiler Welfare*. CABI Publishing, Wallingford, UK, pp. 19-35.

Hocking, P. M. 2006. High-fibre pelleted rations decrease water intake but do not improve physiological indexes of welfare in food-restricted female broiler breeders. *British Poultry Science* 47(1), 19-23. doi:10.1080/00071660500468041.

Hocking, P. M. and Jones, E. K. 2006. On-farm assessment of environmental enrichment for broiler breeders. *British Poultry Science* 47(4), 418–25. doi:10.1080/00071660600825074.

Hocking, P. M., Maxwell, M. H. and Mitchell, M. A. 1993. Welfare assessment of broiler breeder and layer females subjected to food restriction and limited access to water during rearing. *British Poultry Science* 34(3), 443–58. doi:10.1080/00071669308417600.

Hocking, P. M., Maxwell, M. H. and Mitchell, M. A. 1996. Relationships between the degree of food restriction and welfare indices in broiler breeder females. *British Poultry Science* 37(2), 263–78. doi:10.1080/00071669608417858.

Hocking, P. M., Maxwell, M. H., Robertson, G. W. and Mitchell, M. A. 2001. Welfare assessment of modified rearing programmes for broiler breeders. *British Poultry Science* 42(4), 424–32. doi:10.1080/00071660120070677.

Hocking, P. M., Maxwell, M. H., Robertson, G. W. and Mitchell, M. A. 2002. Welfare assessment of broiler breeders that are food restricted after peak rate of lay. *British Poultry Science* 43(1), 5–15. doi:10.1080/00071660120109818.

Hocking, P. M., Channing, C. E., Robertson, G. W., Edmond, A. and Jones, R. B. 2004a. Between breed genetic variation for welfare-related behavioural traits in domestic fowl. *Applied Animal Behaviour Science* 89(1–2), 85–105. doi:10.1016/j.applanim.2004.03.014.

Hocking, P. M., Zaczek, V., Jones, E. K. M. and Macleod, M. G. 2004b. Different concentrations and sources of dietary fibre may improve the welfare of female broiler breeders. *British Poultry Science* 45(1), 9–19. doi:10.1080/00071660410001668806.

Hocking, P. M., Jones, E. K. M. and Picard, M. 2005. Assessing the welfare consequences of providing litter for feed-restricted broiler breeders. *British Poultry Science* 46(5), 545–52. doi:10.1080/00071660500254813.

Horne, P. L. M. V. 2018. *Competitiveness of the EU Poultry Meat Sector, Base Year 2017*. International Comparison of Production Costs, Wageningen, the Netherlands.

Jensen, P. and Toates, F. M. 1993. Who needs 'behavioural needs'? Motivational aspects of the needs of animals. *Applied Animal Behaviour Science* 37(2), 161–81. doi:10.1016/0168-1591(93)90108-2.

Jones, E. K. M. and Prescott, N. B. 2000. Visual cues used in the choice of mate by fowl and their potential importance for the breeder industry. *World's Poultry Science Journal* 56(2), 127–38. doi:10.1079/WPS20000010.

Jones, E. K. M., Prescott, N. B., Cook, P., White, R. P. and Wathes, C. M. 2001. Ultraviolet light and mating behaviour in domestic broiler breeders. *British Poultry Science* 42(1), 23–32. doi:10.1080/713655008.

Jones, E. K., Zaczek, V., Macleod, M. and Hocking, P. M. 2004. Genotype, dietary manipulation and food allocation affect indices of welfare in broiler breeders. *British Poultry Science* 45(6), 725–37. doi:10.1080/00071660400014226.

Kostal, L., Savory, C. J. and Hughes, B. O. 1992. Diurnal and individual variation in behaviour of restricted-fed broiler breeders. *Applied Animal Behaviour Science* 32(4), 361–74. doi:10.1016/S0168-1591(05)80028-0.

Kratzer, D. D. and Craig, J. V. 1980. Mating behaviour of cockerels: effects of social status, group size and group densitiy. *Applied Animal Ethology* 6(1), 49–62. doi:10.1016/0304-3762(80)90093-0.

Kruijt, J. P. 1964. Ontogeny of social behaviour in Burmase red junglefowl (Gallus gallus spadiceus) bonnaterre. *Behaviour* Suppl. XII, 1-201.

Laughlin, K. F. 2009. Breeder management: how did we get here? In: Hocking, P. (Ed.), *Biology of Breeding Poultry. Conference: 29th Poultry Science Symposium*, Edinburgh, Scotland, 23-25 July 2007.

Laurence, A., Lumineau, S., Calandreau, L., Arnould, C., Leterrier, C., Boissy, A. and Houdelier, C. 2014. Short- and long-term effects of unpredictable repeated negative stimuli on Japanese quail's fear of humans. *PLoS ONE* 9(3), e93259. doi:10.1371/journal.pone.0093259.

Lay, D. C., Fulton, R. M., Hester, P. Y., Karcher, D. M., Kjaer, J. B., Mench, J. A., Mullens, B. A., Newberry, R. C., Nicol, C. J., O'Sullivan, N. P. and Porter, R. E. 2011. Hen welfare in different housing systems. *Poultry Science* 90(1), 278-94. doi:10.3382/ps.2010-00962.

Leone, E. H. and Estevez, I. 2008. Economic and welfare benefits of environmental enrichment for broiler breeders. *Poultry Science* 87(1), 14-21. doi:10.3382/ps.2007-00154.

Lindholm, C., Johansson, A., Middelkoop, A., Lees, J. J., Yngwe, N., Berndtson, E., Cooper, G. and Altimiras, J. 2018. The quest for welfare-friendly feeding of broiler breeders: effects of daily vs. 5:2 feed restriction schedules. *Poultry Science* 97(2), 368-77. doi:10.3382/ps/pex326.

Mason, G. J. 1991. Stereotypies: a critical review. *Animal Behaviour* 41(6), 1015-37. doi:10.1016/S0003-3472(05)80640-2.

Mench, J. A. 1988. The development of aggressive behavior in male broiler chicks: a comparison with laying-type males and the effects of feed restriction. *Applied Animal Behaviour Science* 21, 233-42.

Mench, J. A. 2002. Broiler breeders: feed restriction and welfare. *World's Poultry Science Journal* 58(1), 23-9. doi:10.1079/WPS20020004.

Merlet, F., Puterflam, J., Faure, J. M., Hocking, P. M., Magnusson, M. S. and Picard, M. 2005. Detection and comparison of time patterns of behaviours of two broiler breeder genotypes fed ad libitum and two levels of feed restriction. *Applied Animal Behaviour Science* 94(3-4), 255-71. doi:10.1016/j.applanim.2005.02.014.

Millman, S. I. and Duncan, I. J. H 2000. Effect of male-to-male aggressiveness and feed-restriction during rearing on sexual behaviour and aggressiveness towards females by male domestic fowl. *Applied Animal Behaviour Science* 70(1), 63-82. doi:10.1016/s0168-1591(00)00141-6.

Millman, S. T., Duncan, I. J. H. and Widowski, T. M. 2000. Male broiler breeder fowl display high levels of aggression toward females. *Poultry Science* 79(9), 1233-41. doi:10.1093/ps/79.9.1233.

Moradi, S., Zaghari, M., Shivazad, M., Osfoori, R. and Mardi, M. 2013. Response of female broiler breeders to qualitative feed restriction with inclusion of soluble and insoluble fiber sources. *Journal of Applied Poultry Research* 22(3), 370-81. doi:10.3382/japr.2012-00504.

Morrissey, K. L., Widowski, T., Leeson, S., Sandilands, V., Arnone, A. and Torrey, S. 2014. The effect of dietary alterations during rearing on feather condition in broiler breeder females. *Poultry Science* 93(7), 1636-43. doi:10.3382/ps.2013-03822.

Mullens, B. A., Chen, B. L. and Owen, J. P. 2010. Beak condition and cage density determine abundance and spatial distribution of northern fowl mites, Ornithonyssus sylviarum, and chicken body lice, *Menacanthus stramineus*, on caged laying hens. *Poultry Science* 89(12), 2565–72. doi:10.3382/ps.2010-00955.

Nielsen, B. L., Thodberg, K., Malmkvist, J. and Steenfeldt, S. 2011. Proportion of insoluble fibre in the diet affects behaviour and hunger in broiler breeders growing at similar rates. *Animal: an International Journal of Animal Bioscience* 5(8), 1247–58. doi:10.1017/S1751731111000218.

Norman, K. I., Weeks, C. A., Pettersson, I. C. and Nicol, C. J. 2018. The effect of experience of ramps at rear on the subsequent ability of layer pullets to negotiate a ramp transition. *Applied Animal Behaviour Science* 208, 92–9. doi:10.1016/j.applanim.2018.08.007.

OECD/FAO. 2018. World meat projections. *OECDFAO Agricultural Outlook 2018–2027*.

Ordas, B., Vahedi, S., Seidavi, A., Rahati, M., Laudadio, V. and Tufarelli, V. 2015. Effect of testosterone administration and spiking on reproductive success of broiler breeder flocks. *Reproduction in Domestic Animals = Zuchthygiene* 50(5), 820–5. doi:10.1111/rda.12595.

Petitte, J. N., Hawes, R. O. and Gerry, R. W. 1983. The influence of cage versus floor pen management of broiler breeder hens on subsequent performance of cage reared broilers. *Poultry Science* 62(7), 1241–6. doi:10.3382/ps.0621241.

Pollock, D. L. 1999. A geneticist's perspective from within a broiler primary breeder company. *Poultry Science* 78(3), 414–8. doi:10.1093/ps/78.3.414.

Puterflam, J., Merlet, F., Faure, J. M., Hocking, P. M. and Picard, M. 2006. Effects of genotype and feed restriction on the time-budgets of broiler breeders at different ages. *Applied Animal Behaviour Science* 98(1–2), 100–13. doi:10.1016/j.applanim.2005.08.013.

Renema, R. A. and Robinson, F. E. 2004. Defining normal: comparison of feed restriction and full feeding of female broiler breeders. *World's Poultry Science Journal* 60(4), 508–22. doi:10.1079/WPS200434.

Renema, R. A., Rustad, M. E. and Robinson, F. E. 2007. Implications of changes to commercial broiler and broiler breeder body weight targets over the past 30 years. *World's Poultry Science Journal* 63(3), 457–72. doi:10.1017/S0043933907001572.

Riber, A. B., de Jong, I. C., vad de Weerd, H. A. and Steenfeldt, S. 2017. Environmental enrichment for broiler breeders: an undeveloped field. *Frontiers in Veterinary Science* 4, 86. doi:10.3389/fvets.2017.00086.

Rushen, J., Taylor, A. A. and de Passillé, A. M. 1999. Domestic animals' fear of humans and its effect on their welfare. *Applied Animal Behaviour Science* 65(3), 285–303. doi:10.1016/S0168-1591(99)00089-1.

Sandilands, V., Tolkamp, B. J., Savory, C. J. and Kyriazakis, I. 2006. Behaviour and welfare of broiler breeders fed qualitatively restricted diets during rearing: are there viable alternatives to quantitative restriction? *Applied Animal Behaviour Science* 96, 53–67.

Savory, C. J. and Maros, K. 1993. Influence of degree of food restriction, age and time of day on behavior of broiler breeder chickens. *Behavioural Processes* 29(3), 179–89. doi:10.1016/0376-6357(93)90122-8.

Savory, C. J., Hocking, P. M., Mann, J. S. and Maxwell, M. H. 1996. Is broiler breeder welfare improved by using qualitative rather than quantative food restriction to limit growth rate? *Animal Welfare* 5, 105–27.

Shea, M. M., Mench, J. A. and Thomas, O. P. 1990. The effect of dietary tryptophan on aggressive behavior in developing and mature broiler breeder males. *Poultry Science* 69(10), 1664–9. doi:10.3382/ps.0691664.

Siegel, P. B. and Wolford, J. H. 2003. A review of some results of selection for juvenile body weight in chickens. *Journal of Poultry Science* 40(2), 81–91. doi:10.2141/jpsa.40.81.

Steenfeldt, S. and Nielsen, B. L. 2012. Effects of high fibre diets on gut fill, behaviour and productivity in broiler breeders. *2nd JRS Poultry Seminar*, Hannover, Germany.

Stratmann, A., Frohlich, E. K. F., Harlander-Matauschek, A., Schrader, L., Toscano, M. J., Wurbel, H. and Gebhardt-Henrich, S. G. 2015. Soft perches in an aviary system reduce incidence of keel bone damage in laying hens. *PLoS ONE* 10(3), e0122568. doi:10.1371/journal.pone.0122568.

van Emous, R. A. 2010. Quality Time®: an innovative housing concept for broiler breeders. *The 2nd International Nutrition Symposium*, Wageningen, the Netherlands.

van Emous, R. A. and de Jong, I. C. 2013. Promising management measures to solve welfare problems in Broiler Breeders. *2nd International Poultry Meat Congress*, Antalya, Turkey.

van Emous, R. and Gunnink, H. 2011. *Innovative Broiler Breeder Housing System: 'Quality Time' House* [In Dutch: Innovatieve huisvesting voor vleeskuikenouderdieren: "Quality Time" stal). Wageningen UR Livestock Research, Wageningen, The Netherlands.

van Emous, R. A., Kwakkel, R., van Krimpen, M. and Hendriks, W. 2014. Effects of growth pattern and dietary protein level during rearing on feed intake, eating time, eating rate, behavior, plasma corticosterone concentration, and feather cover in broiler breeder females during the rearing and laying period. *Applied Animal Behaviour Science* 150, 44–54. doi:10.1016/j.applanim.2013.10.005.

van Niekerk, T. F. and de Jong, I. 2007. Mutilations in poultry in European poultry production systems. *Lohmann Information* 42, 35–46.

van Niekerk, T. C. G. M. and de Jong, I. 2017. *Stand van zaken achterwege laten van ingrepen bij pluimvee*. Wageningen Livestock Research, Wageningen, The Netherlands.

Vezzoli, G., Mullens, B. A. and Mench, J. A. 2015. Relationships between beak condition, preening behavior and ectoparasite infestation levels in laying hens. *Poultry Science* 94(9), 1997–2007. doi:10.3382/ps/pev171.

William, J. 2012. Ovarian dysfunction in broiler breeder hens. *Medycyna Weterynaryjna* 68, 131–8.

Wood-Gush, D. G. M. 1958. The effect of experience on the mating behaviour of the domestic cock. *Animal Behaviour* 6(1–2), 68–71. doi:10.1016/0003-3472(58)90010-1.

Zuidhof, M. J. 2018. Lifetime productivity of conventionally and precision-fed broiler breeders. *Poultry Science* 97(11), 3921–37. doi:10.3382/ps/pey252.

Zuidhof, M. J., Robinson, F. E., Feddes, J. J. R., Hardin, R. T., Wilson, J. L., McKay, R. I. and Newcombe, M. 1995. The effects of nutrient dilution on the well-being and performance of female broiler breeders. *Poultry Science* 74(3), 441–56. doi:10.3382/ps.0740441.

Zuidhof, M. J., Fedorak, M. V., Ouellette, C. A. and Wenger, I. I. 2017. Precision feeding: innovative management of broiler breeder feed intake and flock uniformity. *Poultry Science* 96(7), 2254–63. doi:10.3382/ps/pex013.

# Chapter 10

# Opportunities to improve the welfare of young chickens

*Elske N. de Haas, Utrecht University, The Netherlands*


1 Introduction
2 Welfare of parental stock and effects on offspring
3 Incubation practices to optimize chick welfare
4 Hatching practices to optimize chick welfare
5 Rearing practices to optimize pullet welfare
6 Conclusions
7 Where to look for further information
8 References


## 1 Introduction

This chapter reviews the range of issues affecting the health and welfare of young chickens. It starts by assessing the welfare of parental stock and its effects on offspring, including nutritional and environmental stressors in breeders. The chapter then reviews research on incubation practices to optimize chick welfare, starting from natural incubation and the development of the embryo to the chick. It moves on to discuss key issues around artificial incubation in a commercial hatchery. Topics include reducing noise around hatching, provision of light during incubation, as well as in-ovo chemosensory learning and temperature programming.

The chapter then reviews research on hatching practices to optimize chick welfare, both in commercial hatcheries and in on-farm hatching. It discusses welfare issues relating to the provision of food and water inside the hatchery, transport in the hatchery and in-ovo sexing. The chapter also discusses ways of promoting chick health via vaccination as well as alternative immune-boosting techniques, as well as the short- and long-term consequences of beak treatments.

Finally, the chapter assesses rearing practices to optimize pullet welfare, starting with the brooding period and early life rearing. The chapter discusses the use of a dark brooder and the importance of light period, wavelength

http://dx.doi.org/10.19103/AS.2020.0078.09

and intensity. It also surveys work on optimal stocking density and group size. Other issues include the importance of learning from older animals as well the importance of olfactory and other forms of enrichment in, for example, reducing the risk of developing injurious pecking behaviours, reducing the risk of injury by improving spatial and navigation skills, as well as reducing feelings of fear and stress.

## 2 Welfare of parental stock and effects on offspring

Commercial layer and broiler chicks hatch from fertilized eggs which are incubated in an artificial incubator inside a hatchery. These fertilized eggs originate from parent flocks; hens and roosters housed together. Parental chickens can greatly affect the behaviour, development and stress sensitivity of their chicks. These effects can be mediated indirectly via the composition of the egg, the level of egg hormones and other inheritable factors apart from genes (Henriksen et al., 2013; Dixon et al., 2016). Epigenetics describes the study of inheritable factors that result in changes to gene activity (i.e. turning a gene on or off). Epigenetic effects can occur as a result of stress in the parental birds (Rodenburg and de Haas, 2016; Jensen, 2014) and can underlie stress resilience in the offspring. High levels of the stress hormone corticosterone in maternal hens in parent stock farms have been negatively related to flock average egg weights (de Haas et al., 2013) and further related to higher levels of anxious and damaging behaviours in their 1-week-old chicks on rearing farms (De Haas et al., 2014b). In the management and housing of parent stock flocks, hardly any quantitative data exists, but they are known to be at risk for welfare impairments by being housed - generally - in rather barren conditions, under high stocking density and low light intensity (EFSA, 2010). Whether these on-farm conditions impact their offspring is yet unknown, but separate housing factors have been found to affect the offspring's welfare, as shown in scientific studies.

### *2.1 Nutritional and environmental stressors in breedings and effects on the welfare of the offspring*

A significant welfare challenge for broiler breeders lies in the feeding regime applied. Broiler breeders are severely feed restricted to maintain fertility (Robinson and Wilson, 1996), which causes chronic hunger and reduces their welfare (see also Chapter 18). A lack of food for the broiler mothers can reduce the hatch weight of her chicks (Van Der Waaij et al., 2011). Further, when their offspring are fed unrestricted rations (as is the normal case in broiler production), their growth curve is hampered and abdominal fat at 6 weeks of age is increased (Van Der Waaij et al., 2011). A mismatch between the dietary condition of the mother and offspring can thus influence the development of the offspring. Having a perch to sit on appears positive for the welfare of broiler breeders but does not affect the welfare of the offspring (Gebhardt-Henrich et al., 2017, 2018)

In layer parental chickens, several studies show that stressful environments and nutritional conditions can impact the welfare of their chicks. Adding fish oil to the diet of maternal hens increases the omega 3:6 (n3:n6) ratio in the yolk (de Haas et al., 2017), and this was predicted to positively affect both brain development and welfare of the offspring, as found in mammals (Bazinet and Layé, 2014). However, the results for the chicks were the total opposite (de Haas et al., 2017). Having a high n3:n6 ratio in the diet-induced nutritional stress in the hens (indicated by raised levels of faecal corticosterone metabolites) and high fear in their chicks (Aigueperse et al., 2013; de Haas et al., 2017). Environmental stressors experienced by parental layers can also influence the welfare of their chicks. Especially, unpredictability in the environment is a strong component of current and long-term stress in animals (Koolhaas et al., 2011). Unpredictable food availability for parental layers increased fear and decreased eating time in the chicks (Janczak et al., 2007). Habituating maternal laying hens to humans, on the other hand, reduces emotional reactivity in the chicks (Bertin et al., 2019). Unpredictable dark-light schedule in maternal layers induced a preference for high energy food in the offspring (Nätt et al., 2009), impaired spatial learning (Lindqvist et al., 2007) and resulted in similar gene expression in the hypothalamus to their parents (Nätt et al., 2009). The hypothalamus is an important brain area for stress regulation in chicken (Cockrem, 2007). These results indicate a programming effect via the environment of the parents to affect the regulation of stress genes in the brain of the offspring and consequently influence behaviour (Jensen, 2014). Recent studies have supported epigenetic programming effects in layers by showing that stressing the bird as a young chick (i.e. during brooding period (Goerlich et al., 2012) or during early rearing (Ericsson et al., 2016)) can influence brain and blood methylation profiles (Pértille et al., 2017) and stress susceptibility of their offspring. This means that the early life conditions in which layer and broiler parental chickens hatch and are kept can have a profound influence on the next generation. The rearing conditions of parental chickens are hardly studied although some very recent studies on broiler breeder pullets show improvements to litter quality and reduction of footpad dermatitis (a local infection of the footpad associated with wet litter) (Kaukonen et al., 2016), adjustments of feeding schedules (Lindholm et al., 2018) and specific fibre-rich diets (De Los Mozos et al., 2017) with positive effects on welfare during rearing (Arrazola et al., 2019a) and lay (Arrazola et al., 2019b). More emphasis should be given to the study and improvement of the welfare of parent birds, as they can pose a transgenerational threat or benefit to the welfare of their offspring.

## 3 Incubation practices to optimize chick welfare

### 3.1 Natural incubation

Naturally, eggs are laid by the hen, and after approximately 8–10 eggs the hen switches to brooding (Romanov et al., 2002). This means that there is a variation in the age of the eggs when brooding commences. From the start of brooding, the hen creates a bald spot of skin, known as a brooding patch, whereby she warms the eggs (Turner, 1997). An adult hen has a body temperature of approximately 41.8°C, but during brooding and fasting her temperature decreases to around 39.6°C (Bolzani et al., 1979). In the period of incubation, brooding hens can lose up to 15% of their body weight due to a 40% reduction in feed intake. During the last days of incubation the hen leaves the eggs more intermittently and in longer bouts to feed, drink, dustbathe and forage (Turner, 1997).

### 3.2 Development of the embryo to chick

Chickens take 21 days to develop from embryo to chick. The development of the sensory systems of the chick develops rapidly, where the auditory system matures faster (i.e. functional at E12 of incubation, E: embryonic day of age) (Rogers, 1995). Already at E11 and E12 chicks respond to sounds (Saunders et al., 1974). The visual system starts development at E2, matures at E17, but is not matured at hatch yet, while the development of the olfactory system starts at E2 and maturation is finalized at E21 when chicks hatch. At E2 the nasal placode forms, where on E4 nasal pits have been differentiated. Here, olfactory nerves grow towards the brain where the olfactory bulb (OB) is formed. On approximately E5 the oronasal grooves form, and on E6 the upper beak with nasal capsules is finalized. On E7 olfactory neurons have reached the OB, and around E10 digestive processes commence with the uptake of amniotic fluids from E12 onwards (Rogers, 1996). Chick embryo's brain electrical responses appear receptive to odour cues already at E6 (Sato et al., 2016), which increase in the intensity of responses from E10 and E15 (Lalloué et al, 2003). This indicates that the olfactory system continues to be more responsive over the course of incubation. Most sensory systems are well developed at hatch, but brain development and maturation continue up until 10 weeks of age (Atkinson et al., 2008, 2010).

### 3.3 Artificial incubation in a commercial hatchery

Commercial chickens are incubated inside an artificial incubator in a hatchery. In the incubator, eggs are first stored. Once incubation commences, the temperature is (generally) fixed at 37.8°C for 14 days, with a set relative humidity (RH) and a turning protocol. During 14 days of incubation, the eggs are kept in trays and turned automatically as would normally be done by the hen. Likely, the hen turns the eggs so as to make sure all eggs are equally warmed (Turner, 1997), as ones at the periphery can be colder than the ones in the centre of the nest. In some hatcheries there is a pre-incubation stage, where eggs are stored and receive a temperature boost, the effect of which on the development or welfare of the chicks is unknown. At E15, generally, the eggs are transferred to a hatcher where they are kept in baskets. Most often candling (i.e. shining a light through the egg) takes place at this time to identify healthy embryos and to exclude non-fertilized or deceased embryos. At the last days of incubation (E15–E21), the developed chick produces heat so that cooling is needed. Cooling takes place by ventilation. Broiler eggs need more cooling than layer eggs as they produce more heat (Nangsuay et al., 2015).

### 3.4 Reducing noise around hatching

Under the mother hen, the developing chick is exposed to the sounds of the environment. Compared to humans, chickens are more sensitive to frequencies below 64 Hz, with optimal sensitivity around 9.1 kHz and 7.2 kHz (Hill et al., 2014). Specifically, the ability to hear low-frequency sounds is likely related to the communication between hen and chicks (Edgar et al., 2016). The chicks and hen respond to each other's vocalizations during incubation and to vocalizations of other chicks in the clutch (Tuculescu and Griswold, 1983). This process could help the chick to learn the calls of conspecifics and facilitates social imprinting (Edgar et al., 2016). In the hatchery the developing chicks are exposed to high levels of noise coming from the ventilators inside the incubator. The noise levels can range from 70 dB to 90 dB (Tong et al., 2015), up to 100 dB (Rodenburg, 2017), which could inhibit social imprinting. Experimental work (Sanyal et al., 2013) showed that chicks exposed to 110 dB of arrhythmic noise at E10 had a delayed motivation to regain closeness to conspecifics. The motivation to move towards conspecifics when repeatedly tested did not change, suggesting sustained high fearfulness in noise-exposed chicks. High sound exposure (90 dB) around hatching does however not cause dramatic effects on hatchability or chick quality but can result in low hatch weight. A low noise treatment (70 db) plus species-specific vocalization playbacks had similar effects of hatching time and chick quality as the high noise treatment (Donofre et al., 2020). Interestingly, adult hens exposed to noise around hatching were more pessimistic than hens which were quietly incubated, as assessed by a judgement bias test (Rodenburg, 2017). Noise around hatching can thus have long-lasting effects on the welfare of chickens. The ventilator is not the only source of noise. The chicks themselves produce a lot of noise, notably high-frequency distress calls. These calls enable the mother hen to reply, reinstate with her chicks and provide comfort by brooding them (Edgar et al., 2016). Stress responses in

15- to 16-day-old chicks have been found to reduce when they hear specific maternal cluck calls in times of stress (Edgar et al., 2015). Exposure to maternal calls has positive effects on memory in the brain of young chicks by the release of noradrenaline (Field et al., 2007). An artificial incubator can contain thousands of, if not a million, chicks (Photo 1). This means that although chicks are exposed to sibling sounds, social imprinting via auditory conditioning is impossible due to the multitude of conspecific sounds they are exposed to. In the hatchery, the high-frequency distress calls expressed by so many chicks can be extremely high (over 100 dB). Workers often wear headphones to reduce the noise they are exposed to (Carvalho et al., 2015). Newly hatched chicks are still exposed to these sounds. High levels of ambient noise cause stress and fear in layers (Campo et al., 2005) and broiler chickens (Voslarova et al., 2011; Chloupek et al., 2009). It is reasonable to assume that high levels of noise will be stressful for young chicks as well. Once the process of conveying, sexing and vaccinating in the hatchery is finished (explained later in this chapter), chicks are placed in transport boxes in another area of the hatchery. These areas are remarkably quieter, as well as when hatching takes place on the farm. More research on reducing excess sounds around hatching, i.e. by playing specific maternal cluck calls, or whether sounds can be reduced by performing hatching practices under dim light should be an avenue to improve the welfare of newly hatched chicks. One strategy might be to have guidelines on the maximum sound level, which could lead to a benchmarking opportunity for welfare-friendly hatcheries.

### 3.5 Provision of light during incubation

Naturally, hens lay eggs in secluded areas under a lower light intensity than in their 'open' habitat. Further, 'the natural forest habitat of chickens is characterized by dimmed light' as described by *Wood-Gush* (1971). When the hen leaves the nest, eggs can be exposed to light, likely of a low intensity. This happens more often during the last period of incubation (Archer and Mench, 2014). The embryo's eye develops rapidly and becomes sensitive to light cues by E4 (Rogers and Krebs, 1996). In contrast, light is not provided in the incubator or hatchery environment, meaning that commercial chicks generally develop in darkness. This can impact their welfare. A summary of the many studies conducted on the effects of light on the welfare of young chicks is given in Table 1. During incubation, the chick's left eye is occluded by the yolk and later the chick's body, whilst the right eye is fully exposed and can receive light that penetrates through the egg shell (Rogers, 1996). The right eye is functionally linked to the left brain hemisphere. The left hemisphere is associated with visual discrimination, food searching, vocalizations and social recognition. The right hemisphere - receiving input via the left eye - is associated with responding to fearful stimuli, avoidance and aggression (Rogers, 1982, 2012). A high level of brain lateralization has been suggested to be positive for welfare (Leliveld et al., 2013; Rogers and Kaplan, 2019). The reason why lateralized chickens benefit from brain lateralization is probably due to an improved ability to continue food searching while being able simultaneously to detect predators (Rogers, 2010). Chicks exposed to light during incubation use *both* spatial and object-specific cues, whereas dark-incubated chickens rely solely on object-specific cues (Rogers, 1990). Chicks - broilers, layers and ducks - that develop in darkness do not benefit from a lateralized brain and are more fearful and stress sensitive (Archer et al., 2017). Today, light-emitting diodes (LEDs) become more popular within the poultry industry due to their energy efficiency, choices in the spectrum and dimming possibilities. However, during incubation, continuous 24 h LED light (van der Pol et al., 2015, 2019) and LED light of high intensity (Shafey et al., 2005) are not recommended. A more natural dark light schedule (12L:12D) when eggs were exposed to light containing red, green or blue colours or of cooler wavelengths and of low intensity (LUX) as opposed to high LUX appears to yield best results on melatonin activity, overall development during incubation, better hatchability (% of hatched eggs) and chick quality parameters, reduced mortality during incubation, less chicks with unhealed navels, better health at hatch and slaughter age and higher welfare parameters (i.e. lower fear and stress responses) (see Table 1). Despite the finding that hens leave the eggs more often later in incubation, light exposure from early on in incubation improves spatial memory in layer chicks (Chiandetti et al., 2005) and dual tasks in foraging and predator vigilance (Dharmaretnam and Rogers, 2005). Light exposure late in incubation can enhance the use of conflict cues and spatial learning in layer chicks (Wichman et al., 2009) and improve monocular sleep (Bobbo et al., 2002). Archer and Mench (2014) found the strongest effects on welfare in broiler chicks when eggs are exposed throughout

**Photo 1** Chicks inside a commercial hatchery upon hatching © www.eggs-incubator.com.

**Table 1** Studies on light during incubation on the welfare of broiler and layer chicks with the main conclusion as stated by the authors

| Animal | Period of treatment | Colour, wavelength and scheme of light | Light intensity at the egg level | Light effects | Conclusion (reference number) |
|---|---|---|---|---|---|
| Broilers | E11–E22[1] | Monochromatic LED[2] in a 12L:12D<br>RED at 632 nm,<br>BLUE at 463 nm,<br>GREEN at 517nm,<br>polychromatic WHITE at 448nm | 0.04 $W/m^2$[3] | RED, WHITE, GREEN vs. DARK:<br>↑ melatonin levels at E20<br>≈ melatonin levels post-hatch | Selective sensitivity of the chick embryo pineal gland to react to different wavelengths of light (1) |
| Broilers | E0–E18 | RED vs. BLUE LEDs<br>12L:12D | 250 LUX irradiance BLUE: 0.8396 $W/m^2$ irradiance RED: 1.265 $W/m^2$ | RED and BLUE vs. DARK:<br>↓ % of chicks with unhealed navels<br>↑ hatchability<br>↑ % of chicks with defects | Either RED or BLUE LEDs or those with similar spectra could be used to improve hatchability and chick quality (2) |
| Layers, broilers and pecking ducks | E0–E18 | WHITE + RED LED<br>12L:12D | 250 LUX | WHITE + RED vs. DARK:<br>↓ mortality<br>↓% of chicks with unhealed navels<br>↓ % of chicks with defects<br>↑ hatchability | Overall, the combined white and red light demonstrated a benefit to the incubation process in both chickens and ducks (3) |
| Broilers | E0–E14 (14d)<br>E15–E21 (7d)<br>E0–E21 (21d) | 12L:12D | 550 LUX | 21d LIGHT:<br>↓ CORT<br>↓ fearfulness<br>↓ immune responses | Providing light only during the last week of incubation is insufficient to improve welfare (4) |
| Broilers | E0–E21 (21d)<br>E7–E21 (14d)<br>E14–E21(7d) | 1L:23D, 6L:18D, 12L:12D | 550 LUX | 12L:12D 21d and 14d:<br>↑ feeding<br>↓ active | 12 h light during the last 2 weeks of incubation influences behaviour (5) |

| | | | | | |
|---|---|---|---|---|---|
| Broilers with eggs of different parental age groups | E0–E18 (18d) | Green 20-watt fluorescent 50-cm tubes | Low: 900 – 1380<br>High: 1430 -2080<br>Light vs. medium vs. dark brown eggs | Low LUX: $\uparrow$ weight at E16<br>Low LUX * light pigmentation: $\uparrow$ hatch of fertile eggs | Light during incubation improved the hatchability of embryos in light brown eggs laid by young hens (6) |
| Broilers | E0–E21 (21d) | Cool white LED light<br>12L:12D<br>24L<br>24D | 500 lux | 12L:12D LIGHT:<br>$\downarrow$ leg bone deformations at hatch and at slaughter<br>$\uparrow$ yolk-free body mass, liver weight, intestine weight at hatch | the light-dark rhythm of 12L: 12D may have a stimulating effect on leg health (7) |
| Layers | E1–E3<br>E18–E21 | 60 W incandescent light bulb | Not specified | LIGHT vs. DARK:<br>$\uparrow$ Left side lateralization<br>No systematic preference for a direction when reaching a goal box | More than one sensitive period seems thus available to light stimulation to trigger brain lateralization. (8) |
| Layers | E18–E21 | Fluorescent incandescent white light | 800 to 1500 LUX<br>1500 to 3000 LUX | LIGHT vs. DARK:<br>No preference for eye use as opposed to dark have preference for left eye use | the effect of light in-vo depends on the amount of light that reaches the embryo (9) |
| Layer | Not specified | Light-incubation vs. dark-incubated<br>details not specified | Not specified | LIGHT vs. DARK:<br>≈ detour abilities | Incubation did not affect spatial behaviour (10) |
| Broilers | E18–E21 | 25 W incandescent light<br>Scheme not specified | 250 LUX | LIGHT vs. DARK:<br>Monocular sleep with right-eye open<br>Dark during incubation:<br>Monocular sleep with left-eye open | Light during incubation eye opening in post-hatching monocular sleep (11) |

(Continued)

**Table 1** Studies on light during incubation on the welfare of broiler and layer chicks with the main conclusion as stated by the authors (*Continued*)

| Animal | Period of treatment | Colour, wavelength and scheme of light | Light intensity at the egg level | Light effects | Conclusion (reference number) |
|---|---|---|---|---|---|
| Broilers | Not specified | 0L:24D, 12L:12D, 24L:0D L:24D, 1L:23D, 6L:18D, and 12L:12D | 550 lUX | LIGHT vs. DARK:<br>12L:12D ↓ fear responses<br>6L:18D and 12L:12D ↑ lateralization | 12 h of light during embryogenesis reduces fearfulness (12) |
| Broilers | E0–E21<br>E14–E21 | 16L:8D vs. 0L:24D | Not specified | LIGHT vs. DARK:<br>↓ CORT 8 h post-hatch | Improved adaptation to the environment (13) |
| Broilers | E0–E21<br>E14–E21 | 16L:8D vs. 0L:24D | Not specified | LIGHT vs. DARK:<br>↑ BW, muscle weight<br>↓ CORT 6 days post-hatch<br>↓ oxidative stress in the brain | Light during incubation and a similar photoperiod improves adaptation (14) |

Footnote: 1: LEDs = light-emitting diode; 2: E = embryonic day of life; 3: $W/m^2$ solar radiation spectrum, LUX ($lm/m^2$); ≈ similar, ↓ lower, ↑ higher.
References: 1: Drozdova et al. (2019); 2: Archer (2018); 3: Archer et al. (2017); 4: Archer and Mench (2014) AABC; 5: Shafey et al. (2005); 6: van der Pol et al. (2015, 2019); 7: Chiandetti et al. (2005); 8: Wichman et al. (2009); 9: Bobbo et al. (2002); 10: Özkan et al. (2012a); 11: Özkan et al. (2012b).

the whole incubation period. Yet dark pigmented eggs absorb more light than less pigmented eggs (Shafey et al., 2005), indicating that the amount of light reaching the embryo may depend on egg colour (Ghatpande et al., 1995). This means that for different egg characteristics light conditions should be adjusted. Further research to optimize lighting practices during incubation is needed. It is known that matching the light conditions during incubation with that during rearing helps adapting chicks to their environment, resulting in better leg health and thereby better welfare in broiler chickens (van der Pol et al., 2017, 2019).

### 3.6 In-ovo chemosensory learning

Chickens have a natural fear of novel items, including food items. This predisposition is named neophobia. In terms of food, it means that chickens are initially unlikely to ingest a novel food item. Especially in young chicks food neophobia can cause risks for survival. Chicks may die from not eating or learning what to eat. Newly hatched chicks contain a large portion of the yolk in their abdomen (yolk-free mass: YFM), which provides some nutrition for the first hours and day(s). Especially for broiler chicks early feeding is very important as, at hatch, broiler chicks have less YFM than layer chicks (Nangsuay et al., 2015). Eating as soon as possible after the hatch is also better for gut development and growth in chickens. Inoculation of feed extracts in the egg can also improve hatchability, chick quality and post-hatch growth while reducing embryo mortality (Bilalissi et al., 2019). Matching maternal diet with chick diet might be another alternative, as especially for fatty acids the composition of the diet reflects that of the yolk in layers (de Haas et al., 2017). During rearing chicks and pullets are given different types of diet, which can vary in particle size, smell, taste and structure. Chicks are often provided with one or more starter diets in the dense mashed condition provided in succession. Point-of-lay pullets are given a pre-lay diet, which could again be different in structure, taste and smell. Provided that a gradual transition occurs during diet changes, the chicks/pullets need time to adapt so that a drop in weight gain is less likely to occur. Food neophobia can be related to the appearance, smell and taste of the food. It is known that olfactory cues provided during incubation may help chicks to adapt to different novel foods. Chicks take in amniotic fluid and are able to discriminate different olfactory cues before hatch (Hagelin et al., 2013). The egg shell contains approximately 150 pores per cm$^2$, and so gas exchange is possible via the porous shell (Onagbesan et al., 2007). Gas exchange occurs via different layers of membrane: the outer shell, which is a porous matrix; then an inner matrix with mammillary cores; and the chorio-allantoic membrane attached to the embryo (Mohammed et al., 2015). Gasses and odour components can enter the nasal cavity of the chick via the shell and bind to at least nine different receptors (Nef et al., 1996). Here lies also a risk as the development and survival of the chick embryo are severely impacted, should toxic gases such as $CO_2$ enter the egg (Onagbesan et al., 2007), especially in bigger eggs (Oro et al., 2019). Information on odour components is taken up by olfactory neurons situated in the olfactory epithelium of the nasal cavity, which are further processed by the OB. Exposing chick embryos from E15 to E20 to strawberry flavoured oil on the shell induced a preference for a strawberry flavoured environment and water in the exposed young chicks (Sneddon et al., 1998). Exposing eggs at the end of incubation to a combination of olfactory cues (0.37% of four pure compounds and two essential oils) also stimulated the ingestion of flavoured food as opposed to non-exposed chicks (Bertin et al., 2011). The intensity of the odour can however yield opposite effects, as a very high level can cause aversion for flavoured food (Bertin et al., 2010). As olfaction plays a role in feeding preferences, in-ovo chemosensory learning can tune in to the role of the OB-determining food preferences. Mabayo et al. (1996) showed that removing the OB or the nerves connecting the OB in 7-day-old male chicks eliminated a preference for different diets. The OB plays an important role in food inhibition, as shown by the removal of the OB in 4-month-old pullets, causing increased food intake (Robinzon et al., 1977). As chicks hatch under the mother, they are also exposed to her specific olfactory cues from the skin, feathers and droppings (Moran, 2019). In some countries, disinfecting eggs and chicks with formaldehyde inside the hatchery removes any olfactory cues, and, further, formaldehyde causes respiratory irritations in broiler chicks (de Gouw et al., 2017). Another more healthy strategy is to disinfect eggs with an ultraviolet light or hydrogen peroxide (Wells et al., 2010; Mcdaniel et al., 2011). Being able to recognize olfactory cues could further help chicks to habituate to their environment and start ingesting food faster with positive effects on gut development, survival and welfare.

### 3.7 In-ovo temperature programming

Chicks, whilst in the egg, do not experience a constant temperature when incubated under a broody hen. In contrast, in a commercial incubator eggs are incubated under a standard temperature scheme, with higher temperatures at the beginning of incubation (37.8°C during E0–E14) and later with lower temperatures and additional cooling. The ambient temperature inside the incubator does not reflect what an embryo experiences, so studies on incubation temperature use eggshell temperature (EST) as a proxy for embryo temperature. The execution of alterations in EST such as low (36.9°C), high (38.6°C) or very high (39.4°C) in comparison to 37.8°C affects chick development. Low temperatures decrease hatchability and chick quality. Chick quality is rated in day-old chick (DOC) by the Tona score: based on qualitative scoring with a max of 100 (Tona et al., 2004) or Pasgar © score with a max of

10 (see Photo 2). These scoring systems combine navel score (black button or leaky navel), yolk sac (large size of the residual yolk sac indicating slow development), red hocks (red or swollen hocks), abnormal beak (red beak or nostrils contaminated with albumen) and low alertness as scientific criteria to as first or second-quality chicks (Van De Ven et al., 2012). Second-quality broiler chicks are less likely to survive as compared to first-quality broiler chicks in the first week of life (62.5% vs. 0.97%), but overall 40% of second-quality chicks can live up to slaughter at 42 days, albeit with a lower bodyweight (33.8 kg) as compared to first-quality chicks. As they represented only 0.04% of the total live weight of the flock, they can be neglected to influence average flock weight as a common reason to discard and cull these chicks. For ethical reasons, a good assessment of the welfare of these second-quality chicks in both broilers and layers could make it justified to cull these chicks when they would otherwise suffer. However, not much is known on the numbers of these chicks at hatching and the choices for culling DOC.

Chickens are kept in various places in the world, in different climates, in different housing conditions, inside/outside and in high or poor climatic conditions. Chicks may experience thermal stress. In light of the projected climate change this could become a bigger issue in tropical extensive chicken facilities. Priming the thermoregulatory capacities of chicks can occur during incubation or in early life. Exposing eggs to high temperature makes broiler chicks more resilient to high temperatures later in life (Morita et al., 2016) and produce better at later ages (Collin et al., 2007). A high incubation temperature can result in hock burns at hatch (Van De Ven et al., 2012), later life ascites and retarded growth (Molenaar et al., 2011), so care is needed when designing a thermo-conditioning protocol during incubation for hot and cold conditions. For layer embryos, exposure for 1 h/day to 27°C at E12–E19 delayed hatching,

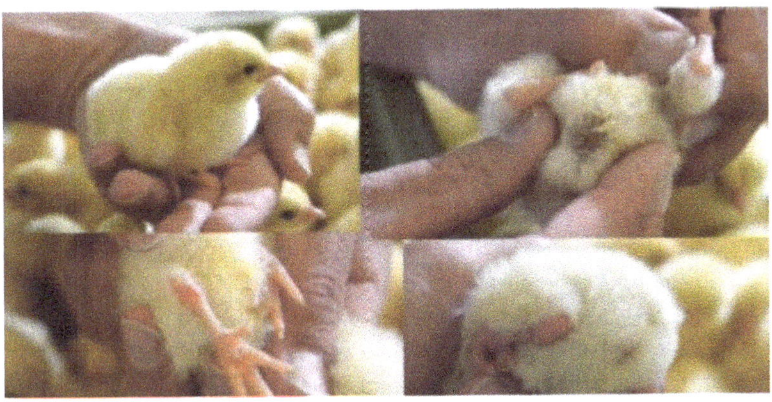

**Photo 2** Four criteria of the Tona and Pascar score for chick quality: alterness, navel conditions, red hocks and full belly; indicative of low yolk absorption © PasReform.

reduced growth rate and resulted in higher emotional reactivity and higher expression of corticotrophin-releasing factor in the amygdala of young layer chicks (Bertin et al., 2018). Temperature treatment during incubation can thus, directly or indirectly via affecting hatching time, prime young chicks' brain and stress resilience. Surprisingly little information exists on the variation in incubation temperature under a broody hen and thus what can be expected from normal deviations in terms of adaptability of chicks to their early post-hatch environment.

## 4 Hatching practices to optimize chick welfare

### *4.1 Process of hatching*

Hatching is an energy-exhausting process for the chick. It can last up to 30 h. First, the chick turns so that the beak with egg tooth is well positioned to peck through the inner membrane of the egg. This moment is called inner pipping. Once inner pipping has been successful, the chick can breathe within the air bubble position at the top of the egg. After some recovery of energy expenditure, the chick tries to peck through the outer shell of the egg. This moment is called outer pipping. Intermittent pecking and resting now occurs, as well as peeping and sleeping (Rogers, 1995). Finally the chick is able to hatch by itself. After hatching, chicks are seen to sleep the majority of the time in order to recover from the energy loss. Helping chicks to hatch is dangerous, as the inner membrane of the shell with its blood vessels is still connected to the body tissue of the chick and thereby can damage the skin tissue and blood vessels. Mothers may peck at the egg but only rarely and are more often seen to respond to the chicks peeping with their maternal calls.

### *4.2 Provision of food and water inside the hatchery*

In general no food, water or light is provided in the hatchery machine. Once the majority of the chicks are hatched, the chicks are taken out of the hatchery machine. In a clutch, chicks can hatch as early as E19, but hatching can extend to E22. The hatch window is generally accepted to be between 476 h and 512 h (Careghi et al., 2005) but can be longer such as from 497 h to 523 h (Løtvedt and Jensen, 2014). Commercial hatcheries mostly open the incubators at different time intervals between 504 h and 516 h, meaning that the first hatched chicks are already without food and water up to 38 h to 50 h (Bergoug et al., 2013). Opening the incubator at certain intervals can cause a decrease in temperature and delay hatch time further, according to Swann and Brake (1990). Staying for a long time in the incubator (>30 h) can cause a loss of body weight up to 14% for the early-hatched chicks (Careghi et al., 2005). These chicks are further at risk of dying from dehydration (Joseph et al., 2006). Time of hatching, i.e. early

(± 490 h), mid (± 503 h) or late (± 510 h), can further influence the development and behaviour of chicks at hatch and thereafter (Bergoug et al., 2013). Early-hatched broiler chicks show to have a higher risk to develop foodpad dermatitis likely due to staying in a too hot environment for an extended time as compared to late-hatched broiler chicks (Careghi et al., 2005). Late-hatched broiler chicks generally have a worse quality score at hatch and lower body weight at day 9 of life as compared to early-hatched chicks (Wang et al., 2014). In layer chicks, hatching time was negatively correlated with growth from day 4 of age up to 8 weeks of age in the females (Løtvedt and Jensen, 2014). For the male layer chicks in this study, early hatching resulted in higher emotional reactivity and increased spatial learning as opposed to hatching midterm. Post-hatch food and water deprivation in broiler chicks can further lead up to lower body weight and mortality up to 6 weeks of life (De Jong et al., 2017). Organ development and physiology are also affected by lack of food and water, but these effects appear short term. Being able to feed and drink shortly after hatch can help overcome the energy dip chicks have from the hatch process. The majority of hatcheries do not provide food and water inside their hatchery machines, although several companies have developed systems which enable food and water inside the hatchery and during transport. In the hatchery chicks are placed in cardboard transport boxes or crates to be transported to the rearing farm. During these transports (which can be several hours up to days) the chicks are generally not provided with food and water or extra heat. The EU council has set that the time of transport should be within 24 h and transport should be completed within 72 h after hatching. These criteria are based on the residual yolk sac which – at best – can last up to 3 days, but there is great variety in this, largely dependent on incubation settings. Obtaining food and water is important for newly hatched commercial chicks, especially as they hatch within an artificial incubator where the air can be dry due to excessive ventilation and cooling, and by exposure to dust or cleaning detergents on the eggs. Making it a standard practice to provide food and water immediately after hatch and during transport will benefit the welfare and optimize growth, and improve immune function for billions of newly hatched chicks.

### 4.3 Optimize the mode of transport in the hatchery

Upon hatching, chicks are removed from their hatching boxes. The boxes are placed on a conveyer system, containing several conveyer belts where the chicks are separated from the remaining egg shells by tilting the box. The chicks are further sorted on quality and sex, further counted and vaccinated and in some cases beak trimmed. A large part of the transport process goes via different conveyer belts under high speed, varying in an acceleration of 4.9 up to 93.9 *g* (Knowles et al., 2004). Passing through the system chicks drop from one conveyer to another at heights up to 55 cm. As a result of the changes in speed and falls, the majority of chicks are not able to stand during transport. At no greater than 0.4 *g* change in speed, only 40% of the chicks are able to maintain standing while experiencing acceleration Knowles et al. (2004). As employees process millions of chicks, desensitization to animal suffering may occur, so chicks that fall or get stuck may not be noticed (or helped) readily. The combined effects of the conveying, sexing, sorting and vaccinating mean that the welfare of the chicks is at risk, but this has been hardly studied. A recent study by Hedlund et al. (2019) showed that conveying and handling upon the hatching of layer chicks causes short- and long-term stress and increases levels of injurious feather pecking (IP) behaviour in later life. Finding better options for transport, reduction of the number of conveyers, conveyer setting and configuration could likely reduce the stress from this process.

### 4.4 In-ovo sexing

During the sexing procedure in the hatchery all layer chicks are sexed, and in some countries broiler chicks are also sexed when the production is specialized to male and female broiler production. These hatching procedures are conducted in a fast manner as thousands of chicks pass the employees specifically trained to determine the sex of the chicks by colour, cloaca or wing sexing (Kaleta and Redmann, 2008). It has been noted that cloaca sexing is not always performed gently, and the rough handling of chicks can cause injuries or even death due to the rupture of cloaca and internal tissues. The speed of the process of sexing and handling of all chicks should thus be very well monitored so as not to cause unnecessary suffering. As only female layer chicks will eventually lay eggs, male chicks are often immediately killed. The methods of killing are either by $CO_2$ exposure or by macerations. Both processes are not very humane and raise ethical questions on the method of killing and the suffering involved. Further, hatching males and killing them after one day of life is another additional question of ethics, as these animals are withheld from living and thriving (Fernyhough et al., 2020). Recent advances in in-ovo gender identification, i.e. determining the sex of the embryo whilst in the egg (Weissmann et al., 2013, 2014), can stop the killing of day-old male chicks, which is likely going to be implemented in France from May 2021. From E8 eggs can be sexed by the assessment of certain biomarkers. From this day onwards, the level of biomarkers that can be taken from the embryo for assay is enough and does not cause a risk for survival. Once the sex has been determined, hatcheries can decide to exclude these eggs for further incubation. In-ovo sexing would enhance efficiency in the hatchery by the use of space in the incubator only for female layer chick. E8 is prior to the formation of the brain stem (Rogers, 1995), an important moment where conscious feeling may occur. Although not optimal, in-ovo sexing might be a better alternative than killing healthy day-old male chicks. Whether or not this technique would be globally implemented is still questionable.

Finding a commercial goal for layer males would be another strategy. Having a double purpose in the breeding strategy for layer chickens, i.e. egg production and weight gain, could lead to heavier breeds where both hens and roosters can serve as a meat source (Giersberg and Kemper, 2018). Layer flocks from a double purpose (*Lohmann Dual*) strain show less behavioural problems such as cannibalism and injurious pecking than layer flocks from an egg-selection strain (Giersberg et al., 2019). Recently, the first double-purpose organic breeding stock has been developed in the Netherlands. Grandparent and parent-stock and production flocks are kept organic to optimize adaptation and production in similar housing situations. This new double-purpose breed (*de Vredelinger*) facilitates keeping males till 18 weeks of age to produce meat and hens to lay up to 255 eggs.

## 4.5 Alternative immune-boosting strategies

In the hatchery and on the rearing farm chickens are vaccinated against viral diseases and some bacterial or parasitic infections. Some vaccinations are given subcutaneously in the hatchery, while others are given on the rearing farm. Some are injected or presented via drops in the eye or sprayed on the chickens. Marek is commonly given subcutaneously, while Newcastle disease, infectious bronchitis and mainly coccidiosis vaccination are commonly given by spray. Spray vaccination takes place in the hatchery for layers or immediately on the rearing farm for broilers, with chicks still in the baskets or transport boxes or on-farm. This spray vaccination is in an attractive colour (pink, purple, green) to stimulate the chicks to peck at the coloured drops of the vaccination liquid on the plumage of their group mates (i.e. see vaccination developers HIPRA, MSD). One can imagine that the first interaction with pecking another chick and ingesting a food particle may cause an initial association between pecking and ingesting. As the chick's reward and avoidance systems are not yet fully developed (Rogers, 1995), the risk of imprinting of edible items found on group mates could lead to a risk for the development of IP later in life. Studies on the potential risk of vaccination strategies for the development of IP have not yet been performed. Other vaccinations can be applied via injection, inside a vaccination carousal or by hand. In a vaccination carousal injection pressure is predetermined and handlers push chicks to the needle. Chicks are taken in pairs and hung by their heads with a pressure device holding them by their head (Photo 3). They are vaccinated and in some cases beak treated at the same time. As the carousal turns, the chicks hang by their neck until they are dropped in

**Photo 3** Chicks being placed by the head in a vaccination or beak-trimming carousel. More info available at: www.fwi.co.uk.

a collection box. This application is performed rapidly to process as many chicks as possible. As injection and handling can be stressful for the chicks, less invasive methods of in-ovo vaccination or alternative strategies such as selection on immune function or prenatal immune programming via the diet would be more optimal in terms of chick welfare. Recent studies show that the selection on high natural antibodies in layers at 16 weeks of age improves general disease resistance and decreased morbidity of chicks at 15 days of age to infections with avian pathogenic *Escherichia coli* (Berghof et al., 2019). Immuno-competence in chicks is already developed prior to hatch, and selection on innate immune genes to Newcastle disease seems another promising route to improve avian innate immune responses (Schilling et al., 2018, 2019). Supplementation of certain vitamins in the diet of the breeders can further improve chick's immune responses (Dixon et al., 2016). Carotenoids (which give the egg yolk its characteristic yellow-orange colour) supplemented to broiler breeders and vitamin E in specific dosages improve immune responses to viral challenges in very young chicks (Haq et al., 1996). Vitamin D3 in broiler breeders also appears to accelerate gut maturity in the chicks, which is important for protection from viral and bacterial infections (Ding et al., 2011). Mass application of in-ovo vaccination or probiotic in-ovo delivery can improve immune programming and resistance to certain viral diseases (Siwek et al., 2018), but not yet all. The technique requires penetrating the egg shell, but when it is safe and applicable could improve welfare of young chicks enormously. It would reduce the handling of chicks to such a big extent that from the hatcher the chicks only need to be placed in boxes and not further conveyed, handled (Bergoug et al., 2013) and transported, which all are very stressful (Hedlund et al., 2019).

## 4.6 The short- and long-term consequences of beak treatments

The beak of a chicken is a highly sensitive organ innervated with a high abundance of nerve endings (Kuenzel et al., 2007), important for foraging, manipulation of food items and preening (McKeegan and Philbey, 2012). In many countries chickens' beaks are shortened at a young age (i.e. from 1 up to 10 days of age). Beak treatment blunts the beak, so as to reduce the impact of IP. The common method was cutting the chick's beak with a hot blade (HB) at around 10 days of age (Glatz, 2000). Infra-red (IR) method - more commonly used nowadays - lasers the beak. Due to the IR the affected area will become necrotic and die off. Both methods are painful for the chick, as shown by a reduction in time spent eating and drinking in the week post-treatment (Marchant-Forde et al., 2008). Although HB-treated chicks experience a short pain-free period lasting up to 26 h (Glatz and Lunam, 1994), there is a general reduction in feeding and pecking behaviour. Feed intake is reduced in the week after, as well as the time spent pecking and absolute peck force at 3 weeks but not 4 or 5 weeks of age (Glatz et al., 1992), with effects stronger when HB trimming occurs at 10 days of age than at 2 days of age. Comparing IR with HB shows that IR treatment causes a reduction in food intake up till 4 weeks of age and a reduction in general activity at 5 weeks of age (Dennis and Cheng, 2012) and a reduction in body weight gain up to 8 weeks of age (Angevaare et al., 2012) and higher mortality than in HB pullets (Honaker and Ruszler, 2004). On the other hand, the beaks of IR trimmed are more consistent in length and morphology (Marchant-Forde et al., 2008), while HB trimming can also lead to chronic deformations of the beak and chronic pain from neuromas which appear less with IR treatment (Marchant-Forde et al., 2008). These neuromas can persist up to 70 weeks of life, causing life-time consequences on pain and feeding behaviour (Lunam et al., 1996). HB trimming performed on older chicks has stronger detrimental welfare effects than on younger chicks, while IR treatment is generally applied in the hatchery on newly hatched chicks (Photo 3). The length of IR trimming should, however, be adapted to the beak length of each DOC, and consideration should be made with regard to the power setting and plate protocol of the IR method (Dennis and Cheng, 2012). This makes it less likely that from small chicks a large portion of their beak is removed. IR-treated chickens do appear less efficient in pecking and had reduced early egg production parameters than non-treated chickens (Angevaare et al., 2012). Beak trimming can have long-lasting effects on meal sizes at 77–80 weeks of age, inhibiting the chickens' daily feeding behaviour and impairing its welfare. And even minor trimming can have long-lasting loss of beak sensitivity (Freire et al., 2008). Instead of beak-trimming young chicks, farmers should use preventive strategies to reduce the development of IP (see Chapter 15). This would greatly enhance the welfare of chicks in the short and long term. In some countries, e.g. in Norway, Switzerland and Denmark, beak treatments do not take place. In Norway, despite having intact beak layer flocks, generally maximum of 2.82% mortality is recorded (Tahamtani et al., 2014). Likely due to smaller flock sizes than in other countries (i.e. maximum of 7.500) and only a few rearing farms, Norway has IP in laying hens well under control.

### 4.7 On-farm hatching

The stress in chicks from hatchery practices can be prevented by on-farm hatching. On-farm hatching is a method by which incubated eggs are placed in the rearing farm at E18. The possibility for chicks to directly eat and drink after hatch increases their survival de Jong et al. (2017). When the chicks hatch they are in a quiet, light and clean environment, with food and water at their disposal and siblings nearby. Broilers which hatch on-farm are less likely to develop footpad dermatitis and have better first-week performance than conventional hatched broilers (De Jong et al., 2017). Adjusting the feeding management of young chicks can also have long-term effects on their health. Having access to food and water early in life for layer chicks has a beneficial effect on immune function by reduced macrophage number in the gut (an indicator of inflammation) (Walstra et al., 2010). The hatching practice of not providing food and water is less optimal for the development of the chicks' gut health. In on-farm hatching systems second-quality chicks remain in the flock. However, when these chicks hatch on-farm in comparison to hatch in a hatchery, their weight is between 50 gm and 200 gm higher (Van De Ven et al., 2012). The immediate access to food and water for on-farm hatched chicks (first- and second-quality chicks) likely stimulates intestinal development as compared to food- and water-deprived hatchery chicks and increases chicks' survival. The importance of on-farm hatching also allows better control over the number of chicks (by counting the number of eggs) placed in the rearing farm.

## 5 Rearing practices to optimize pullet welfare

The rearing period of broiler chickens covers approximately 6 weeks up to their slaughter. The rearing period of layer chickens is until approximately 17–18 weeks of age. For laying hens the rearing period represents more than a quarter of their life. Young point-of-lay hens are generally caught from the rearing farm and transported to the layer farm. On the laying farm, the hens will start laying eggs from around 20 to 21 weeks of age. On some rearing farms a DOC phase is adopted, where brooding conditions are simulated (referred to as brooding period). The duration of this brooding period varies until 4 or up to 6 weeks of age. It has been documented that chicks show distinct behavioural changes within these first 4 weeks of life (Rogers, 1995). These behavioural transitions coincide with changes in brain development and brain specialization. Further, the first 4 weeks of life is the period where most structural brain formation takes place, which primes the brain and sets its orchestra for later life. The conditions in early life can be considered the most important period whereby positive effects on welfare can be achieved, which can last until adulthood.

## 5.1 The brooding period: early life rearing

Newly hatched chicks are poilkilothemic, which means that they are not able to maintain their body temperature. The brooding period is the early life period where chicks need supplementary heat. As they age, chicks become more homeothermic. Under natural brooding, the hen remains on the nest in the first days while taking short bouts of the nest to feed. When this happens the chicks will follow the hen. The chicks remain close to the hen's body and can easily search for cover when they become too cold, scared or tired. As the chicks age, they will take longer bouts of foraging with the hen and gradually adapt to the ambient temperature but also gradually take up more nutrients to maintain body temperature. After approximately 5 days of age, the chicks will leave the nest more often and steadily become independent. Around 10 days of life, the chicks' thermoregulation capacity is fully developed (Rogers, 1995). Chicks' core temperature shows a gradual increase from 38.4°C on day 1 to 40.6°C on day 10, meaning chicks become more tolerant of hypothermia in order to feed (Tzschentke and Nichelmann, 1999). A gradual decrease and stimulated adaptation to ambient temperature initiated by the hen or the chicks themselves helps them adapt to later life ambient temperature. This situation is somewhat different than is generally adopted on the rearing farm, where the ambient temperature is changed gradually, often, weekly, whereby the adaptability of the chicks is not stimulated; neither do the chicks have control over the temperature they prefer. When on a rearing farm spot heating takes place or a brooder guard is present (a secluded heated area) for the first weeks of life, chicks have more possibilities to choose their temperature preference without becoming too cold.

## 5.2 The use of a dark brooder

One option for enabling temperature choice for chicks is a dark brooder. A dark brooder is a dark, separated secluded heated area in the barn (see Photo 4). The benefits of a dark brooder for the chicks are that it provides cover and a warm resting place. A dark brooder may aid in reducing active chicks disrupting inactive or sleeping chicks. Disruption of sleep in young chicks negatively

**Photo 4** Dark brooder installed on commercial rearing farm © T. Decroos (ILVO, Belgium).

affects their welfare, as the brain does not get enough time to recover. Young chicks require a sufficient amount of sleep to help the maturation of the brain at particular stages (Rogers, 1995). Often peaks of sleep coincide with particular transitions in behavioural development and learning. Very young chicks lose the total muscle tone and lie or sit with their head down. For chickens it is possible to sleep with the left eye open (monocular sleep), which allows vigilance for predators. Binocular sleep increases as chicks age, and a typical sleep posture develops (head tucked between wings on a perch). When chickens sleep, they have a higher threshold for arousal. So when they are disturbed, this could initiate fear responses but also restlessness. Sleep helps for memory consolidation and further improves cognition in chicks. Lack of enough sleep, due to excessive light exposure or 24 h light schemes, causes the increase of binocular sleep (catch up sleep) in chicks at the expense of alertness (Boerema et al., 2003). Enabling a proper regulation of rest and sleep by a good daily rhythm for chicks or by social cohesion with the use of dark brooders can improve their welfare. Dark brooders effectively separate active chicks from resting chicks who are often the target of IP (Riber and Forkman, 2007). Thus, chicks reared with dark brooders show decreased IP compared to chicks reared with a heat lamp (Jensen et al., 2006). Commercial flocks of layers which have been reared with a dark brooder were less fearful (Riber and Guzman, 2016) and showed fewer indications of IP both at rear (Gilani et al., 2012) and lay (Riber and Guzman, 2017) compared to non-dark brooded chicks. A dark brooder simulates elements of a mother hen while in addition enables less use of energy resources. Instead of heating the whole barn, only the dark brooders are heated. Chicks reared without a dark brooder are often searching for hiding places, which indicates that the need for cover in dark places is important for the chicks wellbeing (Photo 4).

### 5.3 Light period, wavelength and intensity

When chickens hatch, they have hypermetropia, i.e. far-sightedness, and slowly become emmetropic once they age up to 6 weeks. The photoperiod, colour, flickering and intensity of light which chicks are exposed to in the early life can have consequences on their welfare. On the other hand, chickens have well-developed eyes, large eyes as compared to their body and extra ultraviolet cones which humans do not process and can use mono and binocular vision indicating their sensitive and well-developed vision (Nicol, 2015). Housing broiler chicks under continuous light are executed so as to stimulate food intake; however food intake is not hampered by a more natural photoperiod such as 16L:8D. Indeed the number of chicks eating, drinking, walking, standing and foraging is recorded higher in 24L vs. 16L:8D (Bayram and Özkan, 2010). These results can also be achieved by enhancing synchrony when a natural rhythm 16L:8D is combined with a relatively high light intensity (200 lux vs. 5 and 50 lux) with positive effects on welfare as the chicks show more bouts of uninterrupted resting as compared to 24L (Alvino et al., 2009). Also, when a dark secluded area is not available in the early days of life (see paragraph above), chicks (both layer and broiler) benefit from a dark period, i.e. 4 h light and 2 h dark cycle (4L:2D) for the first week of life, which would roughly occur when brooded by a hen. During the night melatonin production occurs. Under 24L melatonin rhythm becomes disturbed as chicks do not get enough uninterrupted sleep (Sinkalu et al., 2016), while light intensity does not affect melatonin or behavioural rhythms (Deep et al., 2012). Providing melatonin on top of a 12L:12D scheme shows to reduce vigilance and fearfulness in broilers (Sinkalu et al., 2016). Under continuous light, chickens' eyes cannot develop normally, causing larger eyes (Lewis and Gous, 2009). Eye weight positively correlates with increasing photoperiod in broilers (Fidan et al., 2017). It is recommended that 'minimum period of daily darkness required to maintain chicken eye growth within the normal range and show that very short photoperiods, as well as ultra-long photoperiods and continuous illumination, adversely affect ocular development, with potential welfare implications', as stated by Lewis and Gous (2009). A natural rhythm of 16L:8D for broiler chicks causes more comfort behaviour (wing shaking, extensive preening), a reduction in fear responses and a greater degree of sociality without affecting their growth performance at 42 days as compared to 24L light (Bayram and Özkan, 2010). Having a restricted light programme further reduces the risk of leg health problems (Karaarslan and Nazlıgül, 2018). In layer pullets, shorter photoperiods vs. constant light influences gonadal development positively and do not limit food intake or hamper growth (Lewis et al., 2009). When 14-week-old pullets are able to choose which light type they prefer based on their early life rearing conditions, they prefer the light they have been already exposed to (Gunnarsson et al.,

2008a). For organic or free-range layers, rearing with daylight could improve outdoor ranging in daylight as part of their light preference. Similarly, having daylight in an 8L:16D schedule causes an earlier onset of synchronized night-time perching in layer pullets (Gunnarsson et al., 2008a). Daylight combined with enrichment also enhances walking, foraging and exploration in broilers (De Jong and Gunnink, 2019). Housing layer chicks under natural photoperiod (NP) causes a better entrainment of their circadian rhythm displayed by major 'clock' genes in brain and ceca, while chicks under extended period (EP) of light do not have a circadian rhythm (Hieke et al., 2019). Chicks housed under EP of light vs. NP differ in microbiota composition (Fig. 1). Chicks housed under an NP have genera of microbiota resembling a healthy gut, while chicks housed under an EP have genera of microbiota resembling a diseased gut (Hieke et al., 2019). These studies indicate that chickens are sensitive to light period whereby a natural rhythm of light and dark periods promotes adaptation and welfare compared with overly extended photoperiods.

Riber (2015) assessed which LED temperature would be associated with welfare in broilers but found no effects but advised 6065 K for optimal use in terms of productivity. A wavelength of 6065 K is described as 'cold-white' containing mostly blue wavelengths. When using different colours or wavelengths, broilers seem to be more activated and fearful under red and red-yellow light but less active and fearful under blue-green light (Hassan et al., 2013). Red light seems to enhance activity but also aggressive interactions as well as IP, while under blue light Fayoumi chickens are most calm (Hesham et al., 2018). Red light during rearing also reduces stress responses in layers without affecting production (Archer, 2019). Light colour is often combined with intensity in different studies, which are important guidelines for producers to keep so as to ensure optimal growth, welfare and production for broilers and layers. Having light sources exhibiting short wavelengths with LUX greater than 5 after the brooding period stimulates broilers' growth (Arowolo et al., 2019), reduces leg problems and ulcerative footpad lesions (20 LUX vs dim: Fidan et al., 2017) and increases activity (20 LUX vs. 5 LUX: Rault et al., 2017). Similarly, broilers' feed and water intake are greater in high-intensity rearing as compared to medium, low or weak intensity using yellow LEDs (Pan et al., 2019). Enabling the chickens to choose between different light intensity in different zones shows that it can enhance a compensatory regulation of serotonergic and dopaminergic activities (Kang et al., 2020). During rearing of pullets a high light intensity can be feared as a risk factor for the development of IP or cannibalism in flocks kept in high density (Kjaer and Sørensen, 2002). However, when chicks were housed under 5 LUX and 60–80 LUX, it did not influence the incidence of cannibalism (Hartini et al., 2002), likely because chicks were able to search for dim light. Having a varying degree of light intensity stimulates different behaviours, active behaviour in the litter area with

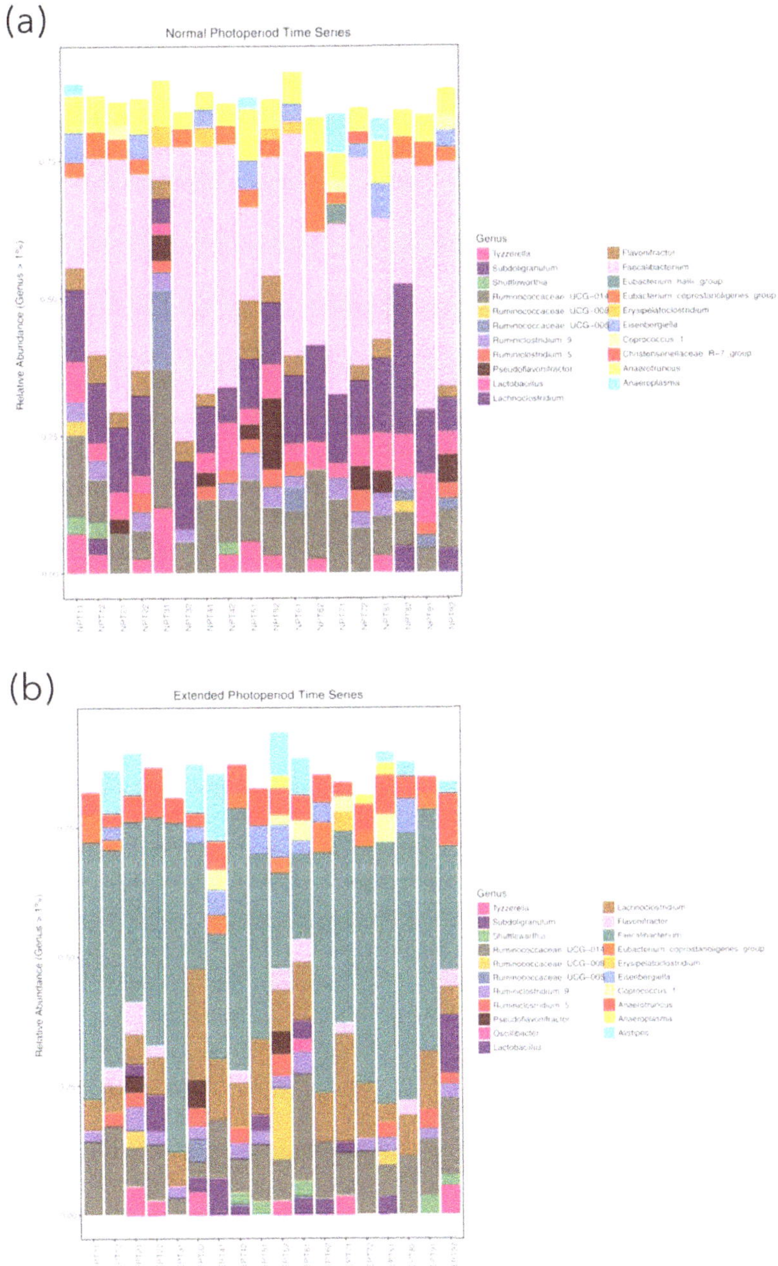


**Figure 1** Gut microbiota of chicks exposed to a natural dark and light period (a) or an extended photoperiod (b) (Hieke et al., 2019). A natural dark/light period stimulates a healthy gut, while extended photoperiod resembles a diseased gut.

high light intensity and a darker area around the perches could help to limit disturbances of resting chicks due to active chicks and reducing the risk of IP or cannibalism development. Generally, management guides for pullets advises high light intensity in the early life (around 40 LUX) and then reducing to 20 LUX; the differences in areas or alterations in light intensity by age are not yet scientifically researched for their impact on the welfare of pullets.

### 5.4 Stocking density and group size

Maximum stocking density for layer pullets is not legally laid down, as it is for adult laying hens in the European Directive (1999/74/EC) and as it is for broilers in the European Directive (2007/43/EC). These standards are based on specific adult body weight for layers (9 hens/m$^2$) or expressed in 33, 39 or 42 live body mass kg/m$^2$ for broilers. Exceeding the stocking densities has consequences for the welfare and growth of broilers and layers. Housing broilers at a greater density than recommended (19 birds/m$^2$) lowers their growth and feed conversion ratio (FCR) while increasing litter moisture and thereby increasing the incidences of footpad lesions (Petek et al., 2014). The EU Platform on Animal Welfare (for further reading see link below this chapter) indicates that that the space chickens need should be calculated in relation to their behavioural demands and their live weight. High stocking density increases risks of IP and feather damage on-farm (> 18 pullets/m$^2$ (Bestman et al., 2009) > 10 pullets/m$^2$ (Huber-Eicher and Audigé, 1999)). Recent research (Liebers et al., 2019) shows that the stocking density of 22–23 pullets/m$^2$ compared to 18 pullets/m$^2$ does not result in lower levels of feather damage or injuries. More space per chick does result in lower anxiety levels and less space per chicks (overcrowding), indicating long-term effect of stress (Eugen et al., 2019). In regard to floor space needed to perform different types of behaviours, Spindler et al. (2013) recommend 11–14 pullets/m$^2$, while Krause and Schrader (2019) define a threshold based on the sitting behaviour of more than 40% as birds age (see Fig. 2). As both broilers and layer pullets undergo a steep growth, providing additional space rather than reducing group size is advisable, as well as choosing a level of which positive effects on welfare have been shown (i.e. <10 pullets/m$^2$; Liebers et al., 2019).

Within the EU, aviary systems or enriched cages are most commonly used. A restricted group size within an enriched cage appears beneficial in terms of adaptation and stability of the social environment, but it does severely limit freedom of movement and thereby impairs welfare (Hartcher and Jones, 2017). Inside an aviary system flock sizes exceed over thousands. Chicks are normally kept within the tiers of the system in the first days of life as they are too small to get back into the system to feed and drink. For 3–4 weeks of age, chicks are either in a small group (ranging from 20 to 80 chicks) or have access to the whole tier

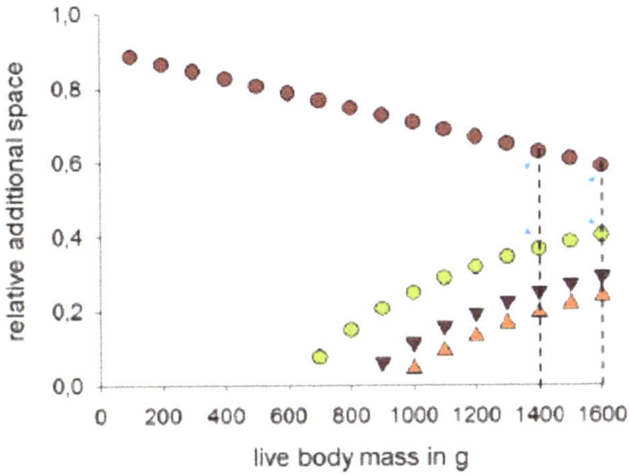


**Figure 2** Stocking density as calculated by relative additional space needed when chickens' age (Krause and Schrader, 2019) for laying hens under EU Directive 199/74/ and for broiler chickens based on EU Directive 2007/43. Red circles: layer EU directive. Green circles: broiler EU directive (33 kg/m$^2$). Brown triangles: broiler EU directive (39 kg/m$^2$). Orange triangles: broiler EU directive (43 kg/m$^2$). The dashed lines at x = 1400 g and x = 1600 g indicate typical live body at sexual maturity. Blue arrows indicate additional space needed at these body weight stages.

and are in a large group (over thousands). A high group size inside these cage-like systems can lead to welfare risks. In this case, group size is confounded with density. More birds in a cage in the first 4 weeks of life increases feed intake and reduces FCR in the first 4 weeks of life and from 4 to 8 weeks of life and body weight gain from 12 to 16 weeks of age (Bozkurt et al., 2006) and up till 20 weeks of age (Thogerson et al., 2009), and increases antibody response to infection (Bozkurt et al., 2008). Certain genotypes are more strongly affected by density, with effects on BW and FCR (Bozkurt et al., 2006), indicating that some breeds need more space and that a recommendation for group size or density should be based on body size and the needed three-dimensional space for their behaviours.

### 5.5 Possibilities to learn from older conspecifics

Chickens are precocial species, meaning they do not require active feeding by their parents. They are mobile and relatively mature at hatch, but they do have to learn what to eat. The mother helps the chicks in learning what they can eat and what not (Edgar et al., 2016). She initiates food and water uptake by the chicks and leads them to these and other resources (i.e. chicks learn by trial and error as well as by copying). Her specific food calls cue the chicks to edible items, while her correction calls discourages chicks from ingesting non-edible particles. Hen vocalizations have shown to increase broiler chicks' body weight and feed conversion in the first 9 days of life (Woodcock et al., 2004). Food distinction is a skill chicks need to improve. Young chicks innately peck at or avoid specific objects. These objects are three-dimensional spherical and of specific colours (Bartashunas and Suboski, 1984). Newly hatched chicks prefer pecking orange over blue (Hess, 1956) or red over yellow (Goodwin and Hess, 1969), moving over static objects and manipulative objects over non-manipulative objects. Chicks often learn and copy the foraging behaviour of their conspecifics. Seeing a companion's aversive reaction to ingestion causes the avoidance of these objects by the chicks (Johnston et al., 1998). Having the ability to learn from slightly older companions has shown to be advantageous. Two-day-old chicks ate more with an eight-day-old chick when foraging alone (Gajdon et al., 2001). A rooster is also a natural guidance for food discrimination and protection, as he will give distinct food calls when finding palatable resources. The lack of a social group with older chicks, a mother hen and/or a rooster requires the chicks to learn from their similar-age chick mates. The advantages for the welfare of brooding by a mother hen have been researched in comparison to brooding without a mother hen. Greater use of space and a higher level of rhythmicity are seen in chicks brooded by a mother hen (Wauters et al., 2002), although another study did not find effects on cognition or fearfulness in chicks brooded by a mother hen as compared to rearing without (Hewlett and Nordquist, 2019). Being with a mother hen for the first 7 weeks of life has shown to change the brain structure of chicks when they are adults. A greater level of positive lateralization effects and neuronal cell mass in the hippocampus (an important brain area involved in learning and cognition) was found at 52 weeks of age (Nordquist et al., 2013), indicating that rearing with a mother can change the brain of chickens positively. These effects are not found with a dark brooder, indicating the specific effects a maternal hen has on the brain of the chicks (Nordquist et al., 2020). The mother hen provides a place for shelter but also protects her chicks. In the first few days of the chick's life, the chicks seek cover under the hen in case of a threat. Under threat, the mother hen actively protects the chicks by taking a threatening posture and by casting specific calls to attract the chicks towards her. Layer chicks brooded by a maternal hen show reduced fear levels as compared to non-brooded chicks (Riber et al., 2007b). Having a safe base from where to explore likely helps with reducing fear levels, which promotes more time to spend on other behaviours. Riber et al. (2007b) showed that brooded layer chicks begin to perch 3 days earlier than non-brooded chicks. Brooded layer chicks show reduced fear, accelerated exploration (Rodenburg et al., 2009) and behavioural synchrony (Riber et al., 2007a and Nielsen et al., 2008). At 3 weeks of age brooded layer chicks are still less fearful as non-brooded chicks but also show more dustbathing and foraging (Shimmura et al., 2010), and these effects pertain up to 8 weeks of age (Roden and Wechsler, 1998) and to young adulthood (14-29 weeks of age (Perré et al., 2002)). A very recent study showed that non-brooded Bantam chicks from 6 till 42 days of age were more aligned and synchronized for a longer time and reacted more strongly to social separation as compared to brooded chicks, indicating that non-brooded chicks are more imprinted on each other Galarut et al. (2020). Having a broody hen inside a rearing farm is, at the moment, not possible due to specialized housing features adapted to rearing thousands of chicks. Overcrowding or negative interactions with the hens may also occur, which scares farmers to adopt this procedure. Despite their lack of guidance from an older chicken, chicks do learn fast where and what to eat, which shows their tremendous adaptability. However, it could be very advantageous for their food intake and welfare to have a social group from which they can learn or to simulate more features of older conspecifics/mother hens, to reduce the requirement to adapt to the unnatural social group conditions as found on farms.

## 5.6 Olfactory enrichment during rearing

Chickens, among birds, have a uropygial gland where preen oil is excreted from and is distributed during preening over their plumage. The composition of this preen oil differs depending on the age, diet, sex, changes in sex hormone levels and whether a chicken receives pecking or not (Sandilands et al., 2004a). Interestingly, hens with chicks have a specific chemical composition of the preen oil, which resembles that of mammals containing pheromones which calm the offspring (Madec et al., 2008). Madec et al. (2006, 2008) used an analogue of this mother hen preen oil (MHUSA: Mother Hens' Uropygial Secretion Analogue) in the environment of broiler chicks (in the air, and by a gelatine block containing the hen scent). In all three studies the MHUSA-exposed birds as compared to controls had an increased mean growth rate (Madec et al., 2006, 2008). MHUSA male birds had lower stress responses (52 days of age), and both MHUSA males and females (at 40 days of age) had higher body weight as compared to no-MHUSA exposed birds, indicating a long-term effect of maternal scent on development and lower stress responsiveness. Apparently, the hens produce this particular preen oil 4 days prior to hatch until chicks become fully independent. Eggs in the nest and the chicks are thus naturally exposed to these odours. Oil from the hens' plumage, however, also constitutes lipids from the litter or other sources in the environment (Sandilands et al., 2004a,b). The developing chicks, as exposed to the mother's feathers, could thus be informed about the future environment based on the fatty acid composition on her feathers and skin. Using odours in the environment can help with stress reduction but also as a kind of enrichment of the environment, with positive effects on animal welfare (Nielsen et al., 2015).

## 5.7 Reducing the risk for development of injurious pecking

Injurious pecking, feather pecking, vent pecking and cannibalism are all pecking behaviours directed to conspecifics, occurring mostly in laying hens. These behaviours have a complex multifactorial causation (see Chapter 15, and Rodenburg et al. 2013), and while there are many factors in early life playing a role, here I focus on the aspect of pecking substrate. Gentle feather pecking or allopecking - as a suggested way to socially recognize group mates (Riedstra and Groothuis, 2002) - occurs at a very young age, with high peak periods around 3–4 weeks of age (van der Eijk et al., 2018; De Haas et al., 2014a; Huber-Eicher and Audigé, 1999). Interestingly, this period coincides with changes in plumage from downy plumage to the development of wing and tail feathers (i.e. high blood innervated roots of tail feather are becoming apparent at the base of the tail and wings). When, instead of pecking to the ground, chicks learn to peck at these areas on their conspecifics, there might be a risk that IP has developed which can pertain until adulthood and may attract other chicks to develop the same behaviour although this has not yet been proven (Zeltner et al., 2000), but anecdotal evidence exists. The risk of IP occuring at rearing appears higher under high stocking density and bright light (Kjaer and Sørensen, 2002). Management in the very early weeks of life could prevent the development of IP at rearing (De Jong et al., 2013; Huber-Eicher and Sebö, 2001; Johnsen et al., 1998). Several large-scale on-farm experiments have shown the various risk factors for IP at rear (Gilani et al., 2013; De Haas et al., 2014a,b; Bestman et al., 2009; Lambton et al., 2010), with a specific provision of continuous floor substrate in the first 3 weeks of life, are important for the reduction of feather pecking at rear (De Haas et al., 2014b) and at lay (Tahamtani et al., 2016a). Further, providing additional fibre sources either separately or in the diet can extend feeding time and reduce the onset of feather pecking (Van Krimpen et al., 2005). Fibre substrate in the form of alfalfa is commonly used as a fibre source for chicken. When the substrate is finished faster than expected, it can be used as an early warning signal for IP. It means that the pullets take up more fibre than normally would be the case, which could also indicate a lack of something in their diet. Similarly the lack of downy feathers on the floor can be an early warning signal for IP (de Haas et al., 2014a). Seeing IP is a direct indicator, but acting before feather damage is occurring, is even more important.

## 5.8 Advantages of environmental enrichment and complexity

The rearing environment where chicks develop can be seen as a kindergarten. Exposure to a variety of different (structural) elements in the early life of chickens can help in adapting to a variety of stressors in later life, such as improved immune responses when faced with pathogens, enhanced skeletal and muscle morphology when needing to jump and fly and skilled behaviour adaptations which help in coping with fear and stress (Campbell et al., 2019).

## 5.9 Improving immune function

Enrichment in the early life of pullets can stimulate the recovery of the gut after infection in layer chicks (Walstra et al., 2010). When chicks are reared in a more complex environment, for example furnished cages versus conventional cages, immune parameters after a viral infection are higher (Matur et al., 2016). Further, providing layer chickens with food and water directly after hatch, and housing them in an enriched housing condition (i.e. various types of floor substrate), yields the most adaptive response to pathogens at 7 and 13 weeks of age (Walstra et al., 2010). Providing young pullets (11 weeks of age) with different types of foraging substrates (straw, wood-shavings, chaff) also improved the immune response (El-Lethey et al., 2003). In cage-free systems, the pathogenic load to which young chickens are exposed is generally higher as compared to cage systems. Exposure to various pathogens likely makes cage-free chickens more adaptive to various types of viral infections (Moe et al., 2010). In support, enriched reared chickens had a higher antibody response compared to non-enriched reared chickens (Matur et al., 2016). On the other hand, cage-free housing is combined with litter provision in the form of wood-shavings or other material. The health risk of being exposed to manure, high levels of ammonia, dust and mycotoxins in the floor substrate has been relatively little studied in terms of welfare in layer pullets or broilers. The quality and quantity of litter material and the continuous distribution hereof are, however, very important for the prevention of IP at rear (De Haas et al., 2014b).

## 5.10 Reducing risk of injuries

Perching behaviour already occurs at 1 week of age when chicks are not reared with a mother hen (Kozak et al., 2016). When perches are provided in cages, chickens show stronger bones of the sternum, tibia and humerus at 12 weeks of age (Enneking et al., 2012) and improved skeletal development (by greater shank width) at 71 weeks of age. Bone strength at lay is further improved when chicks are reared in aviaries as compared to floor rearing (Michel and Huonnic, 2003). An aviary enables pullets to learn to navigate, jump, fly, run and perch up high or on different levels. Provided that pullets have the space to do so (i.e. high stocking density hampers this option), aviary rearing reduces the prevalence of keel bone fractures at rear and lay, likely by the improvement of muscle development and improved bone properties (Casey-Trott et al., 2017a,b). At 16 weeks of age, breast and wing muscle weight was heavier and bone density, bone mineral content and -breaking strength were improved in aviary-reared pullets compared to cage-reared pullets (Casey-Trott et al., 2017c). However, aviary systems do comprise many structural elements which might cause injuries or even increase the risk of keel bone fractures. Therefore, matching rearing and laying systems is important to make pullets adapt to the future laying system conditions (Janczak and Riber, 2015). Inadequate rearing has been associated with the lack of locomotor and flying skills at lay with a high risk of keel bone fractures (Wilkins et al., 2011; Reed et al., 1993; Campbell et al., 2016). To minimize the risk of fractures of having many structural elements, ramps should be provided (Pettersson et al., 2017; LeBlanc et al., 2018; Heerkens et al., 2016). Swinging perches (which reduce the impact of crashes) have also shown to reduce keel bone fractures at lay (Leblanc et al., 2016).

## 5.11 Reducing fear and stress by environmental enrichment

Chickens are a prey species, which likely makes them cautious to potentially harmful stimuli (objects) and predator-like species (humans) (Salmeto et al., 2011). From a few days of age, chicks will show fear responses (Rogers, 1995). A positive association with people and exposure to various stimuli can help chickens to reduce their initial fear level (Barnett et al., 1993; Hemsworth et al., 1994), which thereby positively affects their welfare. A combination of different novel elements, human voices, presence and handling in the first 5 weeks of life reduced chickens' fear reactions and injuries at catching at lay (Reed et al., 1993). Fear of a novel object at 1 and 5 weeks of life has been found to be lower in chicks reared in a complex rearing system (floor based with elevated levels) compared to a closed or partly open aviary system (De Haas et al., 2014b). Fearfulness of young adult chickens at 19 and 23 weeks of age has also been found to be lower when they were reared in an aviary compared to rearing in a cage (Brantsæter et al., 2016a). Even the very early life aviary management, i.e. providing early access to litter or not, can influence fearfulness at adult age when combined with environmental enrichment at lay (Brantsæter et al., 2016b). Stress coping in response to intermitted novel challenges has also been found better when chicks were reared in a complex environment as compared to a simple environment as indicated by lower heterophil/lymphocyte ratio and increased stress recovery shown by resting (Campderrich et al., 2019).

## 5.12 Behaviour and brain effects: improving spatial and navigation skills

Environmental enrichment has proven to be a positive influence on brain development in various species and up to adulthood (van Praag et al., 2000). The first week of life of chicken is very prominent when it comes to hemispheric development and synapse formation, which continues up to 8–10 weeks (Atkinson et al., 2008). Around day 10 of age, chicks' visual processing shifts in hemispheric dominance (Wichman et al., 2009). Environmental enrichment from 10 or 11 days of age has more prominent effects on cognition than enrichment, which starts at day 8 of age (Wichman et al., 2009). But even enrichment later in the rearing period can influence the adult chicken's brain. Having more space to explore and more foraging material to forage during late rearing shows to increase brain hippocampal lateralization at adult age (Patzke et al., 2009). These effects on brain development may relate to the positive results which have been found on cognition. Rearing with structural elements such as perches or A-frames improved chicks' spatial and navigation skills (Gunnarsson et al., 2000) and cognition (Norman et al., 2019). Also rearing in a more complex environment such as an aviary as compared to cages improved spatial skills and memory in adult chicken (Tahamtani et al., 2015), but no effects were found in regard to dopamine involvement (as assessed by tyrosine hydroxylase staining) in the hippocampus (Tahamtani et al., 2016a,b). Environmental enrichment indicates enrichment of the environment, while one could also see it as providing a necessity for the animal to express its full behaviour repertoire. The positive effects reported here support evidence of improved welfare as compared to the standard conditions. The standard condition, however, more likely misses essential elements and rather lead to impaired welfare of young chicks (as seen by the multitude of conditions which affect welfare in young chicks, mentioned earlier).

## 6 Conclusions

Despite advanced knowledge about the factors influencing the welfare of layer and broiler chickens, the conditions that influence their early lives are still little understood. The manner in which their parents are housed as well as their own experience of incubation and rearing practices can have profound effects. Here it is suggested that welfare benefits will arise when commercial practices are better aligned with natural conditions, rather than requiring birds to excessively adapt to environments. Making the right decisions about early life factors has great potential to improve bird welfare because these factors prime the chick's brain and thereby influence behaviour through to adult life.

In this chapter, I mentioned several gaps in the literature by which the welfare of chicks can be influenced, namely, by stress in the paternal birds due to being socially challenged, mutilated, improper housing or feeding regimes. Especially in broiler breeder birds these aspects are hardly studied to impact the behaviour and stress resilience of the future generation(s). In the hatchery, more knowledge on variation in the natural incubation temperature and light exposure on the welfare of the offspring is needed to optimize incubation settings.

Further advancements in the application of in-ovo chemosensory learning and in-ovo temperature programming could aid in adapting chickens to their future life environment. Alternative immune-boosting strategies should be further researched, whether it is essential to provide chicks with over 20 vaccinations in their early life. And specifically whether the use of mass spray vaccination results in risk for the development of injurious pecking.

Recommendations for the improvement of the welfare of chicks are by reducing stress in parental birds, reducing noise levels in hatcheries, hatching with food and water under dim light or darkness to keep them calm and quiet, reducing the use of conveying/transitions over conveying belt/changes in speed of conveying and learning from other countries or examples for control of damaging behaviour and health risk reductions (e.g. Norway, see www.grouphousenet.eu).

Ideally, (layer)roosters should get a better life with optimal welfare or in-ovo sexing should take place so that day-old male chicks are not killed at day 1 of life. This technique also allows both layer and broiler chicks to hatch on-farm with immediate access to food and water, ideally with older conspecifics and with access to a dark secluded and heated area like a dark brooder. The rearing environment for laying hens should provide a complex and variable environment with an abundance of varying foraging material. Their housing system should enable exploration to equip them with the proper cognitive, spatial and coping skills. Pullets should not be hampered in their possibility to perform their whole behavioural repertoire due to high stocking density, so strong regulation and guidelines are important.

## 7 Where to look for further information

- https://www.bruderhahn.de/.
- https://www.pasreform.com/en.
- https://hatchtech.com/hatchery-products/hatchbrood/.
- https://www.vencomaticgroup.com/nl/producten/vleeskuikens/huisvestingsoplossingen/x-treck.
- https://inovo.nl/.
- https://www.farmantibiotics.org/wp-content/uploads/2018/09/commercial-chicken-vaccination-part-1-spray-and-aerosol-usage.pdf.
- https://www.innovatec.com/en/products/chick-room/vaccination-day-old-chicks.
- http://biologischefokkerij.nl.
- https://ec.europa.eu/food/animals/welfare/eu-platform-animal-welfare_en.

## 8 References

Aigueperse, N., Calandreau, L. and Bertin, A. 2013. Maternal diet influences offspring feeding behavior and fearfulness in the precocial chicken. *PLoS ONE* 8(10), e77583 [DOI: 10.1371/journal.pone.0077583].

Alvino, G. M., Blatchford, R. A., Archer, G. S. and Mench, J. A. 2009. Light intensity during rearing affects the behavioural synchrony and resting patterns of broiler chickens. *British Poultry Science*, 50(3), 275-283 [DOI: 10.1080/00071660902942775].

Angevaare, M. J., Prins, S., van der Staay, F. J. and Nordquist, R. E. 2012. The effect of maternal care and infrared beak trimming on development, performance and behavior of Silver Nick hens. *Applied Animal Behaviour Science* 140(1-2), 70-84 [DOI: 10.1016/j.applanim.2012.05.004].

Archer, G. S. 2018. Effect of two different commercially available white light LED fixtures on broiler hatchability and chick quality. *British Poultry Science* 59(3), 251-255 [DOI: 10.1080/00071668.2018.1436160].

Archer, G. S. 2019. Animal well-being and behavior how does red light affect layer production, fear, and stress? *Poultry Science* 98(1), 3-8.

Archer, G. S., Jeffrey, D. and Tucker, C. 2017. Effect of the combination of white and red LED lighting during incubation on layer, broiler, and Pekin duck hatchability. *Poultry Science* 96(8), 2670-2675 [DOI: 10.3382/ps/pex040].

Archer, G. S. and Mench, J. A. 2014. Natural incubation patterns and the effects of exposing eggs to light at various times during incubation on post-hatch fear and stress responses in broiler (meat) chickens. *Applied Animal Behaviour Science* 152, 44-51 [DOI: 10.1016/j.applanim.2013.12.010].

Arowolo, M. A., He, J. H., He, S. P. and Adebowale, T. O. 2019. The implication of lighting programmes in intensive broiler production system. *World's Poultry Science Journal* 75(1), 17-28 [DOI: 10.1017/S0043933918000934].

Arrazola, A., Mosco, E., Widowski, T., Guerin, M., Kiarie, E. and Torrey, S. 2019a. The effect of alternative feeding strategies for broiler breeder pullets: 1. Welfare and performance during rearing. *Poultry Science* 98(9), 3377-3390.

Arrazola, A., Widowski, T., Guerin, M., Kiarie, E. and Torrey, S. 2019b. The effect of alternative feeding strategies for broiler breeder pullets: 2. Welfare and performance during lay. *Poultry Science* 98(12), 6205-6216.

Atkinson, R., Migues, P. V., Cammarota, M., Kavanagh, J. M., Hunter, M. and Rostas, J. A. 2008. Biochemical, behavioural and electrophysiological investigations of brain maturation in chickens. *Brain Research Bulletin* 76(3), 217-223 [DOI: 10.1016/j.brainresbull.2008.02.036].

Atkinson, R., Rostas, J. A. and Hunter, M. 2010. Changes in mid-to-late latency auditory evoked responses in the chicken during neural maturation. *Developmental Psychobiology* 52(1), 24-34 [DOI: 10.1002/dev.20408].

Barnett, J. L., Hemsworth, P. H. and Jones, R. B. 1993. Behavioural responses of commercially farmed laying hens to humans: evidence of stimulus generalization. *Applied Animal Behaviour Science* 37(2), 139-146 [DOI: 10.1016/0168-1591(93)90106-Y].

Bartashunas, C. and Suboski, M. D. 1984. Effects of age of chick on social transmission of pecking preferences from hen to chicks. *Developmental Psychobiology* 17(2), 121-127 [DOI: 10.1002/dev.420170203].

Bayram, A. and Özkan, S. 2010. Effects of a 16-hour light, 8-hour dark lighting schedule on behavioral traits and performance in male broiler chickens. *Journal of Applied Poultry Research* 19(3), 263-273 [DOI: 10.3382/japr.2009-00026].

Bazinet, R. P. and Layé, S. 2014. Polyunsaturated fatty acids and their metabolites in brain function and disease. *Nature Reviews Neuroscience* 15(12):771-785 [DOI: 10.1038/nrn3820. Epub 2014 Nov 12].

Berghof, T. V. L., Matthijs, M. G. R., Arts, J. A. J., Bovenhuis, H., Dwars, R. M., van der Poel, J. J., Visker, M. H. P. W. and Parmentier, H. K. 2019. Selective breeding for high natural antibody level increases resistance to avian pathogenic *Escherichia coli* (APEC) in chickens. *Developmental and Comparative Immunology* 93, 45–57 [DOI: 10.1016/j.dci.2018.12.007].

Bergoug, H., Burel, C., Guinebretière, M., Tong, Q., Roulston, N., Romanini, C. E. B., Exadaktylos, V., Mcgonnell, I. M., Demmers, T. G. M., Verhelst, R., Bahr, C., Berckmans, D. and Eterradossi, N. 2013. Effect of pre-incubation and incubation conditions on hatchability, hatch time and hatch window, and effect of post-hatch handling on chick quality at placement. *World's Poultry Science Journal* 69(2), 313–334 [DOI: 10.1017/S0043933913000329].

Bertin, A., Calandreau, L., Arnould, C. and Levy, F. 2011. The developmental stage of chicken embryos modulates the impact of in Ovo Olfactory Stimulation on Food Preferences. Chemical Senses 37(3), 253–261 [DOI: 10.1093/chemse/bjr101].

Bertin, A., Calandreau, L., Arnould, C., Nowak, R., Levy, F., Noirot, V., Bouvarel, I. and Leterrier, C. 2010. In Ovo Olfactory Experience Influences Post-hatch Feeding Behaviour in Young Chickens. *Ethology* 116(11), 1027–1037 [DOI: 10.1111/j.1439-0310.2010.01820.x].

Bertin, A., Calandreau, L., Meurisse, M., Georgelin, M., Palme, R., Lumineau, S., Houdelier, C., Darmaillacq, A. S., Dickel, L., Colson, V., Cornilleau, F., Rat, C., Delaveau, J.. and Arnould, C. 2018. Incubation temperature affects the expression of young precocial birds' fear-related behaviours and neuroendocrine correlates. *Scientific Reports* 8(1), 1857 [DOI: 10.1038/s41598-018-20319-y].

Bertin, A., Mocz, F., Calandreau, L., Palme, R., Lumineau, S., Darmaillacq, A. S., Dickel, L., Arnould, C. and Houdelier, C. 2019. Human behaviour at the origin of maternal effects on offspring behaviour in laying hens (*Gallus gallus domesticus*). *Physiology and Behavior* 201, 175–183 [DOI: 10.1016/j.physbeh.2019.01.012].

Bestman, M., Koene, P. and Wagenaar, J. P. 2009. Influence of farm factors on the occurrence of feather pecking in organic reared hens and their predictability for feather pecking in the laying period. *Applied Animal Behaviour Science* 121(2), 120–125 [DOI: 10.1016/j.applanim.2009.09.007].

Bilalissi, A., N'nanle, O., Nideou, D., Meteyake, H., Kouame, Y., Decuypere, E., Gbeassor, M., Onagbessan, O. and Tona, K. 2019. The appropriate time to improve day-old chick production and post-hatch growth through *Moringa oleifera* leaf extract inoculation into the hatching egg. *European Poultry Science* 83, 1–12.

Bobbo, D., Galvani, F., Mascetti, G. G. and Vallortigara, G. 2002. Light exposure of the chick embryo influences monocular sleep. *Behavioural Brain Research* 134(1–2), 447–466 [DOI: 10.1016/s0166-4328(02)00059-1].

Boerema, A. S., Riedstra, B. and Strijkstra, A. M. 2003. Decrease in monocular sleep after sleep deprivation in the domestic chicken. *Behaviour* 140(11–12), 1415–1420.

Bolzani, R., Ruggeri,F. and Olivo, O. M. 1979. Average normal temperature of the chicken in the morning and after 1–2 days of fasting. *Bollettino della Societa italiana di biologia sperimentale* 55(16), 1618–1622.

Bozkurt, Z., Byram, I., Bülbül, A. and Aktepe, O. C. 2008. Effects of strain, cage density and position on immune response to vaccines and blood parameters in layer pullets. *Kafkas Universitesi Veteriner Fakultesi Dergisi* 14(2), 191–204 [DOI:10.9775/kvfd.2008.43-A].

Bozkurt, Z., Byram, I., Türkmeno, I. and Aktepe, O. C. 2006. Effects of cage density and cage position on performance of commercial layer pullets from four genotypes. *Turkish Journal of Veterinary and Animal Sciences* 30, 17-28.

Brantsæter, M., Nordgreen, J., Rodenburg, T. B., Tahamtani, F., Popova, A. and Janczak, A. 2016a. Exposure to increased environmental complexity during rearing reduces fearfulness and increases use of three-dimensional space in laying hens (*Gallus gallus domesticus*). *Frontiers in Veterinary Science* 3, 14.

Brantsæter, M., Tahamtani, F., Moe, R., Hansen, T., Orritt, R., Nicol, C. and Janczak, A. 2016b. Rearing laying hens in aviaries reduces fearfulness following transfer to furnished cages. *Frontiers in Veterinary Science* 3, 14.

Campbell, D. L. M., Goodwin, S. L., Makagon, M. M., Swanson, J. C. and Siegford, J. M. 2016. Failed landings after laying hen flight in a commercial aviary over two flock cycles. *Poultry Science* 95(1), 188-197 [DOI: 10.3382/ps/pev270].

Campbell, D. L. M., De Haas, E. N. and Lee, C. 2019. A review of environmental enrichment for laying hens during rearing in relation to their behavioral and physiological development. *Poultry Science* 98(1), 9-28 [DOI: 10.3382/ps/pey319].

Campderrich, I., Nazar, F. N., Wichman, A., Marin, R. H., Estevez, I. and Keeling, L. J. 2019. Environmental complexity: a buffer against stress in the domestic chick. *PLOS ONE* 14(1), e0210270 [DOI: 10.1371/journal.pone.0210270].

Campo, J. L., Garcia-Gil, M. and Davila, S. 2005. Effects of specific noise and music stimuli on stress and fear levels of laying hens of several breeds. *Applied Animal Behaviour Science* 91(1), 75-84 [DOI: 10.1016/j.applanim.2004.08.028].

Careghi, C., Tona, K., Onagbesan, O., Buyse, J., Decuypere, E. and Bruggeman, V. 2005. The effects of the spread of hatch and interaction with delayed feed access after hatch on broiler performance until seven days of age. *Poultry Science* 84(8), 1314-1320 [DOI: 10.1093/ps/84.8.1314].

Carvalho, C., Souza, C., Detinôco, I., Santos, L., Minette, L., Da Silva, E., 2015. Activities and ergonomics of workers in broiler hatcheries. *Brazilian Journal of Poultry Science* 17(2), 123-136.

Casey-Trott, T. M., Guerin, M. T., Sandilands, V., Torrey, S. and Widowski, T. M. 2017a. Rearing system affects prevalence of keel-bone damage in laying hens: a longitudinal study of four consecutive flocks. *Poultry Science* 96(7), 2029-2039 [DOI: 10.3382/ps/pex026].

Casey-Trott, T. M., Korver, D., Guerin, M., Sandilands, V., Torrey, S. and Widowski, T. 2017b. Animal well-being and behavior opportunities for exercise during pullet rearing, Part I: effect on the musculoskeletal characteristics of pullets. *Poultry Science* 96(8), 2509-2517.

Casey-Trott, T. M., Korver, D. R., Guerin, M. T., Sandilands, V., Torrey, S. and Widowski, T. M. 2017c. Opportunities for exercise during pullet rearing, Part II: long-term effects on bone characteristics of adult laying hens at the end-of-lay. *Poultry Science* 96(8), 2518-2527 [DOI: 10.3382/ps/pex060].

Chiandetti, C., Regolin, L., Rogers, L. J. and Vallortigara, G. 2005. Effects of light stimulation of embryos on the use of position-specific and object-specific cues in binocular and monocular domestic chicks (*Gallus gallus*). *Behavioural Brain Research* 163(1), 10-17 [DOI: 10.1016/j.bbr.2005.03.024].

Chloupek, P., Voslářová, E., Chloupek, J., Bedáňová, I., Pištěková, V. and Večerek, V. 2009. Stress in broiler chickens due to acute noise exposure. *Acta Veterinaria Brno* 78(1), 93-98 [DOI: 10.2754/avb200978010093].

Cockrem, J. F. 2007. Stress, corticosterone responses and avian personalities. *Journal of Ornithology* 148, 169–178.

Collin, A., Berri, C., Tesseraud, S., Requena Rodón, F., Skiba-Cassy, S., Crochet, S., Duclos, M., Rideau, N., Tona, K., Buyse, J., Bruggeman, V., Decuypere, E., Picard, M. and Yahav, S. 2007. Effects of thermal manipulation during early and late embryogenesis on thermotolerance and breast muscle characteristics in broiler chickens. *Poultry Science*, 86(5), 795–800.

de Gouw, P., van de Ven, L. J. F., Lourens, S., Kemp, B. and van den Brand, H. 2017. Effects of dust, formaldehyde and delayed feeding on early postnatal development of broiler chickens. *Research in Veterinary Science* 112, 201–207 [DOI: 10.1016/j.rvsc.2017.04.021].

De Haas, E. N., Bolhuis, J. E., de Jong, I. C., Kemp, B., Janczak, A. M. and Rodenburg, T. B. 2014a. Predicting feather damage in laying hens during the laying period. Is it the past or is it the present? *Applied Animal Behaviour Science* 160(1), 75–85 [DOI: 10.1016/j.applanim.2014.08.009].

De Haas, E. N., Bolhuis, J. E., Kemp, B., Groothuis, T. G. and Rodenburg, T. B. 2014b. Parents and early life environment affect behavioral development of laying hen chickens. *PLoS ONE* 9(6), e90577 [DOI: 10.1371/journal.pone.0090577].

De Haas, E. N., Calandreau, L., Baéza, E., Chartrin, P., Palme, R., Darmaillacq, A. S., Dickel, L., Lumineau, S., Houdelier, C., Denis, I., Arnould, C., Meurisse, M. and Bertin, A. 2017. Lipids in maternal diet influence yolk hormone levels and post-hatch neophobia in the domestic chick. *Developmental Psychobiology* 59(3), 400–409 [DOI: 10.1002/dev.21504].

De Haas, E. N., Kemp, B., Bolhuis, J. E., Groothuis, T. and Rodenburg, T. B. 2013. Fear, stress, and feather pecking in commercial white and brown laying hen parent-stock flocks and their relationships with production parameters. *Poultry Science* 92(9), 2259–2269 [DOI: 10.3382/ps.2012-02996].

De Jong, I. C. and Gunnink, H. 2019. Effects of a commercial broiler enrichment programme with or without natural light on behaviour and other welfare indicators. *Animal* 13(2), 384–391 [DOI: 10.1017/S1751731118001805].

De Jong, I. C., Reuvekamp, B. F. and Gunnink, H. 2013. Can substrate in early rearing prevent feather pecking in adult laying hens? *Animal Welfare* 22(3), 305–314 [DOI: 10.7120/09627286.22.3.305].

De Jong, I. C., Van Riel, J., Bracke, M. B. and Brand, H. v. d. 2017. A 'meta-analysis' of effects of post-hatch food and water deprivation on development, performance and welfare of chickens. *PLoS ONE* 12(12): e0189350.

De Los Mozos, J., García-Ruiz, A. I., Den Hartog, L. A. and Villamide, M. J. 2017. Growth curve and diet density affect eating motivation, behavior, and body composition of broiler breeders during rearing. *Poultry Science* 96(8), 2708–2717 [DOI: 10.3382/ps/pex045].

Deep, A., Schwean-Lardner, K., Crowe, T. G., Fancher, B. I. and Classen, H. L. 2012. Effect of light intensity on broiler behaviour and diurnal rhythms. *Applied Animal Behaviour Science* 136(1), 50–56 [DOI: 10.1016/j.applanim.2011.11.002].

Dennis, R. L. and Cheng, H. W. 2012. Effects of different infrared beak treatment protocols on chicken welfare and physiology. *Poultry Science* 91(7), 1499–1505 [DOI: 10.3382/ps.2011-01651].

Dharmaretnam, M. and Rogers, L. J. 2005. Hemispheric specialization and dual processing in strongly versus weakly lateralized chicks. *Behavioural Brain Research* 162(1), 62–70 [DOI: 10.1016/j.bbr.2005.03.012].

Ding, B. A., Pirone, A., Lenzi, C., Baglini, A. and Romboli, I. 2011. Effect of hen diet supplemented with 25-OH-D3 on the development of small intestinal morphology of chick. *Journal of Animal and Feed Sciences* 20, 420–431.

Dixon, L. M., Sparks, N. H. and Rutherford, K. M. 2016. Early experiences matter: a review of the effects of prenatal environment on offspring characteristics in poultry. *Poultry Science* 95(3), 489–499 [DOI: 10.3382/ps/pev343].

Donofre, A. C., da Silva, I. J. O. and Ferreira, I. E. P. 2020. Sound exposure and its beneficial effects on embryonic growth and hatching of broiler chicks. *British Poultry Science* 61(1), 79–85 [DOI: 10.1080/00071668.2019.1673315].

Drozdova, A., Okuliarova, M. and Zeman, M. 2019. The effect of different wavelengths of light during incubation on the development of rhythmic pineal melatonin biosynthesis in chick embryos. 13(8), 1635–1640. [DOI: 10.1017/S1751731118003695].

Edgar, J., Held, S., Jones, C. and Troisi, C. 2016. Influences of maternal care on chicken welfare. *Animals* 6(1), 2 [DOI: 10.3390/ani6010002].

Edgar, J., Held, S., Paul, E., Pettersson, I., I'Anson Price, R. and Nicol, C. 2015. Social buffering in a bird. *Animal Behaviour* 105, 11–19 [DOI: 10.1016/j.anbehav.2015.04.007].

EFSA 2010. Scientific opinion on welfare aspects of the management and housing of the grand-parent and parent stocks raised and kept for breeding purposes. *EFSA Journal* 8(7), 1667.

El-Lethey, H., Huber-Eicher, B. and Jungi, T. W. 2003. Exploration of stress-induced immunosuppression in chickens reveals both stress-resistant and stress-susceptible antigen responses. *Veterinary Immunology and Immunopathology* 95(3–4), 91–101 [DOI: 10.1016/s0165-2427(02)00308-2].

Enneking, S. A., Cheng, H. W., Jefferson-Moore, K. Y., Einstein, M. E., Rubin, D. A. and Hester, P. Y. 2012. Early access to perches in caged White Leghorn pullets. *Poultry Science* 91(9), 2114–2120 [DOI: 10.3382/ps.2012-02328].

Ericsson, M., Henriksen, R., Bélteky, J., Sundman, A. S., Shionoya, K. and Jensen, P. 2016. Long-term and transgenerational effects of stress experienced during different life phases in chickens (*Gallus gallus*). *PLoS ONE* 11(4), e0153879 [DOI: 10.1371/journal.pone.0153879].

Eugen, K. v., Nordquist, R. E., Zeinstra, E. and Staay, F. J. v. d. 2019. Stocking density affects stress and anxious behavior in the laying hen chick during rearing. *Animals* 9(2), 53.

Fernyhough, M., Nicol, C. J., van de Braak, T., Toscano, M. J. and Tønnessen, M. 2020. The ethics of laying hen genetics. *Journal of Agricultural and Environmental Ethics* 33(1), 15–36 [DOI: 10.1007/s10806-019-09810-2].

Fidan, E. D., Nazlıgül, A., Türkyılmaz, M. K., Aypak, S. Ü, Kilimci, F. S., Karaarslan, S. and Kaya, M. 2017. Effect of photoperiod length and light intensity on some welfare criteria, carcass, and meat quality characteristics in broilers. *Revista Brasileira de Zootecnia* 46(3), 202–210 [DOI: 10.1590/s1806-92902017000300004].

Field, S. E., Rickard, N. S., Toukhsati, S. R. and Gibbs, M. E. 2007. Maternal hen calls modulate memory formation in the day-old chick: the role of noradrenaline. *Neurobiology of Learning and Memory* 88(3), 321–330 [DOI: 10.1016/j.nlm.2007.04.001].

Freire, R., Glatz, P. C. and Hinch, G. 2008. Self-administration of an analgesic does not alleviate pain in beak trimmed chickens. *Asian-Australasian Journal of Animal Sciences* 21(3), 443–448 [DOI: 10.5713/ajas.2008.70039].

Gajdon, G. K., Hungerbuhler, N. and Stauffacher, M. 2001. Social influence on early foraging of domestic chicks (Gallus gallus) in a near-to-nature procedure. *Ethology* 107(10), 913–937 [DOI: 10.1046/j.1439-0310.2001.00719.x].

Galuret, S., Lumineau, S., Pouzol, D. and George, I. 2020. Mothering influences domestic chick's laterality. *Animal Behaviour* 159(10), 69-79.

Gebhardt-Henrich, S. G., Toscano, M. J. and Würbel, H. 2017. Perch use by broiler breeders and its implication on health and production. *Poultry Science* 96(10), 3539-3549 [DOI: 10.3382/ps/pex189].

Gebhardt-Henrich, S. G., Toscano, M. J. and Würbel, H. 2018. Use of aerial perches and perches on aviary tiers by broiler breeders. *Applied Animal Behaviour Science* 203, 24-33 [DOI: 10.1016/j.applanim.2018.02.013].

Ghatpande, A., Ghatpande, S. and Khan, M. Z. 1995. Effect of different intensities of fluorescent light on the early development of chick embryos in ovo. *Cellular and Molecular Biology Research* 41(6), 613-621.

Giersberg, M. F. and Kemper, N. 2018. Rearing male layer chickens: a German perspective. *Agriculture* 8(11) [DOI: 10.3390/agriculture8110176].

Giersberg, M. F., Kemper, N. and Spindler, B. 2019. Pecking and piling: the behaviour of conventional layer hybrids and dual-purpose hens in the nest. *Applied Animal Behaviour Science* 214, 50-56 [DOI: 10.1016/j.applanim.2019.02.016].

Gilani, A. M., Knowles, T. G. and Nicol, C. J. 2012. The effect of dark brooders on feather pecking on commercial farms. *Applied Animal Behaviour Science* 142(1-2), 42-50 [DOI: 10.1016/j.applanim.2012.09.006].

Gilani, A. M., Knowles, T. G. and Nicol, C. J. 2013. The effect of rearing environment on feather pecking in young and adult laying hens. *Applied Animal Behaviour Science* 148(1-2), 54-63 [DOI: 10.1016/j.applanim.2013.07.014].

Glatz, P. C. 2000. Beak trimming methods - review. *Asian-Australasian Association of Animal Production Societies* 13(11), 1619-1637 [DOI: 10.5713/ajas.2000.1619].

Glatz, P. C. and Lunam, C. A. 1994. Production and heart rate responses of chickens beak-trimmed at hatch or at 10 or 42 days of age. *Australian Journal of Experimental Agriculture* 34(4), 443-447 [DOI: 10.1071/EA9940443].

Glatz, P. C., Murphy, L. B. and Preston, A. P. 1992. Analgesic therapy of beak-trimmed chickens. *Australian Veterinary Journal* 69(1), 18 [DOI: 10.1111/j.1751-0813.1992.tb09859.x].

Goerlich, V. C., Nätt, D., Elfwing, M., Macdonald, B. and Jensen, P. 2012. Transgenerational effects of early experience on behavioral, hormonal and gene expression responses to acute stress in the precocial chicken. *Hormones and Behavior* 61(5), 711-718 [DOI: 10.1016/j.yhbeh.2012.03.006].

Goodwin, E. B. and Hess, E. H. 1969. Innate visual form preferences in the pecking behavior of young chicks. *Behaviour* 34(4), 223-237 [DOI: 10.1163/156853969x00134].

Gunnarsson, S., Heikkilä, M., Hultgren, J. and Valros, A. 2008a. A note on light preference in layer pullets reared in incandescent or natural light. *Applied Animal Behaviour Science* 112(3-4), 395-399,395-399 [DOI: 10.1016/j.applanim.2007.09.004].

Gunnarsson, S., Heikkilä, M. and Valros, A. 2008b. Effect of day length and natural versus incandescent light on perching and the diurnal rhythm of feeding behaviour in layer chicks (Gallus g. domesticus). *Acta Agriculturae Scandinavica A: Animal Sciences* 6(58(2)), 93-99.

Gunnarsson, S., Yngvesson, J., Keeling, L. J. and Forkman, B. 2000. Rearing without early access to perches impairs the spatial skills of laying hens. *Applied Animal Behaviour Science* 67(3), 217-228 [DOI: 10.1016/s0168-1591(99)00125-2].

Hagelin, J. C., Simonet, J. C. and Lyson, T. R. 2013. Embryonic domestic chickens can detect compounds in an avian chemosignal before breathing air. In: East, M. and

Dehnhard, M. (Eds), *Chemical Signals in Vertebrates 12*. Springer, New York, pp. 363-377 [DOI: 10.1007/978-1-4614-5927-9_29].

Haq, A. -U., Bailey C. A., and Chinnah, A. 1996. Effect of beta-carotene, canthaxanthin, lutein, and vitamine E on neonatal immunity of chicks when supplemented in the broiler breeder diets. *Poultry Science*, 75, 1092-1097.

Hartcher, K. M. and Jones, B. 2017. The welfare of layer hens in cage and cage-free housing systems. *World's Poultry Science Journal* 73(4), 767-782 [DOI: 10.1017/S0043933917000812].

Hartini, S., Choct, M., Hinch, G., Kocher, A. and Nolan, J. V. 2002. Effects of light intensity during rearing and beak trimming and dietary fiber sources on mortality, egg production, and performance of ISA Brown laying hens. *Journal of Applied Poultry Research* 11(1), 104-110 [DOI: 10.1093/japr/11.1.104].

Hassan, M. R., Sultana, S., Choe, H. S. and Ryu, K. S. 2013. Effect of monochromatic and combined light colour on performance, blood parameters, ovarian morphology and reproductive hormones in laying hens. *Italian Journal of Animal Science* 12(3), 359-364 [DOI: 10.4081/ijas.2013.e56].

Hedlund, L., Whittle, R. and Jensen, P. 2019. Effects of commercial hatchery processing on short- and long-term stress responses in laying hens. *Scientific Reports* 9(1) [DOI: 10.1038/s41598-019-38817-y].

Heerkens, J. L., Delezie, E., Ampe, B., Rodenburg, T. B. and Tuyttens, F. A. 2016. Ramps and hybrid effects on keel bone and foot pad disorders in modified aviaries for laying hens. *Poultry Science* 95(11), 2479-2488 [DOI: 10.3382/ps/pew157].

Hemsworth, P. H., Coleman, G. J., Barnett, J. L. and Jones, R. B. 1994. Behavioural responses to humans and the productivity of commercial broiler chickens. *Applied Animal Behaviour Science* 41(1-2), 101-114 [DOI: 10.1016/0168-1591(94)90055-8].

Henriksen, R., Rettenbacher, S. and Groothuis, T. G. G. 2013. Maternal corticosterone elevation during egg formation in chickens (*Gallus gallus domesticus*) influences offspring traits, partly via prenatal undernutrition. *General and Comparative Endocrinology* 191, 83-91 [DOI: 10.1016/j.ygcen.2013.05.028].

Hesham, M. H., El Shereen, A. H. and Enas, S. N. 2018. Impact of different light colors in behavior, welfare parameters and growth performance of Fayoumi broiler chickens strain. *Journal of the Hellenic Veterinary Medical Society* 69(2), 951-958 [DOI: 10.12681/jhvms.18017].

Hess, E. 1956. Natural preferences of chicks and ducklings for objects of different colors. *Psychological Reports* 2(4), 477-483.

Hewlett, S. E. and Nordquist, R. E. 2019. Effects of maternal care during rearing in white leghorn and brown nick layer hens on cognition, sociality and fear. *Animals* 9(7) [DOI: 10.3s390/ani9070454].

Hieke, A. C., Hubert, S. M. and Athrey, G. 2019. Circadian disruption and divergent microbiota acquisition under extended photoperiod regimens in chicken. *PeerJ* 7, e6592 [DOI: 10.7717/peerj.6592].

Hill, E. M., Koay, G., Heffner, R. S. and Heffner, H. E. 2014. Audiogram of the chicken (*Gallus gallus domesticus*) from 2 Hz to 9 kHz. *Journal of Comparative Physiology A: Neuroethology, Sensory, Neural, and Behavioral Physiology* 200(10), 863-870 [DOI: 10.1007/s00359-014-0929-8].

Honaker, C. F. and Ruszler, P. L. 2004. The effect of claw and beak reduction on growth parameters and fearfulness of two leghorn strains. *Poultry Science* 83(6), 873-881 [DOI: 10.1093/ps/83.6.873].

Huber-Eicher, B. and Audigé, L. 1999. Analysis of risk factors for the occurrence of feather pecking in laying hen growers. *British Poultry Science* 40(5), 599–604 [DOI: 10.1080/00071669986963].

Huber-Eicher, B. and Sebö, F. 2001. Reducing feather pecking when raising laying hen chicks in aviary systems. *Applied Animal Behaviour Science* 73(1), 59–68 [DOI: 10.1016/s0168-1591(01)00121-6].

Janczak, A. M. and Riber, A. B. 2015. Review of rearing-related factors affecting the welfare of laying hens. *Poultry Science* 94(7), 1454–1469 [DOI: 10.3382/ps/pev123].

Janczak, A. M., Torjesen, P., Palme, R. and Bakken, M. 2007. Effects of stress in hens on the behaviour of their offspring. *Applied Animal Behaviour Science* 107(1–2), 66–77 [DOI: 10.1016/j.applanim.2006.09.016].

Jensen, A. B., Palme, R. and Forkman, B. 2006. Effect of brooders on feather pecking and cannibalism in domestic fowl (*Gallus gallus domesticus*). *Applied Animal Behaviour Science* 99(3–4), 287–300 [DOI: 10.1016/j.applanim.2005.10.017].

Jensen, P. 2014. Behaviour epigenetics – the connection between environment, stress and welfare. *Applied Animal Behaviour Science* 157, 1–7 [DOI: 10.1016/j.applanim.2014.02.009].

Johnsen, P. F., Vestergaard, K. S. and Nørgaard-Nielsen, G. 1998. Influence of early rearing conditions on the development of feather pecking and cannibalism in domestic fowl. *Applied Animal Behaviour Science* 60(1), 25–41 [DOI: 10.1016/S0168-1591(98)00149-X].

Johnston, A. N. B., Burne, T. H. J. and Rose, S. P. R. 1998. Observation learning in day-old chicks using a one-trial passive avoidance learning paradigm. *Animal Behaviour* 56(6), 1347–1353 [DOI: 10.1006/anbe.1998.0901].

Joseph, N. S., Lourens, A. and Moran, E. T. 2006. The effects of suboptimal eggshell temperature during incubation on broiler chick quality, live performance, and further processing yield. *Poultry Science* 85(5), 932–938 [DOI: 10.1093/ps/85.5.932].

Kaleta, E. F. and Redmann, T. 2008. Approaches to determine the sex prior to and after incubation of chicken eggs and of day-old chicks. *World's Poultry Science Journal* 64(3), 391–399 [DOI: 10.1017/S0043933908000111].

Kang, S. W., Christensen, K. D., Aldridge, D. and Kuenzel, W. J. 2020. Effects of light intensity and dual light intensity choice on plasma corticosterone, central serotonergic and dopaminergic activities in birds, *Gallus gallus*. *General and Comparative Endocrinology* 285, 113289 [DOI: 10.1016/j.ygcen.2019.113289].

Karaarslan, S. and Nazlıgül, A. 2018. Effects of lighting, stocking density, and access to perches on leg health variables as welfare indicators in broiler chickens. *Livestock Science* 218, 31–36 [DOI: 10.1016/j.livsci.2018.10.008].

Kaukonen, E., Norring, M. and Valros, A. 2016. Effect of litter quality on foot pad dermatitis, hock burns and breast blisters in broiler breeders during the production period. *Avian Pathology* 45(6), 667–673 [DOI: 10.1080/03079457.2016.1197377].

Kjaer, J. B. and Sørensen, P. 2002. Feather pecking and cannibalism in free-range laying hens as affected by genotype, dietary level of methionine + cystine, light intensity during rearing and age at first access to the range area. *Applied Animal Behaviour Science* 76(1), 21–39 [DOI: 10.1016/S0168-1591(01)00209-X].

Knowles, T. G., Brown, S. N., Warriss, P. D., Butterworth, A. and Hewitt, L. 2004. Welfare aspects of chick handling in broiler and laying hen hatcheries. *Animal Welfare* 13(4), 409–418(10). https://www.ingentaconnect.com/content/ufaw/aw/2004/00000013/00000004/art00004.

Koolhaas, J. M., Bartolomucci, A., Buwalda, B., de Boer, S. F., Flügge, G., Korte, S. M., Meerlo, P., Murison, R., Olivier, B., Palanza, P., Richter-Levin, G., Sgoifo, A., Steimer, T., Stiedl, O., van Dijk, G., Wöhr, M. and Fuchs, E. 2011. Stress revisited: a critical evaluation of the stress concept. *Neuroscience and Biobehavioral Reviews* 35(5), 1291-1301 [DOI: 10.1016/j.neubiorev.2011.02.003].

Kozak, M., Tobalske, B., Martins, C., Bowley, S., Wuerbel, H. and Harlander-Matauschek, A. 2016. Use of space by domestic chicks housed in complex aviaries. *Applied Animal Behaviour Science* 181, 115-121 [DOI: 10.1016/j.applanim.2016.05.024].

Krause, E. T. and Schrader, L. 2019. Suggestions to derive maximum stocking densities for layer pullets. *Animals* 9(6), 348 [DOI: 10.3390/ani9060348].

Kuenzel, T., Mönig, B., Wagner, H., Mey, J. and Luksch, H. 2007. Neuronal differentiation of the early embryonic auditory hindbrain of the chicken in primary culture. *European Journal of Neuroscience* 25(4), 974-984 [DOI: 10.1111/j.1460-9568.2007.05343.x].

Lalloué, F. L., Ayer-Le Lièvre, C. S. and Sicard, G. 2003. Analysis of the functional maturation of olfactory neurons in chicks before and after birth. *Chemical Senses* 28(8), 729-737 [DOI: 10.1093/chemse/bjg055].

Lambton, S. L., Knowles, T. G., Yorke, C. and Nicol, C. J. 2010. The risk factors affecting the development of gentle and severe feather pecking in loose housed laying hens. *Applied Animal Behaviour Science* 123(1-2), 32-42 [DOI: 10.1016/j.applanim.2009.12.010].

LeBlanc, C., Tobalske, B., Szkotnicki, B. and Harlander-Matauschek, A. 2018. Locomotor behavior of chickens anticipating incline walking. *Frontiers in Veterinary Science* 4, 233.

LeBlanc, S., Tobalske, B., Quinton, M., Springthorpe, D., Szkotnicki, B., Wuerbel, H. and Harlander-Matauschek, A. 2016. Physical health problems and environmental challenges influence balancing behaviour in laying hens. *PLoS ONE* 11(4), e0153477 [DOI: 10.1371/journal.pone.0153477].

Leliveld, L. M. C., Langbein, J. and Puppe, B. 2013. The emergence of emotional lateralization: evidence in non-human vertebrates and implications for farm animals. *Applied Animal Behaviour Science* 145(1-2), 1-14 [DOI: 10.1016/j.applanim.2013.02.002].

Lewis, P. D., Danisman, R. and Gous, R. M. 2009. Photoperiodic responses of broilers. I. Growth, feeding behaviour, breast meat yield, and testicular growth. *British Poultry Science* 50(6), 657-666 [DOI: 10.1080/00071660903338452].

Lewis, P. D. and Gous, R. M. 2009. Photoperiodic responses of broilers. II. Ocular development. *British Poultry Science* 50(6), 667-672 [DOI: 10.1080/00071660903342066].

Liebers, C. J., Schwarzer, A., Erhard, M., Schmidt, P. and Louton, H. 2019. The influence of environmental enrichment and stocking density on the plumage and health conditions of laying hen pullets. *Poultry Science* 98(6), 2474-2488 [DOI: 10.3382/ps/pez024].

Lindholm, C., Johansson, A., Middelkoop, A., Lees, J. J., Yngwe, N., Berndtson, E., Cooper, G. and Altimiras, J. 2018. The quest for welfare-friendly feeding of broiler breeders: effects of daily vs. 5:2 feed restriction schedules. *Poultry Science* 97(2), 368-377 [DOI: 10.3382/ps/pex326].

Lindqvist, C., Janczak, A. M., Nätt, D., Baranowska, I., Lindqvist, N., Wichman, A., Lundeberg, J., Lindberg, J., Torjesen, P. A. and Jensen, P. 2007. Transmission of stress-induced learning impairment and associated brain gene expression from parents to offspring in chickens. *PLoS ONE* 2(4), e364 [DOI: 10.1371/journal.pone.0000364].

Løtvedt, P. and Jensen, P. 2014. Effects of hatching time on behavior and weight development of chickens. *PLoS ONE* 9(7), e103040 [DOI: 10.1371/journal.pone.0103040].

Lunam, C. A., Glatz, P. C. and Hsu, Y. J. 1996. The absence of neuromas in beaks of adult hens after conservative trimming at hatch. *Australian Veterinary Journal* 74(1), 46–49 [DOI: 10.1111/j.1751-0813.1996.tb13734.x].

Mabayo, R. T., Okumura, J., Hirao, A., Sugita, S., Sugahara, K. and Furuse, M. 1996. The role of olfaction in oil preference in the chicken. *Physiology and Behavior* 59(6), 1185-1188 [DOI: 10.1016/0031-9384(95)02152-3].

Madec, I., Gabarrou, J., Guillaumey, D., Lecuelle, C., Bougrat, L. and Pageat, P. 2008. Are thirty-five days enough to observe the stress-reducing effect of a semiochemical analogue on chickens (*Gallus gallus domesticus*) housed under high density? *Poultry Science* 87(2), 222–225 [DOI: 10.3382/ps.2007-00006].

Madec, I., Pageat, P., Bougrat, L., Saffray, D., Falewee, C., Gervasoni, M. A., Bollart, A. and Gabarrou, J. F. 2006. Influence of a semiochemical analogue on growing performances and meat quality of broilers. *Poultry Science* 85(12), 2112–2116 [DOI: 10.1093/ps/85.12.2112].

Marchant-Forde, R. M., Fahey, A. G. and Cheng, H. W. 2008. Environment, well-being, and behavior: comparative effects of infrared and one-third hot-blade trimming on beak topography, behavior, and growth. *Poultry Science* 87(8), 1474–1483.

Matur, E., Akyazi, I., Eraslan, E., Ergul Ekiz, E., Eseceli, H., Keten, M., Metiner, K. and Aktaran Bala, D. 2016. The effects of environmental enrichment and transport stress on the weights of lymphoid organs, cell-mediated immune response, heterophil functions and antibody production in laying hens. *Animal Science Journal* 1 2, 87(2), 284–292.

McDaniel, C. D., Wells, J., Coufal, C., Parker, H., Kiess, A. and Young, K. 2011. Hatchablility of broiler breeder eggs sanitized with a combination of ultraviolet light and hydrogen peroxide. *International Journal of Poultry Science* 10(4), 320-324 [DOI: 10.3923/ijps.2011.320.324].

McKeegan, D. E. and Philbey, A. W. 2012. Chronic neurophysiological and anatomical changes associated with infrared beak treatment and their implications for laying hen welfare. *Animal Welfare* 21(2), 207–217 [DOI: 10.7120/09627286.21.2.207].

Michel, V. and Huonnic, D. 2003. A comparison of welfare, health and production performance of laying hens reared in cages or in aviaries. *British Poultry Science* 44(5), 775–776 [DOI: 10.1080/00071660410001666691].

Moe, R. O., Guémené, D., Bakken, M., Larsen, H. J., Shini, S., Lervik, S., Skjerve, E., Michel, V. and Tauson, R. 2010. Effects of housing conditions during the rearing and laying period on adrenal reactivity, immune response and heterophil to lymphocyte (H/L) ratios in laying hens. *Animal* 4(10), 1709–1715 [DOI: 10.1017/S175173111000100X].

Mohammed, R., Cavallaro, G., Kessels, C. G. and Villamor, E. 2015. Functional differences between the arteries perfusing gas exchange and nutritional membranes in the late chicken embryo. *Journal of Comparative Physiology B: Biochemical, Systemic, and Environmental Physiology* 185(7), 783–796 [DOI: 10.1007/s00360-015-0917-5].

Molenaar, R., Hulet, R., Meijerhof, R., Maatjens, C. M., Kemp, B. and van den Brand, H. 2011. High eggshell temperatures during incubation decrease growth performance and increase the incidence of ascites in broiler chickens. *Poultry Science* 90(3), 624–632 [DOI: 10.3382/ps.2010-00970].

Moran, E. T. 2019. Clutch formation and nest activities by the setting hen synchronize chick emergence with intestinal development to foster viability. *Animal Feed Science and Technology* 250, 69–80 [DOI: 10.1016/j.anifeedsci.2018.09.015].

Morita, V. S., Almeida, V. R., Matos Junior, J. B., Vicentini, T. I., Van Den Brand, H. and Boleli, I. C. 2016. Incubation temperature alters thermal preference and response to heat stress of broiler chickens along the rearing phase. *Poultry Science* 95(8), 1795–1804 [DOI: 10.3382/ps/pew071].

Nangsuay, A., Molenaar, R., Meijerhof, R., Van Den Anker, I., Heetkamp, M. J., Kemp, B. and Van Den Brand, H. 2015. Differences in egg nutrient availability, development, and nutrient metabolism of broiler and layer embryos. *Poultry Science* 94(3), 415–423 [DOI: 10.3382/ps/pev007].

Nätt, D., Lindqvist, N., Stranneheim, H., Lundeberg, J., Torjesen, P. A. and Jensen, P. 2009. Inheritance of acquired behaviour adaptations and brain gene expression in chickens. *PLoS ONE* 4(7), e6405 [DOI: 10.1371/journal.pone.0006405].

Nef, S., Allaman, I., Fiumelli, H., De Castro, E. and Nef, P. 1996. Olfaction in birds: differential embryonic expression of nine putative odorant receptor genes in the avian olfactory system. *Mechanisms of Development* 55(1), 65–77 [DOI: 10.1016/0925-4773(95)00491-2].

Nicol, C. J. 2015. *The Behavioural Biology of Chickens*. CABI International, Egham, UK, pp. 1–192.

Nielsen, B. L., Erhard, H. W., Friggens, N. C. and McLeod, J. E. 2008. Ultradian activity rhythms in large groups of newly hatched chicks (*Gallus gallus domesticus*). *Behavioural Processes* 78(3), 408–415 [DOI: 10.1016/j.beproc.2008.02.010].

Nielsen, B. L., Jezierski, T., Bolhuis, J., Amo, L., Rosell, F., Oostindjer, M., Christensen, J., McKeegan, D., Wells, D. and Hepper, P. 2015. Olfaction: an overlooked sensory modality in applied ethology and animal welfare. *Frontiers in Veterinary Science* 2, 69.

Nordquist, R. E., Zeinstra, E. C., Dougherty, A. and Riber, A. B. 2020. Effects of dark brooder rearing and age on hypothalamic vasotocin and feather corticosterone levels in laying hens. *Frontiers in Veterinary Science* 7, 19 [DOI: 10.3389/fvets.2020.00019].

Nordquist, R. E., Zeinstra, E. C., Rodenburg, T. B. and van der Staay, F. J. 2013. Effects of maternal care and selection for low mortality on tyrosine hydroxylase concentrations and cell soma size in hippocampus and nidopallium caudolaterale in adult laying hen. *Journal of Animal Science* 91(1), 137–146 [DOI: 10.2527/jas.2012-5227].

Norman, K. I., Adriaense, J. E. C. and Nicol, C. J. 2019. The impact of early structural enrichment on spatial cognition in layer chicks. *Behavioural Processes* 164, 167–174 [DOI: 10.1016/j.beproc.2019.05.008].

Onagbesan, O., Bruggeman, V., Smit, L., Debonne, M., Witters, A., Tona, K., Everaert, N. and Decuypere, E. 2007. Gas exchange during storage and incubation of Avian eggs: effects on embryogenesis, hatchability, chick quality and post-hatch growth. *World's Poultry Science Journal* 63(4), 557–573 [DOI: 10.1017/S0043933907001614].

Oro, C. D. S., De Souza, F., Rodrigues, W., Oro, J. and Guirro, E. 2019. Evaluation of the relationship between the breeder's age, breed and the incubator types on quality and animal welfare in the hatchery. *Brazilian Journal of Poultry Science* 21(3) [DOI: 10.1590/1806-9061-2019-1011].

Özkan, S.,Yalçin, S., Babacanoglu, E., Kozanoglu, H., Karadas, F. and Uysal, F. 2012a. Photoperiodic lighting (16 hours of light:8 hours of dark) programs during incubation: 1. Effects on growth and circadian physiological traits of embryos and early stress response of broiler chickens. *Poultry Science* 91(11):2912-2921 [DOI: 10.3382/ps.2012-02426].

Özkan, S., Yalçin, S., Babacanoglu, E., Uysal, S., Karadas, F. and Kozanoglu, H. 2012b. Photoperiodic lighting (16 hours of light:8 hours of dark) programs during incubation: 2. Effects on early posthatching growth, blood physiology, and production performance in broiler chickens in relation to posthatching lighting programs. *Poultry Science* 91(11), 2922-2930 [DOI: 10.3382/ps.2012-02427].

Pan, C., Lu, M., Zhang, Y., Yu, Y., Lu, Q., Yang, Y. and Pan, J. 2019. Unevenly distributed LED light produces distinct behavioral preferences and production performance of broilers. *International Journal of Agricultural and Biological Engineering* 12(2), 49-53 [DOI: 10.25165/j.ijabe.20191202.4689].

Patzke, N., Ocklenburg, S., van der Staay, F. J., Güntürkün, O. and Manns, M. 2009. Consequences of different housing conditions on brain morphology in laying hens. *Journal of Chemical Neuroanatomy* 37(3), 141-148 [DOI: 10.1016/j.jchemneu.2008.12.005].

Perré, Y., Wauters, A. M. and Richard-Yris, M. A. 2002. Influence of mothering on emotional and social reactivity of domestic pullets. *Applied Animal Behaviour Science* 75(2), 133-146 [DOI: 10.1016/S0168-1591(01)00189-7].

Pértille, F., Brantsæter, M., Nordgreen, J., Coutinho, L. L., Janczak, A. M., Jensen, P. and Guerrero-Bosagna, C. 2017. DNA methylation profiles in red blood cells of adult hens correlate with their rearing conditions. *Journal of Experimental Biology* 220(19), 3579-3587 [DOI: 10.1242/jeb.157891].

Petek, M., Ustuner, H. and Yesilbag, D. 2014. Effects of stocking density and litter type on litter quality and growth performance of broiler chicken. *Kafkas Universitesi Veteriner Fakultesi Dergisi* 5, (743-748) [DOI: 10.9775/kvfd.2014.11016].

Pettersson, I. C., Weeks, C. A., Norman, K. I. and Nicol, C. J. 2017. The ability of laying pullets to negotiate two ramp designs as measured by bird preference and behaviour. *PeerJ* 5, e4069 [DOI: 10.7717/peerj.4069].

Rault, J. L., Clark, K., Groves, P. J. and Cronin, G. M. 2017. Light intensity of 5 or 20 lux on broiler behavior, welfare and productivity. *Poultry Science* 96(4), 779-787 [DOI: 10.3382/ps/pew423].

Reed, H. J., Wilkins, L. J., Austin, S. D. and Gregory, N. G. 1993. The effect of environmental enrichment during rearing on fear reactions and depopulation trauma in adult caged hens. *Applied Animal Behaviour Science* 36(1), 39-46 [DOI: 10.1016/0168-1591(93)90097-9].

Riber, A. B. 2015. Effects of color of light on preferences, performance, and welfare in broilers. *Poultry Science* 94(8), 1767-1775 [DOI: 10.3382/ps/pev174].

Riber, A. B. and Forkman, B. 2007. A note on the behaviour of the chicken that receives feather pecks. *Applied Animal Behaviour Science* 108(3-4), 337-341 [DOI: 10.1016/j.applanim.2006.12.008].

Riber, A. B. and Guzman, D. A. 2016. Effects of dark brooders on behavior and fearfulness in layers. *Animals* 6(1), 1.6(1) [DOI: 10.3390/ani6010003].

Riber, A. B. and Guzman, D. A. 2017. Effects of different types of dark brooders on injurious pecking damage and production-related traits at rear and lay in layers. *Poultry Science* 96(10), 3529-3538 [DOI: 10.3382/ps/pex177].

Riber, A. B., Nielsen, B. L., Ritz, C. and Forkman, B. 2007a. Diurnal activity cycles and synchrony in layer hen chicks (*Gallus gallus domesticus*). *Applied Animal Behaviour Science* 108(3-4), 276-287 [DOI: 10.1016/j.applanim.2007.01.001].

Riber, A. B., Wichman, A., Braastad, B. O. and Forkman, B. 2007b. Effects of broody hens on perch use, ground pecking, feather pecking and cannibalism in domestic fowl (*Gallus gallus domesticus*). *Applied Animal Behaviour Science* 106(1-3), 39-51 [DOI: 10.1016/j.applanim.2006.07.012].

Riedstra, B. and Groothuis, T. G. G. 2002. Early feather pecking as a form of social exploration: the effect of group stability on feather pecking and tonic immobility in domestic chicks. *Applied Animal Behaviour Science* 77(2), 127-138 [DOI: 10.1016/S0168-1591(02)00031-X].

Robinson, F. E. and Wilson, J. L. 1996. Reproductive failure in overweight male and female broiler breeders. *Animal Feed Science and Technology* 58(1-2), 143-150 [DOI: 10.1016/0377-8401(95)00880-2].

Robinzon, B., Snapir, N. and Perek, M. 1977. Removal of olfactory bulbs in chickens: consequent changes in food intake and thyroid activity. *Brain Research Bulletin* 2(4), 263-271 [DOI: 10.1016/0361-9230(77)90082-x].

Roden, C. and Wechsler, B. 1998. A comparison of the behaviour of domestic chicks reared with or without a hen in enriched pens. *Applied Animal Behaviour Science* 55(3-4), 317-326 [DOI: 10.1016/S0168-1591(97)00073-7].

Rodenburg, T. B. 2017. Light during incubation and noise around hatching affect cognitive bias in laying hens. Proceedings of the 51st Congress of the International Society for Applied Ethology, Wageningen Academic Publishers.

Rodenburg, T. B. and de Haas, E. N. 2016. Of nature and nurture: the role of genetics and environment in behavioural development of laying hens. *Current Opinion in Behavioral Sciences* 7, 91-94 [DOI: 10.1016/j.cobeha.2015.12.007].

Rodenburg, T. B., Uitdehaag, K. A., Ellen, E. D. and Komen, J. 2009. The effects of selection on low mortality and brooding by a mother hen on open-field response, feather pecking and cannibalism in laying hens. *Animal Welfare* 18(4), 427-432.

Rodenburg, T. B., Van Krimpen, M. M., De Jong, I. C., De Haas, E. N., Kops, M. S., Riedstra, B. J., Nordquist, R. E., Wagenaar, J. P., Bestman, M. and Nicol, C. J. 2013. The prevention and control of feather pecking in laying hens: identifying the underlying principles. *World's Poultry Science Journal* 69(2), 361-374 [DOI: 10.1017/S0043933913000354].

Rogers, L. J. and Kaplan, R. 2019. Does functional lateralization in birds have any implications for their welfare? *Symmetry* 11(8), 1043 [DOI: 10.3390/sym11081043].

Rogers, L. J. 1995. *The Development of Brain and Behaviour in the Chicken*. Cab International, Wallingford.

Rogers, L. J. 1996. Behavioral, structural and neurochemical asymmetries in the avian brain: a model system for studying visual development and processing. *Neuroscience and Biobehavioral Reviews* 20(3), 487-503 [DOI: 10.1016/0149-7634(95)00024-0].

Rogers, L. J. 1982. Light experience and asymmetry of brain function in chickens. *Nature* 297(5863), 223-225 [DOI: 10.1038/297223a0].

Rogers, L. J. 1990. Light input and the reversal of functional lateralization in the chicken brain. *Behavioural Brain Research* 38(3), 211-221 [DOI: 10.1016/0166-4328(90)90176-f].

Rogers, L. J. 2010. Cognition and animal welfare. *Wiley Interdisciplinary Reviews: Cognitive Science* 1(3), 439-445 [DOI: 10.1002/wcs.48].

Rogers, L. J. 2012. The two hemispheres of the avian brain: their differing roles in perceptual processing and the expression of behavior. *Journal of Ornithology* 153(S1), 61–74 [DOI: 10.1007/s10336-011-0769-z].

Rogers, L. J. and Krebs, G. A. 1996. Exposure to different wavelengths of light and the development of structural and functional asymmetrics in the chicken. *Behavioural Brain Research* 80(1–2), 65–73 [DOI: 10.1016/0166-4328(96)00021-6].

Romanov, M. N., Talbot, R. T., Wilson, P. W. and Sharp, P. J. 2002. Genetic control of incubation behavior in the domestic hen. *Poultry Science* 81(7), 928–931.

Salmeto, A. L., Hymel, K. A., Carpenter, E. C., Brilot, B. O., Bateson, M. and Sufka, K. J. 2011. Cognitive bias in the chick anxiety-depression model. *Brain Research* 1373, 124–130 [DOI: 10.1016/j.brainres.2010.12.007].

Sandilands, V., Powell, K., Keeling, L. and Savory, C. J. 2004a. Preen gland function in layer fowls: factors affecting preen oil fatty acid composition. *British Poultry Science* 45(1), 109–115 [DOI: 10.1080/00071660410001668932].

Sandilands, V., Savory, J. and Powell, K. 2004b. Preen gland function in layer fowls: factors affecting morphology and feather lipid levels. *Comparative Biochemistry and Physiology. Part A, Molecular and Integrative Physiology* 137(1), 217–225 [DOI: 10.1016/j.cbpb.2003.10.004].

Sanyal, T., Kumar, V., Nag, T. C., Jain, S., Sreenivas, V. and Wadhwa, S. 2013. Prenatal loud music and noise: differential impact on physiological arousal, hippocampal synaptogenesis and spatial behavior in one day-old chicks. *PLoS ONE* 8(7), e67347 [DOI: 10.1371/journal.pone.0067347].

Sato, T., Kajiwara, R., Takashima, I. and Iijima, T. 2016. A novel method for quantifying similarities between oscillatory neural responses in wavelet time-frequency power profiles. *Brain Research* 1636, 107–117.

Saunders, J. C., Gates, G. R. and Coles, R. B. 1974. Brain-stem evoked responses as an index of hearing thresholds in one-day-old chicks and ducklings. *Journal of Comparative and Physiological Psychology* 86(3), 426–431 [DOI: 10.1037/h0036132].

Schilling, M. A., Katani, R., Memari, S., Cavanaugh, M., Buza, J., Basu, J., Mpenda, F., Deist, M., Lamont, S. and Kapur, V. 2018. Transcriptional innate immune response of the developing chicken embryo to Newcastle disease virus infection. *Frontiers in Genetics* 9, 61.

Schilling, M. A., Memari, S., Cattadori, I. M., Katani, R., Muhairwa, A. P., Buza, J. J. and Kapur, V. 2019. Innate immune genes associated with Newcastle disease virus load in chick embryos from inbred and outbred lines. *Frontiers in Microbiology* 10, 1432 [DOI: 10.3389/fmicb.2019.01432].

Shafey, T. M., Al-Batshan, H. A., Ghannam, M. M. and Al-Ayed, M. S. 2005. Effect of intensity of eggshell pigment and illuminated incubation on hatchability of brown eggs. *British Poultry Science* 46(2), 190–198 [DOI: 10.1080/00071660500065789].

Shimmura, T., Kamimura, E., Azuma, T., Kansaku, N., Uetake, K. and Tanaka, T. 2010. Effect of broody hens on behaviour of chicks. *Applied Animal Behaviour Science* 126(3–4), 125–133 [DOI: 10.1016/j.applanim.2010.06.011].

Sinkalu, V. O., Ayo, J. O., Adelaiye, A. B. and Hambolu, J. O. 2016. Melatonin modulates tonic immobility and vigilance behavioural responses of broiler chickens to lighting regimens during the hot-dry season. *Physiology and Behavior* 165, 195–201 [DOI: 10.1016/j.physbeh.2016.07.019].

Siwek, M., Slawinska, A. Stadnicka, K. et al. 2018. Prebiotics and synbiotics – in ovo delivery for improved lifespan condition in chicken. *BMC Veterinary Research* 14, 402 [DOI: 10.1186/s12917-018-1738-z].

Sneddon, H., Hadden, R. and Hepper, P. G. 1998. Chemosensory learning in the chicken embryo. *Physiology & Behavior* 64, 133–139.

Spindler, B., Clauss, M., Briese, A. and Hartung, J. 2013. Planimetric measurement of floor space covered by pullets. *Berliner und Munchener tierarztliche Wochenschrift* 126(3–4), 156–162.

Swann, G. S. and Brake, J. 1990. Effect of incubation dry-bulb and wet-bulb temperatures on time of hatch and chick weight at hatch. *Poultry Science* 69(6), 887–897 [DOI: 10.3382/ps.0690887].

Tahamtani, F. M., Brantsæter, M., Nordgreen, J., Sandberg, E., Hansen, T. B., Nødtvedt, A., Rodenburg, T. B., Moe, R. O. and Janczak, A. M. 2016a. Effects of litter provision during early rearing and environmental enrichment during the production phase on feather pecking and feather damage in laying hens. *Poultry Science* 95(12), 2747–2756 [DOI: 10.3382/ps/pew265].

Tahamtani, F. M., Hansen, T. B., Orritt, R., Nicol, C., Moe, R. O. and Janczak, A. M. 2014. Does rearing laying hens in aviaries adversely affect long-term welfare following transfer to furnished cages? *PLoS ONE* 9(9), e107357 [DOI: 10.1371/journal.pone.0107357].

Tahamtani, F. M. ,Nordgreen, J., Brantsæter, M., Østby, G., Nordquist, R. and Janczak, A. 2016b. Does early environmental complexity influence tyrosine hydroxylase in the chicken hippocampus and "prefrontal" caudolateral nidopallium? *Frontiers in Veterinary Science* 3, 8.

Tahamtani, F. M., Nordgreen, J., Nordquist, R. E. and Janczak, A. M. 2015. Early life in a barren environment adversely affects spatial cognition in laying hens (Gallus gallus domesticus). *Frontiers in Veterinary Science* 2, 3.

Thogerson, C. M., Hester, P. Y., Mench, J. A., Newberry, R. C., Okura, C. M., Pajor, E. A., Talaty, P. N. and Garner, J. P. 2009. The effect of feeder space allocation on productivity and physiology of Hy-Line W-36 hens housed in conventional cages. *Poultry Science* 88(9), 1793–1799 [DOI: 10.3382/ps.2009-00011].

Tona, K., Onagbesan, O., De Ketelaere, B., Decuypere, E. and Bruggeman, V. 2004. Effects of age of broiler breeders and egg storage on egg quality, hatchability, chick quality, chick weight, and chick posthatch growth to forty-two days. *Journal of Applied Poultry Research* 13(1), 10–18 [DOI: 10.1093/japr/13.1.10].

Tong, Q., McGonnell, I. M., Romanini, C. E., Bergoug, H., Roulston, N., Exadaktylos, V., Berckmans, D., Bahr, C., Guinebretière, M., Eterradossi, N., Garain, P. and Demmers, T. 2015. Effect of species-specific sound stimulation on the development and hatching of broiler chicks. *British Poultry Science* 56(2), 143–148 [DOI: 10.1080/00071668.2014.1000822].

Tuculescu, R. A. and Griswold, J. G. 1983. Prehatching interactions in domestic chickens. *Animal Behaviour* 31(1), 1–10 [DOI: 10.1016/S0003-3472(83)80168-7].

Turner, J. S. 1997. On the thermal capacity of a bird's egg warmed by a brood patch. *Physiological Zoology* 70(4), 470–480 [DOI: 10.1086/515854].

Tzschentke, B. and Nichelmann, M. 1999. Development of avian thermoregulatory system during the early postnatal period: development of the thermoregulatory set-point. *Ornis Fennica* 76, 189–198.

Van De Ven, L. J., Van Wagenberg, A. V., Uitdehaag, K. A., Groot Koerkamp, P. W., Kemp, B. and Van Den Brand, H. 2012. Significance of chick quality score in broiler production. *Animal* 6(10), 1677–1683 [DOI: 10.1017/S1751731112000663].

van der Eijk, J. A. J., Lammers, A., Li, P., Kjaer, J. B. and Rodenburg, T. B. 2018. Feather pecking genotype and phenotype affect behavioural responses of laying hens. *Applied Animal Behaviour Science* 205, 141–150 [DOI: 10.1016/j.applanim.2018.05.027].

van der Pol, C. W., Molenaar, R., Buitink, C., van Roovert-Reijrink, I. A. M., Maatjens, C. M., van den Brand, H. and Kemp, B. 2015. Lighting schedule and dimming period in early life: consequences for broiler chicken leg bone development. *Poultry Science* 94(12) [DOI: 10.3382/ps/pev276].

van der Pol, C. W., van Roovert-Reijrink, I. A. M., Aalbers, G., Kemp, B. and van den Brand, H. 2017. Incubation lighting schedules and their interaction with matched or mismatched post hatch lighting schedules: effects on broiler bone development and leg health at slaughter age. *Research in Veterinary Science* 114, 416–422 [DOI: 10.1016/j.rvsc.2017.07.013].

van der Pol, C. W., van Roovert-Reijrink, I. A. M., Gussekloo, S. W. S., Kranenbarg, S., Leon-Kloosterziel, K. M., van Eijk-Priester, M. H., Zeman, M., Kemp, B. and van den Brand, H. 2019. Effects of lighting schedule during incubation of broiler chicken embryos on leg bone development at hatch and related physiological characteristics. *PLoS ONE* 14(8), e0221083 [DOI: 10.1371/journal.pone.0221083].

Van Der Waaij, E. H., Van Den Brand, H., Van Arendonk, J. A. and Kemp, B. 2011. Effect of match or mismatch of maternal-offspring nutritional environment on the development of offspring in broiler chickens. *Animal* 5(5), 741–748 [DOI: 10.1017/S1751731110002387].

Van Krimpen, M. M., Kwakkel, R. P., Reuvekamp, B. F. J., Van Der Peet-Schwering, C. M. C., Den Hartog, L. A. and Verstegen, M. W. A. 2005. Impact of feeding management on feather pecking in laying hens. *World's Poultry Science Journal* 61(4), 663–686 [DOI: 10.1079/WPS200478].

van Praag, H., Kempermann, G. and Gage, F. H. 2000. Neural consequences of enviromental enrichment. *Nature Reviews Neuroscience* 1(3), 191–198 [DOI: 10.1038/35044558].

Voslarova, E., Chloupek, P., Chloupek, J., Bedanova, I., Pistekova, V. and Vecerek, V. 2011. The effects of chronic intermittent noise exposure on broiler chicken performance. *Animal Science Journal* 82(4), 601–606 [DOI: 10.1111/j.1740-0929.2011.00877.x].

Walstra, I., Ten Napel, J., Kemp, B. and Van Den Brand, H. 2010. Early life experiences affect the adaptive capacity of rearing hens during infectious challenges. *Animal* 5(4(10)), 1688–1696.

Wang, Y., Li, Y., Willems, E., Willemsen, H., Franssens, L., Koppenol, A., Guo, X., Tona, K., Decuypere, E., Buyse, J. and Everaert, N. 2014. Spread of hatch and delayed feed access affect post hatch performance of female broiler chicks up to day 5. *Animal* 8(4), 610–617 [DOI: 10.1017/S175173111400007X].

Wauters, A. M., Perré, Y., Bizeray, D., Leterrier, C. and Richard-Yris, M. A. 2002. Mothering influences the distribution of activity in young domestic chicks. *Chronobiology International* 19(3), 543–559 [DOI: 10.1081/cbi-120004224].

Weissmann, A., Förster, A., Gottschalk, J., Reitemeier, S., Krautwald-Junghanns, M.-E., Preisinger, R. and Einspanier, A. 2014. In ovo-Geschlechtsbestimmung bei Legehybriden: Einfluss auf Schlupf und Produktionsleistung. *European Poultry Science* 78 [DOI: 10.1399/eps.2014.25].

Weissmann, A., Reitemeier, S., Hahn, A., Gottschalk, J. and Einspanier, A. 2013. Sexing domestic chicken before hatch: a new method for in ovo gender identification. *Theriogenology* 80(3), 199–205 [DOI: 10.1016/j.theriogenology.2013.04.014].

Wells, J. B., Coufal, C. D., Parker, H. M. and McDaniel, C. D. 2010. Disinfection of eggshells using ultraviolet light and hydrogen peroxide independently and in combination. *Poultry Science* 89(11), 2499-2505 [DOI: 10.3382/ps.2009-00604].

Wichman, A., Rogers, L. J. and Freire, R. 2009. Visual lateralization and development of spatial and social spacing behaviour of chicks (Gallus gallus domesticus). *Behavioural Processes* 81(1), 14-19 [DOI: 10.1016/j.beproc.2008.12.006].

Wijnen, H. J., Molenaar, R., van Roovert-Reijrink, I. A. M., van der Pol, C. W., Kemp, B. and van den Brand, H. 2020. Effects of incubation temperature pattern on broiler performance. Poultry Science. 99(8), 3897-3907.

Wilkins, L. J., McKinstry, J. L., Avery, N. C., Knowles, T. G., Brown, S. N., Tarlton, J. and Nicol, C. J. 2011. Influence of housing system and design on bone strength and keel bone fractures in laying hens. *Veterinary Record* 169(16), 414 [DOI: 10.1136/vr.d4831].

Woodcock, M. B., Pajor, E. A. and Latour, M. A. 2004. The effects of hen vocalizations on chick feeding behavior. *Poultry Science* 83(12), 1940-1943 [DOI: 10.1093/ps/83.12.1940].

Wood-Gush, D. G. M. 1971. *The Behaviour of the Domestic Fowl*, Issue 7. Nimrod Press.

Zeltner, E., Klein, T. and Huber-Eicher, B. 2000. Is there social transmission of feather pecking in groups of laying hen chicks? *Animal Behaviour* 60(2), 211-216 [DOI: 10.1006/anbe.2000.1453].

# Chapter 11

## Welfare issues in poultry housing and management: broilers

*Ingrid C. de Jong, Wageningen Livestock Research, Wageningen University and Research, The Netherlands*

1 Introduction
2 Broiler behaviour and space use
3 Welfare issues in broiler chicken production: leg health, heat stress and behavioural restrictions
4 The relationship between growth rate and broiler welfare
5 Effects of environment and management on welfare
6 Conclusions and future trends
7 Where to look for further information
8 References

### 1 Introduction

Worldwide, more than 60 billion broiler chickens are reared and slaughtered for meat production on a yearly basis. Poultry meat production even exceeded pork production in 2018, with a worldwide production of more than 123 million metric tons per year (Statistica, 2019). The majority of broiler chickens worldwide are of a conventional, fast-growing breed and are housed in indoor systems with a fully littered floor with either mechanical or natural ventilation (Fig. 1). Feed and water are provided automatically and feeders and drinkers are distributed over the house. The standard broiler house has no windows, but windows can be present in the side walls or the roof. Depending on the local climate open-sided houses can also be used, such as in Brazil (Federici et al., 2016) and southern Europe (de Jong et al., 2012a). In Europe the litter is usually completely removed in between production cycles and houses are cleaned and disinfected, but in other countries, such as the United States, litter can be re-used (Cengiz et al., 2011). In the major producing countries the number of broiler chickens per house is usually between several thousands, around 30 000–50 000 chickens per house. Stocking densities also vary between countries, for example, in Europe in indoor systems a maximum density of 38–42 kg/m$^2$ is a normal

http://dx.doi.org/10.19103/AS.2020.0078.10

**Figure 1** Examples of broiler production systems: indoor system without windows and conventional fast-growing breed (upper picture), and an example of a higher welfare system with a slow-growing breed enrichment (here: Lucerne bales in racks), a reduced stocking density and covered outdoor range (middle and lower picture) (Pictures: Ingrid de Jong, Wageningen Livestock Research).

practice, according to the specifications in the EU Broiler Directive (European Commission, 2007), although in some countries lower maximum densities are allowed such as 30 kg/m$^2$ in Austria and 36 kg/m$^2$ in Sweden (European Commission, 2016). In the United States stocking densities in conventional systems are around 37 kg/m$^2$ and in Brazil between 34 kg/m$^2$ and 38 kg/m$^2$ depending on the type of house (open-sided or tunnel-ventilated) (Vissers et al., 2019). The light intensity in Europe should be at least 20 lux at animal height, as defined in the Broiler Directive (European Commission, 2007), but it can be as low as 5 lux in conventional systems in countries such as the United States and Brazil in windowless houses (Vissers et al., 2019).

Since 2008, in Russia, the Middle East and Asia cage systems are increasingly being used for broiler chickens (e.g. Bracke et al., 2019b). These 'new' cage systems are colony cages with plastic-covered nylon floors and automated systems for depopulation of the chickens using conveyor belts. However, the fact that broiler chickens are still housed in battery cages designed for laying hens, for example, in African countries cannot be ignored (Shields and Greger, 2003).

In contrast, especially in countries in Northwestern Europe, there is an increasing demand for so-called 'higher-welfare' systems, resulting in an increased use of indoor systems with lower stocking densities, environmental enrichment and sometimes a covered veranda (Saatkamp et al., 2019; de Jonge and van Trijp, 2014; RSPCA, 2006; Vissers et al., 2019; Bergmann et al., 2016) (Fig. 1), free-range systems, or broiler chickens kept on pasture (usually organic broiler production systems). Free-range houses are usually similar to indoor production houses, but have pop holes for access to the covered veranda ('free-range indoor') or outdoor range. Pasture systems usually have smaller flock sizes and may include mobile houses (Sossidou et al., 2015). Whereas the majority of broiler chickens in conventional indoor systems are usually of a standard, fast-growing breed, slower-growing broiler strains are usually housed in higher-welfare or pasture systems (Saatkamp et al., 2019; Bergmann et al., 2016; Sossidou et al., 2015). In, for example, Sweden, these have only been introduced recently whereas previously fast-growing broilers were used in organic systems (Wilhelmsson et al., 2019). Slow-growing broiler chickens are usually broiler chickens produced by slow-growing female parent stock that are crossed with a regular or slow-growing male, resulting in a daily growth rate of about 50 g or lower at slaughter weight (2.3-2.5 kg) (Ellen et al., 2012; Saatkamp et al., 2019), whereas fast-growing broiler strains may have an average daily growth rate of more than 65 g at similar slaughter weights (Aviagen, 2019). As a consequence, similar slaughter weights are reached at different ages, that is, around 38 days of age for fast-growing strains, and vary between 46 and 81 days of age for slow-growing strains depending on the daily growth rate.

### 1.1 Genetic selection for growth potential and broiler welfare

There are only three large breeding companies that produce the majority of broiler chickens worldwide. Broiler chickens have been selected intensively for growth, feed efficiency and meat production over the past decades. This means that they have the potential to grow to slaughter weight in a relatively short time with a very low feed conversion rate, provided conditions to do so are (near) optimal. For example, Zuidhof et al. (2014) showed that from 1957 to 2005, broiler growth rates have increased by over 400%, with a concurrent 50% reduction in feed conversion ratio. This means that a body weight of 2.5 kg can be reached in just 38 days (e.g. Aviagen, 2019). Although some of these changes have been attributed to improved housing and management, it was estimated that 85–90% of this achievement could be attributed to genetic selection. The intensive genetic selection, however, has also been shown to lead to concomitant adverse effects on welfare and health, such as leg disorders, metabolic disorders and impaired immune function (Bessei, 2006; EFSA, 2010). This does not mean that breeding companies do not include welfare and health improvement in their selection process. Breeding companies select for optimal leg strength, heart and lung fitness, and against contact dermatitis, and they are culling birds with a family record or signs of genetic abnormalities. For some traits this has already been done for 25 years (Hiemstra and Ten Napel, 2011). For some metabolic disorders (e.g. ascites) and some leg disorders, genetic selection has also been successful in decreasing the prevalence (Hiemstra and Ten Napel, 2011). However, welfare problems linked to genetic selection for efficient growth in broiler chickens are still present, such as leg disorders and health problems in broiler chickens, and the associated feed restriction in the parent stock which is discussed elsewhere in this chapter (de Jong et al., 2012a; Dawkins and Layton, 2012). It has been estimated that the relative weighting of all welfare traits in the combined breeding goal varied from 18% to 33% across breeding companies in 2011 (Hiemstra and Ten Napel, 2011). A more balanced genetic selection with faster progress in welfare traits is possible (e.g. breeding companies do have other breeds available with a lower growth potential), but this may result in a reduction of economically more important traits (Dawkins and Layton, 2012; Hiemstra and Ten Napel, 2011). Only if there is a change in market requirements would breeding companies be expected to change their selection process (Hiemstra and Ten Napel, 2011). In some regions of Europe, the market share of slower-growing strains has substantially increased over the past 5 years because of the change towards higher-welfare systems as a result of increased societal pressure on the negative effects of broiler production on animal welfare (Saatkamp et al., 2019). However, globally, these slower-growing broiler strains in higher-welfare systems account for only a very small part of the total broiler production volumes.

In the next section we will first discuss some aspects of broiler behaviour and space use. Section 3 discusses important welfare problems in broiler production for the on-farm stage, Section 4 discusses the relationship between growth rate and welfare of broiler chickens, and Section 5 discusses environment and management factors affecting broiler welfare.

## 2 Broiler behaviour and space use

Broiler chickens are usually housed in large groups of thousands of birds, and in both mixed and sex-separated groups. Conventional broiler chickens can already be slaughtered at an age of 30 days (usually when thinning a flock, that is, when removing part of the chickens from the house to be slaughtered while the others are staying until a higher slaughter weight), whereas slower-growing strains may reach an age of 49 to over 70 days of age (in case of organic broiler production or Label Rouge production) (e.g. Vissers et al., 2019; Bracke et al., 2019a). This means that these are young birds that will never become adults (which would take around 17–20 weeks), and their behaviour will be juvenile too, that is their sexual and agonistic behaviour will never be fully developed.

Adequate use of space may enable the full performance of natural behaviours. However, space use in commercial broiler houses is often not evenly distributed, with broiler chickens clustering close to the walls from the first day of age onwards (BenSassi et al., 2019a; Leone and Estevez, 2008a). This can be due to their preference to seek protection from perceived predation both from the walls and conspecifics while resting, but it may increase the risk for welfare issues, such as disturbance of resting birds due to movements from the periphery to the central area with feeders and drinkers (Ventura et al., 2012; Buijs et al., 2010), increasing temperature at bird level resulting in heat stress (Ventura et al., 2012), and deteriorating litter quality resulting in contact dermatitis and dirty plumage (BenSassi et al., 2019a). Insight in factors determining space use, neighbour distances and distribution of broilers may help to find ways to promote better use of the available space in commercial houses and thus broiler welfare.

Studies under experimental conditions in fast-growing broiler chickens to determine the effects of group size, pen size and stocking density on the distribution of the chickens in the pen and the use of space showed that pen size had the largest effect on movement and space use patterns, irrespective of group size and stocking density. With increasing pen size, activity, neighbour distances and space use increased, thus chickens used larger space when available (Leone and Estevez, 2008a,b; Leone et al., 2010; Mallapur et al., 2009). Physical barriers of other animals played a role at higher-stocking densities, leading to lower neighbour distances (Leone and Estevez, 2008b). The authors suggested that two motivations play a role in establishing neighbour distances:

on one hand, other chickens offer protection and thus low neighbour distances might be preferred, while on the other hand, resource competition may increase the distance between chickens in a group (Leone et al., 2010). As high-stocking densities have been found to be aversive to broiler chickens, this motivation may also play a role when stocking densities increase (Buijs et al., 2011b), which will be discussed in more detail below. Buijs et al. (2011b) also showed that nearest neighbour distances varied according to the type of behaviour the broilers were performing. Chickens that were eating and drinking showed higher neighbour distances than birds that were foraging, preening or adjusting their sitting or lying posture, or showing other behaviour (which included resting). This is in line with the suggestion of Leone et al. (2010) that the motivation to seek protection plays a role, as chickens are vulnerable to predation when showing behaviours such as resting, dustbathing and preening, which involve periods of time during which the eyes are closed. Apart from seeking social protection during these behaviours (by sitting more close to conspecifics), broilers have also been shown to perform these behaviours more often in areas with artificial protection, such as panels or close to walls (Cornetto and Estevez, 2001). Moreover, providing cover (by using panels) resulted in a better distribution of broiler chickens over the pen (Cornetto et al., 2002), with the largest effects seen in small groups (5 versus 10 or 20 broilers per group) (Leone et al., 2007). This was in line with the hypothesis that broilers in small groups perceive that they are more vulnerable to predation because of reduced social protection.

Similar effects on the spatial distribution of chickens over the pen were found with barrier perches (panels with perches on top) (Ventura et al., 2012). Moreover, both cover panels and barrier perches decreased neighbour disturbances (Ventura et al., 2012; Leone et al., 2007), and also aggression decreased significantly when barrier perches were provided (Ventura et al., 2012). The latter might be due to the better distribution of resting birds over the pen and reducing the open space (Ventura et al., 2012), as it has been shown that aggressive behaviours are mainly observed in the open central space in the pen (Pettit-Riley et al., 2002). Also in commercial flocks clustering of chickens around structures placed in the central area of the house has been observed, for example, when bales of straw or wood shavings were provided (de Jong and Gunnink, 2019; Bailie et al., 2013) although these studies did not find any positive effects on the incidence of disturbance and aggression.

Although interesting from a behavioural point of view, results of experimental studies on effects of enclosure size and group size in broiler chickens on space use are difficult to translate to the commercial situation where broilers are housed in much larger groups. However, providing structures probably improves bird distribution and may reduce negative behaviours, as well as have other potentially positive effects, which will be discussed in the section on environmental enrichment.

## 3 Welfare issues in broiler chicken production: leg health, heat stress and behavioural restrictions

Impaired leg health is considered one of the most severe welfare problems related to the on-farm stage of broiler chickens (EFSA, 2010). Leg health includes lameness (infectious, degenerative and developmental disorders) and contact dermatitis on the feet and hocks (footpad dermatitis and hock burn) (Bradshaw et al., 2002). Other welfare issues are impaired health, including metabolic diseases, which may result in mortality and/or the application of medication, thermal discomfort (mainly heat stress) and so-called behavioural restrictions. The latter refers to environmental and physical restrictions to perform the full range of natural chicken behaviour (EFSA, 2010). For a detailed description of leg disorders and the effect on broiler welfare we refer to elsewhere in this chapter. Here we will only give some information on the prevalence in commercial broiler systems, if known, and discuss the relationship with growth rate.

### *3.1 Prevalence of welfare issues in different broiler production systems*

Walking ability is usually assessed in commercial flocks using the gait score, which assigns a score between 0 and 5 (perfect to unable to walk) (Kestin et al., 1999). In the literature over the past 10 years, the estimated prevalence of a moderate to severe defect, that is a gait score of 3 and higher, may be as high as 30% of the chickens in conventional indoor systems with fast-growing breeds (e.g. Bassler et al., 2013; Sanotra et al., 2001; Knowles et al., 2008) and recent figures illustrate that it can even be higher. In a recent Norwegian study, 24% of the chickens (Ross 308) had a gait score of 3 and higher, but this was at a relatively young age (29 days) due to the low slaughter age in Norway (Kittelsen et al., 2017). A recent study in the United States including various fast-growing breeds showed that an impaired walking ability was found in more than 22% of the chickens in flocks slaughtered at a light body weight (2.2 kg) and more than 34% of the chickens slaughtered at a high body weight (over 3.6 kg) (Opengart et al., 2018). In higher-welfare broiler systems fewer broilers have an impaired walking ability. In Germany, 25% of broilers in a conventional system (fast-growing breed (Ross 308), no enrichment, maximum stocking density of 35 kg/m$^2$) showed an impaired walking ability compared to 1% of broilers in a higher-welfare system (slower-growing breed (Cobb Sasso 175A), enrichment and covered veranda, maximum stocking density of 29 kg/m$^2$) (Bergmann et al., 2016). A Dutch study compared conventional flocks with fast-growing Cobb 500 and Ross 308 breeds (no enrichment, 39–42 kg/m$^2$) with indoor housed slow-growing strains (Hubbard JA757) with environmental enrichment and a lower stocking density (<38 kg/m$^2$) and found more than 60% of the chickens with a moderate to severe defect (gait score 3 and higher) in conventional flocks versus less than 20% in the higher-welfare system (Gerritzen et al., 2019). These figures indicate variation between systems and room for improvement in all systems, but especially in the conventional indoor systems.

Three types of contact dermatitis can be observed in broiler chickens: footpad dermatitis, hock burn and breast burn or breast irritation (Haslam et al., 2007). Usually, prevalence of footpad dermatitis is higher as compared to hock burn and breast burn. In a recent study in the United States, prevalence of 27% and 1.5% for mild and severe footpad dermatitis, respectively, were reported for different breeds of fast-growing broilers slaughtered at light weights (around 2 kg), and 31% and 3.5%, respectively, for fast-growing broiler breeds slaughtered at heavy weights (more than 3.6 kg) (Opengart et al., 2018). Footpad dermatitis is one of the animal-based outcome data routinely collected at the slaughter plant as an indicator of broiler welfare in several European countries (Butterworth et al., 2016). This may have resulted in a decreased prevalence in recent years, for example, in the Netherlands the total footpad lesion score significantly decreased within four years after the start of monitoring from approximately 70 points to a yearly average of around 40 points (on a scale between 0 (no lesions) and 200 (all birds having severe lesions)) (de Jong, personal communication). In Denmark, monitoring started earlier; a decrease of 60 points was achieved in a period of 6 years, and on a yearly basis scores do not exceed 40 points (Kyvsgaard et al., 2013), with a drop in prevalence of severe footpad lesions from 60% to 10% (European Commission, 2016). In higher-welfare broiler systems a significantly lower prevalence of footpad dermatitis is generally found. Bergmann et al. (2016) reported a prevalence of 16.5% and 2.5% around slaughter age for broilers in conventional systems (fast-growing Ross 308 breed, stocking density of 35 kg/m$^2$) and a higher-welfare concept (slower-growing Cobb Sasso 175A breed, covered outdoor and enrichment, stocking density of 29 kg/m$^2$), respectively. Similar differences were found in two Dutch studies (22% and 4.8%, respectively (Ellen et al., 2012), and around 45% and 0%, respectively, in conventional versus higher-welfare systems (Gerritzen et al., 2019)). A comparison of RSPCA freedom food flocks (slow-growing broilers, stocking density maximum 30 kg/m$^2$, environmental enrichment) with the industry standard Red Tractor (indoor housing, fast-growing breeds, no enrichment, stocking density 38 kg/m$^2$) showed that Freedom Food flocks had on average 3.5% footpad dermatitis compared to 6.5% for the conventional flocks (RSPCA, 2006). An exception is the organic broiler farms, where a higher prevalence of footpad dermatitis has been found than on conventional farms (Lund et al., 2017), or a higher variation is observed. For example, in France a prevalence between 4.6% and 39.6% was observed, depending on the farm type (mobile houses had a higher prevalence) and time of visit. This variation might be due to the varying conditions in the outdoor range on organic farms and due to inadequate ventilation in mobile houses causing wet litter (Souillard et al., 2019).

Recent figures on the prevalence of hock burn and breast irritation are scarce. Bassler et al. (2013) reported on average 7.9% of broilers with hock burn on European farms. Bergmann et al. (2016) reported superficial hock lesions in 33% and 20% of the broilers in German conventional (indoor, fast-growing Ross 308 breed, stocking density 35 kg/m$^2$) and higher-welfare systems (slow-growing Cobb Sasso 175A breed, enrichment and veranda, stocking density of 29 kg/m$^2$), respectively, and 1.8% severe hock burn in the conventional system close to slaughter age. When comparing RSPCA freedom food flocks with flocks housed according to the conventional system in the United Kingdom, the RSPCA flocks had on average 3.5% hock burn whereas the conventional flocks had 19% (RSPCA, 2006). A Dutch study reported that 24% of broilers in conventional flocks had hock burn at slaughter, whereas in the higher-welfare concept this was only observed in 1.1% of the chickens (Ellen et al., 2012), and more recently, no severe hock burn was observed in Dutch higher-welfare concept flocks (enrichment, stocking density lower than 38 kg/m2, slower-growing Hubbard JA757 breed) compared to around 10% for the conventional indoor flocks using Ross 308 or Cobb 500 breeds (Gerritzen et al., 2019). Thus, hock burn figures illustrate a similar trend as footpad lesions, with lower levels of hock burn present in higher-welfare systems than in conventional systems. With respect to breast irritation, Gerritzen et al. (2019) found 0% in the higher-welfare system and in less than 5% of the broilers in conventional indoor systems. In an older study of Haslam et al. (2007) breast burn was observed in less than 1% of broilers, but Allain et al. (2009) reported breast burn in 15.8% of broilers with high variation between flocks. They reported an effect of genotype and stocking density with a higher prevalence in the fast-growing broilers and a high stocking density as compared to flocks with slow-growing broilers or a lower stocking density. Taken together, all types of contact dermatitis (footpad dermatitis, hock burn and breast burn) can be present on broiler farms with a lower risk in higher-welfare concepts.

Mortality is related to health, but, not all health issues relevant for welfare are properly captured by mortality figures, although these at least give an indication. With respect to total mortality, the figures show some variation, being sometimes lower in higher-welfare systems, but not always. However, calculated as mortality per day, mortality figures are usually lower in higher-welfare systems. The RSPCA reported figures of 1.8% for their Freedom Food concept versus 5.1% for conventional (RSPCA, 2006). Bassler et al. (2013) reported an average mortality rate of 3.6% on conventional farms in Europe. Bergmann et al. (2016) reported mortality figures that were comparable in indoor conventional and higher-welfare flocks (3.5% and 3.4%, respectively), but in the higher-welfare systems broilers had a longer rearing period. In a Dutch study, the average mortality over the whole rearing period was higher in the higher-welfare system than in the conventional system (3.4 vs. 2.7%), but this was calculated over a rearing period of 56 days for conventional versus 39 days, respectively, for higher-welfare systems (Ellen et al., 2012). However, in a more recent Dutch study, mortality figures of 3.7% and 1.9% were found for conventional indoor and higher-welfare flocks, respectively (Gerritzen et al., 2019). Souillard et al. (2019) reported mortality levels between 2.7% and 3.1% on organic farms in France (excluding mortality caused by predation), depending on farm type, but it should be kept in mind that these were registered over a rearing period of 86 days.

Antibiotics-usage data in the Netherlands in 2018 showed that antibiotic treatments are applied more at farms with fast-growing than with slow-growing breeds, with on average 31% of the farms with fast-growing breeds applying any antibiotic treatment versus 9% of the farms with slow-growing strains, and preventive treatments not being allowed (Avined, 2019). Although antibiotics usage is not necessarily linked to health, these data nevertheless indicate that health problems are less prevalent on farms with slower-growing strains.

With respect to the welfare issue of behavioural restrictions, broilers in a higher-welfare system in Germany (maximum stocking density of 29 kg/m$^2$, slower-growing Cobb Sasso 175A strain and enrichment provided) were significantly more active during the rearing period than fast-growing Ross 308 broilers in an indoor system (maximum stocking density 35 kg/m$^2$) (Bergmann et al., 2017). As far as we know, there are no other publications comparing the behaviour of broiler chickens in conventional versus higher-welfare systems, which indicates that there is a need for studying behaviour of broiler chickens in different production systems.

When comparing conventional indoor-housed broiler flocks with flocks housed in higher-welfare systems, it is often not possible to attribute differences to breed, environmental or management factors, which are normally confounded in such comparative studies. What we know about the separate effects of breed, management and environmental aspects on broiler welfare will be further discussed below.

To date (modern colony) cage systems for broilers have not been studied in great detail. In cage systems broiler welfare may be compromised as compared to conventional indoor floor systems, because of movement restriction, poor bone strength due to lack of exercise, the inability to perform natural behaviours such as dust bathing and scratching, relatively high-stocking densities (which are usually higher as compared to indoor floor systems), and, although depending on the floor type, the risk for skin and leg injuries (Shields and Greger, 2003). Experts assign low welfare scores to these cage systems (Bracke et al., 2019a). In contrast, producers stress that there might be advantages of the plastic floors in the modern colony cages with respect to heat dissipation and the risk for footpad lesions and hock burn, which however requires further study before drawing any conclusion about possible welfare advantages.

# 4 The relationship between growth rate and broiler welfare

## *4.1 Effects of growth rate on leg health, mortality and heat stress*

Few studies are available comparing fast- and slow-growing broiler breeds housed under similar conditions. Van Middelkoop et al. (2002) compared one fast- (Cobb) and one slow-growing (Hubbard) breed that were reared until similar slaughter weight and housed under similar semi-commercial conditions. A significantly lower mortality rate was found in the slow-growing breed than in the fast-growing one (1.5% vs. 5.6%). This was mainly due to heart and circulation problems in the fast-growers and these were nearly absent in the slow growers. In addition, slow growers had a significantly better walking ability (13.1% vs. 43.4% of the birds had a gait score of 3 and higher for the slow vs. fast-growing broiler strain), and significantly better scores for footpad dermatitis, hock burn and breast irritation (proportion of birds with severe footpad dermatitis 0% vs. 9.8%, hock burn 11.5% vs. 44.9%, breast irritation 10.1% vs. 56.6% for slow- vs. fast-growing breed). Similar effects for mortality, gait score, breast irritation, footpad dermatitis and hock burn were found in a second study comparing one slow- (Hubbard) and fast-growing Ross breed, both in two similar semi-commercial housing conditions (with and without an outdoor range) (Rodenburg et al., 2004). For example, severe footpad lesions were found in 6% and 16% of fast-growing broilers (housed indoor and with outdoor access, respectively), compared to 0% and 0.5% in slow-growing birds (housed indoor and with outdoor access, respectively) (Rodenburg et al., 2004). Recently, Dixon (2020) compared three fast-growing strains from the main commercial broiler companies with one slow-growing strain (Hubbard JA757) housed under similar conditions. With respect to mortality, the slow-growing strain and one fast-growing strain had significantly lower mortality levels than the other two fast-growing strains. Moreover, the slow-growing strain had lower proportion of hock burn and significantly better locomotion scores, for example, around 10% of the slow-growing strain had a gait score 3 (moderate defect) as compared to around 30–40% for the three fast-growing strains. Footpad dermatitis scores were low in all strains and did not significantly differ between the strains in that particular study. Slow-growing broilers are thought to be less susceptible to heat stress than conventional fast-growing breeds (de Jong et al., 2012a), but as far as we know there are no data to illustrate this.

Although differences exist between different types of slow-growing strains with respect to the welfare issues discussed here, these seem to be small, especially as compared to the differences between fast- and slower-growing broiler strains (Louton et al., 2019; Rauch et al., 2017). Taken together with the (confounded) system comparisons these few studies indicate that the use of slower-growing broiler strains, indeed, explains part of the lower risk for leg disorders and impaired health in higher-welfare systems.

## 4.2 Effects of growth rate on broiler behaviour

Conventional fast-growing broiler chickens are usually rather inactive with a peak in activity around the fourth week of age (Ventura et al., 2012; Bizeray et al., 2002) and a sharp decrease in activity with increasing age thereafter (Arnould and Faure, 2003; Wallenbeck et al., 2016; Hall, 2001; Shields et al., 2005). This is in part due to their housing conditions, such as the increasing stocking density (kg/m$^2$) with age, preventing the birds from performing undisturbed and active behaviours, and the lack of stimulation from the environment (absence of environmental enrichment, low light intensities, low litter quality) (de Jong et al., 2012a; EFSA, 2010), which will be discussed in the following paragraphs. But the low-activity levels are also due to physical restrictions such as leg health problems and heat stress when these birds try to be active (Bokkers and Koene, 2004). It has been suggested that the high and efficient growth rate prevents fast-growing broilers from becoming active, whereas they are still motivated to perform activities like walking, dust bathing, scratching and jumping onto perches or other elevated structures (Bokkers et al., 2007).

Comparisons of time budgets of broiler strains with different growth rates generally show higher activity levels in slow-growing as compared to fast-growing strains (Nielsen et al., 2004; Branciari et al., 2009; Dal Bosco et al., 2010; Malchow et al., 2019a; Steenfeldt et al., 2019; Wallenbeck et al., 2016; Dixon, 2020). Under similar housing conditions, both fast- and slow-growing strains show a decreased activity and decreased foraging behaviour with increasing age (Bokkers and Koene, 2004; Wallenbeck et al., 2016), although the decline in activity in slow-growing strains has been shown to be more gradual than in the fast-growing strains (Nielsen et al., 2004). Using female broilers in a spacious environment with perches, Bokkers and Koene (2003) observed that fast- and slow-growing Hubbard broilers performed the same behaviours but with different time budgets. Slow-growing broilers perched, walked and scratched more than fast-growing broilers, and fast-growing broilers were sitting more on the floor, eating and drinking. A follow-up experiment indicated that physical ability was the determining factor for activity in heavy broiler chickens, whereas motivation did not seem to be affected. The authors tested broilers weekly in runway tests, where they could receive a food reward. They manipulated motivation by applying short- (3 h) and long-term (24 h) feed deprivation and tested both fast- and slow-growing Hubbard broilers until 12 weeks of age (when both breeds reached a high body weight to include possible physical restrictions). Their results indicated that motivation determined the latency to walk and the walking speed in the runway of broilers with a low body weight, while physical ability determined latency and walking speed in broilers with a high body weight (Bokkers and Koene, 2004). In a later study also Wallenbeck et al. (2016) showed that fast-growing Ross 308 broilers spent more time sitting, drinking and eating, whereas slow-growing Rowan Ranger broilers showed more active behaviours such as foraging and perching up to 12 weeks of age.

There are different slow-growing broiler strains that may potentially differ in their behaviour, and between-breed comparisons of behaviour in slow-growing broiler strains is reported in a few studies. Rayner et al. (2019) showed that, when comparing one fast- and two slow-growing broiler breeds at stocking densities of 30 and 34 kg/m$^2$ under commercial conditions, both slow-growing breeds performed significantly more 'positive behaviours' as compared to the fast-growing breed, also at the higher stocking density. They defined 'positive behaviours' as playful, comfort-related, exploratory-related or safety-related behaviours. Torrey et al. (2019) compared three slow-growing strains and one fast-growing strain with respect to their behavioural time budgets. All strains were housed in enriched pens with a stocking density of 30 kg/m$^2$. The fast-growing strain spent significantly more time sitting and less time standing and walking at a similar age, and at a similar weight the fast-growing strain spent more time sitting and less time walking than the slow-growing strains. They observed no differences in behaviour between the three slow-growing strains.

Thus, slow-growing broiler strains are generally more active as compared to fast-growing strains and there are indications that they also show more species-specific behaviours, but differences between different slow-growing breeds appear to be small or absent.

## 5 Effects of environment and management on welfare

This section discusses the effects of management practices and environmental conditions on broiler welfare. Although single factors are discussed one by one, it should be noted that these factors often interact (Averos and Estevez, 2018) and it is, therefore, sometimes difficult to disentangle the effect of a specific management practice or environmental factor on welfare. In addition, most studies focus on the effect of environmental factors on the welfare of conventional, fast-growing broiler strains. Where available we will include information on slower-growing broiler strains.

### 5.1 Hatching environment

The majority of broiler chicks hatch at the hatchery and are transported to the broiler farm after pulling (being collected from the hatching machine), when the majority of chicks have hatched, and after applying management procedures such as removal of second-grade chicks, sexing and vaccinating. As the chicks do not hatch all at the same time but rather over a period of 24–48 h, the period between hatching and arrival at the broiler farm may last up to 72 h for chicks that hatched first. During this period, broiler chickens usually do not receive food and water (Willemsen et al., 2010). Although yolk sac reserves should, in theory, provide sufficient energy for survival (Mitchell, 2009), it has been suggested that this so-called post-hatch food and water deprivation may have negative effects on broiler welfare (since these birds have been selected for very rapid growth from an early age and may be less well able to deal with a negative energy balance) (Willemsen et al., 2010). A meta-analysis of the literature on the effects of this post-hatch feed and water deprivation indeed showed that a period of post-hatch food and water deprivation of more than 36 h resulted in significantly increased mortality not only at a young age but also up to the slaughter age, suggesting that welfare may be compromised in the longer term too, in addition to thirst and hunger in the initial post-hatch period (de Jong et al., 2017). Apart from the post-hatch feed and water deprivation, hatching conditions in itself (de Gouw et al., 2017) and the resulting handling and transport of day-old chicks are probably stressful (Jacobs et al., 2017; Hedlund et al., 2019). For example, chicks are exposed to high concentrations of dust and pathogens in the hatcher, to a high noise level, darkness and disinfected with formaldehyde (de Gouw et al., 2017).

To overcome these stressful events around hatching and the possible long-term negative consequences on broiler welfare, systems have been developed that enable hatching in the broiler house. In these systems, eggs that have been incubated for 18 days (out of the 21 needed for hatching) are transported to the broiler farm, where the eggs are placed in trays, boxes or in the litter. The chicks are provided with feed and water immediately after hatching. There are also systems where broilers receive feed and water in the hatchery and are transported to the broiler farm thereafter. Until now, very limited research has been done on these practices, but it has been shown that broiler welfare may benefit from such alternative hatching practices. A comparison of on-farm hatched versus hatchery-hatched chickens showed that on-farm hatched flocks had a better litter quality and less footpad dermatitis as compared to hatchery-hatched flocks (de Jong et al., 2019). Broiler chicks that received feed and water immediately after hatching and subjected to handling and transport were less fearful at 30 days of age than broiler chicks that were post-hatch feed and water deprived and subsequently handled and transported, whereas the response at

3 days of age was the other way round (Hollemans et al., 2018). This suggests that post-hatch feed deprivation may have long-term effects on broiler welfare that merit further study.

## 5.2 Stocking density

The effects of stocking density on broiler welfare were reviewed by de Jong et al. (2012a), and we concluded that there was no critical density from which broiler welfare is decreased, as this seemed to vary between studies and welfare indicators. However, it was also concluded that broilers prefer stocking densities that were lower than the maximum densities allowed in the EU legislation (39-42 kg/m$^2$). Broilers were highly motivated to go to areas with a lower stocking density, showing that they at least avoided commercial stocking densities (Buijs et al., 2011a). Spatial analysis suggested that broilers at stocking densities from 15 kg/m$^2$ and higher start to experience the proximity of conspecifics as aversive at some point during the last 3 weeks of rearing. Nearest-neighbour distance analysis even showed evidence that this aversion occurred earlier in life (Buijs et al., 2011b).

Previous studies were inconsistent with respect to the effect of stocking density on most behaviours, but they all indicated that disturbances are increasing with increased stocking densities (de Jong et al., 2012a; Knierim, 2013). This increase seemed to be gradual over a wide range of stocking densities and led to shorter resting and preening durations (Buijs et al., 2011b; Buijs et al., 2010). Avéros and Estevez (2018) found an interaction between stocking density and group size on resting behaviour. In larger groups and at lower stocking densities relatively more birds were resting, which was suggested to be caused by more available space in larger groups and less disturbances caused by the low stocking density. Further, high stocking densities also resulted in a more uneven use of pen space, with a reduced use of the central area in the pen (see review in De Jong et al. (2012a)).

With increasing stocking densities, environmental temperature, humidity and litter moisture increase, especially towards the end of the growing period. It is therefore often difficult to disentangle the effects of the stocking density per se from the deteriorated environmental conditions on broiler welfare indicators, as these often coincide (de Jong et al., 2012a), especially under commercial conditions. However, it is clear from the literature that with increasing stocking density the risk for welfare problems such as impaired walking, contact dermatitis and injuries increases (de Jong et al., 2012a). This was also confirmed in more recent studies. Bailie et al. (2018b) showed that with increasing stocking density between 30 kg/m$^2$ and 36 kg/m$^2$ the prevalence of footpad lesions increased. Knierim (2013) found more footpad dermatitis at stocking densities of 35 kg/m$^2$ and 40 kg/m$^2$ compared to 18 kg/m$^2$ and 25 kg/m$^2$, and a better gait score at 18 $kg/m^2$ compared to 35 $kg/m^2$ and 40 $kg/m^2$. BenSassi et al. (2019b) showed that a reduced stocking density decreased walking difficulties, resulted in a reduction in rejection due to injuries and in prevalence of total welfare problems ('total welfare index'). In a meta-analysis, Avéros and Estevez (2018) showed that gait scores decreased with decreasing stocking densities and in a recent review, Pedersen and Forkman (2019) also concluded that gait scores decreased and walking ability improved with decreasing stocking density. Few studies measured prevalence of tibial dyschondroplasia at different stocking densities, and no studies reported a significant effect. Many studies found an effect of stocking density on footpad dermatitis and hock burn, with a higher prevalence reported in higher stocking densities. Most of these studies were carried out in fast-growing breeds (Pedersen and Forkman, 2019).

### 5.3 Litter quality

Litter, which is referred to as bedding material mixed with faeces, can have a large impact on broiler welfare, as broiler chickens usually spend their whole life or a large part of their life (when elevated resting places are provided) on the litter floor. Good-quality bedding material absorbs moisture, provides thermal insulation, a comfortable resting place, opportunities to perform natural behaviours such as foraging and dust bathing, and affects air quality by affecting the amount of dust, humidity, endotoxins and ammonia in the air. It depends on the particular housing system and country whether or not broiler chickens start with new and fresh bedding material upon placement, or whether or not litter is re-used during several production rounds (and a top dressing of, for example, wood shavings is added when a new production cycle is started) (de Jong et al., 2012a). Several bedding materials are used in practice, dependent on local traditions and availability. Frequently used bedding materials are wood shavings, peat, sawdust, (chopped) straw, straw pellets, sand and rice hulls (e.g. Shepherd et al., 2017). The bedding material is mixed with faeces during the growing period and close to slaughter age the major part will consist of faecal material (dependent on the amount of substrate provided at the placement of the chickens). The litter quality will depend on the quality of the faeces, climate in the house, activity of the chickens and incidents such as leakage of drinkers.

Usually litter quality deteriorates during the production cycle (e.g. BenSassi et al., 2019a; Wilhelmsson et al., 2019) due to increased amount of faeces and increasing stocking density and due to inactivity of the birds. Hence, farmer management (including nutrition) is very important to keep the litter in a good condition (i.e. dry and loose). Wet or sticky litter increases dirtiness of the chickens (BenSassi et al., 2019a); increases the risk for footpad dermatitis, hock burn (Shepherd et al., 2017) and breast irritation; leads to cold stress (due to reduced insulation capacity) (de Jong et al., 2014); and is related to an increased condemnation rate at slaughter (BenSassi et al., 2019a), probably due to infection with bacteria present in the litter (Shepherd et al., 2017). A farmer has several options to control litter quality. A thin layer of litter has been suggested to reduce the risk for footpad dermatitis, although this may depend on the floor type and presence of underfloor heating (Shepherd et al., 2017). To prevent footpad dermatitis due to a low litter quality, it is advised to increase heating and ventilation to keep the litter dry and loose (De Jong et al., 2013). The prevalence of footpad dermatitis is related to the quality of the bedding material, and most likely to its capacity to absorb moisture. For example, straw bedding increases the risk of footpad dermatitis compared to peat and wood shavings (Kyvsgaard et al., 2013; Shepherd et al., 2017). Also, the quality of the feed is important for litter quality (Swiatkiewicz et al., 2017) and, for example, gut health problems leading to diarrhoea may also lead to a deteriorated litter quality. Scratching the litter will keep it dry and loose, and promote drying. Thus, also bird activity may affect litter quality. Lowering the stocking density in the house will increase litter quality, because it reduces the amount of faeces and increases bird activity (de Jong et al., 2012a).

Broiler chickens in conventional systems usually only have the litter floor to perform behaviours such as dust bathing and foraging. Compacted, wet litter or litter covered with a crust will not be suitable for these behaviours, and in this way may hamper the birds to perform their natural behaviours. Furthermore, chickens show preferences for dust bathing materials and bedding material with small particles such as peat or sand would be better suitable to perform this behaviour as compared to, for example, straw bedding (Shields et al., 2004; Baxter et al., 2018a). While on the one hand, dry and loose litter may have positive effects on broiler welfare, on the other hand, this creates higher dust levels which can be a risk for bird's health especially with low ventilation rates (e.g. early in the production cycle or in the cold season) (de Jong et al., 2012a). This will be discussed further below.

## 5.4 Air quality

Air quality refers to the concentration of dust, endotoxins and gases in the air, of which ammonia is the most important, and to the thermal environment (temperature and humidity). Young chickens are not able to maintain their body temperature at the appropriate level and an appropriate temperature should therefore be provided, whereas a too high environmental temperature in the final stage of the rearing period may cause heat stress, especially in fast-growing chickens (de Jong et al., 2012a; EFSA, 2010). Ventilation is important to keep the humidity at the required level to prevent wet litter and to provide a comfortable environment. In wet and cold seasons low ventilation rates, to keep the temperature at the desired level, may result in wet litter and higher levels of footpad dermatitis (de Jong et al., 2012b, 2013). Misting systems can be used during hot summer days to reduce heat stress, but these also increase the moisture level in the house and may thus cause a deteriorated litter quality. The distribution and behaviour of broilers in the house is indicative of the thermal environment, with broilers showing huddling when it is too cold or draughty, and panting behaviour when the environmental temperature is too high (Welfare Quality, 2009).

Dust and ammonia clearly have negative effects on broiler welfare. High ammonia concentrations irritate the eyes, throat and mucous membranes, leading to eye damage, increased disease prevalence and a reduced body weight gain in broilers. When given a choice, broiler chickens avoided concentrations of above 10 ppm, which are commonly present in poultry houses (Jones et al., 2005). The main source of ammonia is the faeces (David et al., 2015a). Ammonia concentrations in broiler houses is affected by many factors such as ventilation, humidity, temperature, litter quality, moisture and pH, feed composition and stocking density (de Jong et al., 2012b; David et al., 2015a).

Both the environment, for example, the litter and feed, and the broilers themselves produce dust. Dust contains both organic and inorganic matter. The type of bedding material affects the amount of dust produced as well as litter quality and bird activity. Dust clearly has a negative effect on broiler health. Airborne micro-organisms are frequently attached to dust particles and these may be directly pathogenic or they may release toxins, increasing the susceptibility of the chickens to non- or low-pathogenic micro-organisms. Because of the antigenic effects/capacity of dust it may activate the immune system, leading to hypersensitivity and respiratory diseases (David et al., 2015b). Although there is no doubt that dust and the attached micro-organisms have a negative effect on the health of the chickens (David et al., 2015b; de Jong et al., 2012b), it is not clear to what extent broilers' health is affected by dust, as broiler chickens are slaughtered at a very young age. Ventilation and humidity are important to control dust levels in broiler houses (de Jong et al., 2012b).

## 5.5 Light

Light has different aspects that can affect broiler welfare: intensity, photoperiod (the number of light and dark hours within the 24 h cycle), source (artificial or natural light) and spectrum. We will discuss these aspects separately below, but it should be noted that the different aspects are usually confounded in studies as well as under commercial conditions.

### 5.5.1 Light intensity

Minimum light intensity has been regulated by legislation; for example, the EU Broiler Directive (European Commission, 2007) stated that a minimum level of 20 lux at bird height is required at 80% of the floor area in the broiler house. Usually during the first days of life a higher light intensity is provided to enable the chicks to find feed and water, according to recommendations of the breeding companies. As reviewed by de Jong et al. (2012a), broilers seem to prefer brighter light when being active and more dim light when resting, but there is remarkably little effect of different light intensities on the time performed on specific behaviours (reviewed in De Jong et al., 2012a), apart from a higher overall activity level with increasing light intensity (e.g. Blatchford et al., 2012; Rault et al., 2017). More important seems to be the difference between the light and the dark period. A low contrast in light intensity between the light and dark periods dampened behavioural rhythms, with broilers being more active during the photophase and less active during the scotophase when a high contrast between the light and dark periods was provided (Alvino et al., 2009a; Blatchford et al., 2012). In addition, it has been shown that high light intensities improve the levels of synchronisation of activities during the photophase and uninterrupted resting during the scotophase (Alvino et al., 2009b) in broilers. Light intensities lower than 5 lux (in both the photoperiod and scotoperiod) cause an increased eye weight and larger eyes, thus affecting eye morphology (Blatchford et al., 2012; Deep et al., 2013; Rault et al., 2017). The majority of studies showed that light intensity did not affect gait score, immunity and mortality (reviewed in De Jong et al. (2012a) and for leg health in Pedersen and Forkman (2019)) whereas some authors found fewer footpad lesions with increasing light intensity (Pedersen and Forkman, 2019; Deep et al., 2013), which might be related to the increased activity under high light intensities.

### 5.5.2 Light programme

Quite a number of studies focussed on the effect of light programmes on broiler production and health. Usually, during the first days of life broilers are provided with continuous or near-continuous light to enable them to find feed and water, and after a week a light programme is started. This is sometimes regulated by legislation, for example, the EU Broiler Directive states that broiler chickens must have a dark period of 6 h per 24 h, of which 4 h of uninterrupted dark period is mandated (European Commission, 2007). Pedersen and Forkman (2019) recently reviewed the effect of light programmes. The majority of studies reviewed reported a significant effect of light programme on leg health. Although many different light schedules were studied, in general, intermittent lighting schedules promoted leg health as compared to continuous lighting (defined as 23 h L (light):1 h D (dark)). Gait score and leg disorders were most affected by the light programme with better scores when intermittent lighting schedules were applied (Pedersen and Forkman, 2019; Karaarslan and Nazligul, 2018). Broilers provided with sufficiently long dark periods generally have fewer health problems, including metabolic diseases and an abnormal eye development (reviewed in De Jong et al. (2012a)). A limited number of studies examined the effect of light programmes on broiler behaviour. Schwean-Lardner et al. (2012) studied uninterrupted dark periods of 10 h, 7 h, 4 h and 1 h per 24 h at 27–28 and 42–43 days of age. They found that time spent on standing, walking, feeding, preening, stretching, dust bathing and litter pecking decreased with increasing day length at both ages, both during the photoperiod and as the daily average (of 24h). Running was not observed with 1 h darkness at both ages and dust bathing was not observed with 1 h darkness at 42–43 days of age. Furthermore, long photoperiods resulted in a reduced responsiveness to an observer. They concluded that welfare was compromised in near-continuous lighting, whereas long dark periods of 10 h did not show any advantage, based on behavioural observations, compared to dark periods of 7 h. A recent study of Sun et al. (2017) confirmed that comfort behaviours were less prevalent under near-continuous lighting (23 h L:1 h D) versus 16 h L 8 h D and an interrupted schedule of 6L:2D:2L:4D. Splitting the scotoperiod, that is, two periods of 4 h darkness per 24 h versus an uninterrupted dark period of 8 h, resulted in adjusted feeding behaviours but no indications of one schedule being better than the other (Duve et al., 2011).

### 5.5.3 Light spectrum

The recent developments in the use of light emitting diode (LED) lighting in broiler houses enables farmers to provide a more complex light spectrum and adjust light spectra according to the preferences of the chickens (Archer, 2018). However, relatively little is currently known on the actual preferences of broiler chickens in relation to light composition. Chickens have a greater spectral sensitivity than humans, especially with respect to the red, blue-green and ultraviolet wavelengths (Prescott and Wathes, 1999). Huth and Archer (2015) compared broiler behaviour and health of two commercially available LED lights with compact fluorescent (CFL) lighting, and found that birds in the LED treatments showed less fearful responses and had better plumage condition, foot and hock scores than in the CFL treatment. Their LED treatments differed in output over the whole spectrum as compared to the CFL bulbs, but had especially higher outputs in the ultraviolet and infrared wavelengths. Others compared welfare aspects of broilers housed under different light spectra using LED lights. Mendes et al. (2013) found that broilers from three weeks of age onwards had a higher feed intake in white than in yellow LED light. In another study broilers housed under cold white light (5000 K) for the whole production period were less anxious and grew faster than broilers housed under warm white light (2700 K) (Archer, 2018). Using a preference test, Riber (2015) compared cold white light (6065 K) with warmer white light (4100 K) and found that from three weeks of age onwards broilers spent more time in the compartment with cold white light than the compartment with warm white light. Furthermore, she observed that resting behaviour was more prevalent in cold white light. These studies thus indicate that cold white light is preferable with respect to broiler welfare. However, the limited amount of studies as well as the number of spectra applied in these studies indicate that more research is necessary to determine light spectrum preferences for broiler chickens in relation to age.

### 5.5.4 Light source

Higher-welfare programmes for broiler chickens often include the provision of natural light (e.g. Saatkamp et al., 2019; Bailie et al., 2013) in broiler houses. However, little is known about the effects of natural light on broiler welfare. With natural light the whole light spectrum can be provided to the chicken and this may affect the broiler's perception of the environment. However, when natural light is provided by windows in the roofs or side-walls, the ultraviolet range may be filtered and is thus not available to/visible to the chickens (Van Niekerk et al., 2016). Broiler chickens provided with natural light show increased activity levels (Bailie et al., 2013; de Jong and Gunnink, 2019) and make better use of the environmental enrichment (de Jong and Gunnink, 2019). Bailie et al. (2013) also found better gait scores when natural light was provided. The (generally) higher light intensities provided with natural light as compared to artificial light, as well as the variation in intensity during the day, may have a large effect on broiler behaviour and potentially stimulate behavioural synchronisation (Bailie et al., 2013; de Jong and Gunnink, 2019). However, the latter aspect needs to be investigated further.

## 5.6 Environmental enrichment

The effect of environmental enrichment on broiler chicken welfare has recently been the topic of a series of studies, in addition to the ones that have been reviewed by Riber et al. (2018). Here we summarise the new findings in addition to the review of Riber et al. (2018). We also refer to the other chapters in this book where the effect of elevated resting places on poultry welfare is reviewed.

Environmental enrichment has been defined as a modification of the environment in which the animals are kept, allowing the animals to perform more of their species-specific behavioural repertoire and thus accommodating a larger range of behavioural choices (Estevez and Newberry, 2017). Environmental enrichments may relate to specific behavioural requirements of chickens, for example, they may meet the need for elevated resting, or for exploration pecking, and sometimes combine these different functions. In their review, Riber et al. (2018) stated that environmental enrichment can prevent abnormal behaviour, increase the positive utilisation of the environment, and increase the ability of the animals to handle behavioural and physiological challenges and thus have a positive effect on welfare. They emphasised that environmental enrichment should be biologically relevant to be effective. The majority of broiler chickens, especially fast-growing breeds in conventional systems, are housed in relatively barren environments, and despite the increased number of scientific studies reporting potentially positive effects of environmental enrichment on broiler welfare, environmental enrichment is only applied on a limited scale in practice. One reason might be that farmers and integrations often think that environmental enrichment increases production costs and thus reduces economic competitiveness (Estevez and Newberry, 2017). However, this does not take into account potential benefits of environmental enrichment on broiler health. For example, effective enrichment may reduce crowding and smothering in a flock by reducing fearfulness (Riber et al., 2018), and as a result decrease carcass condemnations (BenSassi et al., 2019b). Enrichment may also improve leg strength in broiler chickens, reducing the number of culled chickens and application of antibiotics (Kaukonen et al., 2017). And although it may increase broiler activity, most studies do not report a negative effect on feed conversion rate or growth (Riber et al., 2018). Finally, the societal demand for higher-welfare systems may lead to compensation for increased production costs, as, for example, has happened in The Netherlands, where the new retail standards involve new broiler housing concepts that require environmental enrichment (Saatkamp et al., 2019). Most studies on environmental enrichment in broiler chickens include the effect on behaviour and leg health in conventional fast-growing strains. To stimulate implementation in practice, studies on commercial farms including the effect on health, antibiotic treatments, carcass condemnations, feed intake and growth, and thus on economic return of production are therefore necessary, and studies should include slower-growing strains as well.

Riber et al. (2018) discriminated between 'point source enrichments', which they categorised as objects that are generally limited in size and applied on a few locations in the house, and 'complex enriched environments' that differ more radically from conventional systems, usually involving multiple enrichments or point source enrichments in combination with covered verandas or an outdoor range. Before concluding this paragraph, we discuss these two types of enrichments briefly below.

### 5.6.1 Point-source enrichment

To meet the need for an elevated resting place for broiler chickens, perches or elevated platforms can be provided. Studies usually applied different materials for perching, such as metal, wooden or PVC pipes, being oval or round, and used different heights and shapes (e.g. A-shape, X-shaped perches). As the actual use of the perches and their effects on behaviour and other welfare indicators may differ according to the layout of the perches, it is not easy to draw conclusions on the actual use of perches and the benefits for the broilers. However, Riber et al. (2018) concluded that perch use was generally low, especially in fast-growing broilers, and this was confirmed in more recent experiments (de Jong and Gunnink, 2019; Bailie et al., 2018a; Bench et al., 2017; Malchow et al., 2019a; Bergmann et al., 2017). Only a few studies tested perches in slower-growing strains with generally low use (de Jong and Van Wijhe-Kiezebrink, 2014; Bokkers and Koene, 2003; Rodriguez-Aurrekoetxea et al., 2015; Bergmann et al., 2017; Malchow et al., 2019a), although this merits further study, as perch design may have played a role. Perches may potentially provide elevated resting space for slower-growing breeds as these are not hampered as much by physical limitations, but it is important to include the effect of different shapes, design and materials on perch use. Positive effects of perches are increased activity and a better distribution of the broilers in the pen, but in general, no positive effects on foot and leg health were found (reviewed in Riber et al. (2018)). Because body weight and body conformation of broiler chickens, especially in fast-growing strains, may have a negative effect on perch use, platforms have been applied to meet the need for elevated resting places. Platforms are usually made of plastic mesh and provided at various heights with a ramp to promote platform access. Platforms were well used by both fast- and slower growing broiler strains (Tahamtani et al., 2018; Malchow et al., 2019a,b; Kaukonen et al., 2017; de Jong and Van Wijhe-Kiezebrink, 2014; Vasdal et al., 2019), and it was suggested that platforms decrease fearfulness in broiler chickens (Tahamtani et al., 2018). The frequent use of platforms may indicate that these are better suited for broiler chickens, although platform use may also be driven by high animal densities, as they provide additional space. On the other hand, it has been shown that fast-growing breeds at 25 kg/m$^2$ made better use of platforms than broilers at 35 kg/m$^2$, probably because of the 'barrier' effect of broilers on the floor at higher densities (de Jong and Goertz, 2017).

Bales of straw or wood shavings are often used as environmental enrichment in commercial practice, and studies indicate that these are generally well used (see e.g. de Jong and Gunnink, 2019; Baxter et al., 2018b, 2019; Vasdal et al., 2019; Bergmann et al., 2017). Bales may serve different functions, for example, stimulating explorative pecking, dust bathing behaviour because of the provision of fresh litter, resting on an elevated area as long as the bales are intact and providing protection. Panels and barriers, as indicated in the section on broiler behaviour above, provide a quiet resting area and prevent disturbance by other birds and are generally well used (Riber et al., 2018). Dark brooders are not frequently used in broiler chickens but may also provide a quiet resting area in the first two weeks (Stadig et al., 2018). Wheat, maize or other substrates provided in the litter are used to stimulate foraging behaviour, but studies did not show significant effects on leg health and activity (Vasdal et al., 2019).

Since the review of Riber et al. (2018) on several studies focussed on providing dust bathing substrates for broiler chickens, for example, in steel rings or boxes (Baxter et al., 2018a,b, 2019; Vasdal et al., 2019; Baxter and O'Connell, 2019). Different types of substrate have been tested such as oat hulls, wood shavings, straw pellets and peat moss. Such an enrichment is potentially most interesting under conditions where litter quality is insufficient to allow dust bathing and foraging behaviour, that is, in conventional systems with fast-growing breeds. Systems with slower-growing broiler strains usually apply lower stocking densities, and litter quality is much better as compared to intensive indoor systems (Ellen et al., 2012), although the effect of additional dust bathing substrates has not been tested in slow-growing broiler flocks. Both peat moss and oat hulls provided at different locations in the broiler house were attractive for broiler chickens and stimulated dust bathing and foraging behaviour (Baxter et al., 2018a,b), and oat hulls stimulated also leg health in fast-growing chickens (Baxter et al., 2018b).

Finally, pecking objects such as metal chains, strings and pecking stones are used as environmental enrichment (see Baxter and O'Connell, 2019; de Jong and Gunnink, 2019; de Jong and Van Wijhe-Kiezebrink, 2014; Bergmann et al., 2017). The extent to which these are used varied between studies and the effects on activity and other welfare indicators is unclear (Riber et al., 2018). For example, Baxter et al. (2019) found a higher use of pecking chains in their final study, combining different enrichment types, as compared to their previous study (not comparing different types of enrichment), suggesting that further study under commercial conditions is necessary.

### 5.6.2 Complex environmentally enriched environments (including outdoor ranges)

No studies have been published since the review of Riber et al. (2018) on the effects of outdoor range or indoor systems with a covered veranda on broiler welfare. Systems including an outdoor range usually include slower-growing broiler strains (e.g. Stadig et al., 2017; Stadig et al., 2018; Rodriguez-Aurrekoetxea et al., 2015), although fast-growing strains can also be housed in outdoor systems (e.g. Taylor et al., 2017a,b). Studies showed that slower-growing strains are better capable of making use of the extra space and a more diverse environment than fast-growing broiler strains (Wilhelmsson et al., 2019; Lindholm et al., 2016; Nielsen et al., 2003). This may be related to their physical inability to make use of the additional space and resources' variability, and does not exclude that they may still be motivated to use extra space and resources and the ability to perform (certain) natural behaviours.

### 5.6.3 Combining and grouping enrichments

For commercial broiler welfare programmes, different enrichments can be combined (e.g. de Jong and Gunnink, 2019; Vasdal et al., 2019). Baxter et al. (2019) found no clear benefits of grouping enrichments, as compared to spreading the various enrichments, on behaviour and use of the various enrichment types. BenSassi et al. (2019b) compared houses with one, two or more enrichments and found that a greater environmental complexity was associated with a reduction in skin wounds, lower mortality, fewer carcass rejections due to wounds and low body weight, and fewer carcass rejections overall. More research is needed to confirm that combinations of environmental enrichment promote welfare as compared to single enrichment, as well as to determine the minimum number of enrichments that should be provided and the distribution over the house in commercial flocks.

## 6 Conclusions and future trends

In this chapter we provided an overview of welfare issues in broiler chicken production, and how these are affected by genetic, management and environmental factors. Most studies have been performed in the conventional, fast-growing breeds and are related to conditions present in intensive, conventional production systems. Studies in commercial flocks clearly indicate that the prevalence of welfare problems can be high, but variation between flocks and between production systems indicates that there are ways to reduce the prevalence of welfare problems. The increasing number of scientific studies on the effects of environmental enrichment on broiler welfare indicate that, at least in Europe and the United States (CIWF, 2018), there is a trend towards higher-welfare systems for broiler chickens. In Europe, this is also reflected by the increase in the use of slow-growing strains in these systems, such as in The Netherlands and Germany (Saatkamp et al., 2019; Bergmann et al., 2016, 2017). It is also clear that aspects such as societal pressure and market requirements will be important to improve the current situation with respect to broiler welfare (Saatkamp et al., 2019; Bracke et al., 2019a).

The knowledge on the needs and preferences of slower-growing strains as compared to the standard, fast-growing strains is limited and clearly merits further research. With respect to broiler welfare in different production systems, experts recently assigned good welfare scores to organic systems (EU), free range (indoor and outdoor) and Label Rouge, whereas they considered broiler welfare in conventional floor systems and cage systems insufficient. Other systems like organic (US) or middle-segment concepts in The Netherlands scored in-between and were not considered as providing good welfare (Bracke et al., 2019a). However, it would be good to collect more data on system comparisons to determine the actual welfare performance of broilers in various production systems, as these are currently very scarce.

With respect to environmental and management factors, the different aspects of light, especially wavelength preferences and effects of natural light on broiler chicken welfare merit further study. Finally, little is known about the interactions of the different genetic, environmental and management aspects, which is clearly also an area for future research.

## 7 Where to look for further information

Although not very recent, the EFSA report and the update of this report (EFSA, 2010; de Jong et al., 2012a) provide a good starting point for reading. With respect to genetic selection and the relation with broiler welfare, Hiemstra and Ten Napel (2011) provide a lot of background information. The review of Bessei (2006) summarises the main welfare issues in broiler chickens. More specifically, the review by Riber et al. (2018) is a good starting point for the environmental enrichment effects on broiler welfare and the review by Pedersen and Forkman (2019) is a good starting point on the relationship between environmental enrichment and leg health. With respect to societal pressure aimed at higher welfare for broiler chickens, information on requirements in the NGO's European Chicken Commitment and US and Canada Better Chicken Commitment can be found in Anonymous (n.d.) and Anonymous, 2019, and NGO's websites report about companies stating that they will meet these requirements within a couple of years.

## 8 References

Allain, V., Mirabito, L., Arnould, C., Colas, M., Le Bouquin, S., Lupo, C. and Michel, V. 2009. Skin lesions in broiler chickens measured at the slaughterhouse: relationships between lesions and between their prevalence and rearing factors. *British Poultry Science* 50(4), 407–417.

Alvino, G. M., Archer, G. S. and Mench, J. A. 2009a. Behavioural time budgets of broiler chickens reared in varying light intensities. *Applied Animal Behaviour Science* 118(1–2), 54–61.

Alvino, G. M., Blatchford, R. A., Archer, G. S. and Mench, J. A. 2009b. Light intensity during rearing affects the behavioural synchrony and resting patterns of broiler chickens. *British Poultry Science* 50(3), 275–283.

Anonymous n.d. European Chicken Commitment. Available at: https://welfarecommitments.com/europeletter/.

Anonymous 2019. Better Chicken Commitment United States & Canada. Available at: https://welfarecommitments.com/letter.pdf.

Archer, G. S. 2018. Color temperature of light-emitting diode lighting matters for optimum growth and welfare of broiler chickens. *Animal* 12, 1015–1021.

Arnould, C. and Faure, J. M. 2003. Use of pen space and activity of broiler chickens reared at two different densities. *Applied Animal Behaviour Science* 84(4), 281–296.

Averos, X. and Estevez, I. 2018. Meta-analysis of the effects of intensive rearing environments on the performance and welfare of broiler chickens. *Poultry Science* 97(11), 3767–3785.

Aviagen 2019. Ross308 performance objectives. Available at: http://en.aviagen.com/assets/Tech_Center/Ross_Broiler/Ross308-308FF-BroilerPO2019-EN.pdf. Accessed 09/10/2019.

Avined 2019. Antibioticumgebruik pluimveesector in 2018 en de trends van afgelopen jarenen de trends van afgelopen jaren. Available at: https://www.avined .nl/sites/www.avined.nl/files/antibioticagebruik_-_sectorrapportage_2018.pdf. Accessed on 09/10/2019.

Bailie, C. L., Ball, M. E. E. and O'Connell, N. E. 2013. Influence of the provision of natural light and straw bales on activity levels and leg health in commercial broiler chickens. *Animal* 7(4), 618–626.

Bailie, C. L., Baxter, M. and O'Connell, N. E. 2018a. Exploring perch provision options for commercial broiler chickens. *Applied Animal Behaviour Science* 200, 114–122.

Bailie, C. L., Ijichi, C. and O'Connell, N. E. 2018b. Effects of stocking density and string provision on welfare-related measures in commercial broiler chickens in windowed houses. *Poultry Science* 97(5), 1503–1510.

Bassler, A. W., Arnould, C., Butterworth, A., Colin, L., De Jong, I. C., Ferrante, V., Ferrari, P., Haslam, S., Wemelsfelder, F. and Blokhuis, H. J. 2013. Potential risk factors associated with contact dermatitis, lameness, negative emotional state, and fear of humans in broiler chicken flocks. *Poultry Science* 92(11), 2811–2826.

Baxter, M., Bailie, C. L. and O'Connell, N. E. 2018a. An evaluation of potential dustbathing substrates for commercial broiler chickens. *Animal* 12(9), 1933–1941.

Baxter, M., Bailie, C. L. and O'Connell, N. E. 2018b. Evalution of dustbathing substrate and straw bales as environmental enrichments in commercial broiler housing. *Applied Animal Behaviour Science* 200, 78–85.

Baxter, M., Bailie, C. L. and O'Connell, N. E. 2019. Play behaviour, fear responses and activity levels in commercial broiler chickens provided with preferred environmental enrichments. *Animal* 13(1), 171–179.

Baxter, M. and O'Connell, N. E. 2019. Does grouping environmental enrichments together affect the way they are used by commercially housed broiler chickens? *Applied Animal Behaviour Science* 210, 52–59.

Bench, C. J., Oryschak, M. A., Korver, D. R. and Beltranena, E. 2017. Behaviour, growth performance, foot pad quality, bone density, and carcass traits of broiler chickens reared with barrier perches and fed different dietary crude protein levels. *Canadian Journal of Animal Science* 97, 268–280.

BenSassi, N., Averos, X. and Estevez, I. 2019a. The potential of the transect method for early detection of welfare problems in broiler chickens. *Poultry Science* 98(2), 522–532.

BenSassi, N., Vas, J., Vasdal, G., Averos, X., Estevez, I. and Newberry, R. C. 2019b. On-farm broiler chicken welfare assessment using transect sampling reflects environmental inputs and production outcomes. *PLoS ONE* 14, e0214070.

Bergmann, S., Louton, H., Westermaier, C., Wilutzky, K., Bender, A., Bachmeier, J., Erhard, M. H. and Rauch, E. 2016. Field trial on animal-based measures for animal welfare in slow growing broilers reared under an alternative concept suitable for the German market. *Berliner Und Munchener Tierarztliche Wochenschrift* 129, 453–461.

Bergmann, S., Schwarzer, A., Wilutzky, K., Louton, H., Bachmeier, J., Schmidt, P., Erhard, M. and Rauch, E. 2017. Behavior as welfare indicator for the rearing of broilers in an enriched husbandry environmentd-A field study. *Journal of Veterinary Behavior* 19, 90–101.

Bessei, W. 2006. Welfare of broilers: a review. *World's Poultry Science Journal* 62(3), 455–466.

Bizeray, D., Estevez, I., Leterrier, C. and Faure, J. M. 2002. Effects of increasing environmental complexity on the physical activity of broiler chickens. *Applied Animal Behaviour Science* 79(1), 27–41.

Blatchford, R. A., Archer, G. S. and Mench, J. A. 2012. Contrast in light intensity, rather than day length, influences the behavior and health of broiler chickens. *Poultry Science* 91(8), 1768–1774.

Bokkers, E. A. M. and Koene, P. 2003. Behaviour of fast- and slow growing broilers to 12 weeks of age and the physical consequences. *Applied Animal Behaviour Science* 81(1), 59–72.

Bokkers, E. A. M. and Koene, P. 2004. Motivation and ability to walk for a food reward in fast- and slow-growing broilers to 12 weeks of age. *Behavioural Processes* 67(2), 121–130.

Bokkers, E. A. M., Zimmerman, P. H., Rodenburg, T. B. and Koene, P. 2007. Walking behaviour of heavy and light broilers in an operant runway test with varying durations of feed deprivation and feed access. *Applied Animal Behaviour Science* 108(1–2), 129–142.

Bracke, M. B. M., Koene, P., Estevez, I., Butterworth, A. and de Jong, I. C. 2019a. Broiler welfare trade-off: a semi-quantitative assessment of prospects for optimised welfare improvement based on an expert survey. *PLoS ONE* 14(10), e0222955.

Bracke, M. B. M., Vermeer, H. M. and Van Emous, R. A. 2019b. *Animal Welfare Regulations and Practices in 7 (Potential) Trade-Agreement Partners of the EU with a Focus on Laying Hens, Broilers and Pigs: Mexico, Chile, Indonesia, Australia, New Zealand, Turkey and the Philippines*. Wageningen Livestock Research, Wageningen.

Bradshaw, R. H., Kirkden, R. D. and Broom, D. M. 2002. A review of the aetiology and pathology of leg weakness in broilers in relation to welfare. *Avian and Poultry Biology Reviews* 13(2), 45–103.

Branciari, R., Mugnai, C., Mammoli, R., Miraglia, D., Ranucci, D., Dal Bosco, A. and Castellini, C. 2009. Effect of genotype and rearing system on chicken behavior and muscle fiber characteristics. *Journal of Animal Science* 87(12), 4109–4117.

Buijs, S., Keeling, L. J. and Tuyttens, F. A. M. 2011a. Using motivation to feed as a way to assess the importance of space for broiler chickens. *Animal Behaviour* 81(1), 145–151.

Buijs, S., Keeling, L. J., Vangestel, C., Baert, J. and Tuyttens, F. A. M. 2011b. Neighbourhood analysis as an indicator of spatial requirements of broiler chickens. *Applied Animal Behaviour Science* 129(2-4), 111-120.

Buijs, S., Keeling, L. J., Vangestel, C., Baert, J., Vangeyte, J. and Tuyttens, F. A. M. 2010. Resting or hiding? Why broiler chickens stay near walls and how density affects this. *Applied Animal Behaviour Science* 124(3-4), 97-103.

Butterworth, A., de Jong, I. C., Keppler, C., Knierim, U., Stadig, L. and Lambton, S. 2016. What is being measured, and by whom? Facilitation of communication on technical measures amongst competent authorities in the implementation of the European Union Broiler Directive (2007/43/EC). *Animal* 10(2), 302-308.

Cengiz, Ö, Hess, J. B. and Bilgili, S. F. 2011. Effect of bedding type and transient wetness on footpad dermatitis in broiler chickens. *Journal of Applied Poultry Research* 20(4), 554-560.

CIWF 2018. What businesses have pledged to improve broiler welfare? Available at: https://www.compassioninfoodbusiness.com/our-news/2018/11/what-businesses-have-pledged-to-improve-broiler-welfare. Accessed on 13/10/2019.

Cornetto, T. and Estevez, I. 2001. Influence of vertical panels on use of space by domestic fowl. *Applied Animal Behaviour Science* 71(2), 141-153.

Cornetto, T., Estevez, I. and Douglass, L. W. 2002. Using artificial cover to reduce aggression and disturbances in domestic fowl. *Applied Animal Behaviour Science* 75(4), 325-336.

Dal Bosco, A., Mugnai, C., Sirri, F., Zamparini, C. and Castellini, C. 2010. Assessment of a global positioning system to evaluate activities of organic chickens at pasture. *Journal of Applied Poultry Research* 19(3), 213-218.

David, B., Mejdell, C., Michel, V., Lund, V. and Moe, R. O. 2015a. Air quality in alternative housing systems may have an impact on laying hen welfare. Part II: ammonia. *Animals* 5(3), 886-896.

David, B., Moe, R. O., Michel, V., Lund, V. and Mejdell, C. 2015b. Air quality in alternative housing systems may have an impact on laying hen welfare. Part I: Dust. *Animals* 5, 495-511.

Dawkins, M. S. and Layton, R. 2012. Breeding for better welfare: genetic goals for broiler chickens and their parents. *Animal Welfare* 21(2), 147-155.

de Gouw, P., van de Ven, L. J. F., Lourens, S., Kemp, B. and van den Brand, H. 2017. Effects of dust, formaldehyde and delayed feeding on early postnatal development of broiler chickens. *Research in Veterinary Science* 112, 201-207.

de Jong, I. C., Berg, C., Butterworth, A. and Estevez, I. 2012a. *Scientific Report Updating the EFSA Opinions on the Welfare of Broilers and Broiler Breeders*. Supporting Publications 2012:EN-295, [116pp]. Available at: www.efsa.europa.eu/publications.

de Jong, I. C. and Goertz, M. 2017. Broiler chicken stocking density affects use of environmental enrichment objects. Xth European Symposium on Poultry Welfare, Xth European Symposium on Poultry Welfare Ploufragan, France, p. 69.

de Jong, I. C. and Gunnink, H. 2019. Effects of a commercial broiler enrichment programme with or without natural light on behaviour and other welfare indicators. *Animal* 13(2), 384-391.

de Jong, I. C., Gunnink, H. and van Harn, J. 2014. Wet litter not only induces footpad dermatitis but also reduces overall welfare, technical performance, and carcass yield in broiler chickens. *Journal of Applied Poultry Research* 23(1), 51-58.

de Jong, I. C., Gunnink, H., van Hattum, T., van Riel, J. W., Raaijmakers, M. P., Zoet, E. S. and Van den Brand, H. 2019. Comparison of performance, health and welfare aspects between commercially housed hatchery-hatched and on-farm hatched broiler flocks. *Animal* 13, 1269–1277.

de Jong, I. C., van Harn, J., Gunnink, H., Hindle, V. A. and Lourens, A. 2012b. Footpad dermatitis in Dutch broiler flocks: prevalence and factors of influence. *Poultry Science* 91(7), 1569–1574.

de Jong, I. C., van Riel, J., Bracke, M. B. M. and van den Brand, H. 2017. A 'meta-analysis' of effects of post-hatch food and water deprivation on development, performance and welfare of chickens. *PLoS ONE* 12(12), e0189350.

de Jong, I. C. and Van Wijhe-Kiezebrink, M. C. 2014. Use of different types of enrichment in slower growing broilers: a pilot study. Wageningen Livestock Research, Report 810, Wageningen, p. 39. Available at: https://edepot.wur.nl/318861.

de Jong, I. C., Veldkamp, T. and Van Harn, J. 2013. Management tools to reduce footpad dermatitis in broiler chickens, 19th European Symposium on Poultry Nutrition, German Branch of the WPSA, Potsdam, Germany, pp. 78–83.

de Jonge, J. and van Trijp, H. 2014. Heterogeneity in consumer perceptions of the animal friendliness of broiler production systems. *Food Policy* 49, 174–185.

Deep, A., Raginski, C., Schwean-Lardner, K., Fancher, B. I. and Classen, H. L. 2013. Minimum light intensity threshold to prevent negative effects on broiler production and welfare. *British Poultry Science* 54(6), 686–694.

Dixon, L. M. 2020. Slow and steady wins the race: the behaviour and welfare of commercial faster growing broiler breeds compared to a commercial slower growing breed. *PLoS ONE* 15(4), e0231006. doi:10.1371/journal. pone.0231006.

Duve, L. R., Steenfeldt, S., Thodberg, K. and Nielsen, B. L. 2011. Splitting the scotoperiod: effects on feeding behaviour, intestinal fill and digestive transit time in broiler chickens. *British Poultry Science* 52(1), 1–10.

EFSA 2010. Scientific Opinion on the influence of genetic parameters on the welfare and the resistance to stress of commercial broilers. *EFSA Journal* 8(7), 1666.

Ellen, H. H., Leenstra, F., Emous, R. A., Groenestein, K., Van Harn, J., Van Horne, P., de Jong, I. C., Kense, M., Mevius, D. and Wagenaar, J. A. 2012. Broiler Production Systems in the Netherlands, Wageningen Livestock Research, Wageningen, Report 819, 53 pp. Available at: http://library.wur.nl/WebQuery/wurpubs/fulltext/222179.

Estevez, I. and Newberry, R. C. 2017. The contribution of environmental enrichment to sustainable poultry production. In: Applegate, T. (Ed.), *Achieving Sustainable Production of Poultry Meat*. Burleigh Dodds Science Publishing Limited, Cambridge, UK, pp. 247–279.

European Commission 2007. The EU broiler directive 2010/43/EC. Available at: https://eur-lex.europa.eu/eli/dir/2007/43/oj. Accessed on 13/10/19.

European Commission 2016. Use of Slaughterhouse Data to Monitor Welfare of Broilers on Farm. doi:10.2772/57892.

Federici, J. F., Vanderhasselt, R., Sans, E. C. O., Tuyttens, F. A. M., Souza, A. P. O. and Molento, C. F. M. 2016. Assessment of broiler chicken welfare in Southern Brazil. *Revista Brasileira de Ciência Avícola* 18(1), 133–140.

Gerritzen, M., Verkaik, J., Reimert, H., Gunnink, H., Van Hattum, T. and de Jong, I. C. 2019. Letsel en schade bij vleeskuikens als gevolg van vangen, transport en handelingen aan de slachtlijn. Wageningen Livestock Research, Wageningen, Report 1107, 53 pp. doi:10.18174/467902.

Hall, A. L. 2001. The effect of stocking density on the welfare and behaviour of broiler chickens reared commercially. *Animal Welfare* 10, 23–40.

Haslam, S. M., Knowles, T. G., Brown, S. N., Wilkins, L. J., Kestin, S. C., Warriss, P. D. and Nicol, C. J. 2007. Factors affecting the prevalence of foot pad dermatitis, hock burn and breast burn in broiler chicken. *British Poultry Science* 48(3), 264–275.

Hedlund, L., Whittle, R. and Jensen, P. 2019. Effects of commercial hatchery processing on short- and long-term stress responses in laying hens. *Scientific Reports* 9, 2367. doi:10.1038/s41598-019-38817-y.

Hiemstra, S. and Ten Napel, J. 2011. Study of the Impact of Genetic Selection on the Welfare of Chickens Bred and Kept for Meat Production. DG SANCO Letter of Contract N° SANCO/2011/12254 Final Report, 118 pp. https://ec.europa.eu/food/sites/food/files/animals/docs/aw:practice_farm_broilers_653020_final-report_en.pdf

Hollemans, M. S., de Vries, S., Lammers, A. and Clouard, C. 2018. Effects of early nutrition and transport of 1-day-old chickens on production performance and fear response. *Poultry Science* 97(7), 2534–2542.

Huth, J. C. and Archer, G. S. 2015. Comparison of two LED light bulbs to a dimmable CFL and their effects on broiler chicken growth, stress, and fear. *Poultry Science* 94(9), 2027–2036.

Jacobs, L., Delezie, E., Duchateau, L., Goethals, K., Ampe, B., Buyse, J. and Tuyttens, F. A. M. 2017. Impact of transportation duration on stress responses in day-old chicks from young and old breeders. *Research in Veterinary Science* 112, 172–176.

Jones, E. K. M., Wathes, C. M. and Webster, A. J. F. 2005. Avoidance of atmospheric ammonia by domestic fowl and the effect of early experience. *Applied Animal Behaviour Science* 90(3–4), 293–308.

Karaarslan, S. and Nazligul, A. 2018. Effects of lighting, stocking density, and access to perches on leg health variables as welfare indicators in broiler chickens. *Livestock Science* 218, 31–36.

Kaukonen, E., Norring, M. and Valros, A. 2017. Perches and elevated platforms in commercial broiler farms: use and effect on walking ability, incidence of tibial dyschondroplasia and bone mineral content. *Animal* 11(5), 864–871.

Kestin, S. C., Su, G. and Sorensen, P. 1999. Different commercial broiler crosses have different susceptibilities to leg weakness. *Poultry Science* 78(8), 1085–1090.

Kittelsen, K. E., David, B., Moe, R. O., Poulsen, H. D., Young, J. F. and Granquist, E. G. 2017. Associations among gait score, production data, abattoir registrations, and postmortem tibia measurements in broiler chickens. *Poultry Science* 96(5), 1033–1040.

Knierim, U. 2013. Effects of stocking density on the behaviour and bodily state of broilers fattened with a target live weight of 2 kg. *Berliner Und Munchener Tierarztliche Wochenschrift* 126(3–4), 149–155.

Knowles, T. G., Kestin, S. C., Haslam, S. M., Brown, S. N., Green, L. E., Butterworth, A., Pope, S. J., Pfeiffer, D. and Nicol, C. J. 2008. Leg disorders in broiler chickens: prevalence, risk factors and prevention. *PLOS ONE* 3, e1545

Kyvsgaard, N. C., Jensen, H. B., Ambrosen, T. and Toft, N. 2013. Temporal changes and risk factors for foot-pad dermatitis in Danish broilers. *Poultry Science* 92(1), 26–32.

Leone, E. H., Christman, M. C., Douglass, L. and Estevez, I. 2010. Separating the impact of group size, density, and enclosure size on broiler movement and space use at a decreasing perimeter to area ratio. *Behavioural Processes* 83(1), 16–22.

Leone, E. H. and Estevez, I. 2008a. Space use according to the distribution of resources and level of competition. *Poultry Science* 87(1), 3-13.

Leone, E. H. and Estevez, I. 2008b. Use of space in the domestic fowl: separating the effects of enclosure size, group size and density. *Animal Behaviour* 76(5), 1673-1682.

Leone, E. H., Estevez, I. and Christman, M. C. 2007. Environmental complexity and group size: immediate effects on use of space by domestic fowl. *Applied Animal Behaviour Science* 102(1-2), 39-52.

Lindholm, C., Karlsson, L., Johansson, A. and Altimiras, J. 2016. Higher fear of predators does not decrease outdoor range use in free-range rowan ranger broiler chickens. *Acta Agriculturae Scandinavica, Section A - Animal Science* 66(4), 231-238.

Louton, H., Keppler, C., Erhard, M., van Tuijl, O., Bachmeier, J., Damme, K., Reese, S. and Rauch, E. 2019. Animal-based welfare indicators of 4 slow-growing broiler genotypes for the approval in an animal welfare label program. *Poultry Science* 98(6), 2326-2337.

Lund, V. P., Nielsen, L. R., Oliveira, A. R. S. and Christensen, J. P. 2017. Evaluation of the Danish footpad lesion surveillance in conventional and organic broilers: misclassification of scoring. *Poultry Science* 96(7), 2018-2028.

Malchow, J., Berk, J., Puppe, B. and Schrader, L. 2019a. Perches or grids? What do rearing chickens differing in growth performance prefer for roosting? *Poultry Science* 98(1), 29-38.

Malchow, J., Puppe, B., Berk, J. and Schrader, L. 2019b. Effects of elevated grids on growing male chickens differing in growth performance. *Frontiers in Veterinary Science* 6:203. doi:10.3389/fvets.2019.00203

Mallapur, A., Miller, C., Christman, M. C. and Estevez, I. 2009. Short-term and long-term movement patterns in confined environments by domestic fowl: influence of group size and enclosure size. *Applied Animal Behaviour Science* 117(1-2), 28-34.

Mendes, A. S., Paixao, S. J., Restelatto, R., Morello, G. M., de Moura, D. J. and Possenti, J. C. 2013. Performance and preference of broiler chickens exposed to different lighting sources. *Journal of Applied Poultry Research* 22(1), 62-70.

Mitchell, M. A. 2009. Chick transport and welfare. *Avian Biology Research* 2(1-2), 99-105.

Nielsen, B. L., Kjaer, J. B. and Friggens, N. C. 2004. Temporal changes in activity measured by passive infrared detection (PID) of broiler strains growing at different rates. *Archiv fur Geflugelkunde* 68, 106-110.

Nielsen, B. L., Thomsen, M. G., Soerensen, J. P. and Young, J. F. 2003. Feed and strain effects on the use of outdoor areas by broilers. *British Poultry Science* 44(2), 161-196.

Opengart, K., Bilgili, S. F., Warren, G. L., Baker, K. T., Moore, J. D. and Dougherty, S. 2018. Incidence, severity, and relationship of broiler footpad lesions and gait scores of market-age broilers raised under commercial conditions in the southeastern United States. *Journal of Applied Poultry Research* 27(3), 424-432.

Pedersen, I. J. and Forkman, B. 2019. Improving leg health in broiler chickens: a systematic review of the effect of environmental enrichment. *Animal Welfare* 28(2), 215-230.

Pettit-Riley, R., Estevez, I. and Russek-Cohen, E. 2002. Effects of crowding and access to perches on aggressive behaviour in broilers. *Applied Animal Behaviour Science* 79(1), 11-25.

Prescott, N. B. and Wathes, C. M. 1999. Spectral sensitivity of the domestic fowl (Gallus g. domesticus). *British Poultry Science* 40(3), 332-339.

Rauch, E., Keppler, C., Damme, K., Hausleitner, M., Bachmeier, J., Hartmann, J. and Louton, H. 2017. Test of different premium broiler genotypes under Animal Welfare

Label conditions. Part II: animal health. *European Poultry Science* 81. doi:10.1399/eps.2017.169.

Rault, J. L., Clark, K., Groves, P. J. and Cronin, G. M. 2017. Light intensity of 5 or 20 lux on broiler behavior, welfare and productivity. *Poultry Science* 96(4), 779–787.

Rayner, A., Newberry, R. C., Vas, J., Butterworth, A. and Mullan, S. 2019. *Growing Slowly with More Space: Effects on "Positive Behaviours" in Broiler Chickens*. ISAE, Bergen.

Riber, A. B. 2015. Effects of color of light on preferences, performance, and welfare in broilers. *Poultry Science* 94(8), 1767–1775.

Riber, A. B., van de Weerd, H. A., de Jong, I. C. and Steenfeldt, S. 2018. Review of environmental enrichment for broiler chickens. *Poultry Science* 97(2), 378–396.

Rodenburg, T. B., Coenen, E. and Van Harn, J. 2004. Goed gebruik overdekte uitloop. *Pluimveehouderij* 34, 10–11.

Rodriguez-Aurrekoetxea, A., Leone, E. H. and Estevez, I. 2015. Effects of panels and perches on the behaviour of commercial slow-growing free-range meat chickens. *Applied Animal Behaviour Science* 165, 103–111.

RSPCA 2006. *Everyone's a Winner. How Rearing Chickens to Higher Welfare Standards Can Benefit the Chicken, Producer, Retailer and Consumer*. RSPCA, Horsam, West-Sussex, UK, pp. 26.

Saatkamp, H. W., Vissers, L. S. M., Van Horne, P. L. M. and de Jong, I. C. 2019. Transition from conventional broiler meat to meat from production concepts with higher animal welfare: experiences from the Netherlands. *Animals* 9(8), 483.

Sanotra, G. S., Lund, J. D., Ersboll, A. K., Petersen, J. S. and Vestergaard, K. S. 2001. Monitoring leg problems in broilers: a survey of commercial broiler production in Denmark. *World's Poultry Science Journal* 57(1), 55–69.

Schwean-Lardner, K., Fancher, B. I. and Classen, H. L. 2012. Impact of daylength on behavioural output in commercial broilers. *Applied Animal Behaviour Science* 137(1–2), 43–52.

Shepherd, E. M., Fairchild, B. D. and Ritz, C. W. 2017. Alternative bedding materials and litter depth impact litter moisture and footpad dermatitis. *Journal of Applied Poultry Research* 26(4), 518–528.

Shields, S. and Greger, M. 2003. Animal welfare and food safety aspects of confining broiler chickens to cages. *Animals* 3(2), 386–400.

Shields, S. J., Garner, J. P. and Mench, J. A. 2004. Dustbathing by broiler chickens: a comparison of preference for four different substrates. *Applied Animal Behaviour Science* 87(1–2), 69–82.

Shields, S. J., Garner, J. P. and Mench, J. A. 2005. Effect of sand and wood-shavings bedding on the behavior of broiler chickens. *Poultry Science* 84(12), 1816–1824.

Sossidou, E. N., Dal Bosco, A., Castellini, C. and Grashorn, M. A. 2015. Effects of pasture management on poultry welfare and meat quality in organic poultry production systems. *World's Poultry Science Journal* 71(2), 375–384.

Souillard, R., Reperant, J. M., Experton, C., Huneau-Salaun, A., Coton, J., Balaine, L. and Le Bouquin, S. 2019. Husbandry practices, health, and welfare status of organic broilers in France. *Animals* 9, 97. doi:10.3390/ani9030097.

Stadig, L. M., Rodenburg, T. B., Ampe, B., Reubens, B. and Tuyttens, F. A. M. 2017. Effects of shelter type, early environmental enrichment and weather conditions on free-range behaviour of slow-growing broiler chickens. *Animal* 11(6), 1046–1053.

Stadig, L. M., Rodenburg, T. B., Reubens, B., Ampe, B. and Tuyttens, F. A. M. 2018. Effects of dark brooders and overhangs on free-range use and behaviour of slow-growing broilers. *Animal* 12(8), 1621–1630.

Statistica 2019. Production of meat worldwide from 2016 to 2019, by type. Available at: https://www.statista.com/statistics/237632/production-of-meat-worldwide-since-1990/. Accessed on 05/08/2019.

Steenfeldt, S., Sorensen, P. and Nielsen, B. L. 2019. Effects of choice feeding and lower ambient temperature on feed intake, growth, foot health, and panting of fast- and slow-growing broiler strains. *Poultry Science* 98(2), 503–513.

Sun, Y. Y., Li, Y. L., Li, D. L., Chen, C., Bai, H., Xue, F. G. and Chen, J. L. 2017. Responses of broilers to the near-continuous lighting, constant 16-h lighting, and constant 16-h lighting with a 2-h night interruption. *Livestock Science* 206, 135–140.

Swiatkiewicz, S., Arczewska-Wlosek, A. and Jozefiak, D. 2017. The nutrition of poultry as a factor affecting litter quality and foot pad dermatitis – an updated review. *Journal of Animal Physiology and Animal Nutrition* 101(5), e14–e20.

Tahamtani, F. M., Pedersen, I. J., Toinon, C. and Riber, A. B. 2018. Effects of environmental complexity on fearfulness and learning ability in fast growing broiler chickens. *Applied Animal Behaviour Science* 207, 49–56.

Taylor, P. S., Hemsworth, P. H., Groves, P. J., Gebhardt-Henrich, S. G. and Rault, J. L. 2017a. Ranging behaviour of commercial free-range broiler chickens 1: factors related to flock variability. *Animals* 7, 54. doi:10.3390/ani7070054.

Taylor, P. S., Hemsworth, P. H., Groves, P. J., Gebhardt-Henrich, S. G. and Rault, J. L. 2017b. Ranging behaviour of commercial free-range broiler chickens 2: individual variation. *Animals* 7, 55. doi:10.3390/ani7070055.

Torrey, S., Liu, Z., Caston, L., Nascimento dos Santos, M., Rothschild, D., Kiarie, E., Mandell, I. and Widowski, T. 2019. Differences in behavioural time budget between conventional and slow growing broiler chickens. Poultry Science Association Meeting, Poultry Science Association, Montreal.

Van Middelkoop, K., Van Harn, J., Wiers, W. and Van Horne, P. 2002. Slower growing broilers pose lower welfare risks. *World Poultry* 18, 20–21.

Van Niekerk, T., Ellen, H. H. and Winkel, A. 2016. Licht op licht: licht en verlichting in de pluimveehouderij in relatie tot beschadigend pikgedrag. Wageningen Livestock Research, Wageningen, Report 922, 33 pp.

Vasdal, G., Vas, J., Newberry, R. C. and Moe, R. O. 2019. Effects of environmental enrichment on activity and lameness in commercial broiler production. *Journal of Applied Animal Welfare Science* 22(2), 197–205.

Ventura, B. A., Siewerdt, F. and Estevez, I. 2012. Access to barrier perches improves behavior repertoire in broilers. *PLoS ONE* 7(1), e29826.

Vissers, L. S. M., de Jong, I. C., van Horne, P. L. M. and Saatkamp, H. W. 2019. Global prospects of the cost-efficiency of broiler welfare in middle-segment production systems. *Animals* 9, 473. doi:10.3390/ani9070473.

Wallenbeck, A., Wilhelmsson, S., Jonsson, L., Gunnarsson, S. and Yngvesson, J. 2016. Behaviour in one fast-growing and one slower-growing broiler (Gallus gallus domesticus) hybrid fed a high-or low-protein diet during a 10-week rearing period. *Acta Agriculturae Scandinavica, Section A – Animal Science* 66(3), 168–176.

Welfare Quality 2009, *The Welfare Quality Assessment Protocol for Broiler Chickens and Laying Hens*. The Welfare Quality Consortium, Lelystad.

Wilhelmsson, S., Yngvesson, J., Jonsson, L., Gunnarsson, S. and Wallenbeck, A. 2019. Welfare Quality (R) assessment of a fast-growing and a slower-growing broiler hybrid, reared until 10 weeks and fed a low-protein, high-protein or mussel-meal diet. *Livestock Science* 219, 71–79.

Willemsen, H., Debonne, M., Swennen, Q., Everaert, N., Careghi, C., Han, H., Bruggeman, V., Tona, K. and Decuypere, E. 2010. Delay in feed access and spread of hatch: importance of early nutrition. *World's Poultry Science Journal* 66(2), 177–188.

Zuidhof, M. J., Schneider, B. L., Carney, V. L., Korver, D. R. and Robinson, F. E. 2014. Growth, efficiency, and yield of commercial broilers from 1957, 1978, and 2005. *Poultry Science* 93(12), 2970–2982.

# Chapter 12

# Welfare issues in poultry housing and management: laying hens

*Victoria Sandilands, Scotland's Rural College (SRUC), UK*

1 Introduction
2 Conventional cages
3 Enriched cages
4 Cage-free housing
5 Conclusion
6 Where to look for further information
7 References

## 1 Introduction

Laying hens are one of the workhorses of agriculture: unlike meat animals, whose production is based on slaughter output, laying hens produce eggs on an almost daily basis, for a year or more, once they have reached sexual maturity. In 2016, worldwide production of hens' eggs was 73 890 000 metric tons, with China being the largest egg producer (530 billion eggs), followed by the USA (approximately 96 billion eggs), India and Mexico (WATT Global Media, 2018). In 2018, the UK flock was estimated at 40 million hens, which produced approximately 11 139 million eggs (British Egg Information Service, 2019).

Types of permitted housing systems in the EU are set out in Council Directive 1999/74/EC (European Union, 1999), and these systems relate to the way that eggs can be marketed, either as 'eggs from caged hens' or in one of the alternative systems (also known as cage-free, non-cage, floor, and loose-housed systems), which are 'barn eggs' or 'free-range eggs' (European Union, 2008a), the difference being that hens kept in free-range systems must be given access to range (pasture) whereas hens kept in barn systems are not. In addition, free-range eggs can be marketed as organic eggs, providing that they meet additional requirements (European Union, 2008b). Furthermore, on top of any legal requirements, various marketing schemes may impose higher standards above the basic legal minimum (e.g. in the UK, all eggs produced

http://dx.doi.org/10.19103/AS.2020.0078.11

**Table 1** Percent of eggs produced in the UK by housing type (Hewson, 1986; Defra, 2019)

| | 1951 | 1965 | 1980 | 2018 |
|---|---|---|---|---|
| Cage | 8 | 53 | 95 | 44 |
| Cage-free systems | | | | |
| Barn | 12 | 37 | 4 | 1 |
| Free range | 80 | 10 | 1 | 52 |
| Organic | | | | 3 |

under the British Egg Industry Council's Lion Code of Practice must come from hens vaccinated against *Salmonella enteritidis* and *Salmonella typhimurium* (British Egg Information Service, 2018b).

Trends in laying hen housing have changed rapidly in Europe and elsewhere in recent decades. In the UK, for example, 95% of eggs were produced from hens housed in (conventional) cages in 1980 and only 5% in cage-free systems (Table 1). Less than 40 years later, in 2018, the majority of eggs (56%) were produced in cage-free systems, to a situation not dissimilar to the figures of 1965 (when 47% of eggs were produced in cage-free systems). Similarly in Switzerland, although all cages have been banned there since 1992 (Hane et al., 2000), Tauson (2000) described Swiss egg consumption as being 47% from cage-free systems but 53% from imported cage eggs (demonstrating that animal welfare intentions are not always reflected in consumers' buying power). UK caged egg production will probably decline further in the next decade, as the major retailers have pledged to sell only cage-free eggs from 2025 (Black, 2016). The USA is also experiencing a similar move: in 2015, for example, McDonald's USA announced its aim to use only cage-free eggs by 2025, and as of 2019 it had achieved 33% cage-free egg use (McDonald's, 2019).

Each system has unique features which affect the welfare of laying hens and how hens are managed. This chapter sets out the details of each system and some benefits and drawbacks.

## 2 Conventional cages

A conventional cage consists of a wire mesh box with at least one nipple drinker at the back and a feeder at the front, with a sloping floor so that laid eggs roll forward onto an egg collection area, where eggs can be collected by hand or are automatically carried on a moving egg belt out of the hen house and into an egg sorting room. These cages, also known as battery cages, were developed in the 1930s to hold single birds but were later adapted to hold four to five birds (Whittle, 1998). Conventional cages were banned in the EU in 2012 (European Union, 1999) but are still commonly used elsewhere around the world (e.g. China, USA, India, South Africa). Many countries do not set down legal minimum requirements for floor space per hen, but between 2003 and 2012, when the EU still permitted conventional cages, the minimum permissible floor allowance was increased from 450 to 550 cm$^2$/hen (and feed trough space was 10 cm/hen) (European Union, 1999). Space allowances in conventional cages elsewhere in the world are often unregulated and can be as little as 350 cm$^2$/hen (van Horne and Achterbosch, 2008). In the USA, the United Egg Producers guidelines (not a legal requirement) recommend between 67 and 86 in$^2$/hen, equivalent to 432–554 cm$^2$/hen (United Egg Producers, 2016).

The advantages of conventional cages over floor systems in the mid-20th century were several. Egg production tended to be better in caged systems, due to better peak egg production (Appleby et al., 1988) but also fewer egg losses. Eggs were more likely to remain clean since faeces fell away through the wire, and the separation of hens from their faeces (and in the case of free-range birds from that of wild birds also) reduced hens' exposure to parasites and bacteria (Duncan, 2001; Lister and van Nijhuis, 2012). Cages were the first mechanised system of housing layers, which reduced labour (particularly once a system for removing manure was developed) and made it simpler to increase flock sizes to increase production (Whittle, 1998; Duncan, 2001). Another benefit to mechanisation (and indeed, being indoors) was the improved environmental conditions for both birds and people, including reduction in ammonia and dust, which was better for hen health (Duncan, 2001). The smaller group sizes of cages were typically socially stable (Sherwin and Nicol, 1992) and more reflective of natural group sizes of junglefowl (McBride et al., 1969) although possibly smaller than hen preference tests would indicate is desirable (Lindberg and Nicol, 1996). When sick birds need to be caught, or at depopulation, it is easier to catch birds from the contained conventional cage space (this gets considerably harder in enriched cages, where the hens find it easier to escape from an arm's length). Conventional cages were also the most economic: in the mid-1980s Elson (1985) estimated that barn housing was 5–18%, and free range was 35–50%, more expensive than producing eggs in conventional cages (of 450 cm$^2$/hen). The increased costs in non-cage systems were attributed to greater mortality, feed consumption and egg losses, and greater outbreaks of feather pecking and cannibalism (Appleby et al., 1992). Furthermore, hens housed in conventional cages tend to have fewer keel bone fractures than non-cage systems (Gregory and Wilkins, 1991; Fleming et al., 2004), most likely due to limited opportunities for movement and lack of furniture. (For a summary of keel bone damage prevalence, see Table 2.)

One disadvantage of conventional cages is that, while preventing some diseases, they are associated with other health problems. Hens from cages are more likely to suffer from fatty liver haemorrhagic syndrome than those from other housing system types (Rönchen et al., 2008; Shini et al., 2019), probably in combination due to high energy diets and low levels of exercise

**Table 2** Summary of literature on keel bone fracture prevalence

| Authors | Study scope | Type of measure | Outcome |
|---|---|---|---|
| Abrahamsson and Tauson (1993) | Housing systems | KBD | Perches in CCs* improve bone strength but increase KBD |
| Abrahamsson et al. (1996) | Housing systems | KBD | Perches in CCs improve bone strength but increase KBD<br>Perches in ECs associated with KBD |
| Fleming et al. (2004) | Housing systems | KBF | CC < CF |
| Gregory and Wilkins (1991) | Housing systems | KBF | CC < CF |
| Heerkens et al. (2016a) | Ramps in multi-tier | KBF (measured at several places)<br>KBDev | Ramps < no ramps for both total KBF and KBDev at 29, 39, 49 weeks of age |
| Heerkens et al. (2016b) | MT designs | KBF (of two areas of the KB - medial and caudal tip) | Medial KBF: portal 38%, row 13%<br>Caudal KBF: plastic slats 37%, wire mesh 50%<br>Total KBF: plastic slats 76%, wire mesh 86% |
| Hinrichsen et al. (2016) | Housing systems | KBD | MT 53%,<br>ST 20% |
| Riber and Hinrichsen (2016) | Housing systems | KBF | Mean over two collection periods: ST 4%, MT 8% |
| Richards et al. (2012) | Pop hole use | KBF | Use of pop holes: 54% no keel fracture, 35% minor keel fracture, 11% severe keel bone fracture |
| Rodenburg et al. (2008) | Housing systems | KBF | EC 62%, FR 82-97% |
| Sherwin et al. (2010) | Housing systems | KBF | CC 18%, EC 32%, B 69%, FR 60%<br>New KBF CC 25%, EC 4%, B 1%, FR 1% |
| Stratmann et al. (2015a) | Ramps in multi-tier | KBF | No difference at 66 weeks |
| Stratmann et al. (2015b) | Perch design | KBF | Soft perches 15%<br>Hard perches 22% |
| Wilkins et al. (2011) | Housing systems | KBD | EC < CF<br>Raised perches, a source of KBD<br>EC 36%, B 63%, FR 67-86%, organic 45-84%<br>(in FR and organic, prevalence increased with increased perch height) |

*CC, conventional cage; EC, enriched cage; CF, cage free; FR, free range; MT, multi-tier; ST, single tier. Methods of classification are keel bone fracture (KBF), keel bone deviation (KBDev), keel bone damage (KBD, a combination of fracture and deviation). Damage is old (i.e. during the lifetime of the bird in the flock) unless specifically identified as new (i.e. no sign of healing; usually associated with depopulation).

(Butler, 1976). Conventionally caged hens also have poorer bone strength (see later). However, the overwhelming drawback of conventional cages is that they prevent hens from expressing natural behaviour. The first elucidation to the public of living conditions for laying hens in cages (and other farm animals) was probably the publication of Ruth Harrison's seminal book *Animal Machines* in 1964 (Harrison, 1964). From her own farm visits, she was dismayed by the practices that she witnessed and the continuous effort to obtain ever-greater production at whatever cost to the animals. Conversely, Savory (2004) argued that despite the many benefits of cages (with respect to the Five Freedoms; FAWC, 2009), legislation to ban conventional cages was influenced by the general public, who placed great weight on the ability of animals to perform natural behaviours. But it cannot be denied that the constrictions on behaviours are many.

The lack of opportunities for hens in cages to show foraging, dustbathing, perching, and nesting behaviours are well documented. In part, this is due to the lack of resources available for such behaviours - a hen cannot forage or dustbathe without litter, for example. Foraging/ground pecking takes up a large proportion of the time budget of wild (but captive-housed) and domestic (but free-living) chickens, ranging from 48% to 61% of time observed (Savory et al., 1978; Dawkins, 1989). With concentrated feed provided in front of hens and no litter, the opportunities to show foraging in cages are limited to expression at the feed trough, which may explain why conventionally caged hens may spend almost three times longer (65% of observations) at the feeder than hens in cages with litter (23% of observations at feeder, 18% pecking litter) (Bareham, 1976). Litter also enables functional dustbathing (i.e. the removal of stale preen oil) to take place (van Liere and Bokma, 1987; van Liere, 1992). Conventional cages lack perches, yet hens are highly motivated to perch, particularly at night, when almost all hens will perch if enough space is provided (Tauson, 1984; Duncan et al., 1992). Perches are additionally used as a refuge for subordinate birds, to escape the unwanted attention of others (Cordiner and Savory, 2001). Furthermore, hens housed in conventional cages have reduced bone strength compared to those housed in enriched cages and non-cage systems (Hughes and Appleby, 1989; Regmi et al., 2017) due to the lack of exercise they can express, but perches have been shown to increase bone strength (for a review, see Sandilands et al., 2009). Notably, though, the addition of perches into conventional cages not only can improve bone strength but also increases keel bone damage (Abrahamsson and Tauson, 1993; Abrahamsson et al., 1996). Conventional cages also lack nest boxes. Given that egg laying occurs almost daily and nesting behaviour is triggered by hormones by the ruptured egg follicle (Wood-Gush and Gilbert, 1964), the inability to perform nesting behaviours is considered to be a major source of frustration for laying hens (Appleby et al., 1992).

Also, conventional cages lack the space that non-cage systems, or even enriched cages, offer – to enable some behaviours to take place. Dawkins and Hardie (1989) demonstrated that hens need between 540 and 1980 $cm^2$/hen to perform turning, wing stretching, wing flapping, feather ruffling, preening, and ground-scratching behaviours, although the space used has been shown to depend on the space provided (Nicol, 1990), and, of course, hens require much more space to perform a behaviour than the amount of space it takes up, if they are to avoid hitting the enclosure sides or ceiling (Baxter, 1994). Comfort activities such as wing flapping and head scratching increase up to three-fold after living under spatial restriction compared to when space is not restricted (Nicol, 1987). In conventional cages, hens can only stand on wire (cf. enriched cages where they can also perch on plastic or metal poles) and, as a result, often develop hyperkeratosis (Abrahamsson and Tauson, 1995; Abrahamsson et al., 1996). Caged birds are more difficult to inspect at the same rate as those in non-cage systems, and thus staff have to rely more on egg production and feed and water intake than on hen behaviour, which gives delayed information when something goes wrong (Duncan, 2001).

Feather pecking is a misappropriately redirected foraging behaviour, where the motivation to forage for food is directed towards another bird's feathers (Huber-Eicher and Wechsler, 1998). Feather pecking can be gentle or severe and can include feather removal and feather eating (Blokhuis and van der Haar, 1992; Savory, 1995). It can also lead to bleeding skin wounds, vent pecking, cannibalism, and death (Blokhuis and van der Haar, 1992). There is also a heritable component to the behaviour (Cuthbertson, 1980; Buitenhuis et al., 2004). It is a common problem in the housing of laying hens, where compound feed is provided without the need to food search. However, along with other management techniques (Lambton et al., 2010), feather pecking can be reduced by encouraging foraging behaviour via the provision of loose litter to peck at, such as wood shavings or straw, and the provision of grain in the litter (Blokhuis and van der Haar, 1992; Huber-Eicher and Wechsler, 1998; Huber-Eicher and Sebö, 2001; Nicol et al., 2001). Caged hens lack the opportunities to show foraging behaviour, which may result in feather pecking of cage mates. However, where outbreaks occur, caged hens are housed in small groups, meaning the peckers have fewer victims than in non-cage systems, and, as a result, reports of feather pecking and cannibalism tend to be higher in non-cage systems (for a review, see Nicol et al., 2013). In all, conventional cages may be good for the producer but offer little in the way of the quality of life for the hen in terms of behavioural freedom.

## 3 Enriched cages

Enriched cages (also known as furnished, or colony, cages) were designed to maintain the benefits of conventional cages while improving hen welfare and opportunities to express natural behaviour (Hughes, 1994; Elson, 2011). The first prototypes of enriched cages were developed in the mid-1970s, with the Bareham experimental cage and the Elson getaway cage, followed by designs such as the Edinburgh modified cage (Hughes, 1994; Appleby, 1998). Many years of development and modification followed; however, in 1999 the minimum requirement of enriched cages used in the EU is to provide a nest, perches (15 cm/hen), litter so that pecking and scratching are possible, claw shorteners, and more floor space (750 $cm^2$/hen, 600 $cm^2$ of which must be usable - see next) and feeder space (12 cm/hen) than conventional cages. 'Usable area' refers to areas in the cage (or other system) that are at least 30 cm wide, with a floor slope $\leq$ 14%, and headroom of $\geq$ 45 cm, and does not include nest boxes/areas (European Commission, 1999). Interestingly, the legislation does not stipulate how much nest space or litter must be provided, and as a result these resources can vary from manufacturer to manufacturer. However, commonly the litter used is layer's mash, which is dispensed onto a scratch mat. Note that there is no requirement for a dustbath per se (although hens may show dustbathing on the litter for pecking and scratching), most likely because early prototypes of enriched cages that included dustbaths resulted in eggs laid there, resulting in downgrading, or that they were used more frequently for pecking and scratching (Appleby et al., 2002). Colony sizes of enriched cages are not regulated on an EU level and vary by country: in Sweden, for example, the maximum permitted group size per cage is 16 (SVA Animal Health Institute, 2016), whereas no such limit exists in the UK, where typical commercial flocks will be housed in groups of 40-80 hen cages (British Egg Information Service, 2018a). Flock sizes range into hundreds of thousands, because as long as the mechanisation of ventilation, manure removal, feeding, water provision, and egg collecting can cope with it, the sky (or shed height, or environmental application!) is the limit. This ten-tier hen house in the UK holds 123 200 hens and is one of four houses on one farm site (Fig. 1).

Undoubtedly, enriched cages have improved hen welfare over conventional cages. Greater floor space per hen can result in cleaner feathers and less feather loss (Widowski et al., 2017a). Increased food trough space may or may not affect feeding behaviour: reducing bird density in one size of enriched cage, thus giving more feed space/hen, resulted in fewer unsuccessful attempts to reach the feeder and fewer sideways and backward displacements (Albentosa et al., 2007), whereas at 8.9 or 12.8 cm feeder space/hen, no differences in the percentage of birds that were seen feeding simultaneously were observed (Widowski et al., 2017b). Prototype cages that incorporated nests (Sherwin and Nicol, 1992; Abrahamsson et al., 1996) and modern enriched cages (Onbaşılar et al., 2015) showed good use of nest boxes (although some eggs may be laid at the scratch mats, Wall, 2011), suggesting that the design of nests is appropriate

**Figure 1** A ten-tier enriched cage unit in the UK (photo credit: Chicken Behaviour and Welfare MOOC, © University of Edinburgh and Scotland's Rural College 2016 CC BY).

to a degree in fulfilling nesting motivation. Small hen groups (28–40 birds) in enriched cages may lay a greater proportion of eggs in the nest box area than large hen groups (55–80 birds), but space allowance does not appear to affect egg location (Widowski et al., 2017a). Meanwhile, these improvements are gained while maintaining the low mortality rates of conventional cages (Weeks et al., 2016).

Enriched cages have reduced hyperkeratosis of the toe pads associated with sloping wire floors in conventional cages, but perches (in any system) can cause greater incidences of bumblefoot (Abrahamsson and Tauson, 1993) although this can be altered with perch design (for reviews of designs, see Sandilands et al., 2009; Tuyttens, 2009). Claw shorteners have been successful at reducing long and broken claws commonly seen in conventional cages, providing they are suitably positioned and are replaced when worn. Perches are well used by hens in enriched cages but are associated with keel bone damage (Abrahamsson et al., 1996). The provision of litter for pecking and scratching enables hens to show appropriate foraging behaviour, which should reduce the incidence of feather pecking; however, some studies show that feather pecking is still a problem: in a study surveying 79 enriched cage and 80 free-range flocks, severe feather pecking estimated by the farmers in enriched cages was at 33% (Coton et al., 2019).

Enriched cages also have some benefits over non-cage systems. Hens housed in enriched cages experience less keel bone damage compared to those housed in cage-free systems (Wilkins et al., 2011), but keel bone damage occurs none the less. Keel bone fracture affects the behaviour in enriched cages: in a study by Casey-Trott and Widowski (2016), hens with multiple keel fractures perched more than hens with no keel fractures and stood less than hens with no or one keel fracture, although whether behaviour differences are cause or effect is unknown. Like conventional cages, enriched cages maintain some economic advantage over non-cage systems, although to a lesser extent: in 2008, production costs in EU conventional cages were about 83% the cost of non-cage systems, but enriched cages were about 89% (van Horne and Achterbosch, 2008).

Enriched-cage litter is generally easy to manage where layer's mash is used on the mat, as the feed is automatically distributed through separate auger tubes to the mats or via holes in the feed trough. However, the mash is quickly depleted by foraging behaviour or dustbathing. Any scattered litter falls onto the manure belt. Enriched cages which use friable materials require careful management to ensure that the supply remains topped up. Designs of cages using wood shavings or sand need care also, as scattered shavings or sand can interfere with mechanised feeders or egg belts.

## 4 Cage-free housing

As previously described, cage-free housing of laying hens in the EU can be barn or free range, but free-range systems can further be split into standard free range and organic free range. For the rest of this chapter, 'free range' means standard free range and 'organic' means organic free range. Also, most of the housing structures used within the three systems are similar, so they will first be described here for all systems, and then differences for free range and organic will be described later.

### *4.1 Barn and other cage-free systems*

All cage-free housing systems require particular resources, such as 10 cm/hen linear feed space or 4 cm/hen circular feed space; 2.5 cm/hen linear drinking trough, 1 cm/hen circular drinking trough (Fig. 2), or 1 nipple (or cup)/hen; 1 nest/7 hens or 1 m$^2$ nest space/120 hens for group nests; one-third of the floor area that is covered in litter (at least 250 cm$^2$/hen); a maximum stocking density of 9 hens/m$^2$ of usable area (thus, nest boxes, for example, are not included in the m$^2$ calculation for stocking density); and 15 cm perch space/hen (European Commission, 1999).

However, in England and Wales, the governments interpret the raised slatted floor area (at 30 cm intervals) to count as perch space so that they can avoid providing raised perches, an opinion that is frowned upon by the European Commission but so far has brought no prosecution. The argument is that raised perches are a potential source of keel bone injury to laying hens (e.g. Wilkins et al., 2011), which is exacerbated by poor design, inadequate lighting, high stocking density, and poor descending capabilities of laying hens (for a review, see Sandilands et al., 2009). But given the acceptance of other welfare insults to

**Figure 2** Differences in spacing with liner versus circular feeders or drinkers (overhead view). Hens are spread out when clustered around a circular provision, meaning that they need less space at the resource itself.

hens in cage-free housing (i.e. on average greater mortality with free range), the suspicion is that producers see no financial benefit to perches (cf. expense of providing pasture for free-range hens, whose eggs are sold at a premium) but instead a costly investment which takes more time to manage (i.e. is slower to catch birds from and clean). Welfare benefits of perches in cage-free housing include providing a source of refuge from other birds (Cordiner and Savory, 2001), reduced arousal and greater novel object interaction (Whay et al., 2007), reduced aggression and fearfulness (Donaldson and O'Connell, 2012), reduced feather loss (Whay et al., 2007), reduced risk of developing gentle feather pecking (Lambton et al., 2010), and a suitable resting place for roosting. Nonetheless, keel bone damage (fractures and deviations) can be painful and affect production and welfare (Lay et al., 2011; Nasr et al., 2012a,b), but keel bone injury is not solely the domain of cage-free housing. Studies of commercial flocks by Rodenburg et al. (2008), Sherwin et al. (2010), and Wilkins et al. (2011) all found keel bone damage present in all systems, albeit highest in cage-free systems.

Cage-free housing can also result in greater mortality than cage systems (Michel and Huonnic, 2003; Rodenburg et al., 2008; Sherwin et al., 2010; Wilkins et al., 2011). A study of 13 laying flocks from furnished cages (n = 6) and free-range systems (n = 7) found, on average, about 3% mortality in furnished cages but about 8% with single-tier and multi-tier housing at 60 weeks of age (Rodenburg et al., 2008). Similarly, Sherwin et al. (2010) found approximately double mortality rates in commercial cage-free compared to cage systems. An assessment of 3851 flocks showed that model predictions for cumulative mortality at 72 weeks of age was 9.3% in a free-range flock compared to 5.7% in a conventionally caged flock (Weeks et al., 2016). Cage-free housing is also dustier (Rodenburg et al., 2008), carries greater levels of air-borne bacteria and fungi (Vucemilo et al., 2010), and can result in greater foot pad problems (Tauson, 2005; Yilmaz Dikmen et al., 2016) although this has also been shown not to differ (Rodenburg et al., 2008; Sherwin et al., 2010). Furthermore, because hens are no longer separated from their droppings (cf. cages), the presence of intestinal parasites (such as coccidiosis and roundworms) can spread easily amongst the flock, as hens peck at litter which contains faeces that carries the parasite's eggs (Lister and van Nijhuis, 2012).

The requirement for litter in cage-free systems is beneficial for hens to show natural foraging and dustbathing behaviour but requires greater staff time and attention to ensure that the litter is of suitable quality (i.e. dry and friable). In particular, in cold and/or wet countries, ventilation systems (and in the case of free range and organic, open pop holes) will draw in cold/wet air, which makes the litter harder to manipulate by the hens (Duncan, 2001), and wet litter is not ideal or attractive for dustbathing. Furthermore, wet litter is associated with a greater prevalence of foot pad lesions (Wang et al., 1998).

Flock sizes in commercial non-cage systems tend to be of several thousand birds, in order to be economically advantageous. There are large variations, though: flocks of 2450–4200 hens (Zimmerman et al., 2006), 2000–16 000 hens (Gebhardt-Henrich et al., 2014), and 5875–32 000 hens (Rodenburg et al., 2008) are some examples of the variation seen. This can, in part, be explained by the use of two different house designs: mobile sheds are small (holding up to a few thousand birds) but can be moved regularly (useful for free-range and organic egg production in order to locate to fresh pasture) or fixed sheds which are built into foundations (and are larger). An advantage of larger groups sizes may be reduced aggression, although this is probably lost once group sizes exceed 100 birds or so: D'Eath and Keeling (2003) found that aggression between familiar and unfamiliar birds is lower in large (n = 120 hens) compared to small (n = 10 hens) groups, possibly because hens can no longer discriminate between them. In a study of 25 commercial free-range flocks (flock size range, 3000–16 000 hens), observed aggression was low in all flocks, although observations were brief (3 min × 3 locations × 3 areas × 4 ages) (Whay et al., 2007). Disadvantages of large group sizes may include an increased risk of feather pecking and cannibalism, although studies tend to focus on relatively small group sizes compared to commercial flock sizes, and the risk may plateau at some level (for a review, see Rodenburg and Koene, 2007).

Advantages of cage-free housing include freedom of movement and behaviour and better bone strength. A small-scale comparison of the behaviour of hens in pens or cages showed greater wing flapping, dustbathing, and ground scratching in pens (Black and Hughes, 1974). In a study of six enriched-cage flocks and seven cage-free flocks, cage-free birds were more active and made better use of resources, had stronger bones, and were less fearful (Rodenburg et al., 2008). A comparison of 26 commercial flocks from enriched-cage, barn, and free-range systems showed lower proportions of hens with red mite and new keel bone fractures (typically associated with depopulation) in cage-free and enriched-cage systems than with conventional cages (Sherwin et al., 2010). In a study of hens reared in a single-tier floor system and then housed as adults in a multi-tier floor system had stronger humeri and tibiae (based on larger cortical area, greater cortical thickness, and greater bone strength in loading tests) than hens reared in conventional cages and then housed in either conventional or enriched cages as adults (Regmi et al., 2017). Similarly, hens housed in multi-tier systems or cages showed stronger humeri and tibiae when housed in multi-tier systems (Michel and Huonnic, 2003).

Within all three cage-free housing systems, there are two different internal housing structures that can be used to provide these resources: single tier (where there is a litter floor plus one raised level of slats that contains perches, feeders, drinkers, and nest boxes) or multi-tier (where there is a litter floor plus up to four raised levels that contain the other resources, with manure belts under each level) (European Commission, 1999) (Fig. 3).

Single-tier structures, also known as 'flat deck', are an older style of non-cage housing for hens in which the majority of faeces collects under the raised slatted area, typically for the entire laying cycle, which can result in higher ammonia, nitrous oxide, and methane emissions compared to multi-tier structures (Dekker et al., 2011). An advantage to single-tier structures is that they are relatively easy for staff to inspect, because almost all birds can be seen (apart from those in nest boxes) when the flock is walked through. In contrast, multi-tier housing structures, also known as aviaries, provide hens with opportunities to make greater use of the vertical space in the house, not to mention enable the farmer to stock more birds per house footprint, because parts of each tier count as usable space (i.e. where they meet the requirements for area, floor slope, and headroom, and exclude nest boxes). Multi-tier structures have been associated with better plumage condition than single-tier structures (Hinrichsen et al., 2016), possibly because hens can more easily escape unwanted attention by moving through the tiers; however, other studies have found no effect of single- or multi-tier system on plumage damage (Hane et al., 2000) or indeed a tendency for better plumage in single- over multi-tier structures (Riber and Hinrichsen, 2016). It should also be noted that, while opportunities to move through multi-tier structures exist in theory, in fact birds may not move throughout the hen house very much (Pettersson et al., 2018).

(a)

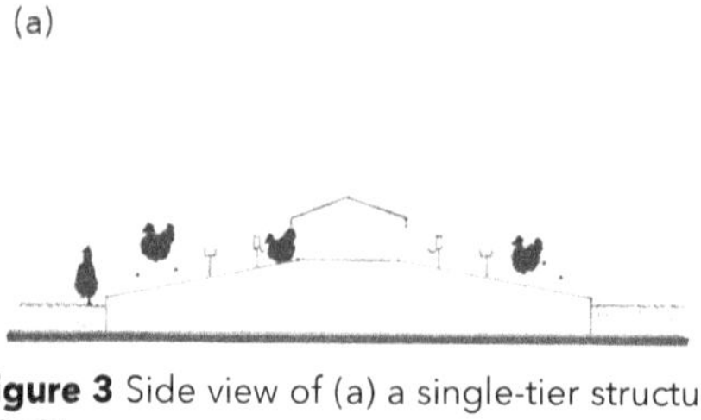

(b)

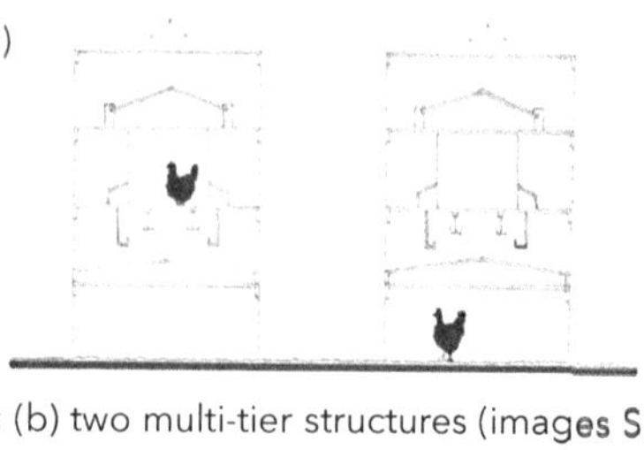

**Figure 3** Side view of (a) a single-tier structure; (b) two multi-tier structures (images SB, 2019).

A disadvantage of multi-tier structures is that hens housed here tend to have greater keel bone damage compared to those housed in single-tier structures (Hinrichsen et al., 2016; Riber and Hinrichsen, 2016; see Chapter 16 for more on this topic). Improved designs to reduce keel bone damage are possible with multi-tier structures. Adding ramps between tier levels to aid hens' descent results in more controlled movements, fewer falls, and fewer collisions compared to aviaries without ramps, in hens observed to 43 weeks of age, although no differences in keel bone fractures were found on dissection at 66 weeks of age (Stratmann et al., 2015a). Heerkens et al. (2016a) also added ramps between each level of an experimental offset multi-tier system (i.e. portal-style) and found a significant reduction in keel bone fractures and foot pad disorders in hens housed with ramps from 17 to 52 weeks of age. Alternative perch designs may also help: Pickel et al. (2011) found that soft perch covers with air cells between the cover and the hard core reduced pressure (N/cm$^2$) on hens' keel bones and foot pads during tests with 36 hens. Stratmann et al. (2015b) found that hens housed in multi-tier systems over a laying cycle and given either metal perches or perches covered in 3.4 cm of soft polyurethane had significantly fewer keel bone fractures with soft perches compared to those with hard perches. Different multi-tier designs may offer varying benefits: multi-tier structure (i.e. portal versus row type) and use of plastic slats versus wire mesh have both been shown to affect keel bone damage, foot health (Heerkens et al., 2016b), plumage condition, mortality, and performance (Heerkens et al., 2015).

### 4.2 Free-range systems

The EU requirements for free-range systems is that hens must have at least 2 m of pophole access to the range per 1000 hens and shelter from weather and predators (European Commission, 1999). The maximum stocking density on the range is 2500 hens/hectare or 1 hen/4 m$^2$ (European Union, 2008a) although national regulations may reduce that further.

The LAYWEL report (2006) dealt eloquently with the conflict that arises within free-range production, between greater freedom, but greater risk, and how to balance welfare between the animal and its environment:

> ...the questions arises, how the risk of diseases, damages and mortality is related to welfare. In many cases the hens do not perceive risk-bearing conditions as adverse experiences. They may even show preferences for these conditions. If there are established relationships between the risk factors and the occurrence of welfare problems, the potential hazard of the birds' welfare has to be balanced against the strength of preference. (LayWel, 2006, p. 4)

Although range access undoubtedly provides greater freedom of behaviour in hens, who can interact with the natural environment (e.g. making and using dustbaths in the earth, foraging for insects and grass, sunbathing), it also comes with additional risks, such as predation and disease. Hens with range access are exposed not only to their own faeces (as with barn systems) but that of wild birds and other animals. Free-range hens have greater exposure to serious diseases including Newcastle disease, mycoplasma, and avian influenza (Lister and van Nijhuis, 2012). Furthermore, poorly drained areas of the range can result in pools of water which hens may drink from, increasing their exposure to bacteria such as *E. coli* and *Brachyspira* (Lister and van Nijhuis, 2012). Such pathogens can lead to significant mortality in a flock: in a study of 18 organic flocks, 4 suffered outbreaks of Pasteurella and had total morality rates of 24–62% (Hegelund et al., 2006).

In free-range systems, there are more variables that can change (e.g. litter quality), which in turn can affect hen welfare (and production). Therefore, in order for the system to work well (for both hens and producers), greater management input is required of producers. For example, the risk of feather pecking developing is higher where only one person inspects the flock, if there is a lack of loose litter at the end of lay, and lack of range use (Green et al., 2000; Lambton et al., 2010). The risk of vent pecking developing has been associated with dim lighting to encourage nest box use (Pötzsch et al., 2001). In all cases, reducing those risk factors takes greater time and effort on the part of staff. Free-range eggs also have a greater tendency to be dirtier than other methods of production (Yilmaz Dikmen et al., 2016), affecting the saleability of eggs.

Access to (and use of) a range area can be beneficial in reducing the risk of feather pecking and cannibalism (Green et al., 2000; Lambton et al., 2010). Hens with access to an outdoor range area have been shown to have a better feather condition than hens without in both commercial systems (Heerkens et al., 2015) and in small-scale research trials (Yilmaz Dikmen et al., 2016), although other work in which all flocks have access to range, but flocks that vary in their use of it show no correlation between range use and plumage condition (Hegelund et al., 2006). However, free-range hens with keel bone fractures may enter the range less than those without (Richards et al., 2012).

While free-range housing systems offer an additional freedom to hens, it may be that only a small proportion of hens go outside: a review of 14 studies on commercial flocks showed that commonly less than half of birds range at any one time, and quite often it is less than 10% of the flock (Pettersson et al., 2016). Evidence suggests that hens which enter the range are the same birds, rather than many birds going outside infrequently, and that range use per day is brief: a study of 12 commercial free-range flocks, in which 5–10% of the hens were fitted with RFID tags and were monitored for 18–21 days over 6 months for range use, showed that 47–90% of tagged hens used the range at least once, but of those hens that used the range, mean time on range varied from 10 to 102 min and 40–85% of days monitored (Gebhardt-Henrich et al., 2014).

There is some argument that large flock sizes reduce the proportion of hens seen on the range (e.g. Hirt et al., 2000; Hegelund et al., 2005). However, in a study of 12 commercial free-range flocks ranging from 2000 to 18 000 hens, there was no correlation between flock size and the percentage of RFID-tagged hens on the range (Gebhardt-Henrich et al., 2014). Also, in a study with 18 organic commercial flocks, all with relatively small flock sizes (1200–5000 hens), the mean percentage of the flocks seen outside was 7–38%, with no correlation between flock size and flock ranging (Hegelund et al., 2006).

Some free-range hen houses are fitted with verandas. Verandas, or winter gardens, are covered areas (solid roof, wholly or partly mesh walls) with litter floors that sit between the hen house and outdoor area. Hens access the veranda through the house pop holes and can then progress to the range through pop holes in the veranda wall. Verandas allow hens access to fresh air, litter for dustbathing and foraging, and natural light while providing some degree of protection from the elements, predators, and wild birds (Gebhardt-Henrich et al., 2014). They also aid in reducing the stocking density inside the shed (Steenfeldt and Nielsen, 2015). If hens are given permanent access to verandas, and the area is lit in the same manner as the inner part of the house, the area can be included as part of the usable space (Sandilands, 2014; Steenfeldt and Nielsen, 2015). Verandas are popular in many northern Europe countries that experience cold winters. Like the range area, there is some evidence that veranda use is lower with larger flock sizes (Gebhardt-Henrich et al., 2014).

## 4.3 Organic egg production

Organic egg production follows more holistic farming methods than other systems, where livestock production should be done (at least in part) on the land, and not solely inside a shed. In the EU, organic egg production is similar to free range (e.g. 2500 hens/ha on the range; one-third floor as litter), but differences include a maximum stocking density of 6 hens/m$^2$; maximum flock size of 3000 birds; 18 cm perch space/hen, and natural lighting, supplemented to a maximum of 16 h light/day (European Union, 2008b). Hens must not be beak-trimmed routinely, but it may be permitted if there is a problem within the flock. Furthermore, pullets should be organically reared (although they can be 'converted' to organic if housed suitably before they are 3 days old and are considered 'organic' after 6 weeks), hens must range on organic land for at least one third of their life, and the diet must contain at least 95% organic ingredients, although this may change from 2025 (European Union, 2018). The use of the EU organic food label (Fig. 4) is mandatory for organic pre-packaged food produced in the EU. Additional requirements for producing organic eggs can be imposed by certification schemes, for example the Soil Association (UK)

**Figure 4** The EU's organic food label (European Commission, 2018).

restricts its flock sizes to 2000 laying hens per shed and 1000 hens/ha (Soil Association, 2019).

Because organic systems tend to use the same housing structures as barn and free-range systems, they often experience similar welfare issues. Keel bone damage, for example, is comparable (Riber and Hinrichsen, 2016). Organic egg production also suffers from a greater proportion of dirty eggs (Ferrante et al., 2009), like free-range egg production. Additionally, organic hens show a similar association between ranging behaviour and better plumage condition in various studies (for a review, see van De Weerd et al., 2009). The lack of routine beak trimming in organic flocks is likely to affect the prevalence of feather pecking, with varying but often high prevalence seen in flocks from various EU countries (van De Weerd et al., 2009). Hen flocks with intact beaks also have a greater cumulative mortality than beak-trimmed flocks, although hen strain is also an important factor (Craig and Lee, 1990; Weeks et al., 2016), but despite this, many EU countries (e.g. Germany, the Netherlands, Sweden, Austria) do not permit (or there are disincentives to perform) routine beak trimming. It is commendable to not routinely mutilate hens; however, severe outbreaks of pecking and cannibalism are major welfare insults to flocks, so the decision to trim or not is a difficult one.

A unique housing structure that came into production in 2010 in the Netherlands is the Rondeel®. This has been classified as three-star based on the welfare level of the animals (three stars is normally granted to organic production methods) by the Dutch Society for Protection of Animals, 'Dierenbescherming' (van Niekerk and Reuvekamp, 2011). The system is circular and consists of multi-tier structures in five radial 'arms', whose walls roll up entirely to create a seamless transition from the shed to the pecking and scratching area. The pecking and scratching areas sit between each of the arms and is covered in Astroturf. The roof of the pecking and scratching area lets

**Figure 5** The Rondeel® system for laying hens. (a) Overhead view showing the radial arms and service area. (b) Inside the laying hen area (Photo credit: Vencomatic Group).

in daylight. In the outer ring of the circle, there is an outdoor run containing trees and covered with mesh on the top and sides which helps to act as a wind-break. The mesh improves the biosecurity of the outdoor area: during the avian influenza outbreak in the Netherlands, Rondeel® hens did not have to be kept inside (Egg Farmers of Alberta, 2017). So far, production and welfare outcomes appear positive, and flocks have been successfully housed in this system without beak trimming, suggesting that novel ways of housing hens may be able to contribute significantly to moving towards more populations of non-beak-trimmed hens (Fig. 5).

## 5 Conclusion

Trends for laying hen housing have changed rapidly in a short period of time in the UK and the EU and are in the process of changing elsewhere, such as the USA. Animal welfare research has contributed significantly to these changes, particularly in highlighting the behavioural drawbacks of conventional cages. However, enriched-cage and non-cage systems are not without their disadvantages, and welfare challenges in these systems still exist, particularly freedom of behaviour in enriched cages, which is difficult to rectify given the limitation in space and the wire floor. Much of the work showcased here, particularly in the last two decades, reports work that is conducted by scientists alongside commercial egg producers, who grant access to their flocks in order to provide meaningful results (cf. small-scale studies that cannot easily be translated into practice). This is to be commended, as by working together, we will more easily return solutions to improve hen welfare further, while still being commercially viable.

As the use of cage-free systems continues to grow, major concerns are the higher levels of keel bone damage and mortality in laying hens. Efforts to address these, by changes in management and genetic selection, are the source of many current research aims. Welfare problems in cage-free systems are potentially reducible, if not altogether solvable, and we should be striving towards systems that give hens the best welfare that we can offer.

## 6 Where to look for further information

Further reading on the pros and cons of cages can be found in Duncan (2001), while a comprehensive look at the effects of laying hen housing on production and welfare is found in Sandilands and Hocking (2012).

There are two particular organisations that host relevant conferences on this subject area:

- ISAE (International Society for Applied Ethology) holds an annual congress, with many talks on farm animal welfare https://www.applied-ethology.org/.
- WPSA (World's Poultry Science Association) has a working group that hosts a European Symposium on Poultry Welfare conference every 4 years http://www.wpsa.com/index.php.

There are a number of current or recently finished international projects, whose websites are a good source of information, including:

- EU COST Action on keel bone damage in laying hens provides guidance in many different languages http://www.keelbonedamage.eu/.
- EU COST Action on preventing damaging behaviour in chickens (and pigs) https://www.grouphousenet.eu/.
- FeatherWel promotes guidance on how to reduce the risk of injurious pecking in cage-free hens http://www.featherwel.org/.

## 7 References

Abrahamsson, P. and Tauson, R. 1993. Effect of perches at different positions in conventional cages for laying hens of 2 different strains. *Acta Agriculturae Scandinavica, Section A - Animal Science* 43(4), 228–35. doi:10.1080/09064709309410171.

Abrahamsson, P. and Tauson, R. 1995. Aviary systems and conventional cages for laying hens - effects on production, egg quality, health and bird location in 3 hybrids. *Acta Agriculturae Scandinavica, Section A - Animal Science* 45(3), 191–203. doi:10.1080/09064709509415851.

Abrahamsson, P., Tauson, R. and Appleby, M. C. 1996. Behaviour, health and integument of four hybrids of laying hens in modified and conventional cages. *British Poultry Science* 37(3), 521–40. doi:10.1080/00071669608417882.

Albentosa, M. J., Cooper, J. J., Luddem, T., Redgate, S. E., Elson, H. A. and Walker, A. W. 2007. Evaluation of the effects of cage height and stocking density on the behaviour of laying hens in furnished cages. *British Poultry Science* 48(1), 1–11. doi:10.1080/00071660601156479.

Appleby, M. C. 1998. The Edinburgh modified cage: effects of groups size and space allowance on brown laying hens. *Journal of Applied Poultry Research* 7(2), 152–61. doi:10.1093/japr/7.2.152.

Appleby, M. C., Hogarth, G. S., Anderson, J. A., Hughes, B. O. and Whittemore, C. T. 1988. Performance of a deep litter system for egg production. *British Poultry Science* 29(4), 735–51. doi:10.1080/00071668808417102.

Appleby, M. C., Hughes, B. O. and Elson, H. A. 1992. *Poultry Production Systems: Behaviour, Management and Welfare*. CAB International, Wallingford, UK.

Appleby, M. C., Walker, A. W., Nicol, C. J., Lindberg, A. C., Freire, R., Hughes, B. O. and Elson, H. A. 2002. Development of furnished cages for laying hens. *British Poultry Science* 43(4), 489–500. doi:10.1080/0007166022000004390.

Bareham, J. R. 1976. A comparison of the behaviour and production of laying hens in experimental and conventional battery cages. *Applied Animal Ethology* 2(4), 291–303. doi:10.1016/0304-3762(76)90064-X.

Baxter, M. R. 1994. The welfare problems of laying hens in battery cages. *Veterinary Record* 134(24), 614–9. doi:10.1136/vr.134.24.614.

Black, A. 2016. All major UK supermarkets' eggs to be 'cage-free' by 2025. *Farmer's Guardian*. Available at: https://www.fginsight.com/news/news/all-major-uk-supermarkets-eggs-to-be-cage-free-by-2025-13961 (accessed on 19 June 2019).

Black, A. J. and Hughes, B. O. 1974. Patterns of comfort behaviour and activity in domestic fowls: a comparison between cages and pens. *British Veterinary Journal* 130(1), 23–33. doi:10.1016/s0007-1935(17)35987-0.

Blokhuis, H. J. and van der Haar, J. W. 1992. Effects of pecking incentives during rearing on feather pecking of laying hens. *British Poultry Science* 33(1), 17–24. doi:10.1080/00071669208417440.

British Egg Information Service. 2018a. Laying cage systems. Available at: https://www.egginfo.co.uk/egg-facts-and-figures/production/cage-egg (accessed on 13 June 2019).

British Egg Information Service. 2018b. Lion code of practice. Available at: https://www.egginfo.co.uk/british-lion-eggs/lion-code-practice (accessed on 11 June 2019).

British Egg Information Service. 2019. Industy data, official egg info – British lion eggs. Available at: https://www.egginfo.co.uk/egg-facts-and-figures/industry-information/data (accessed on 11 June 2019).

Buitenhuis, A. J., Rodenburg, T. B., Wissink, P. H., Visscher, J., Koene, P., Bovenhuis, H., Ducro, B. J. and van der Poel, J. J. 2004. Genetic and phenotypic correlations between feather pecking behavior, stress response, immune response and egg quality traits in laying hens. *Poultry Science* 83(7), 1077–82. doi:10.1093/ps/83.7.1077.

Butler, E. J. 1976. Fatty liver diseases in the domestic fowl a review. *Avian Pathology* 5(1), 1–14. doi:10.1080/03079457608418164.

Casey-Trott, T. M. and Widowski, T. M. 2016. Behavioral differences of laying hens with gractured keel bones within furnished cages. *Frontiers in Veterinary Science* 3, 42. doi:10.3389/fvets.2016.00042.

Cordiner, L. S. and Savory, C. J. 2001. Use of perches and nestboxes by laying hens in relation to social status, based on examination of consistency of ranking orders and frequency of interaction. *Applied Animal Behaviour Science* 71(4), 305–17. doi:10.1016/s0168-1591(00)00186-6.

Coton, J., Guinebretière, M., Guesdon, V., Chiron, G., Mindus, C., Laravoire, A., Pauthier, G., Balaine, L., Descamps, M., Bignon, L., Huneau-Salaün, A. and Michel, V. 2019. Feather pecking in laying hens housed in free-range or furnished-cage systems on French farms. *British Poultry Science* 60(6), 617–27. doi:10.1080/00071668.2019.1639137.

Craig, J. V. and Lee, H. -Y. 1990. Beak trimming and genetic stock effects on behavior and mortality from cannibalism in White Leghorn-type pullets. *Applied Animal Behaviour Science* 25(1–2), 107–23. doi:10.1016/0168-1591(90)90074-N.

Cuthbertson, G. J. 1980. Genetic variation in feather-pecking behaviuor. *British Poultry Science* 21(6), 447–50. doi:10.1080/00071668008416695.

Dawkins, M. S. 1989. Time budgets in Red Junglefowl as a baseline for the assessment of welfare in domestic fowl. *Applied Animal Behaviour Science* 24(1), 77–80. doi:10.1016/0168-1591(89)90126-3.

Dawkins, M. S. and Hardie, S. 1989. Space needs of laying hens. *British Poultry Science* 30(2), 413–6. doi:10.1080/00071668908417163.

D'Eath, R. B. and Keeling, L. J. 2003. Social discrimination and aggression by laying hens in large groups: from peck orders to social tolerance. *Applied Animal Behaviour Science* 84(3), 197–212. doi:10.1016/j.applanim.2003.08.010.

Defra. 2019. United Kingdom egg statistics – quarter 4, 2014, national statistics. Available at: https://assets.publishing.service.gov.uk/government/uploads/system/uploads/attachment_data/file/799166/eggs-statsnotice-31jan19.pdf (accessed on 21 June 2019).

Dekker, S. E. M., Aarnink, A. J. A., de Boer, I. J. M. and Koerkamp, P. W. G. G. 2011. Emissions of ammonia, nitrous oxide, and methane from aviaries with organic laying hen husbandry. *Biosystems Engineering* 110(2), 123–33. doi:10.1016/j.biosystemseng.2011.07.006.

Donaldson, C. J. and O'Connell, N. E. 2012. The influence of access to aerial perches on fearfulness, social behaviour and production parameters in free-range laying hens. *Applied Animal Behaviour Science* 142(1–2), 51–60. doi:10.1016/j.applanim.2012.08.003.

Duncan, I. J. H. 2001. The pros and cons of cages. *World's Poultry Science Journal* 57(4), 381–90. doi:10.1079/WPS20010027.

Duncan, E. T., Appleby, M. C. and Hughes, B. O. 1992. Effect of perches in laying cages on welfare and production of hens. *British Poultry Science* 33(1), 25–35. doi:10.1080/00071669208417441.

Egg Farmers of Alberta. 2017. The Rondeel story. Available at: http://eggs.ab.ca/about/eggnotes/dec17-rondeel/(accessed on 21 June 2019).

Elson, H. A. 1985. The economics of poultry welfare. In: Wegner, R.-M. (Ed.), *Proceedings of the Second European Symposium on Poultry Welfare*. World's Poulty Science Association – German Branch, Celle, Germany, pp. 244–53.

Elson, H. A. 2011. Housing and husbandry of laying hens: past, present and future. *Lohmann Information* 46(2), 16–24. Available at: http://lohmann-information.com/content/l_i_46_artikel11.pdf.

European Commission. 1999. Council Directive 1999/74/EC of 19 July 1999 laying down minimum standards for the protection of laying hens. Available at: http://eur-lex.europa.eu/LexUriServ/LexUriServ.do?uri=CELEX:31999L0074:EN:HTML.

European Commission. 2018. The organic logo, Policies, information and services. Available at: https://ec.europa.eu/info/food-farming-fisheries/farming/organic-farming/organics-glance/organic-logo_en (accessed on 21 June 2019).

European Union. 1999. COUNCIL DIRECTIVE 1999/74/EC of 19 July 1999 laying down minimum standards for the protection of laying hens.

European Union. 2008a. COMMISSION REGULATION (EC) No 589/2008 of 23 June 2008 laying down detailed rules for implementing Council Regulation (EC) No 1234/2007 as regards marketing standards for eggs.

European Union. 2008b. COMMISSION REGULATION (EC) No 889/2008 of 5 September 2008 laying down detailed rules for the implementation of Council Regulation (EC) No 834/2007 on organic production and labelling of organic products with regard to organic production, labelling and co. Available at: https://eur-lex.europa.eu/legal-content/EN/TXT/PDF/?uri=CELEX:32008R0889&from=en.

European Union. 2018. REGULATION (EU) 2018/848 OF THE EUROPEAN PARLIAMENT AND OF THE COUNCIL of 30 May 2018 on organic production and labelling of organic products and repealing Council Regulation (EC) No 834/2007.

FAWC. 2009. The five freedoms. Available at: https://webarchive.nationalarchives.gov.uk/20121010012427/http://www.fawc.org.uk/freedoms.htm (accessed on 21 June 2019).

Ferrante, V., Lolli, S., Vezzoli, G. and Cavalchini, L. G. 2009. Effects of two different rearing systems (organic and barn) on production performance, animal welfare traits and egg quality characteristics in laying hens. *Italian Journal of Animal Science* 8(2), 165–74. doi:10.4081/ijas.2009.165.

Fleming, R. H., McCormack, H. A., McTeir, L. and Whitehead, C. C. 2004. Incidence, pathology and prevention of keel bone deformities in the laying hen. *British Poultry Science* 45(3), 320–30. doi:10.1080/00071660410001730815.

Gebhardt-Henrich, S. G., Toscano, M. J. and Fröhlich, E. K. F. 2014. Use of outdoor ranges by laying hens in different sized flocks. *Applied Animal Behaviour Science* 155, 74–81. doi:10.1016/j.applanim.2014.03.010.

Green, L. E., Lewis, K., Kimpton, A. and Nicol, C. J. 2000. Cross-sectional study of the prevalence of feather pecking in laying hens in alternative systems and its associations with management and disease. *The Veterinary Record* 147(9), 233–8. doi:10.1136/vr.147.9.233.

Gregory, N. G. and Wilkins, L. J. 1991. Broken bones in hens. *Veterinary Record* 129(25–26), 559. doi:10.1136/vr.129.25-26.559-b.

Hane, M., Huber-Eicher, B. and Frohlich, E. 2000. Survey of laying hen husbandry in Switzerland. *World's Poultry Science Journal* 56(1), 21–31. doi:10.1079/WPS20000003.

Harrison, R. 1964. *Animal Machines: the New Factory Farming Industry*. Robinson & Watkins, London, UK.

Heerkens, J. L. T., Delezie, E., Kempen, I., Zoons, J., Ampe, B., Rodenburg, T. B. and Tuyttens, F. A. 2015. Specific characteristics of the aviary housing system affect plumage condition, mortality and production in laying hens. *Poultry Science* 94(9), 2008–17. doi:10.3382/ps/pev187.

Heerkens, J. L. T., Delezie, E., Ampe, B., Rodenburg, T. B. and Tuyttens, F. A. 2016a. Ramps and hybrid effects on keel bone and foot pad disorders in modified aviaries for laying hens. *Poultry Science* 95(11), 2479–88. doi:10.3382/ps/pew157.

Heerkens, J. L. T., Delezie, E., Rodenburg, T. B., Kempen, I., Zoons, J., Ampe, B. and Tuyttens, F. A. 2016b. Risk factors associated with keel bone and foot pad disorders in laying hens housed in aviary systems. *Poultry Science* 95(3), 482–8. doi:10.3382/ps/pev339.

Hegelund, L., Sørensen, J. T., Kjaer, J. B. and Kristensen, I. S. 2005. Use of the range area in organic egg production systems: effect of climatic factors, flock size, age and artificial cover. *British Poultry Science* 46(1), 1–8. doi:10.1080/00071660400023813.

Hegelund, L., Sørensen, J. T. and Hermansen, J. E. 2006. Welfare and productivity of laying hens in commercial organic egg production systems in Denmark. *NJAS - Wageningen Journal of Life Sciences* 54(2), 147–55. doi:10.1016/S1573-5214(06)80018-7.

Hewson, P. 1986. Origin and development of the British poultry industry: the first hundred years. *British Poultry Science* 27(4), 525–39. doi:10.1080/00071668608416912.

Hinrichsen, L. K., Riber, A. B. and Labouriau, R. 2016. Associations between and development of welfare indicators in organic layers. *Animal* 10(6), 953–60. doi:10.1017/S1751731115003018.

Hirt, H., Hordegen, P. and Zeltner, E. 2000. Laying hen husbandry: group size and use of hen-runs. In: Alfoldi, T., Lockeretz, W. and Niggli, U. (Eds), *Proceedings of the 13th International IFOAM Scientific Conference*, Basel, p. 363.

Huber-Eicher, B. and Sebö, F. 2001. Reducing feather pecking when raising laying hen chicks in aviary systems. *Applied Animal Behaviour Science* 73(1), 59–68. doi:10.1016/s0168-1591(01)00121-6.

Huber-Eicher, B. and Wechsler, B. 1998. The effect of quality and availability of foraging materials on feather pecking in laying hen chicks. *Animal Behaviour* 55(4), 861–73. doi:10.1006/anbe.1997.0715.

Hughes, B. O. 1994. Origins and development of modified cages for laying hens. In: Sherwin, C. M. (Ed.), *Modified Cages for Laying Hens*. Universities Press Federation for Animal Welfare (UFAW), Potters Bar, UK, pp. 1–9.

Hughes, B. O. and Appleby, M. C. 1989. Increase in bone strength of spent laying hens housed in modified cages with perches. *Veterinary Record* 124(18), 483–4. doi:10.1136/vr.124.18.483.

Lambton, S. L., Knowles, T. G., Yorke, C. and Nicol, C. J. 2010. The risk factors affecting the development of gentle and severe feather pecking in loose housed laying hens. *Applied Animal Behaviour Science* 123(1–2), 32–42. doi:10.1016/j.applanim.2009.12.010.

Lay, D. C., Fulton, R. M., Hester, P. Y., Karcher, D. M., Kjaer, J. B., Mench, J. A., Mullens, B. A., Newberry, R. C., Nicol, C. J., O'Sullivan, N. P. and Porter, R. E. 2011. Hen welfare in different housing systems. *Poultry Science* 90(1), 278–94. doi:10.3382/ps.2010-00962.

LayWel. 2006. Welfare implications of changes in production systems for laying hens. Deliverable 12: report with consensual version of welfare definition and welfare indicators. Available at: https://www.laywel.eu/web/xmlappservlet12d7.html?action=ProcessSelection&REDIRECT_TEMPLATE=ShowPage&SAVE_PARAMETER_SAV_SOURCE_DATABASE_NAME=/flexyz/projects/wur/fx_cm_laywel.nsf&SAVE_PARAMETER_SAV_DESIGN_CHOICE=interzorg/default&SAVE_PARAMETER_SAV_TEMPLATE_NAME=frontpage&SAVE_PARAMETER_SAV_SOURCE_DOCUMENT_NAME=wp1&SAVE_PARAMETER_SAV_NO_CACHE=TRUE%0A%09.

Lindberg, A. C. and Nicol, C. J. 1996. Effects of social and environmental familiarity on group preferences and spacing behaviour in laying hens. *Applied Animal Behaviour Science* 49(2), 109–23. doi:10.1016/0168-1591(96)01046-5.

Lister, S. and van Nijhuis, B. 2012. The effects of alternative systems on disease and health of poultry. In: Sandilands, V. and Hocking, P. M. (Eds), *Alternative Systems for Poultry: Health, Welfare and Productivity*. CABI Publishing, pp. 62–76.

McBride, G., Parer, I. P. and Foenander, F. 1969. The social organization and behaviour of the feral domestic fowl. *Animal Behaviour Monographs* 2, 125–81. doi:10.1016/S0066-1856(69)80003-8.

McDonald's. 2019. Sharing progress on our cage-free egg commitment. *McDonald's Newsroom*. Available at: https://news.mcdonalds.com/stories/using-our-scale-for-good/cage-free-farmer (accessed on 28 August 2019).

Michel, V. and Huonnic, D. 2003. A comparison of welfare, health and production performance of laying hens reared in cages or in aviaries. *British Poultry Science* 44(5), 775–6. doi:10.1080/00071660410001666691.

Nasr, M., Murrell, J., Wilkins, L. J. and Nicol, C. J. 2012a. The effect of keel fractures on egg-production parameters, mobility and behaviour in individual laying hens. *Animal Welfare* 21(1), 127–35. doi:10.7120/09627281279912937б.

Nasr, M. A. F., Nicol, C. J. and Murrell, J. C. 2012b. Do laying hens with keel bone fractures experience pain? *PLoS ONE* 7(8), e42420. doi:10.1371/journal.pone.0042420.

Nicol, C. J. 1987. Behavioural responses of laying hens following a period of spatial restriction. *Animal Behaviour* 35(6), 1709–19. doi:10.1016/S0003-3472(87)80063-5.

Nicol, C. 1990. Behaviour requirements within a cage environment. *World's Poultry Science Journal* 46(1), 31–3. doi:10.1079/WPS19900005.

Nicol, C. J., Lindberg, A. C., Phillips, A. J., Pope, S. J., Wilkins, L. J. and Green, L. E. 2001. Influence of prior exposure to wood shavings on feather pecking, dustbathing and foraging in adult laying hens. *Applied Animal Behaviour Science* 73(2), 141–55. doi:10.1016/s0168-1591(01)00126-5.

Nicol, C. J., Bestman, M., Gilani, A., De Haas, E. N., De Jong, I. C., Lambton, S., Wagenaar, J. P., Weeks, C. A. and Rodenburg, T. B. 2013. The prevention and control of feather pecking: application to commercial systems. *World's Poultry Science Journal* 69(4), 775–88. doi:10.1017/S0043933913000809. Available at: http://journals.cambridge.org/action/displayFulltext?type=6&fid=9122564&jid=WPS&volumeId=69&issueId=04&aid=9122563&fulltextType=RV&fileId=S0043933913000809.

Onbaşılar, E. E., Ünal, N., Erdem, E., Kocakaya, A. and Yaranoğlu, B. 2015. Production performance, use of nest box, and external appearance of two strains of laying hens kept in conventional and enriched cages. *Poultry Science* 94(4), 559–64. doi:10.3382/ps/pev009.

Pettersson, I. C., Freire, R. and Nicol, C. J. 2016. Factors affecting ranging behaviour in commercial free-range hens. *World's Poultry Science Journal* 72(1), 137–50. doi:10.1017/S0043933915002664.

Pettersson, I. C., Weeks, C. A., Norman, K. I., Knowles, T. G. and Nicol, C. J. 2018. Internal roosting location is associated with differential use of the outdoor range by free-range laying hens. *British Poultry Science* 59(2), 135–40. doi:10.1080/00071668.2017.1404007.

Pickel, T., Schrader, L. and Scholz, B. 2011. Pressure load on keel bone and foot pads in perching laying hens in relation to perch design. *Poultry Science* 90(4), 715–24. doi:10.3382/ps.2010-01025.

Pötzsch, C. J., Lewis, K., Nicol, C. J. and Green, L. E. 2001. A cross-sectional study of the prevalence of vent pecking in laying hens in alternative systems and its associations with feather pecking, management and disease. *Applied Animal Behaviour Science* 74(4), 259–72. doi:10.1016/S0168-1591(01)00167-8.

Regmi, P., Nelson, N., Haut, R. C., Orth, M. W. and Karcher, D. M. 2017. Influence of age and housing systems on properties of tibia and humerus of Lohmann White hens1: bone properties of laying hens in commercial housing systems. *Poultry Science* 96(10), 3755–62. doi:10.3382/ps/pex194.

Riber, A. B. and Hinrichsen, L. K. 2016. Keel-bone damage and foot injuries in commercial laying hens in Denmark. *Animal Welfare* 25(2), 179–84. doi:10.7120/09627286.25.2.179.

Richards, G. J., Wilkins, L. J., Knowles, T. G., Booth, F., Toscano, M. J., Nicol, C. J. and Brown, S. N. 2012. Pop hole use by hens with different keel fracture status monitored throughout the laying period. *Veterinary Record* 170(19), 494. doi:10.1136/vr.100489.

Rodenburg, T. B. and Koene, P. 2007. The impact of group size on damaging behaviours, aggression, fear and stress in farm animals. *Applied Animal Behaviour Science* 103(3-4), 205-14. doi:10.1016/j.applanim.2006.05.024.

Rodenburg, T. B., Tuyttens, F. A. M., de Reu, K., Herman, L., Zoons, J. and Sonck, B. 2008. Welfare assessment of laying hens in furnished cages and non-cage systems: an on-farm comparison. *Animal Welfare* 17, 363-73.

Rönchen, S., Scholz, B., Hamann, H. and Distl, O. 2008. Fat status in Lohmann Silver and Lohmann tradition laying hens kept in modified small group housing systems, small group housing systems, furnished cages and an aviary system. *Berliner und Munchener Tierarztliche Wochenschrift* 121(1-2), 11-8.

Sandilands, V. 2014. Laying hens: supplement to the codes of practice. Available at: https://www.sruc.ac.uk/download/downloads/id/2148/tn662_laying_hens_supplement_to_the_codes_of_practice.pdf.

Sandilands, V. and Hocking, P. M. (Eds). 2012. *Alternative Systems for Poultry: Health, Welfare and Productivity*. CABI Publishing, Wallingford, UK.

Sandilands, V., Moinard, C. and Sparks, N. H. C. 2009. Providing laying hens with perches: fulfilling behavioural needs but causing injury? *British Poultry Science* 50(4), 395-406. doi:10.1080/00071660903110844.

Savory, C. J. 1995. Feather pecking and cannibalism. *World's Poultry Science Journal* 51(2), 215-9. doi:10.1079/WPS19950016.

Savory, C. J. 2004. Laying hen welfare standards: a classic case of 'power to the people'. *Animal Welfare* 13, S153-158.

Savory, C. J., Wood-Gush, D. G. M. and Duncan, I. J. H. 1978. Feeding behaviour in a population of domestic fowls in the wild. *Applied Animal Ethology* 4(1), 13-27. doi:10.1016/0304-3762(78)90090-1.

Sherwin, C. M. and Nicol, C. J. 1992. Behaviour and production of laying hens in three prototypes of cages incorporating nests. *Applied Animal Behaviour Science* 35(1), 41-54. doi:10.1016/0168-1591(92)90015-4.

Sherwin, C. M., Richards, G. J. and Nicol, C. J. 2010. Comparison of the welfare of layer hens in 4 housing systems in the UK. *British Poultry Science* 51(4), 488-99. doi:10.1080/00071668.2010.502518.

Shini, A., Shini, S. and Bryden, W. L. 2019. Fatty liver haemorrhagic syndrome occurrence in laying hens: impact of production system. *Avian Pathology* 48(1), 25-34. Available at: http://www.tandfonline.com/doi/abs/10.1080/03079457.2018.1538550. doi:10.1080/03079457.2018.1538550.

Soil Association. 2019. *Soil Association Standards: Farming and Growing*. Soil Association, Bristol, UK, pp. 1-180.

Steenfeldt, S. and Nielsen, B. L. 2015. Welfare of organic laying hens kept at different indoor stocking densities in a multi-tier aviary system. I: Egg laying, and use of veranda and outdoor area. *Animal* 9(9), 1509-17. doi:10.1017/S1751731115000713.

Stratmann, A., Fröhlich, E. K. F., Gebhardt-Henrich, S. G., Harlander-Matauschek, A., Würbel, H. and Toscano, M. J. 2015a. Modification of aviary design reduces incidence of falls, collisions and keel bone damage in laying hens. *Applied Animal Behaviour*

*Science* 165, 112–23. doi:10.1016/j.applanim.2015.01.012. Available at: https://www.sciencedirect.com/science/article/pii/S0168159115000301?via%3Dihub.

Stratmann, A., Fröhlich, E. K. F., Harlander-Matauschek, A., Schrader, L., Toscano, M. J., Würbel, H. and Gebhardt-Henrich, S. G. 2015b. Soft perches in an aviary system reduce incidence of keel bone damage in laying hens. *PLoS ONE* 10(3), e0122568. doi:10.1371/journal.pone.0122568.

SVA Animal Health Institute. 2016. Animal health: poultry. Available at: https://www.sva.se/en/animal-health/poultry (accessed on 13 June 2019).

Tauson, R. 1984. Effects of a perch in conventional cages for laying hens. *Acta Agriculturae Scandinavica* 34(2), 193–209. doi:10.1080/00015128409435389.

Tauson, R. 2000. European alternative housing systems for layers: health, production and environmental considerations. *Zootecnica International* 23, 14–9.

Tauson, R. 2005. Management and housing systems for layers – effects on welfare and production. *World's Poultry Science Journal* 61(3), 477–90. doi:10.1079/WPS200569.

Tuyttens, F. A. M. 2009. Effects of perch design on behaviour and health of laying hens. *Animal Welfare* 18(4), 533–8.

United Egg Producers. 2016. *Animal Husbandry Guidelines for U.S. Egg Laying Flocks 2016 Edition*. Available at: https://www.google.com/url?sa=t&rct=j&q=&esrc=s&source=web&cd=1&cad=rja&uact=8&ved=2ahUKEwjPmcbasf3kAhVMQxUIHST1CqwQFjAAegQIAhAC&url=https%3A%2F%2Fuepcertified.com%2Fwp-content%2Fuploads%2F2015%2F08%2FUEP-Animal-Welfare-Guidelines-20141.pdf&usg=AOvVaw08kr4M6L2CoURMFuHHqFDv.

van De Weerd, H. A., Keatinge, R. and Roderick, S. 2009. A review of key health-related welfare issues in organic poultry production. *World's Poultry Science Journal* 65(4), 649–84. doi:10.1017/S0043933909000464.

van Horne, P. L. M. and Achterbosch, T. J. 2008. Animal welfare in poultry production systems: impact of EU standards on world trade. *World's Poultry Science Journal* 64(1), 40–52. doi:10.1017/S0043933907001705.

van Liere, D. W. 1992. The significance of fowls bathing in dust. *Animal Welfare* 1, 187–202.

van Liere, D. W. and Bokma, S. 1987. Short-term feather maintenance as a function of dust-bathing in laying hens. *Applied Animal Behaviour Science* 18(2), 197–204. doi:10.1016/0168-1591(87)90193-6.

van Niekerk, T. and Reuvekamp, B. 2011. The Rondeel, a new housing design for laying hens. *Lohmann Information* 46(2), 25–31.

Vucemilo, M., Vinkoviæ, B., Matkoviæ, K., Štokoviæ, I., Jakšiæ, S., Radoviæ, S. and Graniæ, K. 2010. The influence of housing systems on the air quality and bacterial eggshell contamination of table eggs. *Czech Journal of Animal Science* 55(6), 243–9. doi:10.17221/64/2009-CJAS.

Wall, H. 2011. Production performance and proportion of nest eggs in layer hybrids housed in different designs of furnished cages. *Poultry Science* 90(10), 2153–61. doi:10.3382/ps.2011-01495.

Wang, G., Ekstrand, C. and Svedberg, J. 1998. Wet litter and perches as risk factors for the development of foot pad dermatitis in floor-housed hens. *British Poultry Science* 39(2), 191–7. doi:10.1080/00071669889114.

WATT Global Media. 2018. WATT executive guide to world poultry trends. *WATT Poultry Trends*. Available at: http://www.poultrytrends.com/201811/(accessed on 21 June 2019).

Weeks, C. A., Lambton, S. L. and Williams, A. G. 2016. Implications for welfare, productivity and sustainability of the variation in reported levels of mortality for laying hen flocks kept in different housing systems: a meta-analysis of ten studies. *PLoS ONE* 11(1), e0146394. doi:10.1371/journal.pone.0146394.

Whay, H. R., Main, D. C., Green, L. E., Heaven, G., Howell, H., Morgan, M., Pearson, A. and Webster, A. J. 2007. Assessment of the behaviour and welfare of laying hens on free-range units. *Veterinary Record* 161(4), 119–28. doi:10.1136/VR.161.4.119.

Whittle, T. E. 1998. *A Triumph of Science: a 70 Year History of the UK Poultry Industry*. Poultry World Publications, pp. 1-91.

Widowski, T. M., Caston, L. J., Hunniford, M. E., Cooley, L. and Torrey, S. 2017a. Effect of space allowance and cage size on laying hens housed in furnished cages, Part I: performance and well-being. *Poultry Science* 96(11), 3805–15. doi:10.3382/ps/pex197.

Widowski, T. M., Caston, L. J., Casey-Trott, T. M. and Hunniford, M. E. 2017b. The effect of space allowance and cage size on laying hens housed in furnished cages, Part II: behavior at the feeder. *Poultry Science* 96(11), 3816–23. doi:10.3382/ps/pex198.

Wilkins, L. J., McKinstry, J. L., Avery, N. C., Knowles, T. G., Brown, S. N., Tarlton, J. and Nicol, C. J. 2011. Influence of housing system and design on bone strength and keel bone fractures in laying hens. *Veterinary Record* 169(16), 414. doi:10.1136/vr.d4831.

Wood-Gush, D. G. M. and Gilbert, A. B. 1964. The control of the nesting behaviour of the domestic hen. II. The role of the ovary. *Animal Behaviour* 12(4), 451-3. doi:10.1016/0003-3472(64)90065-X.

Yilmaz Dikmen, B., İpek, A., Şahan, Ü, Petek, M. and Sözcü, A. 2016. Egg production and welfare of laying hens kept in different housing systems (conventional, enriched cage, and free range). *Poultry Science* 95(7), 1564–72. doi:10.3382/ps/pew082.

Zimmerman, P. H., Lindberg, A. C., Pope, S. J., Glen, E., Bolhuis, J. E. and Nicol, C. J. 2006. The effect of stocking density, flock size and modified management on laying hen behaviour and welfare in a non-cage system. *Applied Animal Behaviour Science* 101(1-2), 111–24. doi:10.1016/j.applanim.2006.01.005. Available at: wos:000242352400009.

# Chapter 13

# The role of perches in chicken welfare

*Lars Schrader and Julia Malchow, Institute of Animal Welfare and Animal Husbandry – Friedrich-Loeffler-Institut, Germany*

1 Introduction
2 Why do chickens perch?
3 Ontogeny of perching
4 Anatomic prerequisites for perching
5 Properties of perches: the chicken's view
6 Elevated structures in pullet and layer housing
7 Elevated structures in broiler housing
8 Conclusions
9 Future trends in research
10 References

## 1 Introduction

Like most other birds, junglefowls (*Gallus gallus*), i.e. the ancestor of the domestic chicken (*Gallus gallus* forma domestica), live in a three-dimensional habitat and they use their wings to occupy this habitat. Elevated structures such as branches of trees are most often used for resting both during the night and at daytime (Fig. 1). Resting at night most often is termed roosting. However, both during night and daytime elevated structures are used by fowls also for other behaviours (see below). For example, junglefowls occasionally may access trees to eat fruits (Collias and Collias, 1967). This usage of elevated structures still can be observed in the domestic chicken (Wood-Gush and Duncan, 1976). Wood-Gush and Duncan (1976) observed that:

> the 6-7-week-old chicks joined their dam on a branch about 10 m above ground by flying up to a branch approximately 1.5 m above ground, and then working their way from branch to branch until they joined her, whereupon she covered them with her wings. A hen with a stunted 7-8-week-old chick joined the other adults for roosting and left her chick, which could not fly, to roost alone on the ground.
>
> (Wood-Gush and Duncan, 1976)

http://dx.doi.org/10.19103/AS.2020.0078.12

**Figure 1** Male junglefowl perching in an aviary.

This observation documents not only that chickens roost already at young ages but also that they seem to be highly motivated to roost on elevated structures; otherwise the aforementioned hen possibly may have stayed on the ground with her physically compromised chick. As long ago as 1819 the Brothers Grimm mention a possible function of roosting on elevated structures in their tale 'The Bremen Town Musicians': 'and the cock flew up to the top (of the tree), where it was safest for him' (Brüder Grimm, 1819, p. 142, translated by LS). Likewise, Wood-Gush and Duncan (1976) suggested that tree roosting protects the fowl from predators at night.

It is likely that the domestic chicken has a poorer ability to fly than the junglefowl as the relative wing loading of laying hens (Lohmann Brown) is about twice that of junglefowl (Sandilands et al., 2009; Moinard et al., 2004a). In other words, due to a higher body weight related to the wing area there is a twofold force effect per unit area on the lifting surface, i.e. the wings, in the domestic chicken compared to the junglefowl. However, it seems that even junglefowl prefer to move on the ground by walking instead of flying. Walking is the predominant type of locomotion used for foraging and to change location in the junglefowl (Dawkins, 1989; Collias and Collias, 1967) as well as in the domestic chicken (Campbell et al., 2016; Klein et al., 2000). Like most galliforms, chickens spend only a little time in the air, rarely fly long distances, and will fly mainly to escape from predators (Drovetski, 1996). Dial (2003) classified galliforms as 'cursorial ground birds that do fly' (Dial, 2003, p. 941). Thus, there might be no need to offer elevated structures in the housings of modern layer hybrids. At least in enclosed cage-free barns the birds do not have to protect themselves from predators by accessing elevated structures, food is delivered at a sufficient amount at the ground of the barn, and also all other resources such as water or a nest site can be reached by walking, i.e. by the preferred type of locomotion of chickens. Moreover, the use of elevated structures poses a serious risk for the chickens' health and welfare such as for keel bone damages. Including elevated structures also results in higher costs for the equipment of hen houses and more building space is needed for such housing systems. In the following sections, we therefore will address whether roosting and daytime elevated resting are important for chickens. We also will review recent literature on the effects of different types of elevated structures on usage and on the welfare of chickens to derive suggestions for practical implementation.

Chickens have undergone intensive selection for productivity traits, starting from the end of the 19th century and the beginning of the 20th century (Nicol, 2015). This selection is continuing in two directions: on the one side to increase growth rates, i.e. muscle mass, and on the other side to increase reproduction, i.e. egg yield. This has resulted in distinct genetic lines for meat production, i.e. broiler chickens, and in genetic lines for egg production, i.e. layers. Broiler chickens are slaughtered already at a young age between 4 and 6 weeks of life when they reach slaughter weight. Thus, they do not reach maturity and under natural conditions they still would be guarded by their dam (Wood-Gush and Duncan, 1976, see above). This might be the reason for the scarce amount of research on the use of elevated structures in broiler chickens compared to the large amount of research on perching in layers. However, in this chapter we will address both perching in layers and in broiler chickens.

We use the term 'perching' to summarize behaviours shown by chickens on any type of elevated structure at both night and daytime. Most studies investigated perching in chickens by offering rods of different materials and shapes that have a diameter chickens can grasp with their feet rather than ledges, platforms or grids. Thus, in the following sections, we will use the term 'perch' for elevated structures that can be grasped by chickens with their feet. However, whether grasping is a crucial feature for perching will be addressed in a later section.

## 2 Why do chickens perch?

### 2.1 Perching amounts and behaviours

As already mentioned in the introduction of this chapter domestic chickens use elevated structures at both night and daytime (Wood-Gush and Duncan, 1976). Under commercial housing conditions such as in aviary systems, laying hens use perches and elevated structures at night to an approximately fourfold higher frequency than during daytime (Brendler and Schrader, 2016). Comparable differences in perching between night and daytime have been observed in floor housing systems (Giersberg et al., 2019; Campbell et al., 2016; Wichman et al., 2007; Odén et al., 2002; Blokhuis, 1984) and in enriched cages (Louton et al., 2016; Struelens et al., 2008; Appleby et al., 1992; Duncan et al., 1992; Tauson, 1984). Depending on the type of aviary system, the use of perching structures at daytime may become even lower (Steenfeldt and Nielsen, 2015). In enriched cages the use of perches is affected by the distance between the perches and the cage roof and is lowered during both the day and the night if this distance is too small. Perches are used only if there is a distance of at least 19–24 cm between a perch and the cage roof (Struelens et al., 2008). In contrast to adult laying hens, young chickens of both a layer line and a dual-purpose line (Fig. 2) seem to use elevated structures more frequently during daytime than at night. In addition, during daytime they start to use elevated structures at younger ages compared to nighttime (Malchow et al., 2019a; Malchow et al., 2018). The behaviours shown while perching may reflect possible differences in the motivation to perch at day and nighttime.

Accessing perches or elevated structures for night is clearly triggered by a fall in light levels (Kent et al., 1997). Collias and Collias (1967, p. 370) observed: 'They started for their roost at sunset, and it was remarkable to observe the speed and steady rate with which they began to walk, covering some 300 yards within 1.5 minutes and travelling across a branch of a rao [= wadi tree] to an island where they went to roost about 20 feet up in a thorny tree (Zizyphus) less than half an hour before dark'. The predominant behaviour chickens show at night on perches is resting (Blokhuis, 1984). Resting at night is performed in different body postures and positions such as standing or sitting (Pickel et al., 2010;

**Figure 2** Chickens of two genetic lines: Lohmann Tradition (left) is a commercial hybrid line selected for high egg production. Lohmann Dual is a dual-purpose line with moderate egg production of the hens and moderate growth rate of the male chickens. Due to morphological differences (e.g. length of legs, muscle mass in relation the skeleton) the ability of birds to walk, to fly and to access elevated structures is likely to differ.

Blokhuis, 1984). The predominant resting position is sitting but both the body posture, e.g. standing or sitting with the head forward or tucked backwards into the feathers, and the position seems to be related to thermoregulatory needs (Pickel et al., 2011a; Blokhuis, 1984). Besides resting, laying hens show other behaviours such as preening and other comfort behaviours while perching, but such behaviours are observed only to a minor frequency at night (Pickel et al., 2010).

Also during daytime, chickens use perches for resting. However, in contrast to nighttime, other behaviours such as preening, standing, locomotion, and exploration are shown on perches more often than resting (Struelens et al., 2008; Newberry et al., 2001; Braastad, 1990; Blokhuis, 1984). For example, Newberry et al. (2001) observed in 3-6-week-old pullets 32% preening on perches. This decreased to 22% with increasing age. Junglefowl even seem to prefer other sites than perches for resting during daytime (Blokhuis, 1984).

In particular, when kept in aviary systems, laying hens also use perches to change between tiers and to reach food and water facilities (Giersberg et al., 2019; Campbell et al., 2016).

## 2.2 Perching as anti-predator behaviour

As already suggested in the tale 'The Bremen Town Musicians', seeking for shelter against predators is the main underlying motivation for the use of elevated structures in fowl. In the wild, fowls are chased by both aerial and ground predators such as cats, civets, jackals and hawks, eagles, or owls (Collias and Collias, 1967). They react against predators by seeking for shelter on trees, where at the same time they often are covered by leafy branches (Collias and Collias, 1967; Johnson, 1963).

Almost certainly, resting on high places developed during evolution of fowl because this offers shelter against predators. This behavioural preference can still be observed in the domestic chicken. Although they may not be at direct risk of predation when kept in enclosed housings, they still seem to perceive that they are better protected against potential predators when resting on higher or more inaccessible elevated structures. This assumption is supported by findings on the use of perches differing in height and/or covering.

At nighttime laying hens clearly prefer higher perches for roosting than lower ones in commercial aviary systems in which the observed height of lower perches was below 100 cm and that of higher perches was between 200 cm and 260 cm (Giersberg et al., 2019; Brendler and Schrader, 2016; Campbell et al., 2016; Steenfeldt and Nielsen, 2015; Odén et al., 2002). Moreover, if perches are offered at different heights within the tiers of an aviary, hens prefer the highest perches compared to the lowest perches within tiers (Brendler and Schrader, 2016). This can even result in an overcrowding of high perches while space on lower perches remains available (Ali et al., 2019; Brendler and Schrader, 2016). However, laying hens of specific breeds such as dual-purpose hens have been observed to prefer lower perches for nighttime roosting but this seems to result from difficulties of these short-legged birds to reach the higher perches (Giersberg et al., 2019). In addition, at least in aviary systems with a complex three-dimensional structure there also seem to be strain differences in the height preferences for roosting (Ali et al., 2019).

During daytime, however, hens do not prefer the high tiers in multi-tier systems. Instead, within the low tiers the low perches and within the high tier both the lowest and the highest perches are preferred (Brendler and Schrader, 2016). Depending on the design of the aviary systems low perches may be used more often because they provide access to feeders (Giersberg et al., 2019) or they are used to move within the tiers. In experimental studies, preferences for perch heights at daytime are less clear. Wichman et al. (2007) also observed more laying hens (week 18 of life) on low- and medium-high perches during daytime than on high perches. In contrast, Newberry et al. (2001) found that pullets (5–42-days old) preferred the highest perches. However, in addition to the age of chickens the height of the respective highest perch differed in these experimental studies (130 cm in Wichman et al. (2007) and 60 cm in Newberry et al. (2001)).

In enriched cages with perches offered at different heights, hens also prefer high perches over low ones for roosting at nighttime (Struelens et al., 2008). At daytime, however, the lower perches generally are preferred (Roenchen et al., 2010), but height preferences differ for different behaviours. On lower perches, standing, walking and preening are most often shown, but on higher perches, sitting, sleeping and comfort behaviours (Chen et al., 2014; Struelens et al., 2008). However, in enriched cages the use of perches differing in height is strongly affected by the distance between the perches and the cage roof (Struelens et al., 2008), i.e. whether the space on the highest perch is sufficient and how easily the high perches can be reached. On the other hand, if offered with sufficient space, the roof may constitute a cover for chickens and, thus, represents a further possibility for shelter against predators (Fig. 3).

Pullets of domestic fowl (Brown Leghorn) prefer covered areas when performing vulnerable behaviours such as resting and preening (Newberry and Shackleton, 1997). When performing vulnerable behaviours, domestic fowl also reduce distances between each other, which will reduce the risk of an individual becoming a victim (Keeling and Duncan, 1991). In that study adult birds of an old breed (bantam) showed smaller nearest-neighbour distances compared to birds of a hybrid laying strain (Isa Brown). Thus, hybrid chickens still show certain anti-predator behaviours although the degree or intensity may be lowered for some of the anti-predator behaviours.

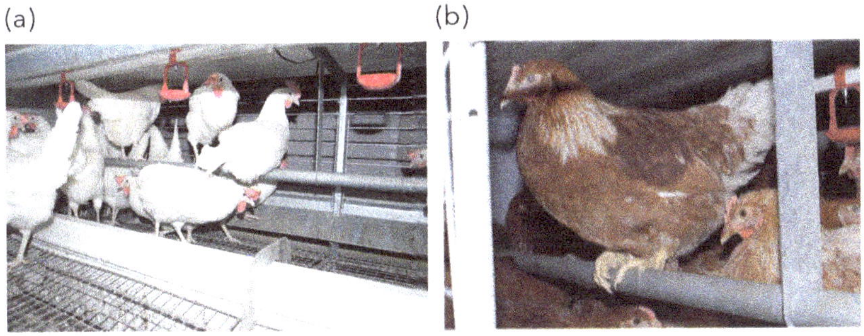

Figure 3 (a) and (b) In enriched cages the distance between perches and the cage roof should be more than 24 cm to allow access. If perches are low, hens on the floor can easily reach the vent of perching hens and, thus, possibly increasing the risk for feather pecking.

So far, the preference of chickens to use high perches in particular while nighttime roosting supports the idea that seeking shelter against predators is the most important motivation for perching in both the ancestral and the domesticated fowl.

If seeking for shelter against predators is the most important motivation for perching, perching will also covary with other behaviours related to anti-predator behaviour such as vigilance behaviour and fearfulness. An individual's predation risk tends to decline with increasing group size (Roberts, 1996). In accordance with this general rule, Newberry et al. (2001) tested the hypothesis that during daytime with increasing group size pullets (White Leghorn) will spend less time on perches, will be less vigilant while perching, and will spend more time on the floor when engaged in vulnerable behaviours such as preening. In compliance with their hypotheses, the proportion of birds that rested on perches decreased from 41% in groups of 15 birds to 33% in groups of 120 birds. In addition, the proportion of pullets being vigilant on the high perches decreased and the proportion of birds preening on the floor increased with increasing group size. They concluded that, despite being protected from predators in enclosed housings, domesticated chickens continue to show anti-predator behaviour. Similar results were obtained by Chen et al. (2014), who offered groups of one to eight laying hens (Hyline Brown) perches at heights of 10–40 cm but in cages of 65 cm height. Usage of perches ranged from 87% in single housed hens to 68% in groups of eight hens. Moreover, hens in smaller groups prefered higher perches (20 cm) and hens in groups of eight birds preferred the lowest perches (10 cm). In contrast, at daytime Guo et al. (2012) found a higher usage of perches in groups of 48 birds compared to groups of 21 hens kept in furnished cages, but perch use was less than 10% in both group sizes.

Based on the hypothesis that perching is an anti-predator behaviour it also can be expected that the reaction towards alerting stimuli will be lower if chickens rest on high compared to low perches. In line with this assumption, laying hens while perching on low perches during daytime reacted with shorter latencies towards a model predator compared to hens on a higher perch (Keeling, 1997). Brendler et al. (2014) tested the reaction of laying hens (Lohmann Selected Leghorn) towards an acoustic disturbance during nighttime. They predicted that hens would show a longer latency to react towards an acoustic disturbance if they roosted on a high compared to a low perch. However, reaction latencies towards the disturbance given in the middle of the night ranged between 0.71 sec and 1.2 sec and did not depend on perch height. One possible explanation for this result is that perch height may be crucial only when hens are searching for an appropriate roosting place (in another trial of this study hens clearly preferred high perches for roosting), but may have limited influence on alertness while roosting. However, on the low perch more hens roosted with their head forward, i.e. with the eyes not covered by feathers, than on high perches. In addition, hens roosting in this head position reacted more slowly (0.87 s) to the acoustic stimulus than hens roosting with their head covered by a wing (1.2 s). The initial reaction of hens roosting with their head forward might have been to open their eyes. This reaction could not be identified from video recordings used in this study. Thus, roosting with head forward may have improved the perception of hens and possibly may indicate a higher level of alertness of hens roosting on low perches.

Perching behaviour itself, i.e. accessing a location that provides a good overview of possible predators, may increase the risk of easy detection by a predator. However, this risk probably is lowered by fowl simultanously seeking covered areas when perching.

So far, there are only very few studies directly addressing the effect of perching on the reaction towards aversive stimuli such as predators. In addition, results are difficult to interpret due to methodological issues. Further evidence that perching is an anti-predator behaviour in fowl may derive from studies on the effect of perches on fearfulness as described in the next section.

## 2.3 Effects of perches on fearfulness

If perching in fowl serves to provide shelter against predators it can be assumed that access to perches will reduce their fearfulness. Indeed, commercial free-range laying hens showed a lower flight distance from the observer and less resistance to being handled if their housing was equipped with (compared to without) aerial perches (Donaldson and O'Connell, 2012). In addition, in the housing with aerial perches hens tended to range farther on the free-range compared to hens from housing without perches. Brantsæter et al. (2016) raised pullets up to an age of 16 weeks either in an aviary system or in cages. At 19 and 23 weeks of age, a human approach and a novel object test was conducted. At both ages, the hens raised in the aviary system showed lower levels of fearfulness than hens reared in cages, suggesting that increased environmental complexity during rearing reduces fearfulness of adult laying hens.

A well-established test for fear of predators in fowl is the tonic immobility test (Gallup et al., 1971). The test bird is restraint in a lateral or supine position. The duration of tonic immobility, i.e. the latency until the test birds will rise from this position positively correlates with its fearfulness. Brake et al. (1994) used the tonic immobility test in broiler breeder pullets and observed shorter durations of tonic immobility, i.e. a lower fear response, when the pullets were raised with (compared to without) perches. Likewise, broiler chickens housed with elevated platforms were less fearful (tested by tonic immobility test) than chickens kept without elevated platforms (Tahamtani et al., 2018). However, Campo et al. (2005) did not find differences in the duration of tonic immobility in adult hens (Spanish ornamental breeds) kept in pens with or without perches.

In the already mentioned study of Brendler et al. (2014), more hens rested with their head forward on low perches compared to high perches. Resting with head forward may improve the perception while roosting and thus indicates a higher alertness or fearfulness of these hens.

An indirect link between fearfulness and perching was found by Roden and Wechsler (1998). They raised chickens (Lohman Selected Leghorn) with or without a hen and observed the pullets in enriched cages equipped with perches. Chickens raised with hens spent less time on the perches compared to chickens raised without a hen. The authors concluded that the increased perch use of chickens raised without a hen may indicate that these chickens felt less secure on the floor; i.e. they may have used the perches more often to increase their feeling of savety.

Adult hens (Spanish ornamental breeds) kept in pens with perches showed a lower heterophil to lymphocyte ratio, indicating that these hens were less stressed than hens kept in pens without perches (Campo et al. 2005).

## 2.4 Perching as a social escape behaviour

Chickens also use perches to escape from or to avoid agonistic interactions with group members. In commercial free-range laying hens, access to aerial perches reduced the level of aggression between hens (Donaldson and O'Connell, 2012). Likewise, in an experimental approach Cordiner and Savory (2001) observed in adult laying hens (ISA Brown) that the presence of perches reduced the number of aggressive pecks. During daytime, lower-ranking birds tended to use perches more often than higher-ranking hens, but at night higher-ranking hens tended to use the perches a little more more often. However, in particular during nighttime, perch usage was comparatively low (18.4% of time spend on perches) in this study. The authors suggested that at daytime provision of perches reduces bird density on the floor, where most social interactions occur. Perching at daytime, thus, may allow lower-ranking hens to avoid agonistic interactions with higher-ranking hens.

Besides agonistic interactions, access to perches can also affect feather pecking.

## 2.5 Measuring motivation to perch

So far, we have presented studies showing that chickens will perch to a considerable amount if perches are offered and that they use perches for different behaviours. Perch use and behaviour while perching differs between day and nighttime. The intensive use of perches already suggests strong underlying motivations that may differ between times of day. However, both during daytime and at night perches seem to be sought by chickens for shelter against possible predators. During daytime, perching additionally may enable subdominant chickens to escape from and to avoid agonistic interactions with dominant conspecifics. Thus, the presence of perches seems to increase the feeling of safety in chickens and, consequently, chickens housed with access to perches show lower signs of fearfulness.

There are also some studies addressing the strength of motivation to perch more directly.

Bubier (1996) did a cost/no-cost multi-choice test with pairs of laying hens (Ross Brown) during daytime. In this test hens had free access (no cost) to small boxes with (1) food and water, (2) woodchips, (3) grass, (4) a perch at a height of 32 cm, (5) a nestbox, and (6) a pen facing another pen with familiar hens. In the cost multi-choice test, they had to squeeze through two dowels in order to access a pen. In both the cost and the no-cost test, the pen with the perch was visited less often and for a shorter total time compared to all other resources tested. This result indicates a rather low motivation for perching during daytime. Based on the anti-predator hypothesis it could have been expected that the test hens would use the perch more often and for a longer duration as small groups of two birds should increase perching and vigilant behaviour. However, the box with grass was most attractive to the test hens in the no-cost condition (on average, hens spent 37 min out of 120 min in the grass box) and still quite attractive in the cost condition (17 min), probably because hens did not have access to grass outside the test arena. This most likely increased their motivation for foraging behaviour resulting in a decrease of resting, i.e. perching, behaviour. Moreover, agonistic behaviour in pairs of hens probably is lower compared to larger groups, and, thus, this may have lowered usage of the perches in order to escape from agonistic interactions. Alternatively, the perches may have been less attractive to the hens due to the low height of 32 cm.

During nighttime, chickens' motivation to perch seems to be notably higher. Olsson and Keeling (2000) prevented groups of three laying hens (Lohmann Selected Leghorn) from accessing perches at nighttime either by removing perches or by covering perches with plexiglass. In that situation hens showed increased signs of agitation such as moving and attempts to take off, which was interpreted as signs of frustration. The same authors also did a demand experiment in which laying hens (Lohmann Selected Leghorn) were trained to push a weighted door in order to gain access to a perch (Olsson and Keeling, 2002). The resistance of the push-door varied between 25% and 100% of each bird's maximum capacity. When tested individually, the hens worked harder; i.e. they overcame doors with heavier weights for access to the perch (75% of individual capacity in median) compared for access to a sham perch that they could not use for perching (0% of individual capacity in median). In a second experiment, a companion hen was roosting either on the perch or on the floor behind the push-door. In this experiment, only four of eight test hens in either of the two treatments opened the push door during the observation time one hour after lights-off. However, the four hens that did not push through the door during observation time were found on the perch the next morning and three of them also opened the door at least at 25% of individual capacity. As a possible explanation for this difference to the results of the first experiment the authors argued that the respective test hens may have not pushed the door in order to avoid aggressive interactions with the companion bird. From their results the authors concluded that hens with prior experience of perches have a strong motivation for nighttime perching. However, so far it is not known whether also chickens without prior experience to perches are motivated to perch at night. In addition, it remains open whether the motivation to roost on a perch at night also is present in the absence of the possibility to perch, i.e. if there is no cue that may trigger perching motivation.

Altogether, the effects of perch height on the usage of perches, on the chickens' behaviour and affective states, and the results on the motivation to perch indicate that perching and, in particular, roosting at night are a behavioural priority for chickens even if chickens are not confronted with any predators in indoor housing systems (Weeks and Nicol, 2006).

## 3 Ontogeny of perching

Perching behaviour and the usage of elevated structures of chickens develop within the first weeks after hatch. The junglefowl and feral domestic chickens lay their eggs and brood in nests built on the ground (McBride et al., 1969; Johnson, 1963). During the first days of life chickens will stay on the ground, at or near the original nest site during the night (McBride et al., 1969; Kent et al., 1997). At an age of 6–14 weeks, chickens will be guided by their hens to roost into trees at night (Wood-Gush and Duncan, 1976; McBride et al., 1969).

Under artificial rearing conditions, i.e. in the absence of a hen or other adult conspecifics, nighttime roosting on perches or elevated structures starts from the third week of life in domestic chickens (Malchow et al., 2018; Riber et al., 2007; Heikkilä et al., 2006). The preference for high roosting sites also develops in the first weeks of life and can differ between different genetic strains. Interestingly, layer chickens seem to start later to roost on higher sites compared to chickens of faster-growing strains (Malchow et al., 2019a; Malchow et al., 2018).

During daytime, chickens will start to use perches beginning from the first week of life (Malchow et al., 2019a; Malchow et al., 2018; Kozak et al., 2016; Wichman et al., 2007; Heikkilä et al., 2006; Newberry et al., 2001). In the presence of their hens, chickens will start some days earlier to use perches during daytime (Riber et al., 2007). During the first weeks of life, chickens prefer low perching sites (Kozak et al., 2016; Gunnarsson et al., 2000) and perch use increases with increasing age. Also daytime-perching chickens of different strains may differ in their use of elevated perching sites (Malchow et al., 2018; Kozak et al., 2016). According to the different development in the usage of elevated structures at night and daytime, during the first weeks of life, growing chickens show a higher usage of elevated structures during the day than at night (Malchow et al., 2019a; Malchow et al., 2018; Norring et al., 2016; Heikkilä et al., 2006).

There are various morphological reasons that influence the use of the third dimension over time in growing chickens (Kozak et al., 2016). After hatching, the chickens' tail and feathers are still developing, and their centre of mass is different from that of adults, hampering balance while climbing in young chickens. Chickens are not able to fly before seven to nine days of life due to feather development (Provine et al., 1984). In addition, the morphology of their toes makes climbing difficult. Although hatchlings are precocial, even walking is less efficient during the first few days after hatching compared to an age of 14 days (Muir et al., 1996).

The development of perception also seems to affect the use of three-dimensional structures. Workman and Andrew (1989) investigated the development of visual lateralization of chickens by recording which eye the chickens were fixating on an observer. On day 10 of life, when chickens started to move independently from their hens, they showed a peak in perching at daytime that coincided with a shift from the right to the left eye, i.e. to the right hemisphere that is dominantly used for spatial processing. After these first days of life chickens will explore the third dimension, and exploration of novel objects peaked in the fifth week of life (Newberry, 1999).

A strong effect of age or development stage on spatial navigation and physical ability in young layer chickens was also found by Norman et al. (2019). Interestingly, also cognition seems to affect the spatial skills of chickens. Gunnarsson et al. (2000) raised chickens for the first 8 weeks of life either with or without perches. At 15 weeks of life, all birds used perches for roosting at night, although chickens raised without perches had to be trained by lifting them onto the perches before nighttime. At 16 weeks of life all chickens were tested to reach food offered at platforms at different heights whereby the highest platforms were above eye-level to the birds and could only be reached by firstly jumping on a lower platform. Both groups of chicken were able to reach the food at the low platforms, indicating that there was no difference in the physical ability of chickens raised with and without perches. However, chickens raised without perches had problems reaching the high platforms. The authors concluded that, in addition to physical training, cognitive training is needed in order to navigate in complex three-dimensional environments.

So far, we have seen that the use of the third dimension depends on the development of the chickens' morphological, perceptive and cognitive abilities. Likewise, the complexity of the chickens' surroundings affects their spatial skills, their working memory (Tahamtani et al., 2015), and their ability to move within the environment (Colson et al., 2008). Thus, the usage of a three-dimensional environment is strongly affected by the spatial complexity of the rearing environment (Brantsæter et al., 2016), and it is strongly recommended to raise chickens in an environment that is comparable in its spatial complexity to that in which the birds will be housed as laying hens (Kozak et al., 2016).

However, at least during early ontogeny Norman et al. (2019) only found minor effects on spatial navigation and physical ability assessed by a detour, a jump and a rotated floor test when they reared layer chickens for first 28–29 days of life with or without elevated structures. They concluded that during early ontogeny, age and development have stronger effects on spatial navigation and physical ability of chickens that may be mediated by experience of a three-dimensional environment.

With respect to the anti-predator hypothesis, suggesting that fowl feel safer on elevated sites (Newberry et al., 2001), it seems surprising that hens brood in ground nests, because this makes the eggs and the chickens vulnerable to predation from ground predators. However, we already have seen that usage of elevated structures is still developing through the first days of life. Thus, elevated nest sites on trees may increase the chickens' risk for injuries due to falling. Interestingly, laying hens of different breeds still seem to prefer ground nests for egg laying but prefer elevated nest sites if they are familiar to them (Krause and Schrader, 2018). In that study, laying hens of four breeds differing in phylogenetic origin (white vs. brown layers) and in productivity (300 vs 200 eggs per year) were offered three nests at 0 cm, 39 cm, 78 cm, and 117 cm above the floor, respectively. During tests of one week, only two hens (brown and white) were kept in a compartment, respectively. Before start of tests, for at least 35 weeks, hens were kept in larger groups in identical compartments but with only one nest on top of a manure pit at a height of 75 cm above ground. Hens of three breeds preferred the ground-level nests for laying and hens of one breed showed an equal preference for the nests at ground level and at 78 cm, i.e. the nest height comparable to that in their home compartments before tests. The results of the first experiment indicate that in a new environment the preference for ground nests is still preserved in the domestic chicken. However, results of the second experiment suggest that chickens also can learn a certain nest site that they will prefer in long-term, irrespective of whether this may increase the risk for potential hatchlings by falling down.

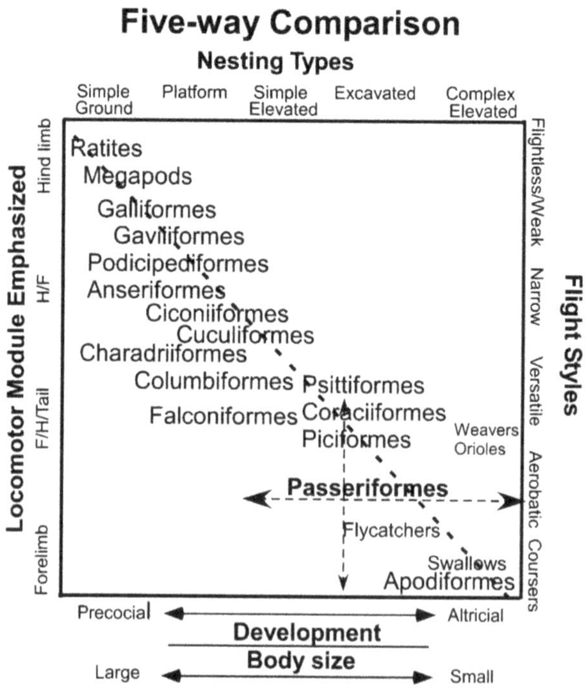


**Figure 4** Concept of Dial (2003) on the relationship nesting types, flight styles, body size, development of hatchlings, and locomotor module emphasized in avian taxa. Chickens belong to the taxa Galliformes and are characterized by having simple ground nests, a weak flight style, a large body size, precocial hatchlings, and a locomotion dominated by the hind limbs.

Dial (2003) suggested a general relationship between nesting strategy, flight ability, body size, developmental precocial-altricial trajectories, and the origin of flight in avian species. Based on these relationships precocial birds such as galliformes species most often have simple ground nests, stronger hindlimbs compared to wings, a relatively large body size and only weak flight abilities (Fig. 4).

## 4 Anatomic prerequisites for perching

So far, we have seen that perching, i.e. the use of elevated structures, is an innate behaviour of chickens serving different functions during night and daytime. Chickens are strongly motivated to perch and they start perching within the first weeks of life. In the wild, chickens most often will perch on the branches of trees, as these are the most common elevated resting structures in their habitat.

In order to understand the perching behaviour of fowl and its basic mechanisms it is helpful to look at the morphology of the feet and the toes. Basically, avian taxa including chickens are counted as theropods (Middleton, 2001), i.e. having four anisodactyl toes (three toes directed forward and one toe directed backwards) and moving bipedally. Due to the arrangement of the four toes, their shape and the resulting functions, the theropods are classified into three major groups among the avian taxa based on biological functions such as locomotion, feeding and perching (Sustaita et al., 2013). The structure of the chickens' toes is 'less curved and relatively dorsoventrally deep' (Birn-Jeffery et al., 2012). Thus, chickens are counted as 'ground-dwellers' (Birn-Jeffery et al., 2012; Pike and Maitland, 2004). In contrast, avian groups from the feeding or perching category have far more curved claws than do ground-dwellers (Feduccia, 1993). Consequently, this anatomy of toes in chickens reflects that they mainly live on the terrestrial ground and to a limited extent in arboreal areas. However, almost all birds have the special feature of enclosing objects with their feet by the special arrangement of digit I, which is located opposite the other three digits II–IV (Middleton, 2001). This allows chickens to grasp objects such as branches.

Moreover, it has been suggested that birds using branches for perching have the ability to passively grasp branches, i.e. that there are certain anatomic adaptations enabling birds to stay in their normal resting position with a minimum of energetic effort. However, there still is a debate on whether grasping can be performed passively or by active synergies of nerves, muscles and tendons.

Grasping is explained by the avian automatic perching mechanism (APM), which is composed of the automatic digital flexor mechanism (ADFM) and the automatic tendon-locking mechanism (ATLM) (Galton and Shepherd, 2012). The prerequisite for these mechanisms is the flexor tendon on the rear side of the heel towards the toe bone. Bending of the knee and of the heel (flexion of the intertarsal leg joint, ADFM) will lead to a tension of the tendon, which in turn attracts the toes. By that, the toes passively will grasp around a branch (ATLM). Thus, due to the low muscle occupancy of the hind limbs and supported by the tendon-locking mechanism, birds are assumed to be able to perch with few muscle expenditure and minimal physical effort.

However, birds do not only grasp while resting on branches. Grasping also serves other biologic functions such as scratching, wading, prey-grasping, tree-climbing and other (Quinn and Baumel, 1990). These functions of grasping presuppose that grasping is actively performed (Backus et al., 2015). Galton and Shepherd (2012) investigated the automatic perching mechanism in anaesthetised starlings. In this state of unconsciousness, the starlings did not show ADFM or ATLM and, consequently, no grasping was seen (Galton and Shepherd, 2012).

Irrespective of whether grasping is an active or passive mechanism, in chickens these complex mechanisms seem to result in a relaxed position of the birds during roosting, and, for example, chickens do not have to correct the foot position over long periods during roosting (Quinn and Baumel, 1983 in Galton and Shepherd, 2012). However, as stated above chickens are mainly active on the ground and they walk while foraging and to change from one place to another. The anatomy of the chickens' feet is adapted to this predominant use.

## 5 Properties of perches: the chicken's view

### 5.1 What is a perch?

A question of immense practical significance is if the welfare of chickens is affected by whether elevated structures can be grasped or not. Most often it is implied that perches - just like branches of trees - should have a diameter and a cross-section enabling the chickens to grasp with their feet around the perch. Moreover, also in many languages 'perch' is translated to 'rod' or 'pole', implying that a perch always has a more or less round shape and a relatively small diameter such as a branch of a tree. However, in a wider sense, 'perch' also means 'any place or object, as a sill, fence, branch, or twig, for a bird, animal, or person to alight or rest upon' (www.dictionary.com) or to put it even more simply: a perch is a secure high seat/position (www.langenscheidt.com). Thus, there may be a need to define what a perch is from the view of chickens.

If there is a special motivation for grasping while perching it can be hypothezised that resting chickens will strongly prefer a small perch that can be clasped compared to any other surface. However, Faure and Jones (1982a,b) offered different resting sites in a small number of groups of naive

15-week-old chickens of both sexes and of different strains. Heights of resting sites ranged from 5 cm to 46 cm, and observations were done during daytime. Chickens did not show a preference for perches they could grasp and a wire meshed resting site was at least equally attractive. Newberry and Shackleton (1997) tested in 3–8-week-old chickens of two layer strains (one experimental and one commercial) the effect of a cover for concealment, whereby the cover varied from a clear cover (0% cover), a different number of 10-cm-wide stripes (33% and 67% cover) to a solid (100%) cover. During daytime, the birds not only used the area under the cover but also rested on the cover except from the clear cover. Interestingly, the chickens tended to use the solid cover even more often for resting compared to the two covers with the 'perch-like' stripes. Schrader and Müller (2009) directly tested whether laying hens of two strains (Lohmann Brown, Lohmann Selected Leghorn) preferred perches compared to a grid (mesh size of 2.2 cm × 3.9 cm, thickness of the grids 0.8 cm) for nighttime roosting. In addition, both resting sites were offered at different heights (15 cm and 60 cm). When offered at the same height, hens preferred the perches over the grids. However, when perches were offered at the low, and grids at the elevated height, the hens clearly preferred the grids for nighttime roosting.

Ali et al. (2019) observed nighttime roosting of laying hens of two white and two brown strains in an aviary system equipped with perches, non-perforated metal ledges and wire floors. A higher proportion of hens of the white strains roosted on the metal ledges compared to the hens of brown strains. In contrast, brown hens used the perches more often than the white hens. However, in both strains a high proportion of hens used the wire floors even if there still was space on the perches and ledges. Regarding the height, hens of the two white strains preferred the highest tier of the aviary whereas hens of the two brown strains were more evenly distributed between the tiers for nighttime roosting. These results suggest that at least for the hens of the white strains height seemed to be more important than the shape of the resting site but preferences for resting sites at nighttime might differ between strains.

Malchow et al. (2018) offered growing chickens of a layer strain, a dual-purpose strain, and a broiler strain mushroom-shaped perches and grids in parallel. Chickens of all strains preferred the grids compared to the perches for resting at daytime and night, except for chickens of the layer strain where the difference between use of grids and perches did not reach significance during nighttime.

In a preliminary study, Schrader et al. (2016) offered laying hens either a metal perch or a metal grid (both at 90 cm). After one week, access to the resting site was denied before dusk by Perspex walls. There was no difference in the usage of resting sites. However, when access was denied hens showed more attempts to reach the grid compared to the perch.

Taken together, these results show no clear preference for resting sites which enable chickens to grasp (i.e. perches) compared to non-graspable resting sites (i.e. grid or non-perforated sledges) while nighttime roosting. Thus, so far there is no indication that chickens are highly motivated to grasp while perching although this might differ between certain strains. However, other features of the resting site such as its height seem to be more important from the chickens' view.

## 5.2 How do chickens perceive height?

We have already discussed that chickens prefer to rest on high sites and that the height of a resting site even seems to be more important for a chicken compared to other features. However, we do not know how chickens perceive the height of their resting site. Intuitively one would say that height is the distance between the resting site and the ground floor or the distance between the resting site and the next visible floor below. But height also could be perceived related to the energy cost necessary to conquer a vertical distance. Resting on a high site even may not relate to the height, but to the distance between the resting site and a protective cover such as the canopy of a tree. For example, Newberry and Shackleton (1997) found that domestic fowl showed more vulnerable behaviours such as resting and preening in covered areas. Already, Johnson (1963) observed that junglefowl roosting in clumps of bamboo are well screened above and below. This may suggest that it is not the height, but being hidden such as by the canopy of a tree or by other vegetation that is the crucial feature that enables birds to feel safe. However, if the distance to a cover is most important it can be expected that in multi-tier systems laying hens will equally distribute between the different tiers as in these systems each tier is covered by a manure belt except for the highest tier. However, this is not the case. Instead, in commercial multi-tier hen housings most hens will rest on the highest tier at nighttime (Brendler and Schrader, 2016; Campbell et al., 2016; Steenfeldt and Nielsen, 2015; Odén et al., 2002). Interestingly, if perches of different heights are offered within tiers laying hens prefer the respective highest perch (Brendler and Schrader, 2016). Thus, the distance between the ground floor and the resting site seems to play the crucial role for resting site preferences although other factors may contribute. In a complex aviary system in which the floor cannot be seen by the hens they nevertheless preferred the highest resting sites (hens of white strains), suggesting that hens most likely perceive the height of resting site by the total distance from ground floor (Ali et al., 2019). Growing chickens of strains differing in growth rate also preferred to roost on the highest resting site when resting sites were offered at a height of 10 cm, 30 cm and 50 cm above floor (Malchow et al., 2018).

Even if we know that chickens prefer high resting sites it remains an open question which height will be high enough to satisfy this behavioural preference. Brendler et al. (2014) offered laying hens (Lohmann Selected Leghorn) perches of different height ranging from 20 cm to 180 cm for one week, respectively, in a random order. Perch usage increased up to a perch height of 90 cm and then remained stable, indicating that a perch should be at least 90 cm high in order to be perceived as high from the chickens' view.

Although the question of how chickens perceive height may appear to be quite academic and very difficult to scientifically investigate, it is noteworthy to address the question in order to design housing systems that meet the chickens' preferences as well as possible.

Another aspect regarding the height of elevated structures is that if these are offered too low or too close to each other, this enables other hens to feather peck as the victims easily can be reached while resting (Wechsler and Huber-Eicher, 1998).

So far, we have seen that perches or elevated structures are important elements that should be offered in order to fulfil their behavioural preferences for a resting site. Under commercial housing conditions, however, the arrangement and properties of perches or elevated structures must also be considered to enable appropriate use and to prevent problems (Struelens and Tuyttens, 2009). These aspects affect different welfare measures in layers and in broiler chickens that also differ in their behaviour, anatomical skills and length of housing period. In the following, we therefore will address these topics separately for layers and broiler chickens.

## 6 Elevated structures in pullet and layer housing

Enrichment of housing by elevated structures can affect welfare measures in pullets. Feather pecking and cannibalism can be major problems in layer housing systems. When pullets have early access (within the first 4 weeks of age) to elevated perches this can decrease the risk of feather pecking four-fold (Huber-Eicher and Audigé, 1999). Also the risk of cloacal cannibalism in the laying period can be decreased by access to perches in the rearing period (Gunnarsson et al., 1999). Pullets kept in an enriched and complex experimental environment including perches showed an improved pro-inflammatory response and coped better with cold stress than pullets from non-enriched environments without perches (Campderrich et al., 2019). When pullets are reared in cages, access to perches can increase their muscle deposition and bone mineralization in the laying period (Hester et al., 2013). Adult laying hens kept in enriched cages equipped with perches (whether cooled or not) can cope better with long periods of heat stress than hens in cages without perches (Hu et al., 2019a). The prevalence of foot pad lesions was lower in cages enriched with perches compared to cage without perches, and foot pad lesions negatively correlated with the time hens spent on perches (Appleby et al., 1992). However, in a study on Get-away cages including perches bumble foot and keel bone damages appeared only in Get-away cages but not in conventional cages (Tauson and Abrahamsson, 1996). It is also noteworthy that elevated structures such as perches significantly increase the risk of keel bone damages.

### 6.1 Arrangement of elevated structures

In order to enable layers to use high resting sites, which they particularly prefer for nighttime roosting, sufficient space should be offered, for example, at the upper tiers of an aviary system. However, this cannot be fully achieved in commercial housing conditions, as most often the stocking density will be too high to allow enough space for all hens at the upper tier. As a result, in commercial housing overcrowding is often observed at the upper tiers, whereas at the lower tiers roosting sites remain vacant (Ali et al., 2019; Brendler and Schrader, 2016; Campbell et al., 2016). Overcrowding may result in hens displaced by other hens from the upper tiers, in particular during dusk. As a result, displaced hens may fall from the upper tier, increasing the risk of keel bone damages (Stratmann et al., 2015a; see chapter 16).

Despite sufficient space for preferred resting sites, the arrangement of three-dimensional structures should also allow the layers to move safely between the different parts of the housing system (Harlander-Matauschek et al., 2015). As discussed in the introduction, chickens prefer to move by walking rather than by flying. Thus, provision of elements such as ramps (Stratmann et al., 2015a) can support the chickens to move through the different parts of the housing system in their preferred manner and additionally can reduce keel bone damages (Heerkens et al., 2016). At the same time, ramps may help laying hens affected by keel bone fractures to move between tiers as hens with fractures tend to reduce their movements between tiers (Rufener et al., 2019) and, thus may have limited access to certain resources. Moreover, ramps can support movement between tiers in hens suffering from poor foot health or poor plumage condition, which impairs their balancing behaviour (LeBlanc et al., 2016). LeBlanc et al. (2018) investigated the force needed by laying hens to climb upwards at different slopes. The steeper the slope (>40°) of a ramp the more force a hen had to invest to climbing up. Consequently, ramps with a slope >40° are less used compared to ramps with lower angles (Zheng et al., 2019). Ramps can be offered in the form of ladders, plastic grids or even wire grid. However, grids seemed to be more suitable for laying hens for ascending and descending between different tiers compared to ladders (Norman et al., 2018; Pettersson et al., 2017; Zheng et al., 2019).

With respect to the poor flying ability of chickens, the distances and the angles between the different parts of the housing system are crucial to consider. By jumping, chickens can conquer gaps of shorter horizontal distances easier than those of longer distances. In laying hens (ISA Brown) Scott and Parker (1994) found that hens move more easily between perches at horizontal distances of less than 1 m than at larger distances, a result confirmed subsequently in other studies (Taylor et al., 2003; Scott et al., 1997, 1999).

Moving upwards is easier for chickens than moving downwards (Moinard et al., 2004a,b). Laying hens (ISA Brown) needed longer to move upwards between perches at an angle of 60° compared to smaller angles (Scott et al., 1997). In this study, the time to move downwards and the number of movement failures increased with increasing angle. Lambe et al. (1997) found in laying hens (ISA Brown) that downward jumps were more difficult at an angle between perches of 45° and 60° compared to 30°. Moinard et al. (2004a) concluded that the problems of landing from greater distances results from the high wing load of hens and that the problems in landing after downward jumping result from the hens' difficulty in controlling their height in relation to approach of the landing perch. Taking these results together, the risk of poor landings, e.g. collisions, increases if chickens have to conquer distances between perches or other elevated structures that are greater than 1 m or have angles greater than 45°.

It is likely that the light intensity in a housing also will affect the hens' ability to navigate three-dimensional structures. Taylor et al. (2003) tested the ability of laying hens to jump at light intensities ranging from 0.8 lux to 40 lux and found that hens had a longer latency to jump between perches at low light intensities. However, Moinard et al. (2004b) did not find an effect of light intensity (5 lux, 10 lux or 20 lux) on the behaviour of hens trained to jump between perches.

It also should be mentioned that layers need sufficient space on the perches both for starting and for landing. A space of 15 cm seems to be enough although at this space the hens have to adapt their take-off and landing behaviour (Moinard et al., 2005).

With respect to the ontogeny of perching it is crucial to offer already pullets a housing condition that is as far as possible comparable to that they will be kept in as laying hens. Pullets can be assisted in discovering three-dimensional housings by ramps or ladders that they can use to easily move between tiers of the system.

Different strains can differ in their use of three-dimensional structures such as in aviary systems (Ali et al., 2019; Giersberg et al., 2019). This may result from strain specific differences in behaviour or in physical and anatomical differences such as the ability to jump or the length of their legs (Fig. 2). Thus, certain features of a housing system such as the distance between elevated structures should be adapted to the strains to achieve a good use of the system.

Although in general access to perches can reduce the risk of feather pecking and cannibalism in particular in enriched cages the risk may increase due to a poor arrangement and height of perches resulting from the relative narrow spatial conditions in this housing system (Sandilands et al., 2009; Struelens and Tuyttens, 2009). When laying hens had access to either low (45 cm) or higher (70 cm) perches, in the experimental pens with the higher perches, feather damage particularly at the vent was poorer in the pens with low compared to higher perches even if observed feather pecking interactions did not differ between treatments (Wechsler and Huber-Eicher, 1998). At low perches such as in enriched cages hens on the floor easily can reach the vent of perching hens and, thus, possibly increasing the risk for feather pecking. In addition, vent and cloacal cannibalism may increase in enriched cages when hens lay eggs from low perches (Moinard et al., 1998; Duncan et al., 1992) as the prolapsed cloaca while and after egg lay attracts other hens to peck.

Most studies on the design and material of elevated structures offered to laying hens have been done with rods, i.e. perches with a relatively small diameter in contrast to plain areas such as grids or non-perforated areas. The design of perches, i.e. the shape and width of perches, their material, colour and cleanliness will affect the usage of perches, the perching behaviour and, in addition, certain aspects of chickens' health.

## 6.2 Perch material

Most perches in layer housings are made of metal, plastic or wood. In an experimental choice test, laying hens did not show any preferences for perches made from wood, metal or plastic (Lambe and Scott, 1998). In contrast, in a study by Chen et al. (2014) laying hens kept in enriched cages preferred wooden perches compared to steel or plastic perches.

In a study comparing enriched cages equipped with perches of different materials (hardwood, textured metal, smooth plastic, softwood and padded vinyl) laying hens spent most time on softwood perches during daytime (Appleby et al., 1992). Perch material does not seem to affect the behaviour while perching (Chen et al., 2014; Pickel et al., 2010; Faure and Jones, 1982a). However, in the study by Pickel et al. (2010) laying hens showed fewer balance movements on rubber perches compared to perches of wood and steel. In addition, laying hens took longer to jump from plastic perches compared to wood and metal perches (Scott and MacAngus, 2004) and showed more safe landings on a round soft perch (experimental air cushion below a soft polyurethane surface) compared to a mushroom-shaped plastic and a round steel perch (Scholz et al., 2014). These findings on the effect of perch material on the behaviour of hens suggest that in particular the surface of a perch, which normally closely is linked to its material, is important, i.e. the slip resistance of the material (Struelens and Tuyttens, 2009). For example, wood or rubber are less slippery compared to steel or plastic and, thus, improve grip and footing stability of birds on the perches (Pickel et al., 2010; Appleby et al., 1992). In addition to the material, dirt can also reduce the slip resistance of perch surface resulting in difficulties to jump from perches (Scott and MacAngus, 2004).

The material of perches is also linked to the temperature of perches by the thermal conductivity of the respective material. Metal perches have a higher thermal conductivity compared to plastic or wooden perches. Materials with a higher thermal conductivity have a higher capacity for heat transfer and this may affect the thermal comfort of birds resting on perches of the respective material. Pickel et al. (2010) observed that under the same climatic conditions laying hens showed less standing and more resting with head tucked backwards into the feathers while perching on steel perches compared to wooden and rubber perches during nighttime roosting. They concluded that the hens' resting position reflected their thermoregulatory behaviour, i.e. reducing heat loss on metal perches. In another study these authors tested this conclusion by observing laying hens roosting on metal perches (steel pipes) which differed in temperature (Pickel et al., 2011a). Different temperatures were achieved by passing water through the perches with 15°C, 18°C (room temperature) and 28°C. The amount of roosting did not differ between temperatures but on the warmest perch, hens roosted more in a standing position with their heads forward. In contrast, on the cooler perches they roosted more often in a standing position with their head covered by feathers.

Cooled perches can also be used in order to control or to reduce the impact of heat stress episodes. In laying hens kept in furnished cages the effects of cooled (10°C), non-cooled and no perches on a variety of measures were investigated. When hens were subjected to short term heat stress (33°C for 4 h) cooled perches resulted in reduced thermoregulatory behaviours (panting, wing spreading, drinking; Hu et al., 2016) and in increased H/L ratios, usually taken as an indicator of raised physiological stress (Strong et al., 2015), but egg production, other immunological, health and stress measures were not affected (Hu et al., 2016; Strong et al., 2015). When exposed to long-term heat stress (35°C at daytime and 28°C during night for several weeks), hens kept with cooled perches showed a lower rectal temperature and, in part, a lower H/L ratio (Hu et al., 2019a). In addition, production measures of hens with cooled perches were improved (Hu et al., 2019b).

These results suggest that perch material can affect the thermoregulation of laying hens due to differences in thermal conductivity. Moreover, cooling of perches seems to be a possibility to reduce negative effects of heat stress on both welfare and productivity in laying hens.

Keel bone damages seem to be related to the hardness of perch material (Kappeli et al., 2011). Hard material such as metal may increase the risk of trauma resulting from falls and collisions in particular if combined with slippery surfaces, impairing transitions between perches. Consequently, in aviary systems keel bone damages could be reduced by covering perches with a soft and comparable thick material (here: polyurethane; Stratmann et al., 2015b). In Get-away cages, however, a soft rubber cover (here: 4 mm rubber layer) did not reduce keel bone damages (Tauson and Abrahamsson, 1996).

In particular, keel bone deviations may also result from high mechanical pressure loading while perching. Pickel et al. (2011b) measured the peak force and contact area of laying hens while sitting and standing on different perches. While sitting the peak forces on the keel bone were about five times higher than on a single foot pad, suggesting that most of the load of body weight is on the keel bone while perching in a sitting position. This high pressure load is likely to contribute to keel bone deviations and also may pose the risk of keel bone fractures. A soft rubber perch (experimental air cushion below a soft polyurethane surface) significantly reduced peak force and increased contact area of keel bone on the perch.

It is also likely that perch material affects foot pad lesions in laying hens. In Get-away cages foot pad lesions (bumble foot) were observed more often when cages were equipped with plastic perches compared to wooden perches (hardwood) but not in all genetic strains of laying hens used in this study (Tauson and Abrahamsson, 1996). Beside the material of perches, ramps (between perches and nest boxes) also seem to effectively reduce the risk of foot pad lesions (Heerkens et al., 2016).

The material of perches often is confounded with their colour. In particular, at low light intensities (<1 lux) bright (e.g. white) perches can be seen better by chickens compared to dark perches (e.g. wooden or black), resulting in shorter latencies to jump on these perches (Taylor et al., 2003). In enriched cages, however, no preference of laying hens for a particular perch colour (white, black or brown) for nighttime resting was found (Chen and Bao, 2012).

### *6.3 Perch shape*

Perches can have different shapes, ranging from round, oval, rectangular, square to mushroom-shaped cross-section. In addition, the wire mesh of tiers is offered for perching.

When presented at the same height (15 cm or 60 cm) laying hens (Lohmann Brown, Lohmann Selected Leghorn) preferred round wooden perches compared to plastic grids for nighttime roosting in an experimental pen. However, if the grid was offered at 60 cm and the perches at 15 cm the hens preferred the grids for roosting indicating that the height of roosting site is more important compared to the shape (Schrader and Müller, 2009). In another experimental study laying hens preferred rectangular (3.5 × 3.5 cm) compared to round (3.5 cm) wooden perches at night but not during daytime (Chen et al., 2014).

In some studies the usage of differently shaped perches was recorded. In cages, laying hens tended to use rectangular (5.0 × 2.2 cm) compared to round (3.5 cm) wooden perches more often during daytime (Duncan et al., 1992). In aviary systems, no difference in the usage between mushroom-shaped perches and oval perches was found (Tauson and Abrahamsson, 1996). In an experimental study laying hens used a wooden rectangular perch (5 × 2 cm) with a height of 45 cm less often during daytime than platforms differing in height (5 cm or 46 cm) and surface (wire mesh or hardboard) (Faure and Jones, 1982a).

Comfort behaviours were observed more often on rectangular (3.5 × 3.5 cm) compared to round (3.5 cm) wooden perches but other behaviours did not differ between perch shapes (Chen et al., 2014). Laying hens showed more safe landings on a round soft perch (experimental air cushion below a soft polyurethane surface), followed by a mushroom-shaped plastic and less often on a round steel perch, but in this study perch shape was confounded with perch material and, thus, the slip resistance of perches (Scholz et al., 2014).

With respect to health measures, Duncan et al. (1992) found fewer foot pad lesions in enriched cages equipped with rectangular (5.0 × 2.2 cm) compared to round (3.5 cm) wooden perches.

Tauson and Abrahamsson (1996) investigated the effect of a variety of perches on the prevalence of bumble foot and keel bone damages. Perches differed in shapes and material: rectangular perches (5.3 × 3.3 cm, hardwood), round perches (3.6 cm, softwood) and a wire mesh perch (10 cm wide). They simultaneously offered these perches in the following combinations: rectangular/rectangular, rectangular/wire, round/rectangular and round/wire. The incidence of bumble foot was highest in the rectangular/rectangular setting and lowest in the round/wire setting. Keel bone lesions were worst in the round/rectangular setting compared with the other combinations. In another experiment of this study, wire mesh perches (10 cm wide) and, in addition, either mushroom-shaped perches (4.8 × 6.8 cm), narrow perches of hardwood (3.5 × 5.0 cm), rectangular perches of hardwood (5.3 × 3.3 cm), or oval perches made of beech wood were offered. Bumble foot scores were poorer in get-away cages with mushroom-shaped perches than in those with oval and narrow perches.

In a further experiment, either the mushroom perches or the oval perches were installed at the top tier of aviary systems. Bumble foot scores were better with the oval perches than with the plastic mushroom perches.

Pickel et al. (2011b) measured the peak force and the contact area on the keel bone and foot pads of hens resting on round, square and oval perches.

With increasing flatness of perch shape (i.e. from round over square to oval shaped perches) the contact area under both the keel bone and the foot pads of sitting hens increased and the peak force decreased.

Taken together, these results on effects of perch shape on behavioural and health measures are inconsistent. However, there are only a few studies dealing with perch shape. In addition, studies differ in their methods and settings and often the perch shape is confounded with other factors. With respect to the risk of keel bone damages, perches or perching sites with a flat surface allowing a large contact area seem to be beneficial.

### 6.4 Perch width

In an experimental study, laying hens preferred smaller (3 x 3 cm) compared to wider (5 x 5 cm) rectangular perches for nighttime roosting, but perch use did not differ during daytime (Chen et al., 2014). In enriched cages, Struelens et al. (2009) provided laying hens with access to perches of widths ranging from 1.5 cm to 10.5 cm (in 1.5 cm steps). During the night, hens did not show a preference for a certain perch width. During the day, hens spent more time on the wider perches (9.0 cm and 10.5 cm) when perch width was arranged stepwise but not when perches of different widths were arranged in parallel.

Pickel et al. (2010) offered laying hens round wooden perches of different diameters (2.7 cm, 3.4 cm or 4.5 cm) in an experimental setting. During nighttime, perch usage, resting and comfort behaviours performed on the perches did not differ between the perch widths. However, balance movements were lowest on the perch with the largest width (4.5 cm). Although perch use was high the width of rectangular perches (3.8 cm, 5.0 cm or 6.0 cm) did not affect their usage by laying hens in enriched cages during daytime (Appleby, 1995).

In sitting laying hens, the contact area under the keel bone was larger and the peak force on the foot pads was lowest on the widest (6.0 cm) compared to smaller (3.4 cm and 4.4 cm) perches (Pickel et al., 2011b). The peak force on keel bones, however, was not affected by perch width in this study. But on the widest perches the contact area under the food pads was larger while hens were standing on the perches.

Perch width does not seem to affect the behaviour of laying hens in a consistent manner. However, wider perches seem to provide a more stable footing for hens and reduce pressure load on the foot pads, both possibly improving the hens' comfort while perching.

## 7 Elevated structures in broiler housing

Although pullets and accordingly also broiler chickens use elevated structures from a young age both during day and nighttime (see Section 3) such structures are hardly ever offered in commercial broiler housings. However, there are several reasons to include elevated structures even in broiler housings. Elevated structures may improve the welfare of broiler chickens by reduction of health problems and by enabling perching behaviour (Riber et al., 2018). Moreover, additional space on elevated structures will increase the space per bird without reduction of stocking density per area of the ground floor.

In general, access to elevated structures has the potential to improve the foot pad health (Zhao et al., 2012; Ventura et al., 2010) of broiler chickens. Foot pad lesions result from the contact to the litter which often is moisty due to faeces and/or malfunctioning drinkers (Ekstrand et al., 1998). By using perches or other elevated structures, contact time with the litter can be reduced and foot pads can dry. However, access to perches may also negatively affect health aspects. Nielsen (2004) found an increased prevalence of severe breast blisters in broiler chickens with access to perches but concluded that genetic strain and sex of chickens seem to have a larger effect on this problem.

Due to their fast growing resulting from a high metabolic rate broiler chickens are at high risk of suffering from high ambient temperatures. Cooled perches can support thermoregulation of broiler chickens (Estevez et al., 2002). At high ambient temperatures cooled perches were used more often compared to non-cooled perches up to an ambient temperature of 28°C (LeVan et al., 2000). At high temperatures, however, heavy broilers in particular become less active and may reduce the usage of elevated structures counteracting this possible positive effect.

Elevated structures such as grids can also have a positive effect on the walking ability of slow- (Malchow et al., 2019a) and fast-growing broiler chickens (Kaukonen et al., 2016). In addition, access to elevated structures may positively affect locomotor activity of broiler chickens by a certain training effect. However, Norring et al. (2016) did not find an effect of elevated structures (plastic platforms, wooden perches) on the general activity of fast growing broiler chickens on a commercial farm. In an experimental study, the locomotor activity of moderate and fast-growing but not of slow-growing chickens tended to be higher in pens with (compared to pens without) elevated structures (Malchow et al., 2019a). 'Barrier perches' (10 cm above litter) installed in experimental pens reduced aggression and disturbances of resting broiler chickens by others compared to control pens without perches (Ventura et al., 2012).

The impaired walking ability of fast-growing broiler chickens generally results in lower usage of elevated structures compared to slower-growing chickens (Malchow et al., 2019a; Malchow et al., 2018; Yngvesson et al., 2016; Oester et al., 2005; Bokkers and Koene, 2003).

This clearly indicates that access to elevated structures should be designed differently for fast- and slower-growing chickens.

As addressed in Section 3 chickens will start exploring and using low structures within the first weeks of life. However, already at this age their growth rate plays an important role regarding their usage of elevated structures (Riber et al., 2018). Fast-growing chickens show a higher level of activity (Malchow et al., 2019a; Malchow et al., 2018) and poorer walking abilities (Malchow et al., 2019c), and are more restricted in their movement due to their rapid muscle growth and skeletal disorder (Bessei, 2006) compared to slow growing chickens. As a result, the body balance of fast-growing broiler chickens is less central and directed to cranio-ventral (Duggan et al., 2017) and vertical back movement (Caplen et al., 2012). Furthermore, fast-growing chickens show a more oscillating walking movement compared to slower-growing chickens (Corr et al., 2003). Due to these impairments of motor abilities, it is likely that especially fast-growing chickens will have problems using perches with a small diameter (Riber et al., 2018), and that platforms or grids will be more appropriate. Indeed, perches are used to a very low extent by fast-growing broiler chickens when offered in commercial housings (Norring et al., 2016). In contrast, grids are used to a much higher proportion (Norring et al., 2016; Oester et al., 2005) and, thus, are more suitable for fast-growing chickens. Independent from growth rate chickens preferred grids compared to perches during daytime (Fig. 5). At night, however, slow growing chickens did not show a preference for either of the elevated structures in this study (Malchow et al., 2018).

Bales of straw are also used as elevated structures and as additional enrichment or bedding material after chickens dispersed the bales (Baxter et al., 2017; Bailie et al., 2013; Riber et al., 2018). In general, elevated structures should enable a stable footing, i.e. should offer sufficient grip.

The effect of height on the usage of elevated structures was investigated by Malchow et al. (2018). In experimental pens they offered chickens of three different breeds (slow, moderate and fast growing, Fig. 6) elevated structures at 10 cm, 30 cm, and 50 cm height. Even the fast growing chickens preferred the highest level of 50 cm at the dark period in the last weeks of life. In this study, a ramp gave the chickens' access to the elevated structures of each height.

In fast-growing chickens, perches can be reached best if they are offered at a low height (Pettit-Riley and Estevez, 2001; LeVan et al., 2000). In order to reach higher elevated structures fast-growing broiler chickens need assistance by ramps or other devices such as ladders especially in the last weeks of their life. A ramp or a ladder should not be too steep and not too long. Da Silva and de Jong (2019) gave an overview on studies in which fast-growing broiler chickens were given access to grids or perches by ramps. The height of the elevated structures in the studies reported varied between 9 cm and 66 cm and the inclination angles of ramps between 10° and 35°. Most of the ramps had an angle below 25° and ramps of and below this angle were assessed as very suitable for broiler chickens. In addition to its slope, the width of a ramp is

**Figure 5** Experimental setup of a study by Malchow et al. (2018) in which they investigated the usage of plastic grids and perches at different heights. The picture shows that even fast growing broiler chickens (Ross 308) are able to access elevated structures at an age of 29 days. A ramp (between the perches and grids) help the chickens accessing the structures.

**Figure 6** Male chickens of three different genetic lines as tested in a study by Malchow et al. (2018): Lohmann Brown Plus (layer line), Lohmann Dual (dual-purpose line), and Ross 308 (fast-growing broiler line) (from left to right). The three chickens had the same age of 42 days.

crucial to consider. A ramp should be wide enough to enable the birds to pass each other and to avoid possible agonistic interactions.

Regarding the design of elevated structures, it is crucial to estimate the space that should be offered on elevated structures. This space depends on the number of birds that will use the elevated structures and the size of birds. Both will vary with the respective strain, i.e. with the growth rate of chickens.

As already indicated for layer pullets, the dimension measured by planimetric methods varies between strains and increases with age (Spindler et al., 2013). While sitting, i.e. the most frequent behaviour on elevated structures, chickens need more space than while standing. For example, at a weight of 2.5 kg broiler chickens are covering an area of about 400 cm² while sitting and about 370 cm² while standing (Giersberg et al., 2016). Interestingly, due to longer fattening periods (and higher weights at slaughter) for slow-growing chickens, at the end of fattening they may cover a larger area compared to fast-growing broiler chickens (Malchow et al., 2019b). Thus, for slow-growing broiler chickens larger areas on elevated structures are necessary than for fast-growing broiler chickens due to a higher usage of elevated structures and a greater body size at slaughtering.

Concerning an implementation of elevated structures in commercial housings, the limited space has to be considered. Such devices can be installed only between the drinking and feeding lines. Perches are easy to install and inspection and control of chickens by the farmer is easy because chickens cannot hide under the small perches. In contrast, chickens can hide under elevated grids or platforms depending on their width and, consequently, be overlooked by the farmers. Grids of a height of 50 cm are well used by all broiler strains. At this height it is also possible for farmers to inspect the birds.

## 8 Conclusions

This chapter reveals that perching, i.e. the usage of elevated structures, is a highly prioritized and innate behaviour of all domestic chickens. The usage of elevated structures differs significantly between night and daytime and chickens perform approximately four times more perching at night compared to daytime. During night chickens rest for most of the time on elevated structures while during daytime behaviours such as preening, standing, locomotion, and exploration are also shown while perching. These differences may reflect different motivations to perch during day and nighttime. However, seeking for shelter against predators seems to be the main underlying motivation for the use of elevated structures in chickens during both night and daytime, and vulnerable behaviours are often shown more while perching. Whereas at night the highest resting sites are clearly preferred, during daytime lower perches are also used, possibly because in commercial housings low perches often provide access to feeders and drinkers or are used to move within tiers.

Studies directly addressing the motivation indicate a high motivation to perch at night but a rather low motivation to perch at daytime. It remains an open question whether prior experience and the presence of elevated structures are necessary to trigger the motivation to perch. Studies directly addressing the anti-predator hypothesis, i.e. testing the effect of elevated structures on fearfulness are scarce and results are difficult to interpret, but some studies suggest that access to elevated structures or a higher environmental complexity reduce fearfulness. In addition, access to elevated structures can reduce aggression between laying hens as lower ranking hens can avoid interactions with higher-ranking hens.

Perching behaviour develops within the first weeks after hatch. The ontogeny of perching behaviour depends on the development of the chickens' morphological, perceptive and cognitive abilities, and is mediated and facilitated by experience of a three-dimensional environment.

Whether there is a separate motivation of chickens to grasp while perching is an open question, which needs clarification as it is of immense practical significance with respect to the design of perching opportunities offered in chicken housings. Based on the anatomy of their feet chickens are 'ground-dwellers' reflecting their original terrestrial and to limited extent arboreal habitat. Indeed, walking is the predominant type of locomotion in the junglefowl and the domestic chicken has an even poorer ability to fly. Consequently, the feet of chickens are adapted to moving and standing on the ground. Thus, a flat surface could be suitable or even better for perching than perches that can be grasped by their feet. So far, studies do not show clear preferences for resting sites which enable chickens to grasp (i.e. perches) compared to non-graspable resting sites (i.e. grids or non-perforated ledges) although this might differ between strains. Furthermore, the height of a resting site at night is more important than whether the surface of the resting site enables grasping or not. This question should be addressed further in future studies, as the results will have enormous impact on the equipment of housing systems. In addition, the risk for keel bone damages may be lower on elevated grids compared to graspable perches.

Another crucial question for the design of housing is how chickens perceive height. Here, studies suggest that the distance between the ground floor and the resting site seems to play the crucial role for resting site preferences although other factors may contribute.

With respect to commercial housings for layers there is a huge conflict between the strong preference for perching and the high risk for keel bone damages, both significantly related to the welfare of birds.

To meet the chickens' preference for nighttime roosting, sites should be offered high and with sufficient space for nearly all birds, e.g. at the upper tiers of an aviary system. But, due to limited space or high density of birds this is not feasible commercial housing conditions. Often, the preferred upper resting sites are overcrowded resulting in hens displaced by other hens and falls in particular during dusk, increasing the risk for keel bone damages.

Because layers have a relatively poor ability to fly the arrangement of three-dimensional structures should allow them to safely move between the different parts of the housing system. This can be supported by ramps if their slope is less than 45° and if they offer a good grip such as grids. In addition, distances between structures should not exceed 1 m to avoid poor landings and collisions. The surface of perching sites should have sufficient slip resistance to ensure safe standing, locomotion and landing. A soft surface additionally is likely to reduce the risk for keel bone fractures resulting from falls and collisions and may reduce the risk of keel bone deformations by reducing mechanical pressure loading while perching. Pressure loading on the keel bone can also be reduced by a larger contact area between the keel bone and the resting area.

As a conclusion, it is worthwhile to reconsider the design of perching areas for layers that could reduce the risk of keel bone damage. Elevated, flat or perforated areas (e.g. grids with sufficient slat width) might be better than perches with a small diameter. Such areas may offer more space, may allow safer standing, locomotion and grids especially may allow safer landing and they may reduce pressure loading on the keel bone while sitting. Although such areas would not permit grasping, this would be unlikely to greatly reduce welfare. However, the suitability and the design of such resting areas must be validated in future research.

Irrespective of considerations on the furnishing of housings, knowledge about the ontogeny of the usage of three-dimensional structures by chickens strongly suggests that chickens should be raised in housings comparable in their structure and spatial complexity to that in which the birds will be housed as laying hens. This will improve their skills to move through complex three-dimensional systems resulting in a decreasing risk for falls and collisions and, thus, may lower the risk for keel bone damages. Likewise, rearing conditions will improve the birds' acceptance of resources such as troughs, drinkers and nests as adult hens.

Different genetic lines of layers can differ in their use of three-dimensional structures such as in aviary systems. This may result from genetic differences in behaviour or differences in physical and anatomical abilities to jump or to fly. These differences appear even more important if alternative layer lines or dual-purpose lines are used. Thus, it might be necessary to adapt the respective housing systems to the hens of certain genetic lines.

Regarding commercial housings for broiler chickens there is increasing evidence that growing chickens show a substantial usage of elevated structures during both night and daytime from just a few days of age. Fast-growing broiler chickens use elevated structures less than slower-growing chickens due to their impaired ability to walk and to balance.

Probably for this reason, fast-growing broiler chickens prefer elevated areas such as grids or platforms for perching compared to perches with a small diameter. But slow-growing chickens also seem to prefer elevated areas for perching during daytime, whereas at night, they use elevated areas and perches to a similar extent. Thus, for both fast- and slower-growing chickens platforms or grids seem to be suitable to enable elevated perching. In addition to their positive effects on the chickens' behaviour, elevated structures may reduce health problems such as foot pad lesions and walking ability. At high ambient temperatures, grids might have a further advantage, offering the chickens a possibility to rest outside the insulating litter, improving ventilation and increasing the dissipation of metabolic heat. However, more research is needed to validate possible effects on the chickens' health.

At a height of about 50 cm, caretakers are able to inspect birds under the elevated platform or grid. Due to their impaired walking ability, fast-growing chickens in particular should be provided with ramps (of angle < 25°) to access elevated structures, and wide enough to enable birds to pass.

If the elevated structures are offered in addition to the ground area this will reduce the stocking density on the ground floor. The size of area that should be offered on the elevated structures is still an open question. Ideally, access should be provided for all birds that are motivated to perch. The area required additionally depends on the number and size of birds that will use the elevated structures. Both will vary with the respective strain, i.e. with the growth rate of chickens, their age and weight at slaughter. The main constraint is that elevated structures can generally be installed only between the drinking and feeding lines.

## 9 Future trends in research

Measures to improve animal welfare are most likely to be implemented in practice if they also have an advantage for the farmer, including economic benefits. These benefits can result from higher production but also from improved animal health, reduced mortality, higher resilience against stressful conditions, or improvements for management.

Elevated structures potentially may contribute to such benefits, in particular for the health of broiler chickens. In future, additional active cooling devices included in the elevated structures may further improve heat dissipation in broiler chickens susceptible to heat stress. In experimental studies, this already has been proven for both broiler chickens and laying hens. Elevated structures may also be integrated in concepts of Precision Livestock Farming, enabling farmers to supplement their control of animals by automatic devices. For example, it may be possible to record the usage of elevated structures based on the weight of birds. These data can give the farmers continuous information on the growth of their birds. Moreover, significant changes in the usage of elevated structures, detected by suitable algorithms, can indicate upcoming health problems. This would allow early detection and interventions. In layer housings it might become possible for perches to integrate technical measures against or detectors for early warning of infestation by the red mite. Thus, combining elevated structures with sensors and other functional properties may lead to an improvement of both the birds' welfare and the farmers' benefit. Such 'functional perching techniques' may become a promising field of research in the future.

However, as stated above there are other fields of research related to perching behaviour of chickens with urgent need for future research: Further knowledge is needed on the motivation and external cues triggering perching behaviour in chickens - in particular, whether there is a motivation to grasp while perching and whether perching areas that do not allow grasping have detrimental effects on the behaviour and health of chickens. For broiler chickens, to date we do not have sufficient information on the area of elevated structures that should be offered. These researches will have significant impacts on the furnishing of future housing systems for both layers and broiler chickens. In addition, it will become more important to take into account that different genetic lines of both layers and broiler chickens can differ in perching behaviour and demands for a housing environment ensuring their welfare.

## 10 References

Ali, A. B. A., Campbell, D. L. M., Karcher, D. M. and Siegford, J. M. (2019). Nighttime roosting substrate type and height among 4 strains of laying hens in an aviary system. *Poult. Sci.* 98(5), 1935-1946. http://doi.org/10.3382/ps/pey574.

Appleby, M. C. (1995). Perch length in cages for medium hybrid laying hens. *Br. Poult. Sci.* 39(1), 186-190. https://doi.org/10.1080/00071669508417749.

Appleby, M. C., Smith, S. F. and Hughes, B. O. (1992). Individual perching behaviour of laying hens and its effects in cages. *Br. Poult. Sci.* 33(2), 227-238.

Backus, S. B., Sustaita, D., Odhner, L. U. and Dollar, A. M. (2015). Mechanical analysis of avian feet: multiarticular muscles in grasping and perching. *J. R. Soc. Open Sci.* 2(2), 140350. https://doi.org/10.1098/rsos.140350.

Bailie, C. L., Ball, M. E. E. and O'Connell, N. E. (2013). Influence of the provision of natural light and straw bales on activity levels and leg health in commercial broiler chickens. *Animal* 7(4), 618-626. https://doi.org/10.1017/S1751731112002108.

Baxter, M., Bailie, C. L. and O'Connell, N. E. (2017). Evaluation of a dustbathing substrate and straw bales as environmental enrichments in commercial broiler housing. *Appl. Anim. Behav. Sci.* 200, 75-85. https://doi.org/10.1016/j.applanim.2017.11.010.

Bessei, W. (2006). Welfare of broilers: a review. *World Poult. Sci. J.* 62(3), 455-466. https://doi.org/10.1017/S0043933906001085.

Birn-Jeffery, A. V., Miller, C. E., Naish, D., Rayfield, E. J. and Hone, D. W. E. (2012). Pedal claw curvature in birds, lizards and Mesozoic dinosaurs-complicated categories and compensating for mass-specific and phylogenetic control. *PLoS ONE* 7(12), e50555. https://doi.org/10.1371/journal.pone.0050555.

Blokhuis, H. J. (1984). Rest in poultry. *Appl. Anim. Behav. Sci.* 12(3), 289-303.

Bokkers, E. A. M. and Koene, P. (2003). Behaviour of fast- and slow growing broilers to 12 weeks of age and the physical consequences. *Appl. Anim. Behav. Sci.* 81(1), 59-72.

Braastad, B. O. (1990). Effects on behaviour and plumage of a key-stimuli floor and a perch in triple cages for laying hens. *Appl. Anim. Behav. Sci.* 27(1-2), 127-139. https://doi.org/10.1016/0168-1591(90)90012-3.

Brake, J., Keeley, T. P. and Jones, R. B. (1994). Effect of age and presence of perches during rearing on tonic immobility fear reactions of broiler breeder pullets. *Poult. Sci.* 73(9), 1470–1474.

Brantsæter, M., Nordgreen, J., Rodenburg, T. B., Tahamtani, F. M., Popova, A. and Janczak, A. M. (2016). Exposure to increased environmental complexity during rearing reduces fearfulness and increases use of three-dimensional space in laying hens (Gallus gallus domesticus). *Front. Vet. Sci.* 3, 14. https://doi.org/10.3389/fvets.2016.00014.

Brendler, C., Kipper, S. and Schrader, L. (2014). Vigilance and roosting behaviour of laying hens on different perch heights. *Appl. Anim. Behav. Sci.* 157, 93–99. https://doi.org/10.1016/j.applanim.2014.06.004.

Brendler, C. and Schrader, L. (2016). Perch use by laying hens in aviary systems. *Appl. Anim. Behav. Sci.* 182, 9–14. https://doi.org/10.1016/j.applanim.2016.06.002.

Brüder Grimm (1819). Die Bremer Stadtmusikanten. In: *Kinder- und Haus-Märchen Band 1. Große Ausgabe, 2. Auflage*, Reimer, G. and Berlin, S., 141–145.

Bubier, N. E. (1996). The behavioural priorities of laying hens: the effect of cost/no cost multi-choice tests on time budgets. *Behav. Process.* 37(2–3), 225–238. https://doi.org/10.1016/0376-6357(96)00019-8.

Campbell, D. L. M., Makagon, M. M., Swanson, J. C. and Siegford, J. M. (2016). Laying hen movement in a commercial aviary: enclosure to floor and back again. *Poult. Sci.* 95(1), 176–187.

Campderrich, I., Nazar, F. N., Wichman, A., Marin, R. H., Estevez, I. and Keeling, L. J. (2019). Environmental complexity: a buffer against stress in the domestic chick. *PLoS ONE* 14(1), e0210270. https://doi.org/10.1371/journal.pone.0210270.

Campo, J. L., Gil, M. G., Davila, S. G. and Munoz, I. (2005). Influence of perches and footpad dermatitis on tonic immobility and heterophil to lymphocyte ratio of chickens. *Poult. Sci.* 84(7), 1004–1009.

Caplen, G., Hothersall, B., Murrell, J. C., Nicol, C. J., Waterman-Pearson, A. E., Weeks, C. A. and Colborne, G. R. (2012). Kinematic analysis quantifies gait abnormalities associated with lameness in broiler chickens and identifies evolutionary gait differences. *PLoS ONE* 7(7), e40800. https://doi.org/10.1371/journal.pone.0040800.

Chen, D. H. and Bao, J. (2012). General behaviors and perching behaviors of laying hens in cages with different colored perches. *Asian-Austral. J. Anim. Sci.* 25(5), 717–724.

Chen, D., Bao, J., Meng, F. and Wei, C. (2014). Choice of perch characteristics by laying hens in cages with different group size and perching behaviours. *Appl. Anim. Behav. Sci.* 150, 37–43. https://doi.org/10.1016/j.applanim.2013.10.007.

Collias, N. E. and Collias, E. C. (1967). A field study of red jungle fowl in North-Central India. *Condor* 69(4), 360–386.

Colson, S., Arnould, C. and Michel, V. (2008). Influence of rearing conditions of pullets on space use and performance of hens placed in aviaries at the beginning of the laying period. *Appl. Anim. Behav. Sci.* 111(3–4), 286–300. https://doi.org/10.1016/j.applanim.2007.06.012.

Cordiner, L. S. and Savory, C. J. (2001). Use of perches and nestboxes by laying hens in relation to social status, based on examination of consistency of ranking orders and frequency of interaction. *Appl. Anim. Behav. Sci.* 71(4), 305–317.

Corr, S. A., Gentle, M. J., McCorquodale, C. C. and Bennett, D. (2003). The effect of morphology on walking ability in the modern broiler: a gait analysis study. *Anim. Welf.* 12(2), 159–171.

da Silva, C. S. and de Jong, I. C. (2019). Literature update on effective environmental enrichment and light provision in broiler chickens. Wageningen Livestock Research, Report 1204. https://doi.org/10.18174/504630.

Dawkins, M. S. (1989). Time budgets in Red Junglefowl as a baseline for the assessment of welfare in domestic fowl. *Appl. Anim. Behav. Sci.* 24(1), 77-80.

Dial, K. P. (2003). Evolution of avian locomotion: correlates of flight, style, locomotor modules, nesting biology, body size, development, and the origin off lapping flight. *Auk* 120(4), 941-952.

Donaldson, C. J. and O'Connell, N. E. (2012). The influence of access to aerial perches on fearfulness, social behaviour and production parameters in free-range laying hens. *Appl. Anim. Behav. Sci.* 142(1-2), 51-60.

Drovetski, S. V. (1996). Influence of the trailing-edge notch on flight performance of galliforms. *Auk* 113(4), 802-810.

Duggan, B. M., Rae, A. M., Clements, D. N. and Hocking, P. M. (2017). Higher heritabilities for gait components than for overall gait scores may improve mobility in ducks. *Genet. Sel. Evol.* 49(1), 42. https://doi.org/10.1186/s12711-017-0317-2.

Duncan, E. T., Appleby, M. C. and Hughes, B. O. (1992). Effect of perches in laying cages on welfare and production of hens. *Br. Poult. Sci.* 33(1), 25-35. https://doi.org/10.1080/00071669208417441.

EFSA AHAW Panel (EFSA Panel on Animal Health and Animal Welfare) (2015). Scientific opinion on welfare aspects of the use of perches for laying hens. *EFSA J.* 13(6), 70pp. https://doi.org/10.2903/j.efsa.2015.4131.

Ekstrand, C., Carpenter, T. E., Andersson, I. and Algers, B. (1998). Prevalence and control of foot-pad dermatitis in broilers in Sweden. *Br. Poult. Sci.* 39(3), 318-324.

Estevez, I., Tablante, N., Pettit-Riley, R. L. and Carr, L. (2002). Use of cool perches by broiler chickens. *Poult. Sci.* 81(1), 62-69.

Faure, J. M. and Jones, R. B. (1982a). Effects of sex, strain and type of perch on perching behaviour in the domestic fowl. *Appl. Anim. Ethol.* 8(3), 281-293.

Faure, J. M. and Jones, R. B. (1982b). Effects of age, access and time of day on perching behaviour in the domestic fowl. *Appl. Anim. Ethol.* 8(4), 357-364.

Feduccia, A. (1993). Evidence from claw geometry indicating arboreal habits of Archaeopteryx. *Science* 259(5096), 790-793.

Gallup, G. G., Nash, R. F. and Ellison, A. L. (1971). Tonic immobility as a reaction to predation: artificial eyes as a fear stimulus for chickens. *Psychon. Sci.* 23(1), 79-80.

Galton, P. M. and Shepherd, J. D. (2012). Experimental analysis of perching in the European starling (Sturnus vulgaris: Passeriformes; Passeres), and the automatic perching mechanism of birds. *J. Exp. Zool. A Ecol. Genet. Physiol.* 317(4), 205-215.

Giersberg, M. F., Hartung, J., Kemper, N. and Spindler, B. (2016). Floor space covered by broiler chickens kept at stocking densities according to Council Directive 2007/43/EC. *Vet. Rec.* 179(5), 124-124.

Giersberg, M. F., Spindler, B. and Kemper, N. (2019). Linear space requirements and perch use of conventional layer hybrids and dual-purpose hens in an aviary system. *Front. Vet. Sci.* 6, 231. https://doi.org/10.3389/fvets.2019.00231.

Gunnarsson, S., Keeling, L. J. and Svedberg, J. (1999). Effect of rearing factors on the prevalence of floor eggs, cloacal cannibalism and feather pecking in commercial flocks of loose housed laying hens. *Br. Poult. Sci.* 40(1), 12-18.

Gunnarsson, S., Yngvesson, J., Keeling, L. J. and Forkman, B. (2000). Rearing without early access to perches impairs the spatial skills of laying hens. *Appl. Anim. Behav. Sci.* 67(3), 217-228. https://doi.org/10.1016/S0168-1591(99)00125-2.

Guo, Y. Y., Song, Z. G., Jiao, H. C., Song, Q. Q. and Lin, H. (2012). The effect of group size and stocking density on the welfare and performance of hens housed in furnished cages during summer. *Anim. Welf.* 21(1), 41-49.

Harlander-Matauschek, A., Rodenburg, T. B., Sandilands, V., Tobalske, B. W. and Toscano, M. J. (2015). Causes of keel bone damage and their solutions in laying hens. *Worlds. Poult. Sci. J.* 71(3), 461-472. https://doi.org/10.1017/S0043933915002135.

Heerkens, J. L. T., Delezie, E., Ampe, B., Rodenburg, T. B. and Tuyttens, F. A. M. (2016). Ramps and hybrid effects on keel bone and foot pad disorders in modified aviaries for laying hens. *Poult. Sci.* 95(11), 2479-2488. https://doi.org/10.3382/ps/pew157.

Heikkilä, M., Wichman, A., Gunnarsson, S., Keeling, L. J. and Valros, A. (2006). Development of perching behaviour in chicks–role of environmental enrichment. *Appl. Anim. Behav. Sci.* 99(1-2), 145-156.

Hester, P. Y., Enneking, S. A., Haley, B. K., Cheng, H. W., Einstein, M. E. and Rubin, D. A. (2013). The effect of perch availability during pullet rearing and egg laying on musculoskeletal health of caged White Leghorn hens. *Poult. Sci.* 92(8), 1972-1980.

Hu, J. Y., Hester, P. Y., Makagon, M. M., Vezzoli, G., Gates, R. S., Xiong, Y. J. and Cheng, H. W. (2016). Cooled perch effects on performance and well-being traits in caged White Leghorn hens. *Poult. Sci.* 95(12), 2737-2746. https://doi.org/10.3382/ps/pew248.

Hu, J. Y., Hester, P. Y., Makagon, M. M., Xiong, Y., Gates, R. S. and Cheng, H. W. (2019a). Effect of cooled perches on physiological parameters of caged White Leghorn hens exposed to cyclic heat. *Poult. Sci.* 98(6), 2317-2325. https://doi.org/10.3382/ps/pez012.

Hu, J. Y., Hester, P. Y., Makagon, M. M., Xiong, Y., Gates, R. S. and Cheng, H. W. (2019b). Effect of cooled perches on performance, plumage condition, and foot health of caged White Leghorn hens exposed to cyclic heat. *Poult. Sci.* 98(7), 2705-2718.

Huber Eicher, B. and Audige, L. (1999). Analysis of risk factors for the occurrence of feather pecking in laying hen growers. *Br. Poult. Sci.* 40(5), 599-604.

Johnson, R. A. (1963). Habitat preference and behavior of breeding jungle fowl in Central Western Thailand. *Wilson Bull.* 75, 270-272.

Kappeli, S., Gebhardt-Henrich, S. G., Frohlich, E., Pfulg, A., Schaublin, H. and Stoffel, M. H. (2011). Effects of housing, perches, genetics, and 25-hydroxycholecalciferol on keel bone deformities in laying hens. *Poult. Sci.* 90(8), 1637-1644.

Kaukonen, E., Norring, M. and Valros, A. (2016). Perches and elevated platforms in commercial broiler farms: use and effect on walking ability, incidence of tibial dyschondrplasia and bone mineral content. *Animal* 11(5), 864-871.

Keeling, L. J. (1997). A comparison of two basic characteristics of a perch for laying hens. In: Proceedings of the 31st International Congress of the International Society of Applied Ethology, Prague, 13-16.

Keeling, L. J. and Duncan, I. J. H. (1991). Social spacing in domestic fowl under seminatural conditions: the effect of behavioural activity and activity transitions. *Appl. Anim. Behav. Sci.* 32(2-3), 205-217.

Kent, J. P., McElligott, A. G. and Budgey, H. V. (1997). Ground-roosting in domestic fowl (Gallus gallus domesticus) in the Gambia: the anticipation of night. *Behav. Process.* 39(3), 271-278.

Klein, T., Teltner, E. and Huber-Eicher, B. (2000). Are genetic differences in foraging behaviour of laying hen chicks paralleled by hybrid-specific differences in feather pecking? *Appl. Anim. Behav. Sci.* 70(2), 143–155.

Kozak, M., Tobalske, B., Springthorpe, D., Szkotnicki, B. and Harlander-Matauschek, A. (2016). Development of physical activity levels in laying hens in three-dimensional aviaries. *Appl. Anim. Behav. Sci.* 185, 66–72.

Krause, E. T. and Schrader, L. (2018). 'High, low, or familiar? Laying hens' nest site preferences of experienced laying hens. *Brit. Poult. Sci.* 59(4), 359–364.

Lambe, N. R. and Scott, G. B. (1998). Perching behaviour and preferences for different perch designs among laying hens. *Anim. Welf.* 7(2), 203–216.

Lambe, N. R., Scott, G. B. and Hitchcock, D. (1997). Behaviour of laying hens negotiating perches at different heights. *Anim. Welf.* 6, 29–41.

LeBlanc, S., Tobalske, B., Quinton, M., Springthorpe, D., Szkotnicki, B., Wuerbel, H. and Harlander-Matauschek, A. (2016). Physical health problems and environmental challenges influence balancing behaviour in laying hens. *PLoS ONE* 11(4), e0153477. https://doi.org/10.1371/journal.pone.0153477.

LeBlanc, C., Tobalske, B., Szkotnicki, B. and Harlander-Matauschek, A. (2018). Locomotor Behaviour of chickens anticipating incline walking. *Front. Vet. Sci.* 4, 233. https://doi.org/10.3389/fvets.2017.00233.

LeVan, N. F., Estevez, I. and Stricklin, W. R. (2000). Use of horizontal and angled perches by broiler chickens. *Appl. Anim. Behav. Sci.* 65(4), 349–365.

Louton, H., Rauch, E., Reese, S., Erhard, M. and Bergmann, S. (2016). Effect of perch height and position on the usage in enriched colony housing systems for laying. *J. Vet. Sci. Technol.* 7(4), http://doi.org/10.4172/2157-7579.1000339.

Malchow, J., Berk, J., Puppe, B.and Schrader, L. (2018). Perches or grids? What do rearing chickens differing in growth performance prefer for roosting? *Poult. Sci.* 98(1), 29–38. http://doi.org/10.3382/ps/pey320.

Malchow, J., Puppe, B., Berk, J. and Schrader, L. (2019a). Effects of elevated grids on growing male chickens differing in growth performance. *Front. Vet. Sci.* 6, 203, http://doi.org/10.3389/fvets.2019.00203.

Malchow, J., Berk, J., Heindorff, S. and Schrader, L. (2019b). Space requirements for fast and slow growing chickens on elevated grids. In: 53rd Congress of the International Society for Applied Ethology, Norway, 343.

Malchow, J., Dudde, A., Berk, J., Krause, E. T., Sanders, O., Puppe, B. and Schrader, L. (2019c). Is the rotarod test an objective alternative to the gait score for evaluating walking ability in chickens? *Anim. Welf.* 28(3), 261–269. http://doi.org/10.7120/09627286.28.3.261.

McBride, G., Parer, I. P. and Foenander, F. (1969). The social organization and behaviour of the feral domestic fowl. *Anim. Behav. Monogr.* 2, 126–181.

Middleton, K. M. (2001). The morphological basis of hallucal orientation in extant birds. *J. Morphol.* 250(1), 51–60.

Moinard, C., Morisse, J. P. and Faure, J. M. (1998). Effect of cage area, cage height and perches on feather condition, bone breakage and mortality of laying hens. *Br. Poult. Sci.* 39(2), 198–202.

Moinard, C., Statham, P. and Green, P. R. (2004a). Control of landing flight by laying hens: implications for the design of extensive housing systems. *Br. Poult. Sci.* 45(5), 578–584.

Moinard, C., Statham, P., Haskell, M. J., McCorquodale, C., Jones, R. B. and Green, P. R. (2004b). Accuracy of laying hens in jumping upwards and downwards between perches in different light environments. *Appl. Anim. Behav. Sci.* 85(1-2), 77-92.

Moinard, C., Rutherford, K. M. D., Statham, P. and Green, P. R. (2005). Visual fixation of a landing perch by chickens. *Exp. Brain Res.* 162(2), 165-171.

Muir, G. D., Gosline, J. M. and Steeves, J. D. (1996). Ontogeny of bipedal locomotion: walking and running in the chick. *J. Physiol. (Lond.)* 493(2), 589-601.

Newberry, R. C. (1999). Exploratory behaviour of young domestic fowl. *Appl. Anim. Behav. Sci.* 63(4), 311-321.

Newberry, R. C., Estevez, I. and Keeling, L. J. (2001). Group size and perching behaviour in young domestic fowl. *Appl. Anim. Behav. Sci.* 73(2), 117-129.

Newberry, R. C. and Shackleton, D. M. (1997). Use of visual cover by domestic fowl: a venetian blind effect? *Anim. Behav.* 54(2), 387-395.

Nicol, C. J. (2015). *The Behavioural Biology of Chickens.* CABI Publishing, Boston, MA.

Nielsen, B. L. (2004). Breast blisters in groups of slow-growing broilers in relation to strain and the availability and use of perches. *Br. Poult. Sci.* 45(3), 306-315.

Norman, K. I., Weeks, C. A., Pettersson, I. C. and Nicol, C. J. (2018). The effect of experience of ramps at rear on the subsequent ability of layer pullets to negotiate a ramp transition. *Appl. Anim. Behav. Sci.* 208, 92-99.

Norman, K. I., Adriaense, J. E. C. and Nicol, C. J. (2019). The impact of early structural enrichment on spatial cognition in layer chicks. *Behav. Processes* 164, 167-74.

Norring, M., Kaukonen, E. and Valros, A. (2016). The use of perches and platforms by broiler chickens. *Appl. Anim. Behav. Sci.* 184, 91-96, http://doi.org/10.1016/j.applanim.2016.07.012.

Odén, K., Keeling, L. J. and Algers, B. (2002). Behaviour of laying hens in two types of aviary systems on 25 commercial farms in Sweden. *Br. Poult. Sci.* 43(2), 169-181.

Oester, H., Wiedmer, H. and Witkowski, A. (2005). Evaluation of elevated surfaces and perches for broilers. *Anim. Sci. Pap. Rep.* 23(1), 231-240.

Olsson, I. A. S. and Keeling, L. J. (2002). The push-door for measuring motivation in hens: laying hens are motivated to perch at night. *Anim. Welf.* 11, 11-19.

Olsson, I. A. S. and Keeling, L. J. (2000). Night-time roosting in laying hens and the effect of thwarting access to perches. *Appl. Anim. Behav. Sci.* 68(3), 243-256.

Pettersson, I. C., Weeks, C. A., Norman, K. I. and Nicol, C. J. (2017). The ability of laying pullets to negotiate two ramp designs as measured by bird preference and behavior. *PeerJ* 5, e4069. https://doi.org/10.7717/peerj.4069.

Pettit-Riley, R. and Estevez, I. (2001). Effects of density on perching behaviour of broiler chickens. *Appl. Anim. Behav. Sci.* 71(2), 127-140.

Pickel, T., Scholz, B. and Schrader, L. (2010). Perch material and diameter affects particular perching behaviours in laying hens. *Appl. Anim. Behav. Sci.* 127(1-2), 37-42.

Pickel, T., Scholz, B. and Schrader, L. (2011a). Roosting behaviour in laying hens on perches of different temperatures: a trade-off between thermoregulation, energy budget, vigilance and recreational resting? *Appl. Anim. Behav. Sci.* 134(3-4), 164-169.

Pickel, T., Schrader, L. and Scholz, B. (2011b). Pressure load on keel bone and foot pads in perching laying hens in relation to perch design. *Poult. Sci.* 90(4), 715-724.

Pike, A. V. L. and Maitland, D. P. (2004). Scaling of bird claws. *J. Zool.* 262(1), 73-81.

Provine, R. R., Strawbridge, C. L. and Harrison, B. J. (1984). Comparative analysis of the development of wing-flapping and flight in the fowl. *Dev. Psychobiol.* 17(1), 1-10.

Quinn, T. H. and Baumel, J. J. (1990). The digital tendon locking mechanism of the avian foot (Aves). *Zoomorphology* 109(5), 281-293.

Riber, A. B., van de Weerd, H. A., de Jong, I. C. and Steenfeldt, S. (2018). Review of environmental enrichment for broiler chickens. *Poult. Sci.* 97(2), 378-396.

Riber, A. B., Wichman, A., Braastad, B. O. and Forkman, B. (2007). Effects of broody hens on perch use, ground pecking and cannibalsim in domestic fowl (Gallus gallus domesticus). *Appl. Anim. Behav. Sci.* 106(1), 39-51.

Roberts, G. (1996). Why individual vigilance declines as group size increases. *Anim. Behav.* 51(5), 1077-1086.

Roden, C. and Wechsler, B. (1998). A comparison of the behavior of domestic chicks reared with or without a hen in enriched pens. *Appl. Anim. Behav. Sci.* 55(3-4), 317-326.

Roenchen, S., Scholz, B., Hamann, H. and Distl, O. (2010). Use of functional areas, perch acceptance and selected behavioural traits in three different layer strains kept in furnished cages, small group systems and modified small group systems with elevated perches. *Arch. Geflugelkd.* 74, 256-264.

Rufener, C., Abreu, Y., Asher, L., Berezowski, J. A., Maximiano Sousa, F., Stratmann, A. and Toscano, M. J. (2019). Keel bone fractures are associated with individual mobility of laying hens in an aviary system. *Appl. Anim. Behav. Sci.* 217, 48-56.

Sandilands, V., Moinard, C. and Sparks, N. H. C. (2009). Providing laying hens with perches: fulfilling behavioural needs but causing injury? *Br. Poult. Sci.* 50(4), 395-406, http://doi.org/10.1080/00071660903110844.

Scholz, B., Kjaer, J. B. and Schrader, L. (2014). Analysis of landing behaviour of three layer lines on different perch designs. *Br. Poult. Sci.* 1668, 37-41. https://doi.org/10.1080/00071668.2014.933175.

Schrader, L., Dippel, S. and Nicol, C. J. (2016). Do laying hens have a motivation to grasp while nighttime roosting? In: Proceedings of the 50th Congress of the International Society for Applied Ethology, UK, 340.

Schrader, L. and Müller, B. (2009). Night-time roosting in the domestic fowl: the height matters. *Appl. Anim. Behav. Sci.* 121(3-4), 179-183.

Scott, G. B., Hughes, B. O., Lambe, N. R. and Waddington, D. (1999). Ability of laying hens to jump between perches: individual variation and the effects of perch separation and motivation on behaviour. *Br. Poult. Sci.* 40(2), 177-184.

Scott, G. B., Lambe, N. R. and Hitchcock, D. (1997). Ability of laying hens to negotiate horizontal perches at different heights, separated by different angles. *Br. Poult. Sci.* 38(1), 48-54.

Scott, G. B. and MacAngus, G. (2004). The ability of laying hens to negotiate perches of different materials with clean or dirty surfaces. *Anim. Welf.* 13(3), 361-365.

Scott, G. B. and Parker, C. A. L. (1994). The ability of laying hens to negotiate between horizontal perches. *Appl. Anim. Behav. Sci.* 42(2), 121-127. https://doi.org/10.1016/0168-1591(94)90152-X.

Spindler, B., Clauß, M., Briese, A. and Hartung, J. (2013). Planimetrische Ermittlung des Mindestplatzbedarfs von Junghennen. *Berl. Munch. Tierarztl. Wochenschr.* 126, 156-162.

Steenfeldt, S. and Nielsen, B. L. (2015). Welfare of organic laying hens kept at different indoor stocking densities in a multi-tier aviary system. II: live weight health measures and perching. *Animal* 9(9), 1518-1528.

Stratmann, A., Fröhlich, E. K. F., Gebhardt-Henrich, S. G., Harlander-Matauschek, A., Würbel, H. and Toscano, M. J. (2015a). Modification of aviary design reduces incidence of falls, collisions and keel bone damage in laying hens. *Appl. Anim. Behav. Sci.* 165, 112–123. https://doi.org/10.1016/j.applanim.2015.01.012.

Stratmann, A., Fröhlich, E. K., Harlander-Matauschek, A., Schrader, L., Toscano, M. J., Würbel, H. and Gebhardt-Henrich, S. G. (2015b). Soft perches in an aviary system reduce incidence of keel bone damage in laying hens. *PLoS ONE* 10(3), e0122568, https://doi.org/10.1371/journal.pone.0122568.

Strong, R. A., Hester, P. Y., Eicher, S. D., Hu, J. and Cheng, H. W. (2015). The effect of cooled perches on immunological parameters of caged white leghorn hens during the hot summer months. *PLoS ONE* 10(10), e0141215. https://doi.org/10.1371/journal.pone.0141215.

Struelens, E. and Tuyttens, F. A. M. (2009). Effects of perch design on behaviour and health of laying hens. *Anim. Welf.* 18, 533–538.

Struelens, E., Tuyttens, F. A. M., Ampe, B., Odberg, F., Sonck, B. and Duchateau, L. (2009). Perch width preferences of laying hens. *Br. Poult. Sci.* 50(4), 418–423.

Struelens, E., Tuyttens, F. A. M., Duchateau, L., Leroy, T., Cox, M., Vranken, E., Buyse, J., Zoons, J., Berckmans, D., Odberg, F. and Sonck, B. (2008). Perching behaviour and perch height preference of laying hens in furnished cages varying in height. *Br. Poult. Sci.* 49(4), 381–389.

Sustaita, D., Pouydebat, E., Manzano, A., Abdala, V., Hertel, F. and Herrel, A. (2013). Getting a grip on tetrapod grasping: form, function, and evolution. *Biol. Rev.* 88(2), 380–405.

Tahamtani, F. M., Nordgreen, J., Nordquist, R. E. and Janczak, A. M. (2015). Early life in a barren environment adversely affects spatial cognition in laying hens (Gallus gallus domesticus). *Front. Vet. Sci.* 2, 1–12. https://doi.org/10.3389/fvets.2015.00003.

Tahamtani, F. M., Pedersen, I. J., Toinon, C. and Riber, A. B. (2018). Effects of environmental complexity on fearfulness and learning ability in fast growing chickens. *Appl. Anim. Behav. Sci.* 207, 49–56.

Tauson, R. (1984). Effects of a perch in conventional cages for laying hens. *Acta Agric. Scand.* 34(2), 193–209.

Tauson, R. and Abrahamsson, P. (1996). Foot and keel bone disorders in laying hens: effects of artificial perch material and hybrid. *Acta Agric. Scand. A* 46(4), 239–246.

Taylor, P. E., Scott, G. B. and Rose, P. (2003). The ability of domestic hens to jump between horizontal perches: effects of light intensity and perch colour. *Appl. Anim. Behav. Sci.* 83(2), 99–108.

Ventura, B. A., Siewerdt, F. and Estevez, I. (2010). Effects of barrier perches and density on broiler leg health, fear, and performance. *Poult. Sci.* 89(8), 1574–1583.

Ventura, B. A., Siewerdt, F. and Estevez, I. (2012). Access to barrier perches improves behaviour repertoire in broilers. *PLoS ONE* 7(1), e29826. https://doi.org/10.1371/journal.pone.0029826.

Wechsler, B. and Huber-Eicher, B. (1998). The effect of foraging material and perch height on feather pecking and feather damage in laying hens. *Appl. Anim. Behav. Sci.* 58(1–2), 131–141.

Weeks, C. A. and Nicol, C. J. (2006). Behavioural needs, priorities and preferences of laying hens. *World Poult. Sci. J.* 62(2), 296–307.

Wichman, A., Heikkila, M., Valros, A., Forkman, B. and Keeling, L. J. (2007). Perching behaviour in chickens and its relation to spatial ability. *Appl. Anim. Behav. Sci.* 105(1–3), 165–179.

Wood-Gush, D. G. M. and Duncan, I. J. H. (1976). Some behavioural observations on domestic fowl in the wild. *Appl. Anim. Ethol.* 2(3), 255–260.

Workman, L. and Andrew, R. J. (1989). Simultaneous changes in behaviour and in lateralization during the development of male and female domestic chicks. *Anim. Behav.* 38(4), 596–605.

Yngvesson, J., Wallenbeck, A., Jönsson, L. and Gunnarsson, S. (2016). Behaviour of broilers in semi-commercial organic rearing - behaviour and mortality of hybrids with rapid or slow growth rate. In: Proceedings of the 50th Congress of the International Society of Applied Ethology, UK, 239.

Zhao, J. P., Jiao, H. C., Jiang, Y. B., Song, Z. G., Wang, X. J. and Lin, H. (2012). Cool perch availability improves the performance and welfare status of broiler chickens in hot weather. *Poult. Sci.* 91(8), 1775–1784.

Zheng, H. Y., Li, B. M., Tong, Q., Chen, G. and Li, X. Y. (2019). Modification of perchery system: preference for ramps rather than ladders during early adaptation period for cage-reared pullets. *Int. J. Agric. Biol. Eng.* 12(4), 34–42.

# Chapter 14

# Improving welfare in catching and transport of chickens

*Leonie Jacobs, Virginia Tech, USA; and Frank A. M. Tuyttens, Institute for Agricultural and Fisheries Research (ILVO) and Ghent University, Belgium*

1 Introduction
2 Broiler chicken pre-slaughter phase and associated welfare concerns
3 Improving pre-slaughter welfare for broiler chickens
4 The pre-slaughter phase for laying hens
5 Conclusions and future trends
6 Where to look for further information
7 References

## 1 Introduction

The primary focus of this chapter is on broiler chickens, because (1) these animals are involved in a large industry, involving billions of birds, therefore they are of high economic value, and (2) more scientific investigation of the pre-slaughter phase has been conducted on broilers as compared to other poultry, such as laying hens, turkeys, or quail. The main body of the chapter describes the pre-slaughter phase of broilers, associated welfare risks, and animal responses. Figure 1 summarizes the risk factors that were identified per pre-slaughter stage and the animal outcomes associated with those. In addition, this chapter includes a brief mention of laying hen welfare during the pre-slaughter phase.

## 2 Broiler chicken pre-slaughter phase and associated welfare concerns

The conventional broiler chicken industry typically requires the birds to be transported twice during their lifetime. The first one, concerns the transportation of chicks from the hatchery to the grower facility. The second one – which is the focus of this chapter – takes place at the end of rearing when the birds are transported from the grower facility to the slaughter plant. These phases

http://dx.doi.org/10.19103/AS.2020.0078.13

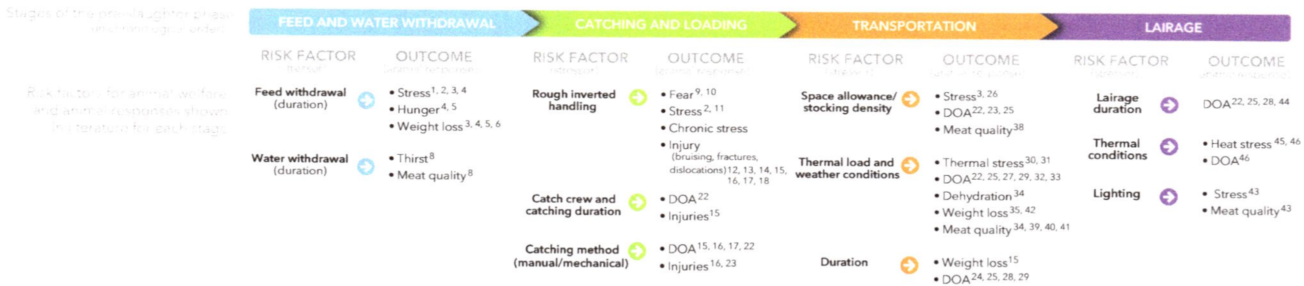


**Figure 1** Summary of risk factors associated with separate stages within the pre-slaughter phase for broilers. These risk factors are shown in scientific literature to have an association with a welfare concern for broilers. Numbers indicate references. 1. Najafi et al., 2016; 2. Kannan and Mench, 1996; 3. Delezie et al., 2007; 4. Nijdam et al., 2005b; 5. Northcutt, Savage and Vest, 1997; 6. Knowles et al., 1995; 7. Buhr et al., 1998; 8. Vanderhasselt et al., 2010; 9. Jones, 1992; 10. Zulkifli et al., 2000; 11. Bedanova et al., 2007; 12. Jacobs et al., 2017c; 13. Nijdam et al., 2005a; 14. Gouveia, Vaz-Pires and Martins da Costa, 2009; 15. Jacobs et al., 2017c; 16. Ekstrand, 1998; 17. Delezie et al., 2006; 18. Farsaie, Carr and Wabeck, 2013; 19. Gregory and Austin, 1992; 20. Nijdam et al., 2006; 21. Kittelsen et al., 2015a; 22. Chauvin et al., 2011; 23. Knierim and Gocke, 2003; 24. Warriss et al., 1992; 25. Nijdam et al., 2004; 26. Bedáňová et al., 2005; 27. Warriss, Pagazaurtundua and Brown, 2005; 28. Caffrey, Dohoo and Cockram, 2017; 29. Vecerek et al., 2006; 30. Knezacek et al., 2010; 31. Burlinguette et al., 2012; 32. Petracci et al., 2006; 33. Haslam et al., 2008; 34. Mitchell and Kettlewell, 2009b; 35. EFSA, 2011; 36. Kotula and Wang, 1994 37. Smith, Lyon and Lyon, 2002; 38. Petracci, Bianchi and Cavani, 2005; 39. Sandercock et al., 2006; 40. Akşit et al., 2006; 41. Sandercock et al., 2001; 42. Petracci, Fletcher and Northcutt, 2001; 43. Barbosa et al., 2013; 44. Villarroel et al., 2018; 45. Warriss et al., 1999; 46. Jacobs et al., 2017a.

of transportation include catching and loading, holding (at the hatchery or at the slaughter plant - lairage) and travel in crates, drawers or boxes - over the road in most cases. Both transportation to the producer (Mitchell, 2009; Jacobs et al., 2016, 2017d) and to the slaughter plant can be intense, stressful events in the birds' lives associated with thermal discomfort, injuries, fatigue, mortality, fear, and production losses (EFSA, 2011). On some large farms, however, live transportation is minimized by having a slaughter plant on site, for example, a large integration in Ukraine (Myronivsky Hliboproduct (MHP), 2018; Manufacturing Journal, 2019). Depopulation and transportation from the production site to the slaughter plant can even be largely automated, such as in the Vencomatic 'Patio' system in which birds are reared in large multi-tier colony cages with a moveable belt as a floor, covered with litter. During depopulation, the belt can be used to move the broilers from the colony cages directly onto a conveyor to the adjacent slaughter plant. In this chapter the focus is, however, on conventional pre-slaughter handling and transportation practices (hereafter the pre-slaughter phase).

Although the pre-slaughter phase is relatively short, many different actors are involved and bear specific responsibilities. Conventionally, the main actors are the producer, the catching crew, the transporter, and the slaughter plant personnel. Broiler producers are responsible for delivering healthy, high-quality broilers that are fit for transport to the slaughter plant. In order to achieve this, they will partly rely on receiving high-quality chicks from the hatchery and partly on their own stockmanship skills and animal facilities. The catching crew is responsible for humane handling and loading of birds under appropriate stocking densities in transportation containers; the transporter for appropriate, safe driving, without stops (depending on local legislation), maintaining a thermal microclimate that is appropriate for bird welfare; and the slaughter plant for appropriate lairage facilities, durations, and handling prior to slaughter. In order to reduce pre-slaughter production losses and animal suffering, clear communication between these actors about their engagements and responsibilities is essential. Qualitative interviews in Flanders (Belgium) showed that producers felt that birds were no longer their responsibility once they have left the premises. Yet in Flanders, producers used to be penalized (thus held responsible) when pre-slaughter mortality exceeded a certain limit (1%; Lambrecht et al., 2020). In turn, transporters and slaughter plant personnel have stated that catchers are responsible for major animal welfare issues, and that transportation in itself does not pose risks to broiler welfare (Lambrecht et al., 2020).

In the European Union there is specific legislation for the transportation of animals, including poultry (EU Council Regulation no 1/2005). This European legislation includes general and specific stipulations for the transport of poultry, including minimum space allowance and duration of transport without access to feed or water (12 h). Poultry transport is legally regulated in other parts of the world as well. For example, Canadian legislation on transportation of all animals is enforced by the Canadian Food Inspection Agency, with the help of other federal, provincial, and territorial authorities (Government of Canada, 2019). In the United States, the sole legislation regarding animal welfare, the Animal Welfare Act (USDA, 1966), does not include poultry (or other birds), allowing, for example, day-old chicks to be sent through regular postal services (Centers for Disease Control and Prevention, 2012).

## 2.1 Age of depopulation

Broiler chickens kept under conventional farming conditions are reared for 5-8 weeks, depending on market demands. For instance, the increasing demand for cuts and processed products has resulted in higher slaughter weights (Baéza et al., 2012). In the United States, average age at slaughter is 48.8 days, with live weights of 2.98 kg (Agri Stats Inc., 2018). Approximately 20% of broilers produced in the United States in 2011 were over 3.5 kg, slaughtered at on average 61 days of age (USDA, 2011). In addition, live weights of approximately 4.54 kg have been reported (anonymous poultry industry veterinarian, 2018). In Europe, birds are slaughtered between 30 and 55 days of age at a live weight of approximately 1.9-3.5 kg (Chauvin et al., 2011; Tuyttens, Vanhonacker and Verbeke, 2014; Grilli et al., 2015; Kittelsen et al., 2017). High (and low) slaughter weight, compared to flock averages, was identified as a 'risk factor' for welfare issues such as mortality during the pre-slaughter phase (Nijdam et al., 2004; Chauvin et al., 2011).

## 2.2 Thinning versus whole flock removal

In EU countries where the legal maximum stocking density is expressed in kg, live weight per $m^2$, it is common practice to thin the flock around 35 days of age with slaughter weights of approximately 1.7-2 kg (Sani, 2006; Tuyttens, Vanhonacker and Verbeke, 2014). Thinning refers to the removal of a part of the flock from the poultry house prior to complete depopulation at final slaughter age, and is sometimes referred to as partial depopulation. Results of a quantitative survey of Flemish broiler producers revealed that thinning was practiced on 79% of the flocks at a median age of 35 days and a median weight of 2 kg (Tuyttens, Vanhonacker and Verbeke, 2014). On average 20% of the flock was thinned. A survey of 27 Australian producers indicated that 90% thin their flocks multiple times around 30-37 days of age, and some indicated they thin flocks as early as 21 days of age (Scott et al., 2017). This practice allows producers to keep more birds in a single house up until thinning and still comply with stocking density requirements (in maximum kg bird per $m^2$) when - remaining - birds reach the preferred slaughter weight.

Some welfare advantages can be identified for birds slaughtered at an earlier age, as a number of welfare concerns associated with commercial broiler production are positively associated with age and/or weight. Young birds may have a better welfare status on-farm due to the limited impact of body weight on behavior, leg health, and dermatitis. As mentioned, high slaughter weight was identified as a risk factor for pre-slaughter mortality (dead on arrivals, recorded after removal from the crates, before birds are shackled; DOA) (Nijdam et al., 2004; Chauvin et al., 2011).

A disadvantage for remaining birds can be reduced biosecurity, with thinning linked to increased Campylobacter jejuni count and prevalence on-farm (Koolman, Whyte and Bolton, 2014; Higham et al., 2018). Other disadvantages that have been suggested include multiple experiences of stressful human contact, especially during dark periods when birds would be resting, associated fear, and feed withdrawal during thinning (Robins and Phillips, 2011).

## *2.3 Feed and water withdrawal*

Broiler chickens are taken off feed prior to catching for transportation to the slaughter plant. The reasoning behind it is from a food-safety perspective. Birds are fasted to avoid fecal contamination at the slaughter plant during processing (in line with EU Council Regulation 853/2004), for which the producer is held responsible in Europe. A gastrointestinal tract containing feed could represent a greater risk for contents to come in contact with the meat when the carcass is being processed, especially when cutting the intestine during automated evisceration, which is considered as contamination and a human health risk. Birds are feed-withdrawn several hours before the catching and loading procedure starts, which may lead to a time without feed of up to 24 h (Jacobs et al., 2017a). Prior research indicated that an 8–12 h feed withdrawal resulted in a sufficiently lowered risk of contamination (Warriss et al., 2004). Furthermore, water provision is stopped shortly before catching, which can lead to birds going without water for up to 15 h (Jacobs et al., 2017c). Broilers showed increased voluntary water consumption after water deprivation of 12 h compared to no deprivation, indicating thirst (Vanderhasselt et al., 2010).

European legislation for animals that undergo transport (EU Council Regulation no 1/2005) prescribes a maximum duration of 12 h for birds to go without water or feed. However, it only regards transportation duration, disregarding the stages that come before and after transport. As broiler chickens are genetically selected for their growth potential, their appetite has also increased (Siegel and Wisman, 1966; Emmerson, 1997), which would mean that fasting would be more stressful for broilers than birds from other genetic lines (for instance laying hens).

The duration of feed withdrawal is the major determinant for the impact on hunger. Fasting broilers for 10 h prior to transportation resulted in a stronger stress response compared to full-fed birds (Nijdam et al., 2005b). This was shown by the increase in plasma thyroxine concentration and decrease in triiodothyronine, triglyceride, glucose, and lactate concentrations in fasted birds compared to the fed birds, although no differences in corticosterone levels were found (Nijdam et al., 2005b). Other studies did report differences in corticosterone due to feed withdrawal. Twenty-one-day-old broiler chickens exposed to feed withdrawal of 24 h (and no other stressors) resulted in a significant increase in plasma corticosterone compared to ad libitum feeding, but no differences in concentrations for shorter feed-withdrawal times were found (Najafi et al., 2016). Kannan and Mench (1996) found that feed-deprived broilers showed increased plasma corticosterone compared to full-fed bird. In that and another study, the impact of feed withdrawal was overruled by being placed in crates, or being exposed to varying stocking densities (Kannan and Mench, 1996; Delezie et al., 2007). However, feed withdrawal could result in coprophagy, indicating a motivation to feed. Gizzards contained litter after 9 h of feed withdrawal, and litter and feces after 16 h of feed withdrawal (Northcutt, Savage and Vest, 1997), indicating that birds ingested or fed on fecal and litter material. The same group recorded that 14–18 h of fasting resulted in heavy sloughing of the intestines (loss of intestinal mucosal lining), which they associated with an increased chance of contamination during processing. Furthermore, fasting was linked to increased *Salmonella* colonization of the crop and intestine (reviewed by Burkholder et al., 2008), with intestinal tissue of fasted broilers being more susceptible to pathogen attachment compared to full-fed broilers (Burkholder et al., 2008). Based on marginal predictions (estimate of effect calculated at specified values of DOA rates, and averaged over the remaining covariates), 8 h of feed withdrawal before loading was associated with lowest DOA rates, with a higher risk for mortality as duration either decreased to 0 h or increased to 19 h (Caffrey, Dohoo and Cockram, 2017).

Feed withdrawal could reduce body temperature, which would be beneficial when transported under hot conditions to avoid heat stress, but a concern when transportation takes place in cold weather conditions. Feed restriction when housed at 24°C or 32°C led to lower body temperatures (40.4°C and 41.0°C) compared to full-fed broilers (40.8°C and 41.9°C) (Deeb and Cahaner, 1999). Similarly, 2-h feed-withdrawn broilers exposed to 35–40°C ambient temperatures showed a slower rise in body temperatures compared to a full-fed control group (Francis, Macleod and Anderson, 1991). Heat stress is one of the major risk factors for broiler DOA during the pre-slaughter phase (Bayliss and Hinton, 1990; Gregory and Austin, 1992). With feed withdrawal being common practice, other strategies need to be applied to address heat stress.

## 2.4 Catching and loading

In most cases, broiler chickens are caught manually by a contracted group of professional catchers. For instance, roughly 75% of flocks in Belgium are caught manually (Jacobs, 2016), the majority of Canadian flocks are caught manually too (Cockram and Dulal, 2018). Manual catching involves the broilers being lifted up from the floor, inverted, carrying birds by one or two legs, with 3-4 birds per hand (Bayliss and Hinton, 1990; Cockram and Dulal, 2018) by a team of 3-12 specialized people (Jacobs et al., 2017c). In other cases, catching is automated by the use of catching equipment, such as the Apollo chicken harvesters (Fig. 2A, Ciemmecalabria S.r.l., Cazzago S. M. (BS), Italy) or the Chicken Cat (Fig. 2B, JTT Conveying, Bredsten, Denmark). Mechanical catching involves the use of large machinery with a sweeping system, where rubber fingers rotate to collect birds and place them on a conveyer belt to then move them to the container (Knierim and Gocke, 2003). Alternatively, one conveyer belt is used to lift birds up from the litter, transferring them to another conveyor belt, into the container with drawers or crates. Catching generally occurs with dimmed lights, or at night to keep birds calm.

European legislation (EU Council Regulation no 1/2005 annex I) stipulates that animal transportation is allowed when 'the animals are fit for the journey' and the animals are deemed unfit for transport when they are injured (with an open wound), unable to move independently, or show physiological weakness. Furthermore, transports should take place 'in conditions guaranteed not to cause [animals] injury or unnecessary suffering'. Although likely initially intended for large livestock, this legislation applies to poultry too. Thus, this legislation requires individual health checks prior to transportation, possibly during the

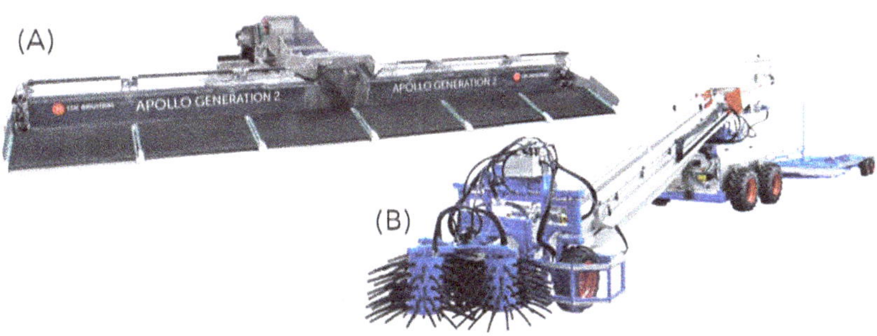


**Figure 2** Poultry catching equipment for mechanical catching. (A) The Apollo harvester from CMC industries (http://poultry.cmcindustries.com/apollo-generation-2-2123746787.html) using conveyor belts to pick up birds, and place them in crates. (B) The Chicken Cat from JJ Conveying (adapted from http://www.chickencat.eu/uk/features.aspx). The harvester picks up birds using the rubber 'fingers', transfers birds onto a conveyor belt, and places the birds in transportation crates (not depicted).

catching stage. Yet, current procedure includes a more practical approach in which the producer or veterinarian signs a form ensuring flock health, even though individual variation within a flock is expected (Jacobs et al., 2017b).

In addition to possible preexisting health conditions, aspects of the catching and loading stages have been linked to major welfare concerns such as injuries and mortality. Throughout their life, broiler chickens are rarely handled by people, usually only briefly as chicks when placed in transportation crates or when placed on farm, and on their last day of life during the pre-slaughter phase (during catching and often during shackling at slaughter). This limited contact can play a role in the acute stress experienced during human-animal interactions. Some stressors related to catching and loading are crate-stocking density/space restriction (Nijdam et al., 2004; Kopecsnik, 2008; Chauvin et al., 2011), loud noises (Chloupek et al., 2009), rough, improper, inverted handling (Jones, 1992; Kannan and Mench, 1996; Zulkifli et al., 2000; Bedanova et al., 2007), and social disruption (Mitchell and Kettlewell, 2009b).

Broilers at 6 weeks of age show increased fearfulness after rough, inverted handling compared to gentle, upright handling, with fearfulness indicated by longer tonic immobility durations (Jones, 1992; Zulkifli et al., 2000). Tonic immobility is induced by a short period of manual restraint and is regarded as an anti-predator reaction, with longer durations indicative of a more fearful affective state (Jones, 1992). Increased levels of plasma corticosterone have been found in male broilers that were handled inverted compared to upright handling, indicative of an acute stress response (Kannan and Mench, 1996). Furthermore, heterophil:lymphocyte ratios peaked at 20 h after catching and crating broilers at slaughter age, which indicates that, in addition to the acute stress response illustrated by corticosterone increases, a more slow-responding mechanism is activated by the stressor (Zulkifli et al., 2000).

While inverted handling may improve the efficiency or speed, suspension by the legs can cause distress in itself illustrated by circulating corticosterone increases and prolonged tonic immobility when assessed after handling. Inverted handling while birds are conscious would occur during catching and loading, and at the slaughter plant when shacking the birds (with the exception of slaughter plants that apply gas stunning, in which birds are shackled after rendered unconscious). Birds may flap their wings and vocalize during inverted handling, or it could induce a tonic immobility response while being carried. The latter animal response plus the relatively small size of poultry compared to other livestock species may be some of the reasons why inversion of conscious poultry has been widely accepted in the industry (Main and Hewson, 2012). Inverted handling presents another welfare concern: the lack of a diaphragm in birds is especially problematic for heavy broilers when handled upside down. The weight of their muscles puts pressure on the heart and respiratory system, which is increasingly uncomfortable. Inverted shackling, similar to inverted handling during catching, had a significant effect on plasma corticosterone, heterophil:lymphocyte ratios, and plasma concentrations of glucose and lactate, with longer shackling durations being perceived to be more stressful (Bedanova et al., 2007). Based on their findings, the authors recommended to keep shackling durations under 60 seconds, which could be similar for inversion during catching.

In addition to fear and distress, the catching and loading stage can result in physical trauma. Examples include fractures, dislocations, and subcutaneous hemorrhaging (Ekstrand, 1998; Delezie et al., 2006; Farsaie, Carr and Wabeck, 2013; Jacobs et al., 2017c). One study found a significant increase in wing fracture and dislocation prevalence after catching of 41-day-old broilers compared to before, with no further increases thereafter (from 0.1% to 1.9%, Jacobs et al., 2017c). Conversely, Kittelsen et al. (2015b) found in 31-day-old broilers, more fractures caused by handling at the slaughter plant than by catching and transportation (2.4% vs. 0.8%, respectively). Bruises are one of the more common injuries found in 3.7–8.4% of birds assessed (Nijdam et al., 2005a; Gouveia, Vaz-Pires and Martins da Costa, 2009; Jacobs et al., 2017c).

In Europe, minimum in-crate space allowances are stipulated by legislation. For broilers between 1.6 kg and 3 kg, birds should have 160 $cm^2$ per kg body weight. However, a contingency is included mentioning that body size, physical condition, weather conditions, and estimated transportation duration could be factors to deviate from the standard (European Commission, 2005). High stocking density (or low space allowance) was identified as a potential stressor during the pre-slaughter phase, with increased risk for mortality compared to lower stocking densities (Warriss et al., 1992; Nijdam et al., 2004; Chauvin et al., 2011). Furthermore, higher levels of plasma corticosterone were found when broilers were crated and transported at high density (350 $cm^2$/bird) compared to the low density (575 $cm^2$/bird; Delezie et al., 2007) and increased heterophil:lymphocyte ratios were found in 3 kg broilers when stocked at 105 $cm^2$/kg for 2 h (Bedáňová et al., 2005).

Restricted space allowance is likely a stressor because birds are severely limited in their ability to express motivated behaviors, such as hide, move away from conspecifics, or express behaviors to maintain thermal homeostasis. The thermal load to which birds are exposed to is multifactorial and depends on environmental factors including ambient temperature, relative humidity, space allowance, active ventilation, curtain configuration on trucks, and duration of trucks being stationary. Furthermore, it depends on animal-related factors, such as body weight, feather coverage, and cleanliness. Especially heat stress is a major risk for broiler chicken welfare, and a cause for DOAs while crated (Bayliss and Hinton, 1990; Ritz, Webster and Czarick, 2005). An ambient temperature of 18°C was identified as a critical upper threshold, above which higher DOA rates were found (Warriss, Pagazaurtundua and Brown, 2005).

Often related to or caused by injuries, a number of studies have found a link between DOA and the catching and loading stages, with choice of catching crew, catching method, and duration affecting DOA percentage (Bayliss and Hinton, 1990; Ekstrand, 1998; Nijdam et al., 2004; Chauvin et al., 2011). Necropsies to assess the cause of death have shown that between 25% and 35% of DOA birds likely died due to trauma inflicted during catching and loading, such as fractures, dislocations, liver ruptures, and trauma to the head (Gregory and Austin, 1992; Nijdam et al., 2006; Kittelsen et al., 2015a). As mentioned earlier, inverted broiler chickens are likely to experience major discomfort from the pressure of their (breast) muscle on the internal organs. In line, inappropriate bird placement in crates during catching was linked to mortality, with the number of supine birds showing a strong positive association with DOA prevalence (Jacobs et al., 2017c).

Besides the impact on animal welfare, the catching and loading procedure is demanding for people involved. The catching crew is involved in tough manual labor, requiring physical strength, exposed to dust in a dark environment, with safety risks due to forklifts driving in and out of the facility. Fatigue in relation to the physical strain may exacerbate rough handling (Delezie, 2006). Accordingly, prolonged catching was linked to increased wing fracture prevalence (Jacobs et al., 2017c) and DOA prevalence (Chauvin et al., 2011).

## 2.5 Transportation

After loading transportation containers onto a specialized poultry transport truck, the birds are transported - over the road - to the slaughter plant. Transportation duration can vary greatly depending on the country considered. Birds may be exposed to a number of different stressors, such as an unfavorable microclimate in the crates, high stocking density, unfamiliar environment, acceleration, vibration, motion, impacts, prolonged feed and water withdrawal, social disruption and noise (Nicol and Scott, 1990; Mitchell and Kettlewell, 1998; Mitchell and Kettlewell, 2009b). Yet, other authors concluded that transportation of circa 2 h itself did not cause physical stress based on plasma CPK activity, which increases when muscles are fatigued or damaged (Warriss et al., 1999). Although most studies have focused on the birds' stress response to pre-slaughter conditions, stress is not the only welfare measure, and certain stimuli can be aversive even if not acutely stressful (Abeyesinghe et al., 2001).

Transportation duration has a positive relationship with body weight loss, with 41 g lost per hour, in transports of on average 113 minutes (Jacobs et al., 2017c), although this relationship was not found for transports of on average 129 minutes (Warriss et al., 1999). Initial weight loss is a consequence of defecation (May, Lott and Deaton, 1990; Warriss et al., 2004). Thereafter, birds will lose muscle tissue (Lyon, Papa and Wilson, 2012). When transported birds experience heat stress, they may lose excess weight because of the increased demand for evaporative heat loss (EFSA, 2011). In turn, birds may experience dehydration, which can affect meat quality (reviewed by Mitchell and Kettlewell, 2009b). Ambient temperature multiplied by transportation duration was identified as an important predictor for DOA prevalence (Nijdam et al., 2004). Long transportation in itself was also identified as a risk factor for bird welfare, with transportation of >4 h or transportation of >300 km increasing the risk for DOAs (Warriss et al., 1992; Vecerek et al., 2006). A Canadian study showed that the risk for mortality increased by 0.045% per hour of transport (Caffrey, Dohoo and Cockram, 2017) and a review concluded that with longer transport journeys, there is a larger risk for birds to die from 'a chronic disease that decreased the ability of the bird to cope with the transport conditions, from an injury sustained during catching and loading, or from environmental extremes possibly aggravated by the period without access to food and water' (Cockram and Dulal, 2018).

Regardless of whether heat stress results in mortality, it is a welfare concern. In addition to heat stress, cold stress can be experienced mostly during transportation rather than when trucks are stationary and it is of additional concern (Knezacek et al., 2010; Burlinguette et al., 2012). Multiple studies have identified season or month to be a risk factor for DOAs, illustrating the impact of weather conditions on broiler welfare. More DOAs were found when transportation occurred between October and April in France (Chauvin et al., 2011), during summer in Italy, the Czech Republic, and the United Kingdom (Petracci et al., 2006; Vecerek et al., 2006; Haslam et al., 2008) and during winter in the Czech Republic (Vecerek et al., 2006).

In most commercial poultry transportation vehicles around the world, there is no active on-board climate control, exposing birds to wind, sun, precipitation, and ambient temperatures. Heat stress may be exacerbated when the truck is stationary on-farm or in traffic, due to lack of ventilation. Often efforts are made, using passive ventilation, to try to avoid thermal stress, for instance, by using tarps on the sides and back of the truck in winter months. Yet, enclosing birds reduces ventilation, increases humidity, and can result in heat stress due to the existence of a 'thermal core' in the truck (Knezacek et al., 2010).

## 2.6 Lairage

Upon arrival at the slaughter plant, containers are sometimes unloaded and stacked outside, in a warehouse or under partial cover. Birds are kept in lairage for a few minutes up to 9 h (Warriss et al., 1999; Jacobs et al., 2017a), although longer lairage periods were observed too (anecdotal observation Jacobs, 2015). Active ventilation via industrial fans is applied if deemed necessary and water may be sprayed in extreme heat. In some cases, birds are kept in dark, or blue-lit areas to keep them calm. Exposure to blue lighting during shackling did result in a 14% reduction of pale, soft, exudative (PSE) meat compared to white lighting, suggesting a reduced stress response during handling (Barbosa et al., 2013).

The thought-process behind lairage is to provide birds with recovery time prior to slaughter. Lairage has been shown to be beneficial in other species. For instance, 3 h of lairage was shown to result in similar blood parameter profiles in pigs compared to no lairage, but better than pigs laired for 9 h (Perez et al., 2002). However, the same authors found elevated cortisol levels in all groups (0-, 3-, 9-h lairage) compared to values in literature, with highest values in the 0-hour group, indicating that lairage may result in some recovery (calmer), but not similar to on-farm values. Furthermore, overnight lairage of pigs was shown to lower cortisol levels compared to shorter lairage times, interpreted as reduced stress levels (Warriss, 2003). The same review concluded that short lairage times were associated with reduced meat quality (pale, soft exudative meat) and long lairage times (>3 h) with increased fighting, associated skin lesions, reduced carcass yield, and reduced meat quality (dark, firm, dry meat) (Warriss, 2003). Possible benefits of lairage could be diminished in poultry as they are not unloaded from transport crates, while large livestock species are laired in pens.

Lairage allows for a 'buffer' of birds to be present at the plant, ensuring a continuous supply of birds for the slaughter line, (Ljungberg, Gebresenbet and Aradom, 2007). Yet, lairage is not a common concept in the United States. Prolonged lairage was identified to increase the risk of DOAs (Nijdam et al., 2004; >4.3 h Chauvin et al., 2011). The risk for mortality increased by 0.044% per hour when kept in lairage after transportation (Caffrey, Dohoo and Cockram, 2017). In Spain, DOA prevalence increased by 0.0021% per hour in lairage (Villarroel et al., 2018). Panting prevalence (an indicator of heat stress) during lairage was positively associated with DOA rates (Jacobs et al., 2017a). Based on liver glycogen depletion and increasing body temperatures, Warriss et al. (1999) recommended to have no lairage or limit lairage durations to 1 h. However, they state that with appropriate active ventilation, lairage duration of 2 h (maximum) could be justified.

## 2.7 Economic impact

Welfare issues during the pre-slaughter phase can have a direct economic impact, including mortality, carcass rejections or downgrades, and weight loss (or live shrink). Additionally, welfare concerns can result in inferior meat quality and negative consumer perception. An estimation of the direct economic impact is shown in Table 1, in which monetary loss due to DOAs, carcass rejections, parts condemnation, and weight loss are calculated based on

**Table 1** Estimated annual economic impact (in millions of dollars, columns i and j) of several pre-slaughter broiler chicken welfare issues in US dollars. US and global impact (column i and j) are extrapolated from information coming from a range of countries, thus represent an approximate estimate on the potential impact, as body weight, economic value, and flock size can vary greatly. Bird value is based on 2018 US Agri Stats data, with an average slaughter weight of 2.96 kg (column b), and average $ value per bird of $0.9877 (column f). Percentages of birds affected come from studies listed in column a. All calculations are based on average flock size of 90 000 birds (Macdonald, 2014). For columns b through j the calculation is given. Calculations were reviewed by Dr. Clinton Neill (agricultural economist, Virginia Tech, 2019)

| | a | b | c | d | e | f | g | h | i | j |
|---|---|---|---|---|---|---|---|---|---|---|
| | Scenarios (country and reference) | Loss in kg/ bird[a] (y=2.96 OR y=2.96*(a/100)) | % of birds affected (y=a OR y=whole flock=100%) | No of birds affected per flock (y=90,000*c) | Kgs loss per flock (y=b*d) | $ value per kg (y=0.988 OR y=0.988/2) | $ loss per flock (y=((e/2)*f)+ (e/2*0.12)[d] OR y=e*f) | $ gain through Cat 3 sales (y=e/2*0.104)[e] | Estimated annual impact (United States) (million $/ year)[b] (y=(100 000*g)-(100 000*h) OR y=100 000*g)) | Estimated annual global impact (million $/ year)[c] (y=(733 333*g)-(733 333*h) OR y=733 333*g) |
| Mortality | United Kingdom (Haslam et al., 2008) | 2.96 | 0.12 | 108 | 320 | 0.988 | 177 | 17 | 16 | 118 |
| | United States (Agri Stats Inc., 2018) | 2.96 | 0.22 | 200.7 | 594 | 0.988 | 329 | 31 | 30 | 219 |
| | Italy (Petracci et al., 2006) | 2.96 | 0.35 | 315 | 932 | 0.988 | 516 | 48 | 47 | 343 |
| | The Netherlands (Nijdam et al., 2004) | 2.96 | 0.46 | 414 | 1 225 | 0.988 | 679 | 64 | 61 | 451 |
| Carcass rejections | United States (Agri Stats Inc., 2018) | 2.96 | 0.27 | 239.4 | 709 | 0.988 | 392 | 37 | 36 | 261 |
| | Belgium (Jacobs et al., 2017a) | 2.96 | 0.75 | 675 | 1 998 | 0.988 | 1 107 | 104 | 100 | 735 |

(Continued)

**Table 1** (Continued)

| | a | b | c | d | e | f | g | h | i | j |
|---|---|---|---|---|---|---|---|---|---|---|
| | Scenarios (country and reference) | Loss in kg/ bird[a] (y=2.96 OR y=2.96*(a/100)) | % of birds affected (y=a OR y=whole flock=100%) | No of birds affected per flock (y=90,000*c) | Kgs loss per flock (y=b*d) | $ value per kg (y=0.988 OR y=0.988/2) | $ loss per flock (y=((e/2)*f)+ (e/2*0.12)[d] OR y=e*f) | $ gain through Cat 3 sales (y=e/2*0.104)[e] | Estimated annual impact (United States) (million $/ year)[b] (y=(100 000*g)-(100 000*h) OR y=100 000*g)) | Estimated annual global impact (million $/ year)[c] (y=(733 333*g)-(733 333*h) OR y=733 333*g) |
| | United Kingdom (Haslam et al., 2008) | 2.96 | 1.23 | 1 107 | 3 277 | 0.988 | 1 815 | 170 | 164 | 1 206 |
| | Norway (Kittelsen et al., 2017)[f] | 2.96 | 1.40 | 1 260 | 3 730 | 0.988 | 2 066 | 194 | 187 | 1,373 |
| Parts condemnation/ rejection[g] | United States (Agri Stats Inc., 2018) | 2.96 | 0.23 | 202.5 | 599 | 0.494 | 296 | | 30 | 217 |
| | Belgium (Jacobs et al., 2017a)[h] | 2.96 | 5.60 | 5 040 | 14 918 | 0.494 | 7 367 | | 737 | 5 403 |
| Weight loss (%) | United States - 4.4% (Rasmussen and Mast, 1989) | 0.13 | 100 | 90 000 | 11 722 | 0.988 | 11 577 | | 1 158 | 8 490 |
| | United States - 3.5% (Buhr et al., 1998) | 0.10 | 100 | 90 000 | 9 324 | 0.988 | 9 209 | | 921 | 6 753 |
| | United Kingdom - 7.3% (Warriss et al., 2004) | 0.22 | 100 | 90 000 | 19 447 | 0.988 | 19 207 | | 1 921 | 14 085 |

| a | b | c | d | e | f | g | h | i | j |
|---|---|---|---|---|---|---|---|---|---|
| Scenarios (country and reference) | Loss in kg/ bird[a] (y=2.96 OR y=2.96*(a/100)) | % of birds affected (y=a OR y=whole flock=100%) | No of birds affected per flock (y=90,000*c) | Kgs loss per flock (y=b*d) | $ value per kg (y=0.988 OR y=0.988/2) | $ loss per flock (y=((e/2)*f)+ (e/2*0.12)[d] OR y=e*f) | $ gain through Cat 3 sales (y=e/2*0.104)[e] | Estimated annual impact (United States) (million $/ year)[b] (y=(100 000*g)-(100 000*h) OR y=100 000*g)) | Estimated annual global impact (million $/ year)[c] (y=(733 333*g)-(733 333*h) OR y=733 333*g) |
| Belgium - 5% (Delezie et al., 2007) | 0.15 | 100 | 90 000 | 13 320 | 0.988 | 13 156 | | 1 316 | 9,648 |
| Belgium - 5.3% (Jacobs et al., 2017c) | 0.16 | 100 | 90 000 | 14 119 | 0.988 | 13 945 | | 1 395 | 10 226 |

[a] Average bird weights and $ value per kg from US data from 2018 for this calculation (Agri Stats Inc., 2018), slaughter weights may differ between countries.
[b] Prevalences and associated economic losses are extrapolated to national US broiler production of 9 billion birds annually (9 billion/90 000 birds per flock=100 000 flocks) (National Chicken Council, 2019) to obtain an estimate of national losses in US dollars.
[c] Prevalences and associated economic losses are extrapolated to global broiler production of 66 billion birds annually (66 billion/90 000 birds per flock=733 333 flocks) (Food and Agriculture Organization of the United Nations (FAO), 2017) to obtain an estimate of international losses in US dollars.
[d] Here we assumed that 50% of the kgs will be designated as Category 2 by-product, with $0.12 as the estimated loss due to cost of rendering. Amount is based on $116/ton cost to send to rendering, which is the average of $77–155/ton (anonymous slaughter plant representative in Flanders, Belgium, 2019).
[e] Here we assume that 50% of the kgs will be designated as Category 3 by-product, with $0.104 income per kg. Amount is based on $105/ton of economic gain, which is the average of $88–121/ton (anonymous slaughter plant representative in Flanders, Belgium, 2019).
[f] Mean rejection % of 'normal' DOA group (Kittelsen et al., 2017).
[g] Loss of parts condemnation in kgs based on 50% loss in carcass value (see $ value per kg) due to cut up (anonymous slaughter plant representative in Flanders, Belgium, 2019).
[h] Percentage based on assumption that all wings and breasts with fractures and/or bruising will be rejected as a part.

reported prevalences. DOAs, rejected parts, and rejected carcasses are either designated as Category 2 animal by-products, which cost the slaughter plant money to process, or as Category 3 by-products (fit for human consumption, used for further processed products), which still provide some income for the plant (anonymous slaughter plant representative in Flanders, Belgium, 2019). Category 2 by-products will go to rendering, at a cost of $77–155 per 1000 kg. Category 3 by-products will provide some income for the plant, estimated at $88–121 per 1000 kg . Seemingly small differences in prevalences of welfare concerns can have a profound impact on financial outcomes because of the vast number of birds involved.

DOAs represent a large economic loss, estimated to have a global impact between $117 and $450 million annually, assuming that half of the DOA birds are rendered, and half sold as Category 3 by-products (Table 1). Birds dying during the pre-slaughter phase can be a complete loss of economic value if designated as Category 2 by-products, because these birds can no longer be sold for human consumption (Petracci et al., 2006). As prevalences of DOAs and other welfare concerns vary greatly (Table 1), it is difficult to precisely estimate the economic impact. In the United States, any condemned parts or carcasses including DOA birds can be rendered for non-human consumption, that is, animal consumption, with the exception of products condemned for biological residue such as growth promoters, antibiotics, or hormones (United States Federal Government, 2019). In contrast, EU member states may not use DOA carcasses for animal consumption. Yet, use for compost or biogas is permitted (European Commission, 2009), so DOA birds and condemned parts are a considerable, but not complete loss of income. The losses are greatest at the pre-slaughter phase compared to earlier in the production process, with all feed, labor, and energy invested in the animal. DOA rates can vary considerably between studies, with values ranging from 0% to 19.4% per transport (Caffrey, Dohoo and Cockram, 2017; Jacobs et al., 2017a). A long-term trend of DOA rate increases during the last two decades from 0.14% to 0.40% was identified in the Czech Republic (Vecerek et al., 2006, 2016). In contrast, DOA rates have been declining in Canada from nearly 0.5% in 1999 to 0.2% in 2016 (Cockram and Dulal, 2018), in Norway from 0.20% in 2009 to 0.10% in 2015 (Animalia, 2015; Kittelsen et al., 2017), and in the United States from 0.37% in 2001 to 0.22% in 2017 (Agri Stats Inc., 2018).

Carcass rejections can either be Category 2 or 3 by-products, meaning they either cost the plant money to send to rendering, or provide some income. Global annual losses are estimated between $260 million and $1.3 billion under the assumption that half of the rejected carcasses go to rendering, and half sold as Category 3 by-product (Table 1). Catching trauma can increase the risk of whole carcass condemnation (Lupo et al., 2010), which can result in another complete economic loss together with DOAs.

Fractures, bruises (Gouveia, Vaz-Pires and Martins da Costa, 2009), and skin lesions can reduce the slaughter yield because of partial carcass rejections or downgrades (Jones, Satterlee and Cadd, 1998). Carcass value is decreased by 50–60% because of cut up (anonymous slaughter plant representative in Flanders, Belgium, 2019). Under the assumption that carcass value decreased by 50%, estimated global annual losses due to parts condemnation is between $217 million and $5.4 billion (Table 1). Pre-slaughter stressors can also reduce meat quality. For instance, high-crate stocking density (> 67 kg/m$^2$) resulted in more downgraded legs (reduced meat quality and reduced income) compared to low-density crating (< 55 kg/m$^2$) (1.75% vs. 1.43%; Petracci, Bianchi and Cavani, 2005).

Fasting is aimed to reduce the risk of contamination of meat. Contaminated meat would result in complete or partial carcass rejection. Prolonged feed withdrawal, although aimed at improving meat quality (avoid contamination), may have a negative economic impact, estimated between $6 and $14 billion dollars on a global, annual basis due to weight loss (Table 1). Feed deprivation has been positively associated with weight loss, regardless of a transportation event (Knowles et al., 1995; Buhr et al., 1998; Nijdam et al., 2005b; Delezie et al., 2007). Initial weight loss is mainly due to emptying of the gastrointestinal tract, which peaks after 4 h of deprivation (Warriss et al., 2004), thereafter it affects slaughter yield.

Some studies have shown that feed withdrawal can affect meat quality, such as decreased tenderness of the breast muscle after 6 h of fasting (Kotula and Wang, 1994) or color of breast muscle after 8 h of fasting, with breasts being lighter and yellower but less red (Smith, Lyon and Lyon, 2002). Although, other studies did not see an effect (Savenije et al., 2002; Delezie et al., 2007). Water deprivation of 36 h and longer resulted in increased breast muscle pH 24 h postmortem, although a shorter (24 h) water deprivation period did not (Vanderhasselt et al., 2010). Muscle pH plays a role in tenderness, water-holding capacity, color, juiciness, and shelf life (Mir et al., 2017), and high pH (>6.0) indicates dark, firm, dry meat (DFD), which is an undesirable trait (Vanderhasselt et al., 2010).

Transporting birds at high-stocking densities is economically attractive, because it reduces the number of trucks needed per flock. However, under certain conditions it could result in economic losses due to increased weight loss, carcass rejections, and mortality (Bedáňová et al., 2005; Delezie et al., 2007). High in-crate stocking densities, especially under hot conditions, can affect meat quality because of heat stress. Hyperthermia and panting-induced respiratory alkalosis could result in myopathies (carcass or parts condemnations) during the pre-slaughter phase (Sandercock et al., 2006), although results differ in literature, as other studies found that high-ambient temperature is of little impact on breast meat quality (Petracci, Fletcher and

Northcutt, 2001). High-ambient temperatures (34°C) when birds are in-crate resulted in decreased meat quality, with lower pH and higher color values (lightness, L*; redness, a*) indicative of reduced meat quality (Pale, Soft, Exudative (PSE)) compared to birds kept at 22°C (Akşit et al., 2006). Similarly, an ambient temperature of 32°C resulted in increased drip loss in 35 day-old birds crated for 2 h (Sandercock et al., 2001). Myopathies can reduce meat quality, with white striping and wooden breast as two examples (affecting muscle pH, color, cooking loss, and shear force values), although these specific myopathies are not related to the pre-slaughter phase, but rather to feeding regime and sex (Trocino et al., 2015). Besides meat quality, meat 'quantity' can be reduced, and birds lost 5.7% of their body weight when kept at 34°C compared to 3.9% and 3.2% when kept at 29.5°C or 25°C for 12 h (Petracci, Fletcher and Northcutt, 2001).

## 3 Improving pre-slaughter welfare for broiler chickens

Despite evidence of considerable animal welfare issues, end-of-life stages are neglected in many livestock welfare-monitoring tools, resulting in incomplete assessments and improvement strategies. Moreover, any welfare-related production losses at this final stage of the production chain (hence having used maximal resource inputs) will have maximal impact on the economic returns and environmental impact implying great relevancy in implementing strategies to prevent such losses.

Improved efficiency of livestock production will inevitably result in improved environmental outcomes. For example, it has been demonstrated that the efficiency gains already achieved in poultry production make poultry farming the most efficient livestock production system in terms of greenhouse gas production, water, and energy-use efficiency (Dunkley, 2014). It is therefore logical that any reduction of production losses (due to mortality, meat condemnation, loss of body weight) during the pre-slaughter phase will result in positive environmental and economical outcomes. A reduced environmental impact contributes to improved sustainability of the sector.

Solutions for welfare concerns need ethical consideration. Some welfare concerns may not have a negative economic consequence, and some welfare improvements may have a negative economic impact. Still, avoiding pain and suffering for any individual animal is preferred from an animal welfare perspective. In addition, benefits to the majority of the birds may be achieved at the cost of an individual or minority. The large number of sentient animals involved in this part of livestock production provides an ethical argument to prioritize broilers over other livestock species, and with poultry production expected to grow as an industry, broiler chicken welfare should be prioritized.

### 3.1 Fitness assessment for broilers

To improve broiler welfare during the pre-slaughter phase, a first step would be to ensure that transported birds are fit and robust to endure the associated stressors. Unfit birds with previously existing pathologies may have those exacerbated by stress during this phase (EFSA, 2011). Furthermore, it is likely that unfit birds are at a greater risk of being injured, rejected for human consumption, or die. In line, existing pathologies (lung congestion, infectious diseases, heart disorders) were more commonly found in DOA birds than birds that died on-farm or birds that were slaughtered (Nijdam et al., 2006; Lund et al., 2013; Kittelsen et al., 2015a). Furthermore, bird sex, age, and weight have been identified as risk factors for mortality (Nijdam et al., 2004; Haslam et al., 2008; Chauvin et al., 2011; Caffrey, Dohoo and Cockram, 2017). These findings support the notion that the physical status of a bird prior to the pre-slaughter phase affects how it experiences this phase. Birds that were deemed unfit for transport showed a stronger stress response toward the pre-slaughter phase compared to birds that were deemed clinically fit (Jacobs et al., 2017b).

Practical guidelines have been developed for such fitness-for-transport assessments (Poultry Industry Council, 2017), although these have not been evaluated scientifically. One study showed that birds that were identified as lame and/or cachexic (growth retardation) were experiencing more stress from the pre-slaughter phase than clinically fit birds, indicated by increased plasma corticosterone levels for birds transported at low- and high-stocking densities (220 $cm^2$ per bird or 160 $cm^2$ per bird; Jacobs et al., 2017d). This suggests that both indicators, of which one is covered in EU legislation, could be included in a fitness assessment, as birds with these conditions are more likely to suffer from transport.

The large number of broilers in a flock will make manual inspection of all animals difficult if not impossible. Furthermore, the consequences for birds that are possibly deemed unfit for transport need to be considered. For instance, if gait would be an indicator of fitness for transport, it is possible that 14–30% of a flock would be deemed unfit to travel due to lameness, as reviewed by Kittelsen et al. (2016).

The only tested method in a research setting is not feasible to apply in practice, as it includes gait scoring as an important determinant for fitness which would be too time-consuming prior to or during the catching process (Jacobs et al., 2019). Both producers and catching crews could perform a fitness-for-transport assessment during daily checks as the transportation date approaches for the former and during catching for the latter. Furthermore, automation of assessment could be considered in combination with mechanical catching. Cachexic individuals would have a significantly lower body weight than flock mates, which could be detected if digital scales were used on-farm in combination with sensor technology. If mechanical harvesters are used during catching and loading, incorporation of scales could also be effective to detect these birds. During manual handling, a quick assessment could consist of obvious signs of poor welfare such as extremely cachexic, ill, or injured birds.

## 3.2 Catching

Most broilers are caught manually by a team of professional catchers. However, mechanization of the catching procedure is possible with equipment commercially available and fewer people needed (see description of the method at lines 219–231). Mechanical catching can provide some improvements for the welfare of both people and birds. Mechanical catching is associated with lower costs and improved working conditions (Lacy and Czarick, 1998). From an animal welfare point of view, fewer bruises on the wings have been reported when birds were caught mechanically (4.2% vs.7.7% when caught manually), although no differences were found for bruises on breasts and legs (CIEMME Super Apollo L harvester; Delezie et al., 2006). In line with those findings, fewer bruises (2.0% vs. 3.0%), fractures (0.7% vs. 0.9%), and dislocations (0.5% vs. 0.6%) were found after mechanical catching when assessing 83 broiler flocks at the slaughter plant compared to manual catching (Chicken Cat harvester; Knierim and Gocke, 2003). However, not all studies indicate an improvement, with mechanical catching resulting in more wing fractures and bruises compared to manual catching (Finnish harvester manufactured by AR Tekniikka Oy; Ekstrand, 1998). Yet, their prevalences between 0.021% and 0.041% are much lower than more recent industry practices show, which may have resulted from some methodological differences in the study. They state that bruising as a reason for rejection of downgrading was recorded, possibly limiting their sample to severe cases only (Ekstrand, 1998).

Furthermore, more DOAs or a greater risk for DOAs were found in mechanically caught flocks compared to manual caught flocks (Ekstrand, 1998; Delezie et al., 2006; Chauvin et al., 2011). An experiment on eight broiler farms indicated that experienced acute stress was similar in both mechanical and manual catching based on corticosterone levels (Nijdam et al., 2005a). Another study on commercial flocks also found comparable corticosterone levels for mechanical or manual catching (Wolff et al., 2019). Although mechanical catching may not be feasible for all farms, it does show some benefits for both people and animals. Variation between equipment, use of equipment (e.g. conveyor speed), and the people involved could be a cause for inconsistencies in injuries, DOAs, and stress caused during catching, thus generalizing outcomes from one study using a single type of equipment, a single 'setting' of the equipment, or a single group of catchers should be avoided. Further development of commercially available systems may improve the catching method.

Inverted handling is common when catching chickens manually, yet inherently stressful for broilers and for other livestock species, conscious inversion by the leg(s) would likely not even be considered. However, if inverted handling is done, handling birds by two rather than one leg would be preferred. Numerous researchers and animal welfare organizations recommend a manual catching method in which the bird is held by the abdomen, with a grip over the wings, with one or two birds at a time, while keeping the bird in an upright position (Eilers et al., 2009; Humane Slaughter Association (HSA), 2016; Consortium of the Animal Transport Guides Project, 2017; Eyes on Animals, 2018).

Upright handling can decrease plasma corticosterone levels compared to inverted manual carrying (Kannan and Mench, 1996). A pilot study showed that when catching 1000 birds per method, upright catching by the abdomen tended to result in fewer wing fractures (1 vs. 7 fractures recorded), and resulted in quicker loading times (21 sec difference; mean of 231 sec vs. 252 sec per transport module) and more consistent crate-stocking densities (range of 26–33 birds vs. 25–36 birds per drawer) compared to conventional catching by two legs (Kittelsen et al., 2018). This upright handling method may not be quicker than conventional inverted catching where birds are held by one leg only (Langkabel et al., 2015). Even though in their study upright catching was faster than inverted catching, the catching crew did perceive the upright catching as more of an effort, as it required bending over more frequently. Application of this upright, abdomen catching would result in broiler welfare improvements (injuries and pain) and decreased mortality (Nijdam et al., 2004; Chauvin et al., 2011; Jacobs et al., 2017c; Kittelsen et al., 2018). The numerous benefits of this methodology and the lack of other manual catching alternatives make the upright handling method the best alternative to inverted catching.

An amendment to manual or mechanical catching of broilers could be to herd birds toward a conveyer belt at the front of the building, similar to how broiler turkeys are caught and loaded. Depending on the system, groups of turkeys are separated from the flock and herded through a race, manually or mechanically placed into the container modules (Prescott et al., 2000). However, in turkeys, this method does come with welfare risks affecting up to 66% of turkeys in some cases, with birds suffering from impacted heads, caught tails, bruising or bloody skin lesions (Prescott et al., 2000). Alternatively, birds can be herded onto a conveyer belt, which leads to the container and requires no manual handling (Erasmus, 2018). The herding approach eliminates the need to catch birds by their legs and invert them during carrying and loading (Fig. 3). However, the impact on welfare of the herding method has received limited attention for broilers and has not yet been scientifically evaluated in

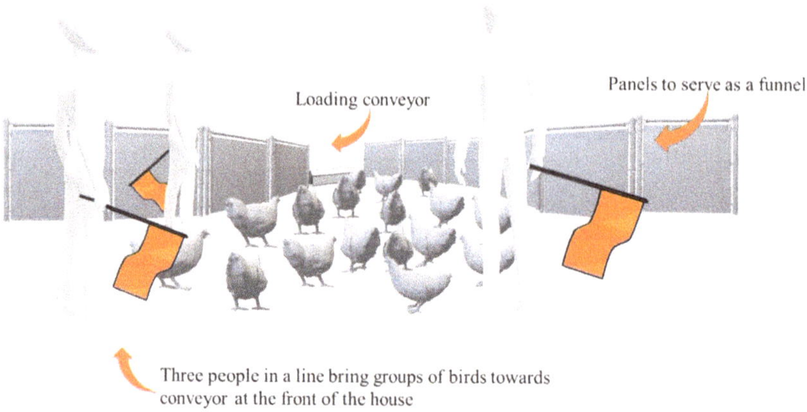


**Figure 3** Herding broiler chickens toward a conveyor belt that loads birds into the transportation container, as done with turkeys prior to transportation. Catchers may use flags, garbage bags, or something otherwise 'noisy' to herd birds toward the conveyor (illustration based on Fig. 20.4, Weeks, 2014).

turkeys. One review described herding birds onto a conveyor, and found that it required a relatively high light intensity of 15 lux, and commented that the method was slow and cumbersome (Kettlewell and Turner, 1985).

So, the potential disadvantage could be that the method may be more time-consuming than conventional manual catching procedures, especially with broiler chickens whose walking ability is often impaired (Silvera et al., 2017). However, according to guidelines and legislation in the European Union, animals that are unable to walk are not allowed to be transported. Thus, this herding method could be an effective way to select out birds with poor gait. An advantage to the herding technique is the limited need for carrying and lifting broilers, which would improve both bird and human welfare (no crouching, no heavy lifting).

### 3.3 Transport and lairage

Even though multiple studies have identified weather conditions, ambient temperatures, and seasons as risk factors for broiler welfare, especially for DOAs, nearly all broiler chickens are transported without active control of the on-truck climate. Thermal comfort could be improved by protecting birds from adverse weather conditions (Jacobs et al., 2017c). A recent review concluded that future developments should focus on increasing control over the thermal environment within trucks (Cockram and Dulal, 2018). Earlier studies have identified thermal stress due to extreme environmental conditions as one of three major risk factors for DOAs (Bayliss and Hinton, 1990; Gregory and

Austin, 1992). Thus, it is clear that broiler welfare during transportation could benefit from control over the thermal environment.

To the authors' knowledge, only one company has developed a climate-conditioned transportation system for broilers in which birds are not exposed to weather conditions. In this commercial system, birds are loaded into 10 separate layers of the truck, rather than in containers. Within the truck, climate is controlled through mechanical ventilation and temperature is monitored (Peer System, 2016). Further developments of controlled-climate transportation trucks could benefit bird welfare during the pre-slaughter phase by limiting thermal stress and plumage soiling.

A number of stressors related to the pre-slaughter phase could be reduced by avoiding the live transportation stage, using on-site slaughter by either mobile or permanent (in case of very large production plants) abattoirs. Current EU legislation (93/119 EC) does allow mobile abattoirs, as in the United States. In the European Union, however, mobile abattoirs, or processing units are a considerable investment (Mancinelli et al., 2018). In the United States approximately nine USDA-inspected mobile poultry units are available, solely used by small-scale operations (USDA, 2010). On-farm slaughter would eliminate the stressors related to transportation and lairage, and would avoid some exposure to unfamiliar surroundings, weather conditions, and road travel. Disadvantages are less control over hygiene, reduced slaughter capacity, fewer advantages related to slaughtering at large economic scale on a centralized location, making the slaughter process more costly (possibly needing more personnel and equipment) (Kettlewell and Turner, 1985). Additionally, catching is still required unless using a whole-house stunning approach or an automated system with conveyer belts moving the birds toward or into the stunning system. Yet for spent layers, economically viable methods (using gas stunning) have been developed for depopulation of laying hens, which could be adapted for broiler slaughter (Webster, Fletcher and Savage, 1996; Berg et al., 2014).

Alternatively, only part of the slaughter process could be performed on-farm. To ensure hygienic conditions and limit bacterial growth, the birds would need to be stunned, exsanguinated, and quickly chilled to below 10°C (Kettlewell and Turner, 1985). More recent results show that carcass holding temperature had a minimal impact on carcass microbiology. Carcasses were stored for 4 h at 4°C, 27°C, or 40°C prior to scalding and further processing, and no significant differences were found in aerobic plate counts, Enterobacteriaceae, Salmonella, or Campylobacter (Bourassa et al., 2019). Furthermore, they found that high holding temperature resulted in similar defeathering success compared to when scalded, which suggests that scalding could be eliminated entirely. However, depending on duration of holding, defeathering can be more difficult due to rigor mortis (holding for longer than 2 h), which requires modification of picking protocols to achieve acceptable defeathering (Harris et al., 2015). Thus, some studies do suggest that on-farm slaughter can be a viable option with regard to food safety, although some procedures may need modification to achieve good carcass quality.

### 3.4 Training, benchmarking, data monitoring and incentives

Training of employees involved with handling birds can reduce pre-slaughter suffering and production losses, as rough handling of broiler chickens during inverted catching was associated with bruising, fractures, acute stress, and fear. Catching crews are generally paid 'per bird', so they do not receive more income if they take their time. Rather, a job done quickly means that they get relatively more money per hour worked. This likely incentivizes catchers to work fast and possibly roughly, rather than slower and gentler. A potential solution for inappropriate, rough handling could lie in training. The incidence of back scratches decreased by 33% after training Brazilian broiler catching crews compared to before training (Pilecco et al., 2013). They trained four catching crews weekly for four weeks, focusing on methodology for upright catching, avoidance of fear-inducing (human) behavior, and environmental aspects (active ventilation and dimmed lighting). Their results show a decrease in back scratch prevalence over time (from approximately 15–10% prevalence) (Pilecco et al., 2013). Training of slaughter plant personnel decreased the prevalence of birds flapping at shackling, birds receiving pre-stun shocks, ineffective neck cuts, broken wings, and red pygostyles (Wigham et al., 2019).

To modify industry stakeholder behavior, cognitive behavior training may be effective, as a person's attitudes and beliefs affect their behavior (theory of planned behavior). This training method aims to achieve attitudinal and behavioral change by targeting both the attitudes underpinning the target behaviors (those that risk animal welfare) and the behaviors themselves. Cognitive behavior training has been effective in changing how stock people handle dairy cattle and pigs (Hemsworth, Coleman and Barnett, 1994; Coleman et al., 2000; Hemsworth et al., 2002). Training improved worker attitudes, reduced fearfulness in pigs, and improved breeding success compared to control farms (Hemsworth, Coleman and Barnett, 1994). Training resulted in declined cattle fearfulness, declined and increased milk yield compared to control farms (Hemsworth et al., 2002). A similar approach could be applied to catching crew training, in which key attitudes will be addressed and in turn behaviors modified.

The training could possibly improve worker satisfaction for catching crews and for producers (better treatment for their animals). Commercial application may reduce common welfare and economic problems associated with the poultry pre-slaughter phase, such as injuries, mortality, carcass contamination, reduced meat quality, and slaughter yield.

To improve accountability, closed-circuit television (CCTV) and person-mounted cameras could be used. CCTV was more common in US slaughter plants than elsewhere. Yet, in 2018, English legislation came into force for mandatory CCTV cameras in slaughter plants (Department for Environment Food & Rural Affairs, 2018; Humane Slaughter Association (HSA), 2018). Either (plant) supervisors or third-party auditors could review video material and act when adjustments are needed.

Recently the Flemish poultry industry agreed on an action plan to reduce pre-slaughter mortality (De Paepe K., pers. comm., 2019). This plan includes detailed engagements and responsibilities of the various actors. It states, for example, that producers ought to give correct information to the transporter about the number of broilers and their mean weight at least two working days beforehand. In turn, the transporter determines the number of vehicles, the number of containers, and the number of broilers per container depending on the transport distance and weather forecast, and informs the producer about these loading instructions. The producer informs the catching crew about these instructions and monitors compliance. If mortality is >1% the animal welfare officer investigates the likely cause and responsible factor. Such an approach could be applied in other countries with similar integrations.

The human component of this production phase cannot be ignored. Intermediary buyers of broilers, for instance slaughter plants or retailers, may put pressure on producers to keep costs low in comparison with competitors, thus if improved welfare conditions would be associated with increased costs, those would likely fall on producers rather than other stakeholders (Appleby, 2005). Labor turnover can be an issue too. As an example, labor turnover in meat and poultry plants can exceed 100% per year at some sites (United States Government Accountability Office (GAO), 2005). Therefore, investing in extensive training for catching crew or slaughter plant employees may not be prioritized. Furthermore, the predominately foreign-born composition of the workforce (United States Government Accountability Office (GAO), 2005) can cause a language barrier (Marín et al., 2009). The use of family, friends, or acquaintances during catching may provide another hurdle with lack of formal training on the topic.

Routine assessment of key welfare indicators combined with incentives to take corrective actions can be very effective in improving broiler welfare during the pre-slaughter phase. Some aspects of welfare are already routinely monitored, such as DOAs and wing fractures (Butterworth et al., 2016; European Union, 2016).

Development of a central database (either within the slaughter plant or on a larger scale) could aid in continuing improvement of animal welfare. Key data could include information on involved parties (producer, catching crew, and transporter) and animal-based welfare indicators. Benchmarking could be used to pinpoint problematic areas and opportunities. Routinely collecting data would provide valuable insights on responsible parties when issues occur. For instance, recording farm of origin, catching crew, transporter, and slaughter plant, in combination with welfare indicators such as DOA prevalence for each flock, will provide details on what farms, catchers, and transporters are more or less likely to be associated with high DOA prevalences.

The next step would be to record additional welfare indicators, either routinely on all flocks, or on flocks that surpass a previously determined threshold level for DOAs. A protocol and a web-based tool have been developed to evaluate and quantify welfare of broilers during the pre-slaughter phase (Jacobs et al., 2019). These tools aim to assist the benchmarking process and are accessible online (https://shiny.ilvo.be/Welltrans/). Incentives or 'premiums' could be provided for producers, catching crews, or transporters with fewest animal welfare problems, and penalties could go to those with the highest level of welfare issues. This has worked well in Denmark, where footpad dermatitis levels improved on-farm, with severe dermatitis prevalence reduced from 60% in 2002 to 10% in 2012 (European Union, 2016). Their slaughter plants routinely apply evaluations since 2002, with economic incentives for low footpad dermatitis levels. In addition, they mandate reduction of stocking density on-farm when annual average level of dermatitis exceeds the trigger level, and they request information from producers if levels reach a specific trigger level, requiring action if the subsequent flock also exceeds the threshold.

Throughout the pre-slaughter phase we recommend a systematic approach of monitoring broiler welfare indicators, with feedback to relevant stakeholders, and encouragement to act with incentives and penalizations when certain thresholds are (repeatedly) surpassed. A continuous cycle of monitoring, followed by action when adjustments are needed, with responsibility taken by the industry where appropriate, will lead to continuous improvement of the system. Initiative can be taken either by the industry, governmental agencies, or other actors such as retailers.

## 4 The pre-slaughter phase for laying hens

The pre-slaughter phase of laying hens is briefly covered in this section. Laying hen transportation is similar to broiler transportation, in that birds are fasted, caught and loaded, transported over the road, laired and slaughtered at a plant. However, laying hen transportation comes with its own set of challenges. Spent layers are used for further processed products, such as soups, or can be rendered to obtain animal by-product (e.g. for poultry feed). Therefore, spent laying hens have little economic value and the economic incentive to handle birds carefully is minimal (Webster, Fletcher and Savage, 1996; Mitchell and Kettlewell, 2009a). It also proves more of a challenge to catch birds, as they are housed in more complex systems such as wire cages or aviary systems. Laying hens are placed in crates rather than container systems, which often results in forceful placement of one or more birds through a small crate opening (to avoid placed birds getting out). Laying hens are more active compared to the more docile broiler chickens, with more fragile bones, which could make handling more difficult. In the United States, manual catching and loading is avoided by using pullet carts. As described by Newberry et al. (1999) 'folding doors are lifted to place hens inside the compartments of the cart. The carts, each of which holds about 210 hens, are rolled along the rows of cages, allowing rapid transfer of hens from the cages into the carts. With no further handling of the hens, the filled carts are rolled directly onto the transport trailer'.

Furthermore, few poultry slaughter plants will process laying hens, resulting, on average, in longer transportation distances compared to broilers. Similar to broilers, laying hens show an acute stress response after conventional, inverted handling with plasma corticosterone levels consistently higher than after gentle handling (Knowles and Broom, 1990).

Laying hens' physical condition can exacerbate the stressors of transportation (Knowles and Broom, 1990) such as age, body weight, and plumage condition. Injuries occur when spent laying hens are caught and loaded into transport crates, although more so in conventionally caged laying hens than in birds with access to tiers and perches during production (Whitehead and Fleming, 2000). Laying hens are particularly sensitive to bone fractures due to osteoporosis, which is a consequence of prolonged and intensive egg production and limited exercise (Whitehead and Fleming, 2000; Kim et al., 2005). Twenty-nine percent of previously caged laying hens obtained one or more bone fractures during the pre-slaughter phase (Gregory and Wilkins, 1989). In contrast, a more recent study on 24 flocks showed only 0.13% damage prevalence, including fractures (van Niekerk, Gunnink and Reuvekamp, 2017). Besides the welfare concern, reducing bone fractures during the pre-slaughter phase can benefit processors, as it would reduce bone fragments in meat (Farm Animal Welfare Council (FAWC), 2010).

Laying hens are prone to experience cold stress (Richards et al., 2012). Poor plumage condition and low metabolic activity due to feed withdrawal make spent layers especially vulnerable to cold stress when transported in cold and wet conditions (Mitchell and Kettlewell, 2009a). This is illustrated by the thermal comfort zone that was modeled for poorly feathered hens, which was between 22°C and 28°C compared to the comfort zone between 10°C and 15°C for well-feathered broilers (Weeks, Webster and Wyld, 1997).

DOAs can be more prevalent in layers compared to broilers, with an average of 1.22% reported from a sample of 54 million spent hens commercially transported to Italian slaughter plants (Petracci et al., 2006). After

2 h of transportation, on average, and 8 h of lairage, 0.28% DOA prevalence was found (van Niekerk, Gunnink and Reuvekamp, 2017) and a survey within Great Britain showed a mean of 0.27% DOAs in 2009 (Weeks et al., 2012). Transportation distance and low ambient temperatures were identified as major risks for DOAs (Voslarova et al., 2007; Weeks et al., 2012), with amean DOA of 0.59% in transports under 50 km, and 1.64% in transports up to 300 km (Voslarova et al., 2007). For these spent hens, discussion is ongoing whether they should be transported at all, with on-farm depopulation as an alternative (Nielsen, Dybkjr and Herskin, 2011), for instance, by $CO_2$ gassing birds in small groups (Newberry et al., 1999). Depopulation for rendering (used for oil, bone meal, and other by-products) is a common method of disposal in the United States with 47% of birds rendered compared to 39% of the birds slaughtered in 2010 (USDA - APHIS, 2011). The disposal for rendering occurs on-farm without transportation (Newberry et al., 1999).

Low economic value of laying hens makes it difficult to incentivize people to handle laying hens more carefully, if that would require more time or money. One approach to make it more feasible, would be to market laying hen meat to consumers, for instance, with a sustainability argument (consumption of hens at end of lay is more sustainable than discarding the carcass for rendering, with more waste). 'Selling' the consumption of hen meat to consumers was identified as a challenge or need from the industry in the EU Hennovation project (EU hennovation, 2017).

Guidelines developed in that project provide practical approaches to improve welfare during the pre-slaughter phase of laying hens, focusing on fitness-for-transport, manual handling, thermal stress while in-crate (Temple et al., 2017). The need for alternatives to conventional catching was identified by stakeholders, especially for hens kept in non-cage systems (EU hennovation, 2017). Upright, abdominal manual catching as an alternative to inverted conventional catching could reduce injury prevalence in laying hens as described for broilers and has been recommended for laying hens by animal welfare organizations and scientific consortia (Humane Slaughter Association (HSA), 2016; Temple et al., 2017; Eyes on Animals, 2018).

Furthermore, knowledge transfer on best practices and guidelines for end of laying hen welfare was prioritized by industry stakeholders (EU hennovation, 2017) and could stimulate continuous improvements for animal welfare due to 'peer pressure'.

## 5 Conclusions and future trends

Broiler chicken welfare can be severely impaired during the pre-slaughter phase, with injuries and mortality as major welfare concerns. In addition, these welfare issues put a strain on production incomes, with global economic losses estimated between $118 million and $14 billion annually. The large, and increasing, number of sentient animals involved in this industry emphasizes the need for improvements in this production phase. The industry should ensure that animals are fit for transport with research needed on valid indicators and feasible evaluation methods. Distress, fear, injury and DOAs due to catching and loading need to be addressed, with research needed on the impact of inversion of birds in itself, on the economics of upright, abdominal catching, on refining mechanical catching methods, and on other alternatives to manual catching than those currently available. Furthermore, the feasibility of air-conditioned transportation needs to be examined. Routine data collection and data storage at the plant could benefit benchmarking and animal welfare through incentives or penalties. There is no data published on the reliability of DOA and rejection rate numbers collected by slaughter plant personnel, especially on inter- and intra-rater reliability.

Human-related challenges, including but not limited to economic pressure to keep costs low, large turnover in the work force, and stakeholder attitudes toward animals may be major hurdles that need to be overcome. Routine monitoring, and actions following when needed, will continuously improve welfare for commercial broiler chickens.

The low economic value of spent laying hens, the birds' fitness, and the complex housing systems provide challenges to the pre-slaughter phase of laying hens. On-farm processing for consumption or rendering could limit the impact of some of these challenges. Future research on alternative catching and loading methods, and data monitoring were prioritized by industry stakeholders.

## 6 Where to look for further information

A recent and comprehensive review on DOAs and injuries during the pre-slaughter phase of broilers can be found here:

- Cockram, M. S. and Dulal, K. J. (2018) 'Injury and mortality in broilers during handling and transport to slaughter', Canadian Journal of Animal Science, 98, pp. 416–432. doi: 10.1139/cjas-2017-0076.

The Hennovation project was completed in 2017. Project findings and resources related to end-of-lay transport can be found on the project website:

- http://hennovation.eu/index.html.

Researchers at ILVO and Ghent University have developed a freely available manual (for now, only in Dutch) and an online tool to evaluate broiler chicken welfare during the pre-slaughter phase. The welfare score calculations are based on expert opinions. This online tool is available in multiple languages and can be used for flock benchmarking:

- https://shiny.ilvo.be/Welltrans/.
- https://www.ilvo.vlaanderen.be/Portals/68/documents/Mediatheek/Mededelingen/248_Vleeskippenwelzijn_tijdens_pre_slachtfase_UPDATE.pdf.

Further information can be available from the following:

- EFSA Panel on Animal Health and Welfare (AHAW); Scientific Opinion concerning the welfare of animals during transport. EFSA Journal 2011; 9(1):1966 [125 pp.].doi:10.2903/j.efsa.2011.1966. Available online: www.efsa.europa.eu/efsajournal.htm.
- http://www.animaltransportguides.eu/ (and more specifically http://animaltransportguides.eu/wp-content/uploads/2016/05/EN-Guides-Poultry-final.pdf).
- Livestock Handling & Transport, 5th edition (ed. T. Grandin), CABI 2019 and in particular Ch. 20 Poultry handling & transport by C. Weeks, F. A. M. Tuyttens & T. Grandin.

## 7 References

Abeyesinghe, S. M., Wathes, C. M., Nicol, C. J. and Randall, J. M. (2001). The aversion of broiler chickens to concurrent vibrational and thermal stressors, *Applied Animal Behaviour Science* 73(3), 199–215. doi: 10.1016/S0168-1591(01)00142-3.

Agri Stats Inc. (2018). *Broiler Chicken Production in the United States*. https://www.agristats.com/.

Akşit, M., Yalçin, S., Ozkan, S., Metin, K. and Ozdemir, D. (2006). Effects of temperature during rearing and crating on stress parameters and meat quality of broilers, *Poultry Science* 85(11), 1867–1874.

Animalia (2015). *Status report on Norwegian meat and egg production*. Available at: http://flashbook.no/animalia/kjottetstilstand15/pubData/source/148314-Animalia-KT15-Flashbook.pdf (Accessed: 30 July 2019).

Appleby, M. C. (2005). The relationship between food prices and animal welfare, *Journal of Animal Science*. American Society of Animal Science, 83(13), E9–E12. doi: 10.2527/2005.8313_SUPPLE9X.

Baéza, E., Arnould, C., Jlali, M., Chartrin, P., Gigaud, V., Mercerand, F., Durand, C., Méteau, K., Le Bihan-Duval, E. and Berri, C. (2012). Influence of increasing slaughter age of chickens on meat quality, welfare, and technical and economic results, *Journal of Animal Science*. Oxford University Press 90(6), 2003–2013. doi: 10.2527/jas.2011-4192.

Barbosa, C. F., Carvalho, R. Hd, Rossa, A., Soares, A. L., Coró, F. A. G., Shimokomaki, M. and Ida, E. I. (2013). Commercial preslaughter blue light ambience for controlling broiler stress and meat qualities, *Brazilian Archives of Biology and Technology* 56(5), 817-821. doi: 10.1590/S1516-89132013000500013.

Bayliss, P. A. and Hinton, M. H. (1990). Transportation of broilers with special reference to mortality rates, *Applied Animal Behaviour Science* 28(1-2), 93-118. doi: 10.1016/0168-1591(90)90048-I.

Bedáňová, I., Voslárová, E., Chloupek, P., Pistěková, V., Suchy, P., Blahova, J., Dobsikova, R. and Vecerek, V. (2007). Stress in broilers resulting from shackling, *Poultry Science* 86(6), 1065-1069. doi: 10.1093/ps/86.6.1065.

Bedáňová, I., Voslárová, E., Vecerek, V., Pistěková, V. and Chloupek, P. (2005). Effects of reduction in floor space during crating on haematological indices in broilers, *Berliner und Munchener Tierarztliche Wochenschrift* 119(1-2), 17-21.

Berg, C., Yngvesson, J., Nimmermark, S., Sandström, V. and Algers, B. (2014). Killing of spent laying hens using CO2 in poultry barns, *Animal Welfare* 23(4), 445-457. doi: 10.7120/09627286.23.4.445.

Bourassa, D. V., Harris, C. E., Bartenfeld, L. N., Richter, S., Daley, W., Wilson, K. M. and Buhr, R. J. (2019). Impact of postmortem holding temperature on feather retention force and broiler carcass microbiology. In: *International Poultry Scientific Forum*. Atlanta, GA.

Buhr, R. J., Northcutt, J. K., Lyon, C. E. and Rowland, G. N. (1998). Influence of time off feed on broiler viscera weight, diameter, and shear, *Poultry Science*. Oxford University Press 77(5), 758-764. doi: 10.1093/ps/77.5.758.

Burkholder, K. M., Thompson, K. L., Einstein, M. E., Applegate, T. J. and Patterson, J. A. (2008). Influence of stressors on normal intestinal microbiota, intestinal morphology, and susceptibility to Salmonella enteritidis colonization in broilers, *Poultry Science*. Oxford University Press 87(9), 1734-1741. doi: 10.3382/ps.2008-00107.

Burlinguette, N. A., Strawford, M. L., Watts, J. M., Classen, H. L., Shand, P. J. and Crowe, T. G. (2012). Broiler trailer thermal conditions during cold climate transport, *Canadian Journal of Animal Science* 92(2), 109-122. doi: 10.4141/cjas2011-027.

Butterworth, A., de Jong, I. C., Keppler, C., Knierim, U., Stadig, L. and Lambton, S. (2016). What is being measured, and by whom? Facilitation of communication on technical measures amongst competent authorities in the implementation of the European Union Broiler Directive (2007/43/EC), *Animal* 10(2), 302-308. doi: 10.1017/S1751731115001615.

Caffrey, N. P. P., Dohoo, I. R. R. and Cockram, M. S. S. (2017). Factors affecting mortality risk during transportation of broiler chickens for slaughter in Atlantic Canada, *Preventive Veterinary Medicine*. Elsevier 147(September), 199-208. doi: 10.1016/j.prevetmed.2017.09.011.

Centers for Disease Control and Prevention (2012). Notes from the field: multistate outbreak of Salmonella Altona and Johannesburg infections linked to chicks and ducklings from a mail-order hatchery - United States, February-October 2011, *MMWR. Morbidity and Mortality Weekly Report* 61(11), 195. Available at: http://www.ncbi.nlm.nih.gov/pubmed/22437914 (Accessed: 5 June 2019).

Chauvin, C., Hillion, S., Balaine, L., Michel, V., Peraste, J., Petetin, I., Lupo, C. and Le Bouquin, S. (2011). Factors associated with mortality of broilers during transport to slaughterhouse, *Animal* 5(2), 287-293. doi: 10.1017/S1751731110001916.

Chloupek, P., Voslářová, E., Chloupek, J., Bedáňová, I., Pištěková, V. and Večerek, V. (2009). Stress in broiler chickens due to acute noise exposure, *Acta Veterinaria Brno* 78(1), 93-98.

Cockram, M. S. and Dulal, K. J. (2018). Injury and mortality in broilers during handling and transport to slaughter, *Canadian Journal of Animal Science* 98(3), 416–432. doi: 10.1139/cjas-2017-0076.

Coleman, G. J., Hemsworth, P. H., Hay, M. and Cox, M. (2000). Modifying stockperson attitudes and behaviour towards pigs at a large commercial farm, *Applied Animal Behaviour Science* 66(1–2), 11–20. doi: 10.1016/S0168-1591(99)00073-8.

Consortium of the Animal Transport Guides Project (2017). *Good practices for animal transport in the EU: poultry*. Available at: www.animaltransportguides.eu (Accessed: 7 June 2019).

Deeb, N. and Cahaner, A. (1999). The effects of naked neck genotypes, ambient temperature, and feeding status and their interactions on body temperature and performance of broilers, *Poultry Science* 78(10), 1341–1346. doi: 10.1093/ps/78.10.1341.

Delezie, E. (2006). *Manual and mechanical catching and transport of broilers: implications for welfare, physiology and product quality and ethical considerations*, Doctoral Thesis. Faculteit Bio-ingenieurswetenschappen. Available at: http://hdl.handle.net/1979/426.

Delezie, E., Lips, D., Lips, R. and Decuypere, E. (2006). Is the mechanisation of catching broilers a welfare improvement? *Animal Welfare* 15(2), 141–147.

Delezie, E., Swennen, Q., Buyse, J. and Decuypere, E. (2007). The effect of feed withdrawal and crating density in transit on metabolism and meat quality of broilers at slaughter weight, *Poultry Science* 86(7), 1414–1423. doi: 10.1093/ps/86.7.1414.

Department for Environment Food & Rural Affairs (2018). *CCTV in slaughterhouses: rules for operators, GOV.uk*. Available at: https://www.gov.uk/government/publications/cctv-in-slaughterhouses-rules-for-operators (Accessed: 24 September 2019).

Dunkley, C. S. (2014). *Global warming: how does it relate to poultry?* Available at: https://extension.uga.edu/publications/detail.html?number=B1382&title=Global Warming, *How Does it Relate to Poultry?* (Accessed: 23 September 2019).

EFSA (2011). Scientific opinion concerning the welfare of animals during transport, *EFSA Journal* 9(1), 1–44. doi: 10.2903/j.efsa.2011.1966.

Eilers, K., Bokkers, E., Mourits, M. and van Kooten, M. (2009). 'Kip, ik heb je!' Het vangen van vleeskuikens – een economische verkenning van verschillende potentieel dierenwelzijnsverbeterende alternatieven. Report 256. Wageningen University, Wageningen, the Netherlands. Available at: www.wetenschapswinkel.wur.nl (Accessed: 4 Sept 2019).

Ekstrand, C. (1998). An observational cohort study of the effects of catching method on carcase rejection rates in broilers, *Animal Welfare* 7(1), 87–96.

Emmerson, D. A. (1997). Commercial approaches to genetic selection for growth and feed conversion in domestic poultry, *Poultry Science*. Oxford University Press 76(8), 1121–1125.

Erasmus, M. A. (2018). Welfare issues in turkey production, *Advances in Poultry Welfare*. Woodhead Publishing, pp. 263–291. doi: 10.1016/B978-0-08-100915-4.00013-0.

EU Hennovation – HORIZON 2020 ISIB-02-2014 project, Grant no. 652638 (2017) *Horizon 2020 EU Research and Innovation programme*. Available at: http://hennovation.eu/index.html (Accessed: 4 September 2019).

European Commission (2005). Council Regulation (EC) No. 1/2005 on the Protection of Animals during Transport and Related Operations and Amending Directives 64/432/EEC and 93/119/ EC and Regulation (EC) No. 1255/97. Available at: https://eur

-lex.europa.eu/legal-content/EN/TXT/PDF/?uri=CELEX:32005R0001andfrom=en (Accessed: 23 Sept 2019).

European Commission (2009). *EC Regulation no 1069/2009*. Available at: https://eur-lex.europa.eu/legal-content/EN/TXT/?uri=CELEX:32009R1069 (Accessed: 23 September 2019).

European Union (2016). *Use of slaughterhouse data to monitor welfare of broilers on farm - Publications Office of the EU*. Luxembourg. Available at: https://publications.europa.eu/en/publication-detail/-/publication/9fbf913d-de15-11e6-ad7c-01aa75ed71a1/language-en/format-PDF (Accessed: 12 June 2019).

Eyes on Animals (2018). *KIP, IK HEB JE! Gids voor het zorgvuldig vangen, laden en transporteren van legkippen*. Available at: https://www.eyesonanimals.com/wp-content/uploads/2018/05/2018-02-22-Kippenvangersbrochure.pdf (Accessed: 30 July 2019).

Farm Animal Welfare Council (FAWC) (2010). *Opinion on osteoporosis and bone fractures in laying hens*. London. Available at: www.fawc.org.uk (Accessed: 13 June 2019).

Farsaie, A., Carr, L. E. and Wabeck, C. J. (2013). Mechanical harvest of broilers, *Transactions of the ASAE*. American Society of Agricultural and Biological Engineers, 26(6), 1650-1653. doi: 10.13031/2013.33818.

Food and Agriculture Organization of the United Nations (FAO) (2017). FAOSTAT. Available at: http://www.fao.org/faostat/en/#data/QL (Accessed: 20 November 2018).

Francis, C. A., Macleod, M. G. and Anderson, J. E. M. (1991). Alleviation of acute heat stress by food withdrawal or darkness, *British Poultry Science*. Taylor & Francis Group 32(1), 219-225. doi: 10.1080/00071669108417343.

Gouveia, K. G., Vaz-Pires, P. and Martins da Costa, P. (2009). Welfare assessment of broilers through examination of haematomas, foot-pad dermatitis, scratches and breast blisters at processing, *Animal Welfare* 18(1), 43-48.

Government of Canada (2019). *Health of animals regulations - Part XII*. Available at: https://laws-lois.justice.gc.ca/eng/regulations/c.r.c.,_c._296/page-15.html#h-548075 (Accessed: 5 June 2019).

Gregory, N. G. and Austin, S. D. (1992). Causes of trauma in broilers arriving dead at poultry processing plants, *The Veterinary Record*. British Medical Journal Publishing Group 131(22), 501-503. doi: 10.1136/VR.131.22.501.

Gregory, N. G. and Wilkins, L. J. (1989). Broken bones in domestic fowl: handling and processing damage in end-of-lay battery hens, *British Poultry Science* 30(3), 555-562. doi: 10.1080/00071668908417179.

Grilli, C., Loschi, A. R., Rea, S., Stocchi, R., Leoni, L. and Conti, F. (2015). Welfare indicators during broiler slaughtering, *British Poultry Science*. Taylor & Francis 56(1), 1-5. doi: 10.1080/00071668.2014.991274.

Harris, C. E., Wilson, K. M., Bourassa, D. V. and Buhr, R. J. (2015). Defeathering of broiler carcasses subjected to delayed scalding 1, 2, 4, and 8 hours after slaughter. In: *International Poultry Scientific Forum*. Atlanta, GA.

Haslam, S. M., Knowles, T. G., Brown, S. N., Wilkins, L. J., Kestin, S. C., Warriss, P. D. and Nicol, C. J. (2008). Prevalence and factors associated with it, of birds dead on arrival at the slaughterhouse and other rejection conditions in broiler chickens, *British Poultry Science*. Taylor & Francis 49(6), 685-696.

Hemsworth, P. H., Coleman, G. J. and Barnett, J. L. (1994). Improving the attitude and behaviour of stockpersons towards pigs and the consequences on the behaviour and reproductive performance of commercial pigs, *Applied Animal Behaviour Science*. Elsevier 39(3–4), 349–362. doi: 10.1016/0168-1591(94)90168-6.

Hemsworth, P. H., Coleman, G. J., Barnett, J. L., Borg, S. and Dowling, S. (2002). The effects of cognitive behavioral intervention on the attitude and behavior of stockpersons and the behavior and productivity of commercial dairy cows, *Journal of Animal Science* 80(1), 68–78. Available at: http://www.ncbi.nlm.nih.gov/pubmed/11831530 (Accessed: 19 March 2018).

Higham, L. E., Scott, C., Akehurst, K., Dring, D., Parnham, A., Waterman, M. and Bright, A. (2018). Effects of financial incentives and cessation of thinning on prevalence of Campylobacter: a longitudinal monitoring study on commercial broiler farms in the UK, *Veterinary Record*. British Veterinary Association 183(19), 595. doi: 10.1136/vr.104823.

Humane Slaughter Association (HSA) (2016). *Practical slaughter of poultry*. Available at: http://www.hsa.org.uk/shop/publications-1/product/practical-slaughter-of-poultry.

Humane Slaughter Association (HSA) (2018). *Media release - monitoring animal welfare at slaughter- new CCTV legislation*. Available at: https://www.gov.uk/government/news/cctv-in-slaughterhouses-legislation-laid-in- (Accessed: 24 September 2019).

Jacobs, L. (2016). Road to better welfare - Welfare of broiler chickens during transportation, Doctoral Thesis. Faculty of Veterinary Medicine, Ghent University.

Jacobs, L., Delezie, E., Duchateau, L., Goethals, K., Ampe, B., Lambrecht, E., Gellynck, X. and Tuyttens, F. A. (2016). Effect of post-hatch transportation duration and parental age on broiler chicken quality, welfare, and productivity, *Poultry Science* 95(9), 1973-1979. doi: 10.3382/ps/pew155.

Jacobs, L., Delezie, E., Duchateau, L., Goethals, K. and Tuyttens, F. A. M. (2017a). Broiler chickens dead on arrival: associated risk factors and welfare indicators, *Poultry Science* 96(2), 259–265. doi: 10.3382/ps/pew353.

Jacobs, L., Delezie, E., Duchateau, L., Goethals, K., Vermeulen, D., Buyse, J. and Tuyttens, F. (2017b). Fit for transport? broiler chicken fitness assessment for transportation to slaughter, *Animal Welfare* 26(3), 335–343. doi: 10.7120/09627286.26.3.335.

Jacobs, L., Delezie, E., Duchateau, L., Goethals, K. and Tuyttens, F. A. M. (2017c). Impact of the separate pre-slaughter stages on broiler chicken welfare, *Poultry Science* 96(2), 266–273. doi: 10.3382/ps/pew361.

Jacobs, L., Delezie, E., Duchateau, L., Goethals, K., Ampe, B., Buyse, J. and Tuyttens, F. A. M. (2017d). Impact of transportation duration on stress responses in day-old chicks from young and old breeders, *Research in Veterinary Science*. Elsevier 112, 172–176. doi: 10.1016/j.rvsc.2017.04.015.

Jacobs, L., Delezie, E., Goethals, K., Ampe, B., Duchateau, L., Watteyn, A. and Tuyttens, F. A. M. (2019). *Vleeskippenwelzijn tijdens de pre-slachtfase - Update - 1.2: evaluatieprotocol en online integratie-tool (ILVO Mededeling 248)*. Melle, Belgium. Available at: https://www.ilvo.vlaanderen.be/Portals/68/documents/Mediatheek/Mededelingen/248_Vleeskippenwelzijn_tijdens_pre_slachtfase_UPDATE.pdf.

Jones, R. B. (1992). The nature of handling immediately prior to test affects tonic immobility fear reactions in laying hens and broilers, *Applied Animal Behaviour Science* 34(3), 247–254. doi: 10.1016/S0168-1591(05)80119-4.

Jones, R. B., Satterlee, D. G. and Cadd, G. G. (1998). Struggling responses of broiler chickens shackled in groups on a moving line: effects of light intensity, hoods, and curtains, *Applied Animal Behaviour Science*. Elsevier 58(3–4), 341–352.

Kannan, G. and Mench, J. A. (1996). Influence of different handling methods and crating periods on plasma corticosterone concentrations in broilers, *British Poultry Science* 37(1), 21–31. doi: 10.1080/00071669608417833.

Kettlewell, P. J. and Turner, M. J. B. (1985). A review of broiler chicken catching and transport systems, *Journal of Agricultural Engineering Research*. Academic Press 31(2), 93–114. doi: 10.1016/0021-8634(85)90064-2.

Kim, W. K., Donalson, L. M., Herrera, P., Kubena, L. F., Nisbet, D. J. and Ricke, S. C. (2005). Comparisons of molting diets on skeletal quality and eggshell parameters in hens at the end of the second egg-laying cycle, *Poultry Science* 84(4), 522-527. doi: 10.1093/ps/84.4.522.

Kittelsen, K. E., Granquist, E. G., Aunsmo, A. L., Moe, R. O. and Tolo, E. (2018). An evaluation of two different broiler catching methods, *Animals : an Open Access Journal from MDPI*. Multidisciplinary Digital Publishing Institute 8(8), 8. doi: 10.3390/ANI8080141.

Kittelsen, K. E., Granquist, E. G., Kolbjørnsen, Ø., Nafstad, O. and Moe, R. O. (2015a). A comparison of post-mortem findings in broilers dead-on-farm and broilers dead-on-arrival at the abattoir, *Poultry Science* 94(11), 2622–2629. doi: 10.3382/ps/pev294.

Kittelsen, K. E., Granquist, E. G., Vasdal, G., Tolo, E. and Moe, R. (2015b). Effects of catching and transportation versus pre-slaughter handling at the abattoir on the prevalence of wing fractures in broilers, *Animal Welfare* 24(4), 387–389. doi: 10.7120/09627286.24.4.387.

Kittelsen, K. E., David, B., Moe, R. O., Poulsen, H. D., Young, J. F. and Granquist, E. G. (2016). Associations among gait score, production data, abattoir registrations, and postmortem tibia measurements in broiler chickens, *Poultry Science* 96(5), 1033-1040, (0513). doi: 10.3382/ps/pew433.

Kittelsen, K. E., Moe, R. O., Hoel, K., Kolbjørnsen, Ø, Nafstad, O. and Granquist, E. G. (2017). Comparison of flock characteristics, journey duration and pathology between flocks with a normal and a high percentage of broilers "dead-on-Arrival" at abattoirs, *Animal* 11(12), 2301-2308. doi: 10.1017/S1751731117001161.

Knezacek, T. D., Olkowski, A. A., Kettlewell, P. J., Mitchell, M. A. and Classen, H. L. (2010). Temperature gradients in trailers and changes in broiler rectal and core body temperature during winter transportation in Saskatchewan, *Canadian Journal of Animal Science*. Ottawa, Canada: NRC Research Press 90(3), 321-330. doi: 10.4141/CJAS09083.

Knierim, U. and Gocke, A. (2003). Effect of catching broilers by hand or machine on rates of injuries and dead-on-arrivals, *Animal Welfare*. Universities Federation for Animal Welfare 12(1), 63–73.

Knowles, T. G. and Broom, D. M. (1990). The handling and transport of broilers and spent hens, *Applied Animal Behaviour Science*. Elsevier 28(1-2), 75–91. doi: 10.1016/0168-1591(90)90047-H.

Knowles, T. G., Warriss, P. D., Brown, S. N., Edwards, J. E. and Mitchell, M. A. (1995). Response of broilers to deprivation of food and water for 24 hours, *British Veterinary Journal*. Elsevier 151(2), 197–202.

Koolman, L., Whyte, P. and Bolton, D. J. (2014). An investigation of broiler caecal Campylobacter counts at first and second thinning, *Journal of Applied Microbiology*. John Wiley & Sons, Ltd 117(3), 876–881. doi: 10.1111/jam.12580.

Kopecsnik, M. (2008). Effect of transport conditions on the losses during transport of broiler chickens, *Magyar Allatorvosok Lapja*. Springer Science and Media 130(7), 391-395 [abstract only]. Available at: http://www.uni-vet.hu/mal (Accessed July 2019).

Kotula, K. L. and Wang, Y. (1994). Characterization of broiler meat quality factors as influence by feed withdrawal time, *Journal of Applied Poultry Research* 3(2), 103-110. doi: 10.1093/japr/3.2.103.

Lacy, M. P. and Czarick, M. (1998). Mechanical harvesting of broilers, *Poultry Science* 77(12), 1794-1797. Available at: https://academic.oup.com/ps/article-abstract/77/12/1794/1567678 (Accessed: 12 July 2019).

Lambrecht, E., Jacobs, L., Delezie, E., De Steur, H., Gellynck, X. and Tuyttens, F. (2020). Stakeholder perceptions on broiler chicken welfare during first-day processing and the pre-slaughter phase: a case study in Belgium. *World's Poultry Science Journal*. doi:10.1080/00439339.2020.1790329

Langkabel, N., Baumann, M. P., Feiler, A., Sanguankiat, A. and Fries, R. (2015). Influence of two catching methods on the occurrence of lesions in broilers, *Poultry Science* 94(8), 1735-1741. doi: 10.3382/ps/pev164.

Ljungberg, D., Gebresenbet, G. and Aradom, S. (2007). Logistics chain of animal transport and abattoir operations, *Biosystems Engineering*. Academic Press 96(2), 267-277. doi: 10.1016/J.BIOSYSTEMSENG.2006.11.003.

Lund, V. P., Kyvsgaard, N. C., Christensen, J. P. and Bisgaard, M. (2013). Pathological manifestations observed in dead-on-arrival broilers at a Danish abattoir, *British Poultry Science*. Taylor & Francis 54(4), 430-440. doi: 10.1080/00071668.2013.804173.

Lupo, C., Bougeard, S., Balaine, L., Michel, V., Petetin, I., Colin, P., LeBouquin, S. and Chauvin, C. (2010). Risk factors for sanitary condemnation in broiler chickens and their relative impact: application of an original multiblock approach, *Epidemiology and Infection*. Cambridge University Press 138(3), 364-375. doi: 10.1017/S0950268809990549.

Lyon, C. E., Papa, C. M. and Wilson, R. L. (2012). Effect of feed withdrawal on yields, muscle pH, and texture of broiler breast meat, *Poultry Science*. Oxford University Press 70(4), 1020-1025. doi: 10.3382/ps.0701020.

Macdonald, J. M. (2014). *Technology, Organization, and Financial Performance in U.S. Broiler Production*. United States Department of Agriculture. Economic Research Service, Economic Information Bulletin, Number 126, EIB-126. June 2014. www.ers.usda.gov/publications/eib-economic-information-bulletin/eib126.aspx (Accessed July 2019).

Main, D. C. J. and Hewson, C. J. (2012). Slaughter and killing of animals for disease control. *World Animal Protection*. Available at: https://www.globalanimalnetwork.org/concepts-animal-welfare-16-livestock-slaughter-and-killing-animals-disease-control.

Mancinelli, A. C., Dal Bosco, A., Mattioli, S., Ranucci, D. and Castellini, C. (2018). Mobile poultry processing unit as a resource for small poultry farms: planning and economic efficiency, animal welfare, meat quality and sanitary implications, *Animals* 8(12), 1-16. doi: 10.3390/ani8120229.

Manufacturing Journal (2019). Innovation and drive for success - Myronivsky Hliboproduct, *Manufacturing Journal*. Available at: http://manufacturing-journal.net/company-profiles-home/50-agriculture/1869-innovation-and-drive-for-success-myronivsky-hliboproduct (Accessed: 23 September 2019).

Marín, A. J., Grzywacz, J. G., Arcury, T. A., Carrillo, L., Coates, M. L. and Quandt, S. A. (2009). Evidence of organizational injustice in poultry processing plants: possible effects on occupational health and safety among latino workers in North Carolina, *American Journal of Industrial Medicine*. John Wiley & Sons, Ltd 52(1), 37–48. doi: 10.1002/ajim.20643.

May, J. D., Lott, B. D. and Deaton, J. W. (1990). The effect of light and environmental temperature on broiler digestive tract contents after feed withdrawal, *Poultry Science*. Oxford University Press 69(10), 1681–1684.

Mir, N. A., Rafiq, A., Kumar, F., Singh, V. and Shukla, V. (2017). Determinants of broiler chicken meat quality and factors affecting them: a review, *Journal of Food Science and Technology*. Springer 54(10), 2997–3009. doi: 10.1007/s13197-017-2789-z.

Mitchell, M. A. (2009). Chick transport and welfare, *Avian Biology Research* 2(1–2), 99–105. doi: 10.3184/175815509X431894.

Mitchell, M. A. and Kettlewell, P. J. (1998). Physiological stress and welfare of broiler chickens in transit: solutions not problems!, *Poultry Science* 77(12), 1803–1814. doi: 10.1093/ps/77.12.1803.

Mitchell, M. A. and Kettlewell, P. J. (2009a). Transport of chicks, pullets and spent hens. In: G.C. Perry (Ed.) *Welfare of the Laying Hen*. Oxon, UK: CAB International, pp. 361–374. doi: 10.1079/9780851998138.0361.

Mitchell, M. A. and Kettlewell, P. J. (2009b). Welfare of poultry during transport – a review. In: *Poultry Welfare Symposium*, Cervia, Italy, (May), pp. 90–100.

Myronivsky Hliboproduct (MHP) Myronivsky (2018). *Our operations*. Available at: https://www.mhp.com.ua/en/operations/op-vinnitskaja-ptitsefabrika-oao-mkhp (Accessed: 23 September 2019).

Najafi, P., Zulkifli, I., Soleimani, A. F. and Goh, Y. M. (2016). Acute phase proteins response to feed deprivation in broiler chickens, *Poultry Science* 95(4), 760–763. doi: 10.3382/ps/pew001.

National Chicken Council (2019). *Broiler chicken industry key Facts 2019*. Available at: https://www.nationalchickencouncil.org/about-the-industry/statistics/broiler-chicken-industry-key-facts/ (Accessed: 8 July 2019).

Newberry, R. C., Webster, A. B., Lewis, N. J. and Van Arnam, C. (1999). Management of spent hens, *Journal of Applied Animal Welfare Science* 2(1), 13–29. doi: 10.1207/s15327604jaws0201_2.

Nicol, C. J. and Scott, G. B. (1990). Pre-slaughter handling and transport of broiler chickens, *Applied Animal Behaviour Science*. Elsevier 28(1–2), 57–73.

van Niekerk, T. G. C. M., Gunnink, H. and Reuvekamp, B. F. J. (2017). Welfare of end-of-lay hens during catching and transport in the winter period. In: *Xth European Symposium on Poultry Welfare*. Ploufragan, France: World's Poultry Science Association, p. 91. Available at: https://www.wpsa.fr/images/publications/Abstract-Book-ESPWelfare.pdf#page=208 (Accessed: 13 June 2019).

Nielsen, B. L., Dybkjr, L. and Herskin, M. S. (2011). Road transport of farm animals: effects of journey duration on animal welfare, *Animal* 5(3), 415–427. doi: 10.1017/S1751731110001989.

Nijdam, E., Arens, P., Lambooij, E., Decuypere, E. and Stegeman, J. A. (2004). Factors influencing bruises and mortality of broilers during catching, transport, and lairage, *Poultry Science* 83(9), 1610–1615. doi: 10.1093/ps/83.9.1610.

Nijdam, E., Delezie, E., Lambooij, E., Nabuurs, M. J., Decuypere, E. and Stegeman, J. A. (2005a). Comparison of bruises and mortality, stress parameters, and meat quality in manually and mechanically caught broilers, *Poultry Science* 84(3), 467–474. doi: 10.1093/ps/84.3.467.

Nijdam, E., Delezie, E., Lambooij, E., Nabuurs, M. J., Decuypere, E. and Stegeman, J. A. (2005b). Feed withdrawal of broilers before transport changes plasma hormone and metabolite concentrations, *Poultry Science* 84(7), 1146–1152. doi: 10.1093/ps/84.7.1146.

Nijdam, E., Zailan, A. R. M., Van Eck, J. H. H., Decuypere, E. and Stegeman, J. A. (2006). Pathological features in dead on arrival broilers with special reference to heart disorders, *Poultry Science* 85(7)(1986), 1303–1308.

Northcutt, J. K., Savage, S. I. and Vest, L. R. (1997). Relationship between feed withdrawal and viscera condition of broilers, *Poultry Science* 76(2), 410–414. doi: 10.1093/ps/76.2.410.

Peer System (2016), *Peer System – Transport*. Available at: http://www.peersystem.nl/en/products/transport (Accessed: 12 June 2019).

Perez, M. P., Palacio, J., Santolaria, M. P., del Aceña, M. C., Chacón, G., Verde, M. T., Calvo, J. H., Zaragoza, M. P., Gascón, M. and García-Belenguer, S. (2002). Influence of lairage time on some welfare and meat quality parameters in pigs, *Veterinary Research* 33(3), 239–250. doi: 10.1051/vetres:2002012.

Petracci, M., Bianchi, M. and Cavani, C. (2005). Pre-slaughter factors affecting mortality, live weight loss, and carcass quality in broiler chickens. In: *Proceedings of the XVII European Symposium on the Quality of Poultry Meat and XI European Symposium on Therapeutic Quality of Eggs and Egg Products*, Doorwerth, The Netherlands, pp. 251–255.

Petracci, M., Bianchi, M., Cavani, C., Gaspari, P. and Lavazza, A. (2006). Preslaughter mortality in broiler chickens, turkeys, and spent hens Under commercial slaughtering, *Poultry Science* 85(9), 1660–1664. doi: 10.1093/ps/85.9.1660.

Petracci, M., Fletcher, D. L. and Northcutt, J. K. (2001). The effect of holding temperature on live shrink, processing yield, and breast meat quality of broiler chickens, *Poultry Science* 80(5), 670–675. doi: 10.1093/PS/80.5.670.

Pilecco, M., Almeida Paz, I., Tabaldi, L., Nääs, I., Garcia, R., Caldara, F. and Francisco, N. (2013). Training of catching teams and reduction of back scratches in broilers, *Revista Brasileira de Ciência Avícola* 15(3), 283–286. doi: 10.1590/S1516-635X2013000300017.

Poultry Industry Council (2017). Should this bird be loaded? A guide for preparing, loading, and transporting poultry. Available at: http://www.poultryindustrycouncil.ca/wp-content/uploads/2017/03/DT-Handbook-final.compressed.pdf.

Prescott, N. B., Berry, P. S., Haslam, S. and Tinker, D. B. (2000). Catching and crating turkeys: effects on carcass damage, heart rate, and other welfare parameters, *Journal of Applied Poultry Research* 9(3), 424–432. doi: 10.1093/japr/9.3.424.

Richards, G. J., Wilkins, L. J., Weeks, C. A., Knowles, T. G. and Brown, S. N. (2012). Evaluation of the microclimate in poultry transport module drawers during the marketing process of end-of-lay hens from farm to slaughter, *The Veterinary Record*. British Medical Journal Publishing Group 171(19), 474. doi: 10.1136/vr.100844.

Ritz, C. W., Webster, A. B. and Czarick, M. (2005). Evaluation of hot weather thermal environment and incidence of mortality associated with broiler live haul, *Journal of Applied Poultry Research*. Oxford University Press 14(3), 594–602. doi: 10.1093/japr/14.3.594.

Robins, A. and Phillips, C. J. C. (2011). International approaches to the welfare of meat chickens, *World's Poultry Science Journal* 67(2), 351–369. doi: 10.1017/S0043933911000341.

Sandercock, D. A., Hunter, R. R., Nute, G. R., Mitchell, M. A. and Hocking, P. M. (2001). Acute heat stress-induced alterations in blood acid-base status and skeletal muscle membrane integrity in broiler chickens at two ages: implications for meat quality, *Poultry Science*. Oxford University Press 80(4), 418–425. doi: 10.1093/ps/80.4.418.

Sandercock, D. A., Hunter, R. R., Mitchell, M. A. and Hocking, P. M. (2006). Thermoregulatory capacity and muscle membrane integrity are compromised in broilers compared with layers at the same age or body weight, *British Poultry Science* 47(3), 322–329. doi: 10.1080/00071660600732346.

Sani, P. (2006). *Thinning flocks to fatten profits, the Poultry site*. Available at: http://www.thepoultrysite.com/cocciforum/issue12a/3/thinning-flocks-to-fatten-profits/ (Accessed: 21 February 2019).

Savenije, B., Schreurs, F. J., Winkelman-Goedhart, H. A., Gerritzen, M. A., Korf, J. and Lambooij, E. (2002). Effects of feed deprivation and electrical, gas, and captive needle stunning on early postmortem muscle metabolism and subsequent meat quality, *Poultry Science* 81(4), 561–571. doi: 10.1093/ps/81.4.561.

Scott, A. B., Singh, M., Toribio, J. A., Hernandez-Jover, M., Barnes, B., Glass, K., Moloney, B., Lee, A. and Groves, P. (2017). Comparisons of management practices and farm design on Australian commercial layer and meat chicken farms: cage, barn and free range, *PLOS ONE*. Public Library of Science 12(11), e0188505. doi: 10.1371/journal.pone.0188505.

Siegel, P. B. and Wisman, E. L. (1966). Selection for body weight at eight weeks of age: 6. Changes in appetite and feed utilization. *Poultry Science* 45(6), 1391–1397. doi: 10.3382/ps.0451391.

Silvera, A. M., Knowles, T. G., Butterworth, A., Berckmans, D., Vranken, E. and Blokhuis, H. J. (2017). Lameness assessment with automatic monitoring of activity in commercial broiler flocks, *Poultry Science* 96(7), 2013–2017. doi: 10.3382/ps/pex023.

Smith, D. P., Lyon, C. E. and Lyon, B. G. (2002). The effect of age, dietary carbohydrate source, and feed withdrawal on broiler breast fillet color, *Poultry Science* 81(10), 1584–1588. doi: 10.1093/ps/81.10.1584.

Temple, D., Weeks, C., van Niekerk, T. and Manteca, X. (2017). *Guidelines end of Lay Hennovation*. Available at: https://docs.google.com/viewerng/viewer?url=http://hennovation.eu/onewebmedia/Guidelines%2520EoL%2520Hennovation%2520v3_FINAL%2520(1).pdf (Accessed: 4 September 2019).

Trocino, A., Piccirillo, A., Birolo, M., Radaelli, G., Bertotto, D., Filiou, E., Petracci, M. and Xiccato, G. (2015). Effect of genotype, gender and feed restriction on growth, meat quality and the occurrence of white striping and wooden breast in broiler chickens, *Poultry Science*. Oxford University Press 94(12), 2996–3004. doi: 10.3382/ps/pev296.

Tuyttens, F. A. M., Vanhonacker, F. and Verbeke, W. (2014). Broiler production in Flanders, Belgium: current situation and producers' opinions about animal welfare, *World's Poultry Science Journal* 70(2), 343–354. doi: 10.1017/S004393391400035X.

United States Federal Government (2019) *9 CFR § 381.95 - Disposal of condemned poultry products*. Available at: https://www.govinfo.gov/app/details/CFR-2019-title9-vol2/CFR-2019-title9-vol2-sec381-95 (Accessed: 23 September 2019).

United States Government Accountability Office (GAO) (2005). *Safety in the meat and poultry industry, while improving, could be further strengthened*. Available at: www.gao.gov/cgi-bin/getrpt?GAO-05-96 (Accessed: 5 June 2019).

USDA (1966). *Animal welfare act*. Available at: https://www.nal.usda.gov/awic/animal-welfare-act (Accessed: 12 June 2019).

USDA (2010). *An introduction to mobile slaughter units*. Available at: https://www.usda.gov/media/blog/2010/08/30/introduction-mobile-slaughter-units (Accessed: 23 September 2019).

USDA (2011). *Broiler industry highlights, ARMS*. Available at: https://www.nass.usda.gov/Surveys/Guide_to_NASS_Surveys/Ag_Resource_Management/ARMS_Broiler_Factsheet/#Broiler_weight (Accessed: 3 July 2019).

USDA APHIS (2011). *Poultry 2010 - structure of the U.S. Poultry industry*. Available at: http://nahms.aphis.usda.gov (Accessed: 8 July 2019).

Vanderhasselt, R., Sprenger, M., Tuyttens, F. A. M., De Brabander, D., Cangar, O., Bahr, C., Berckmans, D., Vranken, E., Decuypere, E., Buyse, J., Lips, D., Cox, M., De Baere, K. and Zoons, J. (2010). *Welzijnsnormen bij Vleeskippen (WELBROIL), FOD Volksgezondheid*. Available at: https://pure.ilvo.be/portal/files/910722/synthese_rapport.pdf.

Vecerek, V., Grbalova, S., Voslarova, E., Janackova, B. and Malena, M. (2006). Effects of travel distance and the season of the year on death rates of broilers transported to poultry processing plants, *Poultry Science* 85(11), 1881–1884. doi: 10.1093/ps/85.11.1881.

Vecerek, V., Voslarova, E., Conte, F., Vecerkova, L. and Bedanova, I. (2016). Negative trends in transport-related mortality rates in broiler chickens, *Asian-Australasian Journal of Animal Sciences*. Asian-Australasian Association of Animal Production Societies (AAAP) 29(12), 1796–1804. doi: 10.5713/ajas.15.0996.

Villarroel, M., Pomares, F., Ibáñez, M. A., Lage, A., Martínez-Guijarro, P., Méndez, J. and De Blas, C. (2018). Rearing, bird type and pre-slaughter transport conditions I. Effect on dead on arrival, *Spanish Journal of Agricultural Research* 16(2), e0503. doi: 10.5424/sjar/2018162-12013.

Voslarova, E., Janackova B, B., Vitula, F., Kozak, A. and Vecerek, V. (2007). Effects of transport distance and the season of the year on death rates among hens and roosters in transport to poultry processing plants in the Czech Republic in the period from 1997 to 2004, *Veterinarni Medicina* 52(6), 262–266. Available at: https://www.researchgate.net/profile/Eva_Voslarova/publication/242545279_Effects_of_transport_distance_and_the_season_of_the_year_on_death_among_hens_and_roosters_in_transport_to_poultry_processing_plants_in_Czech_Republic_in_the_period_from_1997_to_2004/ (Accessed: 13 June 2019).

Warriss, P. D., Bevis, E. A., Brown, S. N. and Edwards, J. E. (1992). Longer journeys to processing plants are associated with higher mortality in broiler chickens, *British Poultry Science* 33(1), 201–206. doi: 10.1080/00071669208417458.

Warriss, P. D., Knowles, T. G., Brown, S. N., Edwards, J. E., Kettlewell, P. J., Mitchell, M. A. and Baxter, C. A. (1999). Effects of lairage time on body temperature and glycogen reserves of broiler chickens held in transport modules, *Veterinary Record* 145(8), 218–222. doi: 10.1136/vr.145.8.218.

Warriss, P. D. (2003). Optimal lairage times and conditions for slaughter pigs: a review, *Veterinary Record* 153(6), 170–176. doi: 10.1136/vr.153.6.170.

Warriss, P. D., Wilkins, L. J., Brown, S. N., Phillips, A. J. and Allen, V. (2004). Defaecation and weight of the gastrointestinal tract contents after feed and water withdrawal in broilers, *British Poultry Science*. Taylor & Francis 45(1), 61–66. doi: 10.1080/0007166041668879.

Warriss, P. D., Pagazaurtundua, A. and Brown, S. N. (2005). Relationship between maximum daily temperature and mortality of broiler chickens during transport and lairage, *British Poultry Science* 46(6), 647–651. doi: 10.1080/00071660500393868.

Webster, A. B., Fletcher, D. L. and Savage, S. I. (1996). Humane on-farm killing of spent hens, *Journal of Applied Poultry Research* 5(2), 191–200. doi: 10.1093/japr/5.2.191.

Weeks, C. (2014). Poultry handling and transport. In: Grandin, T. (ed.) *Livestock Handling and Transport: Theories and Applications* (4th edn). Available at: https://books.google.com/books?hl=en&lr=&id=O8eWBAAAQBAJ&oi=fnd&pg=PA378&dq=pre-slaughter+herding+turkeys&ots=o2IfltjHUh&sig=CqkqIwLOkrU1OcSq_hfjIy7_M4M#v=onepage&q&f=false (Accessed: 1 July 2019).

Weeks, C. A., Brown, S. N., Richards, G. J., Wilkins, L. J. and Knowles, T. G. (2012). Levels of mortality in hens by end of lay on farm and in transit to slaughter in Great Britain, *Veterinary Record* 170(25), 647–647. doi: 10.1136/vr.100728.

Weeks, C. A., Webster, A. J. F. and Wyld, H. M. (1997). Vehicle design and thermal comfort of poultry in transit, *British Poultry Science*. Taylor & Francis Group 38(5), 464–474. doi: 10.1080/00071669708418023.

Whitehead, C. C. and Fleming, R. H. (2000). Osteoporosis in cage layers, *Poultry Science* 79(7), 1033–1041. doi: 10.1093/ps/79.7.1033.

Wigham, E., Grist, A., Mullan, S., Wotton, S. and Butterworth, A. (2019). The influence of welfare training on bird welfare and carcass quality in two commercial poultry primary processing plants, *Animals*. MDPI AG 9(8), 584. doi: 10.3390/ani9080584.

Wolff, I., Klein, S., Rauch, E., Erhard, M., Mönch, J., Härtle, S., Schmidt, P. and Louton, H. (2019). Harvesting-induced stress in broilers: comparison of a manual and a mechanical harvesting method under field conditions, *Applied Animal Behaviour Science*. Elsevier B.V. 221. doi: 10.1016/j.applanim.2019.104877.

Zulkifli, I., Che Norma, M. T., Chong, C. H. and Loh, T. C. (2000). Heterophil to lymphocyte ratio and tonic immobility reactions to preslaughter handling in broiler chickens treated with ascorbic acid, *Poultry Science* 79(3), 402–406. doi: 10.1093/ps/79.3.402.

# Chapter 15

# Improving welfare in poultry slaughter

*Dorothy McKeegan, Institute of Biodiversity, Animal Health and Comparative Medicine, University of Glasgow, UK; and Jessica Martin, The Royal (Dick) School of Veterinary Studies and The Roslin Institute, University of Edinburgh, UK*

1 Introduction
2 Lairage
3 Stunning methods
4 Conclusions
5 Where to look for further information
6 References

## 1 Introduction

In 2018, 68.7 billion chickens were slaughtered for food globally. Based on current trends, this number is predicted to increase to 72.4 billion by 2020, based on an average yearly 2.6% increase (Food and Agriculture Organization of the United Nations (FAO), 2020). In 2018, in the United Kingdom, 1.14 billion broilers were slaughtered across 50 operating slaughter houses and 57 million spent hens were processed in 9 plants (Defra, 2019; Food and Agriculture Organization of the United Nations (FAO), 2020). In comparison for the same year, just over 9 billion broilers were slaughtered in the United States, across approximately 320 plants (United States Department of Agriculture, 2019). Excluding farmed and wild-caught fish, poultry production represents, by a large margin, the greatest number of individual animals killed by humans for food production. This reflects their status as the most numerous terrestrial production animals. Because of the number of sentient animals involved, the manner of their deaths is extremely important, since it may be argued that our ethical responsibilities to animals under our care extend to providing them with good lives and good deaths (Singer, 2016).

For clarity, it is helpful to explain the commonly used terminology in relation to the slaughter of chickens. Slaughter may be defined as the killing of animals intended for human consumption, usually involving exsanguination (bleeding out). During slaughter, animals are usually stunned prior to exsanguination to render them unconscious (though certain types of religious slaughter do

http://dx.doi.org/10.19103/AS.2020.0078.14

not apply stunning, and we will discuss the welfare implications and scale of religious slaughter later in this chapter). Stunning refers to any intentionally induced process which causes loss of consciousness and sensibility with the minimum associated stress and pain, including any process resulting in instantaneous death (European Commission, 2017; Nielsen et al., 2019a). Stunning improves welfare at slaughter because its aim is to prevent the animal from experiencing pain from the neck cut and its own death via blood loss and associated brain hypoxia. Stunning methods can result in a recoverable state (if exsanguination does not occur), referred to as a simple stun or can induce a 'non recovery state', such that the animal would not recover, even in the absence of exsanguination.

The term 'killing' refers to any intentionally induced process which causes the death of an animal (EC, 2009). Emergency killing means the killing of animals which are injured or have a disease associated with severe pain or suffering and where there is no other practical possibility to alleviate this pain or suffering (EC, 2009). Killing methods do not have to include a pre-kill stun step. Therefore the animals may be conscious (if not already unconscious as are result of sickness or injury) when the killing technique is applied and will potentially be able to experience negative emotions (e.g. pain and stress), if the killing method does not render them immediately unconscious. The term 'culling' is used technically to describe identifying and removing individuals from a group (Fetrow et al., 2006), but it has become common to use this term also to describe the removal and killing (usually limited to non-slaughter purposes) of individuals. Euthanasia is defined literally as providing a 'good death' and/or providing a death that is in the animal's interest (usually to relieve suffering). In practice this usually means killing an animal as painlessly as possible with an overdose of anaesthetic. Strictly, the term is only applicable to sick or injured animals (veterinary patients) or animals used in scientific research (Leary et al., 2020, 2007). Euthanasia is often misused as a term for describing killing of animals in other contexts. Loss of consciousness describes a transition in which the animal moves from a state of conscious awareness (connectedness to the environment and responsiveness to stimuli, for example, painful events) to unconsciousness which occurs when the ability to maintain an awareness of self and environment is lost (involving a complete or near-complete lack of responsiveness to environmental stimuli). The existence of an inducible state of unconsciousness and associated changes in brain state and resulting electroencephalogram (EEG) characteristics appear to be common across vertebrates (Pierre et al., 2018; Verhoeven et al., 2014).

In many jurisdictions, animals can only be exsanguinated after stunning, and in accordance with certain allowed methods, with some exceptions authorised for religious rites. For effective stunning, the induced loss of consciousness and sensibility must be maintained until the death of the animal.

In addition to slaughter, large numbers of farmed poultry are killed for other reasons, primarily on-farm due to individual ill health, non-competitive growth, emergency killing for disease control or depopulation (killing of low-value poultry at the end of their productive life) (Gerritzen et al., 2000; Martin et al., 2019b; McKeegan et al., 2011). These activities are associated with related but also distinct welfare challenges and are reviewed elsewhere (McKeegan, 2018; Nielsen et al., 2019b; Sparrey et al., 2014). In this chapter, we will focus on the welfare challenges presented by routine slaughter of chickens for food production. This includes chickens reared for meat (broilers) and also laying hens (so-called spent hens, which are usually slaughtered at the end of the egg production cycle), though it should be noted that the vast majority of available research relates to broilers. We will examine the various methods available and in current use, discussing their welfare costs and benefits. Given the large literature already available on some types of stunning and slaughter, and existing thorough reviews of the physiological basis for stunning methods (Berg and Raj, 2015; Blokhuis et al., 2004; Bøtner et al., 2012; Nielsen et al., 2019a; Raj, 1998; Shields and Raj, 2010), we will focus on outstanding welfare issues, possible opportunities to address these and examine the potential of emerging approaches. The slaughter procedure necessarily follows catching and transportation, processes which also have specific welfare impacts. Again, these are reviewed elsewhere (see Bøtner et al., 2011; Kettlewell and Mitchell, 1994; Knowles and Broom, 1990; Mitchell and Kettlewell, 1994) and will not be included here. We discuss welfare issues related to lairage (the holding period prior to slaughter, discussed briefly in Chapter 13 of this volume) and pre-slaughter handling, when directly relevant to the experience of birds. We will begin by briefly outlining some relevant regulatory frameworks, with a focus on the European Union which is widely recognised to have the most stringent legal protection for animals at the time of killing (EC, 2009). We will then discuss current and emerging methods, concluding with prospects for improvement of welfare based on available systems and identification of knowledge gaps for research.

## 1.1 Regulatory framework

Most countries have regulations that relate to animal welfare and/or meat hygiene at slaughter. Description of the full range of international regulation relevant to this issue is beyond the scope of this chapter, but we will briefly outline some relevant regulatory features, with a focus on the European Union (EU) which imposes detailed regulatory requirements on member states (EC, 2009). In the EU, Council Regulation (EC) No 1099/2009 on the protection of animals at the time of killing regulates the slaughter of animals. As a regulation it automatically became law in member states, with no transposition or scope for local interpretation. However, some member states retain national legislation which may provide additional protection at slaughter. Regulation 1099/2009 requires the use of approved stunning methods for poultry – these include electrical water bath stunning (with minimum electrical requirements) and controlled-atmosphere stunning (including a defined range of allowed gas mixtures and Low Atmospheric Pressure Stunning (LAPS) parameters). The regulation also imposes requirements for the layout, construction and equipment within slaughter houses, the appointment of an Animal Welfare Officer (accountable for implementing animal welfare measures) and certificates of competence for training of staff who handle live animals. Each member state has the responsibility for enforcement of legislation within its territory, but the DG for Health and Food Safety has an overall role for carrying out audits and inspections within member states. The requirement for allowable stunning methods extends to slaughter houses in third countries that export meat to the EU. The contents of Regulation 1099/2009 are evidence based, primarily supported by Opinions of the European Food Safety Authority (EFSA) on welfare at slaughter in 2004 and 2013 (Authie et al., 2013; Blokhuis et al., 2004; More et al., 2017) and the 'precautionary principle' – a strategy of caution when approaching issues of potential harm when extensive scientific knowledge on the matter is lacking.

Following EU exit, the UK will preserve the existing legislative framework initially via a general Withdrawal Bill, and it has been noted that EU exit could present opportunities to review and improve welfare standards (McCulloch, 2018). However, increased costs associated with more stringent welfare standards will need to be balanced with a desire to protect the competitiveness of UK producers in the EU and global markets (European Union, 2017). On exit from the EU, the UK will remain a member of the Council of Europe, an international organisation with 47 member states, which is distinct from the EU. The Council of Europe cannot make binding laws, but it has produced six conventions on animal welfare (including one on slaughter (European Council, 1979)) that act as a framework for animal welfare standards.

Although the United States is the world's largest poultry meat producer with 18% of global output (followed by China, Brazil and the Russian Federation), there are currently no federal regulations to control or safeguard the welfare of animals used in agriculture. An Animal Welfare Act (1966) is in place, but it applies only to animals kept for non-farming purposes. State laws regulate animal welfare in some parts of the country, but no such legislation applies to poultry in the three major poultry-producing states (Georgia, Alabama and Arkansas). Instead, voluntary welfare guidelines (those of the US National Chicken Council) are adopted in the absence of legislation, but these refer primarily to rearing conditions for chickens (e.g. stocking density limits), and not slaughter conditions (National Chicken Council (NCC), 2011). Poultry are not included in the Humane Slaughter Act (US Department of Agriculture, 1958) and electrical stunning (which is used almost universally in the United States) is not subject to stringent technical requirements (as in the EU). Oversight of slaughter is provided by a Poultry Health Veterinarian or other authorised personnel who are required to perform a routine inspection of slaughter procedures to ensure that slaughter is being undertaken according to legislative requirements and good commercial practices. Similarly, China (a key poultry meat producer and the biggest egg producer with 42% of global production), has no legislation with regard to poultry killing and slaughter. Instead, province-specific guidelines are in development. For example, in 2016, Shandong Province issued China's first technical manual for the more humane killing of chickens, which included new standard requirements such as birds being stunned by either gas or electrical methods prior to killing. Where laws are not present, guidelines may inform the basis of welfare standards and practical implementation. For example, the World Organisation for Animal Health (OIE) has developed in its Terrestrial Animal Health Code a chapter on the slaughter of animals (chapter 7.5) in order to ensure welfare during pre-slaughter and slaughter processes for the majority of livestock species (OIE (World Animal Health Organization), 2019).

## 2 Lairage

Lairage is a holding period at the processing plant prior to slaughter (Gregory, 2008; Warriss et al., 2005), lasting from arrival until animals enter the slaughter process. Lairage serves multiple functions – it ensures a continuous and smooth supply of animals to the slaughter line, it provides for ante-mortem inspection to take place for public health and welfare assessment purposes, and it allows birds to rest following transport. The lairage period is usually a maximum of a few hours (often a similar length to the transport journey time), but is occasionally longer (Caffrey et al., 2017; Gregory, 2008; Warriss et al., 2005). Grilli et al. (2015) reported lairage times for 233 different batches of broilers as ranging between 0.2 and 9.4 h with a mean of 4 h, while another study (Jacobs et al., 2017) based on six processing plants found lairage durations from 15 min to 9 h, with a mean of approximately 4.5 h. There is evidence that lengthy lairage durations (more than 6 h) lead to heightened stress and associated metabolic disorders due to prolonged feed and water withdrawal, with significant welfare consequences (Rodrigues et al., 2017; Vieira et al., 2011b).

During lairage, birds are held in their transport containers which are stacked, having been unloaded from transport vehicles, usually by forklift trucks. The condition of the birds at lairage is dependent on multiple factors including rearing conditions, catching practices and transport circumstances. Lairage, therefore, represents the culmination of a series of welfare challenges, with transport the most severe and best characterised of these. Transport imposes multiple stressors (e.g. acceleration, motion, vibration, withdrawal of food and water, social disruption and noise) as well as potentially extreme thermal microenvironments (Mitchell and Kettlewell, 1994, 1998) and its welfare impact may be exacerbated by injuries sustained at catching (Warriss et al., 1992). Spent laying hens endure longer transport distances than broilers (due to fewer processing sites) and are more prone to skeletal injury from handling than broilers (Knowles and Broom, 1990; Knowles and Wilkins, 1998). The most basic index of welfare in lairage is the number of birds dead on arrival (DOA). The number of birds DOA is of interest primarily economically and various studies have reported DOA rates and attempted to determine causative factors. Most studies report DOA rates up to 0.5% (e.g. Bayliss and Hinton, 1990; Ekstrand, 1998; Nijdam et al., 2004; Warriss et al., 1992) with occasionally higher rates, for example, 2% in broilers and 6.6% in spent hens (Petracci et al., 2006). The influences on rates of DOA are complex and include transport duration, waiting time, bird sex (and therefore weight, heavier birds are at greater risk), stocking density, ambient temperature, whether flocks had been previously been thinned, season of the year and feed withdrawal times exceeding 6 h (Cockram et al., 2019; Nijdam et al., 2004; Villarroel et al., 2018; Warriss et al., 1992). Although dead birds may be identified at ante-mortem inspection, this involves only a small sample of animals, so the true extent of DOA is only known at the point of manual shackling either before electrical stunning or after controlled atmosphere stunning (CAS). At least one known available slaughter system has dispensed with lairage and unloads modules directly on to the slaughter line from the vehicle (Meyn, 2020). There is no published research on this practice, and its welfare consequences presumably represent a balance between the negative impact of omitting a rest period versus the positive aspects of reducing feed withdrawal duration and exposure to other challenges in lairage.

Deaths detected at lairage and at subsequent handling most likely reflect mortality during transport, but it has been noted that holding in transport crates at lairage may also generate stressful thermal microclimates (Hunter et al., 1998; Quinn et al., 1998). Ideally, lairage facilities should be designed to protect bird welfare (Grilli et al., 2015), but many lairage areas are open to the outside and therefore reflect ambient conditions which can present thermal challenges. Additionally, limited ventilation in closely stacked containers may be worsened compared to transport by the lack of passive air movement in the absence of forward motion of a vehicle. The same point applies to delayed unloading from the transport vehicle, during which temperature can rise rapidly (Warriss et al., 1999, 2005). Given the practice of vertical stacking of transport containers, solid crate floors are preferred as they prevent birds from being exposed to the droppings of those above; however, perforated floors promote air movement. Forced ventilation and/or climate control is recommended and increasingly utilised in lairages to ensure comfortable thermal conditions, especially in warmer climates, and has been shown to reduce heat stress and mortality (Vieira et al., 2010, 2011a). It is also recommended that transport and lairage activities are reactive to weather conditions, avoiding the hottest part of the day and providing shade from the sun (Warriss et al., 2005). Some processing plants mist transport modules with water in lairage to promote evaporative heat loss, reducing heat stress (Jiang et al., 2015). Space between stacked transport crates is important and should be increased in hot conditions to promote ventilation. Conversely, cold stress in lairage can be reduced by providing shelter from wind and stacking crates closer together. Dimmed or blue lighting (which is perceived by birds as a lower light intensity) is routinely used in lairage to calm the birds and promote resting behaviour (Fig. 1). There is no published evidence of the effect of this practice in lairage, but evidence from rearing environments supports the notion that modified lighting may reduce stress following disturbance and reduces activity (Lewis and Morris, 2000; Mohamed et al., 2014; Prayitno et al., 1997a,b).

The transition from lairage to the slaughter line also presents welfare challenges. Manipulation of containers should avoid tilting, dropping or shaking them (this also applies to unloading from the transport vehicle), and container design is crucial to minimise injury to birds from body parts being caught between containers or in crate openings during stacking and unstacking

**Figure 1** Poultry lairage facility showing stacked transport modules and the use of dimmed blue lighting. Image courtesy of Henny Reimert and Marien Gerritzen, Wageningen Livestock Research.

(Nielsen et al., 2019a). Various methods are used to remove birds from transport crates, and most relevant to welfare are those that take place before stunning. Manual removal of birds from crates (e.g. to allow shackling for electrical stunning, see below) has the potential to cause pain and fear, particularly if birds are handled roughly or held by inappropriate body parts leading to skin lesions, wing fractures and bruising (Jacobs et al., 2017; Sparrey and Kettlewell, 1994). Crate design has a major impact on the success of removal of birds and can contribute to injury if the opening is narrow, and modular systems with larger openings reduce this risk (Tinker et al., 2004). Kittelsen et al. (2015) reported that more wing fracture injuries occurred during pre-slaughter handling than at catching and transport (increasing from 0.8% in lairage to 2.9% after shackling, based on broilers undergoing electrical stunning). Some stunning systems involve tipping of the container such that birds are released and fall onto a flat sprung surface, for delivery onto a conveyer belt (Raj and Tserveni-Gousi, 2000). This practice is likely to cause fear and risks injury, especially if birds pile up, though overcrowding can be reduced with correct conveyor speeds.

It is clear that there are multiple opportunities to improve welfare at lairage including optimising transport conditions, use of controlled temperature lairage environments, minimising the duration of lairage (and thus the food and water withdrawal period), optimised container design and careful bird handling. Appropriate handling in particular is recognised as a key factor in protecting welfare prior to slaughter (OIE (World Animal Health Organization), 2019) and there is evidence that a positive management attitude and staff training is crucial to raise awareness (Nielsen et al., 2019a).

## 3 Stunning methods

In this section the major methods of stunning applied to chickens at slaughter will be discussed. Essentially, these are electrical methods and those that modify the atmosphere. In the UK and Europe, CAS (gas stunning) is most common, while electrical stunning is used in the rest of the world (Defra, 2019; Nielsen et al., 2019a). Mechanical methods are used only as a backup measure for commercial slaughter and will be discussed briefly. Different stunning and slaughter methods have strengths and weaknesses with regard to welfare outcomes, and the relevant risks and hazards for each method were recently reviewed by EFSA (Nielsen et al., 2019a). They noted that two categories of hazards are apparent during stunning – those leading to negative welfare during induction of unconsciousness, and those resulting from a delay or failure to achieve loss of consciousness (Nielsen et al., 2019a). This second category relates to the risk of birds being conscious and therefore exposed to the harms of further processing, since in all cases stunning is followed by neck cutting and exsanguination. Blood loss is the cause of death if the stun is reversible, or may occur after death with stun-kill methods. For chickens, neck cutting is usually done automatically and ideally involves the severing of both carotid arteries (Gregory and Wotton, 1986); however, occasionally only one or partial cuts are made. Whether this is a welfare issue depends on the type of stunning used - if reversible, stunning duration should be sufficient to allow death from blood loss without recovery of consciousness. Whatever method is used, the imperative must be to avoid the severely painful and fearful situations of ineffective stunning, leading to persistence of consciousness during neck cutting, or recovery of consciousness during bleeding.

### 3.1 Electrical stunning

Electrical stunning involves application of an electrical current to the brain, with the intention of inducing unconsciousness immediately prior to exsanguination. This approach is based on the principle of electrical stimulation causing immediate generalised epileptiform activity in the brain and associated unconsciousness and insensibility, protecting welfare during the neck cut and bleeding-out period. Epileptiform activity represents a temporary disruption of normal brain function, consisting of a tonic (e.g. whole body stiffening) and clonic (e.g. convulsions) phase, followed by a period of neuronal exhaustion and then recovery (Raj and Tserveni-Gousi, 2000). There is evidence that chickens do not always show typical epileptiform activity in the brain after electrical stunning (Gregory and Wotton, 1987), but application of adequate current is followed by a period of quiescent or supressed activity (as measured by the EEG, a global measure of brain activity), indicating effective stunning (Raj, 1998). In research settings, various criteria have been used to determine the effectiveness of electrical stunning, including monitoring of EEG activity, abolition of somatosensory evoked potentials in the brain and induction of cardiac arrest (Berg and Raj, 2015; Raj, 1998). Depending on whether cardiac arrest is caused, electrical stunning (electronarcosis) can be a reversible stunning method, so a killing method must be rapidly applied to abolish the risk of recovery of consciousness.

In poultry, electrical stunning is normally applied using a water bath stunner (Fig. 2), whereby the birds are hung upside down on a moving line with their feet held in grounded metal shackles. Shackling involves manually unloading birds from transport containers, inversion of the body by operators and insertion of the legs into parallel metal slots (termed shackles) on a moving conveyor, which transports the birds to the water bath. The birds' heads are then dipped into an electrically charged water bath, closing the circuit and causing current to flow through the head and body to the shackle. As such, birds are potentially stunned as soon as their heads enter the water (Bilgili, 1999; Devos et al., 2018; Prinz et al., 2010a), which allows for automation and high throughput, but shackling is

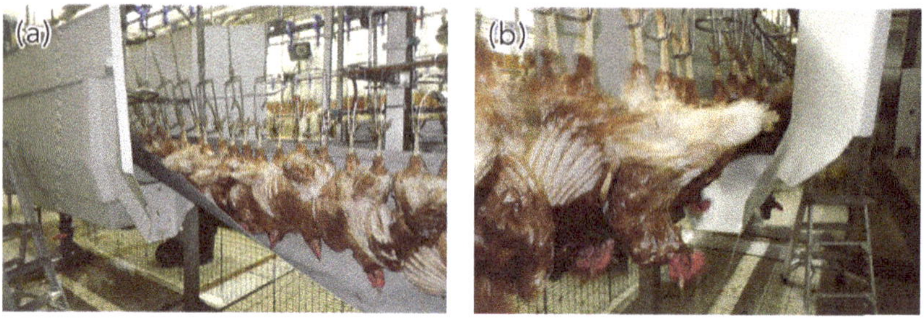


**Figure 2** Water bath stunning of spent hens demonstrating entry to the water bath (a) and exit in an unconscious tonic state (b) image courtesy of Henny Reimert and Marien Gerritzen, Wageningen Livestock Research.

still done manually. While electrical water bath stunning can, in theory, result in immediate loss of consciousness, several major welfare concerns are associated with this approach. These include the risk of ineffective stunning (caused by inadequate current), the painful and aversive nature of shackling, and the risk of pre-stun shocks (e.g. electrically 'live' water on the ramp leading to the water bath coming into contact with the bird's body or wing tips). Other methods of electrical stunning (head only and head to body) are available but are not widely used in commercial slaughter plants, and much less research on these is available.

### *3.1.1 Determination of electrical stunning parameters for poultry*

Because the effectiveness of electrical stunning depends on an adequate delivery of electrical current to the brain, it is affected by electrical variables such as voltage, current, frequency and waveform (Blokhuis et al., 2004; Kranen et al., 1996; Novoa et al., 2019; Raj et al., 2006c). A large body of research has been devoted to determining appropriate parameters for effective electrical stunning of poultry, and previous reviews are available (Berg and Raj, 2015; Blokhuis et al., 2004; Nielsen et al., 2019a; Raj, 2003, 2006; Shields and Raj, 2010). Various studies have determined the effectiveness of stunning parameters at a range of combinations of current, frequency and waveform, most commonly using either EEG output (e.g. Prinz et al., 2012) or reflex responses (e.g. Girasole et al., 2016) as outcome measures. Most of these have been based on 50 Hz alternating current (AC) with a sinusoidal waveform. The findings have underpinned evidence-based policy, and the stipulation of minimum currents according to different frequency ranges in law (e.g. EU Council Regulation (EC) No 1099/2009) (EC, 2009) and global guidelines (OIE (World Animal Health Organization), 2019).

An electrical stun is considered to be effective if it induces unconsciousness rapidly (in less than 1 s EU 1099/2009) and this is sustained for a minimum of 45 s (Gregory and Wotton, 1990b; Raj, 2006). It has been argued that use of measures such as muscle tone or convulsions to determine whether a stun has been successful are problematic (Shields and Raj, 2010), with EEG monitoring considered to be the most informative. In commercial conditions EEG measurements are not possible and other indicators of insensibility (e.g. loss of eye reflexes) are useful (Erasmus et al., 2010c; Prinz et al., 2010b). Signs of consciousness (e.g. wing flapping, responses to painful stimuli, vocalisations) may also indicate an ineffective stun (Grilli et al., 2015; Hindle et al., 2010).

In early work, it was determined that a minimum current of 100 mA per bird was required to ensure supressed EEG associated with a loss of consciousness in broilers, and 120 mA when using a 50 Hz sine wave AC (Gregory et al., 1991; Gregory and Wotton, 1990b, 1991; 1994). Raj et al. (2006c) reported that effective electrical stunning of broilers with a minimum constant current of 100, 150 and 200 mA could be achieved with electrical frequencies of up to 200, 600 and 800 Hz, respectively. Girasole et al. (2016) showed, under slaughterhouse conditions, that the minimum current necessary to achieve effective stunning in 90% of birds was 150 mA for 200 Hz, 200 mA for 400 Hz and 250 mA for 600 Hz. These and other studies show that higher stunning frequencies require greater current intensities to provide an effective stun, and this is reflected in current EU legislation (see Table 1) where allowable frequency/current combinations are defined (and some are not permitted). It is recommended that sine wave AC of 600 Hz maximum is delivered with an average current of 100–400 mA, depending on the species and the frequency (Council Regulation (EC) No 1099/2009) (EC, 2009). Raj et al. (2006a,b) investigated the application of different electrical waveforms, and concluded that sine wave AC is more effective than pulsed direct current (DC), and that a pulse width of at least 30% of current cycle is necessary to reliably induce epileptiform activity. Additionally, some electrical parameters, for example, low-frequency pulsed DC, are undesirable because they may induce cardiac arrest without an EEG pattern that indicates unconsciousness (Raj et al., 2006a). Although cardiac arrest is not necessary for an effective electrical stun, it is considered to be desirable, as it ensures non-recovery (Farm Animal Welfare Council, 2009). The duration of exposure to current is crucial to ensure adequate stunning and

**Table 1** Minimum average current and frequency combinations as defined in the EU Council Regulation (EC) No 1099/2009 (EC, 2009) and OIE Terrestrial Animal Health Code (2005) (OIE (World Animal Health Organization), 2019)

| Frequency (Hz) | Chickens (layers and broilers) (mA) | Turkeys (mA) | Ducks and geese (mA) |
|---|---|---|---|
| <200 | 100 | 250 | 130 |
| 200–400 | 150 | 400 | Not permitted |
| 400–1500 | 200 | 400 | Not permitted |

maintenance of unconsciousness during bleeding. A minimum of 4 s exposure to sine wave AC at a maximum of 600 Hz with a current of 100–400 mA is required by law in the EU but in practice, exposure times are routinely longer than this, in an attempt to diminish the possibility that birds will miss the water bath entirely (EC, 2009).

The meat quality implications of different electrical stunning parameters are routinely assessed because an inherent conflict exists between effective electrical stunning and economically relevant carcass quality outcomes. It is well known that while lower frequency current increases the likelihood of an effective stun, it increases intense muscle contractions leading to carcass defects associated with downgrading (Wilkins et al., 1999). Recent work indicated that a frequency of 750 Hz and current intensity of 200 mA resulted in effective stunning without adverse carcass lesions (Girasole et al., 2015), but others have reported that application of high current at high frequency also results in carcass defects (Ali et al., 2007). The opposition between carcass quality and humane electrical stunning appears to be unavoidable. While allowable electrical current and frequency parameters are set by law in the EU, the desire to promote meat quality has led to the global use of less effective currents that do not protect bird welfare (Bilgili, 1999; Hindle et al., 2010). For example, in the United States, electrical stunning for broilers routinely involves low current (10–45 mA per bird), high frequency (350–500 Hz), pulsed DC, which is unlikely to result in effective stunning (Shields and Raj, 2010). There are no regulations in place to control specific electrical stunning settings in the United States, and indeed AVMA guidelines (2020) allow electrical stunning with pulsed DC, low current (25–45 mA/bird average), low voltage (10–25 V) and high frequency (500 Hz), despite scientific evidence that these parameters are not effective (Leary et al., 2020).

### 3.1.2 Likelihood of effective electrical stunning

Even though suitable electrical parameters for stunning poultry have been determined, there are numerous factors that affect the likelihood of any individual bird being effectively stunned in a water bath system, and hence the reliability of the overall approach is uncertain. A significant welfare issue is the risk of inadequate current delivery to some birds, resulting in conscious birds experiencing pain and fear during subsequent processes. As outlined above, inadequate current could be caused by unsuitable underlying settings (voltage, frequency and waveform), but even when settings are conducive to an effective stun, it is well known that the amount of current applied to individual birds in a water bath stunner varies widely (Sparrey et al., 1992) and requires monitoring and reactive operational adjustment (Humane Slaughter Association, 2016a). Only a small proportion of applied current may flow through the brain, with the majority being distributed across the body (Woolley et al., 1986a,b), and resistance will vary between individuals according to body composition and contact with the shackles. Poor electrical contact with the shackle (caused by dryness, corrosion and debris; (Sparrey et al., 1992; Woolley et al., 1986b)) is therefore a risk, and may be reduced by maintaining shackle condition, cleanliness and wetting (Perez-Palacios and Wotton, 2006; Sparrey et al., 1992).

Due to the complexity of multi-bird water bath stunners, an ongoing issue is the difficulty in determining the actual current received by each bird (Hindle et al., 2010). As outlined above, multiple causes combine to result in a system that is not able to compensate for individual variation and therefore reliably deliver adequate current to all the birds, even if the settings are correct. A further complicating factor is that the visual appearance of inadequately stunned birds can be indistinguishable from effective stunning, since seizures induced by lower current applications effectively cause electro-immobilisation while birds remain conscious (Schutt-Abraham et al., 1983; Shields and Raj, 2010). Although there are some effective, commercially viable means to determine whether a bird is effectively stunned, routine use of rapid line speeds (such as 8000 birds per hour) limits the possibility to check this on an individual basis, and to apply backup interventions (Blokhuis et al., 2004). This is of grave concern because given the very large numbers of birds involved in commercial slaughter, even small percentage failure rates in stunning translate to an appreciable number of birds experiencing severely negative welfare consequences. Prevention is therefore the best approach, and one attempt to overcome the issue of inadequate current delivery is the use of a constant-current stunner, which delivers a pre-set constant current using variable voltages (compensating for the electrical impedance of individual birds). Such equipment has been developed (Sparrey et al., 1993) and tested with positive results (Lambooij et al., 2010), and its implementation would significantly improve the reliability and welfare outcomes associated with electrical stunning of poultry.

There are a couple of scenarios when birds completely avoid being electrically stunned in the water bath system: when they are too small for the head to reach the water, or when a bird lifts its head to avoid contacting the water, therefore entering the neck cutter while conscious. Live birds may also enter the scald tank (a post-death process to allow defeathering) because they miss the neck cutter after being stunned (and recovery ensues) or when both stunning and neck cutting are inadequate (Shields and Raj, 2010). The number of animals not stunned at all is not clear since no reliable figures are available. To avoid the poor welfare outcomes associated with these situations, EU legislation requires manual backup neck cutting, but again fast line speeds make this challenging to actually employ (Nielsen et al., 2019a).

### 3.1.3 Shackling

Concerns about inadequate current leading to ineffective stunning notwithstanding, a major welfare issue associated with electrical water bath stunning is the shackling of conscious birds. There is good evidence that this process causes pain, based on nociceptive activation by the compression forces generated by the weight of a 2–4 kg bird on the small surface area of the leg in contact with the shackle (Gentle and Tilston, 2000). As such, the size of the metal slot in relation to the bird's leg is important, with male broilers at greater risk of increased leg compression compared to smaller females (Satterlee et al., 2000; Sparrey and Kettlewell, 1994). In addition, inversion (hanging upside down) is a potent cause of stress and fear (Bedanova et al., 2007; Korte et al., 1997; Shields and Raj, 2010) and most birds struggle and wing flap immediately following shackling (Kannan et al., 1997). For birds that have leg abnormalities, injuries sustained during catching or inherent fragility (e.g. lame broilers or osteoporotic spent laying hens), the welfare consequences of shackling are likely to be more severe (Butterworth, 1999). Lambooij and Hindle (2018) reported that shackling increased rates of broken bones in spent hens, with 4–20% of freshly broken bones associated with handling immediately prior to stunning. Apart from the issues directly associated with shackling itself, poor manual handling to deliver birds to the shackles can also result in stress, injury, carcase downgrading or even mortality (Knowles and Broom, 1990).

The ongoing welfare costs associated with shackling mean that minimising its duration is a priority. Sparrey and Kettlewell (1994) recommended an optimal shackling time of between 12 and 60 seconds, which allows birds to 'relax' (i.e. stop wing flapping) to ensure an effective stun (Gregory and Bell, 1987) while avoiding longer than necessary shackling durations. The durations actually applied depend on line speed and plant design, which are widely variable. EU regulations stipulate that the shackling durations for chickens must not exceed 1 min (EC, 2009), but the extent of enforcement is not clear. Plant design considerations are also relevant to bird welfare with regard to avoiding sharp curves, steep inclinations or sudden drops in the shackle line, all of which can increase the mechanical forces on the legs and hence pain and fear responses (Nielsen et al., 2019a). Good electrical contact between the bird and the shackle is necessary for effective stunning, which introduces a tension between the degree of leg compression and a desire to use tight-fitting shackles. Compliant shackles have been developed; these are designed to adapt to bird leg size, ensuring good contact while minimising compression (Anastasov and Wotton, 2012; Lines et al., 2012), but are currently not widely used (Humane Slaughter Association, 2016b). A further modification to improve welfare is the introduction of breast support – this can be an almost vertical surface against which birds lean while hanging or take the form of a horizontal conveyor belt which bears weight. Horizontal supports have been shown to decrease wing flapping and should reduce pain and fear, but these are not widely used and measures must be taken to ensure that the birds are not able to remove their legs from the shackles (Lines et al., 2011a, 2012).

Given the inherent welfare insults associated with inversion and leg compression that cannot be dissociated from shackling of conscious birds, the best mitigation is to avoid the process entirely. This has been the starting point for alternative methods of stunning such as CAS. When shackling is used, harms can be somewhat minimised by staff training on good handling technique, ensuring both legs are shackled, not shackling injured or very small birds, minimising shackling duration, plant design that involves smooth and level transit of the shackle line, and use of compliant/well-fitting shackles and/or breast support to reduce weight bearing on the legs (Fig. 3).

### 3.1.4 Pre-stun shocks

A further concern related to water bath stunning is the risk of pre-stun shocks, whereby an electric shock is applied to the bird before loss of consciousness. This can happen when a part of the bird (usually the wing) touches electrified

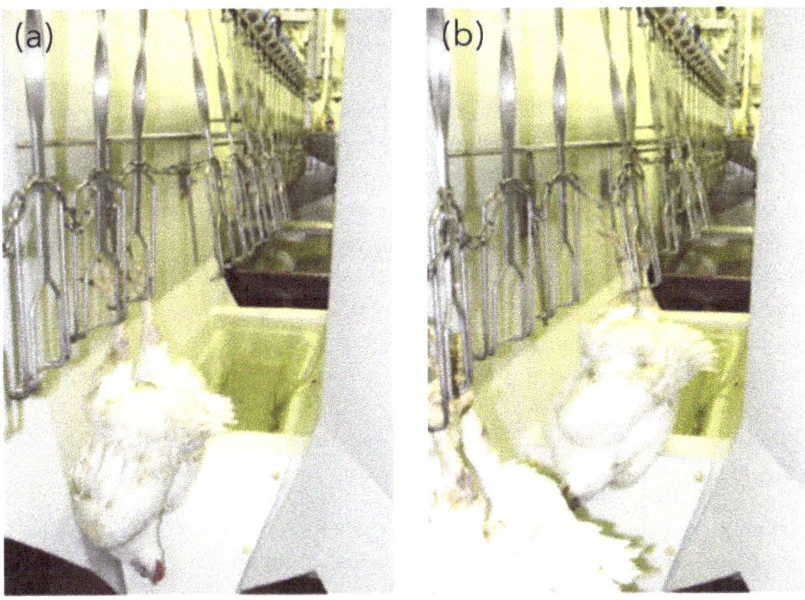


**Figure 3** Broiler chicken approaching the water bath stunner (a) and then at the point of entry (b), where the angled ramp facilitates the swift and gentle swinging of the head into the electrified water in order to minimise the risk of pre-stun shocks. Image credit: Humane Slaughter Association (HSA). Reproduced with permission from the HSA from: Electrical Water bath Stunning of Poultry (Humane Slaughter Association, 2016b).

water before the head (Rao, 2014; Rao et al., 2013). The likelihood of a pre-stun shock is increased by wing flapping at the entrance to the bath or when water levels are too high, causing overflow to come into contact with birds at the entry ramp (Blokhuis et al., 2004). Such shocks are painful and the bird's reaction to them may cause further wing flapping, increasing the chance that the bird's head may miss the water bath completely (Humane Slaughter Association, 2016b; Nielsen et al., 2019a). Mitigation includes measures that promote a smooth and swift entry to the water bath (e.g. subdued lighting, breast support) and ensuring the water level in the bath is sufficient (birds should be immersed up to the base of their wings (Humane Slaughter Association, 2016b; Nielsen et al., 2019a)), while preventing overflow.

### 3.1.5 Head-only stunning

Application of current directly to the head (called 'head-only stunning') involves inversion of the bird into a cone or restraint by hand, and insertion of the head into a device with electrodes to deliver appropriate current (Hillebrand et al., 1996; Lambooij et al., 2010; Raj and O'Callaghan, 2004; Vogel et al., 2011). This approach, then, still involves stressful inversion, since restraint of conscious birds is required for proper positioning of electrodes. However this method is individual to each bird, increasing the chance of adequate current delivery and induction of unconsciousness, though variation resulting from the cleanliness and material of electrodes, feather coverage of head, dryness of skin, and other individual bird parameters remains (Nielsen et al., 2019a). Requirements for head-only electrical stunning equipment are regulated in the EU, being 240 mA for hens and broiler chickens (EC, 2009) and the same signs of unconsciousness are relevant as for water bath stunning. As with water bath stunning, head-only stunning is a simple stun and must be followed with a killing method. Although head-only methods are more likely to guarantee an effective stun, they are not suitable for high throughput and limited to small-scale operations. As such they have limited relevance to commercial poultry slaughter, although some research has investigated the feasibility of head-only stunning using a water bath (Lines et al., 2011b).

### 3.1.6 Novel electrical systems

Novel and emerging electrical stunning systems have been developed in an attempt to combine reliable, instantaneous stunning with welfare-friendly handling and restraint. With high throughput in mind, these have generally involved automated conveyor systems which mechanically locate and control the bird's head to allow the application of head-only stunning and/or head to body stun-killing. An 'upright' system was developed in the UK with initial promise (Tinker et al., 2005), but it has not been commercialised. Head-cloaca stunning has been attempted (Lambooij et al., 2012), which still involves shackling but uses the cloaca instead of the feet as the earthing point to improve electrical contact. It has shown promise in lab settings, but is not currently available. A novel system called 'Dutch Vision Solutions' applies individual head-only stunning. It has been found to be effective though some concerns were raised about prolonged stun to stick intervals and the reliability of backup systems, and it does not eliminate shackling of conscious birds (Gerritzen et al., 2015).

### 3.1.7 Conclusions

The welfare issues associated with electrical water bath stunning are considerable, and they are not easy to overcome. Despite this, electrical water bath systems remain the most commonly used stunning method for poultry worldwide. Even the use of constant current systems does not eliminate the significant harms associated with shackling, or the risk that some birds will miss the stunner completely. The move to alternatives, particularly in Europe, has been driven by a wish to eliminate these concerns. As early as 2004, EFSA recommended the phasing out of electrical methods due to their associated harms, with the Farm Animal Welfare Council (2009) endorsing this in 2009. In the latest EFSA report, there is a specific recommendation that birds should not be shackled while conscious, indirectly indicating that the majority of commercially used electrical stunning methods should cease (Nielsen et al., 2019a). While electrical stunning is still in use in the EU, an interim step has been to ensure it is heavily regulated. Insofar as risks to welfare can be mitigated, this is achieved by specific legislative requirements for water bath design, stipulation of specific currents and frequencies, requirements for water depth, a control box displaying current administered, capacity to adjust voltage, and appropriate back-up procedures for inadequate or missed stuns (Nielsen et al., 2019a). Appropriate inspection and enforcement is provided locally by each EU member state (EC, 2009). The lack of such oversight in other countries slaughtering billions of chickens with electrical stunning is a source of serious welfare concern.

## 3.2 Controlled atmosphere stunning

Controlled atmosphere stunning (CAS) involves exposing animals to modified gas environments, which induce a gradual loss of consciousness followed by a non-recovery state. The term CAS (as opposed to gas stunning) is used by the industry, partly to avoid negative public perceptions and historical connotations of 'gas chambers' (Berg and Raj, 2015). The term controlled atmosphere killing (CAK) is also used (interchangeably with CAS) to refer to this process, because although it is possible to induce a temporary loss of consciousness with gas exposure, all commercial systems are designed to stun-kill since birds tend to rapidly regain consciousness on re-exposure to air leading to a high risk of recovery on the slaughter line (Gerritzen et al., 2000, 2013; McKeegan et al., 2006, 2007; Poole and Fletcher, 1998; Raj, 1998; Raj and Gregory, 1990a). This is because birds have a unique, extremely efficient respiratory system which evolved to meet the demands of flapping flight (Fedde, 2012). Instead of alveoli (which are blind-ended structures in mammalian lungs where gas exchange takes place), parabronchi (a parallel series of tubes) are the primary unit of gas exchange in avian lungs (Fedde, 1998). Both inspiration and expiration require active muscle contractions and there is unidirectional air flow in a tail to head direction during both phases of ventilation which renews gas exchange more frequently than tidal (mammalian) ventilation (Powell and Scheid, 1989). As in mammals, breathing in birds is controlled by motor output from the brainstem and the rate and depth (tidal volume) of breathing is controlled by several reflexes designed to maintain blood gases within acceptable ranges. Central chemoreceptors in the brain are sensitive to changes in arterial carbon dioxide ($CO_2$) and pH, while arterial chemoreceptors (known as carotid bodies) respond to changes in arterial oxygen and pH and facilitate the ventilatory response to hypoxia (low oxygen) (Fedde, 2012; Powell and Scheid, 1989). In birds (but not in mammals), intrapulmonary (lung) chemoreceptors are stimulated by low $CO_2$ levels and inhibited by high $CO_2$ (Milsom et al., 2004). This inhibition of lung receptors when $CO_2$ is high stimulates an increase in ventilation (as in mammals but by a different mechanism).

Commercial systems vary in their design, but CAS is generally applied to birds in their transport containers or on conveyers, involving minimal or no handling of conscious birds. Crucially, shackling is performed after stunning when birds are in a non-recovery state. Correct application of CAS also involves timed exposure which ensures that all birds are stunned, and as such this approach eliminates two major welfare issues associated with electrical stunning - aversive handling and the risk of ineffective or absent stunning. However, unlike effective electrical stunning, induction of unconsciousness with controlled environments is not instantaneous, resulting in a transitional period during which negative experiences are possible. When assessing the welfare impact of CAS it therefore becomes vital to determine the time to loss of consciousness, bearing in mind that vigilance states ranging from alert to deep unconsciousness represent a continuum. As such, the transition from consciousness to unconsciousness should not be thought of as an 'on/off' switch but a gradual process, especially with non-instantaneous stunning approaches such as CAS (Verhoeven et al., 2014). Although it remains challenging to characterise gradually changing brain states, EEG recording and analysis is widely agreed to be the best approach to determine the likelihood of consciousness and there have been attempts to define EEG spectral characteristics associated with different vigilance states in anaesthesia (alert, sedation, surgical plane, deep hypnotic state and brain death) (Benson et al., 2012; Gerritzen et al., 2004; McKeegan et al., 2007; Raj et al., 2006a; Sandercock et al., 2014) so that we can recognise them during stunning.

### 3.2.1 Types of controlled atmosphere stunning

CAS methods can be divided into five main categories which refer to the approach used to modify the environment, usually by displacing air in the system. These are anoxia (inert gases), hypercapnic anoxia (inert gases mixed with $CO_2$), hypercapnic hypoxia (gradual exposure to ultimately high levels of $CO_2$), hypercapnic hyperoxygenation (mixtures of inert gases, $CO_2$ and oxygen such that oxygen levels exceed those naturally found in air) and hypobaric hypoxia (low atmospheric pressure stunning (LAPS), in which birds are subject to gradual decompression). These have all been the basis of commercial CAS systems for poultry at some time, but most current systems employ hypercapnic hypoxia and/or hypercapnic hyperoxygenation, usually in multiple phases (Fig. 4). Systems with multiple steps are designed to first induce loss of consciousness, before application of a second phase (which may involve potentially aversive gas mixtures) to ensure non-recovery. The duration of the conscious phase is reported to be shortest for hypercapnic anoxia, followed by anoxia and hypercapnic hypoxia, and longest with hypercapnic hyperoxygenation (Johnson, 2014).

### 3.2.2 Behavioural responses to controlled atmosphere stunning

Behavioural responses can be used alongside brain measures to assess the stages of loss of consciousness and are the basis of welfare assessment during any stunning or killing method. Mandibulation is a behaviour which resembles a 'tasting' type movement of the bill and has been defined as 'Repetitive and rapid opening and closing of the bill, not associated with inspiration or exhalation' (Martin et al., 2016a,c). Mandibulation is seen in response to both anoxic and hypercapnic CAS mixtures, and LAPS (Abeyesinghe et al., 2007; Martin et al., 2016b). Its function and welfare impact are unclear. Headshaking is a rapid lateral head movement which has been previously associated with disorientation, discomfort and respiratory distress (Webster and Fletcher, 2001) as well as arousal or contexts demanding increased attention (such as the presentation of novel or disturbing stimuli (Hughes, 1983)). Nicol et al. (2011) concluded that head shaking may be a valid indicator of a less-preferred environment and high rates of head shaking are thought to indicate reduced welfare. Headshaking has been interpreted as an aversive reaction to $CO_2$, but it is also seen in response to stunning with inert gases and LAPS (Abeyesinghe et al., 2007; Johnson, 2014; Martin et al., 2016c; McKeegan et al., 2007). Ataxia describes the behavioural responses observed in the initial stages of animals

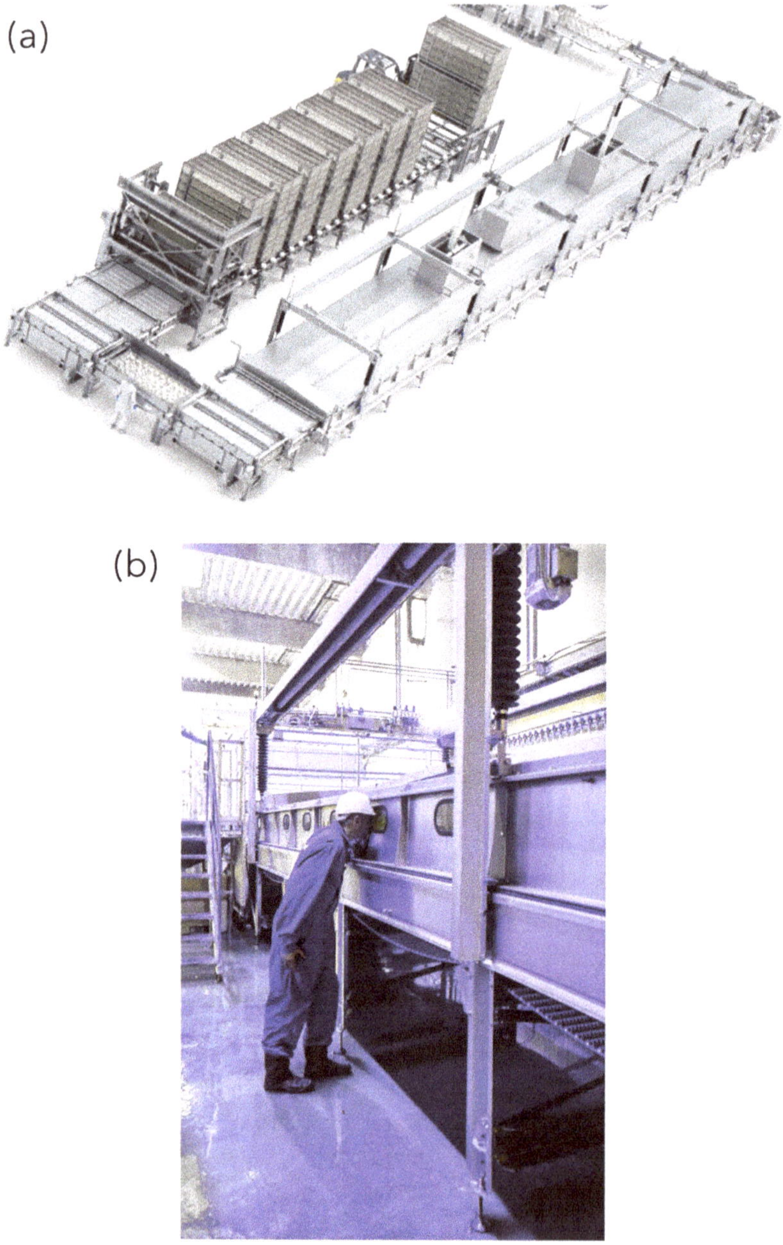

**Figure 4** Marel's ATLAS live bird handling system and controlled atmosphere stunning (CAS) SmoothFlow (a) and the use of windows in the side of the system's hood allow broilers to be observed and monitored (b) images courtesy of Marel.

succumbing to gradual stunning methods such as CAS and it is characterised by progressive loss of voluntary coordination of muscle movement. It results in visible swaying of the body and/or head, attempts to adjust body posture and/or single wing flaps. It may be defined in poultry as 'Apparent dizziness, staggering, swaying of body and/or head, attempts to stand/sit or flaps wings to try and regain balance' (Martin et al., 2016a,b,c).

Rhythmic breathing with an open bill (distinguishable from thermoregulatory panting) is a hyperventilatory response to hypoxia (low oxygen). Deep inhalation (single large gasping breaths, often with neck extension) may also indicate dyspnoea (respiratory discomfort) (Mackie and McKeegan, 2016; Martin et al., 2016c). Open bill and deep breathing have been interpreted as indications of breathlessness in birds (Gerritzen et al., 2004), recently defined in mammals as a negative affective (emotional) experience relating to respiration with multiple qualities (Beausoleil and Mellor, 2015). In humans, these include respiratory effort, air hunger (increased urge to breath) and chest tightness (Beausoleil and Mellor, 2015), though it is not clear which (if any) of these apply to birds, which have a unique respiratory system. Dyspnoea-related behaviours probably reflect reduced welfare during stunning, and in humans there is evidence that air hunger is more unpleasant than pain (Banzett et al., 2008). Deep inhalation has been particularly associated with hyperventilation during $CO_2$ stunning, but it is also seen with inert gases (e.g. McKeegan et al., 2007) and after electrical stunning (Verhoeven et al., 2014). In $CO_2$ stunning, this behaviour continues after loss of posture indicating that consciousness is not required for its performance (Verhoeven et al., 2014). There is evidence that some dyspnoea occurs in all CAS stunning mixtures that have been investigated, including inert gases; however behavioural responses reflecting these have been variously described as 'gasping', 'deep breathing' and 'respiratory disruption' (e.g. Gerritzen et al., 2004; McKeegan et al., 2007) making direct comparison between behavioural categories more difficult.

Loss of posture is the inability of the animal to remain in an initial standing or sitting position and has been widely interpreted as a proxy for loss of consciousness (Authie et al., 2013; Gerritzen et al., 2004) as it reflects loss of muscle control. After loss of posture, convulsions consisting of neuromuscular spasms are observed, usually consisting of a clonic phase (characterized in chickens by vigorous wing flapping) and a tonic phase (characterized by rigidity and twitching of outstretched legs and wings), followed by final leg paddling motions (Erasmus et al., 2010a; Verhoeven et al., 2014). Wing flapping observed before loss of posture may be associated with purposeful escape attempts, indicating poor welfare (Nielsen et al., 2019a). Tonic and clonic convulsions sometimes alternate during CAS. The time at which convulsions cease has been used as an indicator of irreversible brain failure (Erasmus et al., 2010c). Convulsions are never seen before loss of posture, therefore their relevance to welfare is probably limited. However, it has been noted that when birds are killed in groups, it may be possible for birds that are still potentially conscious to be disturbed or even injured by other birds convulsing. Finally, birds will become motionless during CAS, which refers to a limp carcass with the bird being completely still including the cessation of visible breathing movements; it reflects complete and irreversible loss of muscle tone and is considered to confirm a non-recovery state. In this state the bird is no longer breathing and the brain is dead, but the heart (and possibly other body systems) will still be partially functioning (Abeyesinghe et al., 2007).

### 3.2.3 Welfare impacts of different controlled atmosphere stunning types

#### 3.2.3.1 Inert gases

Immersion in inert gases (argon/nitrogen) causes minimal immediate visible and adverse reaction, since these gases cannot be sensed directly by the bird. Ataxia and loss of posture are apparent as the bird succumbs to lack of oxygen (which is sensed by arterial chemoreceptors (Fedde, 2012). While respiratory disruption and presumably a sensation of air hunger is induced by inert gases (in humans (Moosavi et al., 2003)), it is less common and overt compared to hypercapnic gas mixtures in birds (Coenen et al., 2009; Gerritzen et al., 2000; McKeegan et al., 2007). Anoxic CAS is associated with very vigorous convulsions which happen quite early in the process. While the welfare impact of these is unclear (McKeegan et al., 2007; Raj, 2009), they have implications for meat quality. EFSA (2004) opined that this sort of death (by 'pure' anoxia) is humane. However, the use of inert gases is practically problematic due to the difficulty of keeping oxygen below 2% (necessary to cause a non-recovery state in birds) (Raj et al., 1992) in a fast-throughput system. Argon is denser than air which helps to displace oxygen more effectively, but it is prohibitively expensive. As such, anoxic CAS is not common for commercial use.

#### 3.2.3.2 Hypercapnia

Inhalation of $CO_2$ causes hypercapnia, which is an increase in $CO_2$ levels in the blood. This disrupts respiration and normal neuronal function, leading to unconsciousness and eventually to death. $CO_2$ becomes an anaesthetic gas (i.e. causes unconsciousness, and therefore could be used to stun reversibly) at concentrations of around 30% (Raj and Gregory, 1990b). The rate of induction and the depth of anaesthesia achieved depend on gas concentration and exposure time, with higher levels of $CO_2$ requiring shorter exposure times to induce unconsciousness (Raj and Gregory, 1990b). Irreversible stunning with

$CO_2$ is induced by hypoxia and acidification of cerebrospinal fluid, causing brain death (Blokhuis et al., 2004; Raj, 2009; Raj et al., 2006d; Terlouw et al., 2016).

Exposure to $CO_2$ is associated with two welfare concerns: possible activation of mucosal nociceptors, causing pain, and unpleasant respiratory sensations. Either or both of these may underlie aversion to $CO_2$, which has been observed in chickens (McKeegan, 2004; Raj and Gregory, 1991). Based on individual nociceptor recordings, the pain threshold for $CO_2$ exposure in birds is thought to be 45–50% (McKeegan, 2004) and this is reflected in the EU regulation, which stipulates a maximum level of 40% $CO_2$ to which conscious birds may be exposed (EC, 2009). No current CAS systems apply $CO_2$ concentrations higher than 40% to conscious birds. Pain, therefore, is not a major concern during CAS for poultry (unlike in pigs where conscious pigs are exposed to $CO_2$ concentrations in excess of 80% (Llonch et al., 2012; Rodríguez et al., 2008)).

Physiologically, inhalation of $CO_2$ causes three effects - (1) changes in blood $CO_2$ and pH sensed by central chemoreceptors in the brain; (2) changes in blood pH sensed by chemoreceptors in the arteries and (3) inhibition of intrapulmonary (lung) chemoreceptors (IPCs) (Fedde, 2012; Milsom et al., 2004). All of these drive a hypercapnic ventilatory response, resulting in prolonged inspiration, increased tidal volume (bigger breaths) and decreased breathing frequency. This is expressed behaviourally as visible respiratory disruption or dyspnoea. The thresholds for these responses (especially IPC inhibition) are low (<2%) (Milsom et al., 2004) relative to $CO_2$ concentrations that cause unconsciousness and death. The extent to which hypercapnic hypoxia multistage systems present a welfare issue depends on one's view of whether a negative affective (emotional) state is induced by $CO_2$ inhalation at sub-nociceptive concentrations. In humans, air hunger induced by inhalation of $CO_2$ is extremely unpleasant and stressful (Banzett et al., 2008), possibly additionally so since humans can understand the life-threatening implications of inadequate oxygen intake. Since birds have a very different respiratory system in terms of both structure and physiology, it is not clear whether it is valid to directly extrapolate the concept or mechanisms of air hunger from mammals to birds. In mammals, air hunger arises from a mismatch between the automatic motor command to breathe and the degree of lung inflation (Banzett et al., 2008; Beausoleil and Mellor, 2015). Birds have a fixed lung volume, unidirectional airflow and increased dead space (from air sacs), and so they may be better able to accommodate hyperventilation, reducing (or even eliminating) air hunger (Fedde, 2012). As such, behavioural manifestations of hyperventilation might have different welfare consequences in birds compared to mammals. However, if the ventilatory response induced by exposure to $CO_2$ is insufficient (i.e. when the reflex command exceeds normal functional respiratory capacity) it is certainly possible that this could lead to air hunger or a similarly negative sensation. In work on aversion to CAS mixtures, McKeegan et al. (2006) showed that broilers did not withdraw and continued to feed while showing behavioural evidence of dyspnoea in response to $CO_2$. This suggests that respiratory disruption may not in itself be potently aversive, but further study is needed to address this question. If dyspnoea is a potential cause for welfare concern as has been suggested (Beausoleil and Mellor, 2015), it is undesirable to use CAS systems that involve prolonged exposure to concentrations of $CO_2$ that cause dyspnoea, but are not anaesthetic. In other words, there is a tension between inducing unconsciousness very gradually (e.g. with multiple steps of slowly increasing $CO_2$ concentration) and allowing dyspnoea to continue over a long period. There is very little welfare research on multiphase hypercapnic hypoxic CAS, although it is widely used. In one study of a 5-stage $CO_2$ system, transitional EEG waveforms were apparent after 36 s, and the EEG was suppressed (indicating unconsciousness) at 60 s (Gerritzen et al., 2013), suggesting that conscious durations are not excessive and within the range reported for other CAS approaches (Johnson, 2014). Echoing the meat quality/welfare conflict with electrical stunning, multiphase hypercapnic hypoxia systems are popular since wing flapping is reduced, improving product quality (Johnson, 2014).

In hypercapnic hypoxic systems (particularly as $CO_2$ levels rise) or with hypercapnic anoxia, birds are also exposed to hypoxia as the relative amount of $O_2$ in the air decreases by displacement. Arterial chemoreceptors sense hypoxia leading to a hypoxic ventilatory response, also consisting of increased ventilation and tidal volume (Milsom et al., 2004). During hypercapnic hypoxia/anoxia, therefore, both of these reflexes will be activated with possibly greater total dyspnoea (Beausoleil and Mellor, 2015; Steiner et al., 2019). In support of this, there is evidence that hyperoxgenation of hypercapnic gas mixtures causes a smoother induction to unconsciousness (as with anaesthetics), but the stun takes longer since the birds succumb only to $CO_2$ as an anaesthetic, not hypercapnic hypoxia (Johnson, 2014; McKeegan et al., 2007). Hyperoxygenated hypercapnia must be followed by a euthanasia phase (high $CO_2$) since birds will not be killed by such gas mixtures (McKeegan et al., 2007). This approach initially came from work in rats, whereby it was noticed that aversion to $CO_2$ was slightly reduced by supplemental oxygen, and that induction to unconsciousness was smoother (Coenen et al., 1995; Kirkden et al., 2008). However, it should be noted that the presence of supplemental oxygen does not reduce visible respiratory responses to CAS mixtures containing $CO_2$ (e.g. open bill breathing and deep inhalation) (McKeegan et al., 2007)). Whether hyperoxygenation reduces any sensation of 'air hunger' is unknown.

### 3.2.4 Low atmospheric pressure stunning

Low atmospheric pressure stunning (LAPS) involves placing birds in a sealed chamber (Fig. 5) that is gradually decompressed, leading to loss of consciousness and a non-recovery state (McKeegan et al., 2013b; Mackie and McKeegan, 2016; Martin et al., 2016a,b,c, 2019a). Decompression results in the reduction of the ambient atmospheric pressure and a proportional decrease in partial pressure of oxygen, leading to hypobaric hypoxia which is characterised by progressive loss of cognitive and psychomotor skills followed by loss of consciousness (Bouwsema and Lines, 2019; Holloway and Pritchard, 2017). As with other CAS methods, there is no pre-stun handling and birds are stunned in their transport crates. The original concept was developed in the United States in the 1990s following self-reports of a lack of pain or stress during hypobaric hypoxia in humans undergoing high-altitude pilot training (Gradwell, 2016; Gradwell and Macmillan, 2016; US Department of Transportation: Federal Aviation Administration, 2008). Following a positive opinion from EFSA (More et al., 2017), LAPS became legal in the EU for broilers up to 4 kg in May 2018 (Commission Implementing Regulation (EU) 2018/723 amending Annexes I and II to Council Regulation (EC) No 1099/2009 (EC, 2009)). As yet, it has not had significant uptake commercially.

LAPS is applied through a family of six decompression curves (Holloway and Pritchard, 2017; Martin et al., 2016c), which allow the system to automatically

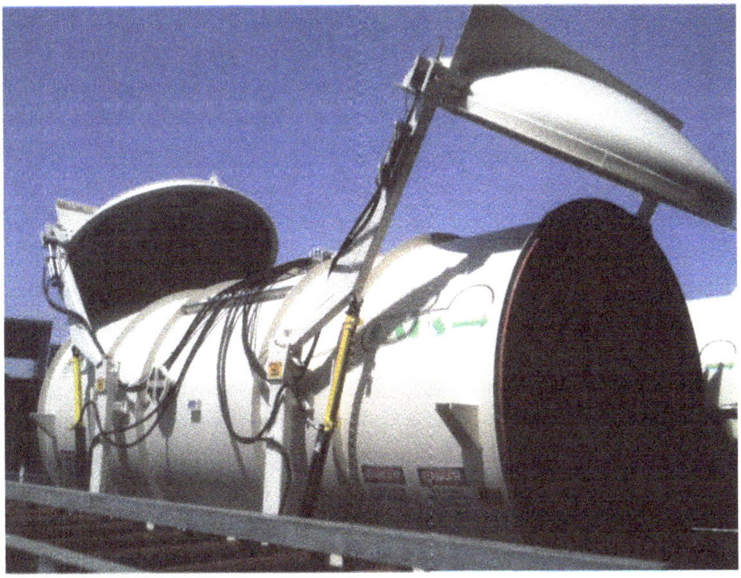

**Figure 5** A chamber forming part of a low atmospheric pressure stunning (LAPS) system. Image courtesy of Randolph Cheek, TechnoCatch LLC.

adjust to variations in ambient temperature (−4°C–30°C) and associated water vapour pressure alterations (Wideman et al., 2013), in order to compensate for the deviation of different gas densities and maintain the same hypoxic effect. These adjustments appear to be effective, albeit resulting in small differences in behavioural latencies (Mackie and McKeegan, 2016; Martin et al., 2016c). In the 280 s LAPS cycle, the most rapid pressure change occurs in the first 67 s, averaging −1.41 kPa $s^{-1}$, before a second phase in which the rate of decompression is greatly reduced (68–180 s), averaging −0.10 kPa $s^{-1}$ (EC, 2009; Holloway and Pritchard, 2017). The final phase, referred to as the 'hold phase' involves a further minor reduction in atmospheric pressure in the final 100 s of the cycle; pressure is then maintained at around 19–20 kPa to ensure non-recovery. For comparison, atmospheric pressure at sea level is approximately 101 kPa and 33 kPa at the summit of Mount Everest (Newcomb, 2010). During the hold phase, the partial pressure of oxygen is approximately 2.7 kPa, which is equivalent to approximately 3.77% of atmospheric oxygen. This is higher than the recommended 2% atmospheric oxygen when using CAS with inert gases (Abeyesinghe et al., 2007; Blokhuis et al., 2004; McKeegan et al., 2013a), but it is sufficient to cause brain death and non-recovery (Martin et al., 2016b,c; McKeegan et al., 2013a). Decompression during LAPS also causes a reduction in temperature (approximately −4°C) and humidity (−50%) inside the chamber (Holloway and Pritchard, 2017).

The welfare concerns specifically associated with LAPS are air hunger responses to hypoxia and potential gas expansion in body cavities leading to the risk of pain and/or barotrauma. Behavioural responses to LAPS are consistent and the sequence of behaviours exhibited closely resembles those seen in normobaric hypoxic environments (Mackie and McKeegan, 2016; Martin et al., 2016a,b,c). The time to loss of consciousness during LAPS (as determined by behavioural and EEG indicators) is approximately 60 s (Martin et al., 2016a,c), which is within the range of other CAS systems (Abeyesinghe et al., 2007; Gerritzen et al., 2013; McKeegan et al., 2007). Respiratory disruption is apparent in LAPS environments (Mackie and McKeegan, 2016; Martin et al., 2016a,b,c), but to a lesser extent (both in counts and proportions of birds) than in hypercapnic or anoxic CAS (McKeegan et al., 2007). Concerns relating to the expansion of gases in internal body cavities have been raised by EFSA (Authie et al., 2014; More et al., 2017), especially with regard to the digestive tract and the sinuses and ears. This reflects locations where pressure-related discomfort has been reported in humans, but more usually on recompression (Bouwsema and Lines, 2019). Determining whether air expansion is causing pain during LAPS is challenging. Thorough pathological examinations of birds following LAPS have shown that organs and cavities are intact (Martin et al., 2019a), but a lack of pathological evidence does not eliminate the possibility of pain or discomfort during decompression. Martin et al. (2016b) investigated by applying an analgesic intervention during LAPS and reported no findings which would suggest colic-like pain or ear barotrauma. Like anoxic CAS, LAPS is associated with severe convulsions after loss of posture, which can have negative meat-quality consequences (such as a higher prevalence of wing damage (Martin et al., 2019a; Vizzier-Thaxton et al., 2010)). The changes in temperature and relative humidity observed during LAPS may have the potential to reduce welfare, for example, dryness of the nasal mucosae (Nielsen et al., 2019a). Though given the limited period of consciousness, this challenge would be brief.

### 3.2.5 Regulations and practical application

In the EU, only certain approaches to CAS are permitted for the routine slaughter of chickens. These are: two or more phases of successive exposure of conscious birds to a gas mixture containing up to 40% of $CO_2$, followed (when animals have lost consciousness) by a higher concentration of $CO_2$; direct or progressive exposure of conscious animals to a gas mixture containing up to 40% of $CO_2$ associated with inert gases; direct or progressive exposure of conscious animals to an inert gas mixture such as argon or nitrogen leading to anoxia. For LAPS an initial phase of a reduction in pressure from 760 Torr (sea level) to 250 Torr within a minimum of 50 s is permitted, and in the second phase a further reduction to 160 Torr within 210 s (Commission Implementing Regulation (EU) 2018/723 amending Annexes I and II to Council Regulation (EC) No 1099/2009 (EC, 2009)).

As with electrical stunning, the effectiveness and welfare impact of CAS in practice depends partly on care with implementation and process control. Nielsen et al. (2019a) identified three key risk areas for gas-based CAS: exposure of birds to painful concentrations of $CO_2$, too short exposure time to ensure stunning or inadequate gas concentration to ensure stunning. Mitigation of these requires proper monitoring of gas concentrations to maintain gas concentrations within legal limits while ensuring adequate exposure. For LAPS, Nielsen et al. (2019a) identified the hazards as too fast decompression, expansion of gases in the body cavity and too short exposure time, all of which are integral to the decompression curve used and its control by operators. Ambient temperature and humidity are also relevant to CAS methods, directly so for LAPS (as evidenced by adjustment of decompression curves (Holloway and Pritchard, 2017)) and because when birds experience heat stress they pant to thermoregulate by evaporative heat loss. Heat stress induces disturbances in blood gases and acid-base balance (called hypocapnic alkalosis - an increase in blood pH due to decreased partial pressure of arterial $CO_2$ (Sandercock et al., 2001)). This phenomenon probably explains anecdotal variation in responses to CAS that seem to be related to ambient temperature conditions, but this has not been formally studied.

### 3.2.6 Conclusions

CAS in poultry was developed to overcome aversive bird handling and improve the reliability of stunning in high throughput systems, and these important welfare benefits apply regardless of the approach. While process control and monitoring remain important, with advanced engineering and considerable progress in systems that can generate highly replicable controlled environments, the identified welfare hazards are less of an issue with CAS than electrical stunning (Authie et al., 2013). This does however mean that CAS systems require considerable capital investment and adequate space for installation in processing plants, both of which are barriers for smaller facilities. The overall perception of regulators and welfare commentators is that CAS represents a high welfare approach to poultry slaughter (Blokhuis et al., 2004; Farm Animal Welfare Council, 2009). Nevertheless, since most CAS systems in use in Europe involve exposure to $CO_2$, the biggest outstanding welfare question is the extent to which hypercapnic exposure induces respiratory responses that cause distress to birds. Research priorities related to this issue were recently outlined for $CO_2$ killing more generally (Steiner et al., 2019) and include correlation of respiratory variables to neurophysiologic and endocrine welfare indicators, further aversion testing and use of pharmacologic interventions to relate observable respiratory patterns to affective states. As noted above, this has been attempted with analgesics to investigate LAPS (Martin et al., 2016b). Use of anxiolytics is likely to be more relevant to air hunger but so far has not been investigated in relation to CAS systems for poultry.

## 3.3 Mechanical methods

Mechanical methods of stunning and killing are not used routinely in large-scale poultry slaughter and are reserved as emergency 'back up' methods, or are used in small-scale slaughtering and on-farm slaughter. Mechanical methods cause unconsciousness and death via destruction of the brain, damage to the brain, concussion or cerebral ischemia (Martin et al., 2018a, 2019b). These methods are not suitable for high throughput and are applied to individual birds (Fig. 6). All require appropriate restraint.

Application of a captive bolt is allowed in EU legislation (EC, 2009) and the bolt can be penetrative or non-penetrative. Captive bolt is stipulated as a reversible method; however, in practice it is irreversible if correctly applied, due to brain destruction (Erasmus et al., 2010b; Martin et al., 2018a). This method tends to be reserved for larger poultry species or individuals for which manual cervical dislocation is not suitable (Sparrey et al., 2014). Effective captive bolt

**Figure 6** Captive bolt (concussive stunning) for chickens. Image credit: Humane Slaughter Association (HSA). Reproduced with permission from the HSA from: Practical Slaughter of Poultry (Humane Slaughter Association, 2016c).

application leads to immediate wing flapping and death can be confirmed by lack of cranial reflexes and absence of rhythmic breathing (Martin et al., 2018a; Raj and O'Callaghan, 2001; Woolcott et al., 2018). Recommended handling is that each bird is placed in a cone (which involves inversion) to access the head and contain wing flapping, and the forces applied (by cartridge, compressed air line pressure or a spring) should be appropriate for the species and size of birds (Nielsen et al., 2019b). As a last resort, EU regulations (EC, 2009) also allow the use of a percussive blow to the head (also known as blunt force trauma), described by Nielsen et al. (2019a,b) as 'holding a bird by its legs, placing its head on a hard surface and delivering a blow to the back of the head with a hard object' or 'holding the bird with both hands around its body and swinging the bird's head towards a hard, stable object such as the rim of a table'. Although there is evidence that such methods can be effective (Erasmus et al., 2010b; Raj and O'Callaghan, 2001), others have reported variable/limited success (Cors et al., 2015; Erasmus et al., 2010a), and there is clearly wide scope for operator error with severe welfare consequences. Better alternative methods are available, and as such use of percussive blows should be avoided.

For broilers of commercial market weight and spent hens, the most common method performed in emergency situations at slaughter is cervical dislocation (Sparrey et al., 2014) (Fig. 7). This can be performed manually (using just the hands of the operator) or mechanically (employing a tool). When performed optimally, cervical dislocation should cause death by cerebral ischaemia and extensive damage to the spinal cord and brainstem (Gregory

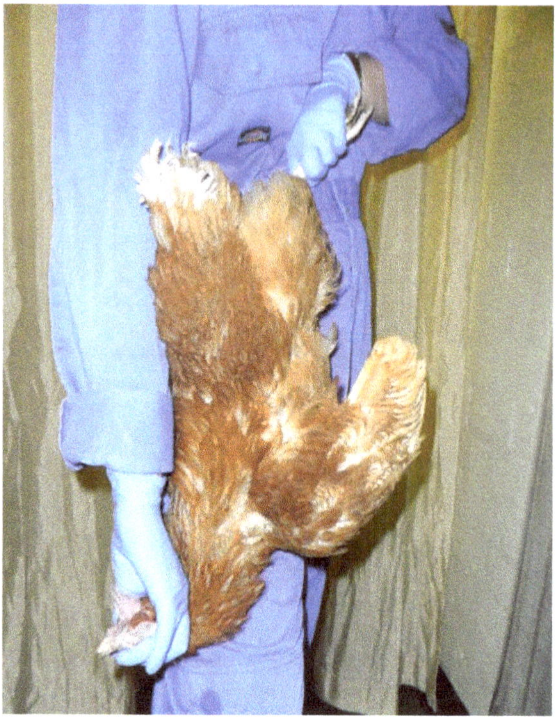

**Figure 7** Demonstration of manual cervical dislocation in chickens. Image courtesy of Jessica Martin, University of Edinburgh.

and Wotton, 1990a; Martin et al., 2019b). Manual cervical dislocation requires training, experience and physical strength and involves firmly stretching and twisting the neck while tilting the head back in a continuous movement (Sparrey et al., 2014). Mechanical cervical dislocation employs devices that dislocate by stretching or crushing, normally after inversion of the conscious bird in a restraint cone. These include crushing at the first cervical vertebra with a pair of pliers such as 'Semark pliers' or the 'Burdizzo' (Erasmus et al., 2010a; Martin et al., 2017; Sparrey et al., 2014). In the EU, manual cervical dislocation can be performed on up to 70 birds/person per day (to minimise operator fatigue), although a recent study found no evidence that fatigue was apparent up to 100 birds (Martin et al., 2018a). Cervical dislocation can be performed manually on birds weighing up to 3 kg, but must be performed mechanically on birds weighing between 3 kg and 5 kg (no cervical dislocation is allowed on birds weighting more than 5 kg) (EC, 2009). A significant advantage of manual cervical dislocation is that it is available at any time without tools, which is important in emergency situations where there is ongoing suffering. However, suboptimal technique remains a concern with the potential to cause painful tissue damage and lack of separation of the brain and spinal cord resulting in delayed death. Mitigation involves rigorous staff training on correct application, which also applies to mechanical cervical dislocation and correct placement of captive bolt shots (Nielsen et al., 2019a,b).

The welfare implications of cervical dislocation, even when performed correctly, relate primarily to concern over how quickly consciousness is lost. Multiple studies have shown that neck crushing may not sever the carotid arteries or spinal cord and does not cause rapid loss of consciousness (Erasmus et al., 2010a; Gregory and Wotton, 1990a; Martin et al., 2018a,b). Recent publications have highlighted distinct differences in performance dependent on method (e.g. manual versus mechanical (Jacobs et al., 2019; Martin et al., 2016d, 2019b) and results have been conflicting regarding periods of reflex persistence or evidence of CNS processing, with wide inter-individual variation. For example, nictitating membrane persistence ranged from 2 s to 106 s in broilers and turkeys following cervical dislocation (Erasmus et al., 2010a; Martin et al., 2016d). Spectral analysis of EEG recordings on lightly anaesthetised broilers and hens subject to either manual or mechanical cervical dislocation indicated that unconsciousness was achieved on average 15 s following method application (Martin et al., 2019b). A limitation of several previous studies is that they conduct neurophysiological and/or reflex assessments when birds are in an anesthetised state, ranging from a general anaesthetic plane (Gregory and Wotton, 1990a) to lighter anaesthesia (Martin et al., 2019b; Woolcott et al., 2018), and lack of consistency between use of mechanical and manual methods. Recent efforts to enhance consistency have resulted in a novel tool, the 'Nex', which has been shown to improve the reliability and effectiveness of manual cervical dislocation (Martin et al., 2019b). This limits the relevance of earlier work to the skilled application of manual cervical dislocation in conscious birds and the lack of a comprehensive welfare assessment of this very common killing method represents a significant gap in knowledge.

### 3.4 Emerging methods

While still in the development stage, there are some methods of stunning that could have potential for future application to poultry. Application of gas mixtures in novel mediums such as high expansion gas-filled foam is a possibility. This approach has been assessed and is considered to be humane (Gerritzen and Sparrey, 2008; McKeegan et al., 2013a), but is currently used only for emergency killing for disease control. Foam is of interest because it facilitates the application of nitrogen (or other inert gases) to produce a rapid anoxic kill, by achieving low oxygen concentrations (< 1%) inside the foam, which are very difficult to achieve with free application of gases. Transcranial magnetic stimulation (TMS) has been trialled in broilers, and involves application of an intense magnetic field to the brain (Lambooij et al., 2011). Although this method appears to have the potential to induce a reversible state of unconsciousness, the period of insensibility was brief (up to 20 s) and further development work is required. Diathermic syncope involves directing microwave energy to the brain to cause a slight increase in brain temperature, which is associated with loss of consciousness (so-called hyperthermic syncope). This has been investigated in sheep and cattle with promising results (Small et al., 2013, 2019), but no research has been conducted into its application to poultry.

Before addition to the EU regulation, any novel method must protect animal welfare to a degree at least equivalent to that ensured by the existing permitted methods, and there is a significant burden of proof required with rigorous evidence standards (Nielsen et al., 2019a). For uptake to be viable, new methods must not be prohibitive economically, and of course ensure acceptable meat quality. For poultry, a requirement to allow the automation associated with high throughput is also a necessity.

## 3.5 Slaughter without stunning and religious requirements

### 3.5.1 Regulatory and religious requirements

The Islamic and Jewish faiths have specific requirements regarding the slaughter of animals for consumption. Both require slaughter methods that are based on scriptural rules, laid down in the Holy Quran and Torah, which require animals to be 'alive', healthy, and not injured at the time of slaughter (Fuseini et al., 2016). The Council of Europe's Convention for the Protection of Animals for Slaughter and EU Council Regulation 1099/2009 requires that animals are stunned before they are slaughtered, but both allow member states to apply derogations to allow meat production that meets the needs of religious communities. Despite this, several European countries do not allow slaughter without stunning including Sweden, Norway, Iceland, Denmark and Slovenia. Some countries stipulate that post-cut stunning must take place, including Austria, Estonia, Greece and Latvia (Global Legal Research Center, 2018).

Slaughter methods related to Islamic rules are termed halal (Arabic for permissible), while the Jewish equivalent is termed shechita (Velarde and Dalmau, 2018). While the Jewish faith rejects any form of stunning, some Islamic authorities allow the use of stunning, provided it does not kill the animals. This is usually defined by the presence of a heart beat at the time of neck cutting (Velarde et al., 2014), but there is no unanimous agreement about this definition or the acceptability of stunning among Islamic scholars (Chandia and Soon, 2018). In practice, the requirement for signs of life after stunning precludes the use of CAS systems (which stun-kill) and leaves electrical stunning (which is potentially reversible) as the only currently available method for halal stun slaughter of poultry. Because halal slaughter may involve reversible stunning or no stunning, 'non stun slaughter' is a more useful term than 'religious slaughter' in discussions about animal welfare consequences. Recent figures from the UK in one surveyed week (Defra, 2019) showed that for broilers 9.7% were non-stunned (9.4% halal and 0.3% shechita) and 12.6% were halal stunned, while for spent hens 1.4% were not stunned (all halal) and 6.4% were halal stunned.

### 3.5.2 Welfare impacts

The purpose of stunning is to render animals unconsciousness so that they do not experience exsanguination, so slaughter without stunning inevitably raises significant welfare concerns. These include aversive pre-slaughter handling, pain induced by the neck cut, distress during bleeding including aspiration of blood, and prolonged times to loss of consciousness (Anil, 2012). To minimize distress in both halal non-stun and shechita methods, the killing must be achieved by an uninterrupted cut from a sharp knife (that severs both jugular veins, carotid arteries, oesophagus and trachea) and by a competent individual using effective equipment (Fuseini et al., 2016; Velarde et al., 2014). This involves manual restraint and the presentation of the neck to a specialist slaughterman who makes the cut, and the bird is then placed into a bleed-out cone (Barnett et al., 2007).

Research is sparse on the welfare impacts of non-stun slaughter. Manual handling and inversion are potential sources of stress and injury for non-stun slaughter, and shackling may be used in electrical halal stunning with its concomitant welfare issues. There have been a few attempts to determine time to loss of consciousness following neck cutting in conscious chickens. Gregory and Wotton (1986) showed that 95% of the electrical-evoked activity in the brain was lost on average by 163 s after the cutting of both carotid arteries and by 349 s after cutting both jugular veins. However, they noted that birds were unlikely to be conscious throughout the entire period when there was measurable electrical activity in their brains. Barnett et al. (2007) reported that chickens undergoing shechita slaughter showed a minimal reaction to the neck cut, but 60% of them showed eye reflex responses 5 s after the neck cut. These were abolished by 15 s, which coincided with loss of muscle control. In a case study, Cranley (2017) described basic behavioural measures during non-stunned slaughter of 250 chickens, and reported that 225 chickens 'died at or before 60 s', but 25 chickens showed prolonged spontaneous behaviour for more than 90 s after the neck cut.

Electrical stunning can meet the needs of halal slaughter because it has the potential to deliver birds to the neck cutter alive and 'undamaged'. However, in systems employing a water bath, it is subject to the same limitations of aversive shackling and unreliability with regard to current delivery, and indeed more so since the stun needs to be demonstrably reversible and so cardiac arrest must not be induced. In a survey of religious slaughter spanning multiple species, Velarde et al. (2014) reported that 15% of the electrically stunned broilers assessed were breathing at the moment of neck cutting and 5% of them were breathing 30 s later. Although breathing does not directly relate to consciousness, it was suggested that the high percentage of rhythmic breathing observed after stunning indicates that the settings of the system were suboptimal. As discussed above, incorrect electrical parameters are likely to result in electro-immobilisation rather than unconsciousness with negative welfare consequences during neck cutting and bleeding. Wotton et al. (2014) suggested that application of 200 mA at 1000 Hz was sufficient to avoid cardiac arrest and so meet halal requirements, but indicators of' consciousness were not measured. Given that electrical stunning remains the only viable method for a reversible stun, novel automated head-only systems (such as 'Dutch Vision Solutions') are in development with the halal slaughter market in mind, and may provide opportunities to improve welfare by ensuring adequate current delivery on an individual basis.

### 3.5.3 Mitigation and conclusions

Various strands of evidence suggest that slaughter to meet religious needs is associated with poor welfare outcomes, even when stunning is used. In non-stun slaughter, opportunities to mitigate these are limited, but meticulous technique and care with handling are desirable, and EU regulations require that where animals are killed without stunning by a religious method the persons responsible for slaughtering must carry out systematic checks to ensure that animals do not present any signs of consciousness or sensibility before further processing (EC, 2009). Further development of electrical stunning systems to meet both welfare and religious needs is urgently required, bearing in mind that the size of these markets is growing (Defra, 2019; Fuseini et al., 2016; Velarde and Dalmau, 2018). Religious slaughter, especially without stunning, remains a controversial issue that is unlikely to be resolved due to the fundamental conflict it presents between animal welfare and human rights with regard to religious freedom (Anil, 2012).

## 4 Conclusions

The stunning and slaughter of chickens is a major commercial enterprise and methods applied must provide a balance between cost, feasibility and bird welfare outcomes. Both electrical stunning and controlled atmosphere methods have strengths and weaknesses in these regards. Electrical stunning has the important advantage of being theoretically instantaneous, but reliability is low in practice and it is associated with highly aversive handling of conscious birds. CAS resolves these issues, but potentially causes welfare harms during gradual induction of unconsciousness. As such, both methods are associated with variation in welfare outcomes based on process aims (in terms of the actual settings applied - electrical parameters or gas mixtures used) in addition to how well these settings are actually implemented in commercial use (i.e. the effectiveness of process control), especially under challenging high throughput conditions. For electrical stunning, suitable parameters to induce unconsciousness of sufficient duration are known, but there is an inherent tension between reliable stunning to protect welfare and poorer meat quality outcomes. In addition, there has been reluctance to take up novel equipment which would improve welfare as part of electrical stunning. The reasons for this are not clear, but are likely related to cost, challenges to practical implementation at high line speeds and in some cases, lack of funding for commercialisation. With CAS, the reliability of stunning is more assured, but uncertainty remains about the welfare implications of exposure to certain controlled atmospheres, especially sub-nociceptive concentrations of $CO_2$. This is a priority area for future research, given that systems which gradually expose birds to progressively increasing concentrations of $CO_2$ are widely used in Europe. An improved understanding of the extent to which avian respiratory systems produce breathlessness, air hunger and distress in hypercapnic and hypoxic environments is needed to assess likely aversion to CAS processes. Welfare outcomes are likely to vary in extent and mechanism depending on gas mixture, and comparisons of hypoxic dyspnoea versus hypercapnic dyspnoea, for example, are needed to underpin the development of improved CAS systems. Following repeated calls to end the use of electrical stunning in the EU, its future use is likely to be confined to small-scale operations for halal markets, which require reversible stunning (which CAS cannot provide). For such a system to be welfare-friendly, it must achieve non-aversive handling and reliable delivery of sufficient current to each bird, and novel systems are making progress towards this, encouraging optimism. It is probably impossible to accomplish stunning and slaughter (especially on a large scale) without imposing some welfare costs on the birds. However, striving to minimise harms to the billions of animals involved is a vital aim, and opportunities to improve welfare at chicken slaughter arise from continued research into better methods, and wider adoption of approaches already known to have acceptable welfare costs. On a global scale and with current knowledge, this leads to a strong recommendation to replace electrical stunning with CAS.

## 5 Where to look for further information

- Authie, E., Berg, C., Bøtner, A., Browman, H., Capua, I., Koeijer, A. de, Depner, K., Domingo, M., Edwards, S., Fourichon, C., Koenen, F., More, S.,

Raj, M., Shivonen, L., Spoolder, H., Stegeman, J. A., Thulke, H.-H., Velarde, A., Vågsholm, I., Willeberg, P., Zientara, S., 2013. Scientific Opinion on monitoring procedures at slaughterhouses for poultry. *EFSA J.* https://doi.org/10.2903/j.efsa.2013.3521.

- Berg, C., Raj, M., 2015. A review of different stunning methods for Poultry–animal welfare aspects (stunning methods for poultry). *Animals* 5, 1207-1219. https://doi.org/10.3390/ani5040407.
- Bøtner, A., Broom, D., Doherr, M. G., Domingo, M., Hartung, J., Keeling, L., Koenen, F., More, S., Morton, D., Oltenacu, P., Salati, F., Salman, M., Sanaa, M., Sharp, J. M., Stegeman, J. A., Szücs, E., Thulke, H.-H., Vannier, P., Webster, J., Wierup, M., 2012. Scientific Opinion on the electrical requirements for waterbath stunning equipment applicable for poultry. *EFSA J.* 10. https://doi.org/10.2903/j.efsa.2012.2757.
- Johnson, C. L. C., 2014. A review of bird welfare during controlled atmosphere and electrical water-bath stunning. *J. Am. Vet. Med. Assoc.* 245, 60–68. https://doi.org/10.2460/javma.245.1.60.
- Leary, S., Underwood, W., Anthony, R., Cartner, S., Grandin, T., Greenacre, C., Gwaltney-Brant, S., McCrackin, M. A., Meyer, R., Miller, D., Shearer, J., Turner, T., Yanong, R., Johnson, C. L., Patterson-Kane, E., 2020. *AVMA Guidelines for the Euthanasia of Animals: 2020 Edition.*
- Nielsen, S. S., Alvarez, J., Bicout, D. J., Calistri, P., Depner, K., Drewe, J. A., Garin-Bastuji, B., Gonzales Rojas, J. L., Gortázar Schmidt, C., Miranda Chueca, M. Á., Roberts, H. C., Sihvonen, L. H., Spoolder, H., Stahl, K., Velarde Calvo, A., Viltrop, A., Winckler, C., Candiani, D., Fabris, C., Van der Stede, Y., Michel, V., 2019a. Slaughter of animals: poultry. *EFSA J.* 17. https://doi.org/10.2903/j.efsa.2019.5849.
- Steiner, A. R., Flammer, S. A., Beausoleil, N. J., Berg, C., Bettschart-Wolfensberger, R., Pinillos, R. G., Golledge, H. D., Marahrens, M., Meyer, R., Schnitzer, T., Toscano, M. J., Turner, P. V., Weary, D. M., Gent, T. C., 2019. Humanely ending the life of animals: research priorities to identify alternatives to carbon dioxide. *Animals* 9, 1-25. https://doi.org/10.3390/ani9110911.
- Velarde, A., Dalmau, A., 2018. Slaughter without stunning, in: *Advances in Agricultural Animal Welfare*. Woodhead Publishing, Duxford, UK, pp. 221-240.

## 6 References

Abeyesinghe, S., McKeegan, D., McLeman, M., Lowe, J., Demmers, T., White, R., Kranen, R., Van Bemmel, H., Lankhaar, J., Wathes, C., Abeyesinghe, S., McLeman, M., Lowe, J., Demmers, T., White, R., Kranen, R., Van Bemmel, H., Lankhaar, J. and Wathes, C. 2007. Controlled atmosphere stunning of broiler chickens. II. Effects on behaviour, physiology and meat quality in a commercial processing plant. *Br. Poult. Sci.* 48(4), 430–442. https://doi.org/10.1080/00071660701543097.

Ali, A. S., Lawson, M. A., Tauson, A. H., Fris Jensen, J. and Chwalibog, A. 2007. Influence of electrical stunning voltages on bleed out and carcass quality in slaughtered broiler chickens. *Arch. Geflugelkd.* 71, 35–40.

Anastasov, M. I. and Wotton, S. B. 2012. Survey of the incidence of post-stun behavioural reflexes in electrically stunned broilers in commercial conditions and the relationship of their incidence with the applied water-bath electrical parameters. *Anim. Welf.* 21(2), 247–256. https://doi.org/10.7120/09627286.21.2.247.

Anil, M. H. 2012. Religious slaughter: A current controversial animal welfare issue. *Anim. Front.* 2(3), 64–67. https://doi.org/10.2527/af.2012-0051.

Authie, E., Berg, C., Bøtner, A., Browman, H., Capua, I., Koeijer, A., Depner, K., Domingo, M., Edwards, S., Fourichon, C., Koenen, F., More, S., Raj, M., Sihvonen, L., Spoolder, H., Stegeman, J., Thulke, H., Vågsholm, I., Velarde, A., Willeberg, P. and Zientara, S. 2014. Scientific Opinion on the use of low atmosphere pressure system (LAPS) for stunning poultry. *Eur. Food Saf. Auth J.* 12, 1-27.

Authie, E., Berg, C., Bøtner, A., Browman, H., Capua, I., Koeijer, A. de, Depner, K., Domingo, M., Edwards, S., Fourichon, C., Koenen, F., More, S., Raj, M., Shivonen, L., Spoolder, H., Stegeman, J. A., Thulke, H.-H., Velarde, A., Vågsholm, I., Willeberg, P. and Zientara, S. 2013. Scientific Opinion on monitoring procedures at slaughterhouses for poultry. *EFSA J.* 11(12) https://doi.org/10.2903/j.efsa.2013.3521.

Banzett, R. B., Pedersen, S. H., Schwartzstein, R. M. and Lansing, R. W. 2008. The affective dimension of laboratory dyspnea: air hunger is more unpleasant than work/effort. *Am. J. Respir. Crit. Care Med.* 177(12), 1384–1390. https://doi.org/10.1164/rccm.200711-1675OC.

Barnett, J. L., Cronin, G. M. and Scott, P. C. 2007. Behavioural responses of poultry during kosher slaughter and their implications for the birds' welfare. *Vet. Rec.* 160(2), 45–49. https://doi.org/10.1136/vr.160.2.45.

Bayliss, P. A. and Hinton, M. H. 1990. Transportation of broilers with special reference to mortality rates. *Appl. Anim. Behav. Sci.* 28(1–2), 93–118. https://doi.org/10.1016/0168-1591(90)90048-I.

Beausoleil, N. J. and Mellor, D. J. 2015. Introducing breathlessness as a significant animal welfare issue. *N. Z. Vet. J.* 63(1), 44–51. https://doi.org/10.1080/00480169.2014.940410.

Bedanova, I., Voslarova, E., Chloupek, P., Pistekova, V., Suchy, P., Blahova, J., Dobsikova, R. and Vecerek, V. 2007. Stress in broilers resulting from shackling. *Poult. Sci.* 86(6), 1065–1069. https://doi.org/10.1093/PS/86.6.1065.

Benson, E. R., Alphin, R. L., Rankin, M. K., Caputo, M. P., Kinney, C. A. and Johnson, A. L. 2012. Evaluation of EEG based determination of unconsciousness vs. loss of posture in broilers. *Res. Vet. Sci.* 93(2), 960–964. https://doi.org/10.1016/j.rvsc.2011.12.008.

Berg, C. and Raj, M. 2015. A review of different stunning methods for Poultry–animal welfare aspects (stunning methods for poultry). *Animals* 5(4), 1207–1219. https://doi.org/10.3390/ani5040407.

Bilgili, S. F. 1999. Recent advances in electrical stunning. *Poult. Sci.* 78(2), 282–286. https://doi.org/10.1093/PS/78.2.282.

Blokhuis, H. J., Roth, B., Holst, S., Kestin, S., Raj, M., Terlouw, C., Calvo, A. V. and Von Wenzlawowicz, M. 2004. Opinion of the Scientific Panel on Animal Health and Welfare on a request from the Commission related to welfare aspects of the main systems of stunning and killing the main commercial species of animals. *EFSA J.* 45, 1–29.

Bøtner, A., Broom, D., Doherr, M., Domingo, M., Hartung, J., Keeling, L., Koenen, F., More, S., Morton, D., Oltenacu, P., Osterhaus, A., Salati, F., Salman, M., Sanaa, M., Sharp, M., Stegeman, J., Szücs, E., Thulke, H., Vannier, P., Webster, J. and Wierup, M. 2011. Scientific opinion concerning the welfare of animals during transport. *EFSA J.* 9(1), 1–125. https://doi.org/10.2903/j.efsa.2011.1966.

Bøtner, A., Broom, D., Doherr, M. G., Domingo, M., Hartung, J., Keeling, L., Koenen, F., More, S., Morton, D., Oltenacu, P., Salati, F., Salman, M., Sanaa, M., Sharp, J. M., Stegeman, J. A., Szücs, E., Thulke, H.-H., Vannier, P., Webster, J. and Wierup, M. 2012. Scientific Opinion on the electrical requirements for waterbath stunning equipment applicable for poultry. *EFSA J.* 10(6). https://doi.org/10.2903/j.efsa.2012.2757.

Bouwsema, J. and Lines, J. 2019. Could low atmospheric pressure stunning (LAPS) be suitable for pig slaughter? A review of available information. *Anim. Welf.* 28(4), 421–432. https://doi.org/10.7120/09627286.28.4.421.

Butterworth, A. 1999. Infectious components of broiler lameness: a review. *World's Poult. Sci. J.* 55(4), 327–352. https://doi.org/10.1079/WPS19990024.

Caffrey, N. P., Dohoo, I. R. and Cockram, M. S. 2017. Factors affecting mortality risk during transportation of broiler chickens for slaughter in Atlantic Canada. *Prev. Vet. Med.* 147, 199–208. https://doi.org/10.1016/j.prevetmed.2017.09.011.

Chandia, M. and Soon, J. M. 2018. The variations in religious and legal understandings on halal slaughter. *Br. Food J.* 120(3), 714–730. https://doi.org/10.1108/BFJ-03-2017-0129.

Cockram, M. S., Dulal, K. J., Mohamed, R. A. and Revie, C. W. 2019. Risk factors for bruising and mortality of broilers during manual handling, module loading, transport, and lairage. *Can. J. Anim. Sci.* 99(1), 50–65. https://doi.org/10.1139/cjas-2018-0032.

Coenen, A. M., Lankhaar, J., Lowe, J. C. and McKeegan, D. E. 2009. Remote monitoring of electroencephalogram, electrocardiogram, and behavior during controlled atmosphere stunning in broilers: implications for welfare. *Poult. Sci.* 88(1), 10–19. https://doi.org/10.3382/ps.2008-00120.

Coenen, A. M., Drinkenburg, W. H., Hoenderken, R. and van Luijtelaar, E. L. 1995. Carbon dioxide euthanasia in rats: oxygen supplementation minimizes signs of agitation and asphyxia. *Lab. Anim.* 29(3), 262–268. https://doi.org/10.1258/002367795781088289.

Cors, J. C., Gruber, A. D., Günther, R., Meyer-Kühling, B., Esser, K. H. and Rautenschlein, S. 2015. Electroencephalographic evaluation of the effectiveness of blunt trauma to induce loss of consciousness for on-farm killing of chickens and turkeys. *Poult. Sci.* 94(2), 147–155. https://doi.org/10.3382/PS/PEU038.

Cranley, J. 2017. Death and prolonged survival in nonstunned poultry: A case study. *J. Vet. Behav. Clin. Appl. Res.* 18, 92–95. https://doi.org/10.1016/j.jveb.2016.09.005.

Defra 2019. *Results of the 2018 FSA Survey into Slaughter Methods in England and Wales*. Defra, London, UK.

Devos, G., Moons, C. P. H. and Houf, K. 2018. Diversity, not uniformity: slaughter and electrical waterbath stunning procedures in Belgian slaughterhouses. *Poult. Sci.* 97(9), 3369–3379. https://doi.org/10.3382/ps/pey181.

Ekstrand, C. 1998. An observational cohort study of the effects of catching method on carcase rejection rates in broilers. *Anim. Welf.* 7, 87–96.

Erasmus, M. A., Lawlis, P., Duncan, I. J. and Widowski, T. M. 2010a. Using time to insensibility and estimated time of death to evaluate a nonpenetrating captive bolt, cervical dislocation, and blunt trauma for on-farm killing of turkeys. *Poult. Sci.* 89(7), 1345–1354. https://doi.org/10.3382/ps.2009-00445.

Erasmus, M. A., Turner, P. V., Nykamp, S. G. and Widowski, T. M. 2010b. Brain and skull lesions resulting from use of percussive bolt, cervical dislocation by stretching, cervical dislocation by crushing and blunt trauma in turkeys. *Vet. Rec.* 167(22), 850–858. https://doi.org/10.1136/vr.c5051.

Erasmus, M. A., Turner, P. V. and Widowski, T. M. 2010c. Measures of insensibility used to determine effective stunning and killing of poultry. *J. Appl. Poult. Res.* 19(3), 288–298. https://doi.org/10.3382/japr.2009-00103.

European Commission 2009. Council regulation (EC) No 1099/2009 of 24 September 2009 on the protection of animals at the time of killing. OJ L 303, 18.11.2009, p. 1-30. Available at: ELI: http://data.europa.eu/eli/reg/2009/1099/oj.

European Commission 2017. Preparation of best practices on the protection of animals at the time of killing. Final Report [WWW Document]. https://doi.org/10.2875/15243.

European Council 1979. European Convention for the protection of animals for slaughter [WWW Document]. Eur. Treaty Ser. Available at: https://www.coe.int/en/web/conventions/full-list/-/conventions/rms/0900001680077d98.

European Food Safety Authority (EFSA). (2004). Opinion of the Scientific Panel on Animal Health and Welfare (AHAW) on a request from the Commission related to welfare aspects of the main systems of stunning and killing the main commercial species of animals. EFSA Journal, 2(7), 45. doi.org/10.2903/j.efsa.2004.45.

European Union 2017. Brexit: farm animal welfare, Chapter 2: Maintaining standards [WWW Document]. Rep. Sess.

Farm Animal Welfare Council 2009. *Report on the Welfare of Farmed Animals at Slaugther or Killing*. Crown Publishing, London, UK.

Fedde, M. 2012. Respiration. In: Sturkie, P. (Ed), *Avian Physiology*. Springer-Verlag New York Inc., New York, pp. 191–221. https://doi.org/10.1007/978-1-4612-4862-0.

Fedde, M. R. 1998. Relationship of structure and function of the avian respiratory system to disease susceptibility. *Poult. Sci.* 77(8), 1130–1138. https://doi.org/10.1093/PS/77.8.1130.

Fetrow, J., Nordlund, K. V. and Norman, H. D. 2006. Invited review: culling: nomenclature, definitions, and recommendations. *J. Dairy Sci.* 89(6), 1896–1905 https://doi.org/10.3168/jds.S0022-0302(06)72257-3.

Food and Agriculture Organization of the United Nations (FAO) 2020. Livestock primary, producing animals/slaughtered [WWW Document]. *FAO Stat.* Available at: http://www.fao.org/faostat/en/#data/QL (Accessed 28 February 2020).

Fuseini, A., Knowles, T. G., Hadley, P. J. and Wotton, S. B. 2016. Halal stunning and slaughter: criteria for the assessment of dead animals. *Meat Sci.* 119, 132–137. https://doi.org/10.1016/J.MEATSCI.2016.04.033.

Gentle, M. J. and Tilston, V. L. 2000. Nociceptors in the legs of poultry: implications for potential pain in pre-slaughter shackling. *Anim. Welf.* 9, 227–236.

Gerritzen, M. A., Lambooij, B., Reimert, H., Stegeman, A. and Spruijt, B. 2004. On-farm euthanasia of broiler chickens: effects of different gas mixtures on behavior and brain activity. *Poult. Sci.* 83(8), 1294–1301. https://doi.org/10.1093/ps/83.8.1294.

Gerritzen, M. and Sparrey, J. 2008. A pilot study to assess whether high expansion CO2-enriched foam is acceptable for on-farm emergency killing of poultry.

Gerritzen, M., van Hattum, T. and Reimert, H. 2015. Efficacy of the Dutch Vision high-low electrical head-only poultry stunner. Livestock Research Report. Wageningen, The Netherlands. https://doi.org/10.13140/RG.2.1.2174.3767.

Gerritzen, M. A., Lambooij, E., Hillebrand, S. J. W., Lankhaar, J. A. C. and Pieterse, C. 2000. Behavioral responses of broilers to different gaseous atmospheres. *Poult. Sci.* 79(6), 928–933. https://doi.org/10.1093/ps/79.6.928.

Gerritzen, M. A., Reimert, H. G. M., Hindle, V. A., Verhoeven, M. T. W. and Veerkamp, W. B. 2013. Multistage carbon dioxide gas stunning of broilers. *Poult. Sci.* 92(1), 41–50. https://doi.org/10.3382/ps.2012-02551.

Girasole, M., Chirollo, C., Ceruso, M., Vollano, L., Chianese, A. and Cortesi, M. L. 2015. Optimization of stunning electrical parameters to improve animal welfare in a poultry slaughterhouse. *Ital. J. Food Saf.* 4(3), 175–178. https://doi.org/10.4081/ijfs.2015.4576.

Girasole, M., Marrone, R., Anastasio, A., Chianese, A., Mercogliano, R. and Cortesi, M. L. 2016. Effect of electrical water bath stunning on physical reflexes of broilers: evaluation of stunning efficacy under field conditions. *Poult. Sci.* 95(5), 1205–1210. https://doi.org/10.3382/PS/PEW017.

Global Legal Research Center 2018. Legal restrictions on religious slaughter in Europe [WWW Document]. Law Libr. Congress. Available at: https://www.loc.gov/law/help/religious-slaughter/europe.php (Accessed 3 February 2020).

Gradwell, D. 2016. Hypoxia and hyperventilation. In: Gradwell, D. and Rainford, D. (Eds), *Ernsting's Aviation and Space Medicine*. CRC Press, Boca Raton, FL, p. 904.

Gradwell, D. and Macmillan, A. 2016. Oxygen systems, pressure cabin and clothing. In: Gradwell, D. and Rainford, D. (Eds), *Ernsting's Aviation and Space Medicine*. CRC Press, Boca Raton, FL.

Gregory, N. G. 2008. Animal welfare at markets and during transport and slaughter. *Meat Sci.* 80(1), 2–11. https://doi.org/10.1016/j.meatsci.2008.05.019.

Gregory, N. G. and Bell, J. C. 1987. Duration of wing flapping in chickens shackled before slaughter. *Vet. Rec.* 121(24), 567–569. https://doi.org/10.1136/VR.121.24.567.

Gregory, N. G., Wilkins, L. J. and Wotton, S. B. 1991. Effect of electrical stunning frequency on ventricular fibrillation, downgrading and broken bones in broilers, hens and quails. *Br. Vet. J.* 147(1), 71–77. https://doi.org/10.1016/0007-1935(91)90069-Y.

Gregory, N. G. and Wotton, S. B. 1986. Effect of slaughter on the spontaneous and evoked activity of the brain. *Br. Poult. Sci.* 27(2), 195–205. https://doi.org/10.1080/00071668608416872.

Gregory, N. G. and Wotton, S. B. 1987. Effect of electrical stunning on the electroencephalogram in chickens. *Br. Vet. J.* 143(2), 175–183. https://doi.org/10.1016/0007-1935(87)90009-1.

Gregory, N. G. and Wotton, S. B. 1990a. Comparison of neck dislocation and percussion of the head on visual evoked responses in the chicken's brain. *Vet. Rec.* 126(23), 570–572.

Gregory, N. G. and Wotton, S. B. 1990b. Effect of stunning on spontaneous physical activity and evoked activity in the brain. *Br. Poult. Sci.* 31(1), 215–220. https://doi.org/10.1080/00071669008417248.

Gregory, N. G. and Wotton, S. B. 1991. Effect of a 350 Hz DC stunning current on evoked responses in the chicken's brain. *Res. Vet. Sci.* 50(2), 250–251. https://doi.org/10.1016/0034-5288(91)90118-8.

Gregory, N. G. and Wotton, S. B. 1994. Effect of electrical stunning current on the duration of insensibility in hens. *Br. Poult. Sci.* 35(3), 463–465. https://doi.org/10.1080/00071669408417711.

Grilli, C., Loschi, A. R., Rea, S., Stocchi, R., Leoni, L. and Conti, F. 2015. Welfare indicators during broiler slaughtering. *Br. Poult. Sci.* 56(1), 1–5. https://doi.org/10.1080/00071668.2014.991274.

Hillebrand, S. J., Lambooy, E. and Veerkamp, C. H. 1996. The effects of alternative electrical and mechanical stunning methods on hemorrhaging and meat quality of broiler breast and thigh muscles. *Poult. Sci.* 75(5), 664–671. https://doi.org/10.3382/ps.0750664.

Hindle, V. A., Lambooij, E., Reimert, H. G. M., Workel, L. D. and Gerritzen, M. A. 2010. Animal welfare concerns during the use of the water bath for stunning broilers, hens, and ducks. *Poult. Sci.* 89(3), 401–412. https://doi.org/10.3382/ps.2009-00297.

Holloway, P. H. and Pritchard, D. G. 2017. Effects of ambient temperature and water vapor on chamber pressure and oxygen level during low atmospheric pressure stunning of poultry. *Poult. Sci.* 96(8), 2528–2539. https://doi.org/10.3382/ps/pex066.

Hughes, B. O. 1983. Headshaking in fowls: the effect of environmental stimuli. *Appl. Anim. Ethol.* 11(1), 45–53. https://doi.org/10.1016/0304-3762(83)90078-0.

Humane Slaughter Association 2016a. Maintaining an uninterrupted electrical circuit and optimising current flow [WWW Document]. HSA online guide – electrical waterbath stunning poultry. Available at: https://www.hsa.org.uk/electrical-waterbath-stunning-of-poultry-introduction/introduction-7 (Accessed 9 March 2020).

Humane Slaughter Association 2016b. *Electrical Waterbath Stunning of Poultry*. Humane Slaughter Association, Wheathampstead, UK.

Humane Slaughter Association 2016c. *Practical Slaughter of Poultry*. Humane Slaughter Association, Wheathampstead, UK.

Hunter, R. R., Mitchell, M. A., Carlisle, A. J., Quinn, A. D., Kettlewell, P. J., Knowles, T. G. and Warriss, P. D. 1998. Physiological responses of broilers to pre-slaughter lairage: effects of the thermal micro-environment? *Br. Poult. Sci.* 39, 53–54. https://doi.org/10.1080/00071669888377.

Jacobs, L., Bourassa, D. V., Harris, C. E. and Buhr, R. J. 2019. Euthanasia: manual Versus mechanical cervical dislocation for broilers. *Animals* 9(2), 47. https://doi.org/10.3390/ani9020047.

Jacobs, L., Delezie, E., Duchateau, L., Goethals, K. and Tuyttens, F. A. M. 2017. Impact of the separate pre-slaughter stages on broiler chicken welfare. *Poult. Sci.* 96(2), 266–273. https://doi.org/10.3382/ps/pew361.

Jiang, N. N., Xing, T., Wang, P., Xie, C. and Xu, X. L. 2015. Effects of water-misting sprays with forced ventilation after transport during summer on meat quality, stress parameters, glycolytic potential and microstructures of muscle in broilers. *Asian-Australas. J. Anim. Sci.* 28(12), 1767–1773. https://doi.org/10.5713/ajas.15.0152.

Johnson, C. L. C. 2014. A review of bird welfare during controlled atmosphere and electrical water-bath stunning. *J. Am. Vet. Med. Assoc.* 245(1), 60–68. https://doi.org/10.2460/javma.245.1.60.

Kannan, G., Heath, J. L., Wabeck, C. J. and Mench, J. A. 1997. Shackling of broilers: effects on stress responses and breast meat quality. *Br. Poult. Sci.* 38(4), 323-332. https://doi.org/10.1080/00071669708417998.

Kettlewell, P. J. and Mitchell, M. A. 1994. Catching, handling and loading of poultry for road transportation. *World's Poult. Sci. J.* 50(1), 54-56. https://doi.org/10.1079/WPS19940005.

Kirkden, R., Niel, L., Stewart, S. and Weary, D. 2008. Gas killing of rats: the effect of supplemental oxygen on aversion to carbon dioxide. *Animal Welfare*, 17(1), 79-87.

Kittelsen, K. E., Granquist, E. G., Vasdal, G., Tolo, E. and Moe, R. O. 2015. Effects of catching and transportation versus pre-slaughter handling at the abattoir on the prevalence of wing fractures in broilers. *Anim. Welf.* 24(4), 387-389. https://doi.org/10.7120/09627286.24.4.387.

Knowles, T. G. and Broom, D. M. 1990. The handling and transport of broilers and spent hens. *Appl. Anim. Behav. Sci.* 28(1-2), 75-91. https://doi.org/10.1016/0168-1591(90)90047-H.

Knowles, T. G. and Wilkins, L. J. 1998. The problem of broken bones during the handling of laying hens - a review. *Poult. Sci.* 77(12), 1798-1802.

Korte, S. M., Beuving, G., Ruesink, W. and Blokhuis, H. J. 1997. Plasma catecholamine and corticosterone levels during manual restraint in chicks from a high and low feather pecking line of laying hens. *Physiol. Behav.* 62(3), 437-441. https://doi.org/10.1016/S0031-9384(97)00149-2.

Kranen, R. W., Veerkamp, C. H., Lambooy, E., Van Kuppevelt, T. H. and Veerkamp, J. H. 1996. Hemorrhages in muscles of broiler chickens: the relationships among blood variables at various rearing temperature regimens. *Poult. Sci.* 75(4), 570-576. https://doi.org/10.3382/PS.0750570.

Lambooij, B. and Hindle, V. 2018. Electrical stunning of poultry. In *Advances in Poultry Welfare*. Woodhead Publishing, UK, pp. 77-98. https://doi.org/10.1016/B978-0-08-100915-4.00004-X.

Lambooij, E., Anil, H., Butler, S. R., Reimert, H., Workel, L., Hindle, V. and Ur, W. 2011. Transcranial magnetic stunning of broilers: a preliminary trial to induce unconsciousness. *Animal Welfare*, 20(3), 407-412.

Lambooij, E., Reimert, H. G. M. and Hindle, V. A. 2010. Evaluation of head-only electrical stunning for practical application: assessment of neural and meat quality parameters. *Poult. Sci.* 89(12), 2551-2558. https://doi.org/10.3382/PS.2010-00815.

Lambooij, E., Reimert, H. G. M., Workel, L. D. and Hindle, V. A. 2012. Head-cloaca controlled current stunning: assessment of brain and heart activity and meat quality. *Br. Poult. Sci.* 53(2), 168-174. https://doi.org/10.1080/00071668.2012.665434.

Leary, S., Underwood, W., Anthony, R., Cartner, S., Grandin, T., Greenacre, C., Gwaltney-Brant, S., McCrackin, M. A., Meyer, R., Miller, D., Shearer, J., Turner, T. and Yanong, R. 2007. AVMA guidelines on euthanasia. *Am. Vet. Med. Assoc. Schaumburg* 218, 1-39. https://doi.org/10.1007/s11634-008-0026-3.

Leary, S., Underwood, W., Anthony, R., Cartner, S., Grandin, T., Greenacre, C., Gwaltney-Brant, S., McCrackin, M. A., Meyer, R., Miller, D., Shearer, J., Turner, T., Yanong, R., Johnson, C. L. and Patterson-Kane, E. 2020. *AVMA Guidelines for the Euthanasia of Animals: 2020 Edition*. AVMA American Veterinary Medical Association, Schaumburg, IL.

Lewis, P. D. and Morris, T. R. 2000. Poultry and coloured light. *Worlds. Poult. Sci. J.* 56(3), 189-207. https://doi.org/10.1079/WPS20000015.

Lines, J. A., Berry, P., Cook, P., Schofield, C. P. and Knowles, T. G. 2012. Improving the poultry shackle line. *Anim. Welf.* 21(1), 69–74. https://doi.org/10.7120/096272812X13353700593608.

Lines, J. A., Jones, T. A., Berry, P. S., Cook, P., Spence, J. and Schofield, C. P. 2011a. Evaluation of a breast support conveyor to improve poultry welfare on the shackle line. *Vet. Rec.* 168(5), 129. https://doi.org/10.1136/vr.c5431.

Lines, J. A., Raj, A. B., Wotton, S. B., O'Callaghan, M. and Knowles, T. G. 2011b. Head-only electrical stunning of poultry using a waterbath: a feasibility study. *Br. Poult. Sci.* 52(4), 432–438. https://doi.org/10.1080/00071668.2011.587180.

Llonch, P., Dalmau, A., Rodríguez, P., Manteca, X. and Velarde, A. 2012. Aversion to nitrogen and carbon dioxide mixtures for stunning pigs. *Anim. Welf.* 21(1), 33–39. https://doi.org/10.7120/096272812799129475.

Mackie, N. and McKeegan, D. E. F. 2016. Behavioural responses of broiler chickens during low atmospheric pressure stunning. *Appl. Anim. Behav. Sci.* 174, 90–98. https://doi.org/10.1016/j.applanim.2015.11.001.

Martin, J. E., Christensen, K., Vizzier-Thaxton, Y. and McKeegan, D. E. 2016a. Effects of light on responses to low atmospheric pressure stunning in broilers. *Br. Poult. Sci.* 57(5), 585–600. https://doi.org/10.1080/00071668.2016.1201200.

Martin, J. E., Christensen, K., Vizzier-Thaxton, Y. and McKeegan, D. E. F. 2016b. Effects of analgesic intervention on behavioural responses to low atmospheric pressure stunning. *Appl. Anim. Behav. Sci.* 180, 157–165. https://doi.org/10.1016/j.applanim.2016.05.007.

Martin, J. E., Christensen, K., Vizzier-Thaxton, Y., Mitchell, M. A. and McKeegan, D. E. F. 2016c. Behavioural, brain and cardiac responses to hypobaric hypoxia in broiler chickens. *Physiol. Behav.* 163, 25–36. https://doi.org/10.1016/j.physbeh.2016.04.038.

Martin, J. E., McKeegan, D. E. F., Magee, D. L., Armour, N. and Pritchard, D. G. 2019a. Pathological consequences of low atmospheric pressure stunning in broiler chickens. *Animal* 14(1), 129–137. https://doi.org/10.1017/S1751731119001411.

Martin, J. E., McKeegan, D. E. F., Sparrey, J. and Sandilands, V. 2017. Evaluation of the potential killing performance of novel percussive and cervical dislocation tools in chicken cadavers. *Br. Poult. Sci.* 58(3), 216–223. https://doi.org/10.1080/00071668.2017.1280724.

Martin, J., McKeegan, D., Sparrey, J. and Sandilands, V. 2016d. Comparison of novel mechanical cervical dislocation and a modified captive bolt for on-farm killing of poultry on behavioural reflex responses and anatomical pathology. *Anim. Welf.* 25(2), 227–241. https://doi.org/10.7120/09627286.25.2.227.

Martin, J. E., Sandercock, D. A., Sandilands, V., Sparrey, J., Baker, L., Sparks, N. H. C. and McKeegan, D. E. F. 2018a. Welfare risks of repeated application of on-farm killing methods for poultry. *Animals* 8(3). https://doi.org/10.3390/ani8030039.

Martin, J. E., Sandilands, V., Sparrey, J., Baker, L., Dixon, L. M. and McKeegan, D. E. F. 2019b. Welfare assessment of novel on-farm killing methods for poultry. *PLoS ONE* 14(2), e0212872. https://doi.org/10.1371/journal.pone.0212872.

Martin, J. E., Sandilands, V., Sparrey, J., Baker, L. and McKeegan, D. E. F. 2018b. On farm evaluation of a novel mechanical cervical dislocation device for poultry. *Animals* 8(1). https://doi.org/10.3390/ani8010010.

McCulloch, S. P. 2018. Brexit and animal protection: legal and political context and a framework to assess impacts on animal welfare. *Animals* 8(11), 213. https://doi.org/10.3390/ani8110213.

McKeegan, D. 2018. Mass depopulation. *Adv. Poult. Welf.*, 351–372. https://doi.org/10.1016/B978-0-08-100915-4.00017-8.

McKeegan, D., McIntyre, J., Demmers, T., Lowe, J., Wathes, C., Van Den Broek, P., Coenen, A. and Gentle, M. 2007. Physiological and behavioural responses of broilers to controlled atmosphere stunning: implications for welfare. *Anim. Welf.* 16, 409–426.

McKeegan, D. E., Reimert, H. G., Hindle, V. A., Boulcott, P., Sparrey, J. M., Wathes, C. M., Demmers, T. G. and Gerritzen, M. A. 2013a. Physiological and behavioral responses of poultry exposed to gas-filled high expansion foam. *Poult. Sci.* 92(5), 1145–1154. https://doi.org/10.3382/PS.2012-02587.

McKeegan, D. E., Sparks, N. H., Sandilands, V., Demmers, T. G., Boulcott, P. and Wathes, C. M. 2011. Physiological responses of laying hens during whole-house killing with carbon dioxide. *Br. Poult. Sci.* 52(6), 645–657. https://doi.org/10.1080/00071668.2011.640307.

McKeegan, D. E. F. 2004. Mechano-chemical nociceptors in the avian trigeminal mucosa. *Brain Res. Rev.* 46(2), 146–154. https://doi.org/10.1016/J.BRAINRESREV.2004.07.012.

McKeegan, D. E. F., McIntyre, J., Demmers, T. G. M., Wathes, C. M. and Jones, R. B. 2006. Behavioural responses of broiler chickens during acute exposure to gaseous stimulation. *Appl. Anim. Behav. Sci.* 99(3–4), 271–286. https://doi.org/10.1016/j.applanim.2005.11.002.

McKeegan, D. E. F., Sandercock, D. A. and Gerritzen, M. A. 2013b. Physiological responses to low atmospheric pressure stunning and the implications for welfare. *Poult. Sci.* 92(4), 858–868. https://doi.org/10.3382/ps.2012-02749.

Meyn 2020. Multistage CO2 stunning system [WWW Document]. Available at: https://www.meyn.com/products/live-bird-handling/multistage-co2-stunning-system (Accessed 2.12.2020).

Milsom, W. K., Abe, A. S., Andradeb, D. V. and Tattersall, G. J. 2004. Evolutionary trends in airway CO2/H+ chemoreception. *Respir. Physiol. Neurobiol.* 144(2-3), 191–202. https://doi.org/10.1016/J.RESP.2004.06.021.

Mitchell, M. A. and Kettlewell, P. J. 1994. Road transportation of broiler chickens: induction of physiological stress. *World's Poult. Sci. J.* 50(1), 57–59. https://doi.org/10.1079/WPS19940006.

Mitchell, M. A. and Kettlewell, P. J. 1998. Physiological stress and welfare of broiler chickens in transit: solutions not problems! *Poult. Sci.* 77(12), 1803–1814.

Mohamed, R. A., Eltholth, M. M. and El-Saidy, N. R. 2014. Rearing broiler chickens under monochromatic blue light improve performance and reduce fear and stress during pre-slaughter handling and transportation. *Biotechnol. Anim. Husb.* 30(3), 457–471. https://doi.org/10.2298/BAH1403457M.

Moosavi, S. H., Golestanian, E., Binks, A. P., Lansing, R. W., Brown, R. and Banzett, R. B. 2003. Hypoxic and hypercapnic drives to breathe generate equivalent levels of air hunger in humans. *J. Appl. Physiol.* 94(1), 141–154. https://doi.org/10.1152/japplphysiol.00594.2002.

More, S., Bicout, D., Bøtner, A., Butterworth, A., Calistri, P., Depner, K., Edwards, S., Garin-Bastuji, B., Good, M., Gortazar Schmidt, C., Miranda, M.A., Nielsen, S.S., Sihvonen, L.,

Spoolder, H., Willeberg, P., Raj, M., Thulke, H., Velarde, A., Vyssotski, A., Winckler, C., Cortiñas Abrahantes, J., Garcia, A., Muñoz Guajardo, I., Zancanaro, G. and Michel, V. 2017. Low atmospheric pressure system for stunning broiler chickens. *EFSA J.* 15(12), e05056. https://doi.org/10.2903/j.efsa.2017.5056.

National Chicken Council (NCC) 2011. *National Chicken Council Animal Welfare Guidelines*. NCC, Washington, DC.

Newcomb, L. 2010. Why use oxygen on everest? The physiological advantages of using supplementary oxygen on the summit day. *BJA Br. J. Anaesth.* 105. https://doi.org/10.1093/bja/el_6439.

Nicol, C., Caplen, G., Edgar, J., Richards, G. and Browne, W. 2011. Relationships between multiple welfare indicators measured in individual chickens across different time periods and environments. *Anim. Welf.* 20, 133–143.

Nielsen, S. S., Alvarez, J., Bicout, D. J., Calistri, P., Depner, K., Drewe, J. A., Garin-Bastuji, B., Gonzales Rojas, J. L., Gortázar Schmidt, C., Miranda Chueca, M. Á., Roberts, H. C., Sihvonen, L. H., Spoolder, H., Stahl, K., Velarde Calvo, A., Viltrop, A., Winckler, C., Candiani, D., Fabris, C., Van der Stede, Y. and Michel, V. 2019a. Slaughter of animals: poultry. *EFSA J.* 17(11), e05849. https://doi.org/10.2903/j.efsa.2019.5849.

Nielsen, S. S., Alvarez, J., Bicout, D. J., Calistri, P., Depner, K., Drewe, J. A., Garin-Bastuji, B., Gonzales Rojas, J. L., Gortázar Schmidt, C., Miranda Chueca, M. Á., Roberts, H. C., Sihvonen, L. H., Spoolder, H., Stahl, K., Velarde Calvo, A., Viltrop, A., Winckler, C., Candiani, D., Fabris, C., Van der Stede, Y. and Michel, V. 2019b. Killing for purposes other than slaughter: poultry. *EFSA J.* 17(11), e05850. https://doi.org/10.2903/j.efsa.2019.5850.

Nijdam, E., Arens, P., Lambooij, E., Decuypere, E. and Stegeman, J. A. 2004. Factors influencing bruises and mortality of broilers during catching, transport, and lairage. *Poult. Sci.* 83(9), 1610–1615. https://doi.org/10.1093/PS/83.9.1610.

Novoa, M., Vázquez, L., Lage, A., González-Torres, I., Pérez-García, L. F., Cobas, N. and Lorenzo, J. M. 2019. Water-bath stunning process in broiler chickens: effects of voltage and intensity. *Span. J. Agric. Res.* 17(2). https://doi.org/10.5424/sjar/2019172-14576.

Perez-Palacios, S. and Wotton, S. B. 2006. Effects of salinity and the use of shackle/leg sprays on the electrical conductivity of a commercial waterbath stunner for broilers. *Vet. Rec.* 158(19), 654–657. https://doi.org/10.1136/vr.158.19.654.

Petracci, M., Bianchi, M., Cavani, C., Gaspari, P. and Lavazza, A. 2006. Preslaughter mortality in broiler chickens, turkeys, and spent hens under commercial slaughtering. *Poult. Sci.* 85(9), 1660–1664. https://doi.org/10.1093/ps/85.9.1660.

Pierre, L. N., Emilie, B., Boissy, A., Boivin, X., Calandreau, L., Delon, N., Deputte, B., Desmoulin-Canselier, S., Dunier, M., Faivre, N., Giurfa, M., Guichet, J.-L. and Terlouw, C. 2018. Animal consciousness. *Anim. Welf.* 27, 87. https://doi.org/10.2903/sp.efsa.2017.EN-1196.

Poole, G. H. and Fletcher, D. L. 1998. Comparison of a modified atmosphere stunning-killing system to conventional electrical stunning and killing on selected broiler breast muscle rigor development and meat quality attributes. *Poult. Sci.* 77(2), 342–347. https://doi.org/10.1093/PS/77.2.342.

Powell, F. and Scheid, P. 1989. Physiology of gas exchange in the avian respiratory system. In: King, A. and McLelland, J. (Eds), *Form and Function in Birds*. Academic Press, London, UK, pp. 393–437.

Prayitno, D. S., Phillips, C. J. and Omed, H. 1997a. The effects of color of lighting on the behavior and production of meat chickens. *Poult. Sci.* 76(3), 452–457. https://doi.org/10.1093/ps/76.3.452.

Prayitno, D. S., Phillips, C. J. and Stokes, D. K. 1997b. The effects of color and intensity of light on behavior and leg disorders in broiler chickens. *Poult. Sci.* 76(12), 1674–1681. https://doi.org/10.1093/ps/76.12.1674.

Prinz, S., van Oijen, G., Ehinger, F., Bessei, W. and Coenen, A. 2010a. Effects of waterbath stunning on the electroencephalograms and physical reflexes of broilers using a pulsed direct current. *Poult. Sci.* 89(6), 1275–1284. https://doi.org/10.3382/ps.2009-00136.

Prinz, S., Van Oijen, G., Ehinger, F., Bessei, W. and Coenen, A. 2012. Electrical waterbath stunning: influence of different waveform and voltage settings on the induction of unconsciousness and death in male and female broiler chickens. *Poult. Sci.* 91(4), 998–1008. https://doi.org/10.3382/PS.2009-00137.

Prinz, S., van Oijen, G., Ehinger, F., Coenen, A. and Bessei, W. 2010b. Electroencephalograms and physical reflexes of broilers after electrical waterbath stunning using an alternating current. *Poult. Sci.* 89(6), 1265–1274. https://doi.org/10.3382/ps.2009-00135.

Quinn, A. D., Kettlewell, P. J., Mitchell, M. A. and Knowles, T. 1998. Air movement and the thermal microclimates observed in poultry lairages. *Br. Poult. Sci.* 39(4), 469–476. https://doi.org/10.1080/00071669888610.

Raj, M. 1998. Welfare during stunning and slaughter of poultry. *Poult. Sci.* 77(12), 1815-1819. https://doi.org/10.1093/ps/77.12.1815.

Raj, M. 2003. A critical appraisal of electrical stunning in chickens. *World's Poult. Sci. J.* 59(1), 89-98. https://doi.org/10.1079/WPS20030005.

Raj, M. 2006. Recent developments in stunning and slaughter of poultry. *World's Poult. Sci. J.* 62(3), 467–484. https://doi.org/10.1079/WPS2005109.

Raj, M. 2009. Stunning and slaughter. In: Duncan, I. and Hawkins, P. (Eds), *The Welfare of Domestic Fowl and Other Captive Birds*. Springer, New York, pp. 259-279. https://doi.org/10.1007/78-90-481-3650-6.

Raj, M., and Gregory, N. G. 1991. Preferential feeding behaviour of hens in different gaseous atmospheres. *Br. Poult. Sci.* 32(1), 57–65. https://doi.org/10.1080/00071669108417327.

Raj, M., and Gregory, N. G. 1990a. Investigation into the batch stunning/killing of chickens using carbon dioxide or argon-induced hypoxia. *Res. Vet. Sci.* 49(3), 364–366. https://doi.org/10.1016/0034-5288(90)90075-F.

Raj, M., and Gregory, N. G. 1990b. Effect of rate of induction of carbon dioxide anaesthesia on the time of onset of unconsciousness and convulsions. *Res. Vet. Sci.* 49(3), 360-363. https://doi.org/10.1016/0034-5288(90)90074-E.

Raj, M. and O'Callaghan, M. 2001. Evaluation of a pneumatically operated captive bolt for stunning/killing broiler chickens. *Br. Poult. Sci.* 42(3), 295–299. https://doi.org/10.1080/00071660120055232.

Raj, M. and O'Callaghan, M. 2004. Effect of amount and frequency of head-only stunning currents on the electroencephalogram and somatosensory evoked potentials in broilers. *Anim. Welf.* 13, 159–170.

Raj, M., O'Callaghan, M. and Hughes, S. 2006a. The effects of amount and frequency of pulsed direct current used in water bath stunning and of slaughter methods on spontaneous electroencephalograms in broilers. *Anim. Welf.* 15, 19–24.

Raj, M., O'Callaghan, M. and Hughes, S. 2006b. The effects of pulse width of a direct current used in water bath stunning and of slaughter methods on spontaneous electroencephalograms in broilers. *Anim. Welf.* 15, 25–30.

Raj, M., O'Callaghan, M. and Knowles, T. 2006c. The effects of amount and frequency of alternating current used in water bath stunning and of slaughter methods on electroencephalograms in broilers. *Anim. Welf.* 15, 7–18.

Raj, M., Sandilands, V. and Sparks, N. H. 2006d. Review of gaseous methods of killing poultry on-farm for disease control purposes. *Vet. Rec.* 159(8), 229–235. https://doi.org/10.1136/vr.159.8.229.

Raj, M. and Tserveni-Gousi, A. 2000. Stunning methods for poultry. *Worlds. Poult. Sci. J.* 56(4), 291–304. https://doi.org/10.1079/WPS20000021.

Raj, M., Wotton, S. B. and Gregory, N. G. 1992. Changes in the somatosensory evoked potentials and spontaneous electroencephalogram of hens during stunning with a carbon dioxide and argon mixture. *Br. Vet. J.* 148(2), 147–156. https://doi.org/10.1016/0007-1935(92)90106-B.

Rao, A. 2014. Prestun shocks and mis-stuns during conventional slaughter. *Vet. Rec.* 174(18), 457–458. https://doi.org/10.1136/vr.g3016.

Rao, M., Knowles, T. and Wotton, S. 2013. The effect of pre-stun shocks in electrical water-bath stunners on carcase and meat quality in broilers. *Anim. Welf.* 22(1), 79–84. https://doi.org/10.7120/09627286.22.1.079.

Rodrigues, D. R., Café, M. B., Jardim Filho, R. M., Oliveira, E., Trentin, T. C., Martins, D. B., Minafra, C. S. and Rodrigues, D. R. 2017. Metabolism of broilers subjected to different lairage times at the abattoir and its relationship with broiler meat quality. *Arq. Bras. Med. Vet. Zootec.* 69, 733–741. https://doi.org/10.1590/1678-4162-9268.

Rodríguez, P., Dalmau, A., Ruiz-De-La-Torre, J., Manteca, X., Litvan, H. and Velarde, A. 2008. Assessment of unconsciousness during carbon dioxide stunning in pigs. *Animal Welfare*, 17(4), 341–349.

Sandercock, D. A., Auckburally, A., Flaherty, D., Sandilands, V. and McKeegan, D. E. F. 2014. Avian reflex and electroencephalogram responses in different states of consciousness. *Physiol. Behav.* 133, 252–259. https://doi.org/10.1016/J.PHYSBEH.2014.05.030.

Sandercock, D. A., Hunter, R. R., Nute, G. R., Mitchell, M. A. and Hocking, P. M. 2001. Acute heat stress-induced alterations in blood acid-base status and skeletal muscle membrane integrity in broiler chickens at two ages: implications for meat quality. *Poult. Sci.* 80(4), 418–425. https://doi.org/10.1093/PS/80.4.418.

Satterlee, D. G., Parker, L. H., Castille, S. A., Cadd, G. G. and Jones, R. B. 2000. Struggling behavior in shackled male and female broiler chickens. *Poult. Sci.* 79(5), 652–655. https://doi.org/10.1093/PS/79.5.652.

Schutt-Abraham, I., Wormuth, H. and Fessel, J. 1983. Electrical stunning of poultry in view of animal welfare and meat production. In: Eikelenboom, G. (Ed.), *Stunning of Animals for Slaughter*. Martinus Nijhoff, The Hague, The Netherlands, p. 154.

Shields, S. J. and Raj, M. 2010. A critical review of electrical water-bath stun systems for poultry slaughter and recent developments in alternative technologies. *J. Appl. Anim. Welf. Sci.* 13(4), 281–299. https://doi.org/10.1080/10888705.2010.507119.

Singer, P. 2016. *The Ethics of Killing Animals*. Oxford University Press, New York.

Small, A., Lea, J., Niemeyer, D., Hughes, J., McLean, D., McLean, J. and Ralph, J. 2019. Development of a microwave stunning system for cattle 2: Preliminary observations on behavioural responses and EEG. *Res. Vet. Sci.* 122, 72-80. https://doi.org/10.1016/J.RVSC.2018.11.010.

Small, A., McLean, D., Keates, H., Owen, J. S. and Ralph, J. 2013. Preliminary investigations into the use of microwave energy for reversible stunning of sheep. *Anim. Welf.* 22(2), 291-296. https://doi.org/10.7120/09627286.22.2.291.

Sparrey, J. M., Paice, M. E. R. and Kettlewell, P. J. 1992. Model of current pathways in electrical water bath stunners used for poultry. *Br. Poult. Sci.* 33(5), 907-916. https://doi.org/10.1080/00071669208417534.

Sparrey, J. M., Sandercock, D. A., Sparks, N. H. C. and Sandilands, V. 2014. Current and novel methods for killing poultry individually on-farm. *World's Poult. Sci. J.* 70(4), 737-758. https://doi.org/10.1017/S0043933914000816.

Sparrey, J. M. and Kettlewell, P. J. 1994. Shackling of poultry: is it a welfare problem? *Worlds. Poult. Sci. J.* 50(2), 167-176. https://doi.org/10.1079/WPS19940014.

Sparrey, J. M., Ketylewell, P. J., Paice, M. E. R. and Whetlor, W. C. 1993. Development of a constant current water bath stunner for poultry processing. *J. Agric. Eng. Res.* 56(4), 267-274. https://doi.org/10.1006/JAER.1993.1078.

Steiner, A. R., Flammer, S. A., Beausoleil, N. J., Berg, C., Bettschart-Wolfensberger, R., Pinillos, R. G., Golledge, H. D. W., Marahrens, M., Meyer, R., Schnitzer, T., Toscano, M. J., Turner, P. V., Weary, D. M. and Gent, T. C. 2019. Humanely ending the life of animals: Research priorities to identify alternatives to carbon dioxide. *Animals* 9(11), 1-25. https://doi.org/10.3390/ani9110911.

Terlouw, C., Bourguet, C. and Deiss, V. 2016. Consciousness, unconsciousness and death in the context of slaughter. Part I. Neurobiological mechanisms underlying stunning and killing. *Meat Sci.* 118, 133-146. https://doi.org/10.1016/J.MEATSCI.2016.03.011.

Tinker, D., Berry, P., Rycroft, J. and Sparks, N. 2004. Handling and catching of hens during depopulation. In: Perry, G. (Ed.), *Welfare of the Laying Hen*. CABI Publishing, Oxford, UK, pp. 345-360.

Tinker, D., Berry, P., White, R., Prescott, N., Welch, S. and Lankhaar, J. 2005. Improvement of the welfare of broilers by changes to a mechanical unloading system. *J. Appl. Poult. Res.* 14(2), 330-337 https://doi.org/10.1093/japr/14.2.330.

United States Department of Agriculture 1958. *Humane Methods of Livestock Slaughter Act*. USDA, Washington, DC.

United States Department of Agriculture 2019. *Poultry Slaughter: 2018 Summary*. USDA, Washington, DC.

US Department of Transportation: Federal Aviation Administration 2008. Aeromedical factors. *Pilot's Handbook of Aeronautical Knowledge* [WWW Document]. Available at: https://www.faa.gov/regulations_policies/handbooks_manuals/aviation/phak/ (Accessed 8 January 2020).

Velarde, A. and Dalmau, A. 2018. Slaughter without stunning. In: Mench, J. A. (Ed.), *Advances in Agricultural Animal Welfare*. Woodhead Publishing, Duxford, UK, pp. 221-240.

Velarde, A., Rodriguez, P., Dalmau, A., Fuentes, C., Llonch, P., von Holleben, K. V. V., Anil, M. H. H., Lambooij, J. B. B., Pleiter, H., Yesildere, T. and Cenci-Goga, B. T. T. 2014. Religious slaughter: evaluation of current practices in selected countries. *Meat Sci.* 96(1), 278-287. https://doi.org/10.1016/j.meatsci.2013.07.013.

Verhoeven, M. T. W., Gerritzen, M. A., Hellebrekers, L. J. and Kemp, B. 2014. Indicators used in livestock to assess unconsciousness after stunning: a review. *Animal* 9(2), 320–330. https://doi.org/10.1017/S1751731114002596.

Vieira, A., Vieira, F., Silva, I. and Filho, J. 2010. Productive losses on broiler preslaughter operations: effects of the distance from farms to abattoirs and of lairage time in a climatized holding area Perdas produtivas nas operações pré-abate de frangos de corte: efeito da distância entre as granjas e os. *Rev. Bras. Zootec.* 39, 2471.

Vieira, F. M., Silva, I. J., Barbosa Filho, J. A., Vieira, A. M. and Broom, D. M. 2011a. Preslaughter mortality of broilers in relation to lairage and season in a subtropical climate. *Poult. Sci.* 90(10), 2127–2133. https://doi.org/10.3382/PS.2010-01170.

Vieira, F. M. C., Silva, I. J. Od, Barbosa Filho, J. A. D., Vieira, A. M. C., Rodrigues-Sarnighausen, V. C. and Garcia, D. 2011b. Thermal stress related with mortality rates on broilers' preslaughter operations: a lairage time effect study. *Cienc. Rural* 41(9), 1639–1644. https://doi.org/10.1590/S0103-84782011005000111.

Villarroel, M., Pomares, F., Ibáñez, M. A., Lage, A., Martínez-Guijarro, P., Méndez, J. and de Blas, C. 2018. Rearing, bird type and pre-slaughter transport conditions. I. Effect on dead on arrival. *Span. J. Agric. Res.* 16(2). https://doi.org/10.5424/sjar/2018162-12013.

Vizzier-Thaxton, Y., Christensen, K. D., Schilling, M. W., Buhr, R. J. and Thaxton, J. P. 2010. A new humane method of stunning broilers using low atmospheric pressure. *J. Appl. Poult. Res.* 19(4), 341–348. https://doi.org/10.3382/japr.2010-00184.

Vogel, K. D., Badtram, G., Claus, J. R., Grandin, T., Turpin, S., Weyker, R. E. and Voogd, E. 2011. Head-only followed by cardiac arrest electrical stunning is an effective alternative to head-only electrical stunning in pigs. *J. Anim. Sci.* 89(5), 1412–1418. https://doi.org/10.2527/jas.2010-2920.

Warriss, P. D., Bevis, E. A., Brown, S. N. and Edwards, J. E. 1992. Longer journeys to processing plants are associated with higher mortality in broiler chickens. *Br. Poult. Sci.* 33(1), 201–206. https://doi.org/10.1080/00071669208417458.

Warriss, P. D., Knowles, T. G., Brown, S. N., Edwards, J. E., Kettlewell, P. J., Mitchell, M. A. and Baxter, C. A. 1999. Effects of lairage time on body temperature and glycogen reserves of broiler chickens held in transport modules. *Vet. Rec.* 145(8), 218–222. https://doi.org/10.1136/vr.145.8.218.

Warriss, P. D., Pagazaurtundua, A. and Brown, S. N. 2005. Relationship between maximum daily temperature and mortality of broiler chickens during transport and lairage. *Br. Poult. Sci.* 46(6), 647–651. https://doi.org/10.1080/00071660500393868.

Webster, A. B. and Fletcher, D. L. 2001. Reactions of laying hens and broilers to different gases used for stunning poultry. *Poult. Sci.* 80(9), 1371–1377. https://doi.org/10.1093/PS/80.9.1371.

Wideman, R. F., Rhoads, D. D., Erf, G. F. and Anthony, N. B. 2013. Pulmonary arterial hypertension (ascites syndrome) in broilers: a review. *Poult. Sci.* 92(1), 64–83. https://doi.org/10.3382/PS.2012-02745.

Wilkins, L. J., Wotton, S. B., Parkman, I. D., Kettlewell, P. J. and Griffiths, P. 1999. Constant current stunning effects on bird welfare and carcass quality. *J. Appl. Poult. Res.* 8(4), 465–471. https://doi.org/10.1093/japr/8.4.465.

Woolcott, C. R., Torrey, S., Turner, P. V., Serpa, L., Schwean-Lardner, K. and Widowski, T. M. 2018. Evaluation of two models of non-penetrating captive bolt devices for on-farm euthanasia of turkeys. *Animals* 8(3), 42. https://doi.org/10.3390/ani8030042.

Woolley, S. C., Borthwick, F. J. and Gentle, M. J. 1986a. Tissue resistivities and current pathways and their importance in pre-slaughter stunning of chickens. *Br. Poult. Sci.* 27(2), 301–306. https://doi.org/10.1080/00071668608416882.

Woolley, S. C., Borthwick, F. J. W. and Gentle, M. J. 1986b. Flow routes of electric currents in domestic hens during pre-slaughter stunning. *Br. Poult. Sci.* 27(3), 403–408. https://doi.org/10.1080/00071668608416896.

World Organisation for Animal Health (OIE) 2019. *Terrestrial Animal Health Code*. OIE, Paris. Available at: www.oie.int/en/international-standard-setting/terrestrial-code/access-online/ (Accessed on 4.1.2020).

Wotton, S., Zhang, X., McKinstry, J., Velarde, A. and Knowles, T. 2014. The effect of the required current/frequency combinations (EC 1099/2009) on the incidence of cardiac arrest in broilers stunned and slaughtered for the halal market. *PeerJ Prepr.* 2, e255v1. https://doi.org/10.7287/peerj.preprints.255v1.

# Chapter 16

## Cause and prevention of injurious pecking in chickens

*Nienke van Staaveren and Alexandra Harlander, University of Guelph, Canada*


1 Introduction
2 Identifying the underlying causes of injurious pecking
3 Practical attempts and management strategies to reduce injurious pecking
4 Conclusion
5 Future trends in research
6 Where to look for further information
7 References


### 1 Introduction

#### *1.1 Why is injurious pecking a problem?*

Birds perform pecking as part of their natural behaviour, and it is an important part of their feeding and social behaviour repertoire (Wood-Gush, 1971). However, when birds apply these pecking movements to the integument and its derivatives, such as feathers, it can lead to a high likelihood of damage, injury and even death. Different forms of bird-to-bird pecking exist (Savory, 1995), and these typically fall under the umbrella term of injurious pecking (IP). IP is reported in a variety of domesticated birds, including pet psittacine birds (Rubinstein and Lightfoot, 2014), laying hens (Savory, 1995; Rodenburg and Koene, 2004), broiler breeders (De Jong and Guemene, 2011), turkeys (Dalton et al., 2013; Erasmus, 2018), quails (Schmid and Wechsler, 1997; Miller and Mench, 2006), pheasants (Hoffmeyer, 1969) and ducks (Raud and Faure, 1994; Gustafson et al., 2007). It is clear that this can have far-reaching implications considering that the global hen population for egg-laying alone is in the billions (IEC, 2015), and that IP prevalence is estimated between 40 and 80% with associated mortality rates between 20% and 40% (reviewed by Blokhuis et al., 2007). Due to the high numbers of birds kept for egg production, the high prevalence of IP and its economic importance, much research has focussed on IP in laying hens. As such, most knowledge on IP

http://dx.doi.org/10.19103/AS.2020.0078.15

originates from research on this group of birds and will be the focus of this chapter.

Focussing on the birds kept for egg production, it is clear that damage of the integument due to IP is a widespread problem in both commercial flocks and backyard chickens (Rodenburg et al., 2013; Elkhoraibi et al., 2014). Reported prevalence of IP depends on the method used and it is important to distinguish between a prevalence that reports on the percentage of flocks affected or the percentage of birds affected within a flock (Nicol et al., 2013). It occurs in all types of housing systems, highlighting the importance of good management. Although non-cage housing systems are associated with an increased risk in IP due to larger group sizes and more difficulty in controlling outbreaks (Nicol et al., 1999; Bilčík and Keeling, 2000), several studies reported no or limited differences between housing systems but rather a large variation within or between flocks (Rodenburg et al., 2008b; Sherwin et al., 2010; Heerkens et al., 2015; Decina et al., 2019a,b; Coton et al., 2019). It is therefore not surprising that while some authors report between 15% and 30% of birds in a flock affected (Green et al., 2000; Sherwin et al., 2010; Decina et al,. 2019a,b) or 8–35% of flocks affected (Comin et al., 2019; Rørvang et al., 2019; Coton et al., 2019), others report higher figures ranging between 45% and 95% of birds in a flock (Gunnarsson et al., 1999; Heerkens et al., 2015; Bestman et al., 2017; Grafl et al., 2017) or 40–65% of the flocks (Green et al., 2000; Bestman and Wagenaar, 2003; Gilani et al., 2013; de Haas et al., 2014a; Heerkens et al., 2015; Bestman et al., 2017). Less information on the prevalence of IP in rearing flocks is available, but it is clear that the behaviour and the percentage of flocks increases with age, for example, from 6% in week 1 to 45% in week 16 (Gilani et al., 2013). Similar to laying hens, reports of IP prevalence in rearing flocks vary and range from 35% to 80% of flocks affected (Huber-Eicher, 1999; Huber-Eicher and Sebö, 2001a; Bestman et al., 2009; de Haas et al., 2014a) with a high likelihood of it persisting during adult life (Bestman et al., 2009).

Comparison of different studies and findings of IP are often complicated due to the different definitions, thresholds and methods used, flock age at the time of recording, breeds and production systems, and whether or not birds were beak trimmed (Nicol et al., 2013). Methods to assess IP include assessments of behaviour or physical appearance of the integument of the bird. Behavioural observations can be used to estimate the rate of IP, or the proportion of flocks or birds within a flock performing IP (Nicol et al., 1999; Lambton et al., 2010). However, directly observing IP can be labour-intensive and time-consuming as well as being dependent on time of day due to the diurnal rhythm of bird behaviour (Kjaer, 2000; Kjaer et al., 2011). This has led to a focus on animal-based outcomes such as feather cover and integument condition as a proxy for IP (EFSA, 2005, 2012; Welfare Quality, 2009). Different scoring systems have been developed to assess injuries caused by aggressive pecking (AP) or tissue pecking (TP) and feather damage caused by feather pecking (FP), varying in definitions and severity scales, the body parts and number of birds to be assessed, and handling method used during the assessment (Bilcik and Keeling, 1999; Tauson et al., 2005; Bright et al., 2006; Welfare Quality, 2009; University of Bristol, 2013). Such methods are often considered vulnerable to subjective scoring or interpretation and require training to be consistent (Bright et al., 2006; Kjaer et al., 2011; Palczynski et al., 2016; Decina et al., 2019c). Advances in technology show promise for the use of infrared thermography as a tool to collect fast and objective measures of feather loss (Cook et al., 2006; Zhao et al., 2013; Pichová et al., 2017); however, its practical application on large-scale commercial farms is yet to be investigated.

IP causes feather damage and injuries and therefore is detrimental to bird health and welfare. Birds rely on an intact feather cover and on integument for different biological processes. The metabolic cost of growing, maintaining and replacing feathers is high (Rubinstein and Lightfoot, 2014), which could explain lower egg production or feed efficiency in birds targeted for IP (Leeson and Morrison, 1978; Glatz, 2001; Su et al., 2006). Furthermore, feathers play an important role in thermoregulation in birds (Tullett et al., 1980; van Zeeland and Schoemaker, 2014) and birds with poor feather cover increase their feed consumption (10-40%) to compensate for this (Tullett et al., 1980; Glatz, 2001; Tauson et al., 2006). Intact feather cover, especially of the wings, is also important for birds to perform balancing behaviour and to navigate the environment, for example, perching and walking-up ramps (LeBlanc et al., 2016), which is especially important in aviary systems. Finally, feather cover also plays a role in social communication, such as displays of aggression or mating behaviour (Scanes, 2015).

IP has many consequences, not only for the birds but also for farmers and society. Economic losses due to mortality, reduced productivity and efficiency are well reported (Leeson and Morrison, 1978; Glatz, 2001). However, there are also more social consequences as farmers receive more satisfaction and take more pride in rearing well-feathered flocks (Palczynski et al., 2016). Additionally, transparency is important, and consumers can lose trust in producers or redefine their animal welfare standards when informed of welfare issues such as IP or beak trimming (Bennett et al., 2016).

With the widespread nature of IP and its potentially disastrous consequences for bird health and welfare, farm productivity and societal implications, scientists, veterinarians, farmers and poultry sector representatives are working to address this issue. With a continual turnover of research, it is critical that IP, which covers a broad range of behaviours and outcomes, should be clearly defined and addressed via a comprehensive approach.

### 1.2 Definitions and descriptions of injurious pecking

IP is an umbrella term used to describe quite distinct forms of injurious bird-to-bird pecking, which lead to distinct categories of damage or injuries to the bird's integument. While this term is widely used in the literature, it is good to take a step back and consider some of the possible ambiguous terminologies in this regard. First of all, the avian integument is adaptive with the skin being thin, elastic and loosely attached to the majority of the body and covered by feathers, yet firmly attached, highly modified and unfeathered on the head, beak, lower legs and feet (Stettenheim, 2000). Apart from feathers as integumentary derivatives, birds show various unfeathered integumentary outgrowths on the head and neck, such as caruncles, wattles and combs (Stettenheim, 2000). IP is named so after the consequences it can have in the form of injuries on the skin, feathers and outgrowths. An injury refers to damage inflicted to the body by an external force (Studdert et al., 2012). However, a proper definition of injury is often lacking and leaves room for interpretation as to what threshold is used to categorize something like an injury. In humans, it is often referred to as harm or hurt and implies the breaking of skin (Dorland, 2007). However, when we speak about IP it often also includes forms that damage the feathers, not necessarily leading to breakage of the skin, and are unlikely to cause harm per se as feathers are dead material (Scanes, 2015). This type of use of the term injury fits more with other definitions of injury that refer to disturbance of structure or function of any part of the body (Youngson, 2005). All in all, it highlights the need for a defined and comprehensive terminology in the field of IP research.

IP can further be classified in a number of ways, for example, by the target area (e.g. head or rest of the body; feathered or unfeathered integument), pecking mechanisms (e.g. single or repetitive movements; gentle or severe pecks), or possible function, cause or motivation (e.g. influence of or on microbiota composition, imbalanced neuromodulators such as serotonin and catecholamines, adding feathers/fibre to the diet). For practicality, a descriptive approach is commonly used to assess symptoms of IP and integument damage. The following descriptive terms are used to better describe the categories of behaviour (Fig. 1) that fall under the IP umbrella term and were first described by Savory (1995):

- Feather pecking (FP): directed at the feathered integument. Can be accompanied with feather removal and consumption.
- Tissue pecking (TP): directed to denuded areas or tissue (also often referred to as cannibalistic pecking). Pecking at particular tissues such as the toes (toe pecking), and the vent or cloaca (vent pecking) are often considered separately.
- Aggressive pecking (AP): mostly directed at the head, including the comb.

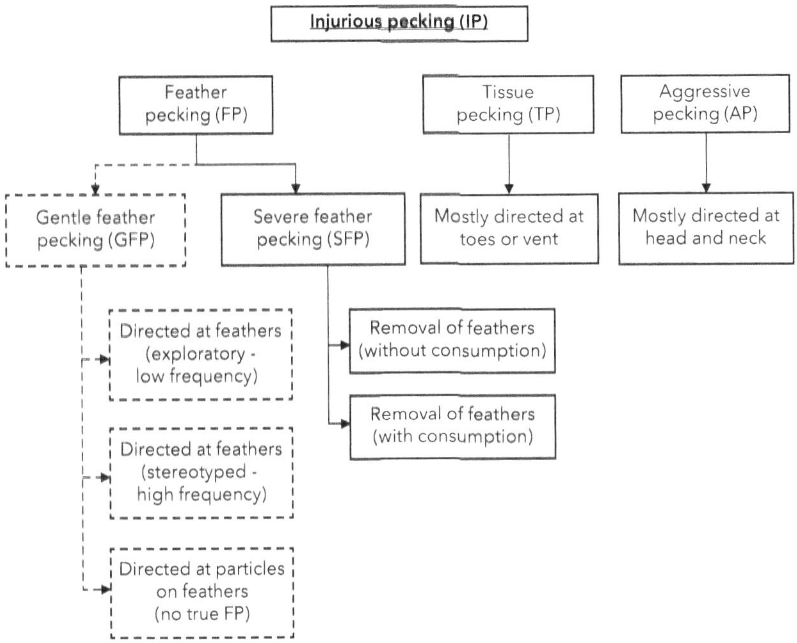


**Figure 1** Schematic overview of injurious pecking (IP) behaviours in chickens. We argue for defined and comprehensive terminology of IP and discuss that gentle feather pecking (GFP) and its sub-forms (dashed lines, -----) should not fall under the IP umbrella term whereas severe feather pecking (SFP), tissue pecking (TP) and aggressive pecking (AP) more clearly fall under the IP umbrella term (solid lines, ----).

Although this descriptive approach is a helpful starting point, it has its limitations. Without necessarily noting the cause or mechanisms to describe the phenomenon, it impedes the biological value to investigate cause and development, or to develop an effective treatment plan given that this cluster of behaviours can result from different biological mechanisms, and is often treated homogenously. As a result, most of the management strategies employ non-specific treatments aimed at addressing all categories of behaviour and which emphasize the role of the environment in IP development.

### 1.2.1 Injurious pecking at the feathered integument

In descriptive terms, pecking against the feathered integument (feather pecking (FP)) can be divided into two intensity categories, each with their own distinct motivational systems (Bilcik and Keeling, 1999; Kjaer and Vestergaard, 1999; McAdie and Keeling, 2002; Rodenburg and Koene, 2004; Newberry et al., 2007; Dixon et al., 2008) and most likely influenced by different, but possibly overlapping, genetic and neurobiological pathways.

The first category is gentle FP, and it is the mildest form consisting of gently repeated pecks at the tips and edges of feathers which can be further sub-divided as exploratory or stereotyped gentle FP (Rodenburg et al., 2004). Birds can also peck at particles in the feather cover which can be mistaken for gentle FP, though this behaviour does not truly fall under FP (Rodenburg et al., 2004). Gentle FP does not include removal of the feathers and causes little damage to or reaction from the recipient. It can be argued whether this gentle FP should fall under the umbrella term of IP due to its low impact (McAdie and Keeling, 2002) and the fact that these pecks are similar to social exploration or allopreening in other birds (Riedstra and Groothuis, 2002; Rodenburg et al., 2004; Dixon et al., 2008). The second category is severe FP and it is the more drastic form which consists of forceful pecking that result in the pulling and removal of feathers, causing feather damage (Savory, 1995; Bilcik and Keeling, 1999). Also, this form is often accompanied by the consumption of feathers (McKeegan and Savory, 1999; Savory and Mann, 1999; Harlander-Matauschek and Bessei, 2005; Ramadan and von Borell, 2008; Harlander-Matauschek and Häusler, 2009). The relationship between gentle and severe FP is controversial. However, it appears that gentle FP is not a good predictor for severe FP later in life, though the two can be related within the same age group (Rodenburg et al., 2004; Newberry et al., 2007). A point of consideration is that the unclear and controversial nature of this relationship between gentle and severe FP might be caused by the subjective, descriptive nature of the two forms (and possible sub-forms of gentle and severe FP) and a lack of a definitive separation between them.

### 1.2.2 Injurious pecking at the unfeathered integument

TP involves pecking at denuded areas or tissues such as the vent or toes, which can lead to serious injuries and result in death (Savory, 1995). It is also frequently referred to as cannibalistic pecking (Savory, 1995). However, cannibalism implies intraspecific predation and while it occurs in many natural populations, it only constitutes a small part of their diet (Fox, 1975). Hunger was mentioned as a possible risk factor for toe pecking (Glatz and Bourke, 2006), but this behaviour is also observed under conditions where hunger is unlikely to play a role (Rodenburg et al., 2009b), and we cannot be certain that this TP is a true form of 'cannibalism' in the case of chickens. TP comes in two particular forms, namely vent/cloaca pecking or toe pecking. Limited research has been done on vent pecking where pecking is specifically directed to the red tissue surrounding the cloaca and often occurs around the onset of lay (Gunnarsson et al., 1999; Pötzsch et al., 2001). Pötzsch et al. (2001) found that nearly 30% of farmers regularly saw vent pecking in their flocks, with a median flock prevalence of 3.5% and a mortality rate of 1.3%

associated with the occurrence of vent pecking. Similarly, a vent pecking prevalence of 2.65% was reported by Gunnarsson et al. (1999). Little is known about toe pecking. Glatz and Bourke (2006) mentioned hunger, excessive temperatures and light intensity as possible factors causing toe pecking, but Rodenburg et al. (2009b) reported a prevalence of toe injuries between 20 and 35% under conditions where these factors were unlikely to play a role. While toe pecking is not often reported (Glatz and Bourke, 2006), it can be an issue in birds and cause wounds (El-Lethey et al., 2000; Buitenhuis et al., 2003; Rodenburg et al., 2009b; Krause et al., 2011) and appears to be receiving increasing attention (*Harlander, personal observation*). Krause et al. (2011) started to observe toe injuries in hens of a particular pen at 35 weeks of age which persisted throughout the experiment, and Rodenburg et al. (2009b) reported a prevalence of toe injuries between 20% and 35% at 42 weeks of age. Others observed that birds with injured toes were more likely to receive toe pecks than those with intact toes (Buitenhuis et al., 2003). The reaction of the recipient bird (e.g. escape, freeze and fight) as well as the presence of damaged feathers or blood were suggested as possible explanations for why a bird might be receiving different forms of pecks (Buitenhuis et al., 2003). Toe pecking can quickly lead to severe injuries, and there are indications of increased physiological stress and fearfulness in birds with injured toes, highlighting its welfare implications (Buitenhuis et al., 2003; Krause et al., 2011).

AP consists of forceful pecks often aimed at the head area, including the comb and possibly the neck. Essentially, AP can be beneficial to some extent to obtain resources, defend territory or establish social rank (Savory, 1995). Once pecks and fights are used to establish the hierarchy, ensuing interactions are often based upon individual recognition as well as threats, submission behaviour and avoidance typically limiting costly injuries (D'Eath and Keeling, 2003; Queiroz and Cromberg, 2006). However, under certain circumstances, AP may escalate and cause significant injuries (Fig. 2) (Birkl et al., 2017a). In commercial conditions where birds are kept in large groups, AP is typically reduced as individual recognition is no longer possible and birds use other markers of status to determine dominance (D'Eath and Keeling, 2003; Rodenburg and Koene, 2007). In these large groups, birds can adopt different social strategies with the majority of birds likely showing low levels of aggression and becoming more tolerant, while a minority might become despotic and target other birds indiscriminately (Estevez et al., 2003). Besides genetic investigations (e.g. Cheng and Muir, 2007) there has been little scientific work on the causes (e.g. Dennis and Cheng, 2011) and prevention of this form of IP behaviour. However, more work has been done on causes and function of aggression in, for example, junglefowl (Kruijt, 1964) or quail (Wechsler and Schmid, 1998), which can provide insights for other types of poultry.

**Figure 2** Severe injury on the head and comb caused by aggressive pecking (AP) in a white laying hen. (Picture: A. Harlander).

### *1.2.3 Relationships between the different forms of injurious pecking*

A major complication with the different forms of IP is that relationships between them are unclear (see also the relationship between gentle and severe FP, Section 1.2.1). Some evidence exists for a relationship between severe FP and TP such as vent or toe pecking (Hughes and Duncan, 1972; Leonard et al., 1995; Cloutier et al., 2000; Pötzsch et al., 2001; Newberry, 2004). During the removal of feathers, the skin can become injured, and this wound may present a pecking target (Cloutier et al., 2000), and birds with more damage to their feathers or skin are more likely to receive toe pecks (Leonard et al., 1995). However, this relationship between severe FP and TP is not always found as TP can also be observed among fully feathered birds (Gunnarsson et al., 1999; Pötzsch et al., 2001; Newberry, 2004). Lastly, severe FP and AP can, but do not necessarily have to, co-occur (Bennewitz et al., 2014; Daigle et al., 2015; Birkl et al., 2017a), which further complicates investigations into IP development. The correlation of these relationships does not reveal the underlying causation of the behaviour. Forms of IP (i.e. severe FP, TP and AP) therefore appear to be aggravated by similar contributing factors (Lambton et al., 2015), though these behaviours do not necessarily occur within the same flock at the same time (Newberry, 2004).

## 2 Identifying the underlying causes of injurious pecking

IP is a multifactorial problem with many contributing factors. These factors can be approached from two broad angles: the ethological view or the dysfunctional view. The ethological view emphasizes the role of the environment and species biology in shaping behaviour. The dysfunctional view emphasizes possible dysfunction due to chronic, unpredictable stress or inappropriate rearing environments. Both approaches may underlie the development of IP, but their relative importance and interactions are unknown. We will illustrate both views on severe FP as this is the most prevalent and investigated form of IP.

## 2.1 The ethological view

### 2.1.1 Determination of behaviour

Behaviour is multifaceted, whereby a multitude of both internal and external factors can influence the final outcome. The combined effect of internal (i.e. physiological stimuli and emotional states) and external stimuli (i.e. sensory input) are integrated into the central nervous system. The result of resolving all these competing stimuli within the context of an animal's intrinsic goals at a given time produces a behavioural output that is deemed to be its most appropriate behaviour under the given circumstances. From an ethological point of view, the summation of the internal and external stimuli can be described as the basis for motivation, which in turn, directly causes the resulting behaviour (Mason and Bateson, 2009). As can be expected, some motivationally led behaviours rely heavily on internal stimuli, while others mostly depend on external cues. It is noteworthy that, in contrast to reflexes, most motivated behaviours are goal-oriented, stimuli-sensitive and dependent on emotional states. Furthermore, they can be learned or changed over time depending on previous positive or negative experiences and feedback loops (Dorman and Gaudiano, 1995). Most of the behavioural problems arise from normal motivation for species-specific behaviour (Mills, 2003). If animals are left in a highly motivated state where they cannot reach a natural goal or achieve homeostasis in response to stimuli, they may become frustrated and redirect that motivation to perform an 'undesirable' (problem) behaviour. It is also possible that an exaggerated external stimulus (superstimulus) may produce such a behavioural response.

### 2.1.2 Ethological view on severe feather pecking

Severe FP is thought to arise from frustration within nutritionally, physically and sensorially deprived environments that prevent birds from performing highly motivated behaviour. It is generally accepted that severe FP is a form of redirected pecking behaviour stemming from frustration at the lack of foraging and feeding opportunities (Wennrich, 1974; Blokhuis, 1986; Huber-Eicher and Wechsler, 1997; Weeks and Nicol, 2006; Dixon et al., 2008). Alternatively, it is suggested that the source of frustration is linked to dust-bathing motivation (Vestergaard and Lisborg, 1993); however, there is comparatively less experimental evidence to support this hypothesis. Foraging behaviour is defined as the behaviour of birds moving in a manner that they are likely to encounter and acquire feed (Broom, 1981). Foraging behaviour consists of two phases: an appetitive phase, also known as the explorative phase, where birds search for feed, followed by a consummatory phase in which birds ingest the feed (Keeling, 2002).

### 2.1.2.1 Unfulfilled motivation to explore and search for feed

In free-living conditions, the explorative phase of foraging in chickens involves walking, ground-scratching, ground-pecking and probing with the beak (Duncan, 1980). The junglefowl spends 34% of its active time ground-scratching and 61% of its active time ground-pecking, even when provided with feed (Dawkins, 1989). This shows the high motivation for expressing foraging behaviour. In laying hen farms, feed is typically presented directly to the bird in a feeder trough which reduces the need to search for feed and means that the appetitive or explorative phase takes a fraction of the time it would in nature. The reduced time spent searching for feed that is readily available in troughs, as well as the reduced foraging activities due to a lack of an appropriate litter substrate in farms (e.g. Blokhuis and Arkes, 1984; Blokhuis, 1986; Decina et al., 2019a), increases the risk of severe FP. This effect is even more pronounced when feed is restricted in quantity or offered as pellets, where again the amount of time and number of pecks required to consume the feed is reduced (Savory and Mann, 1997; van Krimpen et al., 2005). As a result, the laying hens' motivation to perform foraging activities remains unfulfilled. It is assumed that hens redirect pecking behaviour towards their conspecifics, 'misperceiving' feathers as material to forage or explore (Riber, 2007). This suggests that severe FP may be a result of environments that lack sufficient stimuli for normal species-specific behaviours. In other words, FP is potentially a normal behaviour, which is accentuated or overly repeated when hens are exposed to sensorially poor environments.

### 2.1.2.2 Unfulfilled motivation to consume feed

There is much evidence supporting the hypothesis that severe FP develops from the explorative phase of foraging behaviour. Nevertheless, the subsequent consummatory phase cannot be ignored as a contributor to pecking, as it relates to the ultimate goal: foraging-related consumption of feed. Interestingly, the first scientific descriptions of FP behaviour focussed on the consummatory phase, identifying it as feather pulling, eating or plucking (Oettel, 1873; Willimon and Morgan, 1953; Marsboon and Sierens, 1962). In the 1960s and 1970s, motivation to perform this behaviour was identified as a 'pecking drive' (e.g. Hoffmeyer, 1969). This drive would be redirected to feathers in the absence of another suitable substrate. The use of the term 'pecking drive' likely initiated a shift in the terminology to FP when discussing this behaviour. However, in the 1990s, the term 'feather eating' was reintroduced, and several studies showed that some feather peckers consume more feathers than non-peckers (McKeegan and Savory, 1999; Harlander-Matauschek and Bessei, 2005).

Chickens are omnivores (Wood-Gush, 1971), which is evidenced by the variable amounts of vegetation, insects, stones, seeds and worms that are found in the crop contents of birds under natural conditions (Wood et al., 1963). This finding may reflect the patchiness of animal and plant feed sources, or the nutritional need and/or the appetite of chickens for a variety of feed sources. It can also reflect alternating exploration and consumption patterns, where the animal experiences positive (e.g. nutritionally rewarding) and negative (e.g. malaise inducing) consequences of ingesting various items, thereby motivating foraging behaviour. Modern production systems typically provide non-naturalistic, low-fibre (e.g. 2–3%), homogeneous, concentrate-based diets, for example, in the form of pellets (see Section 2.1.1.1), despite evidence that coarse, fibre-rich diets can reduce severe FP, promote satiety and improve the development of chicks' gastrointestinal tract (Hetland et al., 2003; van Krimpen et al., 2005).

The role of feathers as a feed component is not well understood. While feathers are mostly non-nutritive (McCasland and Richardson, 1966), they provide a means to gain required, insoluble fibre which may be lacking in modern, commercial diets (van Krimpen et al., 2005). Indeed, laying hens that exhibit severe FP have been shown to be highly motivated to ingest feathers (McKeegan and Savory, 1999, 2001), and hens studied using operant apparatuses have been shown to exert a significant amount of energy and time to obtain access to feather rewards (Harlander-Matauschek et al., 2006a, 2007). Additionally, chopped feathers in the diet can improve the feather cover of birds (Kriegseis et al., 2012). It is believed that birds have a strong preference for the physical characteristics and visual surface properties of feathers for consumption related to length and part of the feather (Harlander-Matauschek and Feise, 2009). That said, hens do not differ in the number of feathers eaten from birds of different genetic lines (Bögelein et al., 2009), colours (white or brown) or sex (Neumann et al., 2009). Additionally, ingested feathers increase feed passage time and gastrointestinal motility in a manner similar to insoluble fibres (Harlander-Matauschek et al., 2006b). Moreover, it leads to crop (Harlander-Matauschek and Bessei, 2005) and gizzard distension (Benda, 2008), which could create short-term satiety, thereby reducing the motivation to search for feed. In general, feather eating appears to reflect a lack of dietary fibre in the diet, and may have similar functional benefits as insoluble fibres, promoting the movement of feed through the gastrointestinal system.

In summary, severe FP arises from a normal motivation regulating species-specific behaviour, that is, foraging and feeding. Both the explorative searching and the consummatory phase appear connected to the development of FP. However, the extent to which these two phases contribute, separately or combined, is still unclear. Finally, whether a higher contribution from one phase versus the other gives rise to different forms of severe FP requires further investigation. Additionally, it can be debated if corrective management measures for severe FP should be different when the FP originates from an unsatisfied explorative phase compared to FP from an unsatisfied consummatory phase.

### 2.1.3 Challenges and limitations of the ethological view on feather pecking

The ethological view on FP is not without limitations. FP sometimes defies the predictions of the redirected foraging hypothesis. For example, severe FP also occurs in birds in commercial systems having access to pasture where they can perform species-specific foraging behaviour (Fig. 3) (Green et al., 2000; Nicol et al., 2013; Rodenburg et al., 2013; Bestman and Wagenaar, 2014; Elkhoraibi et al., 2014). This suggests that motivational explanations, such as the redirected foraging hypothesis, cannot fully explain severe FP, and that other mechanisms are likely at play. In general, multiple motivational systems contribute to normal activities. Different motivations (e.g. locomotion, foraging, nesting and dust

**Figure 3** Hen with feather damage in the back and wing area due to feather pecking (FP) and in the head area due to aggressive pecking (AP) in an outdoor system with foraging opportunities. (Picture: A. Harlander).

bathing) may trigger the same behavioural outcome (e.g. walking, scratching, pecking and picking up), and if one of these motivations is unfulfilled, it may lead to an FP response.

Additionally, studies where hens that were housed in enriched environments express more pecking behaviour challenge the view that inability to perform foraging behaviour is the main driving force in severe FP (Cuthbertson, 1978). Since a more complex environment elicits more activity in general (Newberry, 1995; Bizeray et al., 2002; Brantsæter et al., 2016), it is conceivable that the same environment may also elicit more pecking. This challenges the notion that once a certain amount of pecking is performed, birds fulfil their motivation for pecking and would then no longer show FP behaviour. However, it is possible that there is an inherent motivation for a basal level of pecking activity. This motivation is directed towards the feathers of conspecifics in a non-enriched environment and towards the enrichment in an enriched environment. In the latter, pecking has a stimulating effect and will elicit additional pecking activity. However, detailed and long-term observation of individual bird pecking behaviour (including severe FP) is missing. As such, this area of study is a fascinating avenue for future research.

Furthermore, the motivation(s) underlying severe FP may not necessarily explain how severe FP is modified into a 'repetitive' behaviour which increases in frequency and duration over time. It also does not address why severe FP behaviour fluctuates over time and why it may vary among individuals in similar environments. Additionally, once severe FP is established, it cannot be completely halted, merely reduced by the addition of foraging opportunities (Lambton et al., 2013) suggesting that repeated pecking at the feather cover might get reinforced through potential beneficial physiological or psychological consequences. These reinforcing consequences might account for the inability of additional foraging opportunities to bring severe FP to a halt when birds get highly motivated to repeatedly experience this positive outcome of FP. However, more information is needed to find out whether reinforcing properties of FP help birds to cope with specific situations.

Plucking and consuming feathers might have useful consequences for gastrointestinal health, such as promoting movement of feed (Harlander-Matauschek et al., 2006b), which could imply it might be a functional behaviour and help birds to cope with discomfort. However, whether rewarding gastrointestinal consequences cause the repetitive nature of severe FP needs to be further investigated. Additionally, some severe FP behaviour does not involve the ingestion of feathers (Riber and Hinrichsen, 2016). Nevertheless, the idea that some forms of severe FP could be linked to gastrointestinal discomfort or possibly gastrointestinal dysfunction, rather than being a response to nutrient deficits or reduced foraging time per se, is interesting.

It is also possible that a superstimulus, such as a large number of loose feathers on the floor (e.g. due to moulting) during rearing, large numbers of feathered birds or densely populated hen houses, produces repetitive pecking and eating as the behavioural response. Ramadan and von Borell (2008) showed that loose feathers on the floor during rearing played a role in the development and severity of FP later in life. However, the social transmission (Zeltner et al., 2000) and learning of FP from primary peckers (Hughes and Duncan, 1972; Dixon, 2008) suggest that foraging motivation or the superstimulus of 'feathers' is unlikely to be the root cause of FP in secondary peckers.

## 2.2 The dysfunctional view

### 2.2.1 When is behaviour considered dysfunctional?

While a universally accepted definition of disorder does not exist, the origins of psychological or psychiatric disorders are attributed to internal mechanisms that cease to function or that fail to perform normally. As such, dysfunction can be defined as the disruption of internal psychological, biological or developmental processes, in such a way that their function deviates from that of healthy individuals (American Psychiatric Association, 2013). Clinically, these disruptions present themselves as disturbances in cognition, emotional regulation and behaviour (American Psychiatric Association, 2013), causing the affected individuals to become disturbing or distressing to others or themselves (Wakefield, 1992). It is important to highlight that the recognition of dysfunction inadvertently constructs an artificial boundary between what is considered as 'normal' and 'abnormal' (i.e. functional vs. dysfunctional) behaviour (Telles-Correia et al., 2018). The term 'abnormal' is often used as a synonym for dysfunction; however, given its negative connotation and its openness to interpretation, the present text will employ the term 'dysfunction' throughout this section to refer to any actions that are considered to lie outside healthy behaviour.

### 2.2.2 Dysfunctional view on severe feather pecking

Severe FP in laying hens has long been thought to be a sign of frustration from failed attempts to forage. However, as previously explained, this view cannot fully explain severe FP (see Section 2.1.3). Neither does it explain why severe FP is so persistent, repetitive-like and involves damaging the feather cover of another bird, resulting in physical harm and distress to others.

It is hypothesized that severe FP behaviour shares similarities with developmental psychological disorders, such as attention-deficit/hyperactivity disorder (ADHD) or autism spectrum disorder (ASD) (Kjaer, 2009; Birkl et al., 2019b). Severe FP has also been compared to obsessive-compulsive disorders (OCD) and related disorders, such as trichotillomania (pulling out one's hair) or excoriation (skin-picking) (van Hierden et al., 2004a). The prevalence of disorders such as ADHD, ASD and OCD ranges between 0% and 10% in human populations (Kessler et al., 2005; Polanczyk et al., 2007; Merikangas et al., 2010; Zablotsky et al., 2019). In contrast, the prevalence of severe FP in laying hen flocks can range between 15% and 95%, and is, therefore, considerably higher (see Section 1.1). It is tempting to speculate that the high prevalence of FP behaviour in laying hens may be a by-product of intensive and controlled breeding, wherein the majority of commercial birds worldwide originates from a few large-breeding companies. This theory is reinforced by the observation that severe FP is a heritable trait (Wysocki et al., 2010) that is closely connected to other genetically determined production and behavioural traits (Korte et al., 1997; Su et al., 2006; Bennewitz et al., 2014; Nicol, 2018). To this end, when genetic lines that have high and low propensity to perform FP are compared, the expression of genes governing the neuroendocrine, the monoaminergic (serotonin (5-HT) and dopamine) and immune systems are distinctly different (Buitenhuis et al., 2004, 2006b; Flisikowski et al., 2009; Biscarini et al., 2010; Brunberg et al., 2011, 2016; Wysocki et al., 2013; van der Eijk et al., 2019a). While the current literature investigating the genetic factors contributing to FP is promising, there is no consensus as to the genes contributing to this behaviour. As severe FP likely results from a complex interplay between genetic and environmental factors, collaboration between geneticists, neuroscientists, physiologists and ethologists is essential to identify the genes and the developmental pathways, whose disruption may cause severe FP.

#### 2.2.2.1 General concepts of the dysfunctional view

While they are more likely to experience higher levels of frustration, wild animals that have been captured perform dysfunctional social or repetitive behaviours at a lower frequency than animals that are born in captivity (Mallapur et al., 2005). It is thought that the former experience more naturalistic early-life experiences that are both socially and physically complex which, in turn, protect them against the development of dysfunctional behaviour (Mason, 2006; Jones et al., 2011a). In contrast, animals born in captivity often experience early social deprivation (e.g. motherless-rearing, isolation or separation from familiar social settings), and they are often raised in standard captive housing, which tend to be physically, nutritionally and sensorially restricted environments. These experiences are powerful stressors that can impair normal development and can induce high rates of dysfunctional behaviour that persist throughout the lifetime of an animal (Lewis, 2004). Interestingly, rearing animals in socially, physically, nutritionally and sensorially rich environments can help attenuate dysfunctional behaviour (Lewis et al., 2007; Jones et al., 2011b; Bechard et al., 2017). Nevertheless, exposure to these complex environments appears to be most effective in young animals and often fails to reduce established dysfunctional behaviour in older animals (Lewis, 2004). It is likely that physically and socially complex environments exert their positive impact by affecting brain development, structure and function (Lewis, 2004), as well as by shaping social learning and by modelling the endocrine-immunological stress response, thereby, helping the organism to cope with stressful conditions (Sandi and Haller, 2015); however, the precise mechanism by which it curtails dysfunctional behaviour is yet to be clearly identified.

In humans, the onset and persistence of psychological disorders are often associated with stress or adverse experiences throughout life (Riboni and Belzung, 2017). Animals experiencing intense or sustained adverse experiences can develop observable behavioural changes, which may result in permanent shifts in neurobiological control (Sandi and Haller, 2015). Although a detailed discussion of stress-related terminology is beyond the scope of this section, it is nonetheless important to discuss the basic concepts to understand why specific neurobiological structures or molecules, discussed herein, are associated with dysfunctional behaviour. While there is no universal definition of stress, it can be characterized as a state of the mind of an individual (McEwen, 2012). The experience of a stressor is associated with arousal and initiation of the physiological stress response, which manifests itself through the autonomic, neuroendocrine, metabolic and immune systems. These reactions to stressors must be terminated as quickly as possible to avoid harm to the organism (McEwen, 2012). A range of responses to various stressors have been described in both animals considered vulnerable or susceptible, and in those described as resilient (Riboni and Belzung, 2017). For instance, resilient animals are capable of terminating stress responses actively by reorganization of neurobiological pathways to cope with environmental stressors/challenges, whereas vulnerable animals are unable to trigger such changes and therefore develop dysfunctional behaviour (Maier and Watkins, 2010; Koolhaas et al., 2011; Riboni and Belzung, 2017). Others hypothesized some changes to biochemical, neurological or genetic pathways may contribute to more efficient stress alleviation by triggering behaviour (e.g. repetitive behaviour in captivity) to function as a coping mechanism to reduce arousal (Cabib and Puglisi-Allegra, 2012). These changes are hypothesized to act as evolutionary advantages for survival in adverse environments (Bateson et al., 2004; Spencer, 2017). Consequently, categorizing all neurobiological alterations as conclusively dysfunctional is difficult, and such classifications can become moot depending on the behavioural outcome in a given individual and context.

The brain is the primary organ of stress adaptation (McEwen, 2012). In addition to the autonomic nervous system, the hypothalamic-pituitary-adrenal (HPA) axis manages the body's response to stressful conditions by regulating stress-related systems. For instance, it impacts neuroendocrine circuits, including molecules such as glucocorticoids, the monoaminergic system (5-HT, dopamine) and inflammatory markers. However, when the HPA axis is activated in response to intense or sustained stress, it can permanently change the downstream effectors and the trigger point of neuroendocrine circuits (Ishiwata et al., 2005; Guilliams and Edwards, 2010; Silverman and Sternberg, 2012). As a result, functional and morphological changes that occur when the HPA axis is engaged may become permanent, thereby altering the neurochemistry and neuroanatomy in areas such as the mammalian limbic system – that includes hippocampus, amygdala, hypothalamus and prefrontal cortex (Ziabreva et al., 2003a,b), neurons between the mammalian cortex and basal ganglia, (i.e. the cortical-basal ganglia circuitry) (Whitehouse and Lewis, 2015) or their functional equivalents in the avian brain (Reiner, 2005; Güntürkün, 2005). For instance, exposure to highly and/or recurrently adverse situations early in life (e.g. prenatal, postnatal and juvenile stressors) can create dysfunctional social behaviour (Sandi and Haller, 2015) or repetitive motor actions that surface at various developmental stages of the individual (Lewis and Kim, 2009). Potentially irreversible changes to these brain areas or circuits can influence the balance between anxiety/fearfulness, mood and motor control, motivation and cognition (Langen et al., 2011; McEwen, 2012). Furthermore, recent studies show stress-induced changes to neural and immune functions (Miura et al., 2008) that coincide with an associated change in the gut microbiome composition and a decrease in microbial diversity, thereby outlining a potential role for gut microbes to further influence behaviour (Bharwani et al., 2016). In sum, all of these systems are, therefore, interconnected and dependent on each other, often creating feedback loops (Hassel et al., 2011).

Pharmacological studies are used to identify neurobiological pathways that contribute to dysfunctional behaviour. For instance, Willemse et al. (1994) found that animals that had recently developed overgrooming behaviour were significantly more sensitive to opioid and dopaminergic antagonists, while animals with established overgrooming behaviour were only sensitive to dopaminergic antagonists. This suggests that different sets of neurobiological pathways control the same behaviour at different stages of life or over time. It is nevertheless important to be mindful that a response to pharmacological treatment does not necessarily correlate with the cause of the dysfunction. Indeed, it is equally possible that the drug sensitivity observed is merely symptomatic (Mills, 2003). The example above does, however, suggest that permanent neurochemical or neuroanatomical changes may explain why it is so difficult to reverse established dysfunctional behaviour (Mason and Rushen, 2006) and why some pharmacological interventions are age-dependent (Mills, 2003).

In summary, social or physical adversities throughout the life span have the potential to permanently alter the neurobiology of an animal, which can lead to dysfunctional behaviour. It is noteworthy that the impact of any given adverse experience is highly dependent on the type of adversity, the duration, the developmental windows in which these stressors are experienced (Mason and Rushen, 2006), as well as the genetic and epigenetic landscape of the individual (Kim et al., 2013; Sandi and Haller, 2015). Importantly, most reported links between neurobiological alterations and behaviours are associations, making causal connections between the two difficult. Most associations between neurobiology and dysfunctional behaviour in human disorders stem from animal research that is carried out under the assumption that research animals have significant genetic, physiological and behavioural similarities to humans. As such, animal models are used in understanding both humans and animals alike.

In the following section, FP studies that report neurobiological outcomes are categorized according to the source of adversity linked to severe FP behaviour: (i) adverse environments and experiences, (ii) pharmacological or nutritional interventions and (iii) laboratory stress tests.

### 2.2.2.2 Environmentally induced neurobiological alterations associated with severe feather pecking

As outlined above, adverse social and physical environments can cause dysfunctional behaviour in mammals. Theoretically, adverse life experiences (see Section 3 for a list of risk factors for IP) can also cause neurobiological changes leading to severe FP in laying hens. The research studies discussed below purposely introduce social and physical stressors in the early life stages of chickens to study the neurobiological changes that precede severe FP.

It is established that mammals subject to early maternal isolation or poor maternal care are at a higher risk of developing dysfunctional behaviour (Martin, 2002). In the context of commercial laying hens, chicks are hatched and reared completely separated from the mother hens. Under a natural setting, the roles of mother hens include, but are not limited to, teaching chicks foraging and feeding habits, indicating rest periods, providing thermoregulation, providing comfort and ensuring protection from predators (Collias and Joos, 1953; Moffatt and Hogan, 1992; Nicol and Pope, 1996; Field et al., 2007; Edgar et al., 2016). These functions fulfilled by the mother reduce the stress responses of the chicks, enhance their memory and improve their learning capacity. It is, therefore, evident that hens' presence can have major effects on chicks' behavioural development (Perré et al., 2002; Rodenburg et al., 2009a; Edgar et al., 2016).

To date, the impact of mother hens on chicks' neurobiology and how maternal presence may prevent severe FP has received limited attention. Several studies comparing chicks reared with and without a hen report conflicting effects. One research group reported that birds reared with a mother exhibited less severe FP (Perré et al., 2002). In contrast, others report that the rate of severe FP between chicks reared with and without hens was not significantly different (Roden and Wechsler, 1998; Riber et al., 2007; Shimmura et al., 2010). It is noteworthy that each study assessed severe FP prevalence at different ages (between 8 and 29 weeks), which may contribute to the inconsistencies observed in the literature between studies. One study suggested that rearing birds with foster hens during the first 7 weeks may reduce the likelihood of FP development as they found differences in peripheral 5-HT concentrations in these birds later in life (Rodenburg et al., 2009a). However, a follow-up study concluded that the plumage condition of brooded laying hens (42 weeks) did not differ significantly from that of unbrooded chicks (Rodenburg et al., 2009b). Yet another study reported that the hippocampal cell sizes of chickens raised with and without a hen was significantly different (Nordquist et al., 2013), an observation that indicates lateralization and vasotocin-positive neurons which are related to the control of social behaviour (Hewlett et al., 2014). Unfortunately, FP behaviour was not monitored in these studies (Rodenburg et al., 2009a,b; Nordquist et al., 2013; Hewlett et al., 2014), and hence, no associations between these brain alterations and FP phenotypes can be drawn. It is also noteworthy that a mother hen's stress level during egg formation can increase chicks' fearfulness and FP activities (de Haas et al., 2013, 2014b), and that routine hatchery practices are associated with higher corticosterone levels, higher fearfulness, stress reactivity and feather damage later in life (Hedlund et al., 2019). In summary, the current literature presents contradicting data regarding the contribution of mother hens to the risk of developing severe FP later in life.

While the studies quoted above presume significant similarities between precocial birds and mammals, it is possible that the degree of parental care required to avoid dysfunctional behaviour is different between the two groups. Therefore, there is a need for validation of this hypothesis via more systematic studies that clearly elucidate the impact of maternal care on long-term neurobiological and behavioural outcomes in birds.

Isolation from conspecifics or the disruption of social bonds by mixing groups of animals has been used to study the neurobiological consequences of social stress as it relates to human behavioural disorder research (Sandi and Haller, 2015). These social stress models are mirrored in the manner in which young chicks are reared in large social groups consisting of thousands of birds of the same age. It is difficult to establish and keep social bonds in this environment (Rodenburg and Koene, 2007). Furthermore, adolescent birds (pullets) can be regrouped when they are transferred from rearing to laying facilities (van Staaveren et al., 2018a). Consequently, recurrent social reorganization can become a chronic stressor (Cheng et al., 2003; Patzke et al., 2009).

Social isolation and disruption have been linked to changes in the monoaminergic pathways in chickens (Gruss and Braun, 1996; Cheng et al., 2003; Cheng and Fahey, 2009). Increased peripheral blood dopamine and corticosterone levels have all been correlated with general social stress, high population density, increased sensitivity to stress, low survival rates and poor feather scores (Craig and Muir, 1996; Cheng et al., 2001a, 2003; Cheng and Muir, 2004; Belujon and Grace, 2015). Additionally, one study concluded that disturbing a social group resulted in lowered 5-HT turnover in the hypothalamus (Cheng and Fahey, 2009) suggestive of deficits in serotonergic neurotransmission. However, much of this research used bird lines selected for improved egg-laying, survival and reduced cannibalism, which were associated with improved feather scores (Craig and Muir, 1996). It is, nevertheless, important to keep in mind that actual FP behaviour was not recorded in most of these studies making associations to severe FP difficult to interpret.

Phenylalanine (PHE) and tyrosine (TYR) make up the two initial steps in the biosynthesis of dopamine, which, in turn, is the metabolic precursor of noradrenaline and adrenaline. A low blood PHE:TYR ratio indicates increased phenylalanine hydroxylase activity, potentially increasing dopamine production. Birkl et al. (2019a) found low PHE:TYR ratios in birds that had experienced social disruption and severe FP. Birkl et al. (2017a) also found low PHE:TYR ratios in birds cohabiting with aggressive peckers, as well as birds experiencing early onset of egg-laying, both experiences that are considered psychologically and physiologically stressful. Furthermore, other groups have also linked aggressive behaviour with high egg production (Cheng and Muir, 2007; Dennis and Cheng, 2011). Interestingly, however, low PHE:TYR ratios can theoretically produce high catecholamine levels (Strasser et al., 2016) that can attenuate the effect of mood lowering adverse environments, potentially aiding the individual to cope with the stress at hand. Notably, Birkl et al. (2019a) and Birkl et al. (2017a) reported the involvement of PHE:TYR ratios in severe FP and AP, illustrating an overlap of neurobiological governance in both IP behaviours. As studies elucidating molecular mechanisms of IP advance, this highlights the importance of careful phenotyping and identification of IP behaviour in each study.

Similarly, tryptophan (TRP) can also influence neurotransmitter levels as it feeds both the 5-HT and kynurenine pathways. A shift in TRP metabolism that favours the kynurenine (KYN) pathway limits 5-HT production and is linked to dysfunctional behaviour in humans (O'Mahony et al., 2015). Surprisingly, the Birkl et al. (2019a) study observed lower KYN:TRP ratios in socially disrupted groups of birds exhibiting high levels of FP. This would indicate that a stress-induced shift to the KYN pathway observed in humans did not occur in laying hens. It is possible that this difference may be attributed to the first and most crucial KYN pathway enzymes: tryptophan-2,3-dioxygenase (TDO) or indoleamine-2,3-dioxygenase (IDO-1 or IDO-2). Indeed, while *Gallus gallus* has orthologs for the TRP-degrading enzymes IDO-2 and TDO, it lacks an ortholog for IDO-1. In humans, IDO-2 only has 3-5% of the enzymatic activity of IDO-1 (Ball et al., 2007). Due to the potential differences in human versus chicken TRP metabolism, more fundamental analyses are required to understand the avian TRP/5-HT metabolism and its role in severe FP in laying hens.

Multiple investigations have compared different housing systems to identify risk factors for FP or feather damage in laying hens (Tahamtani et al., 2015, 2016a; Janczak and Riber, 2015; Brantsæter et al., 2016; Campbell et al., 2019), however, few reports assess the neurobiological changes of hens in different housing conditions and their associations with FP (de Haas et al., 2014b). Tyrosine hydroxylase is a rate-limiting enzyme for dopamine production. One study comparing cage- and aviary-reared hens reported that tyrosine hydroxylase levels were similar in the hippocampus and the avian analogue to the mammalian pre-frontal cortex (nidopallium caudolaterale) of adult hens (Tahamtani et al., 2016b). In contrast, Patzke et al. (2009) compared conventional cages, floor pens and free-range systems and reported that adult laying hens in complex, free-range systems had larger hippocampal soma sizes and more tyrosine hydroxylase-positive fibres in their left hippocampus versus the right and possessed higher serotonergic fibre densities in the nidopallium. This last observation may be due to an altered serotonergic innervation pattern caused by social stress-induced severe FP as the free-range system produced the most crowded social environment accompanied by the most severe FP (Patzke et al., 2009).

### 2.2.2.3 Pharmacological and nutritional modulation of severe feather pecking

In the previous section, we presented examples of how environmental adversity can impact the neurobiology of chickens, and in turn, severe FP behaviour. In this section, we consider examples of pharmacological and nutritional tools that are used to induce or reduce severe FP. These include hormones, neurotransmitters and their respective precursors, as well as agonists and antagonists, each with varying contributions to severe FP behaviour. It is important to highlight that these biomolecules do not function in isolation, and that modulating one may influence other systems. Furthermore, the method used to administer each molecule (e.g. oral vs. injected) and the dosage impacts the outcomes.

El-Lethey et al. (2001) induced FP by feeding pullets a diet with corticosterone (1.5 mg/bird/day). Birds fed this diet were more fearful, had higher serum corticosterone concentrations and developed significantly higher rates of FP than birds that did not receive the corticosterone diet (El-Lethey et al., 2001), indicating a role for the stress response in FP development. Dietary supplementation with 2% TRP was also found to reduce FP damage in 4- and 6-week-old Bantam chicks (Savory et al., 1999). A study using chicks of a high- and low-FP line found that dietary supplementation with 2% TRP reduced gentle, but failed to attenuate severe FP. However, this treatment increased plasma TRP, thereby increasing the large neutral amino acid ratio and forebrain 5-HT turnover in chicks of both lines (van Hierden et al., 2004b). Although toxicity data for oral administration of TRP are limited, a sedative-like effect of such high TRP concentrations cannot be excluded for being responsible for reduced FP damage (Savory et al., 1999).

On the other hand, acute TRP depletion has been used to study the mood-lowering effect of low 5-HT in the mammalian brain concerning psychological disorders (Young, 2013). Birkl et al. (2017b) validated a TRP depletion method in laying hens showing that acute TRP depletion in adolescent birds increased gentle FP in a group setting and showing that peckers were more likely to perform object pecking in an operant chamber (Birkl et al., 2019b). Compared to a bird line bred for low FP, chicks from a high-FP line showed significantly more severe FP when the 5-HT receptor was blocked by an agonist (van Hierden et al., 2004a). This corroborated a previous study that found that high-FP chicks displayed lower brain 5-HT turnover compared to low-FP chicks in response to manual restraint, an acute stressor (van Hierden et al., 2002b). A follow-up study by the same group reported that dopamine receptors of a high-FP line were more sensitive to agonists compared to the receptors of low-FP birds (van Hierden et al., 2005). Furthermore, Kjaer at al. (2004) used the dopamine receptor antagonist haloperidol to demonstrate the possible involvement of dopamine transmission in FP. They observed a decrease in FP in adult birds treated with haloperidol. Indeed, the dopaminergic system has been implicated in reward processes and positive motivation by studies on anticipatory and foraging behaviour in chickens and junglefowl (Moe et al., 2011, 2014). The observed decrease of FP in laying hens in response to haloperidol may be due to pecking being viewed as less-rewarding behaviour.

Yet another intriguing observation is the association between the immune system and FP behaviour. Buitenhuis et al. (2006a) observed a more pronounced specific antibody response following infectious bursal disease vaccination and lower percentage of $CD4^+$ and double-positive $CD4^+$:$CD8^+$ in high-FP birds compared to a low-FP line. These results are in line with a distinct immune response in birds selected for high or low egg-laying and survivability (Cheng et al., 2001b). Birds selected for high egg-laying and survivability (higher feather score) had higher levels of blood lymphocytes and cell-mediated higher $CD4^+$:$CD8^+$ circulating T-cell ratios (indicating a stronger immune system), while those selected for low productivity and survivability showed higher heterophil:lymphocyte ratios (indicator of stress to novel environments) and higher plasma IgG concentrations (Cheng et al., 2001b).

Theoretically, these immunological results could be explained by enhanced corticosterone and blood dopamine in high-FP birds, by having suppressive effects on immunity by suppressing the proliferation of T and B cells (and inhibiting cytokine production) (Cheng et al., 2001a,b). Parmentier et al. (2009) found more feather damage in hens challenged with human serum albumin (stimulating the specific humoral immune response) compared to those challenged with lipopolysaccharide (stimulating the innate immune response). Consequently, the authors suggest that stimulation of the specific immune response can predispose hens to FP later in life; however, they caution that the relationship between immune responsiveness and FP may depend on multiple pathways, including the neuroendocrine system and calls for further investigation (Parmentier et al., 2009).

#### 2.2.2.4 Use of an acute stressor to identify neurobiological differences between severe peckers and non-peckers

The manual restraint test is used to identify neurobiological differences between birds that perform severe FP and birds with low incidences of pecking behaviour. Manual restraint of a bird by a human combines social isolation with the presence of a human predator and is, therefore, considered to be an acute stressor (Korte et al., 1997; Beuving and Blokhuis, 1997).

Social isolation and handling has been linked to changes in the forebrain monoaminergic pathways in chicks (Gruss and Braun, 1996). It is reported to increase a metabolite of 5-HT (5-hydroxyindoleacetic acid, 5-HIAA) in socially isolated chicks, particularly when accompanied by locomotor activities such as running and jumping which is correlated with an increase in distress calls. Additionally, homovanillic acid (HVA), a dopamine metabolite, also increased when birds were handled and lightly chased. HVA also increases to some extent during social isolation. These observations suggest, depending on the stressor, a distinct activation of both the serotonergic and dopaminergic pathways during aversive experiences that involve arousal, distress or fear (Gruss and Braun, 1996).

Rhode Island Red birds were found to be less fearful and show less FP than purebred White Leghorn birds (Uitdehaag et al., 2008, 2009). The former also had higher brain 5-HT and dopamine concentrations, lower 5-HIAA, lower 3,4-dihydroxyphenylacetic acid (DOPAC, a dopamine metabolite), and, consequently, lower 5-HT and dopamine turnover rates after a manual restraint test than White Leghorn birds (Uitdehaag et al., 2011). However, actual differences in severe FP between the two groups were not reported in this study (Uitdehaag et al., 2011). Studies by van Hierden et al. used bird lines selected for high or low FP behaviour and found contradicting results. They report low plasma corticosterone, low 5-HT and low dopamine turnover in the forebrain in high-FP birds relative to low-FP birds, suggesting that these biochemical changes predispose high-FP animals to readily develop FP behaviour (van Hierden et al., 2002a,b). Similarly, Kops et al. (2017) confirmed that high FP birds have lower brain 5-HT and, to a lesser extent, dopamine turnover rates after a manual restraint test than low-FP birds. The high FP birds in this study were found to develop more severe FP, as well as toe pecking, than low FP birds as pullets, a life stage where the brain is still in development (Kops et al., 2017). As adults, high FP birds continued to perform severe FP at a higher rate than low FP birds, but neurobiologically, few differences were observed; for example, high FP birds had higher turnover rates of dopamine and 5-HT, higher DOPA and 5-HIAA in certain brain areas (Kops et al., 2017). If this neurobiological change in adulthood was caused by severe FP or severe FP was caused by the neurobiological change could not be elicited. Serotonergic differences in certain brain areas after manual restraint tests were also observed between severe FP birds (higher 5-HIAA, higher 5-HT turnover) and non-peckers (Kops et al., 2013a). Interestingly, when birds were selected for low mortality instead of FP behaviour, the differences observed were mostly within the dopaminergic system with lower levels of noradrenaline and DOPAC (Kops et al., 2013b); however, this study did not report differences in severe FP between the lines.

Finally, high FP birds showed more pronounced variations in the heart rate and reduced vagal nerve-mediated cardiac control in response to manual restraint, suggesting that they were more susceptible to stress relative to low FP birds (Kjaer and Jørgensen, 2011). As higher dopamine levels were found in stressed FP birds, and peripheral dopamine levels have been used to assess the intensity of the sympathetic response to a stimulus, lower vagal tone in feather peckers fits into this concept. This is in line with findings by two groups that reported lower parasympathetic activity, higher HPA-axis reactivity and increased plasma corticosterone in high FP birds following handling by humans (Korte et al., 1998; Kjaer and Guémené, 2009). Others, however, have reported lower plasma corticosterone following manual restraint in high FP birds (Korte et al., 1997; van Hierden et al., 2002b).

### 2.2.3 Challenges and limitations of the dysfunctional view on severe feather pecking

Adversity is an inevitable part of daily life that shapes the behaviour of any animal. Brief exposure to adversity or stress can be a powerful motivating force driving the individual to obtain a reward or avoid adversity. In contrast, chronic exposure to adverse environments can trigger permanent neurobiological changes, which are known to play a critical role in the etiology and exacerbation of certain human psychological disorders.

Similar to psychological disorders, severe FP-associated neurobiological findings suggest the involvement of the HPA axis, the autonomic and monoaminergic system, and immunological parameters. Nevertheless, studies reporting these effects are inconsistent when describing the degree to which each pathway contributes to severe FP. These differences can be explained by the small number of studies investigating avian neurobiology and constraints in the experimental set-up or subjects. The studies employ prominent and widely accepted approaches that are often used in human psychological disorder research, such as a brief stressor, a pharmacological intervention or a nutritional treatment. However, birds used to study acute stress are often reared under adverse social and physical environments, sustaining intense and chronic stressors before being subject to the treatment under investigation. Moreover, the rearing conditions, and therefore, the chronic stress experienced by the birds in each study are different. This unintended combination of acute and chronic stress, as well as the type of chronic stress, likely has diverging consequences on the neurochemical balance of experimental subjects, making it difficult to tease out the changes that are caused by each treatment. Furthermore, severe FP research often uses genetic lines selected for high or low FP behaviour, for high or low egg production or survival rates. These genetic lines do not reflect commercial flocks. Also, individual birds, even within high or low FP lines, may present extreme phenotypes showing either low, high or no FP behaviour (Piepho et al., 2017; Iffland et al., 2020). The variations in age and experimental settings also are likely to contribute to variation in the observed results. Most studies are also not conducted in commercial farm settings, eliminating thereby potentially important environmental factors that may impact the neurobiological landscape of the birds.

Additionally, most studies measure metabolites of the autonomic and neuroendocrine systems and brain circuits that are involved in a broad range of biological functions and behaviours. Given the scope and crosstalk between the systems under study, it is likely that the levels of the measured biomolecules reflect these additive and interactive effects. As such, this may account for the contradicting associations that have been reported between severe FP and neurochemicals, immune molecules and perceived fearfulness (e.g. Rodenburg et al., 2010; Grams et al., 2015; de Haas and van der Eijk, 2018; van der Eijk et al., 2018).

There is a knowledge gap in understanding avian brain processes and pathways in general. While comparisons to the mammalian system are often made, there are a few established differences in the stress-responsive circuits and metabolic pathways. For example, the avian hippocampus is not the primary target of stress-related hormones, unlike its mammalian counterpart. It is also not clear whether the strong connection between the hippocampus and the prefrontal cortex in mammals is mirrored in birds (Patzke et al., 2009). Further, it is noteworthy that dopamine and 5-HT cannot cross the blood-brain barrier, while their precursors (TRP, PHE and TYR) and glucocorticoids readily permeate it. There is some evidence that peripheral serotonin concentrations mirrored the 5-HT levels in the avian brain after manual restraint (Uitdehaag et al., 2011). Given that each study measures neurochemical levels in different locations of the body, cross-study comparisons become difficult when the relationship between the peripheral and central nervous system metabolite levels is unclear. As such, further avian studies are required to understand this dynamic under both acute and chronic stress. As 95% of TRP is degraded to kynurenine (5% to 5-HT), one rate-limiting enzyme is regulated by glucocorticoids and the other, IDO-1, by cytokines in mammals. This pathway has received attention in human immune dysfunction and central nervous system disorders. Intriguingly, the IDO-1 enzyme is missing in *Gallus gallus* domesticus (Birkl et al., 2019a). The implications of this on the 5-HT balance are yet unclear. There is, therefore, a clear need for avian fundamental research in this area of study that would allow direct and informed comparison with the mammalian system.

A fundamental question that should be addressed is whether the neurobiological differences between severe feather peckers and non-peckers currently presented in the literature are sufficient to even qualify to interpret the former as possessing a dysfunctional biological system. Pursuant to this, there must be a discussion to determine the tipping point at which normal function is perceived to be disrupted by the adverse effects of FP. To date, this is still unclear. If indeed the disruptions to some neurobiological pathways are deemed to be dysfunctional, follow-up work must determine whether severe feather peckers become more repetitive and persistent in various other tasks. This area of research, such as cognitive tasks, has garnered little attention until now (Kjaer et al., 2015; Heinsius et al., 2020).

Regardless of the trigger that initiates FP, the neurobiological markers discussed in this chapter do not provide insight into the molecular mechanisms that induce the final FP behaviour, nor the neurobiological foundation that establishes a persistent behaviour in some birds, while failing to do so in others. Daigle et al. (2015) found that over 50% of birds studied were inconsistent in their behavioural patterns throughout their lifetime. Whether birds exhibiting inconsistent FP behaviour are dysfunctional is yet another topic that needs to be discussed.

Given the limited understanding of laying hen neurobiology and current FP literature, it is maybe not possible to allocate FP behaviour to a dysfunctional biological mechanism. As further research delves into identifying both biochemical and structural changes within the neurobiological systems and the brain, the idea of neuroplasticity will need to be considered as a physical change that is not always indicative of dysfunctions, but maybe a product of adaptation. Finally, neurochemical measurements do not linearly correspond to FP behaviour through a cause-and-effect model due to its complex nature. Consequently, the neurobiological measurements that characterize severe FP may not necessarily cause the behaviour to which it is associated. To conclude, the field of research to better understand severe FP through a biochemical or a neuroscientific lens is rich with opportunities.

## 3 Practical attempts and management strategies to reduce injurious pecking

In attempts to prevent IP, or more specifically the damage caused by IP, beak trimming is widely applied, whereby the sharp upper and lower mandible tips of the beak are removed via either infrared treatment or hot blade trimming depending on the country (Nicol, 2018). However, this practice is under increasing scrutiny due to the ethical and societal implications. Beak trimming can affect bird welfare through short- and long-term pain, the formation of neuromas, loss of normal beak function and consequent changes in behaviour (Hughes and Gentle, 1995; Kuenzel, 2007; Nicol, 2018). Furthermore, ethical concerns play a large role as people prefer to see changes in housing and management, rather than adapting the animals to the environment (Rollin, 2015). Currently, the practice is heavily regulated in countries where it is allowed. In North America, beak trimming is still highly prevalent (Duncan, 2001), while the EU Council Directive 1999/74/EC prohibits beak trimming in the European Union. However, individual member states may allow beak trimming under certain conditions (European Commission, 1999). The practice is banned (e.g. Sweden, Norway, Finland and the Netherlands) or voluntarily phased out through agreements or penalties (e.g. Denmark, Germany and Austria) in several countries. The UK postponed implementing a ban due to the high risk of keeping intact flocks but is monitoring progress in developing other strategies to reduce IP to revisit a possible ban in the future (Nicol, 2018). While beak trimming is effective in reducing the consequences of IP, it is clearly not able to eliminate this behaviour and thus only treats the symptoms rather than the causes or other contributing factors of the behaviour (Nicol et al., 2013).

IP is a multifactorial issue which makes its prevention and control complex (Nicol et al., 2013; Rodenburg et al., 2013). Research based on experimental and epidemiological approaches has identified many interrelated risk factors for IP (with a focus on FP), during the rearing and laying period which should be addressed in a combined manner (Pötzsch et al., 2001; Lambton et al., 2013; Bestman et al., 2017; Jung and Knierim, 2018). Jung and Knierim (2018) nicely summarize results from experimental and epidemiological studies to provide an overview of such factors, the evidence of their protective effects ranked as good, contentious or no confirmed evidence and how they relate to practical recommendations. While they acknowledge that this ranking was somewhat arbitrary, it provides a good overview of the work done to date. They also emphasized that due to the confounding interactions between factors and different methods used in the consulted studies, they cannot conclude that no influence exists for those factors with contentious or no evidence but rather that further investigation is needed (Jung and Knierim, 2018). This also highlights the understandable difficulty for farmers in finding the right approach to prevent and/or control IP. These approaches, which should target bird-related factors and factors in the social environment and the physical environment, should be addressed in a combined manner (Pötzsch et al., 2001; Lambton et al., 2013; Jung and Knierim, 2018).

### 3.1 Bird-related factors

IP has a genetic component with FP, TP and AP, all being heritable traits (Kjaer and Sorensen, 1997; Kjaer and Hocking, 2004; Bennewitz et al., 2014). Selection for low FP and high FP with a consequent difference in plumage condition (Kjaer and Sorensen, 1997; Kjaer et al., 2001) or high- and low-mortality lines is possible (Muir, 1996; Rodenburg et al., 2008a). In other cases, differences in FP occurred due to selection for production traits as they appear correlated (Korte et al., 1997; Su et al., 2006). Furthermore, FP lines also differ in other correlated traits such as their stress response, aggression, fearfulness, locomotor activity, foraging behaviour, serotonin metabolism and gut microbiota (reviewed by Nicol, 2018). This interconnectedness between FP and other traits makes selection against FP difficult (Rodenburg et al., 2013; Nicol, 2018). It is, therefore, not surprising that genetics can play an important role, and that different selection strategies are investigated to influence FP (Rodenburg et al., 2008a). Kjaer et al. (2001) showed that it is easier to select for increased FP than for decreased FP due to an asymmetric response to divergent selection. Recently, the heterogeneous nature of FP was further highlighted by the finding of subpopulations, such as 'extreme' feather peckers that consistently pecked more than the 'regular' feather peckers (Labouriau et al., 2009; Piepho et al., 2017; Iffland et al., 2020). These extreme feather peckers came in the mood for FP twice as often, and within these periods showed a nine times higher pecking frequency than birds from the FP group (Iffland et al., 2020). Furthermore, even in low FP lines extreme FP phenotypes still exist (Piepho et al., 2017). To date, breeding companies often perform indirect, downward selection on IP victims or mortality as continuous behavioural observations to select on IP itself are too labour-intensive and time-consuming and so far automatic recording of IP in groups of birds was not possible. However, current developments in sensor technology and genomic approaches are making it more likely that individual animal behaviour can be monitored in group settings and can be used in breeding programmes in the future to select against damaging behaviours (Ellen et al., 2019).

Different plumage patterns or colours can affect birds' perception and form an attractive pecking substrate. Brown- and white-feathered birds can differ in the amount of IP they perform and the amount of damage they receive (Oden et al., 2002; Dixon and Duncan, 2010; Yamak and Sarica, 2012; de Haas et al., 2013; Decina et al., 2019b), though differences between brown- and white-feathered hens are often confounded with the genetic strain. Additionally, strain differences can be found between birds with the same feather colour, for example, brown-feather genotypes (Coton et al., 2019) as well as differences between different coloured birds within the same strain (Bright, 2007). Bright (2007) found that birds with white plumage had less feather damage than grey or black birds of the same strain. Furthermore, white birds with brown pigmentation are more vulnerable to FP than all-white birds (Keeling et al., 2004) and damaged feathers receive more pecking than undamaged feathers (McAdie and Keeling, 2000). A likely explanation for these differences is the difference in colour contrast when the lighter base of the feathers becomes exposed (McAdie and Keeling, 2000; Keeling et al., 2004; Bright, 2007).

There is a general consensus in literature that FP and feather damage increase as birds get older (Bilcik and Keeling, 1999; Huber-Eicher and Sebö, 2001a; Lambton et al., 2013; Decina et al., 2019a,b). FP can already be observed in very young birds after hatch and affect large proportions of a flock by 4–6 weeks of age (Wechsler et al., 1998; Huber-Eicher and Sebö, 2001a; Riedstra and Groothuis, 2002). Feather damage can also already be observed in pullets (see Section 1.2) though it can be concealed due to moulting and the more subtle nature of the behaviour occurring in the growing phase (Huber-Eicher and Sebö, 2001b; Bestman et al., 2009). TP, such as toe pecking and vent pecking, is typically seen among older birds (Pötzsch et al., 2001; Rodenburg et al., 2009b; Krause et al., 2011), with an early onset of lay being a risk factor for vent pecking (Pötzsch et al., 2001).

## 3.2 Factors in the social environment

The occurrence of IP or integument damage in young birds highlights the importance of the rearing period (van de Weerd and Elson, 2006; Janczak and Riber, 2015), especially considering that feather damage observed during this period is indicative of feather damage observed during the laying period (Bestman et al., 2009; Drake et al., 2010; Gilani et al., 2013). Bestman et al. (2009) found that when flocks presented with feather damage during rear, they also developed feather damage during the laying period in 90% of the cases. As FP can spread through social learning (Zeltner et al., 2000) or the same reaction to environmental conditions in many birds (Wechsler et al., 1998), and damaged feathers are more attractive for further pecking (McAdie and Keeling, 2000), this shows the importance of limiting IP during this crucial rearing period. Furthermore, maternal stress can increase the risk of anxiety or FP in the offspring (de Haas et al., 2014b). The role of epigenetics and the hormonal influence of the parent flock on the development of IP in the offspring require more attention (de Haas et al., 2014a,b; Nicol, 2018).

Other important factors within the social environment that are linked to this social transmission and can play a role in the development of IP are the related concepts of group size and stocking density. Stocking density is often manipulated by changing the group size in experimental studies which complicates untangling which factor contributes the most. Group sizes are larger in non-cage systems and this is often brought up as a potential issue when it comes to controlling IP as it contains a larger number of potential victims and allows for a fast spread of the behaviour. Larger group sizes have been associated with increased risk of severe FP and feather damage (Hughes and Duncan, 1972; Bilcik and Keeling, 1999; Nicol et al., 1999; de Haas et al., 2014a). However, AP was more commonly observed in smaller flocks at lower stocking densities, likely in an effort to establish social hierarchies (Nicol et al., 1999), which is typically avoided in large groups where birds use other cues of social hierarchy to avoid numerous fights (Estevez et al., 2003; D'Eath and Keeling, 2003; Rodenburg and Koene, 2007). High stocking densities are also associated with increased feather damage among chicks reared in aviaries (Zepp et al., 2018), chicks reared under organic conditions (Bestman et al., 2009), hens in furnished cage flocks (Widowski et al., 2017; Coton et al., 2019), free-range flocks (Coton et al., 2019) and hens housed in percheries (Nicol et al., 1999).

### 3.3 Factors in the physical environment

Early-life experience influences the adaptability of pullets to later housing and management practices (Widowski and Torrey, 2018). It is therefore not surprising that the conditions during the rearing period can influence the occurrence of IP later in life. Generally, similar rearing and laying environments are thought to be beneficial (Janczak and Riber, 2015; Widowski and Torrey, 2018), but the number of changes between these two periods per se does not necessarily increase the risk of IP (Drake et al., 2010). However, combining the rearing and laying period on the same farm (Drake et al., 2010) or earlier transfer of birds between farms (Bestman and Wagenaar, 2003) can help reduce FP by allowing pullets to familiarize themselves with and adapt to the new housing environment before the onset of lay (Nicol, 2018). In general, early access to perches is thought to benefit pullet leg health and adaptability (Widowski and Torrey, 2018). Access to perches was associated with a lower risk of FP (Huber-Eicher and Audigé, 1999) and vent pecking (Gunnarsson et al., 1999). However, it is important to note that perches need to be designed appropriately as Rodenburg et al. (2009b) suggested that low perches allow toes of perching birds to be targeted for toe pecking by birds on the ground.

The flooring and provision of litter substrate are some of the most investigated factors in relation to FP due to their role in redirected ground-pecking or dust-bathing theories (Blokhuis, 1986; Vestergaard et al., 1993). Pullets reared without suitable litter material can develop FP and feather damage during rear and lay (Fig. 4) (Huber-Eicher and Wechsler, 1997; Johnsen et al., 1998; Nicol et al., 2001; Bestman et al., 2009; de Haas et al., 2014b); however, the presence or absence of litter is not always a predisposing factor in the development of FP (Gunnarsson et al., 1999). The role of chick paper on behavioural development during rear was recently reviewed and appears to function as foraging material for chicks (reviewed by Schreiter et al., 2019). The presence of suitable litter material or foraging opportunities is important during lay, given that the absence of a scratch area in furnished cages, having all wire/ slatted flooring in non-cage systems, restricted access to the litter during nest box training, and the absence of loose litter at the end of lay all contribute to higher levels of FP or feather damage (Green et al., 2000; Lambton et al., 2010; Decina et al., 2019a,b). Litter quality is thus important and while little research had been conducted regarding strategies to maintain litter in a loose and friable state (Lambton et al., 2013), Bestman et al. (2017) found that topping-up

**Figure 4** Severe feather damage and injuries in pullet housed on wire-mesh flooring. (Picture: A. Harlander).

of litter during lay reduced the percentage of white birds (though not brown birds) with wounds on the back and vent. Additionally, increased foraging opportunities provided by outdoor range access can reduce FP depending on the range use, though it does not completely prevent the behaviour (Green et al., 2000; Bestman and Wagenaar, 2003; Nicol et al., 2003; Lambton et al., 2010). Further attempts to increase foraging opportunities come in the form of enrichment elements, that is, objects which can be manipulated but not used for foraging or dust bathing (e.g. strings), substrates which can be consumed as feed (e.g. straw, grain, alfalfa, pecking stones etc.) and installations with dust bathing substrates (Schreiter et al., 2019). Chickens with access to strings had better feather cover than those that did not have strings in their rearing cages (McAdie et al., 2005). While unable to investigate the efficacy of strings to reduce FP, Jones et al. (2002) found that birds maintained interest in the strings two weeks after their introduction. Other forms of visual, auditory and olfactory enrichment (e.g. video playback, radio, smells and ultraviolet light) have been explored mostly in terms of whether they sustain birds' interest or affect their development, but are often not validated on a commercial scale (Jones, 2004; Campbell et al., 2019). In general, substrates which can be consumed as feed are more effective at reducing severe FP compared to objects or dust-bathing substrates (Dixon et al., 2010). Providing these types of substrates can not only reduce FP but also AP (Zepp et al., 2018). However, the benefit of these enrichment elements is more clear in barren environments (e.g. cage environment), while in littered housing systems conflicting results are reported (for review see Schreiter et al., 2019). Additionally, there is no consensus on whether additional enrichment should be provided as a preventive measure (e.g. permanently) or a curative measure (e.g. added after IP is observed) in flocks (Schreiter et al., 2019). The use of enrichment as a curative measure can explain why in Decina et al. (2019a) higher levels of feather damage were observed in flocks that provided enrichment, as one-third of farmers did not provide enrichment until after they had established problems with FP.

In line with foraging opportunities, feed management (including diet composition and feeding strategies) is another important tool that can be used to reduce IP. Deficiencies in nutrients such as protein, amino acids or minerals can increase FP and cannibalism, as reviewed by van Krimpen et al. (2005). A commonly observed behaviour in conjunction with severe FP is feather eating and this could be a method through which birds attempt to increase the fibre in their diet (McKeegan and Savory, 1999; Harlander-Matauschek and Häusler, 2009). Aversive spraying solutions might deter birds from FP but might raise concerns from health, food, veterinary or environmental agencies (Harlander-Matauschek et al., 2009; Harlander-Matauschek and Rodenburg, 2011). Other strategies which increase the time spent on feeding such as mash (instead of crumbles or pelleted feed), low-energy diets and high (in)-soluble fibre diets or roughage may have further positive effects by decreasing FP (van Krimpen et al., 2005). Pelleted feed is consumed more quickly, which could explain the associations with high occurrences of severe FP and higher levels of feather damage (Aerni et al., 2000; El-Lethey et al., 2000; Lambton et al., 2010). Scattering of grain or straw during rear to promote foraging in the litter was found to reduce feather damage during lay (Blokhuis and Vanderhaar, 1992); however, subsequent research did not find such an effect (Bestman and Wagenaar, 2014). Jung and Knierim (2018) similarly found no evidence of a preventive role, and in fact, found increased risk of FP in some studies. They noted that when recommended, the frequency, amount and location of scattering grain need to be considered to avoid stress and smothering of birds (Jung and Knierim, 2018). The number of diet changes during rear or lay influences severe FP (Green et al., 2000; Gilani et al., 2013) and vent pecking (Pötzsch et al., 2001), though Heerkens et al. (2015) found decreased feather damage in the neck area when more frequent diet changes were occurred. Diet changes can increase competition or temporary aversion to feed resulting in hungry and more peck-prone birds which can thereby affect IP (Green et al., 2000; Dixon et al., 2006; Dixon and Nicol, 2008).

Apart from feeding strategies, the feeder and drinker equipment used is also implicated in the development of FP. Drake et al. (2010) found that FP developed earlier in laying flocks depending on the feeder or drinker type used during rear, though no clear explanation for why this occurred could be provided. Birds developed FP more quickly when reared with chain feeders rather than pan feeders or a combination of different feeder types (Drake et al., 2010). Interestingly, chain feeder systems were associated with increased risk of FP in other studies (Green et al., 2000; Gilani et al., 2013). Gilani et al. (2013) found that the rate of severe FP was lower when a combination of feeder types was used compared to only chain feeders. In contrast, the use of compartmentalized pans increased the rate of severe FP (Gilani et al., 2013). The use of bell and nipple drinker systems during rear compared to only nipple drinkers or nipple/cup drinker systems was also associated with earlier onset of FP during lay (Drake et al., 2010) and bell drinkers were also a risk factor for vent pecking (Pötzsch et al., 2001). The use of bell drinkers can possibly increase the risk of IP due to water spillage leading to poorer litter quality or increased competition around bell drinkers due to bird clustering (Green et al., 2000).

Factors related to house climate such as temperature, humidity, sound, light and dust levels play an important role in the development of FP during rear and lay (Gilani et al., 2013). House climate influences both litter and air quality which are other important factors to consider in the management of IP and can have consequences for bird health (EFSA, 2005). Bird health is important to reduce the risk of FP, and higher levels of FP or feather damage have been observed in flocks with red mite infections or other diseases (Green et al., 2000; Heerkens et al., 2015). Furthermore, apart from high levels of ammonia being aversive to birds (David et al., 2015; Pokharel et al., 2017), higher levels of ammonia and $CO_2$ are linked to an earlier onset of FP (Drake et al., 2010). Decina et al. (2019a) found that flocks where manure was removed only at the end of the production period had a higher prevalence of feather damage than flocks where manure was removed more frequently. While this was mostly the case in barns with all-wire/slatted systems which lacked foraging opportunities, this can further highlight the importance of litter management and air quality. Additionally, litter and air quality are also of importance to farm workers and can influence their health (Donham et al., 2000), possibly leading to reduced care and detection of welfare issues within the flock when staff are reluctant to enter the facility or visibility is reduced.

Dimming of lights is a commonly reported management strategy used to prevent or reduce IP (Green et al., 2000; Kaukonen and Valros, 2019). Vision is a dominant sense in birds and reducing light intensity limits their ability to recognize plumage cover as well as their overall activity (Bright, 2007; O'Connor et al., 2011). In a cross-sectional epidemiological study by Green et al. (2000) dimming of lights was initially indicated as a risk factor; however, they found that this dimming was likely in response to FP. Over 61% of farmers dimmed the lights to reduce FP when it occurred, while 19% of farmers indicated that they dimmed the lights to prevent FP (Green et al., 2000). Similarly, Decina et al. (2019a,b) were unable to evaluate light intensity as a risk factor, as this was further complicated because nearly one-quarter of the farmers could not provide the lux value in the house (van Staaveren et al., 2018a). Longitudinal and experimental studies provide more insight in this regard and found an earlier start of severe FP and feather damage under high light intensities (Kjaer and Vestergaard, 1999; Drake et al., 2010). However, dimming of light to encourage birds to use nest boxes has also been associated with an increased risk in vent pecking (Pötzsch et al., 2001). Furthermore, low light intensities can influence the development of birds' visual abilities and lead to other welfare problems (Prescott et al., 2003; Nicol et al., 2013).

Different light sources or wavelengths can influence bird behaviour, with blue light being associated with higher levels of FP while red light is thought to reduce the appearance of blood or bare skin (Mohammed et al., 2010; Lambton et al., 2013). The use of red light to prevent or reduce FP was reported by farmers in the UK (3% and 9%, respectively), though not as frequently as the dimming of light intensity as a management strategy (Green et al., 2000). The duration of the light period also plays a role, and the amount of daylight during rear was used as a predictor of feather damage during lay (Bestman et al., 2009), and severe FP during rear was reduced when birds had more light hours (Gilani et al., 2013). Apart from duration, the schedule of light and dark periods also impacts the birds' diurnal rhythm (Kjaer, 2000). Decina et al. (2019b) reported a positive association between feather damage and the practice of midnight feeding, which they related to the interruption of the sleeping cycle and increased restlessness and activity of the birds. The ability to retreat from active birds and have a secure resting area in the form of dark brooders was previously reported as being protective against FP in chicks (Jensen et al., 2006; Gilani et al., 2012).

Another aspect which has been somewhat neglected, but plays a vital role, is the farm staff and their perceptions of IP. When more people or more experienced caretakers inspect the birds, the probability of severe FP is reduced (Green et al., 2000; Gilani et al., 2013); however, others found no association between experience and FP (Bestman and Wagenaar, 2003; Heerkens et al., 2015). Additionally, the amount of time spent inspecting the birds was associated with less damage to the vent plumage (Heerkens et al., 2015). Farmers can use different criteria to define when action should be taken and have different thresholds as to what kind of damage is still considered acceptable (Huber-Eicher, 1999; Palczynski et al., 2016; Kaukonen and Valros, 2019). For example, one-third of farmers took action when they observed birds pecking each other's feathers, while half of the farmers did so only when they observed feather damage, nearly 8% took action when they observed naked areas and 12% did not consider FP as a problem until they observed injuries or losses (Huber-Eicher, 1999). In other cases, incidences of up to 5–7% of FP or 1–2% of cannibalism were considered tolerable by farmers (Kaukonen and Valros, 2019). Coton et al. (2019) noted differences in farmer perceptions of FP. A smaller proportion of farmers mentioned 'moderate to substantial' FP problems in furnished cage systems compared to free-range systems while the actual prevalence of severe FP was similar in both. The authors attributed this to differences in farmers' opinions but also less favourable conditions (e.g. light, overview of flock) to observe FP in furnished cage flocks (Coton et al., 2019). Farmers can detect both the behaviour and the consequent damage, but farmers may also tend to underestimate the prevalence of FP (Lambton et al., 2010; van Staaveren et al., 2018b). Lambton et al. (2010) surveyed farmers who indicated that 65% of their flocks showed FP while researchers who observed these flocks found signs of FP in 86% of them. The reason for this underestimation can vary, for example, a lack of recognition or systematic recording (Nicol et al., 2013), but highlights the importance of factors that influence farmers' understanding and management actions (Palczynski et al., 2016). These factors, also termed secondary risk factors, need to be addressed together with the primary risk factors for interventions to be effective (Whay, 2007).

Despite these methods to reduce IP, it is clear that this is still a prevalent problem with many contributing factors. Therefore, understanding the underlying mechanism of IP is necessary to develop more robust methods to deal with IP in commercial settings and has the potential to benefit the welfare of birds as well as farmers and society.

## 4 Conclusion

IP is an umbrella term used to describe quite distinct forms of injurious bird-to-bird pecking, which lead to damage or injuries to the bird's integument. Due to the high number of birds kept for egg-laying, the high prevalence of IP and its animal welfare and economic importance, this is particularly a problem in laying hen production. Over the years, much work has been done in an effort to elucidate the underlying mechanisms of this behaviour with limited results. There is a need for clear definitions of the different forms of IP (feather pecking - FP, aggressive pecking - AP, and tissue pecking - TP), and this is especially important for the heterogeneous forms of FP. Two different views are proposed to explain severe FP, each with their own challenges and limitations. The ethological view focusses on severe FP as a redirected foraging behaviour, while the dysfunctional view looks towards dysfunction within neurobiological pathways induced by environmental adversity. The complexity in addressing severe FP can likely be found in the interconnectedness between different implicated neurobiological systems, as well as the difficulty in translating findings from different experimental lines used in research settings to the commercial lines currently used on-farm. As such, currently, many risk factors and potential management strategies to reduce IP are available; however, without fully understanding the underlying cause of the behaviour, these will have limited success in preventing IP.

## 5 Future trends in research

Poultry research has attempted to alleviate and characterize IP in birds for over 70 years. Despite this, the mechanisms underlying IP remain elusive, and farmers still do not have a reliable and consistent solution to mitigate this disruptive behaviour. Although risk factors and management strategies (see Section 3) have been proposed, it is clear that IP is a complex and multifactorial problem that demands to be better understood, before widely applicable prevention tools can be put forth. Given the long history of IP studies, new etiological theories and treatments that can challenge traditional thinking can bring a fresh perspective and unlikely solutions to the current field. Nevertheless, the validation of the existing reports continues to be important to solidify the foundation upon which future research will build.

While many studies use birds that are bred for high- or low-IP susceptibility, the underlying mechanisms that give rise to these phenotypes are not understood. The fast-paced advances in genomics research will help in the identification of animals susceptible to IP; however, both genetic and environmental factors interact in complex and possibly unknown ways to produce IP behaviour. Moving forward, it is imperative to clearly and narrowly define both the environments that are used to study IP and the phenotyping of experimental subjects (Ellen et al., 2019). This will contribute to validate future studies in strengthening core concepts and, importantly, to disentangle the interactions between the genetic make-up of the individual and its environment. Additionally, exacting definitions will also allow identifying the genetic backgrounds contributing to different IP phenotypes such as severe FP, TP and AP. A better understanding of the interplay between genetic and environmental factors in IP etiology will not only identify its origins, but it may also unveil novel pathways for prevention and treatment. This will certainly require collaboration between geneticists, neuroscientists, physiologists and ethologists.

To date, the ethological view and the dysfunctional view are investigated in isolation. Despite the reality that both approaches may contribute to IP, their relative importance in IP and interactions are unknown. Mirroring the proposed study of genetics versus the environments, new research should strive to integrate both the ethological and the dysfunctional approaches to understand the underlying mechanisms of IP.

Exploring non-pharmacological methods to prevent or reduce IP may be an additional alternative research avenue that has garnered increased attention. This area of study is especially relevant as the use of drug interventions are increasingly viewed as negative by the consumer. For example, individuals suffering from psychological disorders are reported to benefit from dietary interventions such as increasing tryptophan content in the diet relative to the other amino acids or time spent outdoors, exposure to bright light and exercise. Similarly, future research should investigate the impact of similar stimuli on IP behaviour. Green et al. (2000) have already reported improved plumage in birds that had access to an outdoor area. Furthermore, several studies indicate associations between FP and the gut microbiota (Meyer et al., 2012; Birkl et al., 2018; van der Eijk et al., 2019b), its metabolites (Meyer et al., 2013) and the immune system (van der Eijk et al., 2019c). Pioneering work has also investigated the impact of faecal transplants in FP behaviour (van der Eijk et al., 2019d). The bidirectional communication between the gut and the brain is mediated through hormones, nerves (van Staaveren et al. 2020), cytokines and microbial metabolites, among other molecules. Dissecting this intricate network, identifying its contribution to modify behaviour and using this information to develop nutraceuticals are exciting avenues of research. In the short-term, on-farm testing of interventions that are proposed by the current understanding of IP and re-evaluation of management strategies (see Section 3) on commercial farms will help to reduce and prevent pecking behaviour which, in turn, will improve the welfare of the birds.

## 6 Where to look for further information

The following articles provide a good overview of the subject:

- Brunberg et al. (2016). Omnivores going astray: A review and new synthesis of abnormal behavior in pigs and laying hens. *Frontiers in Veterinary Science*, 3: 57.
- de Haas and van der Eijk. (2018). Where in the serotonergic system does it go wrong? Unravelling the route by which the serotonergic system affects feather pecking in chickens. *Neuroscience & Biobehavioral Reviews*, 95: 170–188.
- Jung and Knierim. (2018). Are practice recommendations for the prevention of feather pecking in laying hens in non-cage systems in line with the results of experimental and epidemiological studies? *Applied Animal Behaviour Science*, 200: 1–12.
- Nicol et al. (2013). The prevention and control of feather pecking: application to commercial systems. *World's Poultry Science Journal*, 69(4): 775–788.
- Nicol, C. (2018). Feather pecking and cannibalism – can we really stop beak trimming? In: *Advances in Poultry Welfare – A volume in Food Science, Technology and Nutrition* (Ed. Joy A. Mench). Woodhead Publishing, Elsevier Ltd, UK.
- Rodenburg et al. (2013). The prevention and control of feather pecking in laying hens: identifying the underlying principles. *World's Poultry Science Journal*, 69(2): 361–373.
- Schreiter et al. (2019). Effects of litter and additional enrichment elements on the occurrence of feather pecking in pullets and laying hens – A focused review. *Veterinary Medicince and Science*, doi: 10.1002/vms3.184.
- van Krimpen et al. (2005). Impact of feeding management on feather pecking in laying hens. *World's Poultry Science Journal*, 61(4): 663–685.

## 7 References

Aerni, V., El-Lethey, H. and Wechsler, B. 2000. Effect of foraging material and food form on feather pecking in laying hens. *British Poultry Science* 41(1), 16–21.

American Psychiatric Association 2013. *Diagnostic and Statistical Manual of Mental Disorders*. American Psychiatric Association, Arlington, VA.

Ball, H. J., Sanchez-Perez, A., Weiser, S., Austin, C. J. D., Astelbauer, F., Miu, J., McQuillan, J. A., Stocker, R., Jermiin, L. S. and Hunt, N. H. 2007. Characterization of an indoleamine 2,3-dioxygenase-like protein found in humans and mice. *Gene* 396(1), 203–213.

Bateson, P., Barker, D., Clutton-Brock, T., Deb, D., D'Udine, B., Foley, R. A., Gluckman, P., Godfrey, K., Kirkwood, T., Lahr, M. M., McNamara, J., Metcalfe, N. B., Monaghan, P.,

Spencer, H. G. and Sultan, S. E. 2004. Developmental plasticity and human health. *Nature* 430(6998), 419-421.

Bechard, A. R., Bliznyuk, N. and Lewis, M. H. 2017. The development of repetitive motor behaviors in deer mice: effects of environmental enrichment, repeated testing, and differential mediation by indirect basal ganglia pathway activation. *Developmental Psychobiology* 59(3), 390-399.

Belujon, P. and Grace, A. A. 2015. Regulation of dopamine system responsivity and its adaptive and pathological response to stress. *Proceedings of the Royal Societe B* 282, 20142516.

Benda, I. 2008. *Untersuchungen zu den Beziehungen von Federpicken, Exploration und Nahrungsaufnahme bei Legehennen*. Universitaet Hohenheim, Stuttgart, Germany.

Bennett, R. M., Jones, P. J., Nicol, C. J., Tranter, R. B. and Weeks, C. A. 2016. Consumer attitudes to injurious pecking in free-range egg production. *Animal Welfare* 25(1), 91-100.

Bennewitz, J., Boegelein, S., Stratz, P., Rodehutscord, M., Piepho, H. P., Kjaer, J. B. and Bessei, W. 2014. Genetic parameters for feather pecking and aggressive behavior in a large F-2-cross of laying hens using generalized linear mixed models. *Poultry Science* 93(4), 810-817.

Bestman, M., Koene, P. and Wagenaar, J.-P. 2009. Influence of farm factors on the occurrence of feather pecking in organic reared hens and their predictability for feather pecking in the laying period. *Applied Animal Behaviour Science* 121(2), 120-125.

Bestman, M., Verwer, C., Brenninkmeyer, C., Willett, A., Hinrichsen, L. K., Smajlhodzic, F., Heerkens, J., Gunnarsson, S. and Ferrante, V. 2017. Feather-pecking and injurious pecking in organic laying hens in 107 flocks from eight European countries. *Animal Welfare* 26(3), 355-363.

Bestman, M. W. P. and Wagenaar, J. P. 2003. Farm level factors associated with feather pecking in organic laying hens. *Livestock Production Science* 80(1-2), 133-140.

Bestman, M. and Wagenaar, J. P. 2014. Health and welfare in Dutch organic laying hens. *Animals* 4(2), 374-390.

Beuving, G. and Blokhuis, H. J. 1997. Effect of novelty and restraint on catecholamines in plasma of laying hens. *British Poultry Science* 38(3), 297-300.

Bharwani, A., Mian, M. F., Foster, J. A., Surette, M. G., Bienenstock, J. and Forsythe, P. 2016. Structural & functional consequences of chronic psychosocial stress on the microbiome & host. *Psychoneuroendocrinology* 63, 217-227.

Bilčík, B. and Keeling, L. J. 1999. Changes in feather condition in relation to feather pecking and aggressive behaviour in laying hens. *British Poultry Science* 40(4), 444-451.

Bilčík, B. and Keeling, L. J. 2000. Relationship between feather pecking and ground pecking in laying hens and the effect of group size. *Applied Animal Behaviour Science* 68(1), 55-66.

Birkl, P., Bharwani, A., Kjaer, J. B., Kunze, W., McBride, P., Forsythe, P. and Harlander-Matauschek, A. 2018. Differences in cecal microbiome of selected high and low feather-pecking laying hens. *Poultry Science* 97(9), 3009-3014.

Birkl, P., Chow, J., Forsythe, P., Gostner, J. M., Kjaer, J. B., Kunze, W. A., McBride, P., Fuchs, D. and Harlander-Matauschek, A. 2019a. The role of tryptophan-kynurenine in feather pecking in domestic chicken lines. *Frontiers in Veterinary Science* 6, 209.

Birkl, P., Chow, J., McBride, P., Kjaer, J. B., Kunze, W., Forsythe, P. and Harlander-Matauschek, A. 2019b. Effects of acute tryptophan depletion on repetitive behavior in laying hens. *Frontiers in Veterinary Science* 6, 230.

Birkl, P., Franke, L., Bas Rodenburg, T., Ellen, E. and Harlander-Matauschek, A. 2017a. A role for plasma aromatic amino acids in injurious pecking behavior in laying hens. *Physiology and Behavior* 175, 88–96.

Birkl, P., Kjaer, J. B., Szkotnicki, W., Forsythe, P. and Harlander-Matauschek, A. 2017b. Acute tryptophan depletion : the first method validation in an avian species (Gallus gallus domesticus). *Poultry Science* 96(9), 3021–3025.

Biscarini, F., Bovenhuis, H., van der Poel, J., Rodenburg, T. B., Jungerius, A. P. and van Arendonk, J. A. M. 2010. Across-line SNP association study for direct and associative effects on feather damage in laying hens. *Behavior Genetics* 40(5), 715–727.

Bizeray, D., Estevez, I., Leterrier, C. and Faure, J. M. 2002. Effects of increasing environmental complexity on the physical activity of broiler chickens. *Applied Animal Behaviour Science* 79(1), 27–41.

Blokhuis, H. J. J. 1986. Feather-pecking in poultry: its relation with ground-pecking. *Applied Animal Behaviour Science* 16(1), 63–67.

Blokhuis, H. J. J. and Arkes, J. G. G. 1984. Some observations on the development of feather-pecking in poultry. *Applied Animal Behaviour Science* 12(1–2), 145–157.

Blokhuis, H. J. and Vanderhaar, J. W. 1992. Effects of pecking incentives during rearing on feather pecking of laying hens. *British Food Journal* 33, 17–24.

Blokhuis, H. J., van Niekerk, T. F., Bessei, W., Elson, A., Guémené, D., Kjaer, J. B., Maria Levrino, G. A., Nicol, C. J., Tauson, R., Weeks, C. A. and van De Weerd, H. A. 2007. The LayWel project: welfare implications of changes in production systems for laying hens. *World's Poultry Science Journal* 63(1), 101–114.

Bögelein, S., Bessei, W. and Harlander-Matauschek, A. 2009. Do laying hens differ in their relative preferences for feathers of different genetic lines? In: Proceedings of the 2009 Joint East and West Central Europe ISAE Regional Meeting, p. 24. Universität für Bodenkultur, Vienna, Austria.

Brantsæter, M., Nordgreen, J., Rodenburg, T. B., Tahamtani, F. M., Popova, A. and Janczak, A. M. 2016. Exposure to increased environmental complexity during rearing reduces fearfulness and increases use of three-dimensional space in laying hens (Gallus gallus domesticus). *Frontiers in Veterinary Science* 3, 14.

Bright, A. 2007. Plumage colour and feather pecking in laying hens, a chicken perspective? *British Poultry Science* 48(3), 253–263.

Bright, A., Jones, T. A. and Dawkins, M. S. 2006. A non-intrusive method of assessing plumage condition in commercial flocks of laying hens. *Animal Welfare* 15, 113–118.

Broom, D. M. 1981. *Biology of Behaviour*. Cambridge University Press, Cambridge, UK.

Brunberg, E., Jensen, P., Isaksson, A. and Keeling, L. 2011. Feather pecking behavior in laying hens: hypothalamic gene expression in birds performing and receiving pecks. *Poultry Science* 90(6), 1145–1152.

Brunberg, E. I., Rodenburg, T. B., Rydhmer, L., Kjaer, J. B., Jensen, P. and Keeling, L. J. 2016. Omnivores going astray: a review and new synthesis of abnormal behavior in pigs and laying hens. *Frontiers in Veterinary Science* 3, 57.

Buitenhuis, A. J., Kjaer, J. B., Labouriau, R. and Juul-Madsen, H. R. 2006a. Altered circulating levels of serotonin and immunological changes in laying hens divergently selected for feather pecking behavior. *Poultry Science* 85(10), 1722–1728.

Buitenhuis, A. J., Rodenburg, T. B., van Hierden, Y. M., Siwek, M., Cornelissen, S. J. B., Nieuwland, M. G. B., Crooijmans, R. P., Groenen, M. A. M., Koene, P., Korte, S. M., Bovenhuis, H. and van der Poel, J. J. 2006b. Mapping quantitative trait loci affecting feather pecking behavior and stress response in laying hens. *Poultry Science* 85, 1215–1222.

Buitenhuis, A. J., Rodenburg, T. B., Siwek, M., Cornelissen, S. J. B., Nieuwland, M. G. B., Crooijmans, R. P., Groenen, M. A. M., Koene, P., Bovenhuis, H. and van der Poel, J. J. 2003. Identification of quantitative trait loci for receiving pecks in young and adult laying hens. *Poultry Science* 82(11), 1661–1667.

Buitenhuis, A. J., Rodenburg, T. B., Wissink, P. H., Visscher, J., Koene, P., Bovenhuis, H., Ducro, B. J. and van der Poel, J. J. 2004. Genetic and phenotypic correlations between feather pecking behavior, stress response, immune response, and egg quality traits in laying hens. *Poultry Science* 83(7), 1077–1082.

Cabib, S. and Puglisi-Allegra, S. 2012. The mesoaccumbens dopamine in coping with stress. *Neuroscience and Biobehavioral Reviews* 36(1), 79–89.

Campbell, D. L. M., de Haas, E. N. and Lee, C. 2019. A review of environmental enrichment for laying hens during rearing in relation to their behavioral and physiological development. *Poultry Science* 98(1), 9–28.

Cheng, H. W., Dillworth, G., Singleton, P., Chen, Y. and Muir, W. M. 2001a. Effects of group selection for productivity and longevity on blood concentrations of serotonin, catecholamines, and corticosterone of laying hens. *Poultry Science* 80(9), 1278–1285.

Cheng, H. W., Eicher, S. D., Chen, Y., Singleton, P. and Muir, W. M. 2001b. Effect of genetic selection for group productivity and longevity on immunological and hematological parameters of chickens. *Poultry Science* 80(8), 1079–1086.

Cheng, H. W. and Fahey, A. 2009. Effects of group size and repeated social disruption on the serotonergic and dopaminergic systems in two genetic lines of White Leghorn laying hens. *Poultry Science* 88(10), 2018–2025.

Cheng, H. W. and Muir, W. M. 2004. Chronic social stress differentially regulates neuroendocrine responses in laying hens: II. Genetic basis of adrenal responses under three different social conditions. *Psychoneuroendocrinology* 29(7), 961–971.

Cheng, H. W. and Muir, W. M. 2007. Mechanisms of aggression and production in chickens: genetic variations in the functions of serotonin, catecholamine, and corticosterone. *World's Poultry Science Journal* 63(2), 233–254.

Cheng, H. W. W., Singleton, P. and Muir, W. M. M. 2003. Social stress differentially regulates neuroendocrine responses in laying hens: I. Genetic basis of doparnine responses under three different social conditions. *Psychoneuroendocrinology* 28(5), 597–611.

Cloutier, S., Newberry, R. C., Forster, C. T. and Girsberger, K. M. 2000. Does pecking at inanimate stimuli predict cannibalistic behaviour in domestic fowl? *Applied Animal Behaviour Science* 66(1–2), 119–133.

Collias, N. and Joos, M. 1953. The spectrographic analysis of sound signals of the domestic fowl. *Behaviour* 5(1), 175–188.

Comin, A., Jeremiasson, A., Kratzer, G. and Keeling, L. 2019. Revealing the structure of the associations between housing system, facilities, management and welfare of commercial laying hens using Additive Bayesian Networks. *Preventive Veterinary Medicine* 164, 23–32.

Cook, N. J., Smykot, A. B., Holm, D. E., Fasenko, G. and Church, J. S. 2006. Assessing feather cover of laying hens by infrared thermography. *The Journal of Applied Poultry Research* 15(2), 274–279.

Coton, J., Guinebretiere, M., Guesdon, V., Chiron, G., Mindus, C., Laravoire, A., Pauthier, G., Balaine, L., Descamps, M., Bignon, L., Huneau-Salaun, A. and Michel, V. 2019.

Feather pecking in laying hens housed in free-range or furnished-cage systems on French farms. *British Poultry Science* 60(6), 617–627.

Craig, J. V. and Muir, W. M. 1996. Group selection for adaptation to multiple-hen cages: beak-related mortality, feathering, and body weight responses. *Poultry Science* 75(3), 294–302.

Cuthbertson, G. J. 1978. The development of feather pecking and cannibalism in chicks. *Applied Animal Ethology* 4(3), 292.

Daigle, C. L., Rodenburg, T. B., Bolhuis, J. E., Swanson, J. C. and Siegford, J. M. 2015. Individual consistency of feather pecking behavior in laying hens: once a feather pecker always a feather pecker? *Frontiers in Veterinary Science* 2, 6.

Dalton, H. A., Wood, B. J. and Torrey, S. 2013. Injurious pecking in domestic turkeys: development, causes, and potential solutions. *World's Poultry Science Journal* 69(4), 865–876.

David, B., Mejdell, C., Michel, V., Lund, V. and Moe, R. O. 2015. Air quality in alternative housing systems may have an impact on laying hen welfare. Part II - ammonia. *Animals* 5(3), 886–896.

Dawkins, M. S. 1989. Time budgets in Red Junglefowl as a baseline for the assessment of welfare in domestic fowl. *Applied Animal Behaviour Science* 24(1), 77–80.

D'Eath, R. B., Keeling, L. J. 2003. Social discrimination and aggression by laying hens in large groups: from peck orders to social tolerance. *Applied Animal Behaviour Science* 84(3), 197–212.

Decina, C., Berke, O., van Staaveren, N., Baes, C. F., Widowski, T. M. and Harlander-Matauschek, A. 2019a. A cross-sectional study on feather cover damage in Canadian laying hens in non-cage housing systems. *BMC Veterinary Research* 15, 435.

Decina, C., Berke, O., van Staaveren, N., Baes, C. F., Widowski, T. M. and Harlander-Matauschek, A. 2019b. An investigation of associations between management and feather damage in Canadian laying hens housed in furnished cages. *Animals* 9(4), 135.

Decina, C., Berke, O., van Staaveren, N., Baes, C. F. and Harlander-Matauscheck, A. 2019c. Development of a scoring system to assess feather damage in Canadian laying hen flocks. *Animals* 9(7), 436.

de Haas, E. N., Bolhuis, J. E., de Jong, I. C., Kemp, B., Janczak, A. M. and Rodenburg, T. B. 2014a. Predicting feather damage in laying hens during the laying period. Is it the past or is it the present? *Applied Animal Behaviour Science* 160, 75–85.

de Haas, E. N., Bolhuis, J. E., Kemp, B., Groothuis, T. G. G. and Rodenburg, T. B. 2014b. Parents and early life environment affect behavioral development of laying hen chickens. *PLoS ONE* 9(6), e90577.

de Haas, E. N. and van der Eijk, J. A. J. 2018. Where in the serotonergic system does it go wrong? Unravelling the route by which the serotonergic system affects feather pecking in chickens. *Neuroscience and Biobehavioral Reviews* 95, 170–188.

de Haas, E. N., Kemp, B., Bolhuis, J. E., Groothuis, T. and Rodenburg, T. B. 2013. Fear, stress, and feather pecking in commercial white and brown laying hen parent-stock flocks and their relationships with production parameters. *Poultry Science* 92(9), 2259–2269.

De Jong, I. C. and Guemene, D. 2011. Major welfare issues in broiler breeders. *World's Poultry Science Journal* 67(1), 73–82.

Dennis, R. L. and Cheng, H. W. 2011. The dopaminergic system and aggression in laying hens. *Poultry Science* 90(11), 2440–2448.

Dixon, L. M. 2008. Feather pecking behaviour and associated welfare issues in laying hens. *Avian Biology Research* 1(2), 73–87.

Dixon, L. M. and Duncan, I. J. H. 2010. Changes in substrate access did not affect early feather-pecking behavior in two strains of laying hen chicks. *Journal of Applied Animal Welfare Science* 13(1), 1–14.

Dixon, L. M. M., Duncan, I. J. H. and Mason, G. 2008. What's in a peck? Using fixed action pattern morphology to identify the motivational basis of abnormal feather-pecking behaviour. *Animal Behaviour* 76(3), 1035–1042.

Dixon, L. M., Duncan, I. J. H. and Mason, G. J. 2010. The effects of four types of enrichment on feather-pecking behaviour in laying hens housed in barren environments. *Animal Welfare* 19, 429–435.

Dixon, G., Green, L. E. and Nicol, C. J. 2006. Effect of diet change on the behavior of chicks of an egg-laying strain. *Journal of Applied Animal Welfare Science* 9(1), 41–58.

Dixon, G. and Nicol, C. J. 2008. The effect of diet change on the behaviour of layer pullets. *Animal Welfare* 17, 101–109.

Donham, K. J., Cumro, D., Reynolds, S. J. and Merchant, J. A. 2000. Dose-response relationships between occupational aerosol exposures and cross-shift declines of lung function in poultry workers: recommendations for exposure limits. *Journal of Occupational and Environmental Medicine* 42(3), 260–269.

Dorland, W. A. N. 2007. *Dorland's Illustrated Medical Dictionary*. Saunders, Philadelphia, PA.

Dorman, C. and Gaudiano, P. 1995. Motivation. In: *Handbook of Brain Theory and Neural Networks*, Arbib, M. A. (Ed.), pp. 591–594. MIT Press, Cambridge, MA.

Drake, K. A., Donnelly, C. A. and Dawkins, M. S. 2010. Influence of rearing and lay risk factors on propensity for feather damage in laying hens. *British Poultry Science* 51(6), 725–733.

Duncan, I. J. H. 1980. The ethogram of the domesticated hen. In: *The Laying Hen and Its Environment*, Moss, R. (Ed.), pp. 5–16. Martinus Nijhoff Publishers, The Hague, The Netherlands.

Duncan, I. J. H. 2001. Animal welfare issues in the poultry industry: Is there a lesson to be learned? *Journal of Applied Animal Welfare Science* 4(3), 207–221.

Edgar, J., Held, S., Jones, C. and Troisi, C. 2016. Influences of maternal care on chicken welfare. *Animals* 6(1), 2.

EFSA 2005. Opinion of the Scientific Panel on Animal Health and Welfare on a request from the Commission related to the welfare aspects of various systems of keeping laying hens. *The EFSA Journal* 197, 1–23.

EFSA 2012. Statement on the use of animal-based measures to assess the welfare of animals. *The EFSA Journal* 10, 2767.

Elkhoraibi, C., Blatchford, R. A., Pitesky, M. E. and Mench, J. A. 2014. Backyard chickens in the United States: a survey of flock owners. *Poultry Science* 93(11), 2920–2931.

El-Lethey, H., Aerni, V., Jungi, T. W. and Wechsler, B. 2000. Stress and feather pecking in laying hens in relation to housing conditions. *British Poultry Science* 41(1), 22–28.

El-Lethey, H., Jungi, T. W.. and Huber-Eicher, B. 2001. Effects of feeding corticosterone and housing conditions on feather pecking in laying hens (Gallus gallus domesticus). *Physiology and Behavior* 73(1–2), 243–251.

Ellen, E. D., van der Sluis, M., Siegford, J., Guzhva, O., Toscano, M. J., Bennewitz, J., van der Zande, L. E., van der Eijk, J. A. J., de Haas, E. N., Norton, T., Piette, D., Tetens, J., de Klerk, B., Visser, B. and Rodenburg, T. B. 2019. Review of sensor technologies in animal breeding: phenotyping behaviors of laying hens to select Against feather pecking. *Animals* 9(3), 108.

Erasmus, M. A. 2018. 13 - Welfare issues in turkey production. In: *Advances in Poultry Welfare*, Mench, J. A. (Ed.), *Food Science, Technology and Nutrition*, pp. 263-291. Woodhead Publishing, Duxford, UK.

Estevez, I., Keeling, L. J. and Newberry, R. C. 2003. Decreasing aggression with increasing group size in young domestic fowl. *Applied Animal Behaviour Science* 84(3), 213-218.

European Commission 1999. Council Directive 1999/74/EC of 19 July 1999 laying down minimum standards for the protection of laying hens. *Official Journal of the European Communities* L203, 53-57.

Field, S. E., Rickard, N. S., Toukhsati, S. R. and Gibbs, M. E. 2007. Maternal hen calls modulate memory formation in the day-old chick: the role of noradrenaline. *Neurobiology of Learning and Memory* 88(3), 321-330.

Flisikowski, K., Schwarzenbacher, H., Wysocki, M., Weigend, S., Preisinger, R., Kjaer, J. B. and Fries, R. 2009. Variation in neighbouring genes of the dopaminergic and serotonergic systems affects feather pecking behaviour of laying hens. *Animal Genetics* 40(2), 192-199.

Fox, L. R. 1975. Cannibalism in natural populations. *Annual Review of Ecology and Systematics* 6(1), 87-106.

Gilani, A.-M., Knowles, T. G. and Nicol, C. J. 2012. The effect of dark brooders on feather pecking on commercial farms. *Applied Animal Behaviour Science* 142(1-2), 42-50.

Gilani, A.-M., Knowles, T. G. and Nicol, C. J. 2013. The effect of rearing environment on feather pecking in young and adult laying hens. *Applied Animal Behaviour Science* 148(1-2), 54-63.

Glatz, P. C. 2001. Effect of poor feather cover on feed intake and production of aged laying hens. *Asian-Australasian Journal of Animal Sciences* 14(4), 553-558.

Glatz, P. C. and Bourke, M. 2006. *Beak Trimming Handbook for Egg Producers: Best Practices for Minimising Cannibalism in Poultry*. Landlinks Press, Collingwood, VIC, Australia.

Grafl, B., Polster, S., Sulejmanovic, T., Pürrer, B., Guggenberger, B. and Hess, M. 2017. Assessment of health and welfare of Austrian laying hens at slaughter demonstrates influence of husbandry system and season. *British Poultry Science* 58(3), 209-215.

Grams, V., Boegelein, S., Grashorn, M. A., Bessei, W. and Bennewitz, J. 2015. Quantitative genetic analysis of traits related to fear and feather pecking in laying hens. *Behavior Genetics* 45(2), 228-235.

Green, L. E., Lewis, K., Kimpton, A. and Nicol, C. J. 2000. Cross-sectional study of the prevalence of feather pecking in laying hens in alternative systems and its associations with management and disease. *Veterinary Record* 147(9), 233-238.

Gruss, M. and Braun, K. 1996. Distinct activation of monoaminergic pathways in chick brain in relation to auditory imprinting and stressful situations: a microdialysis study. *Neuroscience* 76(3), 891-899.

Guilliams, T. G. and Edwards, L. 2010. Chronic stress and the HPA axis. *The Standard* 9, 1-2.

Gunnarsson, S., Keeling, L. J. and Svedberg, J. 1999. Effect of rearing factors on the prevalence of floor eggs, cloacal cannibalism and feather pecking in commercial flocks of loose housed laying hens. *British Poultry Science* 40(1), 12-18.

Güntürkün, O. 2005. Avian and mammalian 'prefrontal cortices': limited degrees of freedom in the evolution of the neural mechanisms of goal-state maintenance. *Brain Research Bulletin* 66(4-6), 311-316.

Gustafson, L. A., Cheng, H. W., Garner, J. P., Pajor, E. A. and Mench, J. A. 2007. The effects of different bill-trimming methods on the well-being of Pekin ducks. *Poultry Science* 86(9), 1831-1839.

Harlander-Matauschek, A., Baes, C. and Bessei, W. 2006a. The demand of laying hens for feathers and wood shavings. *Applied Animal Behaviour Science* 101(1-2), 102-110.

Harlander-Matauschek, A., Beck, P. and Piepho, H.-P. 2009. Taste aversion learning to eliminate feather pecking in laying hens, Gallus gallus domesticus. *Animal Behaviour* 78(2), 485-490.

Harlander-Matauschek, A., Benda, I., Lavetti, C., Djukic, M. and Bessei, W. 2007. The relative preferences for wood shavings or feathers in high and low feather pecking birds. *Applied Animal Behaviour Science* 107(1-2), 78-87.

Harlander-Matauschek, A. and Bessei, W. 2005. Feather eating and crop filling in laying hens. *Archiv für Geflügelkunde* 69, 241-244.

Harlander-Matauschek, A. and Feise, U. 2009. Physical characteristics of feathers play a role in feather eating behavior. *Poultry Science* 88(9), 1800-1804.

Harlander-Matauschek, A. and Häusler, K. 2009. Understanding feather eating behaviour in laying hens. *Applied Animal Behaviour Science* 117(1-2), 35-41.

Harlander-Matauschek, A., Piepho, H. P. and Bessei, W. 2006b. The effect of feather eating on feed passage in laying hens. *Poultry Science* 85(1), 21-25.

Harlander-Matauschek, A. and Rodenburg, T. B. 2011. Applying chemical stimuli on feathers to reduce feather pecking in laying hens. *Applied Animal Behaviour Science* 132(3-4), 146-151.

Hassel, S., McKinnon, M. C., Cusi, A. M. and MacQueen, G. M. 2011. An overview of psychological and neurobiological mechanisms by which early negative experiences increase risk of mood disorders. *Journal of the Canadian Academy of Child and Adolescent Psychiatry* 20(4), 277-288.

Hedlund, L., Whittle, R. and Jensen, P. 2019. Effects of commercial hatchery processing on short- and long-term stress responses in laying hens. *Scientific Reports* 9(1), 2367.

Heerkens, J. L. T., Delezie, E., Kempen, I., Zoons, J., Ampe, B., Rodenburg, T. B. and Tuyttens, F. A. M. 2015. Specific characteristics of the aviary housing system affect plumage condition, mortality and production in laying hens. *Poultry Science* 94(9), 2008-2017.

Heinsius, J., van Staaveren, N., Kwon, I., Li, A., Kjaer, J. and Harlander-Matauschek, A. 2020. Chickens selected for feather pecking can inhibit prepotent motor responses in a Go/No-Go task. *Scientific Reports*, 10, 6485. doi: 10.1038/s41598-020-63618-z.

Hetland, H., Svihus, B. and Krogdahl, A. 2003. Effects of oat hulls and wood shavings on digestion in broilers and layers fed diets based on whole or ground wheat. *British Poultry Science* 44(2), 275-282.

Hewlett, S. E., Zeinstra, E. C., van Eerdenburg, F. J. C. M., Rodenburg, T. B., van Kooten, P. J. S., van der Staay, F. J. and Nordquist, R. E. 2014. Hypothalamic vasotocin and tyrosine hydroxylase levels following maternal care and selection for low mortality in laying hens. *BMC Veterinary Research* 10, 167.

Hoffmeyer, I. 1969. Feather-pecking in pheasants an ethological approach to the problem. *Danish Review of Game Biology* 6, 1-36.

Huber-Eicher, B. 1999. A survey of layer-type pullet rearing in Switzerland. *World's Poultry Science Journal* 55(1), 83-91.

Huber-Eicher, B. and Audigé, L. 1999. Analysis of risk factors for the occurrence of feather pecking in laying hen growers. *British Poultry Science* 40(5), 599-604.

Huber-Eicher, B. and Sebö, F. 2001a. The prevalence of feather pecking and development in commercial flocks of laying hens. *Applied Animal Behaviour Science* 74(3), 223-231.

Huber-Eicher, B. and Sebö, F. 2001b. Reducing feather pecking when raising laying hen chicks in aviary systems. *Applied Animal Behaviour Science* 73(1), 59-68.

Huber-Eicher, B. and Wechsler, B. 1997. Feather pecking in domestic chicks: its relation to dustbathing and foraging. *Animal Behaviour* 54(4), 757-768.

Hughes, B. O. and Duncan, I. J. H. 1972. The Influence of strain and environmental factors upon feather pecking and cannibalism in fowls. *British Poultry Science* 13(6), 525-547.

Hughes, B. O. and Gentle, M. J. 1995. Beak trimming of poultry - its implications for welfare. *World's Poultry Science Journal* 51(1), 51-61.

IEC 2015. *Egg Industry Review 2015*. International Egg Commission, Church Stretton, UK.

Iffland, H., Wellmann, R., Preuss, S., Tetens, J., Bessei, W., Piepho, H.-P. and Bennewitz, J. 2020. A novel model to explain extreme feather pecking behavior in laying hens. *Behavior Genetics,* 50, 41-50.

Ishiwata, H., Shiga, T. and Okado, N. 2005. Selective serotonin reuptake inhibitor treatment of early postnatal mice reverses their prenatal stress-induced brain dysfunction. *Neuroscience* 133(4), 893-901.

Janczak, A. M. and Riber, A. B. 2015. Review of rearing-related factors affecting the welfare of laying hens. *Poultry Science* 94(7), 1454-1469.

Jensen, A. B., Palme, R. and Forkman, B. 2006. Effect of brooders on feather pecking and cannibalism in domestic fowl (Gallus gallus domesticus). *Applied Animal Behaviour Science* 99(3-4), 287-300.

Johnsen, P. F., Vestergaard, K. S. and Nørgaard-Nielsen, G. 1998. Influence of early rearing conditions on the development of feather pecking and cannibalism in domestic fowl. *Applied Animal Behaviour Science* 60(1), 25-41.

Jones, R. B. 2004. Environmental enrichment: the need for practical strategies to improve poultry welfare. In: *Welfare of the Laying Hen*, Perry, G. (Ed.), *Poultry Science Symposium Series*, pp. 215-225.World's Poultry Science Association (UK Branch), Bristol, UK.

Jones, M. A., Mason, G. J. and Pillay, N. 2011a. Correlates of birth origin effects on the development of stereotypic behaviour in striped mice, Rhabdomys. *Animal Behaviour* 82(1), 149-159.

Jones, M. A., Mason, G. and Pillay, N. 2011b. Early environmental enrichment protects captive-born striped mice against the later development of stereotypic behaviour. *Applied Animal Behaviour Science* 135(1-2), 138-145.

Jones, R. B., McAdie, T. M., McCorquodale, C. and Keeling, L. J. 2002. Pecking at other birds and at string enrichment devices by adult laying hens. *British Poultry Science* 43(3), 337-343.

Jung, L. and Knierim, U. 2018. Are practice recommendations for the prevention of feather pecking in laying hens in non-cage systems in line with the results of experimental and epidemiological studies? *Applied Animal Behaviour Science* 200, 1-12.

Kaukonen, E. and Valros, A. 2019. Feather pecking and cannibalism in non-beak-trimmed laying hen flocks–farmers' perspectives. *Animals* 9(2)., 43.

Keeling, L. 2002. Behaviour of fowl and other domesticated birds. In: *The Ethology of Domestic Animals: An Introductory Text*, Jensen, P. (Ed.), pp. 107–108. CABI Publishing, Oxon, UK.

Keeling, L., Anderson, L., Schutz, K. E., Kerje, S., Fredriksson, R., Carlborg, O., Cornwallis, C. K., Pizzari, T. and Jensen, P. 2004. Chicken genomics: feather-pecking and victim pigmentation. *Nature* 431(7009), 645–646.

Kessler, R. C., Chiu, W. T., Demler, O., Merikangas, K. R. and Walters, E. E. 2005. Prevalence, severity, and comorbidity of 12-month DSM-IV disorders in the National comorbidity Survey Replication. *Archives of General Psychiatry* 62(6), 617–627.

Kim, S.-J., Lewis, M. and Veenstra-VanderWeele, J. 2013. Chapter 38. The developmental neurobiology of repetitive behavior. In: *Neural Circuit Development and Function in the Brain*, Rubenstein, J. L. R. and Rakic, P. (Eds), pp. 761–782. Academic Press, Oxford.

Kjaer, J. B. 2000. Diurnal rhythm of feather pecking behaviour and condition of integument in four strains of loose housed laying hens. *Applied Animal Behaviour Science* 65(4), 331–347.

Kjaer, J. B. 2009. Feather pecking in domestic fowl is genetically related to locomotor activity levels: implications for a hyperactivity disorder model of feather pecking. *Behavior Genetics* 39(5), 564–570.

Kjaer, J. B., Glawatz, H., Scholz, B., Rettenbacher, S. and Tauson, R. 2011. Reducing stress during welfare inspection: validation of a non-intrusive version of the LayWel plumage scoring system for laying hens. *British Poultry Science* 52(2), 149–154.

Kjaer, J. B. and Guémené, D. 2009. Adrenal reactivity in lines of domestic fowl selected on feather pecking behavior. *Physiology and Behavior* 96(2), 370–373.

Kjaer, J. B., Hjarvard, B. M., Jensen, K. H., Hansen-Møller, J. and Naesbye Larsen, O. 2004. Effects of haloperidol, a dopamine D2 receptor antagonist, on feather pecking behaviour in laying hens. *Applied Animal Behaviour Science* 86(1-2), 77–91.

Kjaer, J. B. and Hocking, P. M. 2004. The genetics of feather pecking and cannibalism. In: *Welfare of the Laying Hen*, Perry, G. C. (Ed.), *Poultry Science Symposium Series*, pp. 109–123.World's Poultry Science Association (UK Branch), Bristol, UK.

Kjaer, J. B. and Jørgensen, H. 2011. Heart rate variability in domestic chicken lines genetically selected on feather pecking behavior. *Genes, Brain and Behavior* 10(7), 747–755.

Kjaer, J. B. and Sorensen, P. 1997. Feather pecking behaviour in White Leghorns, a genetic study. *British Poultry Science* 38(4), 333–341.

Kjaer, J. B., Sørensen, P. and Su, G. 2001. Divergent selection on feather pecking behaviour in laying hens (Gallus gallus domesticus). *Applied Animal Behaviour Science* 71(3), 229–239.

Kjaer, J. B. and Vestergaard, K. S. 1999. Development of feather pecking in relation to light intensity. *Applied Animal Behaviour Science* 62(2–3), 243–254.

Kjaer, J. B., Würbel, H. and Schrader, L. 2015. Perseveration in a guessing task by laying hens selected for high or low levels of feather pecking does not support classification of feather pecking as a stereotypy. *Applied Animal Behaviour Science* 168, 56–60.

Koolhaas, J. M., Bartolomucci, A., Buwalda, B., de Boer, S. F., Flügge, G., Korte, S. M., Meerlo, P., Murison, R., Olivier, B., Palanza, P., Richter-Levin, G., Sgoifo, A., Steimer, T., Stiedl, O., van Dijk, G., Wöhr, M. and Fuchs, E. 2011. Stress revisited: A critical evaluation of the stress concept. *Neuroscience and Biobehavioral Reviews* 35(5), 1291–1301.

Kops, M. S., de Haas, E. N., Rodenburg, T. B., Ellen, E. D., Korte-Bouws, G. A. H., Olivier, B., Güntürkün, O., Bolhuis, J. E. and Korte, S. M. 2013a. Effects of feather pecking phenotype (severe feather peckers, victims and non-peckers) on serotonergic and dopaminergic activity in four brain areas of laying hens (Gallus gallus domesticus). *Physiology and Behavior* 120, 77–82.

Kops, M. S., de Haas, E. N., Rodenburg, T. B., Ellen, E. D., Korte-Bouws, G. A. H., Olivier, B., Güntürkün, O., Korte, S. M. and Bolhuis, J. E. 2013b. Selection for low mortality in laying hens affects catecholamine levels in the arcopallium, a brain area involved in fear and motor regulation. *Behavioural Brain Research* 257, 54–61.

Kops, M. S., Kjaer, J. B., Güntürkün, O., Westphal, K. G. C., Korte-Bouws, G. A. H., Olivier, B., Korte, S. M. and Bolhuis, J. E. 2017. Brain monoamine levels and behaviour of young and adult chickens genetically selected on feather pecking. *Behavioural Brain Research* 327, 11–20.

Korte, S. M., Beuving, G., Ruesink, W. and Blokhuis, H. J. 1997. Plasma catecholamine and corticosterone levels During manual restraint in chicks from a high and low feather pecking line of laying hens. *Physiology and Behavior* 62(3), 437–441.

Korte, S. M. M., Ruesink, W. and Blokhuis, H. J. 1998. Heart rate variability during manual restraint in chicks from high- and low-feather pecking lines of laying hens. *Physiology and Behavior* 65(4–5), 649–652.

Krause, E. T., Petow, S. and Kjaer, J. B. 2011. A note on the physiological and behavioural consequences of cannibalistic toe pecking in laying hens (Gallus gallus domesticus). *Archiv für Geflügelkunde* 75, 140–143.

Kriegseis, I., Bessei, W., Meyer, B., Zentek, J., Wuerbel, H. and Harlander-Matauschek, A. 2012. Feather-pecking response of laying hens to feather and cellulose-based rations fed during rearing. *Poultry Science* 91(7), 1514–1521.

Kruijt, J. P. 1964. Ontogeny of social behaviour in burmese red junglefowl (*Gallus gallus spadiceus*) Bonnaterre. *Behaviour Supplement* no. 12, I-201. Retrieved August 4, 2020, from www.jstor.org/stable/30039152.

Kuenzel, W. J. 2007. Neurobiological basis of sensory perception: welfare implications of beak trimming. *Poultry Science* 86(6), 1273–1282.

Labouriau, R., Kjaer, J. B., Abreu, G. C. G., Hedegaard, J. and Buitenhuis, A. J. 2009. Analysis of severe feather pecking behavior in a high feather pecking selection line. *Poultry Science* 88(10), 2052–2062.

Lambton, S. L., Knowles, T. G., Yorke, C. and Nicol, C. J. 2010. The risk factors affecting the development of gentle and severe feather pecking in loose housed laying hens. *Applied Animal Behaviour Science* 123(1–2), 32–42.

Lambton, S. L., Knowles, T. G., Yorke, C. and Nicol, C. J. 2015. The risk factors affecting the development of vent pecking and cannibalism in free-range and organic laying hens. *Animal Welfare* 24(1), 101–111.

Lambton, S. L., Nicol, C. J., Friel, M., Main, D. C. J., McKinstry, J. L., Sherwin, C. M., Walton, J. and Weeks, C. A. 2013. A bespoke management package can reduce levels of injurious pecking in loose-housed laying hen flocks. *Veterinary Record* 172(16), 423–423.

Langen, M., Kas, M. J. H., Staal, W. G., Van Engeland, H. and Durston, S. 2011. The neurobiology of repetitive behavior: of mice.... *Neuroscience and Biobehavioral Reviews* 35(3), 345–355.

LeBlanc, S., Tobalske, B., Quinton, M., Springthorpe, D., Szkotnicki, B., Wuerbel, H. and Harlander-Matauschek, A. 2016. Physical health problems and environmental challenges influence balancing behaviour in laying hens. *PLoS ONE* 11(4), e0153477.
Leeson, S. and Morrison, W. D. 1978. Effect of feather cover on feed efficiency in laying birds. *Poultry Science* 57(4), 1094–1096.
Leonard, M. L., Horn, A. G. and Fairfull, R. W. 1995. Correlates and consequences of allopecking in White Leghorn chickens. *Applied Animal Behaviour Science* 43(1), 17–26.
Lewis, M. H. 2004. Environmental complexity and central nervous system development and function. *Mental Retardation and Developmental Disabilities Research Reviews* 10(2), 91–95.
Lewis, M. H. and Kim, S. J. 2009. The pathophysiology of restricted repetitive behavior. *Journal of Neurodevelopmental Disorders* 1(2), 114–132.
Lewis, M. H., Tanimura, Y., Lee, L. W. and Bodfish, J. W. 2007. Animal models of restricted repetitive behavior in autism. *Behavioural Brain Research* 176(1), 66–74.
Maier, S. F. and Watkins, L. R. 2010. Role of the medial prefrontal cortex in coping and resilience. *Brain Research* 1355, 52–60.
Mallapur, A., Waran, N. and Sinha, A. 2005. Factors influencing the behaviour and welfare of captive lion-tailed macaques in Indian zoos. *Applied Animal Behaviour Science* 91(3–4), 337–353.
Marsboon, R. and Sierens, G. 1962. Treatment and prophylaxis of cannibalism in poultry with haloanisone-R 2028. *Poultry Science* 41, 776–780.
Martin, J. E. 2002. Early life experiences: activity levels and abnormal behaviours in resocialised chimpanzees. *Animal Welfare* 11, 419–436.
Mason, G. 2006. Are wild-born animals 'protected' from stereotypies? In: *Stereotypic Animal Behaviour: Fundamentals and Applications to Welfare*, Mason, G. and Rushen, J. (Eds). CAB International, Oxford, UK.
Mason, G. and Bateson, M. 2009. Motivation and the organization of behaviour. In: *The Ethology of Domesticated Animals*, Jensen, P. (Ed.), pp. 38–56. CABI Publishing, Wallingford, UK.
Mason, G. and Rushen, J. (Eds.) 2006. *Stereotypic Behaviour in Captive Animals: Fundamentals and Applications to Welfare*. CAB International, Wallingford, UK.
McAdie, T. M. and Keeling, L. J. 2000. Effect of manipulating feathers of laying hens on the incidence of feather pecking and cannibalism. *Applied Animal Behaviour Science* 68(3), 215–229.
McAdie, T. M. and Keeling, L. J. 2002. The social transmission of feather pecking in laying hens: effects of environment and age. *Applied Animal Behaviour Science* 75(2), 147–159.
McAdie, T. M., Keeling, L. J., Blokhuis, H. J. and Jones, R. B. 2005. Reduction in feather pecking and improvement of feather condition with the presentation of a string device to chickens. *Applied Animal Behaviour Science* 93(1–2), 67–80.
McCasland, W. E. and Richardson, L. R. 1966. Methods for determining the nutritive value of feather meals. *Poultry Science* 45(6), 1231–1236.
McEwen, B. S. 2012. Brain on stress: how the social environment gets under the skin. *Proceedings of the National Academy of Sciences of the United States of America* 109(Suppl. 2), 17180–17185.
McKeegan, D. E. F. and Savory, C. J. 1999. Feather eating in layer pullets and its possible role in the aetiology of feather pecking damage. *Applied Animal Behaviour Science* 65(1), 73–85.

McKeegan, D. E.. and Savory, C. J. 2001. Feather eating in individually caged hens which differ in their propensity to feather peck. *Applied Animal Behaviour Science* 73(2), 131–140.

Merikangas, K. R., He, J. P., Burstein, M., Swanson, S. A., Avenevoli, S., Cui, L., Benjet, C., Georgiades, K. and Swendsen, J. 2010. Lifetime prevalence of mental disorders in U.S. Adolescents: results from the National comorbidity survey replication-adolescent supplement (NCS-A). *Journal of the American Academy of Child and Adolescent Psychiatry* 49(10), 980–989.

Meyer, B., Bessei, W., Vahjen, W., Zentek, J. and Harlander-Matauschek, A. 2012. Dietary inclusion of feathers affects intestinal microbiota and microbial metabolites in growing Leghorn-type chickens. *Poultry Science* 91(7), 1506–1513.

Meyer, B., Zentek, J. and Harlander-Matauschek, A. 2013. Differences in intestinal microbial metabolites in laying hens with high and low levels of repetitive feather-pecking behavior. *Physiology and Behavior* 110–111, 96–101.

Miller, K. A. and Mench, J. A. 2006. Differential effects of 4 types of environmental enrichment on aggressive pecking, feather pecking, feather loss, food wastage and productivity in Japanese quail. *British Poultry Science* 47(6), 646–658.

Mills, D. S. 2003. Medical paradigms for the study of problem behaviour: a critical review. *Applied Animal Behaviour Science* 81(3), 265–277.

Miura, H., Ozaki, N., Sawada, M., Isobe, K., Ohta, T. and Nagatsu, T. 2008. A link between stress and depression: shifts in the balance between the kynurenine and serotonin pathways of tryptophan metabolism and the etiology and pathophysiology of depression. *Stress* 11(3), 198–209.

Moe, R. O., Nordgreen, J., Janczak, A. M., Bakken, M., Spruijt, B. M. and Jensen, P. 2014. Anticipatory and foraging behaviors in response to palatable food reward in chickens: effects of dopamine D2 receptor blockade and domestication. *Physiology and Behavior* 133, 170–177.

Moe, R. O., Nordgreen, J., Janczak, A. M., Spruijt, B. M., Kostal, L., Skjerve, E., Zanella, A. J. and Bakken, M. 2011. Effects of haloperidol, a dopamine D2-like receptor antagonist, on reward-related behaviors in laying hens. *Physiology and Behavior* 102(3–4), 400–405.

Moffatt, C. A. and Hogan, J. A. 1992. Ontogeny of chick responses to maternal food calls in the Burmese red junglefowl (Gallus gallus spadiceus). *Journal of Comparative Psychology* 106(1), 92–96.

Mohammed, H. H., Grashorn, M. A. and Bessei, W. 2010. The effects of lighting conditions on the behaviour of laying hens. *Archiv für Geflügelkunde* 74, 197-202.

Muir, W. M. 1996. Group selection for adaptation to multiple-hen cages: selection program and direct responses. *Poultry Science* 75(4), 447–458.

Neumann, M., Bessei, W. and Harlander-Matauschek, A. 2009. Do feather-pecking hens have a preference for different colours or gender of feathers? In: Proceedings of the 2009 Joint East and West Central Europe ISAE Regional Meeting, p. 32. Universität für Bodenkultur, Vienna, Austria.

Newberry, R. C. 1995. Environmental enrichment: increasing the biological relevance of captive environments. *Applied Animal Behaviour Science* 44(2-4), 229-243.

Newberry, R. C. 2004. Cannibalism. In: *Welfare of the Laying Hen*, Perry, G. (Ed.), *Poultry Science Symposium Series*, pp. 239-258.World's Poultry Science Association (UK Branch), Bristol, UK

Newberry, R. C., Keeling, L. J., Estevez, I. and Bilčík, B. 2007. Behaviour when young as a predictor of severe feather pecking in adult laying hens: the redirected foraging hypothesis revisited. *Applied Animal Behaviour Science* 107(3-4), 262-274.

Nicol, C. 2018. 9Feather pecking and cannibalism: can we really stop beak trimming? In: *Advances in Poultry Welfare*, Mench, J. A. (Ed.), *Food Science, Technology and Nutrition*, pp. 175-197. Woodhead Publishing, Duxford, UK.

Nicol, C. J., Bestman, M., Gilani, A.-M., De Haas, E. N., De Jong, I. C., Lambton, S., Wagenaar, J. P., Weeks, C. A. and Rodenburg, T. B. 2013. The prevention and control of feather pecking: application to commercial systems. *World's Poultry Science Journal* 69(4), 775-788.

Nicol, C. J., Gregory, N. G., Knowles, T. G., Parkman, I. D. and Wilkins, L. J. 1999. Differential effects of increased stocking density, mediated by increased flock size, on feather pecking and aggression in laying hens. *Applied Animal Behaviour Science* 65(2), 137-152.

Nicol, C. J., Lindberg, A. C., Phillips, A. J., Pope, S. J., Wilkins, L. J. and Green, L. E. 2001. Influence of prior exposure to wood shavings on feather pecking, dustbathing and foraging in adult laying hens. *Applied Animal Behaviour Science* 73(2), 141-155.

Nicol, C. J. and Pope, S. J. 1996. The maternal feeding display of domestic hens is sensitive to perceived chick error. *Animal Behaviour* 52(4), 767-774.

Nicol, C. J., Pötzsch, C., Lewis, K. and Green, L. E. 2003. Matched concurrent case-control study of risk factors for feather pecking in hens on free-range commercial farms in the UK. *British Poultry Science* 44(4), 515-523.

Nordquist, R. E., Zeinstra, E. C., Rodenburg, T. B. and van der Staay, F. J. 2013. Effects of maternal care and selection for low mortality on tyrosine hydroxylase concentrations and cell soma size in hippocampus and nidopallium caudolaterale in adult laying hen. *Journal of Animal Science* 91(1), 137-146.

O'Connor, E. A., Parker, M. O., Davey, E. L., Grist, H., Owen, R. C., Szladovits, B., Demmers, T. G. M., Wathes, C. M. and Abeyesinghe, S. M. 2011. Effect of low light and high noise on behavioural activity, physiological indicators of stress and production in laying hens. *British Poultry Science* 52(6), 666-674.

Oden, K., Keeling, L. J. and Algers, B. 2002. Behaviour of laying hens in two types of aviary systems on 25 commercial farms in Sweden. *British Poultry Science* 43(2), 169-181.

Oettel, R. 1873. *Der Huehner oder Gefluegelhof*. Bernhardt, Friedrich Boigt, Weimar, Germany.

O'Mahony, S. M., Clarke, G., Borre, Y. E., Dinan, T. G. and Cryan, J. F. 2015. Serotonin, tryptophan metabolism and the brain-gut-microbiome axis. *Behavioural Brain Research* 277, 32-48.

Palczynski, L., Buller, H., Lambton, S. and Weeks, C. 2016. Farmer attitudes to injurious pecking in laying hens and to potential control strategies. *Animal Welfare* 25(1), 29-38.

Parmentier, H. K., Rodenburg, T. B., Reilingh, G. D. V., Beerda, B. and Kemp, B. 2009. Does enhancement of specific immune responses predispose laying hens for feather pecking? *Poultry Science* 88(3), 536-542.

Patzke, N., Ocklenburg, S., van der Staay, F. J. J., Güntürkün, O. and Manns, M. 2009. Consequences of different housing conditions on brain morphology in laying hens. *Journal of Chemical Neuroanatomy* 37(3), 141-148.

Perré, Y., Wauters, A.-M. and Richard-Yris, M.-A. 2002. Influence of mothering on emotional and social reactivity of domestic pullets. *Applied Animal Behaviour Science* 75(2), 133-146.

Pichová, K., Bilčík, B. and Košťál, L. 2017. Assessment of the effect of housing on feather damage in laying hens using IR thermography. *Animal*, 11(4), 661-669.

Piepho, H. P., Lutz, V., Kjaer, J. B., Grashorn, M., Bennewitz, J. and Bessei, W. 2017. The presence of extreme feather peckers in groups of laying hens. *Animal* 11(3), 500-506.

Pokharel, B. B., dos Santos, V. M., Wood, D., van Heyst, B. and Harlander-Matauschek, A. 2017. Laying hens behave differently in artificially and naturally sourced ammoniated environments. *Poultry Science* 96(12), 4151-4157.

Polanczyk, G., de Lima, M. S., Horta, B. L., Biederman, J. and Rohde, L. A. 2007. The worldwide prevalence of ADHD: a systematic review and metaregression analysis. *The American Journal of Psychiatry* 164(6), 942-948.

Pötzsch, C. J., Lewis, K., Nicol, C. J. J., Green, L. E. E., Potzsch, C. J. 2001. A cross-sectional study of the prevalence of vent pecking in laying hens in alternative systems and its associations with feather pecking, management and disease. *Applied Animal Behaviour Science* 74(4), 259-272.

Prescott, N. B., Wathes, C. M. and Jarvis, J. R. 2003. Light , vision and the welfare of poultry. *Animal Welfare* 12, 269-288.

Queiroz, S. A. and Cromberg, V. U. 2006. Aggressive behavior in the genus Gallus sp. *Brazilian Journal of Poultry Science* 8(1), 1-14.

Ramadan, S. G. A. and von Borell, E. 2008. Role of loose feathers on the development of feather pecking in laying hens. *British Poultry Science* 49(3), 250-256.

Raud, H. and Faure, J. M. 1994. Welfare of ducks in intensive units. *Revue Scientifique et Technique*. International Office of Epizootics 13(1), 119-129.

Reiner, A. 2005. A new avian brain nomenclature: why, how and what. *Brain Research Bulletin* 66(4-6), 317-331.

Riber, A. 2007. *Ontogeny of Behaviour in Domestic Fowl–With Emphasis on Feather Pecking*. University of Copenhagen, Denmark.

Riber, A. B. and Hinrichsen, L. K. 2016. Feather eating and its associations with plumage damage and feathers on the floor in commercial farms of laying hens. *Animal* 10(7), 1218-1224.

Riber, A. B., Wichman, A., Braastad, B. O. and Forkman, B. 2007. Effects of broody hens on perch use, ground pecking, feather pecking and cannibalism in domestic fowl (*Gallus gallus domesticus*). *Applied Animal Behaviour Science* 106(1-3), 39-51.

Riboni, F. V. and Belzung, C. 2017. Stress and psychiatric disorders: from categorical to dimensional approaches. *Current Opinion in Behavioral Sciences* 14, 72-77.

Riedstra, B. and Groothuis, T. G. G. 2002. Early feather pecking as a form of social exploration: the effect of group stability on feather pecking and tonic immobility in domestic chicks. *Applied Animal Behaviour Science* 77(2), 127-138.

Roden, C. and Wechsler, B. 1998. A comparison of the behaviour of domestic chicks reared with or without a hen in enriched pens. *Applied Animal Behaviour Science* 55(3-4), 317-326.

Rodenburg, T. B., Bolhuis, J. E., Koopmanschap, R. E., Ellen, E. D. and Decuypere, E. 2009a. Maternal care and selection for low mortality affect post-stress corticosterone and peripheral serotonin in laying hens. *Physiology and Behavior* 98(5), 519-523.

Rodenburg, T. B., de Haas, E. N., Nielsen, B. L. and Buitenhuis, A. J. 2010. Fearfulness and feather damage in laying hens divergently selected for high and low feather pecking. *Applied Animal Behaviour Science* 128(1-4), 91-96.

Rodenburg, T. B., van Hierden, Y. M., Buitenhuis, A. J., Riedstra, B., Koene, P., Korte, S. M., van der Poel, J. J., Groothuis, T. G. G. and Blokhuis, H. J. 2004. Feather pecking in laying hens: new insights and directions for research? *Applied Animal Behaviour Science* 86(3-4), 291-298.

Rodenburg, T. B. and Koene, P. 2004. Feather pecking and feather loss. In: *Welfare of the Laying Hen* Perry, G. (Ed.), *Poultry Science Symposium Series*, pp. 227-238. World's Poultry Science Association (UK Branch), Bristol, UK.

Rodenburg , T. B. and Koene, P. 2007. The impact of group size on damaging behaviours, aggression, fear and stress in farm animals. *Applied Animal Behaviour Science* 103(3-4), 205-214.

Rodenburg, T. B., Komen, H., Ellen, E. D., Uitdehaag, K. A. and van Arendonk, J. A. M. 2008a. Selection method and early-life history affect behavioural development, feather pecking and cannibalism in laying hens: a review. *Applied Animal Behaviour Science* 110(3-4), 217-228.

Rodenburg, T. B., van Krimpen, M. M., de Jong, I. C., de Haas, E. N., Kops, M. S., Riedstra, B. J., Nordquist, R. E., Wagenaar, J. P., Bestman, M. and Nicol, C. J. 2013. The prevention and control of feather pecking in laying hens: identifying the underlying principles. *World's Poultry Science Journal* 69(2), 361-374.

Rodenburg, T. B., Tuyttens, F. A. M., de Reu, K., Herman, L., Zoons, J. and Sonck, B. 2008b. Welfare assessment of laying hens in furnished cages and non-cage systems: an on-farm comparison. *Animal Welfare* 17, 363-373.

Rodenburg, T. B., Uitdehaag, K. A., Ellen, E. D. and Komen, J. 2009b. The effects of selection on low mortality and brooding by a mother hen on open-field response, feather pecking and cannibalism in laying hens. *Animal Welfare* 18, 427-432.

Rollin, B. E. 2015. Telos, conservation of welfare, and ethical issues in genetic engineering of animals. *Current Topics in Behavioral Neurosciences* 19, 99-116.

Rørvang, M. V., Hinrichsen, L. K. and Riber, A. B. 2019. Welfare of layers housed in small furnished cages on Danish commercial farms: the condition of keel bone, feet, plumage and skin. *British Poultry Science* 60(1), 1-7.

Rubinstein, J. and Lightfoot, T. 2014. Feather loss and feather destructive behavior in pet birds. *Veterinary Clinics of North America - Exotic Animal Practice* 17(1), 77-101.

Sandi, C. and Haller, J. 2015. Stress and the social brain: behavioural effects and neurobiological mechanisms. *Nature Reviews Neuroscience* 16(5), 290-304.

Savory, C. J. 1995. Feather pecking and cannibalism. *World's Poultry Science Journal* 51(2), 215-219.

Savory, C. J. and Mann, J. S. 1997. Behavioural development in groups of pen-housed pullets in relation to genetic strain, age and food form. *British Poultry Science* 38(1), 38-47.

Savory, C. J. and Mann, J. S. 1999. Feather pecking in groups of growing bantams in relation to floor litter substrate and plumage colour. *British Poultry Science* 40(5), 565-572.

Savory, C. J., Mann, J. S. and Macleod, M. G. 1999. Incidence of pecking damage in growing bantams in relation to food form, group size, stocking density, dietary tryptophan concentration and dietary protein source. *British Poultry Science* 40(5), 579-584.

Scanes, C. G. (Ed). 2015. *Sturkie's Avian Physiology*. Academic Press, London.

Schmid, I. and Wechsler, B. 1997. Behaviour of Japanese quail (*Coturnix japonica*) kept in semi-natural aviaries. *Applied Animal Behaviour Science* 55(1-2), 103-112.

Schreiter, R., Damme, K., von Borell, E., Vogt, I., Klunker, M. and Freick, M. 2019. Effects of litter and additional enrichment elements on the occurrence of feather pecking in pullets and laying hens - a focused review. *Veterinary Medicine and Science* 5(4), 500-507.

Sherwin, C. M., Richards, G. J. and Nicol, C. J. 2010. Comparison of the welfare of layer hens in 4 housing systems in the UK. *British Poultry Science* 51(4), 488-499.

Shimmura, T., Kamimura, E., Azuma, T., Kansaku, N., Uetake, K. and Tanaka, T. 2010. Effect of broody hens on behaviour of chicks. *Applied Animal Behaviour Science* 126(3-4), 125-133.

Silverman, M. N. and Sternberg, E. M. 2012. Glucocorticoid regulation of inflammation and its functional correlates: from HPA axis to glucocorticoid receptor dysfunction. *Annals of the New York Academy of Sciences* 1261, 55-63.

Spencer, K. A. 2017. Developmental stress and social phenotypes: integrating neuroendocrine, behavioural and evolutionary perspectives. *Philosophical Transactions of the Royal Society of London. Series B, Biological Sciences* 372(1727), 20160242.

Stettenheim, P. R. 2000. The integumentary morphology of modern birds - an overview. *American Zoologist* 40(4), 461-477.

Strasser, B., Geiger, D., Schauer, M., Gatterer, H., Burtscher, M. and Fuchs, D. 2016. Effects of exhaustive aerobic exercise on tryptophan-kynurenine metabolism in trained athletes. *PLoS ONE* 11(4), e0153617.

Studdert, V., Gay, C. and Blood, D. 2012. *Saunders Comprehensive Veterinary Dictionary*. Elsevier Health Sciences, Edinburgh, UK.

Su, G., Kjaer, J. B. and Sorensen, P. 2006. Divergent selection on feather pecking behavior in laying hens has caused differences between lines in egg production, egg quality, and feed efficiency. *Poultry Science* 85(2), 191-197.

Tahamtani, F. M., Brantsaeter, M., Nordgreen, J., Sandberg, E., Hansen, T. B., Nodtvedt, A., Rodenburg, T. B., Moe, R. O. and Janczak, A. M. 2016a. Effects of litter provision during early rearing and environmental enrichment during the production phase on feather pecking and feather damage in laying hens. *Poultry Science* 95(12), 2747-2756.

Tahamtani, F. M., Nordgreen, J., Brantsæter, M., Østby, G. C., Nordquist, R. E. and Janczak, A. M. 2016b. Does early environmental complexity influence tyrosine hydroxylase in the chicken hippocampus and "prefrontal" caudolateral nidopallium? *Frontiers in Veterinary Science* 3, 8.

Tahamtani, F. M., Nordgreen, J., Nordquist, R. E. and Janczak, A. M. 2015. Early life in a barren environment adversely affects spatial cognition in laying hens (*Gallus gallus domesticus*). *Frontiers in Veterinary Science* 2, 1-12.

Tauson, R., Kjaer, J., Maria, G. A., Cepero, R. and Holm, K.-E. 2005. Applied scoring of integument and health in laying hens. *Animal Science Papers and Reports* 23, 153-159.

Tauson, R., Kjaer, J. B., Maria, G. A., Cepero, R., Holm, K.-E., Elwinger, K. and Wall, H. 2006. LayWel: Welfare implications of changes in production systems for laying hens - WP3 - Health, pp. 1-70. Available at: https://www.laywel.eu/web/pdf/deliverables %2031-33%20health-2.pdf.

Telles-Correia, D., Saraiva, S. and Gonçalves, J. 2018. Mental disorder–the need for an accurate definition. *Frontiers in Psychiatry* 9, 64.

Tullett, S. G., Macleod, M. G. and Jewitt, T. R. 1980. The effects of partial defeathering on energy metabolism in the laying fowl. *British Poultry Science* 21(3), 241-245.

Uitdehaag, K. A., Rodenburg, T. B., Bolhuis, J. E., Decuypere, E. and Komen, H. 2009. Mixed housing of different genetic lines of laying hens negatively affects feather pecking and fear related behaviour. *Applied Animal Behaviour Science* 116(1), 58-66.

Uitdehaag, K. A., Rodenburg, T. B., van Hierden, Y. M., Bolhuis, J. E., Toscano, M. J., Nicol, C. J. and Komen, J. 2008. Effects of mixed housing of birds from two genetic lines of laying hens on open field and manual restraint responses. *Behavioural Processes* 79(1), 13-18.

Uitdehaag, K. A., Rodenburg, T. B., van Reenen, C. G., Koopmanschap, R. E., Reilingh, G. D. V., Engel, B., Buist, W. G., Komen, H. and Bolhuis, J. E. 2011. Effects of genetic origin and social environment on behavioral response to manual restraint and monoamine functioning in laying hens. *Poultry Science* 90(8), 1629-1636.

University of Bristol 2013. *AssureWel Laying Hens Assessment Protocol*. University of Bristol, Bristol, UK.

van de Weerd, H. A. and Elson, A. 2006. Rearing factors that influence the propensity for injurious feather pecking in laying hens. *World's Poultry Science Journal* 62, 654-664.

van der Eijk, J. A. J., Lammers, A., Kjaer, J. B. and Rodenburg, T. B. 2019a. Stress response, peripheral serotonin and natural antibodies in feather pecking genotypes and phenotypes and their relation with coping style. *Physiology and Behavior* 199, 1-10.

van der Eijk, J. A. J., Lammers, A., Li, P., Kjaer, J. B. and Rodenburg, T. B. 2018. Feather pecking genotype and phenotype affect behavioural responses of laying hens. *Applied Animal Behaviour Science* 205, 141-150.

van der Eijk, J. A. J., de Vries, H., Kjaer, J. B., Naguib, M., Kemp, B., Smidt, H., Rodenburg, T. B. and Lammers, A. 2019b. Differences in gut microbiota composition of laying hen lines divergently selected on feather pecking. *Poultry Science* 98(12), 7009-7021.

van der Eijk, J. A. J., Rodenburg, T. B., de Vries, H., Kjaer, J. B., Smidt, H., Naguib, M., Kemp, B. and Lammers, A. 2019c. Early-life microbiota transplantation affects behavioural responses, serotonin, and immune characteristics in chickens divergently selected on feather pecking. *Scientific Reports*10, 2750.

van der Eijk, J. A. J., Verwoolde, M. B., de Vries Reilingh, G., Jansen, C. A., Rodenburg, T. B. and Lammers, A. 2019d. Chicken lines divergently selected on feather pecking differ in immune characteristics. *Physiology and Behavior* 212, 112680.

van Hierden, Y. M., de Boer, S. F., Koolhaas, J. M. and Korte, S. M. 2004a. The control of feather pecking by serotonin. *Behavioral Neuroscience* 118(3), 575-583.

van Hierden, Y. M., Koolhaas, J. M. and Korte, S. M. M. 2004b. Chronic increase of dietary L-tryptophan decreases gentle feather pecking behaviour. *Applied Animal Behaviour Science* 89(1-2), 71-84.

van Hierden, Y. M., Koolhaas, J. M., Košťál, L., Výboh, P., Sedlačková, M., Rajman, M., Juráni, M., Mechiel Korte, S. 2005. Chicks from a high and low feather pecking line of laying hens differ in apomorphine sensitivity. *Physiology and Behavior* 84(3), 471-477.

van Hierden, Y. M., Korte, S. M. M., Ruesink, E. W. W., van Reenen, C. G., Engel, B., Koolhaas, J. M. and Blokhuis, H. J. 2002a. The development of feather pecking behaviour and targeting of pecking in chicks from a high and low feather pecking line of laying hens. *Applied Animal Behaviour Science* 77(3), 183-196.

van Hierden, Y. M., Korte, S. M., Ruesink, E. W., van Reenen, C. G., Engel, B., Korte-Bouws, G. A. H., Koolhaas, J. M. and Blokhuis, H. J. 2002b. Adrenocortical reactivity and central serotonin and dopamine turnover in young chicks from a high and low feather-pecking line of laying hens. *Physiology and Behavior* 75(5), 653-659.

van Krimpen, M. M., Kwakkel, R. P., Reuvekamp, B. F. J., van Der Peet-Schwering, C. M. C., Den Hartog, L. A. and Verstegen, M. W. A. 2005 Impact of feeding management on feather pecking in laying hens. *World's Poultry Science Journal* 61(4), 663-686.

van Staaveren, N., Decina, C., Baes, C. F., Widowski, T. M., Berke, O. and Harlander-Matauschek, A. 2018a. A description of laying hen husbandry and management practices in Canada. *Animals* 8(7), 114.

van Staaveren, N., Decina, C., Baes, C. F., Widowski, T., Berke, O. and Harlander-Matauschek, A. 2018b. Animal care by farmers in relation to feather cover damage in laying hens. In: Proceedings of the First Annual Meeting of the European Congress of Behavioural Medicine and Animal Welfare (ECAWBM), Schoning, B., Jonckher-Sheehy, V., Morton, D., Roger, P. and Camps, T. (Eds), pp. 195-196. GTVMT, Berlin, Germany.

van Staaveren, N., Krumma, J., Forsythe, P., Kjaer, J. B., Kwon, I. Y., Mao, Y.-K., West, C., Kunze, W. and Harlander-Matauschek, A. 2020. Cecal motility and the impact of Lactobacillus in feather pecking laying hens. *Scientific Reports* 10, 12978. doi: 10.1038/s41598-020-69928-6.

van Zeeland, Y. R. A. and Schoemaker, N. J. 2014. Plumage disorders in psittacine birds - part 1: feather abnormalities. *European Journal of Companion Animal Practice* 24, 34-47.

Vestergaard, K. S., Kruijt, J. P. and Hogan, J. A. 1993. Feather pecking and chronic fear in groups of red junglefowl: their relations to dustbathing, rearing environment and social status. *Animal Behaviour* 45(6), 1127-1140.

Vestergaard, K. S. and Lisborg, L. 1993. A model of feather pecking development which relates to dustbathing in the fowl. *Behaviour* 126(3-4), 291-308.

Wakefield, J. C. 1992. The concept of mental disorder: on the boundary between biological facts and social values. *American Psychologist* 47(3), 373-388.

Wechsler, B., Huber-Eicher, B. and Nash, D. R. 1998. Feather pecking in growers: a study with individually marked birds. *British Poultry Science* 39(2), 178-185.

Wechsler, B. and Schmid, I. 1998. Aggressive pecking by males in breeding groups of Japanese quail (Coturnix japonica). *British Poultry Science* 39(3), 333-339.

Weeks, C. A. and Nicol, C. J. 2006. Behavioural needs, priorities and preferences of laying hens. *World's Poultry Science Journal* 62(2), 296-307.

Welfare Quality 2009. *Welfare Quality® Assessment Protocol for Poultry*. Welfare Quality, Lelystad, Netherlands.

Wennrich, G. 1974. Ethological studies of domestic chickens (*Gallus domesticus*) of different hybrid origin in floor management with special reference to aggressive behaviour as well as feather pecking and cannibalism. 1st Communication: pecking movements in feeding behaviou. *Archiv für Geflügelkunde* 38, 143-149.

Whay, H. R. 2007. The journey to animal welfare improvement. *Animal Welfare* 16, 117-122.

Whitehouse, C. M. and Lewis, M. H. 2015. Repetitive behavior in neurodevelopmental disorders: clinical and translational findings. *The Behavior Analyst* 38(2), 163-178.

Widowski, T. M., Caston, L. J., Hunniford, M. E., Cooley, L. and Torrey, S. 2017. Effect of space allowance and cage size on laying hens housed in furnished cages, Part I: performance and well-being. *Poultry Science* 96(11), 3805-3815.

Widowski, T. M. and Torrey, S. 2018. Rearing young birds for adaptability. In: *Advances in Poultry Welfare* Mench, J. A. (Ed.), pp. 49-77. Woodhead Publishing Group, Duxford, UK.

Willemse, T., Mudde, M., Josephy, M. and Spruijt, B. M. 1994. The effect of haloperidol and naloxone on excessive grooming behavior of cats. *European Neuropsychopharmacology* 4(1), 39-45.

Willimon, C. P. and Morgan, C. L. 1953. The effect of minor nutrient mineral elements in the diet of chickens on feather pulling and cannibalism. *Poultry Science* 32(2), 309-313.

Wood-Gush, D. G. M. 1971. *The Behaviour of the Domestic Fowl*. Heinemann Educational Publishers, London, UK.

Wood, G. M., Smith, R. T. and Henderson, D. C. 1963. A crop analysis technique for studying the food habits and preferences of chickens on range. *Poultry Science* 42(2), 304-309.

Wysocki, M., Bessei, W., Kjaer, J. B. and Bennewitz, J. 2010. Genetic and physiological factors influencing feather pecking in chickens. *World's Poultry Science Journal* 66(4), 659-672.

Wysocki, M., Preuss, S., Stratz, P. and Bennewitz, J. 2013. Investigating gene expression differences in two chicken groups with variable propensity to feather pecking. *Animal Genetics* 44(6), 773-777.

Yamak, U. S. and Sarica, M. 2012. Relationships between feather score and egg production and feed consumption of different layer hybrids kept in conventional cages. *Archiv fur Geflugelkunde* 76, 31-37.

Young, S. N. 2013. Acute tryptophan depletion in humans: a review of theoretical, practical and ethical aspects. *Journal of Psychiatry and Neuroscience* 38(5), 294-305.

Youngson, R. M. 2005. *Collin's Dictionary of Medicine*. Harper Collins, Glasgow, UK.

Zablotsky, B., Black, L. I., Maenner, M. J., Schieve, L. A., Danielson, M. L., Bitsko, R. H., Blumberg, S. J., Kogan, M. D. and Boyle, C. A. 2019. Prevalence and trends of developmental disabilities among children in the United States: 2009-2017. *Pediatrics* 144(4), e20190811.

Zeltner, E., Klein, T. and Huber-Eicher, B. 2000. Is there social transmission of feather pecking in groups of laying hen chicks? *Animal Behaviour* 60(2), 211-216.

Zepp, M., Louton, H., Erhard, M., Schmidt, P., Helmer, F. and Schwarzer, A. 2018. The influence of stocking density and enrichment on the occurrence of feather pecking and aggressive pecking behavior in laying hen chicks. *Journal of Veterinary Behavior* 24, 9-18.

Zhao, Y., Xin, H. and Dong, B. 2013. Use of infrared thermography to assess laying-hen feather coverage. *Poultry Science* 92(2), 295-302.

Ziabreva, I., Poeggel, G., Schnabel, R. and Braun, K. 2003a. Separation-induced receptor changes in the hippocampus and amygdala of *Octodon degus*: influence of maternal vocalizations. *Journal of Neuroscience* 23(12), 5329-5336.

Ziabreva, I., Schnabel, R., Poeggel, G. and Braun, K. 2003b. Mother's voice 'buffers' separation-induced receptor changes in the prefrontal cortex of *Octodon degus*. *Neuroscience* 119(2), 433-441.

# Chapter 17

# Bone health and associated problems in layer hens

*Christina Rufener, University of California-Davis, USA; and Michael J. Toscano, University of Bern, Switzerland*


1 Introduction
2 Bone development, growth and remodelling
3 Identified bone health problems
4 Contributory factors to poor bone health
5 Influence of poor bone health on productivity and welfare
6 Strategies for improving bone health
7 Future trends in research
8 Where to look for further information
9 References


## 1 Introduction

Housing and management systems for laying hens are variable but must allow for appropriate skeletal and cognitive development during rearing to ensure adult hens can support normal biological functioning while producing a large number of eggs. High egg production requires mobilization of resources including minerals and energy. Likely as a consequence of this mobilization and related factors, laying hens are believed to have weakened skeletal systems leading to specific problems including fractured keels during the laying period and other bone injuries during removal from the barn at the end of lay. To combat these problems, multiple strategies have been undertaken including effective breeding and nutritional strategies as well as adaptations of rearing practices and layer housing. The following chapter provides a summary of the basic skeletal system and its development, specific problems of bone health and efforts to reduce the problem. References to specific types of housing systems will be based on the three broad categories described elsewhere in this book. Within cage-free systems, housing type is further distinguished as single-tier or multi-tier (e.g. aviary) systems.

http://dx.doi.org/10.19103/AS.2020.0078.16

## 2 Bone development, growth and remodelling

Bone development, growth and remodelling of the avian skeleton are highly complex processes - especially as birds have special features including pneumatic (hollow) bones enabling flight and structures such as the keel bone serving as attachment point for the wing muscles (Sisson et al., 1975). To discuss influencing factors and potential solutions for bone health problems in adult laying hens, it is crucial to understand the mechanisms involved in the development and growth of the skeleton in pullets as well as ongoing bone remodelling processes after the onset of lay.

### 2.1 Bone development during rearing

The genesis of structures for cartilage formation is considered to be the earliest stage of skeletal development (Pines and Reshef, 2014) and starts as early as the first day of embryo development (Pourquié, 2004). At hatch and then continuing for the first 17 weeks of life, the structural skeleton becomes fully developed in terms of shape and dimensions, though it is likely to continue undergoing maturation (e.g. strength of internal linkages) until 35 weeks of age (Rath et al., 2000). Depending on the bone type, bone development and ossification are driven by either endochondral or intramembranous processes (Gilbert, 2000). Long bones of the appendicular skeleton such as the tibia or humerus develop by endochondral ossification, where cartilage is replaced by bone tissue. Flat bones such as the skull, on the other hand, are formed directly from connective tissue (mesenchyme) without a cartilage template. The keel bone is a special hybrid of these conditions as it is a rather flat bone (and therefore subject to intramembranous ossification), but has a cartilage template which is progressively ossifying with increasing age. Although the keel has previously been shown to be fully ossified at 40 weeks of age (Buckner et al., 1949), anecdotal reports have found that the keel tip (i.e. the most caudal part) remains mostly cartilaginous until 70 weeks of age.

Skeletal growth of long bones involves two main processes: elongation (i.e. longitudinal growth) and widening of the bones (Whitehead, 2004). Bones grow in length by proliferation (i.e. cellular division), maturation and calcification of cartilage cells at the growth plate located at the ends of the bone. Osteoblasts, the cells responsible for bone formation, then secrete bone tissue on the calcified cartilage to ossify the newly formed bone matrix. The second process of bone growth, widening of bones, involves the resorption of old bone in the medullary cavity by osteoclasts. Osteoblasts deposit new bone tissue on the outer surface of the bone, increasing the diameter of the bone and the medullary cavity. Once the pullet skeleton is fully developed, avian long bones are similar in structure to mammalian bones with an outer, compact cortical and an inner trabecular (or cancellous) bone (Fig. 1). Although both cortical and trabecular bones consist of the lamellar bone, the organization of lamellae differs and affects the function and stability of the bone. Cortical bone with a compact organization of lamellae serves as the principal structure of the skeleton, whereas trabecular bone arranged in rods and plates reduces bone density and allows long bones to compress if stress is applied.

## 2.2 Medullary bone and bone remodelling during lay

At the age of sexual maturity (17–19 weeks of age) and shortly before the onset of lay, the bone biology of laying hens changes dramatically. To meet the calcium demand for egg shell formation, avian females produce a special bone type referred to as medullary bone (Dacke et al., 1993). In contrast to the outer, compact cortical bone and the inner trabecular bone commonly found in all mammals, medullary bone is a woven, unorganized and highly calcified bone type that serves as a labile calcium source to cover high calcium requirements emerging during egg shell formation (Fig. 1; Etches, 1987). Once hens reach sexual maturity, the formation of structural bone, that is, the cortical and trabecular bones, ceases (Hudson et al., 1993) and the medullary bone begins to be deposited in the medullary cavities of long bones (Whitehead, 2004). The shift from the formation of structural bone towards the deposition of the medullary bone is driven by oestrogen, which increases osteoblastic activity and stimulates the formation of the woven bone instead of the lamellar bone (Whitehead and Fleming, 2000).

The medullary bone undergoes constant remodelling within the daily routine of bone depletion during egg shell calcification alternating with bone formation during periods when no egg shell is being produced (Kerschnitzki

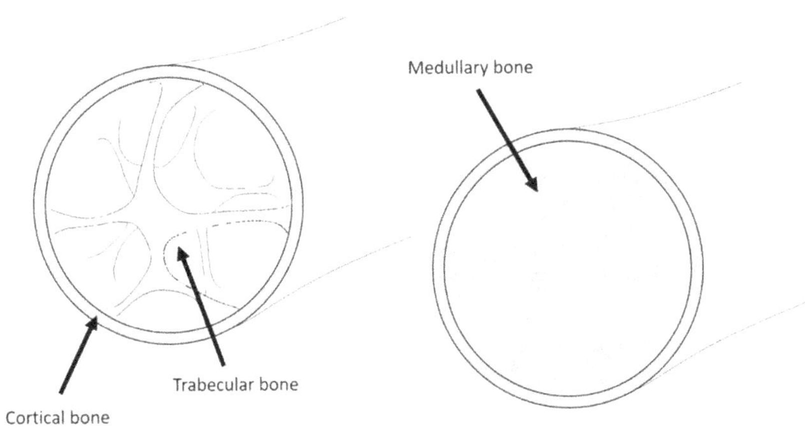


**Figure 1** Schematic representation of cortical, trabecular and medullary bone in an avian long bone. Source: C. Rufener.

et al., 2014; Johnson, 2015). No new structural bone is formed while the hen is in lay, and bone resorption by osteoclastic is not specific for one bone type (Fleming et al., 2006). As a result, calcium is mobilized during egg laying from both medullary and structural bone, but with different rates (Simkiss, 1967) as the medullary bone can be metabolized 10–15 times faster than the cortical bone (Hurwitz, 1965). Junglefowl – the ancestor of all commercial chickens – lay a clutch of 5–8 eggs once or twice a year and incubate them for 19–21 days (Del Hoyo et al., 2001). Once a hen has laid the last egg of a clutch and oestrogen levels decrease, the medullary bone is gradually lost and remodelling of the new cortical bone will be initiated (Whitehead, 2004). In modern layer strains producing 250–310 eggs in the first year after the onset of lay (Karcher and Mench, 2018), hens lay an egg on multiple consecutive days followed by only a few days without oviposition. As a result, modern commercial hens are not able to remodel and/or deposit a new cortical bone between two egg laying periods.

## 3 Identified bone health problems

Within commercial laying hen management, two major issues of bone health exist: keel bone damage (including deviations and fractures) and injuries during depopulations. Both are believed to relate to the underlying poor bone health and manifest in response to interactions with their environment and/or the handling required for the removal of hens at the end of lay (i.e. depopulation).

### 3.1 Keel bone damage: deviations and fractures

For laying hens, damaged keel bones, broadly classified as manifesting either fractures, deviations or a combination of the two, is one of the most important welfare problems facing the commercial laying hen industry as suggested by the UK's Farm Animal Welfare Committee (FAWC, 2010, 2013). The issue of keel bone damage has also been cited as a major problem by the EFSA-AHAW panel (European Food Safety Authority, 2015) and a North American-based consortium of welfare researchers (Lay et al., 2011).

Keel bone fractures can be defined as breaks in the bone that typically manifest as a callus around the fracture site but may also involve sharp, unnatural deviations or bending (Fig. 2; Casey-Trott et al., 2015). While fractures are considered to be a defect in the bone architecture, deviations are generally believed to result from the normal process of bone remodelling (Harlander-Matauschek et al., 2015). Deviations are defined as keels with an abnormally shaped structure that have not resulted from a fracture incident, but contain section(s) that vary from a theoretically perfect two-dimensional straight plane in either the transverse or sagittal planes (Fig. 3; Casey-Trott et al., 2015). Given that fractures are viewed as defects of the bone rather than a consequence of

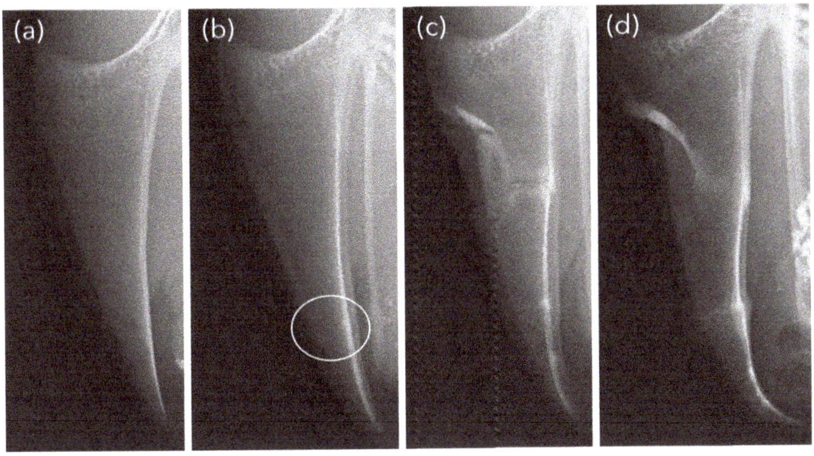


**Figure 2** Latero-lateral X-rays of the same hen without fracture (a), with a minor fracture indicated by a circle (b), with multiple major fractures (c) and partially healed fractures (d). Source: C Rufener.

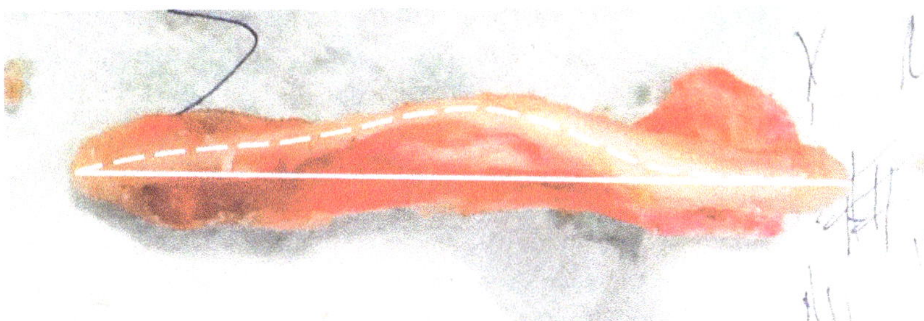

**Figure 3** Deviated keel bone (ventrodorsal view). The solid line indicates the theoretically perfect two-dimensional straight transverse plane and the dashed line represents the deviated shape. Source: C. Rufener.

a normal physiological process (i.e. remodelling), fractures are more critical in terms of welfare and productivity. Deviations remain important to consider as a dramatically altered structure is likely to be weakened and more susceptible to fracture (Harlander-Matauschek et al., 2015).

Damage to the keel bone has been widely reported in multiple countries and levels are exceptionally high, typically in excess of 50% by the end of lay (Canada (Petrik et al., 2015); Belgium (Heerkens et al., 2013); Holland (Rodenburg et al., 2008); Switzerland (Käppeli et al., 2011); UK (Wilkins et al., 2011; Tarlton et al., 2013; Toscano et al., 2015)). In addition, the problem appears to occur in all types of housing systems (Wilkins et al., 2011), including conventional and enriched cages (Petrik et al., 2015) and organic systems (Bestman and Wagenaar, 2014), and across genetic lines (Käppeli et al., 2011;

Candelotto et al., 2017). The problem has also been observed in commercial broiler breeder flocks (Gebhardt-Henrich et al., 2017b). The underlying cause of keel bone fractures is generally considered to be high and sustained egg production leading to weak and brittle bones that are susceptible to fracture during collision, a theory discussed in more detail below.

### 3.2 Injuries during depopulation

At the end of a commercial laying period, all hens are removed from the barn in a systematic fashion to allow cleaning and preparation for the subsequent flock. The removal process, commonly referred to as depopulation, will occur typically between 60 and 90 weeks of age though sometimes up to 100 weeks of age depending on a host of factors including the flock's health, productivity and availability of the final processing system or location.

Despite the increased automation of tasks within the laying hen industry, the depopulation process remains largely a labour-intensive process requiring human workers to manually collect the hens within the barn. Once collected, the hens are normally crated for transport to the slaughter facility though can also be killed on-site for use as an energy source, for example, biogas (Afazeli et al., 2014). Beginning with the initial collection of hens in the barn and continuing to the arrival at the abattoir or death of hens otherwise, numerous occasions exist for injuries to occur including collisions with pen or cage furnishings or loading into the transport crate. Injuries are also likely to happen during the actual transport of the live hens, though the principal danger here is exposure to extreme temperatures, withdrawal from feed and water, and the consequent physiological stress (reviewed by: Mitchell and Kettlewell, 2009; Schwartzkopf-Genswein et al., 2012). Given that the collection of hens is considered the most stressful process as well as the source of injury during depopulation (Schwartzkopf-Genswein et al., 2012), the focus here is on the catching, carrying and crating of hens within the overall depopulation process. Similar to the concerns raised with keel bone fractures, the bones of the hens are believed to be relatively weak by this point making them susceptible to damage.

Focussing on hen injuries during collection, Gregory and Wilkins (1989) reported that an average of 24% of hens (ranging from 13% to 41% of farms surveyed) had a broken bone immediately following commercial depopulation which they attributed to the removal process. The seminal effort by Gregory and Wilkins (Gregory and Wilkins, 1989), as well as the majority of work in the area performed by others (Gregory et al., 1990; Kestin et al., 1992; Reed et al., 1993; Budgell and Silversides, 2004), focussed on laying hens from cage systems, although comparable levels of damage have been reported in aviaries (cage systems: 10.1%, Gregory et al., 1990; aviaries: 8%, Gerpe et al., 2017).

The removal process for aviary and cage systems has many parallel events and is likely to result in similar injuries as, in each case, the bird is typically forcefully pulled from the cage or their resting/perching position in the aviary. In this process, birds can collide with the edge of the cage as well as furnishings (e.g. perches, drinkers, etc.) either in the cage or that exist as part of the non-cage system.

Bones commonly found to be broken during removal from cages include those of the hips and legs: humerus, ischium and pubis (Gregory and Wilkins, 1989; Gregory et al., 1990). A similar category of bone injuries was also found to occur in aviaries, albeit with greater frequency in the keel, pubis and furcular with less to bones of the wings and legs (Gerpe et al., 2017). The reason for the variation in injury rates between housing systems is not clear though likely associated with multiple study-specific factors including genetic line (Budgell and Silversides, 2004; Candelotto et al., 2017), the increased load-bearing activity of hens in non-cage systems (Blokhuis et al., 2005) leading to increased bone strength (Leyendecker et al., 2005), as well as the differences in how hens are collected (i.e. pulled from within a cage in one direction vs. from multiple angles in aviary systems).

Once collected, birds are then bundled into groups of 3-5 hens and carried to a transport crate or other disposal device, for example, on-site gas chamber. While being carried, they are routinely held by a single leg which can result in twisting that can easily result in bone fracture or displacement. During the catching and carrying process, birds are frequently passed from a person performing the initial catching to a second person responsible for carrying and/or bundling, and possibly, a third person who actually loads the birds into the transport crate. Each passing of the birds between persons creates an additional opportunity for injury as the position of the birds is often rotated to accommodate the change in grasp. In support of the notion that improved handling during this stage of removal can lead to reduced injury, removal of birds by two hands versus one reduced injury from 14% to 5% (Gregory et al., 1993). Similarly, Gregory and Wilkins (1989) found that when birds were handled one bird at a time and brought to the transport crate, the rate of injury fell from 24% to 14%. During loading of hens, the keel and wings are especially susceptible to damage by collision with the sides of the transport crate (during placement into the crate) (Van Niekerk and Reuvekamp, 1994; Knowles and Wilkins, 1998). Mainly with broilers, a variety of commercial means have been pursued to reduce injury with moderate success including: legislation and better training of poultry handlers (Kettlewell and Mitchell, 1994) as well as technical improvements in how birds are collected and removed (Lacy and Czarick, 1998; Kristensen et al., 2001). Despite these efforts, laying hen removal remains a labour-intensive process with little or no advances in automation to reduce bird injuries.

## 4 Contributory factors to poor bone health

Skeletal injuries in laying hens are believed to result from compromised skeletal health though the underlying reasons of poor health are not fully understood. High and sustained egg production is considered the dominant factor, though recent evidence suggests the contribution of alternative factors as well.

### 4.1 Excessive calcium demand of egg production

The underlying cause that is most routinely associated with keel bone fractures during lay and injuries during depopulation is high and sustained egg production of modern commercial laying hens. To support the process of egg shell formation and the required calcium demand, bone resorption is used to supplement the exogenous (dietary) calcium required to form the eggshell. As mentioned above, bone resorption continues throughout the lay cycle with high rates of egg production. Over time, the production of eggs is believed to result in hens with a net loss of mineral in structural (cortical) bone leaving the bones weak and brittle (Whitehead and Fleming, 2000; Fleming et al., 2004; Bain et al., 2016). It is this loss of structural cortical bone that is thought to be responsible for the increased frequency of keel bone fracture as the hen struggles to support nearly continuous eggshell formation with diminishing endogenous mineral reserves. From the status of relative bone weakness, the hen is now susceptible to injury from collisions with barn furniture. Wilkins et al. (2011) suggested that the variation in fractures observed between systems was likely due to perches (and other system furnishings) in some systems resulting in more high-energy collisions, a view supported by the increased fracture occurrence and severity seen in systems with greater environmental complexity (i.e. more and higher items available for perching). Most studies have focussed on non-cage systems, though fractures do appear in caged systems (Wilkins et al., 2011) where high-energy collisions are also known to occur (Baker et al., 2020). Regarding multi-tier systems, reports have detailed collisions specifically within an aviary environment (Stratmann et al., 2015a), where hens had collisions as they sought positions at higher elevations, particularly during the dusk/dark transition (Stratmann et al., 2019). During this transition period, sitting hens occasionally get pushed from their positions by conspecifics or fail to land successfully, particularly where local bird density is very high.

The role of egg production as the underlying reason of keel bone fractures is supported directly by work using a synthetic gonadotropin antagonist to experimentally suspend lay resulting in reduced keel fractures (Eusemann et al., 2018b). Similarly, poultry bred for relatively high- and low-bone strength over seven generations found all males to be free of keel bone fractures whereas the proportion in hens ranged from 0% to 40% depending on the genetic line and generation (Fleming et al., 2004). Comparisons between high- and low-producing hybrids (Habig et al., 2017; Eusemann et al., 2018a), including heritage lines (Regmi et al., 2016a), also support the position of egg production as a cause of poor bone health. However, these studies are typically confounded by strain effects, for example, differential behavioural activity between genetic lines.

Although egg production is likely the major contributing factor to keel bone fractures, the problem appears to pre-date modern intensive production programmes with the earliest published observations made by Darwin (1868) reporting that 88% of chickens (both males and females) across 12 commercial breeds had a deformed sternum. Later reports by Warren (1937) and Buckner et al. (1949), though still preceding modern selection programmes, highlighted keel bone damage with the suggestion that the cause was related to genetics and the availability of perches and vitamin D. More recently, within the last 3–5 years, methodological advances have allowed for a more reliable and detailed assessment of keel damage with specific techniques including: improved palpation training (Casey-Trott et al., 2015; Gebhardt-Henrich et al., 2019); routine use of radiographic imaging (Baur, 2017; Eusemann et al., 2018a; Rufener et al., 2018a,b, 2019; Gebhardt-Henrich et al., 2019) and computed tomography (Candelotto et al., n.d.; Regmi et al., 2016b; Chargo et al., 2018a,b).

Improved assessment, in combination with the use of an ex vivo model that used an impact testing apparatus to experimentally create fractures in deceased birds (Toscano et al., 2013, 2018), has suggested alternative or additional influencing factors that may be contributing to the development of keel bone damage. The additional factors may operate in addition to or in synergism with underlying poor bone health in response to high and sustained egg production (Toscano et al., 2020). In support of the role of other factors beyond egg production causing keel bone fractures, the use of an impact-testing protocol identified a post-peak egg production phase where, beyond 49 weeks of age, susceptibility to new fractures appeared to decrease. The finding was surprising given that egg production remains relatively high (> 85% hen daily percentage) and that hens are not believed to produce cortical bone while they remain in lay (Hudson et al., 1993). Nonetheless, the findings parallel to age-related (i.e., longitudinal) keel bone fracture frequencies reported independently in commercial flocks where rates of fracture often stabilize or decrease in the same period (Richards et al., 2012; Petrik et al., 2015; Stratmann et al., 2015a; Toscano et al., 2015). The stabilization or decrease in new fractures could be attributed to altered behaviour or improved navigation abilities, though the impact testing data, in combination with similar patterns in commercial flocks, suggests some qualitative change within the animal such as bone maturation and/or age-related changes in strength (Rath et al., 2000). In other words, if egg production was directly related to bone health and occurrence of fractures, the rates of fracture would continue increasing beyond 49 weeks of age, which does not seem to be the case. Rather, the rate of new fractures appears to level-off or possibly even decrease in both commercial and experimental settings. In further support of the unclear role between egg production as the underlying reason of keel bone fractures, hens showing a decreased susceptibility for fracture had greater egg shell width and strength (Candelotto et al., 2017), a finding in conflict with the expectation that egg production and bone health are in direct competition.

## 4.2 Alternative contributing factors to keel bone fractures

Beyond intensive egg production, the length of bone development in the period preceding puberty, that is, the age at first egg, has been suggested to influence bone quality (Wolc et al., 2010), which in turn is influenced by growth rate, photoperiod and body weight (Dunn and Sharp, 1992). As the keel is understood to ossify late compared to other parts of the skeleton (Buckner et al., 1948) and remains cartilaginous even at puberty (Casey-Trott et al., 2017b), a premature transition to the lay period before completion of bone development (Rath et al., 2000) could prevent the keel from reaching peak strength in terms of its full genetic potential. Given that the keel ossifies cranial to caudal leaving the tip to be the final portion to mature (Buckner et al., 1948), which does not finally ossify until an estimated 35–40 weeks of age, the delay in maturity could explain reports where the tip (i.e. most caudal aspect) of the keel also manifests the highest rates of damage (Riber and Hinrichsen, 2016; Baur, 2017; Baur et al., 2020). A comparison of traditional and commercial genetic lines (Hocking et al., 2003) found the former come into lay an average of 4 weeks later while also possessing improved bone health, though results were confounded by breed differences such as differences in body mass. By manipulating photoperiod with a small but definitive delay of four days, Silversides et al. (2006) was able to induce positive effects on medullary and cortical density and area of the radius and humerus. Hester et al. (Hester et al., 2011) found no effect of pullet lighting on end of lay bone mineral content suggesting the effects may be short-lived, though they were unable to document whether there was a latency to enter lay between treatments. Interestingly, Gebhardt-Henrich and Fröhlich (2015) examined natural variation in fracture occurrence and found that hens without evidence of fracture at end of lay came into lay 17 days later than hens with fractures. In summary, there is evidence to suggest that premature occurrence of lay could impinge on a hen's ability to develop a mature skeletal structure, which could lead to increased susceptibility for keel bone fractures.

Other possibilities to explain keel bone fractures include the existence of underlying pathological conditions that may resemble an injury that would result from a collision. Most work to date assessing fractures has used rather coarse techniques (i.e. palpation and direct examination of excised keels (Casey-Trott et al., 2015)), with the presence of callus material considered as positive identification of fracture. However, a callus-like expansion of bone can also result from other processes such as periosteal irritations, congenital malformations, periostitis, osteomyelitis and bone tumours (Craig et al., 2016). In this scenario, while more severe or traumatic fractures may result from collisions, there is also a possibility for pathological fractures or a predisposing disease resulting in a weakened skeletal architecture (in contrast to the skeleton being weakened by high egg production). These predisposed bones would be expected to break more easily, even with routine daily forces that would generally be considered innocuous. In contrast, traumatic fractures occur only with an increased/massive force acting on normally structured bones (Craig et al., 2016), though this work has yet to be studied in terms of fractures to the keel. Pathological conditions could include diseases of a genetic (e.g. osteogenesis imperfecta, or glass bone disease), metabolic (e.g. osteoporosis) or nutritional (e.g. osteodystrophia fibrosa, rickets) origin, or bone tumours. Each of these pathological states could result from factors independent of egg production while yielding conditions that make bone susceptible to fracture. It is even possible that, as a result of the intensive selection for egg production, linked traits that relate to these pathological conditions could have been inadvertently selected in the process via a form of genetic hitchhiking.

## 5 Influence of poor bone health on productivity and welfare

Keel bone fractures appear to have detrimental effects on hens' productivity and welfare (Riber et al., 2018). In terms of productivity, fracture repair was suggested to involve a diversion of resources where minerals (e.g. calcium) and energy, normally directed towards the reproductive process of egg production, now must be reallocated to the process of healing bone (Thiruvenkadan et al., 2010). In support of this prediction, hens with increasing severity of keel fracture as defined by a radiographic protocol had decreased egg production that became visible from approximately 50 weeks of age and increased in magnitude with age (Rufener et al., 2018a). The greatest modelled difference between the least and most severe manifestations of damage by 61 weeks of age was 16.2%, meaning that 61-week-old hens with the most severe keel bone fractures would be expected to lay 16.2% fewer eggs than 61-week-old hens without fractures. Similar findings have been reported by others who identified an associated increase with feed and water consumption (Nasr et al., 2012a), although this report did not control for individual variation. Interestingly, assessments done at the flock level were unable to identify a relationship between flock production and the percentage of hens with keel fractures (Heerkens et al., 2013). The reason for the lack of a consistent relationship between egg production and fractures is not clear, where the ambiguity is likely worsened by: difficulties in assessment by palpation (Casey-Trott et al., 2015), high-variable rates of fracture incidence at the flock level, as well as the high, individual-level variation of egg production combined with the suspected/predicted relatively small effects of keel fractures on egg production (Candelotto et al., 2017; Rufener et al., 2018a). Farm-level management factors that are difficult to quantify, such as feed quality or efforts to calm hens, are also likely to contribute to variation and handicap efforts at linking egg production and keel bone fractures.

Parallel to the correlations between the presence of fractures and productivity, changes in mobility can also be indicative of pain and compromised welfare (Prunier et al., 2013). Behavioural indicators of pain are normally characterized by reactions, which serve to terminate the causative stimulus or reduce its effects (Mellor et al., 2000). Specifically in regard to keel bone fracture, Nasr et al. (2012b, 2015), using birds trained to descend from a 50 cm high perch for a food reward, found birds with fractures manifested an increased latency to descend in comparison to those without, a response which was attenuated with analgesics suggesting the involvement of pain pathways. In this scenario, the latency to descend from the perch was reasoned to involve the bird's expectation of pain that would result from musculature attached to the keel generating force during motion as well as the keel making contact with the floor during descent. The observed latency may reflect an interplay of conflicting motivations - the first, where the hen's eagerness to consume the reward competes with the second, the desire to avoid the likely pain that would result. If this mechanism is occurring, it reflects an interference of normal locomotive behaviour and raises the possibility of compromised welfare. As an extension of Nasr's training paradigm, Rufener et al. (2019) demonstrated altered movement patterns between defined zones (i.e. aviary tiers and an external wintergarden) with increasing severity of keel bone fracture. Hens with damaged keels spent more time in the uppermost tiers containing resources (feed, water, nest box) for which demand is generally inelastic. In contrast, these hens also made fewer visits to the lower levels containing resources (e.g. litter for dust bathing) for which demand is more elastic, suggesting definitive behavioural changes in response to fracture. Similar responses in terms of nest box usage (Gebhardt-Henrich and Fröhlich, 2015) and accessing the outdoor (i.e. in this case, free range pasture) environment (Richards et al., 2012) have been reported. Movement within cage systems, although associated with less severe keel damage, results in altered behaviour (Casey-Trott and Widowski, 2016). Taken together, these results support the position that hens manifesting keel fractures within cage and cage-free systems are adopting their behaviour to avoid pain-associated with movement.

Interestingly, specific aspects of fractures such as a fracture gap relate to decreased nest and perch access (Rentsch et al., 2019), suggesting particular types or features of fractures may be more critical than others. Breaks (or gaps) in the keel, and by extension the periosteum, which have the highest density of sensory and sympathetic nerves (Mach et al., 2002; Nencini and Ivanusic, 2016), would result in the greatest sensation of pain (Freeman et al., 2008). More work defining the specific features of keel fractures and their associated behavioural responses, ideally in concert with analgesic manipulation, is needed to determine how different types of fractures relate to pain and consequent welfare.

## 6 Strategies for improving bone health

The problem of keel bone damage is a complex and multifactorial issue, thus solutions will need to consider these various causes in detail. Furthermore, it is likely that no single strategy will suffice, but rather a combination of strategies - genetic, nutritional and management - tailored to the specific housing and production environment will be needed.

### 6.1 *Effective breeding strategies*

In considering potential tools for genetic selection against keel fractures, it needs to be determined which traits should be selected where possibilities may include: stronger bones, improved physical ability, increased mineral absorption, rates of bone remodelling and increased docility. Strategies to link traits with keel fractures have suggested that keel and humerus bone mineral density and strength are associated with reduced keel fracture susceptibility (Candelotto et al., 2017, n.d.) as well selection of specific QTLs (Dunn et al., 2007; Podisi et al., 2012). Alternatively, as susceptibility to bone fractures are a relatively complex genetic trait, genomic selection is also an option (Fulton, 2012). Genomic selection involves monitoring the incidence of fracture in a large population of laying hens and comparing the genomic information of the two sub-populations. One of the strengths of this approach is that it does not target a single factor, for example, bone strength, but focusses on the presence or absence of damage. As part of the process, the mechanism associated with the trait of interest is selected. A challenge with the genomic approach is that it requires a large sampling population (>5000) of which a clear phenotype is needed. Given recent emergence of the complexity of damage that actually exists (Baur et al., 2020), selection of a clear fracture-free phenotype may be problematic. However, if successful and a genomic phenotype is identified, no further phenotypic measurements are required for the selection programme (Eggen, 2012). Although breeding for reduced keel bone damage is attractive, efforts need to be made to ensure that only the traits of interest are selected with minimal loss of desired traits, for example, shell quality.

### 6.2 Nutritional strategies

Efforts to improve bone strength have principally involved better calcium delivery for egg production or slowing the process of de-mineralization. One example that combined these strategies was use of omega-3 fatty acids, which can modulate the processes that drive bone de-mineralization by decreasing the activity of osteocytes, the cells responsible for bone resorption (Liu et al., 2003; Sun et al., 2003; Griel et al., 2007; Shen et al., 2007). Omega-3 fatty acids also enhance the formation of bone by osteoblasts (Watkins et al., 2003). Intervention with omega-3-enhanced diets results in reduced fracture incidence (Tarlton et al., 2013; Toscano et al., 2015), but is also known to benefit human health (Calder, 2006, 2009) leading to benefits for human consumers as well. There is also evidence from various animal studies, however, that omega-3 fatty acids can have detrimental effects on health, possibly through interactions with immune function (Anderson and Fritsche, 2002) or production of damaging free radicals (Aruoma, 1998). The detrimental effects may also be relevant for laying hens (Toscano et al., 2012, 2015).

A second strategy is use of larger calcium particles which take longer to break down and thus maintain elevated dietary calcium throughout a 24-h period. Provision of limestone particulate, oyster shell or other forms of calcium with a relatively large particle size and thus requiring a longer duration for digestion, can maintain egg-shell quality and extend the lay period up to 100 weeks of age (Thiele, 2015; Pottgüter, 2016). Egg shell formation occurs in the evening when the hen is sleeping and is associated with a nine-fold increase in resorption (Van De Velde et al., 1985) to provide greater calcium to supplement dietary sources as needed. Larger dietary calcium particles can thus increase the amount of calcium available during this critical night time period. Although published work on this strategy has focussed on egg shell quality, the reduced need for endogenous calcium, for example, from bone, would likely extend to the keel and thus aid in maintaining skeletal integrity. In other words, as greater amounts of calcium would be available in the blood from the diet, less would be required from the bone. Beyond the form of calcium, feeding the daily requirement of calcium towards the end of the day and/or the last feeding has also been suggested as a strategy to increase calcium concentrations at night and reduce the need for bone resorption (Thiele, 2015).

### 6.3 Rearing hens for better cognitive and musculoskeletal development

The rearing phase is crucial to prepare pullets for their life in the layer house, particularly in the case of more complex (e.g. furnished cages, open barn) and multi-tier housing. In terms of bone health, the environment a pullet is exposed to during her early life affects the musculoskeletal development as well as spatial cognition and behavioural experience with three-dimensional space, which in turn relates to both susceptibility to and the actual occurrence of fractures. In other words, the rearing environment affects the musculoskeletal strength, and thus the susceptibility to fractures when exposed to trauma, for example, due to a fall or collision. The likelihood for falls or collisions, on the other hand, relates to the hen's ability to move through vertical space which is modulated by her behavioural and cognitive development during early life.

### 6.3.1 Musculoskeletal development

Physical exercise leads to higher load-bearing activity on the skeleton which in turn enhances bone strength (Lanyon, 1992). It is well established that housing systems offering opportunities for physical activity (e.g. walking, running, flying and jumping) are associated with improved bone health compared to cage systems, which allow comparatively limited exercise and bone loading (reviewed by Webster, 2004).

The link between the physical activity and bone strength has been well established in adult hens. In humans, the effect of physical exercise on bone strength is strongest before the onset of sexual maturity (Bass, 2000). Correspondingly, the rearing phase (i.e. from hatch to approximately 17 weeks of age) is believed to be a particularly crucial phase regarding bone development in laying hens - especially because the formation of structural bone ceases once a hen enters lay (Hudson et al., 1993). Compared to pullets reared in conventional cages, pullets reared in cages containing perches or in aviaries had improved bone health as indicated by a variety of responses including: increased long bone width and keel length (Casey-Trott et al., 2017b), increased bone mineral density or content (Enneking et al., 2012; Casey-Trott et al., 2017b), and higher cortical bone density, ash content, stiffness and strength (Regmi et al., 2015).

Importantly, increased bone strength seen at the end of rearing seems to persist throughout lay even if hens were transferred to cages after rearing, that is, with minimal bone loading during the laying phase (Casey-Trott et al., 2017c). Hester et al. (2013) also found greater bone mineral density at the end of lay in caged hens that had access to perches only during the rearing phase. Michel and Huonnic (2003) reared pullets in aviaries or floor pens with perches before moving them into aviaries and found higher breaking strength of tibia and humerus in aviary-reared hens at the end of lay. Increased bone strength in aviary-reared, but cage-housed hens further resulted in a lower keel bone fracture prevalence at the end of lay (Casey-Trott et al., 2017a). However, as a suggestion of the limits for enriched rearing, increased mineralization of the keel seen as a result of perch access during rearing did not prevent higher keel bone fracture incidence at the end of lay if hens were given access to perches during the laying phase (Hester et al., 2013).

Besides promoting skeletal development, physical exercise increases muscle mass. Casey-Trott et al. (2017c) found heavier breast and wing muscles in aviary-reared pullets compared to pullets reared in conventional cages, though cage-reared pullets had heavier leg muscles. As with bone health, this effect seems to last throughout the laying phase as Hester et al. (2013) showed increased muscle deposition at the end of lay when hens were given access to perches as pullets. In an experimental study using an ex vivo protocol to model bone fractures resulting from collisions (described previously in this chapter), Toscano et al. (2018) found that breast muscle mass was related to a lower risk to obtain major and severe experimental fractures. The benefits of breast muscling could relate to breast muscles serving as cushion if a keel bone is exposed to trauma, for instance, when a hen collides with a perch. As total body mass is associated with keel bone fracture prevalence and severity (with heavier birds having more severe damage at the keel; Gebhardt-Henrich et al., 2017a) the nature of the relationship is not clear. Breast muscle mass could have a cushioning effect resulting in a reduced impact on the bone, or increased muscle mass could link with higher load bearing on the keel and therefore a stronger bone being less susceptible to break.

Overall, multiple studies have shown that the properties of the rearing system - especially the provision of perches and access to vertical space - affect the physical properties of the skeleton and lead to a lower susceptibility for fractures when exposed to trauma. In addition, housing systems promoting exercise and therefore muscle deposition, especially breast muscles around the keel, might affect skeletal integrity positively due to a reduced risk for injury at the keel. Providing access to resources allowing exercise such as flying and perching during the rearing phase is therefore a powerful strategy to enhance bone health with lasting effects throughout the laying period. As there seems to be only a limited association between bone traits and fracture prevalence (Gebhardt-Henrich et al., 2017b), non-skeletal factors such as behaviour and spatial abilities are believed to strongly influence the risk for injuries at the keel bone (Toscano et al., 2018).

### 6.3.2 Spatial abilities: locomotor skills, perception and cognition

In aviaries where resources (e.g. food, water, nest boxes, perches and litter) are distributed over several stacked tiers up to 3.5 m high, a hen must be able to move within her environment to access these resources. Failure to coordinate movement successfully and consequent falls and collisions within these complex housing systems are likely to result in trauma resulting in keel bone fractures (Sandilands et al., 2009; Wilkins et al., 2011; Harlander-Matauschek et al., 2015; Stratmann et al., 2015b). Although the development of spatial abilities is important for all housing systems, the complexities of aviary systems with their multiple vertical tiers combined with the dangers of poor movement coordination and locomotor skill are the focus of this section. Complexity of the housing environment is an abstract term, but here it is intended to convey the relative amount of resources within housing across multiple areas that require memory and skill to move between (horizontally and vertically), typically with a variety of movement types (e.g. jumping, walking, and flying).

To successfully move within three-dimensional space and avoid falls and collisions, pullets need to have appropriate: (visual) perception, cognitive abilities involving processing of visual and spatial information, and locomotor skills and physical strength. Perception involves visual acuity or depth perception, whereas the cognitive abilities refer to processes such as memory, the discrimination of local and global cues, route planning or knowledge of location. Locomotor skills and physical strength are closely related to the musculoskeletal development described earlier, but require further training of specific locomotor patterns. For instance, while a certain range of locomotor skills and physical strength is needed for a pullet to run horizontally in a floor housing system, a different range must be developed for the jumping and flying to a perch required in an aviary system.

The behavioural component of successfully moving within a housing system is intertwined with cognition and perception. As an example, independent of a hen's strength, a hen needs proper visual acuity and depth perception to assess flying distance as well as spatial memory and knowledge of location to orient herself. In turn, a hen needs to gain experience with specific locomotor patterns to develop the corresponding cognitive abilities. For instance, a hen has to access an elevated platform first to develop knowledge of location corresponding to her position in vertical space. The following sections describe the development of perception, cognition and locomotor skills and give an overview of how housing complexity during rearing could affect these aspects of spatial abilities.

To successfully move within structures of a housing system, a hen needs to be able to perceive her environment properly. Perception involves visual acuity, depth perception, spectral and flicker sensitivity, and accommodation (the ability to change focus). Compared to single-tier or cage-systems, living in an aviary system requires greater depth perception and contrast sensitivity, as the ability to assess distance and angles is crucial to move between vertical elements successfully. Exposure to light plays an important role regarding normal eye development (Prescott et al., 2003), but it remains unclear how other environmental factors affect perception. The visual system in chicks seems to be fully developed by two days of age (Over and Moore, 1981). At this age, chicks can avoid obstacles, follow moving objects and peck at objects. As an indication of depth perception at an early age, newly hatched as well as two-day-old chicks discriminate between shallow and deep surfaces on a visual cliff (Walk and Gibson, 1961; Green et al., 1993). The early environment seems to affect depth perception to some degree (Tallarico and Farrell, 1964), but the underlying mechanisms have not been investigated.

Visual and spatial cognition also develops early in life. Aspects of cognitive abilities such as the interpretation of object permanence, amodal completion, use of spatial cues, memory or knowledge of location have been discussed elsewhere in this chapter. In brief, spatial cognition develops with age, and lateralization plays an important role. A left hemisphere bias develops in ovo and is dominant throughout the first day of life. As a result, a chick's visuospatial analysis initially focusses on local features (e.g. to approach objects or discriminate food items; Vallortigara and Rogers, 2005). Around 11 days of age, chicks would begin to move out of sight of the hen, which coincides with a shift of bias towards the right hemisphere (Freire and Cheng, 2004). Accordingly, chicks start to recognize global cues and relational properties of spatial layouts after 11 days of age (Vallortigara et al., 1997).

The degree of spatial information available in the environment affects the cognitive development of chicks. For instance, rearing chicks with visual barriers resulted in better/faster performance in spatial memory tests (visual displacement and detour tests) when tested at 11 days of age (Freire and Cheng, 2004). In more applied settings, pullets reared in aviaries had a better working memory than pullets reared in cages (Tahamtani et al., 2015), and hens not having access to perches within the first eight weeks of their life had longer latency to find a food reward via platforms with increasing difficulty of the task (Gunnarsson et al., 2000). As the Gunnarsson study did not find any differences in locomotor skills for the easiest task (i.e. reaching a platform 40 cm above ground) and because all hens used vertical space via perches in their home pens, the authors concluded that hens with early access to perches had improved spatial cognition rather than just better locomotor skills (compared to hens that had access to perches only after 8 weeks of age). Further, chicks reared with access to perches and elevated structures were faster than floor-reared chicks in a test requiring chicks to navigate a detour in order to reach conspecifics (Norman et al., 2019), a task where differences in physical strength are unlikely to play a major role.

Locomotor skills develop relatively early during the rearing phase and map onto cognitive changes related to the shift in hemisphere bias. Accordingly, if given the opportunity to access three-dimensional space, individual chicks accessed perches of 10–40 cm height as early as 8 days of age (Gunnarsson et al., 2000; Heikkilä et al., 2006; Wichman et al., 2007), though most studies showed that the majority of chicks used elevated structures such as perches and platforms around 14 days of age (Heikkilä et al., 2006; Wichman et al., 2007; Kozak et al., 2016; Habinski et al., 2017).

If pullets or hens have not gained experience accessing vertical structures such as perches or raised tiers at an early age, they struggle to utilize the full area of complex housing systems as adult hens. Hens reared in aviaries are distributed on all tiers in experimental pens, whereas hens reared in cages spent most time on the floor (Brantsæter et al., 2016). Along those lines, Colson et al. (2008) compared floor- and aviary-reared birds that were housed in an aviary system during lay and reported that hens reared on the floor used upper levels less, showed reduced accuracy in long flights and jumps, and preferred to stay on litter and lower levels. Similarly, hens reared on the floor used the highest tier of the aviary in the layer house less frequently than aviary-reared hens (Michel and Huonnic, 2003).

Despite the evidence showing that space use in the adult housing system is affected by the complexity experienced during rearing, it is difficult to disentangle the underlying mechanisms. In other words, differences in space use in response to the rearing system are probably not only related to the locomotor skills and physical strength of a hen, but more likely the result of interactions between locomotor skills, cognitive development and possibly perception.

In addition to the effects of general housing complexity on spatial abilities, the lack of experience with specific housing-system elements (e.g., ramps) affect the ability to negotiate transitions in a novel environment even if pullets had access to vertical space during rearing. Hence, it is important that the rear and lay environments are considered together. Norman et al. (2018) reared pullets with perches and platforms, but half of the birds were additionally provided with ramps (Fig. 4) to facilitate platform access. During testing at 12–14 weeks of age, pullets that had access to ramps were more successful in reaching food rewards on a platform connected with a ramp and took less time to complete these upward transitions. In addition, ramp-reared pullets showed double the number of transitions between the litter and platforms during group-housing observations immediately after being moved to a novel environment. The study by Norman et al. indicates that both general access to vertical space as well as experience with specific structural elements influences the locomotor skills within three-dimensional space.

In summary, spatial abilities obtained from interactions with the rearing system are highly relevant for laying hens being housed in complex environments. Environmental complexity and provision of structures enabling the access to vertical space promote necessary cognitive development and allow training of locomotor skills, which are critical in aiding pullets as they are moved from the rearing facility to the laying environment. The transition from

**Figure 4** Pullet using a ramp within a commercial multi-tier housing system. Source: A. K. Rentsch.

the pullet to the layer house has a myriad of challenges as hens must - again - learn to navigate within their novel environment.

It is the general consensus that differences between the rearing and the laying environment should be reduced (reviewed by Janczak and Riber, 2015), and meeting a pullet's cognitive and behavioural skills obtained during the rearing phase is crucial to maintain good bone health. Ideally, a hen placed in a novel environment will have had previous experience with all elements of the housing system (perches including specific materials and shapes, ramps, platforms), and had the opportunity to learn how to transit between tiers to reach all resources successfully. To prepare pullets optimally for their novel environment as a laying hen, it is therefore important to not only rear for a specific laying environment, but to enhance the laying environment in order to facilitate adaptability to the laying house.

## 6.4 Adjustments to layer housing

Even though the design and management of the rearing environment can contribute to the cognitive, behavioural and musculoskeletal development of pullets, it is unlikely that stronger bones and enhanced navigation abilities obtained during rearing can reduce keel bone fracture frequency and severity on their own. Adaptations of the laying environment are therefore - in addition to the provision of an optimal rearing environment - necessary to improve the skeletal integrity and reduce keel bone fracture prevalence in adult laying hens maintained under commercial conditions.

### *6.4.1 Complexity of modern housing systems and the role of perches*

Modern housing systems such as aviaries can be up to 3.5 m high with several stacked tiers providing different resources via perches or platforms, which require the hens to jump or fly in order to reach the next tier and the associated resource. While aviaries benefit the hen by providing specific resources for highly motivated, species-specific activities (Weeks and Nicol, 2006), their increasing adoption by producers is also related to the economic benefit of being able to place more hens within a given area (Fröhlich and Oester, 2001; Aerni et al., 2005).

Domestic laying hens are descendants from the red junglefowl (Fumihito et al., 1994), a generally ground-dwelling species, and are therefore not particularly well suited to move within these complex environments. More so, increased body weight relative to wing area as a consequence of focussed genetic breeding has most probably led to a hen that lacks the stability of its ancestor and thus may have greater problem moving within aviary systems (Sandilands et al., 2009). Thus, it is not surprising that hens might have difficulty descending within an aviary system. In commercial aviaries, angles and distances between aviary elements often exceed the navigation abilities of laying hens which might explain why Campbell et al. (2015) observed up to 21% of flights in an aviary resulting in failed landings. Failed landings and/or falls can result in crashes and collisions which are assumed to be one reason for the high prevalence of keel bone fractures in laying hens (Campbell et al., 2015; Stratmann et al., 2015a,b). Similarly, a positive relationship was seen between the inter-row distance of aviaries and rates of keel bone fractures, which the authors reasoned resulted from failed attempts to jump between rows (Heerkens et al., 2015). Perches are especially seen as a hazardous element in aviary systems due to their exposure increasing the risk for high impacts after uncontrolled falls and collisions (reviewed by Sandilands et al., 2009).

Even though perches are suspected to be directly related to keel bone fracture occurrence, they are an important element in a layer housing system which is a paradox highlighted previously (Sandilands et al., 2009). Perches not only offer the opportunity to fulfil the behavioural need for roosting in elevated locations, but also serve as a means to access the stacked tiers of an aviary and thus, allow for movement within the system (reviewed by Struelens and Tuyttens, 2009). Perch material and shape are known to mitigate the problem of perches being a hazardous element. Perch shape affects keel bone health, with circular perches causing more damage to the keel than rectangular perches (Tauson and Abrahamsson, 1994). In the same vein, Pickel et al. (2011) measured the peak force experienced at the keel in hens perching on different perch types. Round and oval perches resulted in the highest force, followed by squared and an experimental, cushioned prototype perch. The authors suggested that increased contact area, thus, reduces localized pressure on the keel as well as a soft surface may reduce keel bone problems.

Within more applied settings, use of soft perches and reduced keel bone damage is supported by Stratmann et al. (2015b) who found a reduction in keel bone fracture prevalence and hazardous landings (Scholz et al., 2014). Perch design may also increase landing safety as Pickel et al. (2010) reported fewer balancing movements with increasing perch diameter (27, 34, 45 mm) on rubber compared to wood and steel perches. Perch material, as well as cleanliness, is likely to influence overall ability to move as the latency, pre-jumping behaviour, slips, failures to jump and crashes were influenced by jumping from wooden, metal and PVC perches (Scott and MacAngus, 2004). It is suggested to install non-slippery perches with a large contact area, possibly made from a soft material, to reduce keel bone fracture prevalence. To date, the development of such perches and the application to practical farming have proven difficult due to problems with hygiene and pest management (e.g. infestation of soft surfaces by mites). Thus, not only the shape and material of perches, but also their placement must be considered in order to facilitate bird movement within a given housing system.

### 6.4.2 Facilitating bird movement

Depending on distance and angle between take-off and landing sites in a commercial non-cage system, laying hens may have difficulty moving between resources and tiers successfully. Under experimental conditions, angles steeper than 60° for upward movements (Scott et al., 1997) and steeper than 30° for downward movements (Lambe et al., 1997; Scholz et al., 2014; Scott et al., 1997) have been shown to result in falls at landing, whereas a horizontal distance of more than 50 cm seems to increase the difficulty to navigate (Scott and Parker, 1994). Downward paths appear to be more difficult for laying hens than upward paths (Moinard et al., 2004; Scott et al., 1997), and the presence of conspecifics or obstacles at the landing site resulted in an increased risk for falls (Moinard et al., 2005). Lighting conditions only affect landing accuracy and the latency to jump when the intensity is very low (i.e. below 2 lux; Moinard et al., 2004; Taylor et al., 2003). Even if navigated successfully (i.e. without a fall at landing), increased angles and distances as well as downward compared to upward movements result in higher forces experienced at the keel (Rufener et al., 2020). Accordingly, the European Food Safety Authority (EFSA) reviewed the welfare aspects of perch use for laying hens extensively and stated that the risk of injury increases when hens have to jump a distance of more than 80 cm vertically, horizontally or diagonally, or jump an angle steeper than 45 cm (EFSA Panel on Animal Health and Animal Welfare, 2015). It is therefore recommended to install perches in a way to allow transitions of a short distance and flat angle.

However, angles and distance in commercial aviaries are often steeper and longer than recommended, and rearranging perches in a given housing system can be difficult as they often serve as a supporting structure. Several studies have proposed alternative solutions to facilitate hen movement within aviaries. Stratmann et al. (2015a) modified a commercial aviary setting by adding perches, platforms or ramps to facilitate movements between the tiers of the system, aiming to reduce the risk for falls and collisions. The authors found fewer falls and collisions as well as a lower keel bone fracture prevalence in hens provided with ramps and suggested that the continuous path between the tiers supports safer and more appropriate behaviours such as walking and running instead of flying and jumping. Accordingly, Heerkens et al. (2016) found similar effects in an experimental setting, where ramps were used to connect stacked tiers and perches at different heights resulting in fewer keel bone fractures. As different strains show considerable differences in the spatial distribution in vertical space (Ali et al., 2016; Kozak et al., 2016), it is important to work on solutions facilitating movements irrespective of strain.

Overall, the installation of ramps (Fig. 1) is a relatively inexpensive and straightforward adaptation of most non-cage systems which could improve navigation and consequently reduce fracture prevalence. The opportunity to walk instead of fly or jump might be especially beneficial for hens already having keel bone fractures as the wing muscles are attached to the keel (Sisson et al., 1975) and because fractures have been associated with limited bird movement in aviary systems (Rentsch et al., 2019; Rufener et al., 2019). Flying, wing-assisted running, jumping and other activities involving wing muscles are likely to apply higher forces to the keel presumably resulting in more pain if a fracture is present. Hens can climb ramps without using their wings up to an angle of 40° (LeBlanc, 2016), thus the installation of ramps with an angle up to 40° could allow hens having keel bone fractures to access upper levels and associated resources with less pain. Accordingly, Rentsch et al. (2019) found that hens with fresh fractures were less likely to show vertical locomotion than hens with absent or healed fractures, but demonstrated that hens used ramps irrespective of keel bone health. As hens can walk on grid-type ramps but need to jump from rung to rung on latter-type ramps (Pettersson et al., 2017; Norman et al., 2018), the installation of grid ramps at an angle of <40° can facilitate transitions between tiers and reduce keel bone fracture risk.

### *6.5 Improving hen handling during depopulation*

To improve welfare during hen removal, it is key to reduce the amount and duration of manual handling (Scott, 1993). For short periods or with small flocks (e.g. less than 200 birds), catchers are typically able to perform the task with minimal injury to the birds and themselves. However, once the duration of removal is extended and fatigue does become a factor, it becomes difficult for even the most devoted of caretakers to maintain the appropriate concentration and physicality for such a task resulting in animal suffering and worker hazard (Kettlewell and Mitchell, 1994). For instance, as an example of attempts to reduce the human factor of fatigue in hen removal, Kristensen et al. (2001) reported the development of a modular system on wheels that brought the transport crate directly to the birds and minimized the carrying process, reducing the handling time for 450 hens from 65 min to 4.5 min. Despite the increased speed of removal, the technique resulted in a statistically similar (though numerically less) frequency of bone breakage. As only two farms were assessed and the collectors were aware of the research objectives, this may have impacted observed results. Along the same vein, in the last 60 years, various formats of collection techniques have been developed to reduce the human component including: a recessed conveyor belt that moved the birds directly to the truck, mats that could be laid down prior to collection that would be lifted out of the house with birds contained and large-tined forks that could scoop birds off the floor (Lacy and Czarick, 1998). While each of these mechanical efforts have held some degree of promise, none has been widely adopted by the industry due to a variety of reasons including: excessive capital costs, poor integration into existing systems, discomfort with the new technology, among others (Scott, 1993). Nonetheless, improving the quality of handling by reducing the fatigue of the collectors is key to improving the welfare of removed hens (Gregory and Wilkins, 1989; Kettlewell and Mitchell, 1994).

## 7 Future trends in research

Given the existing knowledge that has been generated, the scientific community is poised to make rapid advancements in understanding and resolving the problem of keel bone damage. One key step has been advances in detection, which can be done rapidly and within commercially relevant settings (e.g. by standardizing palpation and through radiographic images). While not providing direct benefit, improved assessment will allow for accuracy to better evaluate intervention strategies of nutritional, genetic and management natures. Another key area will be greater focus on developmental processes during rearing, particularly with increased use of cage-free systems and the appreciation of how birds must develop to use and benefit from the contained resources.

# 8 Where to look for further information

## 8.1 Further reading

An additional resource focussing on keel bone damage is the chapter 'Skeletal problems in contemporary commercial laying hens', Advances in Poultry Welfare Science, 2018. General principles on bone health.

## 8.2 Major international research projects and conferences

- An EU-COST Action (www.keelbonedamage.eu, CA15224) will continue until November 2020 and is an excellent source to provide the latest information regarding ongoing activities, workshops and data.
- Several large EU-funded projects involving assessment of hen health and welfare include ChickenStress (www.chickenstress.edu) and the 'Poultry and pig low input and organic production systems' Welfare'.

# 9 References

Aerni, V., Brinkhof, M. W. G., Wechsler, B., Oester, H. and Fröhlich, E. (2005). Productivity and mortality of laying hens in aviaries: a systematic review, *World's Poultry Science Journal*. Cambridge University Press on behalf of World's Poultry Science Association 61(1), 130–142.

Afazcli, H., Jafari, A., Rafiee, S. and Nosrati, M. (2014). An investigation of biogas production potential from livestock and slaughterhouse wastes, *Renewable and Sustainable Energy Reviews*. Elsevier 34, 380–386.

Ali, A. B. A., Campbell, D. L., Karcher, D. M. and Siegford, J. M. (2016). Influence of genetic strain and access to litter on spatial distribution of 4 strains of laying hens in an aviary system, *Poultry Science* 95(11), 2489–2502. doi: 10.3382/ps/pew236.

Anderson, M. and Fritsche, K. L. (2002). (n-3) Fatty acids and infectious disease resistance, *Journal of Nutrition*. 132(12), 3566–3576. Available at: http://www.ncbi.nlm.nih.gov/pubmed/12468590.

Aruoma, O. I. (1998). Free radicals, oxidative stress, and antioxidants in human health and disease, *Journal of the American Oil Chemists' Society*. Springer-Verlag 75(2), 199–212. doi: 10.1007/s11746-998-0032-9.

Bain, M. M., Nys, Y. and Dunn, I. C. (2016). Increasing persistency in lay and stabilising egg quality in longer laying cycles. What are the challenges?, *British Poultry Science*. Taylor & Francis 57(3), 330–338.

Baker, S. L., Robison, C. I., Karcher, D. M., Toscano, M. J. and Makagon, M. M. (2020). Keel impacts and associated behaviors in laying hens. *Applied Animal Behaviour Science* 222:104886 Available at: https://linkinghub.elsevier.com/retrieve/pii/S0168159119301467.

Bass, S. L. (2000). The prepubertal years: a uniquely opportune stage of growth when the skeleton is most responsive to exercise?, *Sports Medicine* 30(2), 73–78.

Baur, S. (2017). Radiographic evaluation of keel bone damage. Universität Bern.

Baur, S., Rufener, C., Toscano, M. J. and Geissbühler, U. (2020). Radiographic evaluation of keel bone damage in laying hens - morphologic and temporal observations in a longitudinal study, *Frontiers in Veterinary Science* 7, 129.

Bestman, M. and Wagenaar, J. P. (2014). Health and welfare in Dutch organic laying hens, *Animals*. Multidisciplinary Digital Publishing Institute 4(2), 374–390. doi: 10.3390/ani4020374.

Blokhuis, H., et al. (2005). Welfare aspects of various systems of keeping laying hens, *EFSA Journal* 197, 1–23.

Brantsæter, M., Nordgreen, J., Rodenburg, T. B., Tahamtani, F. M., Popova, A. and Janczak, A. M. (2016). Exposure to increased environmental complexity during rearing reduces fearfulness and increases use of three-dimensional space in laying hens (Gallus gallus domesticus), *Frontiers in Veterinary Science*. Frontiers Media SA 3, 14.

Buckner, G. D., Insko, W. M., Henry, A. H. and Wachs, E. F. (1948). Rate of growth and calcification of the sternum of male and female New Hampshire chickens, *Poultry Science*. Oxford University Press: Oxford, UK 27(4), 430–433.

Buckner, G. D., Insko, W. M., Henry, A. H. and Wachs, E. F. (1949). Rate of growth and calcification of the sternum of male and female New Hampshire chickens having crooked keels, *Poultry Science*. Oxford University Press, Oxford, UK 28(2), 289–292.

Budgell, K. L. and Silversides, F. G. (2004). Bone breakage in three strains of end-of-lay hens, *Canadian Journal of Animal Science*. Agricultural Institute of Canada 84(4), 745–747. doi: 10.4141/A04-040.

Calder, P. C. (2006). Polyunsaturated fatty acids and inflammation, *Prostaglandins, Leukotrienes and Essential Fatty Acids* 75(3), 197–202. doi: 10.1016/j.plefa.2006.05.012.

Calder, P. C. (2009). Polyunsaturated fatty acids and inflammatory processes: new twists in an old tale, *Biochimie*. 91(6), 791–795. Available at: http://www.ncbi.nlm.nih.gov/entrez/query.fcgi?cmd=Retrieve&db=PubMed&dopt=Citation&list_uids=19455748.

Campbell, D. L. M., Makagon, M. M., Swanson, J. C., and Siegford, J. M. (2015). Litter use by laying hens in a commercial aviary: dust bathing and piling. *Poultry Science* 95, 164–175.

Candelotto, L., Stratmann, A., Gebhardt-Henrich, S. G., Rufener, C., van de Braak, T. and Toscano, M. J. (2017). Susceptibility to keel bone fractures in laying hens and the role of genetic variation, *Poultry Science* 96(10), 3517–3528. doi: 10.3382/ps/pex146.

Candelotto, L., et al. (n.d.). Genetic variation of keel and long bone skeletal properties for five lines of laying hens, *Journal of Applied Poultry Research*.

Casey-Trott, T., et al. (2015). Methods for assessment of keel bone damage in poultry, *Poultry Science* 71(3), 461–472. doi: 10.3382/ps/pev223.

Casey-Trott, T. M., Guerin, M. T., Sandilands, V., Torrey, S. and Widowski, T. M. (2017a). Rearing system affects prevalence of keel-bone damage in laying hens: a longitudinal study of four consecutive flocks, *Poultry Science* 96(7), 2029–2039. doi: 10.3382/ps/pex026.

Casey-Trott, T. M., Korver, D. R., Guerin, M. T., Sandilands, V., Torrey, S. and Widowski, T. M. (2017b). Opportunities for exercise during pullet rearing, Part I: effect on the musculoskeletal characteristics of pullets, *Poultry Science* 96(8), 2509–2517. doi: 10.3382/ps/pex059.

Casey-Trott, T. M., Korver, D. R., Guerin, M. T., Sandilands, V., Torrey, S. and Widowski, T. M. (2017c). Opportunities for exercise during pullet rearing, Part II: long-term effects on bone characteristics of adult laying hens at the end-of-lay, *Poultry Science* 96(8), 2518–2527. doi: 10.3382/ps/pex060.

Casey-Trott, T. M. and Widowski, T. M. (2016). Behavioral differences of laying hens with fractured keel bones within furnished cages, *Frontiers in Veterinary Science* 3, 42.

Chargo, N. J., Robison, C. I., Akaeze, H. O., Baker, S. L., Toscano, M. J., Makagon, M. M. and Karcher, D. M. (2018a). Keel bone differences in laying hens housed in enriched colony cages, *Poultry Science*. Poultry Science Association, Inc. 98(2), 1031–1036.

Chargo, N. J., Robison, C. I., Baker, S. L., Toscano, M. J., Makagon, M. M. and Karcher, D. M. (2018b). Keel bone damage assessment: consistency in enriched colony laying hens, *Poultry Science*. Poultry Science Association, Inc. 98(2), 1017–1022.

Colson, S., Arnould, C. and Michel, V. (2008). Influence of rearing conditions of pullets on space use and performance of hens placed in aviaries at the beginning of the laying period, *Applied Animal Behaviour Science* 111(3–4), 286–300. doi: 10.1016/j.applanim.2007.06.012.

Craig, L. E., Dittmer, K. E. and Thompson, K. G. (2016). Bones and joints. In: Maxie, G. (Ed.) *Jubb, Kennedy & Palmer's Pathology of Domestic Animals*, Volume 1, 6th edn. Elsevier, St Louis, MO, pp. 16–163.

Dacke, C. G., et al. (1993). Medullary bone and avian calcium regulation, *Journal of Experimental Biology*. The Company of Biologists Ltd, 184(1), 63–88.

Darwin, C. R. (1868). Variation of plants and animals under domestication.

Del Hoyo, J., Elliott, A. and Sargatal, J. (2001). *Hand Book of the Birds of the World. Volume 2: New World Vultures to Guinea Fowl*. Lynx Editions, Barcelona.

Dunn, I. C. and Sharp, P. J. (1992). The effect of photoperiodic history on egg laying in dwarf broiler hens, *Poultry Science*. Oxford University Press, Oxford, UK 71(12), 2090–2098.

Dunn, I. C., Fleming, R. H., McCormack, H. A., Morrice, D., Burt, D. W., Preisinger, R. and Whitehead, C. C. (2007). A QTL for osteoporosis detected in an F2 population derived from White Leghorn chicken lines divergently selected for bone index, *Animal Genetics*. Wiley Online Library 38(1), 45–49.

EFSA Panel on Animal Health and Animal Welfare (2015). Scientific opinion on welfare aspects of the use of perches for laying hens. *EFSA Journal* 13, 4131.

Eggen, A. (2012). The development and application of genomic selection as a new breeding paradigm, *Animal Frontiers*. American Society of Animal Science 2(1), 10–15.

Enneking, S. A., Cheng, H. W., Jefferson-Moore, K. Y., Einstein, M. E., Rubin, D. A. and Hester, P. Y. (2012). Early access to perches in caged White Leghorn pullets, *Poultry Science*. Oxford University Press 91(9), 2114–2120.

Etches, R. J. (1987). Calcium logistics in the laying hen, *The Journal of Nutrition* 117(3), 619–628.

Eusemann, B. K., Baulain, U., Schrader, L., Thöne-Reineke, C., Patt, A. and Petow, S. (2018a). Radiographic examination of keel bone damage in living laying hens of different strains kept in two housing systems, *PLoS ONE*. Public Library of Science 13(5), e0194974.

Eusemann, B. K., Sharifi, A. R., et al. (2018b). Influence of a sustained release deslorelin acetate implant on reproductive physiology and associated traits in laying hens, *Frontiers in Physiology* 9, 1846.

FAWC (2010). *Opinion on Osteoporosis and Bone Fractures in Laying Hens*. Farm Animal Welfare Council, London. Available at: http://www.fawc.org.uk/pdf/bone-strength-opinion-101208.pdf.

FAWC (2013). *An Open Letter to Great Britain Governments: Keel Bone Fracture in Laying Hens*. Farm Animal Welfare Council, London.

Fleming, R. H., et al. (2004). Assessing bone mineral density in vivo: digitized fluoroscopy and ultrasound, *Poultry Science* 83(2), 207–214. Available at: http://www.ncbi.nlm.nih.gov/pubmed/14979571.

Fleming, R. H., McCormack, H. A., McTeir, L. and Whitehead, C. C. (2006). Relationships between genetic, environmental and nutritional factors influencing osteoporosis in laying hens, *British Poultry Science*. Taylor & Francis 47(6), 742–755.

Freeman, K. T., et al. (2008). A fracture pain model in the rat: adaptation of a closed femur fracture model to study skeletal pain, *Anesthesiology: The Journal of the American Society of Anesthesiologists*. The American Society of Anesthesiologists, 108(3), pp. 473-483.

Freire, R. and Cheng, H. W. (2004). Experience-dependent changes in the hippocampus of domestic chicks: a model for spatial memory, *European Journal of Neuroscience* 20(4), 1065–1068. doi: 10.1111/j.1460-9568.2004.03545.x.

Fröhlich, E. K. F. and Oester, H. (2001). From battery cages to aviaries: 20 years of Swiss experiences, In: Sambraus, M. G. H. and Steiger, A. (Eds.), *Das Wohlergehen von Legehennen in Europa-Berichte, Analysen und Schlussfolgerungen*. Citeseer, 28:176.

Fulton, J. E. (2012). Genomic selection for poultry breeding, *Animal Frontiers*. American Society of Animal Science, 2(1), 30–36.

Fumihito, A., Miyake, T., Sumi, S., Takada, M., Ohno, S. and Kondo, N. (1994). One subspecies of the red junglefowl (Gallus gallus gallus) suffices as the matriarchic ancestor of all domestic breeds, *Proceedings of the National Academy of Sciences of the United States of America* 91(26), 12505–12509. doi: 10.1073/pnas.91.26.12505.

Gebhardt-Henrich, S. G. and Fröhlich, E. K. F. (2015). Early onset of laying and bumblefoot favor keel bone fractures, *Animals* 5(4), 1192–1206. doi: 10.3390/ani5040406.

Gebhardt-Henrich, S. G., Pfulg, A., Fröhlich, E. K. F., Käppeli, S., Guggisberg, D., Liesegang, A. and Stoffel, M. H. (2017a). Limited associations between keel bone damage and bone properties measured with computer tomography, three-point bending test, and analysis of minerals in Swiss laying hens', *Frontiers in Veterinary Science*. Lausanne, 4(128), 128. doi: 10.3389/fvets.2017.00128.

Gebhardt-Henrich, S. G., Toscano, M. J. and Würbel, H. (2017b). Perch use by broiler breeders and its implication on health and production, *Poultry Science* 96(10), 3539–3549. doi: 10.3382/ps/pex189.

Gebhardt-Henrich, S. G., Rufener, C. and Stratmann, A. (2019). Improving intra-and inter-observer repeatability and accuracy of keel bone assessment by training with radiographs, *Poultry Science*. doi: 10.1093/ps/pez410.

Gerpe, C., Stratmann, A. and Toscano., M. J. (2017). Assessment of animal welfare during the collection process of depopulation for end of lay hens. In: de Jong, I. and Koene, P. (Eds) Proceedings of the 7th International Conference on the Assessment of Animal Welfare at Farm and Group Level. Wageningen Academic Publishers, Ede, p. 184.

Gilbert, S. F. (2000). Osteogenesis: the development of bones. In: Gilbert, S. F. (Ed.), *Developmental Biology*. 6th edn. Sinauer Associates, Sunderland, MA.
Green, P. R., Davies, I. B. and Davies, M. N. (1993). 'Interaction of visual and tactile information in the control of chicks' locomotion in the visual cliff, *Perception* 22(11), 1319–1331. doi: 10.1068/p221319.
Gregory, N. G. and Wilkins, L. J. (1989). Broken bones in domestic fowl: handling and processing damage in end-of-lay battery hens, *British Poultry Science* 30(3), 555–562. Available at: http://www.ncbi.nlm.nih.gov/entrez/query.fcgi?cmd=Retrieve&db=PubMed&dopt=Citation&list_uids=2819499.
Gregory, N. G., Wilkins, L. J., Eleperuma, S. D., Ballantyne, A. J. and Overfield, N. D. (1990). Broken bones in domestic fowls: effect of husbandry system and stunning method in end-of-lay hens, *British Poultry Science*. Taylor & Francis 31(1), 59–69.
Gregory, N. G., Wilkins, L. J., Alvey, D. M. and Tucker, S. A. (1993). Effect of catching method and lighting intensity on the prevalence of broken bones and on the ease of handling of end-of-lay hens, *The Veterinary Record* 132(6), 127–129.
Griel, A. E., Kris-Etherton, P. M., Hilpert, K. F., Zhao, G., West, S. G. and Corwin, R. L. (2007). An increase in dietary n-3 fatty acids decreases a marker of bone resorption in humans, *Nutrition Journal* 6, 2. doi: 10.1186/1475-2891-6-2.
Gunnarsson, et al. (2000). Rearing without early access to perches impairs the spatial skills of laying hens, *Applied Animal Behaviour Science*. Netherlands 67(3), 217–228. doi: 10.1016/S0168-1591(99)00125-2.
Habig, C., et al. (2017). How bone stability in laying hens is affected by phylogenetic background and performance level, *European Poultry Science*. Ulmer 81, ISSN 1612-9199, © Verlag Eugen Ulmer, Stuttgart. doi: 10.1399/eps.2017.200.
Habinski, A. M., Caston, L. J., Casey-Trott, T. M., Hunniford, M. E. and Widowski, T. M. (2017). Animal well-being and behavior: development of perching behavior in 3 strains of pullets reared in furnished cages, *Poultry Science* 96(3), 519–529. doi: 10.3382/ps/pew377.
Harlander-Matauschek, A., Rodenburg, T. B., Sandilands, V., Tobalske, B. W. and Toscano, M. J. (2015). Causes of keel bone damage and their solutions in laying hens, *World's Poultry Science Journal* 71(3), 461–472. doi: 10.1017/S0043933915002135.
Heerkens, J., et al. (2013). Do keel bone deformations affect egg production in end-of-lay housing hens housed in aviaries? In: Tauson, R., et al. (Eds) 9th European Poultry Conference, Uppsala, Sweden, p. 127.
Heerkens, J. L. T., Delezie, E., Rodenburg, T. B., Kempen, I., Zoons, J., Ampe, B. and Tuyttens, F. A. (2015). Risk factors associated with keel bone and foot pad disorders in laying hens housed in aviary systems, *Poultry Science*. Oxford University Press 95(3), 482–488. doi: 10.3382/ps/pev339.
Heerkens, J. L. T., Delezie, E., Ampe, B., Rodenburg, T. B. and Tuyttens, F. A. (2016). Ramps and hybrid effects on keel bone and foot pad disorders in modified aviaries for laying hens, *Poultry Science*. Oxford University Press 95(11), 2479–2488. doi: 10.3382/ps/pew157.
Heikkilä, M., Wichman, A., Gunnarsson, S. and Valros, A. (2006). Development of perching behaviour in chicks reared in enriched environment, *Applied Animal Behaviour Science*. Elsevier 99(1–2), 145–156.
Hester, P. Y., Wilson, D. A., Settar, P., Arango, J. A. and O'Sullivan, N. P. (2011). Effect of lighting programs during the pullet phase on skeletal integrity of egg-laying

strains of chickens, *Poultry Science*. Oxford University Press, Oxford, UK 90(8), 1645–1651.

Hester, P. Y., Enneking, S. A., Haley, B. K., Cheng, H. W., Einstein, M. E. and Rubin, D. A. (2013). The effect of perch availability during pullet rearing and egg laying on musculoskeletal health of caged White Leghorn hens, *Poultry Science*. Oxford University Press 92(8), 1972–1980.

Hocking, P. M., Bain, M., Channing, C. E., Fleming, R. and Wilson, S. (2003). Genetic variation for egg production, egg quality and bone strength in selected and traditional breeds of laying fowl, *British Poultry Science*. Taylor & Francis 44(3), 365–373.

Hudson, H. A. et al. (1993) Histomorphometric bone properties of sexually immature and mature White Leghorn hens with evaluation of fluorochrome injection on egg production traits, *Poultry science*. Oxford University Press 72(8), 1537–1547.

Hurwitz, S. (1965). Calcium turnover in different bone segments of laying fowl, *American Journal of Physiology* 208(1), 203–207.

Janczak, A. M. and Riber, A. B. (2015). Review of rearing-related factors affecting the welfare of laying hens, *Poultry Science* 94(7), 1454–1469. doi: 10.3382/ps/pev123.

Johnson, A. L. (2015). Reproduction in the female. In: Scanes, C. G. (Ed.) *Sturkie's Avian Physiology*, 6th edn. Elsevier Academic Press, London, pp. 635–665. doi: 10.1016/B978-0-12-407160-5.00028-2.

Käppeli, S., Gebhardt-Henrich, S. G., Fröhlich, E., Pfulg, A. and Stoffel, M. H. (2011). Prevalence of keel bone deformities in Swiss laying hens, *British Poultry Science*. Lidfors, L., Blokhuis, H. J. and Keeling, L. (Eds). Taylor & Francis, Uppsala, Sweden 52(5), 531–536. doi: 10.1080/00071668.2011.615059.

Karcher, D. M. and Mench, J. A. (2018). Overview of commercial poultry production systems and their main welfare challenges. In: Mench, J. A. (Ed.) *Advances in Poultry Welfare Science*, 1st edn. Woodhead Publishing, Cambridge, pp. 3–25.

Kerschnitzki, M., Zander, T., Zaslansky, P., Fratzl, P., Shahar, R. and Wagermaier, W. (2014). Rapid alterations of avian medullary bone material during the daily egg-laying cycle, *Bone*. Elsevier B.V. 69, 109–117. doi: 10.1016/j.bone.2014.08.019.

Kestin, S. C., et al. (1992). Prevalence of leg weakness in broiler chickens and its relationship with genotype, *Veterinary Record* 131(9), 190–194. Available at: http://www.ncbi.nlm.nih.gov/entrez/query.fcgi?cmd=Retrieve&db=PubMed&dopt=Citation&list_uids=1441174.

Kettlewell, P. J. and Mitchell, M. A. (1994). Catching, handling and loading of poultry for road transportation, *World's Poultry Science Journal*. Cambridge University Press 50(1), 54–56.

Knowles, T. G. and Wilkins, L. J. (1998). The problem of broken bones during the handling of laying hens - a review, *Poultry Science*. Oxford University Press 77(12), 1798–1802.

Kozak, M., Tobalske, B., Martins, C., Bowley, S., Wuerbel, H. and Harlander-Matauschek, A. (2016). Use of space by domestic chicks housed in complex aviaries, *Applied Animal Behaviour Science*. Elsevier B.V. 181, 115–121. doi: 10.1016/j.applanim.2016.05.024.

Kristensen, H. H., Berry, P. S. and Tinker, D. B. (2001). Depopulation systems for spent hens–a preliminary evaluation in the United Kingdom, *The Journal of Applied Poultry Research*. Oxford University Press 10(2), 172–177.

Lacy, M. P. and Czarick, M. (1998). Mechanical harvesting of broilers, *Poultry Science*. Oxford University Press 77(12), 1794–1797.

Lambe, N. R., Scot, G. B., and Hitchcock, D. (1997). Behaviour of laying hens negotiating perches at different heights. *Animal Welfare* 6, 29-41.

Lanyon, L. E. (1992). Control of bone architecture by functional load bearing, *Journal of Bone and Mineral Research* 7 (Suppl. 2), S369-S375. doi: 10.1002/jbmr.5650071403.

Lay, D. C., Fulton, R. M., Hester, P. Y., Karcher, D. M., Kjaer, J. B., Mench, J. A., Mullens, B. A., Newberry, R. C., Nicol, C. J., O'Sullivan, N. P. and Porter, R. E. (2011). Hen welfare in different housing systems, *Poultry Science* 90(1), 278-294. doi: 10.3382/ps.2010-00962.

LeBlanc, C. (2016). Development of domestic fowl locomotion over inclined surfaces and use of anticipation strategies. Thesis. University of Guelph. Available at: http://hdl.handle.net/10214/9670.

Leyendecker, M., Hamann, H., Hartung, J., Kamphues, J., Neumann, U., Sürie, C. and Distl, O. (2005). Keeping laying hens in furnished cages and an aviary housing system enhances their bone stability, *British Poultry Science*. Taylor & Francis 46(5), 536-544.

Liu, D., et al. (2003). Long-term supplementation of various dietary lipids alters bone mineral content, mechanical properties and histological characteristics of Japanese quail, *Poultry Science* 82(5), 831-839. Available at: http://www.ncbi.nlm.nih.gov/entrez/query.fcgi?cmd=Retrieve&db=PubMed&dopt=Citation&list_uids=12762407.

Mach, D. B., Rogers, S. D., Sabino, M. C., Luger, N. M., Schwei, M. J., Pomonis, J. D., Keyser, C. P., Clohisy, D. R., Adams, D. J., O'Leary, P. and Mantyh, P. W. (2002). Origins of skeletal pain: sensory and sympathetic innervation of the mouse femur, *Neuroscience*. Elsevier 113(1), 155-166.

Mellor, D. J., Cook, C. J. and Stafford, K. J. (2000). Quantifying some responses to pain as a stressor, *The Biology of Animal Stress: Basic Principles and Implications for Welfare*. CABI Publishing, Wallingford, Oxon, UK, pp. 171-198.

Michel, V. and Huonnic, D. (2003). A comparison of welfare, health and production performance of laying hens reared in cages or in aviaries, in *Abstracts from the Spring Meeting of the WPSA French Branch, British Poultry Science* 44(5), 769-831, doi: 10.1080/00071668.2003.10871381.

Mitchell, M. A. and Kettlewell, P. J. (2009). Welfare of poultry during transport - a review. In: Poultry Welfare Symposium. Association Proceeding, Cervia, pp. 90-100.

Moinard, C., Rutherford, K. M. D., Haskell, M. J., McCorquodale, C., Jones, R. B. and Green, P. R. (2005). Effects of obstructed take-off and landing perches on the flight accuracy of laying hens. *Applied Animal Behaviour Science* 93, 81-95. Available at: http://www.sciencedirect.com/science/article/pii/S0168159104003077.

Moinard, C., Statham, P., Haskell, M. J., McCorquodale, C., Jones, R. B. and Green, P. R. (2004). Accuracy of laying hens in jumping upwards and downwards between perches in different light environments, *Applied Animal Behaviour Science* 85(1-2), 77-92. doi: 10.1016/j.applanim.2003.08.008.

Nasr, M. A. F., Murrell, J., Wilkins, L. and Nicol, C. (2012a). The effect of keel fractures on egg production parameters, mobility and behaviour in individual laying hens, *Animal Welfare* 21(1), 127-135. Available at: http://www.ufaw.org.uk/documents/nasr.pdf.

Nasr, M. A., Nicol, C. J. and Murrell, J. C. (2012b). Do laying hens with keel bone fractures experience pain?, *PLoS ONE*. Public Library of Science 7(8), e42420. doi: 10.1371/journal.pone.0042420.

Nasr, M. A. F., Nicol, C. J., Wilkins, L. and Murrell, J. C. (2015). The effects of two non-steroidal anti-inflammatory drugs on the mobility of laying hens with keel bone fractures, *Veterinary Anaesthesia and Analgesia*. Wiley Online Library 42(2), 197–204.

Nencini, S. and Ivanusic, J. J. (2016). The physiology of bone pain. How much do we really know?, *Frontiers in Physiology*. Frontiers 7, 157.

Van Niekerk, T. and Reuvekamp, B. (1994). Husbandry factors and bone strength in laying hens. In: Proceedings of the 9th European Symposium on Poultry Welfare, Volume II. Glasgow, Scotland, pp. 133–136.

Norman, K. I., Adriaense, J. E. C. and Nicol, C. J. (2019). The impact of early structural enrichment on spatial cognition in layer chicks, *Behavioural Processes*. Elsevier 164(May), 167–174. doi: 10.1016/j.beproc.2019.05.008.

Norman, K. I., Weeks, C. A., Pettersson, I. C. and Nicol, C. J. (2018). The effect of experience of ramps at rear on the subsequent ability of layer pullets to negotiate a ramp transition, *Applied Animal Behaviour Science* 208, 92–99. doi: 10.1016/j.applanim.2018.08.007.

Over, R. and Moore, D. (1981). Spatial acuity of the chicken, *Brain Research* 211(2), 424–426. doi: 10.1016/0006-8993(81)90967-7.

Petrik, M. T., Guerin, M. T. and Widowski, T. M. (2015). On-farm comparison of keel fracture prevalence and other welfare indicators in conventional cage and floor-housed laying hens in Ontario, Canada, *Poultry Science* 94(4), 579–585. Available at: http://ps.oxfordjournals.org/content/94/4/579.abstract.

Pettersson, I. C., et al. (2017). The ability of laying pullets to negotiate two ramp designs as measured by bird preference and behaviour, *PeerJ*. United States 5, e4069. doi: 10.7717/peerj.4069.

Pickel, T., Scholz, B. and Schrader, L. (2010). Perch material and diameter affects particular perching behaviours in laying hens, *Applied Animal Behaviour Science* 127(1–2), 37–42. Available at: http://www.sciencedirect.com/science/article/pii/S0168159110002078.

Pickel, T., Schrader, L. and Scholz, B. (2011). Pressure load on keel bone and foot pads in perching laying hens in relation to perch design, *Poultry Science*. Oxford University Press 90(4), 715–724. doi: 10.3382/ps.2010-01025.

Pines, M. and Reshef, R. (2014). Poultry bone development and bone disorders. In: Scanes, C. G. (Ed.) *Sturkie's Avian Physiology*, 6th edn. Elsevier, London, pp. 367–364.

Podisi, B. K., Knott, S. A., Dunn, I. C., Burt, D. W. and Hocking, P. M. (2012). Bone mineral density QTL at sexual maturity and end of lay, *British Poultry Science*. Taylor & Francis 53(6), 763–769.

Pottgüter, R. (2016). Feeding laying hens to 100 weeks of age, *Lohmann Information* 501(1), 18–21. Available at: http://www.ltz.de/de-wAssets/docs/lohmann-information/Lohmann-Information_3.pdf.

Pourquié, O. (2004). The chick embryo: a leading model in somitogenesis studies, *Mechanisms of Development* 121(9), 1069–1079. doi: 10.1016/j.mod.2004.05.002.

Prescott, N. B., Wathes, C. M. and Jarvis, J. R. (2003). Light, vision and the welfare of poultry, *Animal Welfare* 12(2), 269–288.

Prunier, A., Mounier, L., Le Neindre, P., Leterrier, C., Mormède, P., Paulmier, V., Prunet, P., Terlouw, C. and Guatteo, R. (2013). Identifying and monitoring pain in farm animals: a review, *Animal*. Cambridge University Press 7(6), 998–1010.

Rath, N. C., Huff, G. R., Huff, W. E. and Balog, J. M. (2000). Factors regulating bone maturity and strength in poultry, *Poultry Science* 79(7), 1024–1032. Available at: http://ps.fass.org/content/79/7/1024.abstract.

Reed, H. J., Wilkins, L. J., Austin, S. D. and Gregory, N. G. (1993). The effect of environmental enrichment during rearing on fear reactions and depopulation trauma in adult caged hens, *Applied Animal Behaviour Science*. Elsevier 36(1), 39–46. doi: 10.1016/0168-1591(93)90097-9.

Regmi, P., Deland, T. S., Steibel, J. P., Robison, C. I., Haut, R. C., Orth, M. W. and Karcher, D. M. (2015). Effect of rearing environment on bone growth of pullets, *Poultry Science* 94(3), 502–511. doi: 10.3382/ps/peu041.

Regmi, P., Nelson, N., Steibel, J. P., Anderson, K. E. and Karcher, D. M. (2016a). Comparisons of bone properties and keel deformities between strains and housing systems in end-of-lay hens, *Poultry Science*. Poultry Science Association, Inc. 95(10), 2225–2234.

Regmi, P., Smith, N., Nelson, N., Haut, R. C., Orth, M. W. and Karcher, D. M. (2016b). Housing conditions alter properties of the tibia and humerus during the laying phase in Lohmann white Leghorn hens, *Poultry Science*. Poultry Science Association, Inc. 95(1), 198–206.

Rentsch, A. K., et al. (2019). Laying hens' mobility is impaired by keel bone fractures though effect is not reversed by paracetamol treatment, *Applied Animal Behaviour Science*. Elsevier 217, 48–56. doi: 10.1016/J.APPLANIM.2019.04.015.

Riber, A. B., Casey-Trott, T. M. and Herskin, M. S. (2018). The influence of keel bone damage on welfare of laying hens', *Frontiers in Veterinary Science*. Frontiers 5, 6.

Riber, A. B. and Hinrichsen, L. K. (2016). Keel-bone damage and foot injuries in commercial laying hens in Denmark, *Animal Welfare*. Universities Federation for Animal Welfare 25(2), 179–184.

Richards, G. J. J., Wilkins, L. J., Knowles, T. G., Booth, F., Toscano, M. J., Nicol, C. J. and Brown, S. N. (2012). Pop hole use by hens with different keel fracture status monitored throughout the laying period, *Veterinary Record* 170(19), 494. doi: 10.1136/vr.100489.

Rodenburg, T. B., et al. (2008). Welfare assessment of laying hens in furnished cages and non-cage systems : an on-farm comparison, *Animal Welfare* 17, 363–373.

Rufener, C., Baur, S., Stratmann, A. and Toscano, M. J. (2018a). Keel bone fractures affect egg laying performance but not egg quality in laying hens housed in a commercial aviary system, *Poultry Science* 98(4), 1589–1600. doi: 10.3382/ps/pey544.

Rufener, C., et al. (2018b). Subjective but reliable: assessing radiographs of keel bone fractures in laying hens with a tagged visual analogue scale, *Frontiers in Veterinary Science* 5, 124. doi: 10.3389%2Ffvets.2018.00124.

Rufener, C., Abreu, Y., Asher, L., Berezowski, J. A., Maximiano Sousa, F., Stratmann, A. and Toscano, M. J. (2019). Keel bone fractures are associated with individual mobiltiy of laying hen in aviary systems, *Applied Animal Behaviour Science* 217, 48–56. doi: 10.1016/j.applanim.2019.05.007.

Rufener, C., Rentsch, A. K., Stratmann, A. and Toscano, M. J. (2020). Perch positioning affects both laying hen locomotion and forces experienced at the keel. *Animals* 10, 1223.

Sandilands, V., Moinard, C. and Sparks, N. H. C. (2009). Providing laying hens with perches: fulfilling behavioural needs but causing injury?, *British Poultry Science*. Taylor & Francis 50(4), 395–406. doi: 10.1080/00071660903110844.

Scholz, B., Kjaer, J. B. and Schrader, L. (2014). Analysis of landing behaviour of three layer lines on different perch designs, *British Poultry Science*. Taylor and Francis 55(4), 419–426. Taylor and Francis. doi: 10.1080/00071668.2014.933175.

Schwartzkopf-Genswein, K. S., Faucitano, L., Dadgar, S., Shand, P., González, L. A. and Crowe, T. G. (2012). Road transport of cattle, swine and poultry in North America and its impact on animal welfare, carcass and meat quality: a review, *Meat Science*. Elsevier 92(3), 227–243.

Scott, G. B. (1993). Poultry handling: a review of mechanical devices and their effect on bird welfare, *World's Poultry Science Journal*. Cambridge University Press 49(1), 44–57.

Scott, G. B. and MacAngus, G. (2004). The ability of laying hens to negotiate perches of different materials with clean or dirty surfaces, *Animal Welfare* 13(3), 361–365.

Scott, G. B. and Parker, C. A. L. (1994). The ability of laying hens to negotiate between horizontal perches. *Applied Animal Behaviour Science* 42, 121–127. Available at: http://www.sciencedirect.com/science/article/pii/016815919490152X.

Scott, G. B., Lambe, N. R. and Hitchcock, D. (1997). Ability of laying hens to negotiate horizontal perches at different heights, separated by different angles. *British Poultry Science* 38(1), 48–54. https://doi.org/10.1080/00071669708417939.

Shen, C. L., Yeh, J. K., Rasty, J., Chyu, M. C., Dunn, D. M., Li, Y. and Watkins, B. A. (2007). Improvement of bone quality in gonad-intact middle-aged male rats by long-chain n-3 polyunsaturated fatty acid, *Calcified Tissue International* 80(4), 286–293. doi: 10.1007/s00223-007-9010-8.

Silversides, F. G., Korver, D. R. and Budgell, K. L. (2006). Effect of strain of layer and age at photostimulation on egg production, egg quality, and bone strength, *Poultry Science*. Oxford University Press, Oxford, UK 85(7), 1136–1144.

Simkiss, K. (1967). *Calcium in Reproductive Phsysiology*. Chapman and Hall Ltd, London.

Sisson, S., Grossman, J. D. and Getty, R. (1975). In: Getty, R. (Ed.) *Sisson and Grossman's the Anatomy of the Domestic Animals*, 5th edn. Saunders, Philadelphia, PA.

Stratmann, A., Fröhlich, E. K. F., Gebhardt-Henrich, S. G., Harlander-Matauschek, A., Würbel, H. and Toscano, M. J. (2015a). Modification of aviary design reduces incidence of falls, collisions and keel bone damage in laying hens, *Applied Animal Behaviour Science* 165, 112–123. doi: 10.1016/j.applanim.2015.01.012.

Stratmann, A., Fröhlich, E. K. F., Harlander-Matauschek, A., Schrader, L., Toscano, M. J., Würbel, H. and Gebhardt-Henrich, S. G. (2015b). Soft perches in an aviary system reduce incidence of keel bone damage in laying hens, *PLOS ONE*. Public Library of Science 10(3), e0122568. doi: 10.1371/journal.pone.0122568.

Stratmann, A., Mühlemann, S., Vögeli, S. and Ringgenberg, N. (2019). Frequency of falls in commercial aviary-housed laying hen flocks and the effects of dusk phase length, *Applied Animal Behaviour Science*. Elsevier 216, 26–32. doi: 10.1016/j.applanim.2019.04.008.

Struelens, E. and Tuyttens, F. A. M. (2009). Effects of perch design on behaviour and health of laying hens, *Animal Welfare* 18, 533–538.

Sun, D., Krishnan, A., Zaman, K., Lawrence, R., Bhattacharya, A. and Fernandes, G. (2003). Dietary n-3 fatty acids decrease osteoclastogenesis and loss of bone mass in ovariectomized mice, *Journal of Bone and Mineral Research* 18(7), 1206–1216. doi: 10.1359/jbmr.2003.18.7.1206.

Tahamtani, F. M., Nordgreen, J., Nordquist, R. E. and Janczak, A. M. (2015). Early life in a barren environment adversely affects spatial cognition in laying hens (Gallus gallus domesticus), *Frontiers in Veterinary Science*. Frontiers Media SA 2, 3.

Tallarico, R. B. and Farrell, W. M. (1964). Studies of visual depth perception: an effect of early experience on chicks on a visual cliff, *Journal of Comparative and Physiological Psychology* 17, 1695-1702. doi: 10.1046/j.1460-9568.2003.02593.x.

Tarlton, J. F. F., Wilkins, L. J., Toscano, M. J., Avery, N. C. and Knott, L. (2013). Reduced bone breakage and increased bone strength in free range laying hens fed omega-3 polyunsaturated fatty acid supplemented diets, *Bone* 52(2), 578-586. doi: 10.1016/j.bone.2012.11.003.

Tauson, R. and Abrahamsson, P. (1994). Foot and skeletal disorders in laying hens: effects of perch design, hybrid, housing system and stocking density, *Acta Agriculturae Scandinavica, Section A - Animal Science*. Taylor & Francis 44(2), 110-119. doi: 10.1080/09064709409410189.

Taylor, P. E., Scott, G. B. and Rose, P. (2003). The ability of domestic hens to jump between horizontal perches: effects of light intensity and perch colour, *Applied Animal Behaviour Science*. Elsevier 83(2), 99-108.

Thiele, H.-H. (2015). Optimal calicum supply for laying hens, *Lohmann Information* 50(2), 32-35. Available at: http://www.ltz.de/de-wAssets/docs/lohmann-information/Lohmann-Information2_2015_Vol.-49-2-October-2015_Thiele.pdf.

Thiruvenkadan, A. K., Panneerselvam, S. and Prabakaran, R. (2010). Layer breeding strategies: an overview, *World's Poultry Science Journal*. Cambridge University Press 66(3), 477-502.

Toscano, M. J., Wilkins, L. J., Millburn, G., Thorpe, K. and Tarlton, J. F. (2013). Development of an ex vivo protocol to model bone fracture in laying hens resulting from collisions, *PLoS ONE* Witten, P. E. (Ed.). Public Library of Science, 8(6), e66215. doi: 10.1371/journal.pone.0066215.

Toscano, M. J., Booth, F., Richards, G., Brown, S., Karcher, D. and Tarlton, J. (2018). Modeling collisions in laying hens as a tool to identify causative factors for keel bone fractures and means to reduce their occurrence and severity, *PLoS ONE*. Yildirim, A. (Ed.) 13(7), e0200025. doi: 10.1371/journal.pone.0200025.

Toscano, M. J., Booth, F., Wilkins, L. J., Avery, N. C., Brown, S. B., Richards, G. and Tarlton, J. F. (2015). The effects of long (C20/22) and short (C18) chain omega-3 fatty acids on keel bone fractures, bone biomechanics, behaviour and egg production in free range laying hens, *Poultry Science* 94(5), 823-835. doi: 10.3382/ps/pev048.

Toscano, M. J., Dunn, I. C., Christensen, J.-P., Petow, S., Kittelsen, K. and Ulrich, R. 2020. Explanations for keel bone fractures in laying hens: considering alternatives other than high egg production, *Poultry Science*. doi:10.1016/j.psj.2020.05.035.

Toscano, M. J. J., Wilkins, L. J. J. and Tarlton, J. F. F. (2012). Impact of a mixed chain length omega-3 fatty acid diet on production variables in a commercial free-range laying hens, *British Poultry Science* 53(3), 360-365. doi: 10.1080/00071668.2012.698726.

Vallortigara, G., Andrew, R. J., Sertori, L. and Regolin, L. (1997). Sharply timed behavioral changes during the first 5 weeks of life in the domestic chick (Gallus gallus), *Bird Behavior* 12(1), 29-40. doi: 10.3727/015613897797141290.

Vallortigara, G. and Rogers, L. J. (2005). Survival with an asymmetrical brain: advantages and disadvantages of cerebral lateralization', *Behavioral and Brain Sciences*. UC Davis, 28(4), 575-589. doi: 10.1017/S0140525X05000105.

Van De Velde, J. P., Vermeiden, J. P. W. and Bloot, A. M. (1985). Medullary bone matrix formation, mineralization, and remodeling related to the daily egg-laying cycle of Japanese quail: a histological and radiological study, *Bone* 6(5), 321-327. doi: 10.1016/8756-3282(85)90322-9.

Walk, R. D. and Gibson, E. J. (1961). A comparative and analytical study of visual depth perception, *Psychological Monographs: General and Applied* 75(15), 1–44. doi: 10.1037/h0093827.

Warren, D. C. D. E. (1937). Physiological and genetic studies of crooked keels in chickens, *Kansas Agricultural Experimental Station Technical Bulletin* 44, 1–32.

Watkins, B. A., Li, Y., Lippman, H. E. and Feng, S. (2003). Modulatory effect of omega-3 polyunsaturated fatty acids on osteoblast function and bone metabolism, *Prostaglandins, Leukotrienes and Essential Fatty Acids* 68(6), 387–398.

Webster, A. B. (2004). Welfare implications of avian osteoporosis, *Poultry Science* 83(2), 184–192. Available at: http://www.ncbi.nlm.nih.gov/entrez/query.fcgi?cmd=Retrieve&db=PubMed&dopt=Citation&list_uids=14979568.

Weeks, C. A. and Nicol, C. J. (2006). Behavioural needs, priorities and preferences of laying hens, *World's Poultry Science Journal*. Cambridge University Press 62(2), 296–307. Available at: http://journals.cambridge.org/abstract_S0043933906000195 (Accessed: 14 March 2014).

Whitehead, C. C. (2004). Overview of bone biology in the egg-laying hen, *Poultry Science* 83(2), 193–199. Available at: http://www.ncbi.nlm.nih.gov/pubmed/14979569.

Whitehead, C. C. and Fleming, R. H. (2000). Osteoporosis in cage layers, *Poultry Science*. Oxford University Press 79(7), 1033–1041.

Wichman, A., Heikkilä, M., Valros, A., Forkman, B. and Keeling, L. J. (2007). Perching behaviour in chickens and its relation to spatial ability, *Applied Animal Behaviour Science*. Elsevier 105(1–3), 165–179.

Wilkins, L. J., McKinstry, J. L., Avery, N. C., Knowles, T. G., Brown, S. N., Tarlton, J. and Nicol, C. J. (2011). Influence of housing system and design on bone strength and keel bone fractures in laying hens, *Veterinary Record* 169(16), 414. doi: 10.1136/vr.d4831.

Wolc, A., Bednarczyk, M., Lisowski, M. and Szwaczkowski, T. (2010). Genetic relationships among time of egg formation, clutch traits and traditional selection traits in laying hens, *Journal of Animal and Feed Sciences* 19(3), 452–459.

# Chapter 18

## Poultry health monitoring and management: bone and skin health in broilers

*Gina Caplen, University of Bristol, UK*

1 Introduction
2 Leg disorders and lameness
3 Contact dermatitis
4 Conclusion and future trends
5 Where to look for further information
6 References

### 1 Introduction

Consumer demand for cheap meat has driven the broiler industry to set ever-increasing production targets, often at the expense of the animals' wellbeing. High levels of lameness and contact dermatitis, an inflammatory skin condition, within intensively reared broiler flocks have long raised welfare concern. Reduced mobility, physical restrictions in behavioural expression and pain experience, associated with leg pathologies and skin lesions, will have negative welfare repercussions. Despite some success with interventions to improve bone and skin conditions, including sustained efforts by the genetic companies to select for improvements in leg health, some of the same problems still persist decades later. One estimate indicated that 12.5 billion broilers experience leg problems annually worldwide (FAO, 2010).

With demand for chicken meat still escalating globally, it is imperative that the risk factors for lameness and contact dermatitis are fully understood so that novel and effective management solutions can be developed and better employed. Not only would the improvement of leg and skin health directly benefit broiler welfare, but there is also potential to confer many economical benefits to the producers. Lameness is associated with higher flock morbidity and mortality (Dinev, 2009) and is one of the primary causes of on-farm culling (especially in older birds). In addition, both lameness and contact dermatitis lower growth performance indices (Knowles et al., 2008; Butterworth

http://dx.doi.org/10.19103/AS.2020.0078.17

and Haslam, 2009) and are responsible for a large proportion of carcass condemnations at the slaughter plant (Kyvsgaard et al., 2013; Hashimoto et al., 2013).

## 2 Leg disorders and lameness

### 2.1 Categories of lameness and impaired walking ability

Leg disorders encompass a broad range of pathologies and can involve any combination of tendons, joints, ligaments and bones (Bradshaw et al., 2002). Lameness can arise due to developmental bone deformities, be of an infectious origin, or be degenerative (e.g. develop as a consequence of a trauma or progressive load-bearing).

#### 2.1.1 Gait and morphology

Over the last 60 years the morphology, walking style and locomotive capacity of commercial broilers have been profoundly altered by genetic selection for increased growth rate and body mass. Selection has specifically targeted the breast muscle, which, as a proportion of the total body mass, is now double that of traditional strains (Schmidt et al., 2009). This 'pectoral hypertrophy', in addition to greater thigh muscle and leg bone mass, and relatively short legs distinguish modern broiler strains from traditional meat breeds.

Broilers display pronounced gait alterations when compared to red junglefowl (the ancestral species), and additional differences are evident between non-lame and moderately lame individuals (Caplen et al., 2012). Compensatory gait adaptations, including the wider stance and more pronounced hip and foot rotations characteristic of broilers, develop in response to increased load (body mass), morphology (unbalanced posture), pathology and injury (Corr et al., 2003a,b). In the short term such adaptations minimise the additional energy required for movement and enable the birds to retain mobility; however, ultimately, the gait adaptations themselves are likely to be fatiguing and increase the risk of modern broilers developing progressive leg pathologies.

#### 2.1.2 Developmental lameness

A rapid growth rate and unnaturally high body mass place extreme loads on immature bones and joints. Frequently observed skeletal abnormalities include rotational (tortional) deformities such as 'twisted legs' (unilateral angulation of tibiotarsal articulation) and 'valgus-varus' deformities (Riddell and Kong, 1992). Valgus or varus angulation is mainly associated with bilateral tibial deviation and rotation (often with gastrocnemius tendon displacement); however, some

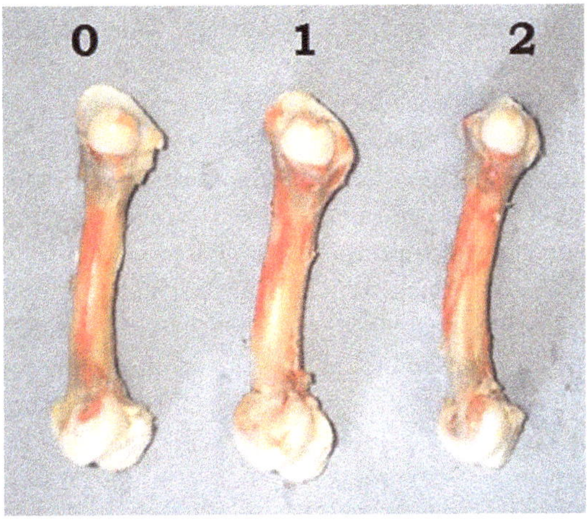


**Figure 1** Examples of femoral curvature: 0 = normal, almost straight; 1 = mild curvature; 2 = moderate curvature (photo courtesy of Dr Andrew Butterworth, University of Bristol).

studies also report femoral deformations (e.g. Duff and Thorp, 1985). Such deviations remain poorly understood but are likely to be a consequence of poor bone mineralisation. Leg angulations and bone deformities can occur over a range of severities; for example valgus has been described as mild (tibia-metatarsus angle between 10° and 25°), intermediate (25–45°), and severe (>45°) (Shim et al., 2012a). A severity scale for femoral curvature is illustrated in Fig. 1.

Tibial dyschondroplasia (TD) is a leg pathology resulting from inadequate ossification and vascularisation of the epiphyseal growth plates. The TD lesion is characterised by an abnormal mass of cartilage occurring, most commonly, in the tibial metaphysis (Orth and Cook, 1994). Severe TD lesions appear to cause lameness, although the effect is less evident in milder forms (Lynch et al., 1992). Rickets is characterised by an enlargement of the epiphysial growth plate and soft rubbery bones, a result of inadequate endochondral ossification and mineralisation (Julian, 1998; Dinev and Kanakov, 2011). Rickets develops as a result of nutrient malabsorption, for example following intestinal disease, or due to inadequate nutrition, and is relatively uncommon within modern broiler flocks.

### *2.1.3 Infectious lameness*

Bacterial chondronecrosis with osteomyelitis (BCO), more specifically known as femoral head necrosis, is recognised as a major cause of lameness in commercial broilers worldwide and is reviewed in detail elsewhere (e.g. Wideman, 2016). BCO is primarily identified by the presence of lesions in the femur (in particular, the metaphysis, femoral head and proximal femoral growth plate) and the tibiotarsus. The condition is associated with a complex range of opportunistic infective organisms, predominantly avian pathogenic *Escherichia coli* and *Staphylococcus* spp. (Dinev, 2009; Mandal et al., 2016; Al-Rubaye et al., 2017; Wijesurendra et al., 2017). Rapid increases in body weight cause mechanical damage to the immature cartilage (osteochondrosis). Bacterial proliferation at the wound sites triggers an immunological response and limits blood flow to the growth plate (Wideman and Prisby, 2013), which, in turn, leads to the development of necrotic abscesses and voids that are characteristic of BCO (Wideman, 2016). Bacteria are thought to be distributed haematologically, via the respiratory or gastrointestinal tract; *S. agnetis*, for example, can be transferred to the blood stream via drinking water (Al-Rubaye et al., 2017). Since broilers favour sitting, reduced blood flow resulting from prolonged compression of the leg arteries may further promote the development of this pathology (Wideman, 2016). Broilers housed on wire flooring are particularly susceptible to developing BCO lesions due to persistent footing instability and physiological stress (Al-Rubaye et al., 2017).

Tenosynovitis (an arthritis caused by avian reoviruses) causes the inflammation of the hock joints, lesions in the gastrocnemius and digital flexor tendons and, ultimately, lameness (Sellers, 2017). The gastrocnemius tendons leave the gastrocnemius muscle and pass over the intertarsal (hock) joint and attach to the posterior surface of the tarsometatarsus, while the digital flexor tendons extend along the tarsometatarsus to the phalanges. Upon dissection the infected synovia of swollen hocks often appears thickened, often in association with an increase in joint fluid, blood or pus (Fig. 2). In China, bacterial arthritis is also common in broilers, the main serovar being *Salmonella pullorum* (Guo et al., 2019).

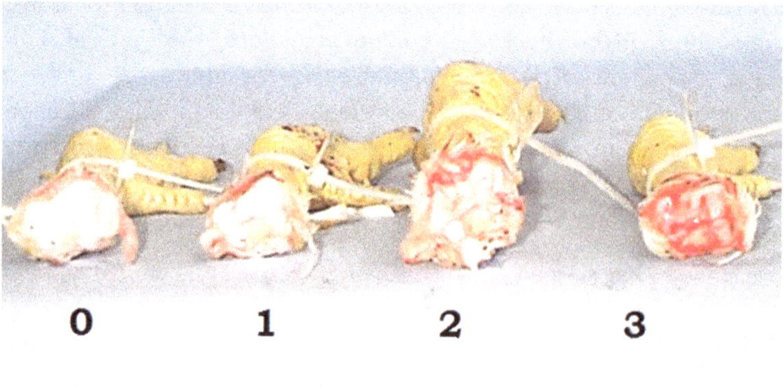


**Figure 2** Inflammation of the joint capsule (proximal tarsometatarsus). Synovitis score scale: 0 = none, 1 = mild, 2 = medium, 3 = severe (photo courtesy of Dr Andrew Butterworth, University of Bristol).

## 2.2 Qualitative and quantitative assessment of leg health

Several gait scoring methods, based upon the visual appraisal of walking ability, have been developed for assessing broiler lameness. The widely used Bristol six-point 'gait score' (GS) scale, developed by Kestin et al. (1992), has now been adopted in a modified form for use within the broiler Welfare Quality® Assessment; scores range between 'GS 0: Normal, dextrous and agile' and 'GS 5: Incapable of walking' (Welfare Quality, 2009). Although this system is suited for on-farm welfare assessment (being relatively quick and requiring no specialised equipment), it lacks the capacity to discriminate between lameness types. A moderately lame bird assigned 'GS 3' could be affected bilaterally (e.g. valgus) or unilaterally (e.g. singular hock inflammation), or lack any obvious pathology.

Objective methodologies for quantitatively assessing gait are available, including kinetic (the measure of forces involved in walking) and kinematic (the study of body motion) systems, but at the current time these technologies remain better suited to experimental work. Several techniques have been employed to collect kinetic data from broilers, with varying measures of success. These include force plates (Corr et al., 2007; Sandilands et al., 2011), which require a constant walking speed and generate significant background 'noise', and piezoelectric pressure-sensing mats (Nääs et al., 2010). Image analysis systems have enabled the collection of three-dimensional kinematic data for the purpose of correlating gait characteristics, such as walking speed, step length and lateral body oscillation (Fig. 3), with defined lameness scores, (Caplen et al., 2012; Aydin, 2017a) and have been used to detect subtle localised changes in gait parameters as part of analgesic drug studies (Caplen et al., 2013a).

The latency-to-lie (LTL) test has been widely employed as a simple index of leg weakness (Weeks et al., 2002; Berg and Sanotra, 2003). The test is based

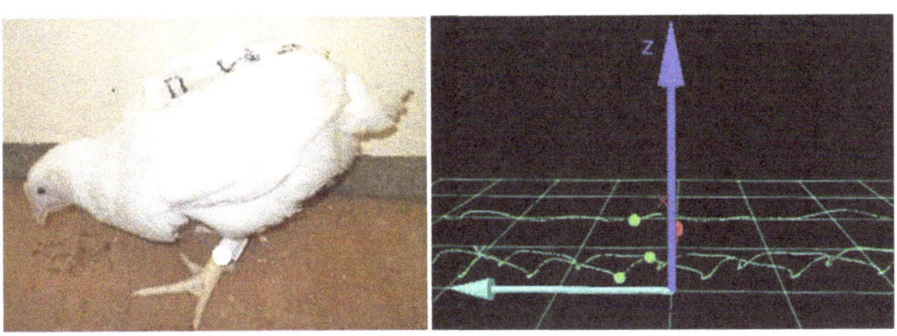


**Figure 3** Broiler fitted with retro-reflective markers (one on the midline of the back, and a further two attached to the posterior aspect of the metatarsal bone, immediately above each foot). An infra-red four-camera motion capturing system was used to collect three-dimensional kinematic data as the bird walked down a runway. Images: Gina Caplen, University of Bristol.

upon the premise that chickens find sitting in water aversive and that birds with poor leg health will sit sooner when they are placed standing in shallow water (Fig. 4). Due to the logistics involved in performing this test, it is better suited to experimental trials rather than on-farm welfare assessments. An automated visual monitoring methodology has been recently developed to assign GS to broilers based upon 'the number of lying events' and 'latency to sit' when birds traverse a test corridor (Aydin et al., 2015; Aydin, 2017b). Although this new system may have the potential for transfer to a farm environment, any methodology that relies upon broilers voluntarily moving along a standard walkway will run the risk of sample bias towards the more mobile individuals (and under sample the lame birds) within a flock.

Significant efforts are currently being made to develop automated farm-based camera systems that remotely monitor flock behaviour and forewarn producers of the development of muscular-skeletal problems, for example when flock activity levels fall below an accepted level. Image analysis systems are reviewed in Chapter 7. In brief, several systems show great promise for incorporation with existing management software and wide-scale employment. Optical flow measures of flock movement have been shown to correlate with lameness (Dawkins et al., 2009, 2013; Roberts et al., 2012), while a different visual system has also been used to correlate bird activity and flock distribution with GS (Van Hertem et al., 2018).

Infrared thermography (IRT) provides a non-invasive means of measuring infrared radiation (surface heat) from an object, and the technology is being

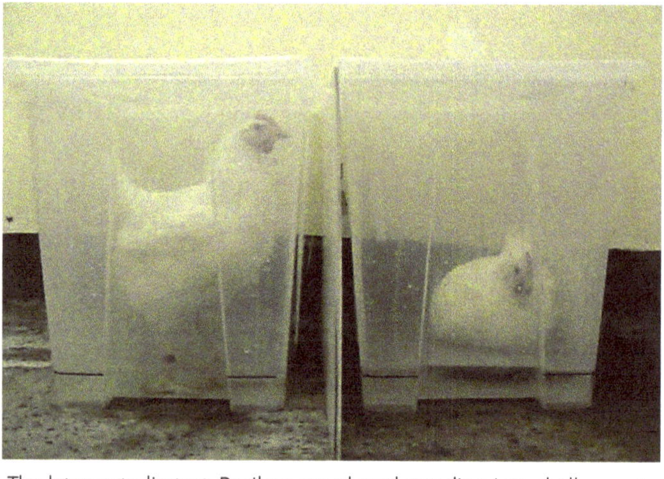

**Figure 4** The latency-to-lie test. Broilers are placed standing into shallow water at room temperature (without visual contact of conspecifics) as per the left-hand bird. The time taken for the birds to sit, as per the right-hand bird, provides a simple index of leg weakness. Birds are removed from the water as soon as they sit. Image: Gina Caplen, University of Bristol.

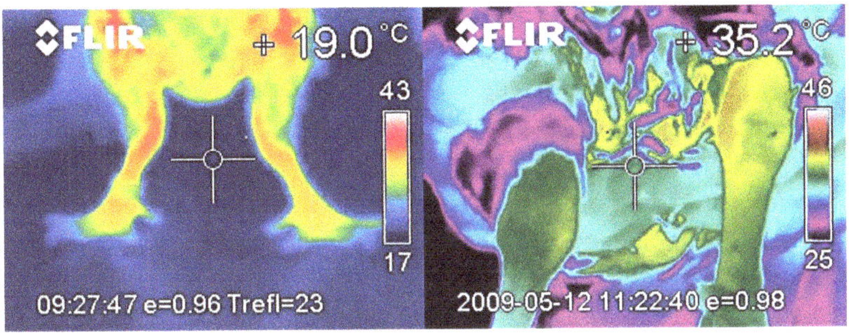


**Figure 5** Thermal images of a broiler with a developmental 'valgus' leg deformity (left image) and a broiler with unilateral inflammation of the right hock joint (right image). Images: Gina Caplen, University of Bristol.

increasingly utilised in the fields of clinical and veterinary health. Although IRT cannot diagnose specific pathologies, it can be used to detect localised areas of increased or decreased heat production, that is due to inflammation or reduced blood flow, accordingly (Fig. 5). Relating to poultry production, IRT may prove to be useful as a non-invasive technique for detecting and monitoring leg pathologies on farm that are currently diagnosable only at post-mortem, for example BCO (Weimer et al., 2019).

## 2.3 Prevalence of lameness and specific leg pathologies

According to a comprehensive UK study, by the end of the rearing period more than a quarter (27.6%) of standard broilers had a moderate to severe gait impediment (GS 3+) and 3.3% were unable to walk (Knowles et al., 2008). More recent studies estimate a slightly lower prevalence of moderate to severe gait impediments in European flocks (Europe-wide: 15.6%, Bassler et al., 2013; Norway: 24.6%, Kittelsen et al., 2017; 19%, Granquist et al., 2019). A Danish survey found 77% of conventional broilers to have at least some form of impaired walking ability (GS 1+), while the prevalence of birds with moderate to severe lameness (GS 2+) was, a modest, 6% (Tahamtani et al., 2018).

Very little information is available regarding the current prevalence of specific leg-health pathologies. Sanotra et al. (2003) report very high levels of TD within Scandinavian flocks (Denmark: 57.1%; Sweden: 45.2–56.3%), yet later studies report much less (Finland: 2.3%, Kaukonen et al., 2017; Denmark: 5%, Tahamtani et al., 2018). BCO appears to retain a persistent presence within broiler flocks. BCO lesions have been recorded in >90% of lameness-related mortality cases (Bulgaria: Dinev, 2009), and in almost 30% of all broiler mortalities and culls (Australia: Wijesurendra et al., 2017). Although joint lesions (including arthritis and tenosynovitis) were reported to increase within UK flocks by 35% between 2011 and 2013, the actual increase was relatively small (2.16

to 2.92 per 10 000 birds slaughtered, Part et al., 2016) and, therefore, perhaps not fully reliable. Valgus and other long-bone deformities are frequently seen within standard UK broiler flocks, but exact estimates are not available. An earlier study reports a very high prevalence of valgus-varus in Scandanavia (Denmark: 36.9%; Sweden: 46.4–52.6%, Sanotra et al., 2003).

A wide variation in the estimates of general walking ability and specific leg disorders is to be expected when you consider the diversity in production systems employed on both regional and global scales (including stocking density, climate, barn design and diet). The different housing systems for broilers are described in Chapter 10. In addition, breeding companies are constantly selecting for improved production parameters and leg health (e.g. valgus/varus and TD), and so the genotype of birds supplied for use in commercial systems is in perpetual flux. In addition, management practices can become out-dated. For example, vaccination had successfully controlled tenosynovitis for decades; however, in recent years there has been a dramatic increase in the number of clinical cases as new genetic virus variants emerge (Sellers, 2017). There is an obvious need for independent and systematic monitoring of specific pathologies to document and better understand such trends.

## 2.4 Lameness risk factors

Key risk factors associated with lameness and poor leg health in modern broilers include those directly associated with the growth rate (including genotype, age, body mass and feeding regimen) and those indicative of sub-optimal environmental management.

### 2.4.1 Sex, age and body mass

Lameness increases with broiler age and body mass (e.g. Henriksen et al., 2016). Males generally have a higher GS than females and are more prone to developing valgus-varus and femoral degenerative joint lesions (Paz et al., 2013), especially if they are heavy. Some studies report a higher TD incidence in male broilers (e.g. Birgul et al., 2012), presumably since TD is also positively associated with body mass (e.g. Shim et al., 2012a). Worryingly, TD lesions have been recorded in broilers as young as 20 days, which may be a direct consequence of selection pressure for increased developmental rates (Dinev et al., 2012).

### 2.4.2 Genotype and system

Slower-growing genotypes demonstrate less lameness than modern fast-growing breeds when reared on the same feeding regimen (Kestin et al., 2001). Fast-growing breeds tend to have lower bone strength, which makes them susceptible to developing leg deformities (Shim et al., 2012b). The choice of system (and appropriate broiler genotype) has an obvious impact upon the leg health of the flock. Organically reared broilers tend to have better leg health than conventional flocks (Tuyttens et al., 2008; Tahamtani et al., 2018); however, flock size and stocking density are usually much lower in the organic farms, and organic systems are required to use slower-growing breeds. Fast-growing breeds do not perform well under extended production and are unsuited for use in extensive systems, as the birds make low use of the outdoor space and have a tendency to develop severe welfare problems, including high cull and mortality rates, impaired mobility, joint inflammation and severe footpad dermatitis (FPD) (Nielsen et al., 2003; Dal Bosco et al., 2014; Castellini et al., 2016).

### 2.4.3 Stocking density

Research consistently indicates that the health and welfare of broilers is compromised above stocking densities of 34–38 kg/m$^2$, dependent upon final body weight (e.g. Estevez, 2007; Sun et al., 2013; Das and Lacin, 2014). Accordingly the European Broiler Directive (2007/43/EC), which lays down the minimum rules for the protection of chickens kept for meat production, specifies that, as a general rule, stocking density should not exceed 33 kg/m$^2$. Although broilers grow more slowly at high stocking densities (Dawkins et al., 2004; Sun et al., 2013), there is evidence that leg health can become compromised at densities as low as 23 kg/m$^2$ (Buijs et al., 2009). Birds present on the barn floor effectively act as obstacles, and this has been termed the 'barrier effect' (e.g. Collins, 2008). Prolonged activity becomes more difficult at higher stocking densities, and behaviour becomes fragmented; locomotion slows and walking bouts decrease in length (e.g. Buijs et al., 2011; Ventura et al., 2012). Although stocking density does not appear to be directly linked with the prevalence of specific pathologies (e.g. valgus–varus: Arnould and Faure, 2004; TD: Das and Lacin, 2014), higher stocking densities are associated with tibial deformities and reduced bone strength (Buijs et al., 2012; Sun et al., 2013; Vargas-Galicia et al., 2017). The apparent effects of stocking density on skin health are due to correlations with deteriorating environment, especially litter quality, and this is associated with poor environmental control (Section 3.3.1). Optimal environmental management is essential for maintaining general flock health status since the combination of wet litter with warm moist air will promote bacterial growth and transmission. Gut infections, such as enteritis, provide one of many indirect mechanisms by which the intestinal microbiota may influence skeletal fitness and bone mass (Charles et al., 2015), while lameness may also arise directly as a result of bone or joint infection.

### 2.4.4 Environmental conditions

Key requirements for achieving and sustaining good leg health include adequate ventilation of the production facility, provision of high-quality litter and the maintenance of temperature and relative humidity (RH) within an optimum range. Higher GSs have been associated with increased hock burn (HB) and FPD scores (see Section 3), and reduced feather cleanliness, suggesting that a sub-optimal physical environment (i.e. poor litter quality) may be detrimental to leg health (Granquist et al., 2019). High temperature, high ammonia concentrations and the percentage of time that temperature and RH remain outside of the breeder's recommended range have also been associated with gait deficiencies and leg deformities (Dawkins et al., 2004; Jones et al., 2005; Tullo et al., 2017).

Rigid environmental control is also very important during incubation; temperature and oxygen concentrations can influence embryonic bone development, post-hatch production parameters and leg health (e.g. Ipek and Sozcu, 2016; Oznurlu et al., 2016). TD incidence has been associated with temperature deviations during the early stages of embryo development (Yalçin et al., 2007). Small changes in egg shell temperature (EST) during incubation could easily occur in practice due to the widespread use of large-capacity incubators and the fast growth rate of broiler embryos. Indeed, Oviedo-Rondón et al. (2009a) observed that broilers hatched from multi-stage incubators went on to develop a higher GS and greater prevalence of valgus leg deformations than those hatched from single-stage incubators.

## 2.5 Welfare impact of lameness

The terms 'lameness' and 'leg health' are not fully interchangeable. Although correlations between body mass, growth rate and lameness are well documented (e.g. Kestin et al., 2001), there is little evidence to conclusively link lameness severity with pathology (Garner et al., 2002; Sandilands et al., 2011; Fernandes et al., 2012). Instead, lameness is more likely to reflect the birds' subjective experience of the pathology, the manifestation of an integrated behavioural response. Investigating the true impact of poor leg health at the bird level is complicated by the gross morphological modifications that define the modern broiler bird and the multiple, often inter-linked, aetiologies and pathologies to which they are susceptible. As already mentioned lameness has been associated with contact dermatitis, and it is likely that, if left untreated, many progressive leg pathologies will confer a greater risk of developing secondary complications. Condemnations (carcasses deemed unfit for human consumption) at post-mortem inspection have been associated with increasing GS (Granquist et al., 2019), indicating that at least a proportion of lame broilers display pathological changes on the carcasses.

### 2.5.1 Dehydration and reduced feeding

Lame broilers may endure dehydration (Butterworth et al., 2002) since decreased mobility can limit their ability to access water, even in systems that otherwise provide appropriate water delivery. Nipple drinkers minimise water spillage and, as such, can help to maintain litter quality; however, their position, above the broilers' heads, can cause some birds to lose balance when stretching to drink (Jones et al., 2005). Very lame birds are often seen to have a lower body mass. Food intake may be reduced if it is difficult to reach; however, appetite may also become suppressed as a result of inflammatory-induced sickness behaviour (e.g. Dantzer, 2009).

### 2.5.2 Pain

The prevalence and severity of pain associated with broiler lameness remains poorly understood due to the impact of biomechanical factors on gait pattern, the heterogeneous nature of leg pathologies, and a requirement for rigorous experimental design (since pain experience must be inferred using indirect non-verbal measures). Strong physiological evidence indicates that broilers have the capacity to experience leg pain. Slowly adapting mechanoreceptors are present within the skin of the chicken tarsometatarsus, and these become sensitised following induced inflammation (Gentle et al., 2001). Inflammatory arthropathy, a condition that can cause pain in humans, has also been identified within the hock joints of spontaneously (naturally) lame broilers (Corr et al., 2003c). Some very lame broilers have also been reported as having a greater relative adrenal mass, which is likely to be indicative of chronic stress (Müller et al., 2015).

Visual identification of a pain state is complicated by the fact that non-lame broilers spend the majority of their time sitting (Weeks et al., 2000). In addition, poultry are a prey species and it is likely that they avoid overt pain-associated behaviour, as any display of weakness could make them more vulnerable to predation. Morphology (via mechanical limitations and a reduced motivation to walk) has a significant influence upon gait and activity levels in modern strains, regardless of any assumed discomfort. Since certain pathologies (e.g. a mild skeletal deformity) are likely to be less painful than others (e.g. inflammatory or necrotic conditions), the welfare implications of failing to quantify and differentiate pain from other causes of gait abnormality are substantial.

The provision of analgesic drugs under experimental conditions can provide indirect evidence for pathological pain if positive changes in behaviour, or improvements in a predefined test performance, are observed post-treatment. The majority of studies to date have utilised non-steroidal anti-inflammatory drugs (NSAIDs), such as carprofen and meloxicam, as they are routinely used to manage pain associated with osteoarthritis in dogs and cats and may have a therapeutic potential in poultry. A self-selection experiment, reporting the preferential selection of food spiked with NSAID, claimed to provide early evidence for the occurrence of pain in lame broilers (Danbury et al., 2000); however, a later study was unable to corroborate these findings (Siegel et al., 2011).

NSAID treatment has been observed to increase walking velocity and modify gait in moderately lame broilers (Nääs et al., 2009; Caplen et al., 2013a) and improve LTL performance (Hothersall et al., 2016). However, such improvements in mobility could be attributable to a reduction in joint inflammation as opposed to any direct analgesic effect. Although the relationship between lameness and pain (and thus welfare) is complicated by confounding 'risk factors' such as sex, strain, bodyweight and pathology, 'lameness' has been identified as the most consistent predictor for several broiler mobility measures. This indicates that there is a constituent of 'lameness' that cannot be explained by any combination of the more obvious bird characteristics (e.g. being a heavy male), and it is this component that may represent pain or discomfort (Caplen et al., 2014).

Hyperalgesia (heightened sensitivity to pain) is prevalent in many disease states as part of an inflammatory response to prevent further tissue damage. Pain-producing chemicals (cytokines and chemokines) trigger primary afferent nociceptor sensitisation. Primary hyperalgesia describes pain sensitivity that occurs directly in the damaged tissues; a lowered nociceptive/pain threshold to both thermal and mechanical stimuli is usual. Caplen et al. (2013b) utilised a specially designed apparatus to detect primary thermal hyperalgesia in broilers with experimentally induced inflammatory arthropathies (an acute pain model) and demonstrated that NSAID treatment could reverse this effect via anti-nociception (Fig. 6). A comparable effect has not yet been demonstrated in broilers with non-induced 'spontaneous' lameness; conversely, a higher baseline thermal threshold was reported in farm-lame broilers, which increased further following NSAID treatment, potentially due to altered nociceptive processing (Hothersall et al., 2014). The fact that hyperalgesia was detected in a group of experimental birds following pain induction under controlled conditions but not in a group of chickens with (potentially mixed) on-farm pathologies does not prove that the latter group had no pain experience. Pain is not always accompanied by hyperalgesia, and, therefore, a lack of response is not an evidence for an absence of pain. Clear difficulties exist in obtaining large groups of birds with comparable pathologies for experimental work and in linking lameness severity with pathology (e.g. Sandilands et al., 2011).

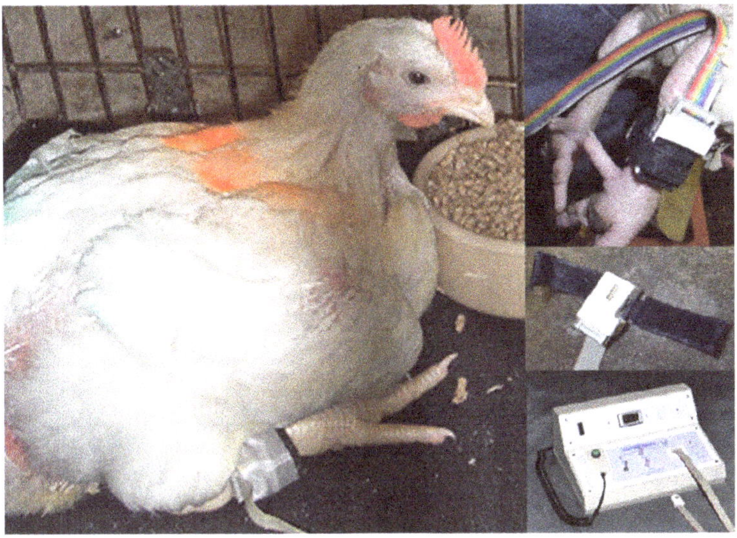

**Figure 6** Apparatus used for thermal threshold testing (Images: Gina Caplen). The leg probe (containing a temperature sensor and a heated element) was attached to the lateral aspect of the tarsometatarsus via a Velcro strap. A ramped heat stimulus was applied until a behavioural endpoint was detected and then immediately removed. The response temperature was held on a digital readout.

### 2.5.3 Limited behavioural expression

Selection for broiler growth has significantly narrowed their ethogram and altered their time budget, compared to other breeds of *Gallus gallus domesticus,* by restricting the range of behaviours that they are physically able to perform. The effect of posture on resting metabolic rate becomes increasingly significant as broilers grow; locomotion becomes very energetically expensive, standing more so than sitting. Since the metabolic scope for exercise decreases throughout their development, the proportion of the overall metabolic rate accounted for by locomotor behaviour also decreases, which corresponds to declining activity levels and low walking speeds (Tickle et al., 2018). This is particularly apparent in the fast-growing breeds, since slower-growing broilers perform less sitting and more perching, walking and ground scratching throughout the production period (Bokkers and Koene, 2003; Reiter and Bessei, 2009).

In welfare terms inherent inactivity is problematic for several reasons. Walking acts to strengthen muscles and bones, and inactivity is thought to be a direct cause of leg weakness. Lameness further compromises mobility, limiting behavioural expression and reducing the value of any enrichment provision. There is evidence that physically impaired individuals retain at least some motivation to perform locomotory behaviour (Rutten et al., 2002; Bokkers and Koene, 2004; Bokkers et al., 2007). When this motivation remains unfulfilled, it is likely to trigger frustration (e.g. as indicated by displacement preening, Bokkers and Koene, 2003), stress and suffering.

## 2.6 Prevention and control of lameness

### 2.6.1 Genetics

Long-term selection for improved broiler leg health (in particular, long-bone deformities and TD) over the last 30 years or so has been partially effective, despite some unfavourable genetic correlations with body mass (Kapell et al., 2012a). A marked reduction in TD within commercial strains over recent years is testimony to what is possible. Estimated heritabilities of non-infectious skeletal disorders (including TD and valgus–varus), and susceptibility to infection (e.g. FHN), indicate that genetic selection and breeding programmes offer a means to further reduce multiple lameness aetiologies and leg pathologies alongside the main selection focus, to further improve production parameters (Akbaş et al., 2009; Wideman et al., 2014). At the producer level, a greater commercial uptake of slower-growing strains (with their lower predisposition for musculo-skeletal health problems) housed within appropriate systems would directly alleviate the main welfare concerns.

### 2.6.2 Incubation conditions

Incubation lighting and heating schedules both appear to have a marked effect upon leg health. Egg exposure to continuous light has been shown to have a detrimental effect upon embryonic leg bone development. The risk of both poor bone strength and developing TD in later life is increased, compared to incubation regimes that include a continuous period (at least 8 h) of darkness (van der Pol et al., 2017, 2019).

The optimum incubation EST for healthy leg development appears to be lower than 37.8°C, the temperature currently recommended for maximising hatchability and chick growth. Embryos incubated under 'slow start' conditions, that is under lower temperatures (EST: 36.9–37.5°C) during the first few weeks of incubation, hatched later, had greater femoral bone ash (Groves and Muir, 2014; Muir and Groves, 2018), grew slower over the first week post-hatch, and had a lower prevalence of TD lesions at day 34 (Groves and Muir, 2017). Hatchery and chick quality issues clearly influence the susceptibility of broilers to BCO. Although it remains widespread within European flocks, the pathology is being successfully addressed via improvements in hatchery hygiene (Dinev, 2009).

### 2.6.3 Barn management

As mentioned previously, it is very important to effectively utilise heating and ventilation systems to stringently maintain the environmental parameters of the barn within the guidelines for the breed. Maintaining dry litter also appears to be important; the use of wheat straw as litter has been associated with a higher prevalence of lameness than (more absorbent) wood-shavings (Su et al., 2000), while the addition of vermiculite to wood-shavings has been found to lower lameness severity further (Yildiz et al., 2014).

### 2.6.4 Diet and feeding regimes

Diet and feeding regimes are utilised to control growth rate and/or increase bone strength. Quantitative or qualitative dietary restriction can be used as a means to slow weight gain and promote healthier anatomical leg development (Kestin et al., 2001; Wijtten et al., 2010). However, feed restriction in any form will trigger overt hunger, especially in the meat breeds that have been selected for high appetite and metabolic growth. Hunger is often associated with frustration, stereotypies and adverse behaviour, occasionally including cannibalism (Eriksson et al., 2010), all of which have obvious adverse impacts upon bird welfare.

A change from ad-libitum feeding to meal feeding (i.e. limiting food availability to 240 min/day split between two, three and four discrete meal times) has been seen to improve walking ability and reduce TD (Su et al., 1999). Responsive nutritional management can also be used to control lameness by slowing down the rate at which young broilers grow and then speeding the growth rate back up to achieve a standard finishing commercial body mass once the leg bones have better developed. This can be achieved via the sequential feeding of two diets, a high-energy/low-protein diet and a low-energy/high-protein diet, over 48-hour feeding cycles (Leterrier et al., 2008); a standard finishing diet is provided from day 29 until the last day of production to compensate for reduced growth. The inclusion of whole wheat within the diet is beneficial as it slows digestion and lowers the feed conversion rate, thereby, reducing both growth rate and lameness (Knowles et al., 2008). All three of these management practices, via health improvements and decreased feed costs, have been identified as having economic potential to realise substantial improvements in gross margin and net return for the farmer (Gocsik et al., 2017).

Broilers fed mashed diet have been shown to have higher bone ash and lower GS than broilers fed the same diet in pellet form (Brickett et al., 2007). Presumably birds eat a greater volume of pelleted food (and achieve a higher corresponding weight gain), while those fed mashed diet benefit physiologically from being able to preferentially select components of their ration and benefit behaviourally from spending more time foraging for their feed. Since deficiencies and imbalances of numerous vitamins and minerals can impact upon broiler bone mineralisation and leg health, the provision of an optimal dietary formulation is extremely important. Due to the ongoing selection for improved production parameters, dietary guidelines require a regular review. Numerous nutritional factors influence broiler leg health, often via complex interactions (e.g. facilitating mineral assimilation), and these are reviewed in detail elsewhere (see Waldenstedt, 2006; Kwiatkowska et al., 2017). Calcium (Ca), phosphorus (P) and vitamin $D_3$ (cholecalciferol) appear to be of key importance.

Dietary Ca supplementation is beneficial for increasing bone quality and reducing TD incidence (Coto et al., 2008; Abdulla et al., 2017); sources such as oyster shell, snail shell and limestone are reported as being most effective (Oso et al., 2011). In addition to quantity, the dietary balance between available Ca and P is extremely important. The optimum feed ratio is 2:1, and an excessive supply of either element can lessen the assimilation of both (Coto et al., 2008; Bradbury et al., 2014). Rickets can occur as a Ca-deficiency or P-deficiency type, resulting from either a direct dietary deficiency or an excessive proportion of either. An increase in TD incidence has also been reported as a consequence of Ca:P imbalance (Waldenstedt, 2006). Vitamin $D_3$ is beneficial as a dietary supplement since it increases Ca and P intestinal absorption, improves bone quality and walking ability (Baracho et al., 2012; Sun et al., 2013) and is valuable in the prevention of TD (Whitehead et al., 2004) and BCO (Wideman et al., 2015). Ascorbic acid (vitamin C) supplementation has been found to benefit both bone quality (Yildiz et al., 2009) and walking ability (Petek et al., 2005).

Phytase is an enzyme that acts to increase the bioavailability of various minerals (e.g. Ca, P, Mg, Zn, Fe), a large proportion of which are present within grains and seeds as insoluble complexes. Dietary phytase has been shown to increase the apparent digestibility of Ca by 4–6% (Saima et al., 2009) and to improve bone mineralisation (Bradbury et al., 2017).

Dietary supplementation of probiotics (live organisms intended to improve the gut microflora), prebiotics (non-digestible feed components that promote the growth of beneficial intestinal microorganisms) and synbiotics (the combination of a probiotic with a prebiotic) benefit bone development and mineralisation by increasing the intestinal absorption and assimilation of nutrients and minerals (Scholz-Ahrens et al., 2007; Yan et al., 2018). The inclusion of a dietary synbiotic (containing a prebiotic and a probiotic mixture of four microbial strains) was found to improve multiple indices of broiler leg health, LTL performance and walking ability (Yan et al., 2019). Probiotics are most effective at reducing BCO incidence if they are given proactively as part of the feed, rather than therapeutically after the onset of lameness (Wideman et al., 2012).

It is also important to recognise that there are environmental contaminants that can negatively impact upon health, even with the provision of an 'optimum diet'. Bacterial, viral and parasitic infections can reduce the ability of the intestinal epithelium to absorb nutrients, while feed contamination with certain mycotoxins can induce or exacerbate skeletal problems due to interference with vitamin D metabolism (Waldenstedt, 2006).

### 2.6.5 Photoperiod and light intensity

Continuous bright light is normally provided during the first 4 days following barn placement post-hatch (the brooding period) to stimulate feeding; however, exposure to intermittent lighting during the same period has been shown to benefit leg health by slowing bone development and increasing leg bone symmetry (van der Pol et al., 2015). Within the EU the Broiler Directive (2007/43/EC) states that lighting must follow a 24-h rhythm and include periods of darkness lasting at least 6 h in total, with at least one uninterrupted period of darkness of at least 4 h, excluding dimming periods (see Chapter 10). The presence of a dark phase (scotophase) throughout the production cycle is conclusively beneficial for broiler leg health. Provision of a scotophase is associated with a reduction in lameness (Brickett et al., 2007; Knowles et al., 2008; Bassler et al., 2013; Schwean-Lardner et al., 2013; Das and Lacin, 2014) and TD (Petek et al., 2005; Karaarslan and Nazligul, 2018) and an increase in tibial strength (Lewis et al., 2009; Yang et al., 2015), compared to exposure to continuous light. Bone mineralisation peaks during the dark period and is sensitive to diurnal rhythm (Russell et al., 1984, as cited by Bassler et al., 2013). In addition, broilers provided with a scotophase are physically more active during the light period than those kept under near-continuous light (Sanotra et al., 2002; Bayram and Özkan, 2010; Schwean-Lardner et al., 2012). A photoperiod of 16L:8D appears optimal for maximising both welfare and feed conversion (Classen, 2004).

Broilers are typically housed under low light intensity during the light phase for the majority of the production cycle. Those reared under very dim lighting (<5 lux) develop a higher body mass, display less activity and have a higher GS than birds reared under brighter (10–320 lux) lighting regimes (Blatchford et al., 2009, 2012; Senaratna et al., 2016; Fidan et al., 2017). Exposure of chicks to low levels of UV-B illumination (to simulate natural daylight) has been trialled as an alternative non-dietary means of enhancing vitamin $D_3$. UV-B treatment has been found to have substantial benefits for bone characteristics and virtually eliminates growth-plate abnormalities caused by Ca and vitamin D deficiencies (Fleming, 2008). Broilers reared under environments including supplementary UV-A light-emitting diode (LED) and UV-B fluorescent lighting had a lower GS than those of a white LED control group (James et al., 2018). The incorporation of UV wavelengths within commercial lighting regimes may, therefore, be beneficial for broiler leg health.

### 2.6.6 Environmental enrichment

Increasing activity levels have known benefits for broiler leg health since mechanical loading is essential for maintaining normal bone formation and remodelling (Hester et al., 2013). Broilers encouraged to walk on treadmills were shown to have better bone density, thickness and fewer leg bone deformities than control birds (Reiter and Bessei, 2009), while cage-reared broilers were observed to have lower leg bone mineralisation than birds reared in open barn systems (Aguado et al., 2015).

Standard broiler breeds are typically barn-reared in low complexity environments with minimal enrichment; however, studies that have provided commercial flocks with elevated structures in an attempt to encourage locomotor activity (and thus improve leg and foot pad health) have met with varying success. Although the provision of platform perches in combination with dust baths was found unsuccessful in reducing lameness (Bailie et al., 2018a), the provision of platforms in a different study improved GS and TD incidence/severity (Kaukonen et al., 2017). Presumably the key to use is in the platform design. A small-scale study using experimental pens reported that slow-growing breeds use perches more frequently than fast-growing strains (Bokkers and Koene, 2003), presumably due to intrinsically higher activity levels and superior leg strength; however, low usage overall suggests perch provision to be inappropriate for broilers. In contrast, the provision of straw and/or wood-shaving bales has reliably been associated with increased physical activity (Bailie and O'Connell, 2015; Ohara et al., 2015; de Jong and Gunnink, 2019) and improved leg health (Bailie et al., 2013; Vasdal et al., 2019) in commercial flocks.

Body mass appears to physically limit the utilisation of elevated structures. Usage undergoes a marked decline towards the end of the production cycle (perches: Bailie and O'Connell, 2015; hay bales: Ohara et al., 2015; platforms: Norring et al., 2016), and females exploit the enrichment more than the (heavier) males (Estevez et al., 2002; Ohara et al., 2015). The design of commercially appropriate feeding regimes (e.g. distribution of the dietary ration on the litter surface and wheat and mealworm provision in addition to the dietary ration) to stimulate activity has largely been unsuccessful (Bizeray et al., 2002a,b; Jordan et al., 2011). The provision of oat hulls has, however, been shown to promote dust-bathing and foraging and reduce lameness (Baxter et al., 2018). The provision of slow-growing breeds with an outdoor range has many fundamental welfare benefits over indoor systems, including the encouragement of increased locomotor activity (Nielsen et al., 2003;

Sosnówka-Czajka et al., 2007) and the potential for lowering lameness levels (Zhao et al., 2014; Ipek and Sozcu, 2017).

Selection for fast early growth rate has made broilers susceptible to developing leg disorders as a consequence of significant morphological alterations and a predisposition to inactivity. This, in turn, produces direct welfare problems and confers difficulties in improving housing design. Although the majority of broiler welfare problems have a genetic basis, many, such as lameness, are further exasperated by interactions with a poor environmental management. The provision and maintenance of high-quality litter is very important for general flock health. Contact dermatitis is a skin condition frequently seen in commercial broiler flocks, being particularly prevalent in lame birds since it affects those skin surfaces that have prolonged contact with poor quality (wet) litter.

## 3 Contact dermatitis

Broilers are particularly susceptible to developing contact dermatitis, due to management practices (particularly litter contamination with faecal ammonia and water) and low activity levels. Contact dermatitis can range in severity from dark skin discolouration and superficial erosions (mild) to pronounced inflammation of subcutaneous tissue and deep necrotic lesions (severe). Most commonly the plantar surface of the feet, including the skin of the central pad and toes, is affected, and this is referred to as footpad dermatitis (FPD) or pododermatitis. Contact dermatitis is also commonly observed to affect the skin of the hock joint and that overlying the sternum, termed 'hock burn' (HB) and breast burn, accordingly.

### *3.1 Prevalence*

A wide variation in the occurrence of moderate to severe FPD has been reported, both between and within studies (e.g. Italy: 0–90%, Meluzzi et al., 2008a,b; USA: 5%, Opengart et al., 2018; Denmark: 13.1%, Lund et al., 2017; Bulgaria: 16%, Dinev et al., 2019; Canada: 28.7%, Hunter et al., 2017; Europe: 37.3%, Bassler et al., 2013; the Netherlands: 38%, de Jong et al., 2012a; UK: 0–48%, Jones et al., 2005; France: 83%, Allain et al., 2009). Although the majority of these studies used a three-point scale, some, for example Allain et al. (2009), used complex scoring systems which can make inter-study comparisons difficult.

Higher prevalence and severity of FPD in European, compared to American, flocks could reflect multiple factors. These include the common European practice of flock thinning (see Section 3.3.1), the type of bedding material (chopped straw litter is frequently used in some parts of Europe while wood-shavings were used in the American study), a more pronounced seasonal (winter) effect in certain parts of Europe, concrete floors in broiler houses (more likely to produce condensation and trap litter moisture compared with the dirt floors typically found in U.S. broiler houses) and the utilisation of a relatively thin layer of bedding, completely replaced between flocks (compared to the deeper build-up and re-use of litter common in the USA) (Opengart et al., 2018).

Although HB develops more slowly than FPD (Skrbic et al., 2015) and is observed less frequently, the two are often associated (e.g. Bassler et al., 2013). As with FPD, a wide variation in HB prevalence has also been reported (e.g. Bulgaria: <1%, Dinev et al., 2019; UK: 0–33%, Haslam et al., 2007; Hepworth et al., 2011; Italy: 3–87%, Melluzzi et al., 2008a,b; Europe: 7.9%, Bassler et al., 2013; France: 59%, Allain et al., 2009), presumably for similar reasons. Although the presence of breast burn and severe HB has been positively correlated (Allain et al., 2009), breast burn is much less frequently seen (e.g. UK: <1%, Haslam et al., 2007; Bulgaria: 5%, Dinev et al., 2019; France: 16%, Allain et al., 2009; Portugal: 18%, Gouveia et al., 2009) and, as a consequence, the condition remains comparatively understudied.

### 3.2 Qualitative assessment

Many different scoring systems have historically been used to study FPD and HB, ranging from simple binary scores to complex severity scales with as many as ten different categories (e.g. Allain et al., 2009). To enable a meaningful assessment of dermatitis prevalence over time and between systems, it is important to establish a standardised measure. Widely used scoring systems for categorising FPD and HB lesion severity in broilers on farm as part of the 'Welfare Quality (2009) Assessment Protocol for Poultry' utilise five-point severity scales and photographic reference images (Fig. 7). To account for the likelihood that FPD lesion severity is unevenly distributed over commercial broiler units, it is recommended that at least five different sampling locations are employed within each unit and a minimum of 100 birds are scored (de Jong et al., 2012b).

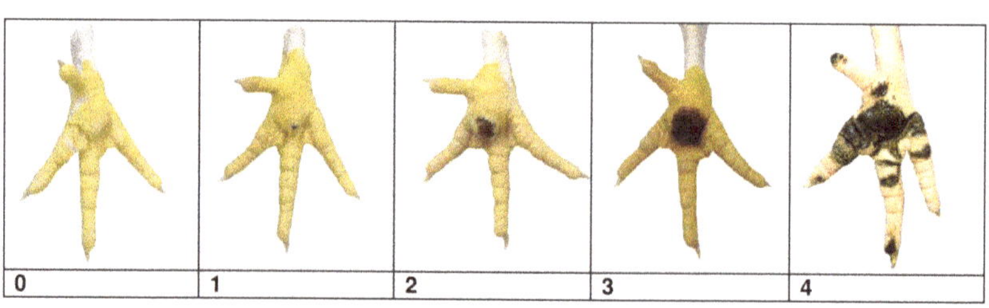


**Figure 7** A five-point scoring scale (0–4) for categorising the severity of foot pad dermatitis lesion in broilers (Welfare Quality®, 2009).

## 3.3 Contact dermatitis risk factors

### 3.3.1 Age and stocking density

Older broilers are heavier, less active (e.g. Alvino et al., 2009), and consequently experience more skin contact with increasingly faecal-loaded litter. Although contact dermatitis is often reported as increasing with age (FPD: Kyvsgaard et al., 2013; Sarica et al., 2014; HB: Kjaer et al., 2006; Haslam et al., 2007), there are exceptions. FPD lesions have been observed as early as 7 days post-hatch when litter moisture is low (Berk, 2009; Hashimoto et al., 2011). This suggests that young broilers may be more vulnerable to acquiring FPD, with or without exposure to poor litter. There is further evidence for an age-related increase in skin resistance to environmental irritation. Early exposure to high-moisture litter has been shown to increase FPD prevalence and severity (at 14 days); however, no further increase was observed following litter wetting at >56 days of age (Cengiz et al., 2011).

A positive relationship between contact dermatitis and stocking density has often been reported (FPD: Haslam et al., 2007; Bailie et al., 2018b; Ventura et al., 2010; Kyvsgaard et al., 2013; HB: Karaarslan and Nazligul, 2018), associated with high litter moisture and pH (Dozier et al., 2006; Petek et al., 2010). However, stocking density is not a definitive indicator of poor welfare as several studies report no direct impact of density on FPD (Dawkins et al., 2004; Haslam et al., 2006; Meluzzi et al., 2008a; Allain et al., 2009).

Some systems stock at high density and incorporate flock thinning into their production cycle. This involves the removal of a proportion of the flock, usually 1 week prior to the end of production, in order to maintain stocking density (body mass per area) within legal limits. More severe FPD has been reported in birds at final depopulation than in younger thinned-out birds, consistent with the theory that the risk increases with age; however, litter improvement following a reduction in stocking density (post-thinning) may actually enable foot pad health to improve with age (de Jong et al., 2012a).

It has been suggested that placement density ($m^2$/bird) may be a more important factor than finishing density (kg/$m^2$), assuming other environmental factors that impact FPD remain constant. FPD lesions have been observed to form relatively early in life (between 12 and 21 days), correlating with deteriorating litter conditions (including increases in ammonia and moisture), and then to persist over the life of the flock (Opengart et al., 2018).

### 3.3.2 Sex and body mass

There is plenty of evidence to indicate that male and heavier birds are more at risk of developing HB (Oviedo-Rondón et al., 2009b; Henriksen et al.,

2016; Louton et al., 2018). Indeed body mass at only 2 weeks of age has been identified as a predictor of those flocks most at risk of developing HB prior to slaughter (Hepworth et al., 2010). Although some consider male birds to be more pre-disposed to developing FPD than females due to a heavier body mass (e.g. Gouveia et al., 2009; Sarica et al., 2014; Villarroel et al., 2018), not all studies are in agreement with this. Conflicting observations may reflect a loss of condition (and thus body mass) in broilers with severe FPD, especially if the lesion becomes the site of a secondary infection (e.g. Kyvsgaard et al., 2013).

### 3.3.3 Production systems

Contact dermatitis lesions are more prevalent/severe in the fast-growing than slower-growing genotypes (FPD: Sarica et al., 2014; Yamak et al., 2016; HB: Haslam et al., 2007; Skrbic et al., 2015), presumably due to different feed conversion rates (and related excreta properties) and activity levels. Unsurprisingly, both FPD prevalence and severity are higher in solid-floor than wire-floor (cage) housing systems (Cengiz et al., 2013; Simsek et al., 2014).

A higher prevalence of FPD has been reported in extensive indoor systems compared to traditional free-range farms (Portugal: Gouveia et al., 2009), and lower HB has been observed in organically reared broilers than in conventional flocks (Broom and Reefmann, 2005; Tuyttens et al., 2008). However, other studies have reported higher FPD scores in flocks with outdoor access than those without (UK: Pagazaurtundua and Warriss, 2006; Turkey: Sarica et al., 2014). The feet of free-range birds will be exposed to ammonia-rich litter in the barns at night, in addition to hard stony ground and pathogens (from wild birds/rodents) while they are out on the range. Broilers with access to a range incorporating both trees and grass have been reported to have better foot condition than those provided with grass alone (Dal Bosco et al., 2014), and this improvement is likely to reflect an increase in range use. Since range conditions, particularly wetness, are an additional risk factor for the pathology (Sans et al., 2014), a well-considered range design, (including the provision of shelter and drainage) should benefit foot health by allowing broilers feet to dry during their time spent outside.

### 3.3.4 Litter quality

The Welfare Quality Assessment Protocol for broilers describes a five-point scale for scoring litter quality ranging from '0: Completely dry and flaky, moves easily with foot' to '4: Sticks to boots once the cap or compacted crust is broken' (Welfare Quality, 2009). The prevalence and severity of contact dermatitis are often correlated with deterioration in litter quality, particularly when the surface of the litter becomes compacted and the moisture content increases (FPD: Kyvsgaard et al., 2013; de Jong et al., 2014; HB: Allain et al., 2009; Bassler et al., 2013). Wet litter and litter containing a high nitrogen content (from animal waste) are generally more alkaline, that is have a higher pH value (Meluzzi et al., 2008b; Abd El-Wahab et al., 2013). Indeed, improvements in litter quality have been shown to reduce FPD lesion severity in market-age broilers (Cengiz et al., 2011).

Litter quality is related to the type of bedding material used, nutrition, gut health and the optimal management of the housing environment (including heating, ventilation and minimising water spillage). The use of misting systems has been linked with higher levels of FPD (Jones et al., 2005), while the provision of more drinkers per unit area leads to higher litter moisture and a greater prevalence of HB (Jones et al., 2005). Increased water consumption has also been associated with higher FPD (Manning et al., 2007) and HB (Hepworth et al., 2010). Litter quality is presumably reduced directly via increased water spillage and indirectly via increased (and loose) excreta.

Several studies have highlighted that, at least under commercial conditions, litter condition is likely to be more critical than stocking density in the development of FPD (Dawkins et al., 2004; Haslam et al., 2007). Stocking density can contribute towards litter moisture levels; however, with appropriate gut health and effective environmental management, the maintenance of litter quality appears to be possible throughout the production cycle, even under high stocking densities.

### 3.3.5 Diet

Nutrition, including diet composition and feeding programs, plays a significant role in the aetiology of contact dermatitis due to feed conversion rate, water intake, properties of excreta and the resulting litter quality (see Section 3.3.4). Links between physiological stress, elevated corticosterone and increased water content of faeces are well established (e.g. Puvadolpirod and Thaxton, 2000; Nicol et al., 2009), most likely due to sub-optimal gut function. Jacob et al. (2016a) observed an increase in FPD prevalence between 19 and 29 days old, which correlated with a dietary switch from starter to grower feed; this abrupt dietary transition was hypothesised to reduce gut health and, therefore, litter quality. Presumably the risk of causing stress to the digestive system could easily be reduced by introducing a novel diet gradually, initially presented as a low proportion of the ration and increased over time.

Studies outlining the optimal dietary levels of crude protein, biotin and electrolytes, and the effects of diet composition and nutrient concentration on litter quality and FPD occurrence in broilers, are reviewed in detail elsewhere (e.g. Swiatkiewicz et al., 2017). Dietary constituents vary between countries, and many studies have trialled novel protein sources. FPD is significantly increased in broilers receiving an entirely vegetable-based protein diet compared to those given a diet containing both vegetable- and animal-based proteins

(Nagaraj et al., 2007; Cengiz et al., 2013), and further FPD reductions can be achieved with the addition of corn gluten meal (Eichner et al., 2007). Barley-based diets have been associated with poor litter quality and FPD increase compared to corn-based diets (Cengiz et al., 2017). On the basis of FPD prevalence, rapeseed meal has also been identified as an unsuitable alternative dietary protein to soyabean meal (Abd El-Wahab et al., 2018).

Dietary additives have also been well-researched. The provision of an increased percentage of dietary wheat has been linked with reduced HB (Haslam et al., 2007) and is also seen to benefit leg health (see Section 2.7.4). The inclusion of reduced levels of 2-hydroxy-4-(methylthio)butanoic acid-chelated trace minerals into broiler diets as an alternative to industry levels of inorganic trace minerals has been found to significantly reduce FPD (Zhao et al., 2010; Manangi et al., 2012; Da Costa et al., 2016). Rations supplemented with additional fat (Fuhrmann and Kamphues, 2016), fatty acids (Khosravinia, 2015), vitamin $D_3$ (Sun et al., 2013), biotin (Abd El-Wahab et al., 2013; Sun et al., 2017), tannic acid (Cengiz et al., 2017) and a directly fed microbe (*Bacillus*) in combination with dietary enzymes (Dersjant-Li et al., 2015) have all been shown to reduce FPD. Dietary enzymes, such as phytase, are beneficial to foot health as they improve nutrient digestibility and reduce litter moisture content (Farhadi et al., 2017).

### 3.3.6 Lighting

Low-level lighting (0.5–5 lux) delivered with a short scotoperiod (period of darkness) appears to be a risk factor for contact dermatitis (Deep et al., 2013; Bassler et al., 2013; Schwean-Lardner et al., 2013; Senaratna et al., 2016) compared to birds reared under brighter lighting (10–320 lux). This is likely to be associated with an increase in body mass in the low-light birds (Lien et al., 2007; Blatchford et al., 2012; Senaratna et al., 2016). The type of lighting may also be important. The use of LED bulbs has been linked with improved FPD and HB scores when compared with compact fluorescent bulbs (Huth and Archer, 2015).

### 3.3.7 Egg quality, incubation and early life

At least some susceptibility to contact dermatitis appears to have its foundations in pre-hatch conditions. Lower HB scores have been reported for broilers originating from young parent breeders (particularly those individuals with a low hatching weight) than those originating from older parents (Henriksen et al., 2016).

Appropriate incubation temperatures also appear to be important. Oviedo-Rondón et al. (2009a) observed HB lesions to be more prevalent in 8-week-old commercial broilers hatched from multi-stage incubators (associated with greater temperature fluctuation) than those hatched from single-stage incubators. Fluctuating incubation temperatures were also seen to increase FPD susceptibility via alterations in foot skin structure (Da Costa et al., 2016). Broilers incubated at temperatures (39–40°C) higher than the optimum recommended range may also have a greater risk of developing severe HB in later life (Ipek and Sozcu, 2016).

Raising brooding temperature may help reduce the risk of HB development in heavy chicks. Henriksen et al. (2016) reported that a high brooding temperature (37°C) during the first week post-hatch increased activity, delayed body weight gain and reduced the prevalence of HB at 5 weeks in both high- and low-hatch-weight chicks (compared to high-hatch-weight chicks brooded at 33°C).

### 3.4 Welfare impact of contact dermatitis

Although there is no evidence for a relationship between contact dermatitis and mortality (de Jong et al., 2012a), skin lesions provide an obvious point of entry for pathogenic microorganisms, which can lead to secondary infections and advanced pathologies such as gangrenous dermatitis and osteomyelitis (Dinev, 2009). In broilers, both FPD and HB are associated with an increased prevalence and severity of *Campylobacter* infection (Bull et al., 2008; Rushton et al., 2009). Hepworth et al. (2011) reported HB to be a useful indicator of flock health since HB was positively associated with the percentage of birds with septicaemia and fever, detected at post-mortem inspection. In addition HB is positively associated with lameness (Sørensen et al., 2000; Haslam et al., 2007); presumably as broilers age and become heavier (and experience deteriorations in leg health), they spend more time lying down, increasing contact time with the litter.

De Jong et al. (2014) observed reductions in production parameters (including body weight) and walking ability (increased GS) in broilers reared on wet litter. As FPD also increased on wet litter the authors hypothesised that birds with deep or infected lesions may have experienced pain and consequently fed and drunk less due to inappetence (de Jong et al., 2014). Sherlock et al. (2012) compared global hepatic gene expression in control birds and in those with experimentally induced FPD and HB lesions. They reported evidence for the inflammatory reaction to impact upon key pathways linked with growth, metabolism and energy utilisation and hypothesised that pain may be the underlying trigger for the up-regulation of genes linked to a pro-inflammatory response and energy metabolism (Sherlock et al., 2012). Further evidence for FPD having a pain component is provided by Hothersall et al. (2016); birds with more severe FPD lesions sat sooner in an LTL test; however, standing ability was improved following NSAID treatment.

### 3.5 Prevention and control of contact dermatitis

#### 3.5.1 Genetic selection

Since fast-growing genotypes have conclusively been shown to demonstrate greater susceptibility to developing contact dermatitis than the slower-growing genotypes (Sarica et al., 2014; Yamak et al., 2016), greater use of slow-growing strains in extended production systems would have direct and obvious benefits to broiler welfare. HB prevalence also varies according to the standard commercial hybrid used, and this diversity has been attributed to differences in both susceptibility and predisposition to generate wet litter (DEFRA, 2010). Breeding programmes have been shown to offer a direct means by which flock susceptibility to FPD and HB can be selected against and reduced without compromising further improvements in production parameters (Kjaer et al., 2006; Akbas et al., 2009; Ask, 2010; Kapell et al., 2012b), and such selection should continue.

#### 3.5.2 Litter properties

The use of appropriate bedding material in floor-based poultry production systems is necessary to meet several health and welfare requirements. In regard to managing dermatitis deep litter systems should be 'deep', with at least a 7.6 cm depth of litter provided (Shepherd et al., 2017). Bedding material should have high absorbance properties (i.e. have a small particle size) as well as the ability to quickly release moisture (Bilgili et al., 2009; Cengiz et al., 2011). Flocks reared on wood-shavings or sawdust exhibit less FPD than those reared on chopped straw (e.g. Kyvsgaard et al., 2013; Skrbic et al., 2015; Villarroel et al., 2018), rice husks (e.g. Jacob et al., 2016b), grass (e.g. Garcia et al., 2012) or corncob litter (Xavier et al., 2010). Wood-shavings also appear to be the most appropriate litter for managing HB. Lower HB scores were observed at the end of the grow-out period in broilers reared on wood-shavings than paper sludge (Villagrá et al., 2011) or chopped straw (Skrbic et al., 2015). Unchopped straw is unsuitable for use as litter because it has low moisture absorbency and is associated with high FPD (Đukić Stojčić et al., 2016).

Although broiler chickens are commonly housed on sawdust and wood-shavings in some parts of Europe and North America, due to the lack of availability elsewhere, alternative litter sources are required in many parts of the world. Other bedding materials with a potential for maintaining healthy skin include pelleted straw (Berk, 2009; Avdalovic et al., 2017), sand (Simsek et al., 2009) and vermiculite (Yildiz et al., 2014). Similar levels of HB and FPD are reported for conventional wood-shaving litter systems and alternative perforated flooring (netting) systems (Li et al., 2017). The use of shredded paper, although cheap, readily available and highly absorbent, has proved to be unsuccessful in the control of contact dermatitis (Villagrá et al., 2011; Kheravii et al., 2017) since it becomes sodden and fails to release moisture.

In some countries, due to costs and availability, litter (at least the lower layer) is often re-used. Although some studies report a higher prevalence of contact dermatitis in broilers reared on re-used litter than in those reared on fresh (HB: Jacob et al., 2016b; FPD: de Oliveira et al., 2015; Shepherd et al., 2017), other studies report the opposite (Jacob et al., 2016b; Yamak et al., 2016). The ability to successfully re-use sawdust litter appears to be limited to production facilities located in hot dry climates, which exploit effective litter management, ventilation and composting systems.

Litter additives have been shown to be useful in improving litter quality, such as the addition of microorganisms and enzymes to increase the metabolisation of organic components. The application of a biological promoter aimed at wastewater treatment (Micropan®Simplex, Eurovix USA) to chopped straw successfully reduced both litter pH and FPD prevalence compared to a control flock (Đukić Stojčić et al., 2016; Zikic et al., 2017). The addition of sodium bisulphate complex (SBS) to litter has also been shown to be beneficial for foot pad health; this treatment also acidifies the litter, thereby reducing the microbial growth of nitrifying bacteria and $NH_3$ formation (Toppel et al., 2019).

### 3.5.3 Managing litter moisture

Litter moisture levels are directly linked to air temperature and RH within the broiler house, which are themselves directly influenced by ambient air temperature and season, the choice of heating system, the ventilation rate, the stocking density and broiler body mass. Deviations in environmental parameters outside of the optimum range recommended by the breeder companies have been identified as a key factor adversely influencing FPD (Jones et al., 2005). A detrimental effect of stocking density on RH has been observed to occur towards the end of the production cycle, indicating that environmental control becomes more difficult at this time (Jones et al., 2005). Presumably the ventilation rate requires to be increased once the birds get larger, less active and cover more of the litter surface, to improve air flow at the litter surface and remove litter moisture. With appropriate environmental management/control and the maintenance of temperature and RH levels within optimum limits, it should be possible to maintain good-quality dry litter throughout the production cycle.

Season has a marked influence upon contact dermatitis, with more lesions observed during colder months (Meluzzi et al., 2008a; Hepworth et al., 2010; Dinev et al., 2019). RH and litter moisture are also usually higher during colder months (e.g. Dawkins et al., 2004; Hermans et al., 2006; Meluzzi et al., 2008a). To save heating costs and maintain shed temperatures, ventilation rates are often reduced at this time. Gas-burning space heaters are a common form of heating in older UK broiler houses (Jones et al., 2005). These systems are located within the sheds themselves and generate a lot of water vapour as a by-product of combustion.

Gut health is linked with litter quality as any increase in faecal water content will be directly transferred to the litter. Causes of diarrhoea (scour) in broilers include coccidiosis, worms, viral or bacterial infection, a diet too high in protein, abrupt dietary alterations or long periods without feed (e.g. caused by feed equipment failures) and biosecurity breaches (Hermans et al., 2006). The use of antibiotics has been shown to improve broiler gut health, indirectly improve litter quality and reduce FPD severity (de Jong et al., 2012a). Global pressure to reduce antibiotic usage in all forms of livestock production has recently sparked much interest in the impact of intestinal microbiota upon broiler productivity (improved feed conversion efficiency) and health (e.g. Stanley et al., 2014; Oviedo-Rondón, 2019). Healthy gut microbiota lowers the risk of enteric diseases due to a reduction in pathogenic load (via competitive exclusion) and improved immune function. A reduction in enteric diseases should, in turn, improve litter quality. The use of probiotics, prebiotics and synbiotics (mixtures of both) to promote beneficial gut microbiota is emerging as a possible alternative to in-feed antibiotics. Results are very encouraging; some studies have even demonstrated prebiotic treatment to be more effective than antibiotics at maintaining broiler intestinal health (e.g. Al-Baadani et al., 2016).

The majority of broiler farms now use nipple and/or cup drinkers as opposed to bell drinkers since they minimise water spillage, and water spillage directly reduces litter quality. Increased water consumption has also been directly associated with FPD (Manning et al., 2007). Daily water consumption is currently monitored in l/bird/day to monitor flock performance and provide an indication of bird health. However, it has been shown that monitoring water consumption as l/m$^2$ floor area/day is a good lag (end of crop) indicator of litter quality (Manning et al., 2007). The use of automatic water meters has been linked with a decrease in HB (Hepworth et al., 2010). An ability to monitor and control water intake may, therefore, facilitate better stockmanship and improve litter management.

### 3.5.4 Environmental enrichment

Although evidence for a positive influence of elevated structures on contact dermatitis is variable and likely to reflect a different uptake in perch use between studies, the provision of straw bales and perches has been reported to reduce contact dermatitis (FPD: Ventura et al., 2010; Hongchao et al., 2013; Ohara et al., 2015; Kiyma et al., 2016; HB: Karaarslan and Nazligul, 2018). Elevated structures encourage better vertical distribution of birds within the available space, decreasing bird density at the floor level and allowing better circulation of air at the litter surface. They also provide a cleaner and dryer surface for the birds to sit and stand on, away from the litter surface.

### 3.5.5 Monitoring and surveillance systems

Camera systems are being developed to continuously model and monitor the movement and activity of broiler flocks throughout the production cycle to automatically detect and trigger alerts for FPD and HB (Dawkins et al., 2017; Fernandez et al., 2018). One system, in particular, shows great potential to inform improved flock management by providing farmers with an early warning of welfare issues. A camera-based optical flow system (OPTICFLOCK, see Chapter 7) has been shown to have a greater capacity to predict the prevalence of FPD and HB at slaughter (even before external signs become visible) than water consumption, bodyweight or mortality data (Dawkins et al., 2017).

FPD is an important welfare indicator. It is encouraging that government surveillance schemes, whereby producers have to reduce their stocking density or correct management deficiencies (e.g. improve litter quality) if their overall flock score (calculated from the prevalence and severity of FPD lesions at slaughter) exceeds certain predefined trigger limits, have proved to be successful at reducing FPD in Sweden, Denmark (e.g. Kyvsgaard et al., 2013) and the UK (Part et al., 2016). It is hoped that other countries will adopt similar national monitoring schemes and that a widespread uptake of automated monitoring systems will soon provide a means for continuous on-farm surveillance.

## 4 Conclusion and future trends

Broiler lameness and contact dermatitis remain commonplace within intensively reared commercial flocks, yet it is difficult to infer whether there have been significant improvements or indeed global declines in broiler welfare over recent years. Drawing any definitive conclusions is problematic since data remain sparse and are collected in different ways, multiple management differences exist both within and between countries and the breeds are undergoing constant change (via genetic selection). Leg problems certainly occur more frequently in heavy fast-growing strains due to muscular tissue accumulation occurring at a faster rate than the developing skeleton can support. Although lameness becomes more apparent at higher stocking densities (likely due to forced inactivity), environmental management, rather than stocking density, appears to be critical in the prevention of wet litter and contact dermatitis. Appropriate ventilation should be provided year-round, and both temperature and RH should remain within the recommended range at all times.

Broilers display low activity levels as a consequence of genetic selection for an unnatural body morphology and an excessive body mass; accordingly, they spend the majority of their time sitting in contact with the litter surface. The provision of elevated structures (such as straw bales and platforms with ramp-access), in combination with appropriate lighting, should facilitate rest and encourage activity, ultimately benefiting bone strength and leg and skin health. The provision of slow-growing genotypes with a suitable well-drained range should potentially encourage broilers to perform more activities outside, away from the litter surface of the barn, and thereby limit dermatitis progression.

Both lameness and contact dermatitis pose serious welfare concerns. Lameness physically limits behavioural expression and compromises mobility; birds may experience difficulty in gaining access to feeders and drinkers. Skin lesions provide a gateway for bacteria which can, in turn, trigger joint inflammation. Although evidence for an association between lameness, pain and underlying pathologies remains inconclusive, it is highly likely that some forms of lameness and severe footpad lesions are painful. More research is required to quantify the level of pain associated with specific pathologies, to enable us to differentiate these from other, morphological, gait abnormalities and to better focus breeding programmes (i.e. provide the genetics companies with targeted pressure).

Contact dermatitis and many leg pathologies are heritable and can be selected against. Although the use of genetic selection has had some success in the control of certain pathologies, the persistence of leg and skin conditions in commercial flocks indicates that more needs to be done. A greater uptake, commercially, of the slower-growing breeds would directly benefit broiler welfare. Non-governmental organisations (NGOs) have been, and will continue to be, fundamental in highlighting the requirement for robust genotypes. The Royal Society for Protection of Cruelty to Animals (RSPCA, UK) independently assesses new breeds according to a broiler breed welfare assessment protocol prior to its acceptance for use within the RSPCA-assured scheme; this includes key welfare outcome measures such as growth rate, leg heath and mortality. A number of animal protection organisations have recently set out an NGO 'Broiler Ask' initiative in the EU and the USA. The EU letter can be found at https://welfarecommitments.com/europeletter/. This scheme directly targets food suppliers and retailers to request that they raise welfare standards for meat chickens including better labelling on meat products, a maximum stocking density of 30 kg/m$^2$, the adoption of approved higher-welfare breeds and improved lighting by 2026.

Lameness, FPD and HB are generally assessed visually using severity scales, either on farm or at the processing plant. All three measures are used as flock welfare indicators and are crucial for setting (and reviewing) targets to progressively improve welfare within the industry. It is hoped that the development of fully automated video surveillance systems for use on farm will soon enable flock behaviour to be continuously monitored and any anomalies detected to be flagged up to provide producers with an early warning prior to the development of potential welfare issues.

Although standard broiler diets have been developed to deliver optimum nutrition, additional supplements have been observed to confer substantial health benefits. Dietary supplementation with prebiotics, probiotics and synbiotics has been shown to improve the gut uptake of beneficial nutrients and minerals, leading to improvements in the leg health. The addition of dietary enzymes improves nutrient digestibility and has been associated with improvements in both litter quality and footpad health. The uptake of increasingly integrated and automated management systems such as Flockman® (a combined lighting and feed control system for broilers) appears to have numerous welfare benefits. Such systems claim to improve digestion, feed conversion, litter quality and leg health via the automation of lighting programmes and provision of meal feeding to broiler flocks.

More needs to be done to help farmers recognise that the implementation of different feeding practices (e.g. whole wheat feeding, meal feeding and sequential feeding) could have a substantial positive impact upon both broiler welfare and farm economics.

The overarching problem preventing fundamental change within the poultry industry for the improvement of broiler skin and leg health is one of financial cost. The majority of factors that increase bird activity (i.e. appropriate environmental enrichment, lighting regimes, diet and gentotype) also improve skin and leg health; however, the producers (and the genetics companies) hesitate to make these changes due to a real or perceived association of increased activity with lowered production (food conversion) parameters. In fact, it is entirely possible that extra production costs resulting from the implementation of certain management practices designed to reduce lameness and/or improve skin health could be compensated for by other gains. For example, improving leg health and increasing flock uniformity have the potential to confer many economical benefits, including better technical performance (and thus reduced production costs), lowered mortality and reduced carcass condemnation rates. Comprehensive economic cost-benefit analyses may therefore be necessary to bolster both consumer pressure and ethical persuasion for change.

## 5 Where to look for further information

### Further reading

- *Measuring and Auditing Broiler Welfare* by Weeks and Butterworth (2004) provides a good introduction to lameness and contact dermatitis.

- For a useful recent review of the applied ethology of broilers, including space use, stocking density, behaviour and welfare, see *The Behavioural Biology of Chickens* by Nicol (2015).
- *Poultry Feathers and Skin: The Poultry Integument in Health and Welfare*, edited by Olukosi et al. (2019), contains chapters on contact dermatitis and the genetics of contact dermatitis.
- Bradshaw, R. H., Kirkden, R. D. and Broom, D. M. 2002. A review of the aetiology and pathology of leg weakness in broilers in relation to their welfare. *Avian and Poultry Biological Reviews* 13(2), 45–103.
- Dinev, I. 2012. Leg weakness pathology in broiler chickens. *Journal of Poultry Science* 49(2), 63–67.
- Kierończyk, B., Rawski, M., Józefiak, D. and Świątkiewicz, S. 2017. Infectious and non-infectious factors associated with leg disorders in poultry – a review. *Annals of Animal Science* 17(3), 645–69.
- Mayne, R. 2005. A review of the aetiology and possible causative factors of foot pad dermatitis in growing turkeys and broilers. *World's Poultry Science Journal* 61(2), 256–67.
- Pedersen, I. J. and Forkman, B. 2019. Improving leg health in broiler chickens: a systematic review of the effect of environmental enrichment. *Animal Welfare* 28(2), 215–30.
- Swiatkiewicz, S., Arczewska-Wlosek, A. and Jozefiak, D. 2017. The nutrition of poultry as a factor affecting litter quality and foot pad dermatitis – an updated review. *Journal of Animal Physiology and Animal Nutrition (Berlin)* 101(5), e14–20. doi:10.1111/jpn.12630.
- Waldenstedt, L. 2006. Nutritional factors of importance for optimal leg health in broilers: a review. *Animal Feed Science and Technology* 126(3–4), 291–307.
- Wideman, R. F. 2016. Bacterial chondronecrosis with osteomyelitis and lameness in broilers: a review. *Poultry Science* 95(2), 325–44.

## Key Journals

- *Animal Welfare*
- *British Poultry Science*
- *Journal of Applied Poultry Research*
- *Poultry Science*
- *Veterinary Record*

## Key Conferences

- European Symposium on Poultry Welfare
- UFAW Animal Welfare Conference: Recent Advances in Animal Welfare Science

- WAFL: International Conference on the Assessment of Animal Welfare at Farm and Group Level

## 6 References


Abd El-Wahab, A., Radko, D. and Kamphues, J. 2013. High dietary levels of biotin and zinc to improve health of foot pads in broilers exposed experimentally to litter with critical moisture content. *Poultry Science* 92(7), 1774–82. doi:10.3382/ps.2013-03054.

Abd El-Wahab, A., Visscher, C. and Kamphues, J. 2018. Impact of different dietary protein sources on performance, litter quality and foot pad dermatitis in broilers. *Journal of Animal and Feed Sciences* 27(2), 148–54. doi:10.22358/jafs/90696/2018.

Abdulla, N. R., Loh, T. C., Akit, H., Sazili, A. Q., Foo, H. L., Kareem, K. Y., Mohamad, R. and Abdul Rahim, R. 2017. Effects of dietary oil sources, calcium and phosphorus levels on growth performance, carcass characteristics and bone quality of broiler chickens. *Journal of Applied Animal Research* 45(1), 423–9. doi:10.1080/09712119.2016.1206903.

Aguado, E., Pascaretti-Grizon, F., Goyenvalle, E., Audran, M. and Chappard, D. 2015. Bone mass and bone quality are altered by hypoactivity in the chicken. *PLoS ONE* 10(1), e0116763. doi:10.1371/journal.pone.0116763.

Akbaş, Y., Yalcin, S., Ozkan, S., Kirkpinar, F., Takma, C., Gevrekçi, Y,, Guller, H. C. and Turkmut, L. 2009. Heritability estimates of tibial dyschondroplasia, valgus-varus, foot-pad dermatitis and hock burn in broiler. *Archiv fur Geflugelkunde* 73(1), 1–6.

Al-Baadani, H. H., Abudabos, A. M., Al-Mufarrej, S. I. and Alzawqari, M. 2016. Effects of dietary inclusion of probiotics, prebiotics and Synbiotics on intestinal histological changes in challenged broiler chickens. *South African Journal of Animal Science* 46(2), 157–65. doi:10.4314/sajas.v46i2.6.

Allain, V., Mirabito, L., Arnould, C., Colas, M., Le Bouquin, S., Lupo, C. and Michel, V. 2009. Skin lesions in broiler chickens measured at the slaughterhouse: relationships between lesions and between their prevalence and rearing factors. *British Poultry Science* 50(4), 407–17. doi:10.1080/00071660903110901.

Al-Rubaye, A. A. K., Ekesi, N. S., Zaki, S., Emami, N. K., Wideman, R. F. and Rhoads, D. D. 2017. Chondronecrosis with osteomyelitis in broilers: further defining a bacterial challenge model using the wire flooring model. *Poultry Science* 96(2), 332–40. doi:10.3382/ps/pew299.

Alvino, G. M., Archer, G. S. and Mench, J. A. 2009. Behavioural time budgets of broiler chickens reared in varying light intensities. *Applied Animal Behaviour Science* 118(1–2), 54–61. doi:10.1016/j.applanim.2009.02.003.

Arnould, C. and Faure, J. M. 2004. Use of pen space and activity of broiler chickens reared at two different densities (vol 84, pg 281, 2003). *Applied Animal Behaviour Science* 87(1–2), 153–70.

Ask, B. 2010. Genetic variation of contact dermatitis in broilers. *Poultry Science* 89(5), 866–75. doi:10.3382/ps.2009-00496.

Avdalovic, V., Vueinic, M., Resanovic, R., Avdalovic, J., Maslic-Strizaks, D. and Vucicevic, M. 2017. Effect of pelleted and chopped wheat straw on the footpad dermatitis in broilers. *Pakistan Journal of Zoology* 49(5), 1639–46. doi:10.17582/journal.pjz/2017.49.5.1639.1646.

Aydin, A. 2017a. Development of an early detection system for lameness of broilers using computer vision. *Computers and Electronics in Agriculture* 136, 140–6. doi:10.1016/j.compag.2017.02.019.

Aydin, A. 2017b. Using 3D vision camera system to automatically assess the level of inactivity in broiler chickens. *Computers and Electronics in Agriculture* 135, 4–10. doi:10.1016/j.compag.2017.01.024.

Aydin, A., Bahr, C. and Berckmans, D. 2015. Automatic classification of measures of lying to assess the lameness of broilers. *Animal Welfare* 24(3), 335–43. doi:10.7120/09627286.24.3.335.

Bailie, C. L. and O'Connell, N. E. 2015. The influence of providing perches and string on activity levels, fearfulness and leg health in commercial broiler chickens. *Animal* 9(4), 660–8. doi:10.1017/S1751731114002821.

Bailie, C. L., Ball, M. E. E. and O'Connell, N. E. 2013. Influence of the provision of natural light and straw bales on activity levels and leg health in commercial broiler chickens. *Animal* 7(4), 618–26. doi:10.1017/S1751731112002108.

Bailie, C. L., Baxter, M. and O'Connell, N. E. 2018a. Exploring perch provision options for commercial broiler chickens. *Applied Animal Behaviour Science* 200, 114–22. doi:10.1016/j.applanim.2017.12.007.

Bailie, C. L., Ijichi, C. and O'Connell, N. E. 2018b. Effects of stocking density and string provision on welfare-related measures in commercial broiler chickens in windowed houses. *Poultry Science* 97(5), 1503–10. doi:10.3382/ps/pey026.

Baracho, M. S., Naas, I. A., Bueno, L. G. F., Nascimento, G. R. and Moura, D. J. 2012. Broiler walking ability and toe asymmetry under harsh rearing conditions. *Revista Brasileira de Ciência Avícola* 14(3), 217–22. doi:10.1590/S1516-635X2012000300009.

Bassler, A. W., Arnould, C., Butterworth, A., Colin, L., De Jong, I. C., Ferrante, V., Ferrari, P., Haslam, S., Wemelsfelder, F. and Blokhuis, H. J. 2013. Potential risk factors associated with contact dermatitis, lameness, negative emotional state, and fear of humans in broiler chicken flocks. *Poultry Science* 92(11), 2811–26. doi:10.3382/ps.2013-03208.

Baxter, M., Bailie, C. L. and O'Connell, N. E. 2018. Evalution of dustbathing substrate and straw bales as environmental enrichments in commercial broiler housing. *Applied Animal Behaviour Science* 200, 78–85. doi:10.1016/j.applanim.2017.11.010.

Bayram, A. and Özkan, S. 2010. Effects of a 16-hour light, 8-hour dark lighting schedule on behavioral traits and performance in male broiler chickens. *Journal of Applied Poultry Research* 19(3), 263–73. doi:10.3382/japr.2009-00026.

Berg, C. and Sanotra, G. S. 2003. Can a modified latency-to-lie test be used to validate gait-scoring results in commercial broiler flocks? *Animal Welfare* 12(4), 655–9.

Berk, J. 2009. Effect of litter type on prevalence and severity of pododermatitis in male broilers. *Berliner Und Munchener Tierarztliche Wochenschrift* 122(7–8), 257–63.

Bilgili, S. F., Hess, J. B., Blake, J. P., Macklin, K. S., Saenmahayak, B. and Sibley, J. L. 2009. Influence of bedding material on footpad dermatitis in broiler chickens. *Journal of Applied Poultry Research* 18(3), 583–9. doi:10.3382/japr.2009-00023.

Birgul, O. B., Mutaf, S. and Alkan, S. 2012. Effects of different angled perches on leg disorders in broilers. *Archiv fur Geflugelkunde* 76(1), 44–8.

Bizeray, D., Estevez, I., Leterrier, C. and Faure, J. M. 2002a. Effects of increasing environmental complexity on the physical activity of broiler chickens. *Applied Animal Behaviour Science* 79(1), 27–41. doi:10.1016/S0168-1591(02)00083-7.

Bizeray, D., Estevez, I., Leterrier, C. and Faure, J. M. 2002b. Influence of increased environmental complexity on leg condition, performance, and level of fearfulness in broilers. *Poultry Science* 81(6), 767–73. doi:10.1093/ps/81.6.767.

Blatchford, R. A., Klasing, K. C., Shivaprasad, H. L., Wakenell, P. S., Archer, G. S. and Mench, J. A. 2009. The effect of light intensity on the behavior, eye and leg health, and immune function of broiler chickens. *Poultry Science* 88(1), 20–8. doi:10.3382/ps.2008-00177.

Blatchford, R. A., Archer, G. S. and Mench, J. A. 2012. Contrast in light intensity, rather than day length, influences the behavior and health of broiler chickens. *Poultry Science* 91(8), 1768–74. doi:10.3382/ps.2011-02051.

Bokkers, E. A. M. and Koene, P. 2003. Behaviour of fast- and slow growing broilers to 12 weeks of age and the physical consequences. *Applied Animal Behaviour Science* 81(1), 59–72. doi:10.1016/S0168-1591(02)00251-4.

Bokkers, E. A. M. and Koene, P. 2004. Motivation and ability to walk for a food reward in fast- and slow-growing broilers to 12 weeks of age. *Behavioural Processes* 67(2), 121–30. doi:10.1016/j.beproc.2004.03.015.

Bokkers, E. A. M., Zimmerman, P. H., Rodenburg, T. B. and Koene, P. 2007. Walking behaviour of heavy and light broilers in an operant runway test with varying durations of feed deprivation and feed access. *Applied Animal Behaviour Science* 108(1-2), 129–42. doi:10.1016/j.applanim.2006.10.011.

Bradbury, E. J., Wilkinson, S. J., Cronin, G. M., Thomson, P. C., Bedford, M. R. and Cowieson, A. J. 2014. Nutritional geometry of calcium and phosphorus nutrition in broiler chicks. Growth performance, skeletal health and intake arrays. *Animal* 8(7), 1071–9. doi:10.1017/S1751731114001037.

Bradbury, E. J., Wilkinson, S. J., Cronin, G. M., Thomson, P., Walk, C. L. and Cowieson, A. J. 2017. Evaluation of the effect of a highly soluble calcium source in broiler diets supplemented with phytase on performance, nutrient digestibility, foot ash, mobility and leg weakness. *Animal Production Science* 57(10), 2016–26. doi:10.1071/AN16142.

Bradshaw, R. H., Kirkden, R. D. and Broom, D. M. 2002. A review of the aetiology and pathology of leg weakness in broilers in relation to welfare. *Avian and Poultry Biology Reviews* 13(2), 45–103. doi:10.3184/147020602783698421.

Brickett, K. E., Dahiya, J. P., Classen, H. L., Annett, C. B. and Gomis, S. 2007. The impact of nutrient density, feed form, and photoperiod on the walking ability and skeletal quality of broiler chickens. *Poultry Science* 86(10), 2117–25. doi:10.1093/ps/86.10.2117.

Broom, D. M. and Reefmann, N. 2005. Chicken welfare as indicated by lesions on carcases in supermarkets. *British Poultry Science* 46(4), 407–14. doi:10.1080/00071660500181149.

Buijs, S., Keeling, L., Rettenbacher, S., Van Poucke, E. and Tuyttens, F. A. M. 2009. Stocking density effects on broiler welfare: identifying sensitive ranges for different indicators. *Poultry Science* 88(8), 1536–43. doi:10.3382/ps.2009-00007.

Buijs, S., Keeling, L. J., Vangestel, C., Baert, J. and Tuyttens, F. A. M. 2011. Neighbourhood analysis as an indicator of spatial requirements of broiler chickens. *Applied Animal Behaviour Science* 129(2–4), 111–20. doi:10.1016/j.applanim.2010.11.017.

Buijs, S., Van Poucke, E., Van Dongen, S., Lens, L., Baert, J. and Tuyttens, F. A. 2012. The influence of stocking density on broiler chicken bone quality and fluctuating asymmetry. *Poultry Science* 91(8), 1759–67. doi:10.3382/ps.2011-01859.

Bull, S. A., Thomas, A., Humphrey, T., Ellis-Iversen, J., Cook, A. J., Lovell, R. and Jorgensen, F. 2008. Flock health indicators and *Campylobacter* spp. in commercial housed broilers reared in Great Britain. *Applied and Environmental Microbiology* 74(17), 5408–13. doi:10.1128/AEM.00462-08.

Butterworth, A. and Haslam, S. M. 2009. A lameness control strategy for broiler fowls. Welfare Quality Reports No. 13. Cardiff School of City and Regional Planning, Cardiff, UK. Available at: http://www.welfarequality.net/media/1119/wqr13.pdf.

Butterworth, A., Weeks, C. A., Crea, P. R. and Kestin, S. C. 2002. Dehydration and lameness in a broiler flock. *Animal Welfare* 11(1), 89–94.

Caplen, G., Hothersall, B., Murrell, J. C., Nicol, C. J., Waterman-Pearson, A. E., Weeks, C. A. and Colborne, G. R. 2012. Kinematic analysis quantifies gait abnormalities associated with lameness in broiler chickens and identifies evolutionary gait differences. *PLoS ONE* 7(7), e40800. doi:10.1371/journal.pone.0040800.

Caplen, G., Colborne, G. R., Hothersall, B., Nicol, C. J., Waterman-Pearson, A. E., Weeks, C. A. and Murrell, J. C. 2013a. Lame broiler chickens respond to non-steroidal anti-inflammatory drugs with objective changes in gait function: a controlled clinical trial. *Veterinary Journal* 196(3), 477–82. doi:10.1016/j.tvjl.2012.12.007.

Caplen, G., Baker, L., Hothersall, B., McKeegan, D. E. F., Sandilands, V., Sparks, N. H., Waterman-Pearson, A. E. and Murrell, J. C. 2013b. Thermal nociception as a measure of non-steroidal anti-inflammatory drug effectiveness in broiler chickens with articular pain. *Veterinary Journal* 198(3), 616–9. doi:10.1016/j.tvjl.2013.09.013.

Caplen, G., Hothersall, B., Nicol, C. J., Parker, R. M. A., Waterman-Pearson, A. E., Weeks, C. and Murrell, J. 2014. Lameness is consistently better at predicting broiler chicken performance in mobility tests than other broiler characteristics. *Animal Welfare* 23(2), 179–87. doi:10.7120/09627286.23.2.179.

Castellini, C., Mugnai, C., Moscati, L., Mattioli, S., Amato, M. G., Cartoni Mancinelli, A. and Dal Bosco, A. 2016. Adaptation to organic rearing system of eight different chicken genotypes: behaviour, welfare and performance. *Italian Journal of Animal Science* 15(1), 37–46. doi:10.1080/1828051X.2015.1131893.

Cengiz, Ö, Hess, J. B. and Bilgili, S. F. 2011. Effect of bedding type and transient wetness on footpad dermatitis in broiler chickens. *Journal of Applied Poultry Research* 20(4), 554–60. doi:10.3382/japr.2011-00368.

Cengiz, O., Hess, J. B. and Bilgili, S. F. 2013. Effect of protein source on the development of footpad dermatitis in broiler chickens reared on different flooring types. *Archiv fur Geflugelkunde* 77(3), 166–70.

Cengiz, Ö, Koksal, B. H., Tatli, O., Sevim, Ö, Ahsan, U., Bilgili, S. F. and Gökhan Önol, A. 2017. Effect of dietary tannic acid supplementation in corn- or barley-based diets on growth performance, intestinal viscosity, litter quality, and incidence and severity of footpad dermatitis in broiler chickens. *Livestock Science* 202, 52–7. doi:10.1016/j.livsci.2017.05.016.

Charles, J. F., Ermann, J. and Aliprantis, A. O. 2015. The intestinal microbiome and skeletal fitness: connecting bugs and bones. *Clinical Immunology* 159(2), 163–9. doi:10.1016/j.clim.2015.03.019.

Classen, H. L. 2004. Day length affects performance, health and condemnations in broiler chickens. *Proceedings of the 16th Australian Poultry Science Symposium*, Sydney, New South Wales, Australia, 9–11 February 2004, pp. 11215. Available at: https://sydney.edu.au/vetscience/apss/documents/2004/APSS2004-classen-pp112-115.pdf.

Collins, L. M. 2008. Non-intrusive tracking of commercial broiler chickens *in situ* at different stocking densities. *Applied Animal Behaviour Science* 112(1-2), 94-105. doi:10.1016/j.applanim.2007.08.009.

Corr, S. A., Gentle, M. J., McCorquodale, C. C. and Bennett, D. 2003a. The effect of morphology on walking ability in the modern broiler: a gait analysis study. *Animal Welfare* 12(2), 159-71.

Corr, S. A., Gentle, M. J., McCorquodale, C. C. and Bennett, D. 2003b. The effect of morphology on the musculoskeletal system of the modern broiler. *Animal Welfare* 12(2), 145-57.

Corr, S. A., Maxwell, M., Gentle, M. J. and Bennett, D. 2003c. Preliminary study of joint disease in poultry by the analysis of synovial fluid. *Veterinary Record* 152(18), 549-54. doi:10.1136/vr.152.18.549.

Corr, S. A., McCorquodale, C., McDonald, J., Gentle, M. and McGovern, R. 2007. A force plate study of avian gait. *Journal of Biomechanics* 40(9), 2037-43. doi:10.1016/j.jbiomech.2006.09.014.

Coto, C., Yan, F., Cerrate, S., Wang, Z., Sacakli, P., Halley, J. T., Wiernusz, C. J., Martinez, A. and Waldroup, P. W. 2008. Effects of dietary levels of calcium and nonphytate P in broiler starter diets on live performance, bone development, and growth plate conditions in male broilers fed a corn-based diet. *International Journal of Poultry Science* 7(7), 638-45. doi:10.3923/ijps.2008.638.645.

Da Costa, M. J., Oviedo-Rondon, E. O., Wineland, M. J., Claassen, K. and Osborne, J. 2016. Effects of incubation temperatures and trace mineral sources on chicken live performance and footpad skin development. *Poultry Science* 95(4), 749-59. doi:10.3382/ps/pev446.

Dal Bosco, A. D., Mugnai, C., Amato, M. G., Piottoli, L., Cartoni, A. and Castellini, C. 2014. Effect of slaughtering age in different commercial chicken genotypes reared according to the organic system: 1. Welfare, carcass and meat traits. *Italian Journal of Animal Science* 13(2), 3308. doi:10.4081/ijas.2014.3308.

Danbury, T. C., Weeks, C. A., Chambers, J. P., Waterman-Pearson, A. E. and Kestin, S. C. 2000. Self-selection of the analgesic drug carprofen by lame broiler chickens. *Veterinary Record* 146(11), 307-11. doi:10.1136/vr.146.11.307.

Dantzer, R. 2009. Cytokine, sickness behavior, and depression. *Immunology and Allergy Clinics of North America* 29(2), 247-64. doi:10.1016/j.iac.2009.02.002.

Das, H. and Lacin, E. 2014. The effect of different photoperiods and stocking densities on fattening performance, carcass and some stress parameters in broilers. *Israel Journal of Veterinary Medicine* 69(4), 211-20.

Dawkins, M. S., Donnelly, C. A. and Jones, T. A. 2004. Chicken welfare is influenced more by housing conditions than by stocking density. *Nature* 427(6972), 342-4. doi:10.1038/nature02226.

Dawkins, M. S., Lee, H. J., Waitt, C. D. and Roberts, S. J. 2009. Optical flow patterns in broiler chicken flocks as automated measures of behaviour and gait. *Applied Animal Behaviour Science* 119(3-4), 203-9. doi:10.1016/j.applanim.2009.04.009.

Dawkins, M. S., Cain, R., Merelie, K. and Roberts, S. J. 2013. In search of the behavioural correlates of optical flow patterns in the automated assessment of broiler chicken welfare. *Applied Animal Behaviour Science* 145(1-2), 44-50. doi:10.1016/j.applanim.2013.02.001.

Dawkins, M. S., Roberts, S. J., Cain, R. J., Nickson, T. and Donnelly, C. A. 2017. Early warning of footpad dermatitis and hockburn in broiler chicken flocks using optical flow, bodyweight and water consumption. *Veterinary Record* 180(20), 499. doi:10.1136/vr.104066.

de Jong, I. C. and Gunnink, H. 2019. Effects of a commercial broiler enrichment programme with or without natural light on behaviour and other welfare indicators. *Animal* 13(2), 384–91. doi:10.1017/S1751731118001805.

de Jong, I. C., van Ham, J., Gunnink, H., Hindle, V. A. and Lourens, A. 2012a. Footpad dermatitis in Dutch broiler flocks: prevalence and factors of influence. *Poultry Science* 91(7), 1569–74. doi:10.3382/ps.2012-02156.

de Jong, I. C., van Harn, J., Gunnink, H., Lourens, A. and van Riel, J. W. 2012b. Measuring foot-pad lesions in commercial broiler houses. Some aspects of methodology. *Animal Welfare* 21(3), 325–30. doi:10.7120/09627286.21.3.325.

de Jong, I. C., Gunnink, H. and van Harn, J. 2014. Wet litter not only induces footpad dermatitis but also reduces overall welfare, technical performance, and carcass yield in broiler chickens. *Journal of Applied Poultry Research* 23(1), 51–8. doi:10.3382/japr.2013-00803.

de Oliveira, M. C., Goncalves, B. N., Padua, G. T., da Silva, V. G., da Silva, D. V. and Freitas, A. I. M. 2015. Treatment of poultry litter does not improve performance or carcass lesions in broilers. *Revista Colombiana De Ciencias Pecuarias* 28(4), 331–8.

Deep, A., Raginski, C., Schwean-Lardner, K., Fancher, B. I. and Classen, H. L. 2013. Minimum light intensity threshold to prevent negative effects on broiler production and welfare. *British Poultry Science* 54(6), 686–94. doi:10.1080/00071668.2013.847526.

Department of the Environment Food and Rural Affairs. 2010. Foot pad dermatitis & hock burn in broilers: risk factors, aetiology & welfare consequences. London, UK. Available at: http://sciencesearch.defra.gov.uk/Document.aspx?Document=AW1137SID5FinalReport.pdf (accessed on 16 June 2019).

Dersjant-Li, Y., van de Belt, K., van der Klis, J. D., Kettunen, H., Rinttila, T. and Awati, A. 2015. Effect of multi-enzymes in combination with a direct-fed microbial on performance and welfare parameters in broilers under commercial production settings. *Journal of Applied Poultry Research* 24(1), 80–90. doi:10.3382/japr/pfv003.

Dinev, I. 2009. Clinical and morphological investigations on the prevalence of lameness associated with femoral head necrosis in broilers. *British Poultry Science* 50(3), 284–90. doi:10.1080/00071660902942783.

Dinev, I. and Kanakov, D. 2011. Deep pectoral myopathy: prevalence in 7 weeks old broiler chickens in Bulgaria. *Revue De Medecine Veterinaire* 162(6), 279–83.

Dinev, I., Denev, S. A. and Edens, F. W. 2012. Comparative clinical and morphological studies on the incidence of tibial dyschondroplasia as a cause of lameness in three commercial lines of broiler chickens. *Journal of Applied Poultry Research* 21(3), 637–44. doi:10.3382/japr.2010-00303.

Dinev, I., Denev, S., Vashin, I., Kanakov, D. and Rusenova, N. 2019. Pathomorphological investigations on the prevalence of contact dermatitis lesions in broiler chickens. *Journal of Applied Animal Research* 47(1), 129–34. doi:10.1080/09712119.2019.1584105.

Dozier, W. A., Thaxton, J. P., Purswell, J. L., Olanrewaju, H. A., Branton, S. L. and Roush, W. B. 2006. Stocking density effects on male broilers grown to 1.8 kilograms of body weight. *Poultry Science* 85(2), 344–51. doi:10.1093/ps/85.2.344.

Duff, S. R. I. and Thorp, B. H. 1985. Abnormal angulation/torsion of the pelvic appendicular skeleton in broiler fowl: morphological and radiological findings. *Research in Veterinary Science* 39(3), 313–9. doi:10.1016/S0034-5288(18)31720-X.

Đukić Stojčić, M., Bjedov, S., Zikic, D., Peric, L. and Milosevic, N. 2016. Effect of straw size and microbial amendment of litter on certain litter quality parameters, ammonia emission, and footpad dermatitis in broilers. *Archives Animal Breeding* 59(1), 131–7. doi:10.5194/aab-59-131-2016.

Eichner, G., Vieira, S. L., Torres, C. A., Coneglian, J. L. B., Freitas, D. M. and Oyarzabal, O. A. 2007. Litter moisture and footpad dermatitis as affected by diets formulated on an all-vegetable basis or having the inclusion of poultry by-product. *Journal of Applied Poultry Research* 16(3), 344–50. doi:10.1093/japr/16.3.344.

Eriksson, M., Waldenstedt, L., Elwinger, K., Engstrom, B. and Fossum, O. 2010. Behaviour, production and health of organically reared fast-growing broilers fed low crude protein diets including different amino acid contents at start. *Acta Agriculturae Scandinavica, Section A - Animal Science* 60(2), 112–24. doi:10.1080/09064702.2010.502243.

Estevez, I. 2007. Density allowances for broilers: where to set the limits? *Poultry Science* 86(6), 1265–72. doi:10.1093/ps/86.6.1265.

Estevez, I., Tablante, N., Pettit-Riley, R. L. and Carr, L. 2002. Use of cool perches by broiler chickens. *Poultry Science* 81(1), 62–9. doi:10.1093/ps/81.1.62.

FAO. 2010. Poultry welfare in developing countries - welfare issues in commercial broiler production, pp. 117–18. Available at: http://www.fao.org/3/a-al723e.pdf.

Farhadi, D., Karimi, A., Sadeghi, G., Rostamzadeh, J. and Bedford, M. R. 2017. Effects of a high dose of microbial phytase and myo-inositol supplementation on growth performance, tibia mineralization, nutrient digestibility, litter moisture content, and foot problems in broiler chickens fed phosphorus-deficient diets. *Poultry Science* 96(10), 3664–75. doi:10.3382/ps/pex186.

Fernandes, B. C. D., Martins, M. R. F. B., Mendes, A. A., Paz, I., Komiyama, C. M., Milbradt, E. L. and Martins, B. B. 2012. Locomotion problems of broiler chickens and its relationship with the gait score. *Revista Brasileira de Zootecnia - Brazilian Journal of Animal Science* 41(8), 1951–5. doi:10.1590/S1516-35982012000800021.

Fernandez, A. P., Norton, T., Tullo, E., van Hertem, T., Youssef, A., Exadaktylos, V., Vranken, E., Guarino, M. and Berckmans, D. 2018. Real-time monitoring of broiler flock's welfare status using camera-based technology. *Biosystems Engineering* 173, 103–14. doi:10.1016/j.biosystemseng.2018.05.008.

Fidan, E. D., Nazligul, A., Turkyilmaz, M. K., Aypak, S. Ü., Kilimci, F. S., Karaarslan, S. and Kaya, M. 2017. Effect of photoperiod length and light intensity on some welfare criteria, carcass, and meat quality characteristics in broilers. *Revista Brasileira de Zootecnia - Brazilian Journal of Animal Science* 46(3), 202–10. doi:10.1590/s1806-92902017000300004.

Fleming, R. H. 2008. Nutritional factors affecting poultry bone health. *Proceedings of the Nutrition Society* 67(2), 177–83. doi:10.1017/S0029665108007015.

Fuhrmann, R. and Kamphues, J. 2016. Effects of fat content and source as well as of calcium and potassium content in the diet on fat excretion and saponification, litter quality and foot pad health in broilers. *European Poultry Science* 80, 118. doi:10.1399/eps.2016.118.

Garcia, R., Almeida Paz, I., Caldara, F., Nääs, I., Bueno, L., Freitas, L., Graciano, J. and Sim, S. 2012. Litter materials and the incidence of carcass lesions in broilers chickens. *Revista Brasileira de Ciência Avícola* 14(1), 27–32. doi:10.1590/S1516-635X2012000100005.

Garner, J. P., Falcone, C., Wakenell, P., Martin, M. and Mench, J. A. 2002. Reliability and validity of a modified gait scoring system and its use in assessing tibial dyschondroplasia in broilers. *British Poultry Science* 43(3), 355–63. doi:10.1080/00071660120103620.

Gentle, M. J., Tilston, V. and McKeegan, D. E. F. 2001. Mechanothermal nociceptors in the scaly skin of the chicken leg. *Neuroscience* 106(3), 643–52. doi:10.1016/s0306-4522(01)00318-9.

Gocsik, É., Silvera, A. M., Hansson, H., Saatkamp, H. W. and Blokhuis, H. J. 2017. Exploring the economic potential of reducing broiler lameness. *British Poultry Science* 58(4), 337–47. doi:10.1080/00071668.2017.1304530.

Gouveia, K. G., Vaz-Pires, P. and da Costa, P. M. 2009. Welfare assessment of broilers through examination of haematomas, foot-pad dermatitis, scratches and breast blisters at processing. *Animal Welfare* 18(1), 43–8.

Granquist, E. G., Vasdal, G., de Jong, I. C. and Moe, R. O. 2019. Lameness and its relationship with health and production measures in broiler chickens. *Animal* 13(10), 2365–72. doi:10.1017/S1751731119000466.

Groves, P. J. and Muir, W. I. 2014. A meta-analysis of experiments linking incubation conditions with subsequent leg weakness in broiler chickens. *PLoS ONE* 9(7), e102682. doi:10.1371/journal.pone.0102682.

Groves, P. J. and Muir, W. I. 2017. Earlier hatching time predisposes Cobb broiler chickens to tibial dyschondroplasia. *Animal* 11(1), 112–20. doi:10.1017/S1751731116001105.

Guo, R. X., Li, Z. Y., Zhou, X. H., Huang, C. Y., Hu, Y. C., Geng, S., Chen, X., Li, Q., Pan, Z. and Jiao, X. 2019. Induction of arthritis in chickens by infection with novel virulent *Salmonella pullorum* strains. *Veterinary Microbiology* 228, 165–72. doi:10.1016/j.vetmic.2018.11.032.

Hashimoto, S., Yamazaki, K., Obi, T. and Takase, K. 2011. Footpad dermatitis in broiler chickens in Japan. *Journal of Veterinary Medical Science* 73(3), 293–7. doi:10.1292/jvms.10-0329.

Hashimoto, S., Yamazaki, K., Obi, T. and Takase, K. 2013. Relationship between severity of footpad dermatitis and carcass performance in broiler chickens. *Journal of Veterinary Medical Science* 75(11), 1547–9. doi:10.1292/jvms.13-0031.

Haslam, S. M., Brown, S. N., Wilkins, L. J., Kestin, S. C., Warriss, P. D. and Nicol, C. J. 2006. Preliminary study to examine the utility of using foot burn or hock burn to assess aspects of housing conditions for broiler chicken. *British Poultry Science* 47(1), 13–8. doi:10.1080/00071660500475046.

Haslam, S. M., Knowles, T. G., Brown, S. N., Wilkins, L. J., Kestin, S. C., Warriss, P. D. and Nicol, C. J. 2007. Factors affecting the prevalence of foot pad dermatitis, hock burn and breast burn in broiler chicken. *British Poultry Science* 48(3), 264–75. doi:10.1080/00071660701371341.

Henriksen, S., Bilde, T. and Riber, A. B. 2016. Effects of post-hatch brooding temperature on broiler behavior, welfare, and growth. *Poultry Science* 95(10), 2235–43. doi:10.3382/ps/pew224.

Hepworth, P. J., Nefedov, A. V., Muchnik, I. B. and Morgan, K. L. 2010. Early warning indicators for hock burn in broiler flocks. *Avian Pathology* 39(5), 405–9. doi:10.1080/03079457.2010.510500.

Hepworth, P. J., Nefedov, A. V., Muchnik, I. B. and Morgan, K. L. 2011. Hock burn: an indicator of broiler flock health. *Veterinary Record* 168(11), 303. doi:10.1136/vr.c6897.

Hermans, P. G., Fradkin, D., Muchnik, I. B. and Morgan, K. L. 2006. Prevalence of wet litter and the associated risk factors in broiler flocks in the United Kingdom. *Veterinary Record* 158(18), 615–22. doi:10.1136/vr.158.18.615.

Hester, P. Y., Enneking, S. A., Haley, B. K., Cheng, H. W., Einstein, M. E. and Rubin, D. A. 2013. The effect of perch availability during pullet rearing and egg laying on musculoskeletal health of caged White Leghorn hens. *Poultry Science* 92(8), 1972-80. doi:10.3382/ps.2013-03008.

Hongchao, J., Jiang, Y., Song, Z., Zhao, J., Wang, X. and Lin, H. 2013. Effect of perch type and stocking density on the behaviour and growth of broilers. *Animal Production Science* 54(7), 930-41. doi:10.1071/AN13184.

Hothersall, B., Caplen, G., Parker, R. M. A., Nicol, C. J., Waterman-Pearson, A. E., Weeks, C. A. and Murrell, J. C. 2014. Thermal nociceptive threshold testing detects altered sensory processing in broiler chickens with spontaneous lameness. *PLoS ONE* 9(5), e97883. doi:10.1371/journal.pone.0097883.

Hothersall, B., Caplen, G., Parker, R. M. A., Nicol, C. J., Waterman-Pearson, A. E., Weeks, C. and Murrell, J. 2016. Effects of carprofen, meloxicam and butorphanol on broiler chickens' performance in mobility tests. *Animal Welfare* 25(1), 55-67. doi:10.7120/09627286.25.1.055.

Hunter, J. M., Anders, S. A., Crowe, T., Korver, D. R. and Bench, C. J. 2017. Practical assessment and management of foot pad dermatitis in commercial broiler chickens: a field study. *Journal of Applied Poultry Research* 26(4), 593-604. doi:10.3382/japr/pfx019.

Huth, J. C. and Archer, G. S. 2015. Comparison of two LED light bulbs to a dimmable CFL and their effects on broiler chicken growth, stress, and fear. *Poultry Science* 94(9), 2027-36. doi:10.3382/ps/pev215.

Ipek, A. and Sozcu, A. 2016. The effects of eggshell temperature fluctuations during incubation on welfare status and gait score of broilers. *Poultry Science* 95(6), 1296-303. doi:10.3382/ps/pew056.

Ipek, A. and Sozcu, A. 2017. The effects of access to pasture on growth performance, behavioural patterns, some blood parameters and carcass yield of a slow-growing broiler genotype. *Journal of Applied Animal Research* 45(1), 464-9. doi:10.1080/09712119.2016.1214136.

Jacob, F. G., Baracho, M. S., Naas, I. A., Salgado, D. A. and Souza, R. 2016a. Incidence of pododermatitis in broiler reared under two types of environment. *Revista Brasileira de Ciência Avícola* 18(2), 247-54. doi:10.1590/1806-9061-2015-0047.

Jacob, F. G., Baracho, M. S., Naas, I. A., Lima, N. S. D., Salgado, D. D. and Souza, R. 2016b. Risk of incidence of hock burn and pododermatitis in broilers reared under commercial conditions. *Revista Brasileira de Ciência Avícola* 18(3), 357-62. doi:10.1590/1806-9061-2015-0183.

James, C., Asher, L., Herborn, K. and Wiseman, J. 2018. The effect of supplementary ultraviolet wavelengths on broiler chicken welfare indicators. *Applied Animal Behaviour Science* 209, 55-64. doi:10.1016/j.applanim.2018.10.002.

Jones, T. A., Donnelly, C. A. and Dawkins, M. S. 2005. Environmental and management factors affecting the welfare of chickens on commercial farms in the United Kingdom and Denmark stocked at five densities. *Poultry Science* 84(8), 1155-65. doi:10.1093/ps/84.8.1155.

Jordan, D., Stuhec, I. and Bessei, W. 2011. Effect of whole wheat and feed pellets distribution in the litter on broilers' activity and performance. *Archiv fur Geflugelkunde* 75(2), 98-103.

Julian, R. J. 1998. Rapid growth problems: ascites and skeletal deformities in broilers. *Poultry Science* 77(12), 1773-80. doi:10.1093/ps/77.12.1773.

Kapell, D. N., Hill, W. G., Neeteson, A. M., McAdam, J., Koerhuis, A. N. M. and Avendaño, S. 2012a. Twenty-five years of selection for improved leg health in purebred broiler lines and underlying genetic parameters. *Poultry Science* 91(12), 3032–43. doi:10.3382/ps.2012-02578.

Kapell, D. N., Hill, W. G., Neeteson, A. M., McAdam, J., Koerhuis, A. N. M. and Avendaño, S. 2012b. Genetic parameters of foot-pad dermatitis and body weight in purebred broiler lines in 2 contrasting environments. *Poultry Science* 91(3), 565–74. doi:10.3382/ps.2011-01934.

Karaarslan, S. and Nazligul, A. 2018. Effects of lighting, stocking density, and access to perches on leg health variables as welfare indicators in broiler chickens. *Livestock Science* 218, 31–6. doi:10.1016/j.livsci.2018.10.008.

Kaukonen, E., Norring, M. and Valros, A. 2017. Perches and elevated platforms in commercial broiler farms: use and effect on walking ability, incidence of tibial dyschondroplasia and bone mineral content. *Animal* 11(5), 864–71. doi:10.1017/S1751731116002160.

Kestin, S. C., Knowles, T. G., Tinch, A. E. and Gregory, N. G. 1992. Prevalence of leg weakness in broiler-chickens and its relationship with genotype. *Veterinary Record* 131(9), 190–4. doi:10.1136/vr.131.9.190.

Kestin, S. C., Gordon, S., Su, G. and Sorensen, P. 2001. Relationships in broiler chickens between lameness, liveweight, growth rate and age. *Veterinary Record* 148(7), 195–7. doi:10.1136/vr.148.7.195.

Kheravii, S. K., Swick, R. A., Choct, M. and Wu, S. B. 2017. Potential of pelleted wheat straw as an alternative bedding material for broilers. *Poultry Science* 96(6), 1641-7. doi:10.3382/ps/pew473.

Khosravinia, H. 2015. Effect of dietary supplementation of medium-chain fatty acids on growth performance and prevalence of carcass defects in broiler chickens raised in different stocking densities. *Journal of Applied Poultry Research* 24(1), 1-9. doi:10.3382/japr/pfu001.

Kittelsen, K. E., David, B., Moe, R. O., Poulsen, H. D., Young, J. F. and Granquist, E. G. 2017. Associations among gait score, production data, abattoir registrations, and postmortem tibia measurements in broiler chickens. *Poultry Science* 96(5), 1033–40. doi:10.3382/ps/pew433.

Kiyma, Z., Kucukyilmaz, K. and Orojpour, A. 2016. Effects of perch availability on performance, carcass characteristics, and footpad lesions in broilers. *Archives Animal Breeding* 59(1), 19–25. doi:10.5194/aab-59-19-2016.

Kjaer, J. B., Su, G., Nielsen, B. L. and Sorensen, P. 2006. Foot pad dermatitis and hock burn in broiler chickens and degree of inheritance. *Poultry Science* 85(8), 1342-8. doi:10.1093/ps/85.8.1342.

Knowles, T. G., Kestin, S. C., Haslam, S. M., Brown, S. N., Green, L. E., Butterworth, A., Pope, S. J., Pfeiffer, D. and Nicol, C. J. 2008. Leg disorders in broiler chickens: prevalence, risk factors and prevention. *PLoS ONE* 3(2), e1545. doi:10.1371/journal.pone.0001545.

Kwiatkowska, K., Winiarska-Mieczan, A. and Kwiecien, M. 2017. Feed additives regulating calcium homeostasis in the bones of poultry - a review. *Annals of Animal Science* 17(2), 303-16. doi:10.1515/aoas-2016-0031.

Kyvsgaard, N. C., Jensen, H. B., Ambrosen, T. and Toft, N. 2013. Temporal changes and risk factors for foot-pad dermatitis in Danish broilers. *Poultry Science* 92(1), 26–32. doi:10.3382/ps.2012-02433.

Leterrier, C., Vallee, C., Constantin, P., Chagneau, A. M., Lessire, M., Lescoat, P., Berri, C., Baéza, E., Bizeray, D. and Bouvarel, I. 2008. Sequential feeding with variations in energy and protein levels improves gait score in meat-type chickens. *Animal* 2(11), 1658–65. doi:10.1017/S1751731108002875.

Lewis, P. D., Danisman, R. and Gous, R. M. 2009. Photoperiodic responses of broilers. III. Tibial breaking strength and ash content. *British Poultry Science* 50(6), 673–9. doi:10.1080/00071660903365612.

Li, H., Wen, X., Alphin, R., Zhu, Z. and Zhou, Z. 2017. Effects of two different broiler flooring systems on production performances, welfare, and environment under commercial production conditions. *Poultry Science* 96(5), 1108–19. doi:10.3382/ps/pew440.

Lien, R. J., Hess, J. B., McKee, S. R., Bilgili, S. F. and Townsend, J. C. 2007. Effect of light intensity and photoperiod on live performance, heterophil-to-lymphocyte ratio, and processing yields of broilers. *Poultry Science* 86(7), 1287–93. doi:10.1093/ps/86.7.1287.

Louton, H., Bergmann, S., Reese, S., Erhard, M., Bachmeier, J., Rösler, B. and Rauch, E. 2018. Animal- and management-based welfare indicators for a conventional broiler strain in 2 barn types (Louisiana barn and closed barn). *Poultry Science* 97(8), 2754–67. doi:10.3382/ps/pey111.

Lund, V. P., Nielsen, L. R., Oliveira, A. R. S. and Christensen, J. P. 2017. Evaluation of the Danish footpad lesion surveillance in conventional and organic broilers: misclassification of scoring. *Poultry Science* 96(7), 2018–28. doi:10.3382/ps/pex024.

Lynch, M., Thorp, B. H. and Whitehead, C. C. 1992. Avian tibial dyschondroplasia as a cause of bone deformity. *Avian Pathology* 21(2), 275–85. doi:10.1080/03079459208418842.

Manangi, M. K., Vazquez-Anon, M., Richards, J. D., Carter, S., Buresh, R. E. and Christensen, K. D. 2012. Impact of feeding lower levels of chelated trace minerals versus industry levels of inorganic trace minerals on broiler performance, yield, footpad health, and litter mineral concentration. *Journal of Applied Poultry Research* 21(4), 881–90. doi:10.3382/japr.2012-00531.

Mandal, R. K., Jiang, T. S., Al-Rubaye, A. A., Rhoads, D. D., Wideman, R. F., Zhao, J., Pevzner, I. and Kwon, Y. M. 2016. An investigation into blood microbiota and its potential association with bacterial chondronecrosis with osteomyelitis (BCO) in broilers. *Scientific Reports* 6, 25882. doi:10.1038/srep25882.

Manning, L., Chadd, S. A. and Baines, R. N. 2007. Water consumption in broiler chicken: a welfare indicator. *World's Poultry Science Journal* 63(1), 63–71. doi:10.1017/S0043933907001274.

Meluzzi, A., Fabbri, C., Folegatti, E. and Sirri, F. 2008a. Survey of chicken rearing conditions in Italy: effects of litter quality and stocking density on productivity, foot dermatitis and carcase injuries. *British Poultry Science* 49(3), 257–64. doi:10.1080/00071660802094156.

Meluzzi, A., Fabbri, C., Folegatti, E. and Sirri, F. 2008b. Effect of less intensive rearing conditions on litter characteristics, growth performance, carcase injuries and meat quality of broilers. *British Poultry Science* 49(5), 509–15. doi:10.1080/00071660802290424.

Muir, W. I. and Groves, P. J. 2018. Incubation and hatch management: consequences for bone mineralization in Cobb 500 meat chickens. *Animal* 12(4), 794–801. doi:10.1017/S1751731117001938.

Müller, B. R., Medeiros, H. A. S., de Sousa, R. S. and Molento, C. F. M. 2015. Chronic welfare restrictions and adrenal gland morphology in broiler chickens. *Poultry Science* 94(4), 574–8. doi:10.3382/ps/pev026.

Nääs, IdA., Paz, I. C. L. A., Baracho, M. S., Menezes, A. G., Bueno, L. G. F., Almeida, I. C. L. and Moura, D. J. 2009. Impact of lameness on broiler well-being. *Journal of Applied Poultry Research* 18(3), 432–9. doi:10.3382/japr.2008-00061.

Nääs, IdA., Paz, I. CdL. A., Baracho, MdS., Menezes, A. Gd, Lima, K. A. Od, Bueno, L. GdF., Mollo Neto, M., Carvalho, V. Cd, Almeida, I. CdL. and Souza, A. Ld 2010. Assessing locomotion deficiency in broiler chicken. *Scientia Agricola* 67(2), 129–35. doi:10.1590/S0103-90162010000200001.

Nagaraj, M., Wilson, C. A. P., Hess, J. B. and Bilgili, S. F. 2007. Effect of high-protein and all-vegetable diets on the incidence and severity of pododermatitis in broiler chickens. *Journal of Applied Poultry Research* 16(3), 304–12. doi:10.1093/japr/16.3.304.

Nicol, C. J. 2015. *The Behavioural Biology of Chickens*. CABI Publishing, Wallingford, UK.

Nicol, C. J., Caplen, G. J., Edgar, J. L. and Browne, W. J. 2009. Associations between welfare indicators and environmental choice in laying hens. *Animal Behaviour* 78(2), 413–24. doi:10.1016/j.anbehav.2009.05.016.

Nielsen, B. L., Thomsen, M. G., Sorensen, J. P. and Young, J. F. 2003. Feed and strain effects on the use of outdoor areas by broilers. *British Poultry Science* 44(2), 161–9. doi:10.1080/0007166031000088389.

Norring, M., Kaukonen, E. and Valros, A. 2016. The use of perches and platforms by broiler chickens. *Applied Animal Behaviour Science* 184, 91–6. doi:10.1016/j.applanim.2016.07.012.

Ohara, A., Oyakawa, C., Yoshihara, Y., Ninomiya, S. and Sato, S. 2015. Effect of environmental enrichment on the behavior and welfare of Japanese broilers at a commercial Farm. *Journal of Poultry Science* 52(4), 323–30. doi:10.2141/jpsa.0150034.

Olukosi, O. A., Olori, V. E., Helmbrecht, A., Lambton, S. and French, N. A. (Eds). 2019. *Poultry Feathers and Skin*. The Poultry Integument in Health and Welfare, CABI Publishing, Wallingford, UK.

Opengart, K., Bilgili, S. F., Warren, G. L., Baker, K. T., Moore, J. D. and Dougherty, S. 2018. Incidence, severity, and relationship of broiler footpad lesions and gait scores of market-age broilers raised under commercial conditions in the southeastern United States. *Journal of Applied Poultry Research* 27(3), 424–32. doi:10.3382/japr/pfy002.

Orth, M. W. and Cook, M. E. 1994. Avian tibial dyschondroplasia - a morphological and biochemical review of the growth-plate lesion and its causes. *Veterinary Pathology* 31(4), 403–4. doi:10.1177/030098589403100401.

Oso, A. O., Idowu, A. A. and Niameh, O. T. 2011. Growth response, nutrient and mineral retention, bone mineralisation and walking ability of broiler chickens fed with dietary inclusion of various unconventional mineral sources. *Journal of Animal Physiology and Animal Nutrition* 95(4), 461–7. doi:10.1111/j.1439-0396.2010.01073.x.

Oviedo-Rondón, E. O. 2019. Holistic view of intestinal health in poultry. *Animal Feed Science and Technology* 250, 1–8. doi:10.1016/j.anifeedsci.2019.01.009.

Oviedo-Rondón, E. O., Wineland, M. J., Funderburk, S., Small, J., Cutchin, H. and Mann, M. 2009a. Incubation conditions affect leg health in large, high-yield broilers. *Journal of Applied Poultry Research* 18(3), 640–6. doi:10.3382/japr.2008-00127.

Oviedo-Rondón, E. O., Wineland, M. J., Small, J., Cutchin, H., McElroy, A., Barri, A. and Martin, S. 2009b. Effect of incubation temperatures and chick transportation conditions on bone development and leg health. *Journal of Applied Poultry Research* 18(4), 671-8. doi:10.3382/japr.2008-00135.

Oznurlu, Y., Sur, E., Ozaydin, T., Celik, I. and Uluisik, D. 2016. Histological and histochemical evaluations on the effects of high incubation temperature on the embryonic development of tibial growth plate in broiler chickens. *Microscopy Research and Technique* 79(2), 106-10. doi:10.1002/jemt.22611.

Pagazaurtundua, A. and Warriss, P. D. 2006. Levels of foot pad dermatitis in broiler chickens reared in 5 different systems. *British Poultry Science* 47(5), 529-32. doi:10.1080/00071660600963024.

Part, C. E., Edwards, P., Hajat, S. and Collins, L. M. 2016. Prevalence rates of health and welfare conditions in broiler chickens change with weather in a temperate climate. *Royal Society Open Science* 3(9), 160197. doi:10.1098/rsos.160197.

Paz, I. C. L. A., Garcia, R. G., Bernardi, R., Seno, LdO., Naas, IdA. and Caldara, F. R. 2013. Locomotor problems in broilers reared on new and re-used litter. *Italian Journal of Animal Science* 12(2). doi:10.4081/ijas.2013.e45.

Petek, M., Sonmez, G., Yildiz, H. and Baspinar, H. 2005. Effects of different management factors on broiler performance and incidence of tibial dyschondroplasia. *British Poultry Science* 46(1), 16-21. doi:10.1080/00071660400023821.

Petek, M., Cibik, R., Yildiz, H., Sonat, F. A., Gezen, S. S., Orman, A. and Aydin, C. 2010. The influence of different lighting programs, stocking densities and litter amounts on the welfare and productivity traits of a commercial broiler line. *Veterinarija ir Zootechnika* 51(73), 36-43.

Puvadolpirod, S. and Thaxton, J. P. 2000. Model of physiological stress in chickens: 4. Digestion and metabolism. *Poultry Science* 79(3), 383-90. doi:10.1093/ps/79.3.383.

Reiter, K. and Bessei, W. 2009. Effect of locomotor activity on leg disorder in fattening chicken. *Berliner Und Munchener Tierarztliche Wochenschrift* 122(7-8), 264-70.

Riddell, C. and Kong, X. M. 1992. The influence of diet on necrotic enteritis in broiler-chickens. *Avian Diseases* 36(3), 499-503. doi:10.2307/1591740.

Roberts, S. J., Cain, R. and Dawkins, M. S. 2012. Prediction of welfare outcomes for broiler chickens using Bayesian regression on continuous optical flow data. *Journal of the Royal Society Interface* 9(77), 3436-43. doi:10.1098/rsif.2012.0594.

Rushton, S. P., Humphrey, T. J., Shirley, M. D. F., Bull, S. and Jorgensen, F. 2009. Campylobacter in housed broiler chickens: a longitudinal study of risk factors. *Epidemiology and Infection* 137(8), 1099-110. doi:10.1017/S095026880800188X.

Rutten, M., Leterrier, C., Constantin, P., Reiter, K. and Bessei, W. 2002. Bone development and activity in chickens in response to reduced weight-load on legs. *Animal Research* 51(4), 327-36. doi:10.1051/animres:2002027.

Saima, M. Z. U., Jabbar, M. A., Ijaz, M. and Qadeer, M. A. 2009. Efficacy of microbial phytase at different levels on growth performance and mineral availability in broiler chicken. *Journal of Animal and Plant Sciences* 19(2), 58-62.

Sandilands, V., Brocklehurst, S., Sparks, N., Baker, L., McGovern, R., Thorp, B. and Pearson, D. 2011. Assessing leg health in chickens using a force plate and gait scoring: how many birds is enough? *Veterinary Record* 168(3), 77. doi:10.1136/vr.c5978.

Sanotra, G. S., Lund, J. D. and Vestergaard, K. S. 2002. Influence of light-dark schedules and stocking density on behaviour, risk of leg problems and occurrence of chronic fear in broilers. *British Poultry Science* 43(3), 344-54. doi:10.1080/000716601201036023611.

Sanotra, G. S., Berg, C. and Lund, J. D. 2003. A comparison between leg problems in Danish and Swedish broiler production. *Animal Welfare* 12(4), 677–83.

Sans, E., Federici, J., Dahlke, F. and Molento, C. 2014. Evaluation of Free-Range broilers using the welfare quality® protocol. *Revista Brasileira de Ciência Avícola* 16(3), 297-306. doi:10.1590/1516-635x1603297-306.

Sarica, M., Yamak, U. S. and Boz, M. A. 2014. Effect of production systems on foot pad dermatitis (FPD) levels among slow-, medium- and fast-growing broilers. *European Poultry Science* 78. doi:10.1399/eps.2014.52.

Schmidt, C. J., Persia, M. E., Feierstein, E., Kingham, B. and Saylor, W. W. 2009. Comparison of a modern broiler line and a heritage line unselected since the 1950s. *Poultry Science* 88(12), 2610–9. doi:10.3382/ps.2009-00055.

Scholz-Ahrens, K. E., Ade, P., Marten, B., Weber, P., Timm, W., Asil, Y., Gluer, C. C. and Schrezenmeir, J. 2007. Prebiotics, probiotics, and synbiotics affect mineral absorption, bone mineral content, and bone structure. *Journal of Nutrition* 137(3), 838S–46S. doi:10.1093/jn/137.3.838S.

Schwean-Lardner, K., Fancher, B. I. and Classen, H. L. 2012. Impact of daylength on behavioural output in commercial broilers. *Applied Animal Behaviour Science* 137(1–2), 43–52. doi:10.1016/j.applanim.2012.01.015.

Schwean-Lardner, K., Fancher, B. I., Gomis, S., Van Kessel, A., Dalal, S. and Classen, H. L. 2013. Effect of day length on cause of mortality, leg health, and ocular health in broilers. *Poultry Science* 92(1), 1–11. doi:10.3382/ps.2011-01967.

Sellers, H. S. 2017. Current limitations in control of viral arthritis and tenosynovitis caused by avian reoviruses in commercial poultry. *Veterinary Microbiology* 206, 152-6. doi:10.1016/j.vetmic.2016.12.014.

Senaratna, D., Samarakone, T. S. and Gunawardena, W. W. 2016. Red color light at different intensities affects the performance, behavioral activities and welfare of broilers. *Asian-Australasian Journal of Animal Sciences* 29(7), 1052–9. doi:10.5713/ajas.15.0757.

Shepherd, E. M., Fairchild, B. D. and Ritz, C. W. 2017. Alternative bedding materials and litter depth impact litter moisture and footpad dermatitis. *Journal of Applied Poultry Research* 26(4), 518–28. doi:10.3382/japr/pfx024.

Sherlock, L., McKeegan, D. E. F., Cheng, Z., Wathes, C. M. and Wathes, D. C. 2012. Effects of contact dermatitis on hepatic gene expression in broilers. *British Poultry Science* 53(4), 439–52. doi:10.1080/00071668.2012.707310.

Shim, M. Y., Karnuah, A. B., Anthony, N. B., Pesti, G. M. and Aggrey, S. E. 2012a. The effects of broiler chicken growth rate on valgus, varus, and tibial dyschondroplasia. *Poultry Science* 91(1), 62–5. doi:10.3382/ps.2011-01599.

Shim, M. Y., Karnuah, A. B., Mitchell, A. D., Anthony, N. B., Pesti, G. M. and Aggrey, S. E. 2012b. The effects of growth rate on leg morphology and tibia breaking strength, mineral density, mineral content, and bone ash in broilers. *Poultry Science* 91(8), 1790–5. doi:10.3382/ps.2011-01968.

Siegel, P. B., Gustin, S. J. and Katanbaf, M. N. 2011. Motor ability and self-selection of an analgesic drug by fast-growing chickens. *Journal of Applied Poultry Research* 20(3), 249–52. doi:10.3382/japr.2009-00118.

Simsek, U. G., Dalkilic, B., Ciftci, M., Cerci, I. H. and Bahsi, M. 2009. Effects of enriched housing design on broiler performance, welfare, chicken meat composition and serum cholesterol. *Acta Veterinaria Brno* 78(1), 67–74. doi:10.2754/avb200978010067.

Simsek, U. G., Erisir, M., Ciftci, M. and Tatli Seven, P. 2014. Effects of cage and floor housing systems on fattening performance, oxidative stress and carcass defects in broiler chicken. *Kafkas Universitesi Veteriner Fakultesi Dergisi* 20(5), 727–33.

Skrbic, Z., Pavlovski, Z., Lukic, M. and Petricevic, V. 2015. Incidence of footpad dermatitis and hock burns in broilers as affected by genotype, lighting program and litter type. *Annals of Animal Science* 15(2), 433–45. doi:10.1515/aoas-2015-0005.

Sørensen, P., Su, G. and Kestin, S. C. 2000. Effects of age and stocking density on leg weakness in broiler chickens. *Poultry Science* 79(6), 864–70. doi:10.1093/ps/79.6.864.

Sosnówka-Czajka, E., Skomorucha, I., Herbut, E. and Muchacka, R. 2007. Effect of management system and flock size on the behaviour of broiler chickens. *Annals of Animal Science* 7(2), 329–35.

Stanley, D., Hughes, R. J. and Moore, R. J. 2014. Microbiota of the chicken gastrointestinal tract: influence on health, productivity and disease. *Applied Microbiology and Biotechnology* 98(10), 4301–10. doi:10.1007/s00253-014-5646-2.

Su, G., Sorensen, P. and Kestin, S. C. 1999. Meal feeding is more effective than early feed restriction at reducing the prevalence of leg weakness in broiler chickens. *Poultry Science* 78(7), 949–55. doi:10.1093/ps/78.7.949.

Su, G., Sorensen, P. and Kestin, S. C. 2000. A note on the effects of perches and litter substrate on leg weakness in broiler chickens. *Poultry Science* 79(9), 1259–63. doi:10.1093/ps/79.9.1259.

Sun, Z. W., Yan, L., G, Y. Y., Zhao, J. P., Lin, H. and Guo, Y. M. 2013. Increasing dietary vitamin D-3 improves the walking ability and welfare status of broiler chickens reared at high stocking densities. *Poultry Science* 92(12), 3071–9. doi:10.3382/ps.2013-03278.

Sun, Z. W., Fan, Q. H., Wang, X. X., Guo, Y. M., Wang, H. J. and Dong, X. 2017. High dietary biotin levels affect the footpad and hock health of broiler chickens reared at different stocking densities and litter conditions. *Journal of Animal Physiology and Animal Nutrition* 101(3), 521–30. doi:10.1111/jpn.12465.

Swiatkiewicz, S., Arczewska-Wlosek, A. and Jozefiak, D. 2017. The nutrition of poultry as a factor affecting litter quality and foot pad dermatitis – an updated review. *Journal of Animal Physiology and Animal Nutrition* 101(5), e14–20. doi:10.1111/jpn.12630.

Tahamtani, F. M., Hinrichsen, L. K. and Riber, A. B. 2018. Welfare assessment of conventional and organic broilers in Denmark, with emphasis on leg health. *Veterinary Record* 183(6), 192. doi:10.1136/vr.104817.

Tickle, P. G., Hutchinson, J. R. and Codd, J. R. 2018. Energy allocation and behaviour in the growing broiler chicken. *Scientific Reports* 8(1), 4562. doi:10.1038/s41598-018-22604-2.

Toppel, K., Kaufmann, F., Schon, H., Gauly, M. and Andersson, R. 2019. Effect of pH-lowering litter amendment on animal-based welfare indicators and litter quality in a European commercial broiler husbandry. *Poultry Science* 98(3), 1181–9. doi:10.3382/ps/pey489.

Tullo, E., Fontana, I., Fernandez, A. P., Vranken, E., Norton, T., Berckmans, D. and Guarino, M. 2017. Association between environmental predisposing risk factors and leg disorders in broiler chickens. *Journal of Animal Science* 95(4), 1512–20. doi:10.2527/jas.2016.1257.

Tuyttens, F., Heyndrickx, M., De Boeck, M., Moreels, A., Van Nuffel, A., Van Poucke, E., Van Coillie, E., Van Dongen, S. and Lens, L. 2008. Broiler chicken health, welfare and fluctuating asymmetry in organic versus conventional production systems. *Livestock Science* 113(2–3), 123–32. doi:10.1016/j.livsci.2007.02.019.

van der Pol, C. W., Molenaar, R., Buitink, C. J., van Roovert-Reijrink, I. A. M., Maatjens, C. M., van den Brand, H. and Kemp, B. 2015. Lighting schedule and dimming period in early life: consequences for broiler chicken leg bone development. *Poultry Science* 94(12), 2980–8. doi:10.3382/ps/pev276.

van der Pol, C. W., van Roovert-Reijrink, I. A. M., Aalbers, G., Kemp, B. and van den Brand, H. 2017. Incubation lighting schedules and their interaction with matched or mismatched post hatch lighting schedules: effects on broiler bone development and leg health at slaughter age. *Research in Veterinary Science* 114, 416–22. doi:10.1016/j.rvsc.2017.07.013.

van der Pol, C. W., van Roovert-Reijrink, I. A. M., Maatjens, C. M., Gussekloo, S. W. S., Kranenbarg, S., Wijnen, J., Pieters, R. P. M., Schipper, H., Kemp, B. and van den Brand, H. 2019. Light-dark rhythms during incubation of broiler chicken embryos and their effects on embryonic and post hatch leg bone development. *PLoS ONE* 14(1), e0210886. doi:10.1371/journal.pone.0210886.

Van Hertem, T., Norton, T., Berckmans, D. and Vranken, E. 2018. Predicting broiler gait scores from activity monitoring and flock data. *Biosystems Engineering* 173, 93–102. doi:10.1016/j.biosystemseng.2018.07.002.

Vargas-Galicia, A. J., Sosa-Montes, E., Rodriguez-Ortega, L. T., Pro-Martinez, A., Ruiz-Feria, C. A., González-Cerón, F., Gallegos-Sánchez, J., Arreola-Enríquez, J. and Bautista-Ortegac, J. 2017. Effect of litter material and stocking density on bone and tendon strength, and productive performance in broilers. *Canadian Journal of Animal Science* 97(4), 673–82.

Vasdal, G., Vas, J., Newberry, R. C. and Moe, R. O. 2019. Effects of environmental enrichment on activity and lameness in commercial broiler production. *Journal of Applied Animal Welfare Science* 22(2), 197–205. doi:10.1080/10888705.2018.1456339.

Ventura, B. A., Siewerdt, F. and Estevez, I. 2010. Effects of barrier perches and density on broiler leg health, fear, and performance. *Poultry Science* 89(8), 1574–83. doi:10.3382/ps.2009-00576.

Ventura, B. A., Siewerdt, F. and Estevez, I. 2012. Access to barrier perches improves behavior repertoire in broilers. *PLoS ONE* 7(1), e29826. doi:10.1371/journal.pone.0029826.

Villagrá, A., Olivas, I., Benitez, V. and Lainez, M. 2011. Evaluation of sludge from paper recycling as bedding material for broilers. *Poultry Science* 90(5), 953–7. doi:10.3382/ps.2010-00935.

Villarroel, M., Francisco, I., Ibanez, M. A., Novoa, M., Martinez-Guijarro, P., Méndez, J. and De Blas, C. 2018. Rearing, bird type and pre-slaughter transport conditions of broilers II. Effect on foot-pad dermatitis and carcass quality. *Spanish Journal of Agricultural Research* 16(2), e0504. Available at: http://revistas.inia.es/index.php/sjar/article/view/12015/4038 (accessed on 19 June 2019). doi:10.5424/sjar/2018162-12015.

Waldenstedt, L. 2006. Nutritional factors of importance for optimal leg health in broilers: a review. *Animal Feed Science and Technology* 126(3–4), 291–307. doi:10.1016/j.anifeedsci.2005.08.008.

Weeks, C. A. and Butterworth, A. 2004. *Measuring and Auditing Broiler Welfare* (vol. 1). CABI Publishing, Wallingford, UK.

Weeks, C. A., Danbury, T. D., Davies, H. C., Hunt, P. and Kestin, S. C. 2000. The behaviour of broiler chickens and its modification by lameness. *Applied Animal Behaviour Science* 67(1–2), 111–25. doi:10.1016/s0168-1591(99)00102-1.

Weeks, C. A., Knowles, T. G., Gordon, R. G., Kerr, A. E., Peyton, S. T. and Tilbrook, N. T. 2002. New method for objectively assessing lameness in broiler chickens. *Veterinary Record* 151(25), 762–4.

Weimer, S. L., Wideman, R. F., Scanes, C. G., Mauromoustakos, A., Christensen, K. D. and Vizzier-Thaxton, Y. 2019. The utility of infrared thermography for evaluating lameness attributable to bacterial chondronecrosis with osteomyelitis. *Poultry Science* 98(4), 1575–88. doi:10.3382/ps/pey538.

Welfare Quality. 2009. Welfare Quality assessment protocol for poultry (broilers, laying hens). Welfare Quality Consortium, Lelystad, the Netherlands. Available at: http://www.welfarequalitynetwork.net/downloadattachment/45627/21652/Poultry%20Protocol.pdf (accessed on 19 June 2019).

Whitehead, C. C., McCormack, H. A., McTeir, L. and Fleming, R. H. 2004. High vitamin D-3 requirements in broilers for bone quality and prevention of tibial dyschondroplasia and interactions with dietary calcium, available phosphorus and vitamin A. *British Poultry Science* 45(3), 425–36. doi:10.1080/00071660410001730941.

Wideman, R. F. 2016. Bacterial chondronecrosis with osteomyelitis and lameness in broilers: a review. *Poultry Science* 95(2), 325–44. doi:10.3382/ps/pev320.

Wideman, R. F. and Prisby, R. D. 2013. Bone circulatory disturbances in the development of spontaneous bacterial chondronecrosis with osteomyelitis: a translational model for the pathogenesis of femoral head necrosis. *Frontiers in Endocrinology* 3, 183. doi:10.3389/fendo.2012.00183.

Wideman, R. F., Hamal, K. R., Stark, J. M., Blankenship, J., Lester, H., Mitchell, K. N., Lorenzoni, G. and Pevzner, I. 2012. A wire-flooring model for inducing lameness in broilers: evaluation of probiotics as a prophylactic treatment. *Poultry Science* 91(4), 870–83. doi:10.3382/ps.2011-01907.

Wideman, R. F., Al-Rubaye, A., Reynolds, D., Yoho, D., Lester, H., Spencer, C., Hughes, J. D. and Pevzner, I. Y. 2014. Bacterial chondronecrosis with osteomyelitis in broilers: influence of sires and straight-run versus sex-separate rearing. *Poultry Science* 93(7), 1675–87. doi:10.3382/ps.2014-03912.

Wideman, R. F., Blankenship, J., Pevzner, I. Y. and Turner, B. J. 2015. Efficacy of 25-OH vitamin D-3 prophylactic administration for reducing lameness in broilers grown on wire flooring. *Poultry Science* 94(8), 1821–7. doi:10.3382/ps/pev160.

Wijesurendra, D. S., Chamings, A. N., Bushell, R. N., Rourke, D. O., Stevenson, M., Marenda, M. S., Noormohammadi, A. H. and Stent, A. 2017. Pathological and microbiological investigations into cases of bacterial chondronecrosis and osteomyelitis in broiler poultry. *Avian Pathology* 46(6), 683–94. doi:10.1080/03079457.2017.1349872.

Wijtten, P. J. A., Hangoor, E., Sparla, J. K. and Verstegen, M. W. A. 2010. Dietary amino acid levels and feed restriction affect small intestinal development, mortality, and weight gain of male broilers. *Poultry Science* 89(7), 1424–39. doi:10.3382/ps.2009-00626.

Xavier, D. B., Broom, D. M., McManus, C. M. P., Torres, C. and Bernal, F. E. M. 2010. Number of flocks on the same litter and carcase condemnations due to cellulitis, arthritis and contact foot-pad dermatitis in broilers. *British Poultry Science* 51(5), 586–91. doi:10.1080/00071668.2010.508487.

Yalçin, S., Molayoglu, H. B., Baka, M., Genin, O. and Pines, M. 2007. Effect of temperature during the incubation period on tibial growth plate chondrocyte differentiation and the incidence of tibial dyschondroplasia. *Poultry Science* 86(8), 1772–83. doi:10.1093/ps/86.8.1772.

Yamak, U. S., Sarica, M., Boz, M. A. and Ucar, A. 2016. Effect of reusing litter on broiler performance, foot-pad dermatitis and litter quality in chickens with different growth rates. *Kafkas Universitesi Veteriner Fakultesi Dergisi* 22(1), 85–91.

Yan, F. F., Wang, W. C. and Cheng, H. W. 2018. Bacillus subtilis based probiotic improved bone mass and altered brain serotoninergic and dopaminergic systems in broiler chickens. *Journal of Functional Foods* 49, 501–9. doi:10.1016/j.jff.2018.09.017.

Yan, F. F., Mohammed, A. A., Murugesan, G. R. and Cheng, H. W. 2019. Effects of a dietary synbiotic inclusion on bone health in broilers subjected to cyclic heat stress episodes. *Poultry Science* 98(3), 1083–9. doi:10.3382/ps/pey508.

Yang, H. M., Xing, H., Wang, Z. Y., Xia, J. L., Wan, Y., Hou, B. and Zhang, J. 2015. Effects of intermittent lighting on broiler growth performance, slaughter performance, serum biochemical parameters and tibia parameters. *Italian Journal of Animal Science* 14(4), 4143. doi:10.4081/ijas.2015.4143.

Yildiz, H., Gunes, N., Cengiz, S., Ozcan, R., Petek, M., Yilmaz, B. and Arican, I. 2009. Effects of ascorbic acid and lighting schedule on tibiotarsus strength and bone characteristics in broilers. *Archiv fur Tierzucht* 52(4), 432–44.

Yildiz, A., Yildiz, K. and Apaydin, B. 2014. The effect of vermiculite as litter material on some health and stress parameters in broilers. *Kafkas Universitesi Veteriner Fakultesi Dergisi* 20(1), 129–34.

Zhao, J., Shirley, R. B., Vazquez-Anon, M., Dibner, J. J., Richards, J. D., Fisher, P., Hampton, T., Christensen, K. D., Allard, J. P. and Giesen, A. F. 2010. Effects of chelated trace minerals on growth performance, breast meat yield, and footpad health in commercial meat broilers. *Journal of Applied Poultry Research* 19(4), 365–72. doi:10.3382/japr.2009-00020.

Zhao, Z. G., Li, J. H., Li, X. and Bao, J. 2014. Effects of housing systems on behaviour, performance and welfare of fast-growing broilers. *Asian-Australasian Journal of Animal Sciences* 27(1), 140–6. doi:10.5713/ajas.2013.13167.

Zikic, D., Djukic-Stojcic, M., Bjedov, S., Peric, L., Stojanovic, S. and Uscebrka, G. 2017. Effect of litter on development and severity of foot pad dermatitis and behavior of broiler chickens. *Revista Brasileira de Ciência Avícola* 19(2), 247–54. doi:10.1590/1806-9061-2016-0396.

# Index

ADHD. *see* Attention-deficit/hyperactivity disorder (ADHD)
Ad-libitum feeding 617
Adult hens 383
Adversity 532
Aggression reducing methods 244
  housing systems separating sexes 247-249
  normal sexual behaviour breeding 249-251
  panel effect 244-245
  stocking density effect 245-247
  ultraviolet light effect 247
Aggressive pecking (AP) 510-512
Air hunger 479
Allopecking 290
Amodal completion 97
Anaesthetic gas 480
Analgesic drugs 613
ANI-35L. *see* Animal Needs Index (ANI-35L)
Animal behaviour 150
Animal Needs Index (ANI-35L) 200
Animal Welfare Act (1966) 420, 462
Anoxia 477
  *see also* Controlled atmosphere stunning (CAS)
Antipredator behaviour 381
AP. *see* Aggressive pecking (AP)
Appetitive phase 518
Apollo chicken harvesters 423
Argon 480
  *see also* Inert gases
Arterial chemoreceptors 476, 482
Artificial insemination 249
Artificial rearing conditions 386
Ascorbic acid (vitamin C) supplementation 618
ASD. *see* Autism spectrum disorder (ASD)
Ataxia 477-478
ATLAS, Marel's 478
Attention-deficit/hyperactivity disorder (ADHD) 522
Autism spectrum disorder (ASD) 522
Automated conveyor systems 474
Automatic digital flexor mechanism (ADFM) 389
Automatic perching mechanism (APM) 389
Avian hippocampus 533
Aviaries 360
Aviary system 378

Bacterial chondronecrosis with osteomyelitis (BCO) 605
Bales 335
Barley-based diets 626
Barrier perches 401
Battery cages 350
BCO. *see* Bacterial chondronecrosis with osteomyelitis (BCO)
Beak trimming 73, 279
Bedding material 628
Behavioural responses 477-480
Behavioural restrictions 319
Biological promoter 629
Biological residue 432
Bird-related factors 536-537
Blunt force trauma 487
Bone development during rearing 568-569
Bone mineralisation 619
Breeding programmes 628
Broiler breeder fowl 77
Broiler breeders
  feed restriction 230-232
  future trends 251-252
  growth potential 230-232
  housing conditions 228-230
  overview 227-228
  welfare issues 232

aggression and associated welfare problems 239-240
mutilation and associated welfare problems 240-251
unfulfilled behavioural and physiological needs 232-239
Broiler chickens 377
elevated structures 400-404
fitness assessment 435-436
herding 438
without elevated platforms 383
Broiler Directive (2007/43/EC) 619
Broiler housings 400-404
Broilers welfare issues
behaviour and space use 317-318
broiler production systems 319-323
environment and management effects 325
air quality 329-330
environmental enrichment 333-337
hatching environment 326-327
light 330-333
litter quality 328-329
stocking density 327-328
future trends 337-338
genetic selection 316-317
growth rate effects
on broiler behaviour 324-325
on leg health, mortality and heat stress 323-324
overview 313-315
Brooding period 281
Brown Leghorn 380
*see also* Pullets
Bumblefoot 73
Bumble foot 398
Burdizzo 488

Cage system 249
CAK. *see* Controlled atmosphere killing (CAK)
*Campylobacter* 181, 182
Canadian Food Inspection Agency 420
Cancellous bone. *see* Inner trabecular bone
Cannibalism 209, 234, 393
Cannibalistic pecking 515
Captive bolt 486
*see also* Mechanical methods 486
Carcass rejections 432
Carcass value 433
Carotenoids 278
Carotid bodies. *see* Arterial chemoreceptors
Caruncles 512
CAS. *see* Controlled atmosphere stunning (CAS)
Catching and loading 423-426
Catching crews 440
Catching trauma 432
Category 2 animal by-products 432
Category 3 by-products 432
Cellular division 568
Cervical dislocation 487
CFL. *see* Compact fluorescent (CFL) lighting
Chemokines. *see* Pain-producing chemicals
Chicken vocalizations 48
Climate conditioned transportation system 439
Cloacal cannibalism 209, 393
Clonic phase 479
*see also* Neuromuscular spasms
Closed-circuit television (CCTV) 441
Cluck calls 92
Coercive matings 138
Cognitive behavior training 440
Cognitive judgement bias test 215
Cold white 284
Combining and grouping enrichments 337
Combs 512
Commercial aviary systems 379
Commercial farm settings 533
Commercial multi-tier housing system 586
Commercial slaughter 471
Compact fluorescent (CFL) lighting 332
Complex environmentally enriched environments 336-337
Concussive stunning. *see* Captive bolt
CONLERN 123
Consummatory phase 518
Contact dermatitis
prevalence 621-622
prevention and control of
environmental enrichment 630-631
genetic selection 628
litter properties 628-629
managing litter moisture 629-630
monitoring and surveillance systems 631
qualitative assessment 622
risk factors
age and stocking density 623
diet 625-626
egg quality/incubation/early life 626-627
lighting 626

litter quality 624-625
production systems 624
sex and body mass 623-624
welfare impact of 627
Contrafreeloading 95
Controlled atmosphere killing (CAK) 475
Controlled-atmosphere stunning 462
Controlled atmosphere stunning (CAS) 463, 475-477
welfare impacts 480-482
Controlled-climate transportation trucks 439
Coprophagy 422
Corn-based diets 626
Corporate social responsibility (CSR) 252
Cortical-basal ganglia circuitry 525
Cortical bone 568
Cover panels 245
Creeping/recurrent smothering 135
CSR. *see* Corporate social responsibility (CSR)
Cytokines. *see* Pain-producing chemicals

Dark brooders 335
Day-old chick (DOC) 272
DDX4 chickens 9
Declarative representations 103
Deep inhalation 479
Defeathering 439
Depopulation, age of 420
Detour test 99
Developmental lameness 604-605
Deviations 570-571
DGE. *see* Direct Genetic Effect (DGE)
Diathermic syncope 490
Dierenbescherming 364
Dietary Ca supplementation 618
Dietary enzymes 626
Different plumage patterns/colours 537
Direct Genetic Effect (DGE) 16
DOC. *see* Day-old chick (DOC)
Domestic chicken 375
Domestic fowl 64, 120
Dual-purpose hens 380
Dual-purpose line 378
Dust-bathing motivation 517
Dutch vision solutions 475
Dysfunctional view 516-517
abnormal behaviour 522
general concepts of 523-526
normal behaviour 522
Dyspnoea 479

E8 eggs 276
ECG. *see* Electrocardiogram (ECG)
EEG. *see* Electroencephalogram (EEG)
Effective breeding strategies
fracture-free phenotype 579
genomic selection 579
EFSA. *see* European Food Safety Authority (EFSA)
eGFPs. *see* Green Fluorescent Proteins (eGFPs)
Egg production 377
Egg shell calcification 569
Egg shell temperature (EST) 272, 612
Egg yield 377
Electrical stunning 467-468
effective, likelihood of 470-471
epileptiform activity 467
parameters, determination of 468-470
in poultry 467
shackling 472-473
somatosensory evoked potentials, abolition of 467
in United States 470
Electrical water bath stunning 462, 472
Electrocardiogram (ECG) 161
Electroencephalogram (EEG) 161
Electro-immobilisation 471
Electronarcosis. *see* Electrical stunning
Elevated resting places and welfare problems 239
Emergence test 63
Emerging methods 489-490
Environmentally induced neurobiological alterations 526-529
EP. *see* Extended period (EP)
Epigenetics 262
EQTL. *see* Expression QTL (eQTL) analysis
EST. *see* Egg shell temperature (EST)
Ethological view 516-517
behaviour, determination of 517
on severe feather pecking 517-518
European Food Safety Authority (EFSA) 462, 589
Exaggerated external stimulus 517
Excoriation 523
Explorative phase 518
Expression QTL (eQTL) analysis 10
Extended period (EP) 284
Extended Welfare Assessment Grid 216
External stimuli 517
EYenamic system 181

Facilitating bird movement 588-590
Farm Animal Welfare Committee (FAWC) 570
Fast-growing breeds 610
FAWC. *see* Farm Animal Welfare Committee (FAWC)
FCR. *see* Feed conversion ratio (FCR)
Feather cover 511
Feather eating 519
Feather pecking (FP) 16, 72, 207, 214, 234, 354, 393, 396, 511-513
Feed and water withdrawal 421-422
Feed conversion ratio (FCR) 286
Feed-deprived broilers 422
Five Domains Model 202
Five Freedoms 179
Flapping flight 476
Flat deck 360
Flemish poultry industry 441
Flock thinning 621
Fluctuating incubation temperatures 627
Food distinction 288
Footpad dermatitis 320, 323
Footpad dermatitis (FPD) 621
Foot pad lesions 393
Foraging 70, 71
Foraging behaviour 518
Forceful pecks 515
FP. *see* Feather pecking (FP)
FP-associated neurobiological findings 533
FPD. *see* Footpad dermatitis (FPD)

Gait and morphology 604
Gait score (GS) scale 607
Galliforms 376
*Gallus gallus*. *see* Junglefowls
*Gallus gallus* forma domestica. *see* Domestic chicken
Gas-burning space heaters 630
Gas chambers 475
Genetic architecture 3
Genetics of poultry behaviour
  aggressive male mating behaviour 18
  anxiety behaviour 11-12
    open field test 12-13
    social reinstatement test 14-15
    tonic immobility 13-14
  bottom-up approaches, mapping genes 9
    global gene expression analysis 10
    transgenic analysis 9-10
  brooding behaviour 15
  commercial aspects and research 21-22
  epigenetics and behaviour 20-21
  feather pecking (FP) behaviour 16
    linkage analyses 17
    transcriptomics 17-18
  general activity, chickens 18
  genetics of behaviour 5-6
  overview 3-5
  pleiotropy and potential for selection 19-20
  top-down and bottom-up approaches 10-11
  top-down approaches, mapping genes
    association mapping 6-8
    heritability studies 6
    pedigree analysis 6
    quantitative trait loci (QTL) 6-8
    selective sweep mapping 8
Genome editors 9
Genome Wide Association Studies (GWAS) 13
Genomic prediction 22
Gentle feather pecking (GFP) 513
Gentle FP 514
Get-away cages 394
GFP. *see* Gentle feather pecking (GFP)
Global annual losses 432
Global hen population 509
Glucocorticoids 525
Green Fluorescent Proteins (eGFPs) 10
Grid-type ramps 589
Ground-dwellers 389
Ground-level nests 388
Group-level technologies
  automated assessment types
    conclusions 183
    group *vs.* individual 179-180
    sound 182
    temperature 182-183
    visual information from CCTV/video 180-182
  automated welfare assessment 184-186
  automated welfare measures, precision farming 183-184
  future trends 186-189
  overview 177-179
Gustation 45
Gut health 630
Gut infections 611
GWAS. *see* Genome Wide Association Studies (GWAS)

Halal slaughter 490
HB trimming 279
Head-cloaca stunning 474-475
Head-only stunning 474
Headshaking 477
Healthy gut microbiota 630
Herding approach 437
Hetergeneous Stock intercross 12
Heterophil lymphocyte ratios 424
High-altitude pilot training 483
Higher-welfare systems 315
High expansion gas-filled foam 489
High roosting sites 386
Homovanillic acid (HVA) 531
Housing system 395-396
Humane Slaughter Act 463
Hunger alleviation and fulfilment, broiler breeders 234-235
- appetite suppressants 238-239
- feed scattering in bedding 238
- number of daily meals 236-238
- qualitative feed restriction 235-236

HVA. *see* Homovanillic acid (HVA)
Hybrid laying strain (Isa Brown) 380
2-hydroxy-4-(methylthio)butanoic acid 626
Hyline Brown 381
Hyperalgesia 614
Hypercapnia 480-482
Hypercapnic anoxia 477
- *see also* Controlled atmosphere stunning (CAS)

Hypercapnic hyperoxygenation 477
Hypercapnic hypoxia 477
- *see also* Controlled atmosphere stunning (CAS)

Hypercapnic hypoxia multistage systems 481
Hypercapnic ventilatory response 481
Hyperthermic syncope 490
Hypobaric hypoxia 477
Hypocapnic alkalosis 485
Hypothalamic-pituitary-adrenal (HPA) axis 524
Hypothalamus 263

Iceberg indicators 214
Identified bone health problems
- depopulation 572-573
- keel bone damage 570-572

IDO-1 enzyme 534
IDO-2, enzymatic activity of 529
IGE. *see* Indirect Genetic Effect (IGE)
Impact-testing protocol 575
Imprinting 92
Improving bone health, strategies for 579-590
Improving hen handling during depopulation 590
Indirect Genetic Effect (IGE) 16
Indoleamine-2,3-dioxygenase (IDO-1 or IDO-2). *see* Kynurenine (KYN) pathway enzymes
Ineffective stun 469
Inert gases 480
Infectious lameness 605-606
Inflammatory arthropathy 613
Inflammatory-induced sickness behaviour 613
Inflammatory markers 525
Infra-red four-camera motion capturing system 607
Infra-red heat treatment (IRBT) 75
Infra-red (IR) method 279
Infra-red thermography 182
Infrared thermography (IRT) 608
Injurious bird-to-bird pecking 512
Injurious pecking (IP)
- animal-based outcomes 510
- backyard chickens 510
- behavioural observations 510
- definitions and descriptions of 512-513
- disastrous consequences 511
- domesticated birds 509
- economic losses 511
- far-reaching implications 509
- at feathered integument 513-514
- flocks 510
- labour-intensive 510
- lower egg production 511
- metabolic cost 511
- non-cage housing systems 510
- practical attempts and management 535-536
- prevalence 509
- reduced productivity and efficiency 511
- relationships between different forms 516
- time-consuming 510
- underlying causes, identifying the 516-517
- at unfeathered integument 514-515

Inner trabecular bone 568
Intermediary buyers 441
Internal stimuli 517

Intrapulmonary (lung) chemoreceptors 476
Intraspecific predation 515
Inverted handling 437
IP. *see* Injurious pecking (IP)
IR. *see* Infra-red (IR) method
IRBT. *see* Infra-red heat treatment (IRBT)
Irreversible brain failure 479
IRT. *see* Infrared thermography (IRT)

Junglefowls 375
  chicks 62
  male 376

Keel bone 568
  deviations 398
  fractures 161
Killing method. *see* Electrical stunning
Kynurenine (KYN) pathway enzymes 528

Labor turnover 441
Lame broilers 613
LAPS. *see* Low atmospheric pressure stunning (LAPS)
Latency-to-lie (LTL) test 607
Layer housing adjustments 586-590
Layer line 378
Laying hens
  acoustic disturbance 382
  commercial free range 382
  different strains 395
  experimental choice test 396
  ISA Brown 395
  landing behaviour 395
  layer strain 391
  light intensity 395
  night time roosting of 391
  reaction latencies 382
  relative wing loading of 376
  resting sites 391
  shorter latencies 382
  take-off 395
Laying hens welfare issues
  cage-free housing 357
    barn and other cage-free systems 357-361
    free-range systems 361-363
    organic egg production 363-365
  conventional cages 350-354
  enriched cages 354-357
  overview 349-350
LayWel scoring system 207
Learned helplessness 95
Learning and cognition abilities of chicken
  associative learning 94-95
  cognition
    affective states 100-101
    communication 102
    memory 102-104
    numerical abilities 98-99
    social learning 95-96
    spatial cognition 99-100
    time perception 97-98
    transitive inference 98
    visual cognition 97
  early learning
    brood companion effects 93-94
    maternal effects 92-93
  habituation 94-95
  improved management 104
    animal welfare, cognition implications 109-110
    cognitive development ensuring 105
    maternal influences, chicken welfare 105-106
    moving in and out of houses 109
    shape correct responses 106-108
    signal events associations 106-108
    social cognition 108-109
    up and down perches 109
  overview 91
LEDs. *see* Light-emitting diodes (LEDs)
Leg disorders and lameness
  categories 604-606
  prevalence of 609-610
  prevention and control
    barn management 617
    diet and feeding regimes 617-619
    environmental enrichment 620-621
    genetics 616
    incubation conditions 616
    photoperiod and light intensity 619-620
  qualitative and quantitative assessment 607-609
  risk factors
    environmental conditions 612
    genotype and system 610-611
    sex/age/body mass 610
    stocking density 611
  welfare impact
    dehydration and reduced feeding 613

limited behavioural expression 615-616
pain 613-614
Leg paddling motions 479
Light-emitting diodes (LEDs) 40, 267, 619
Light intensity 331, 395
Light programme 331-332
Light source 333
Light spectrum 332-333
Litter 328, 353
Lohmann dual 378
Lohmann Selected Leghorn 385, 393
Lohmann tradition 378
Longitudinal growth 568
Loss of posture 479
Low atmospheric pressure stunning (LAPS) 462, 483-485
Low oxygen 476

Mammalian limbic system 525
Mandibulation 477
Manual restraint test 531
Marek 277
Mashed diet 617
Measuring motivation to perch 384-385
Meat production 377
Mechanical catching 435-436
commercial flocks study 436
in turkeys 437
Mechanical methods
death 486
neurophysiological and/or reflex assessments 489
reversible method 486
suboptimal technique 488
unconsciousness 486
Medullary bone and bone remodelling during lay 569-570
Metabolic pathways 533
Methyl anthranilate 45
MHUSA. *see* Mother Hens' Uropygial Secretion Analogue (MHUSA)
Misperceiving feathers 518
Misting systems 330, 625
Mitigation 474
Mobile abattoirs 439
Modern housing systems complexity 587-588
Modern production systems 519
Modified gas environments 475
Monoaminergic system (5-HT, dopamine) 525
Mother Hens' Uropygial Secretion Analogue (MHUSA) 289
Multi-tier housing structures 360
Musculoskeletal development, rearing hens 581-582
Mushroom perches 399

Natural photoperiod (NP) 284
Necropsies 426
Nest box smothering 135
Neuromuscular spasms 479
Neuronal exhaustion 467
Nipple drinkers 613
Nitrogen 480
*see also* inert gases
Nociception 71
Non-infectious skeletal disorders 616
Non-instantaneous stunning approaches. *see* Controlled atmosphere stunning (CAS)
Non-sexual aggression 239
Non-steroidal anti-inflammatory drugs (NSAIDs) 614
Non stun slaughter 490
Non-stun slaughter 492
Normobaric hypoxic environments 484
Novel electrical systems 474-475
NP. *see* Natural photoperiod (NP)
NSAIDs. *see* Non-steroidal anti-inflammatory drugs (NSAIDs)
Nutritional strategies 580

OB. *see* Olfactory bulb (OB)
Obsessive-compulsive disorders (OCD) 522
Old breed (bantam) 380
Olfactory bulb (OB) 264
On-farm welfare assessment 607
Open bill 479
Open field test 63
Operant conditioning 95
Optimal environmental management 611
Optimum incubation EST 616
Osteoblasts 568
Osteochondrosis 606
Oval perches 399
Overcrowding 394

Pain-producing chemicals 614
Parabronchi 476
Pasture systems 315
Pavlovian conditioning 94

Peak force 399, 400
Pebble-floor test 39
Pecking drive 518
Pectoral hypertrophy 604
Pelleted food 617
Perches 353
Perches in chicken welfare, role of
  anatomic prerequisites 389-390
  broiler housing
    elevated structures in 400-404
  chickens perch
    amounts and behaviours 377-379
    anti-predator behaviour 379-382
    measuring motivation 384-385
    ontogeny 385-389
    social escape behaviour 383-384
  conclusions 404-407
  properties
    chickens perceive height 392-393
    perch 390-392
  pullet and layer housing
    arrangement of 394-396
    elevated structures in 393-394
    perch material 396-398
    perch shape 398-400
    perch width 400
  research, future trends in 407-408
Perching
  anatomic prerequisites for
    fowl, perching behaviour of 389
    grasping 389
  behaviour 382
  during daytime 380
  enriched cages 378
  in enriched cages
    on higher perches 380
    on lower perches 380
  floor housing systems 378
  fowls 379
  head roosting 382
  reaction towards aversive stimuli 382
  shelter against predators 381
  as social escape behaviour 383-384
Perch material 396-398
Perch shape 398-400
Perch usage 393
Perspex walls 391
PGCs. *see* Primordial Germ Cells (PGCs)
Pharmacological and nutritional modulation 529-531
Phenylalanine (PHE) 528
Physical environment factors 538-544
Phytase 618
Placement density 623
Planimetric methods 404
Point-source enrichment 335-336
Poor bone health, contributory factors to
  alternative factors 576-577
  egg production, excessive calcium demand of 574-576
  on productivity and welfare, influence of 577-579
Poultry catching equipment 423
Poultry farming 185
Poultry Health Veterinarian 463
Poultry lairage facility 465
Poultry slaughter, improving welfare in
  lairage 463-466
  regulatory framework 461-463
  stunning methods 466-467
Poultry social behaviour
  aggressive interactions, factors management 131-133
  animals social group 119-121
  animal welfare implications 126-129
  breeding flocks welfare implications 135-138
  early-life experiences 121-126
  imprinting 121-126
  managing conflict over feeding 130-131
  other considerations 133-135
  overview 117-119
  poultry flocks management strategies 139-140
Prebiotics 618
Precautionary principle 462
Precision Livestock Farming 407
Precision technology 150
Predominant resting position 378-379
Pre-laying call 7
Pre-slaughter phase and associated welfare concerns
  catching crew 419
  depopulation 419
  economic impact
    animal consumption 432
    monetary loss 428
    welfare concerns 428
  fasting 433
  feed deprivation 433
  lairage 427-428
  for laying hens 442-444
  live transportation 419
  mortality 419

myopathies 433
poultry transport 419
producers 419
prolonged feed withdrawal 433
qualitative interviews 419
stressors 433
transportation 426-427

Pre-slaughter welfare, improving
benchmarking process 442
central database development 441
livestock production 434
poultry production 434
record additional welfare indicators 441
reduced environmental impact 434
welfare-related production losses 434

Pre-stun shocks 473-474
Primordial Germ Cells (PGCs) 9
Probiotics 618
Procedural representations 103
Proximal tarsometatarsus 606
Pullets 380
Pulsed direct current (DC) 469

QTG. *see* Quantitative Trait Gene (QTG)
QTL. *see* Quantitative trait loci (QTL)
QTN. *see* Quantitative Trait Nucleotide (QTN)
Quality Time Concept® 247
Quantitative/qualitative dietary restriction 617
Quantitative Trait Gene (QTG) 3
Quantitative trait loci (QTL) 3, 6-8, 13
Quantitative Trait Nucleotide (QTN) 3

Radio-frequency identification (RFID) 151
Raising brooding temperature 627
Ramps 394
Rearing environment 387
Referential communication 102
Regulatory and religious requirements 490-491
Religious slaughter 490, 492
Reproductive behaviour 119, 135
Respiratory disruption 479, 484
Responsive nutritional management 617
Restricted space allowance 425
Restrictive feeding, welfare problems 233-234
RFID. *see* Radio-frequency identification (RFID)
Rhode Island Red birds 531
Rhythmic breathing 479
Rickets 605
Rigid environmental control 612
Rondeel® 364
Rooster 288
Roosting 375
Rotated floor test 387
Royal Society for Protection of Cruelty to Animals (RSPCA) 321, 632

*Salmonella* colonization 422
SBS. *see* Sodium bisulphate complex (SBS)
Scald tank 471
Scoring cleanliness 209
Scotophase 619
Semark pliers 488
Sensory perception, chickens
hearing 47-49
olfaction 42-45
overview 37-38
taste 45-47
touch 49-50
vision 38-42
welfare implications 50-51

Sequential feeding 617
Severe FP 514
Severe peckers and non-peckers, neurobiological differences between 531-532
Sexual aggression 240
Shackles 467
Shechita 490
Single Nucleotide Polymorphism (SNP) 6
Single-tier structures 360
Sinusoidal waveform 468
Slower-growing chickens 401
longer fattening periods 404

Slower-growing genotypes 610
Slowly adapting mechanoreceptors 613
Smothering 187
SNP. *see* Single Nucleotide Polymorphism (SNP)
Social environment factors 537-538
Social plasticity 128
Sodium bisulphate complex (SBS) 629
Solid crate floors 464
Spanish ornamental breeds 383
Spatial abilities 582-586
Spatial navigation 387
Spatial processing 386
Spiking 137
Spray vaccination 277
Starveout 106

States of suffering, improved management of poultry
- behavioural restrictions 78-79
- fear responses 61
  - intense and sudden stimuli 62-63
  - novelty 63-64
  - special evolutionary dangers 64-67
- frustration 67-71
- future trends 80
- hunger 77-78
- overview 59-61
- pain 71-72
  - artificial environment 73-75
  - breeding practices 76
  - pecking 72-73
  - surgical procedures 75-76
- social stresses 79-80

Stress-related terminology 524
Stress-responsive circuits 533
Strong physiological evidence 613
Subjective affective states 60
Sub-optimal physical environment 612
Superstimulus. *see* Exaggerated external stimulus
Synbiotics 618
Synovitis score scale 606

Tasting type movement 477
TD. *see* Tibial dyschondroplasia (TD)
Tendon-locking mechanism 389
Tenosynovitis 606
Terrestrial Animal Health Code 463
Test bird 383
*The Bremen Town Musicians*, Brothers Grimm 376
Thermal core 427
Thermal imaging 183
Theropods 389
Thinning *vs.* whole flock removal 420-421
Thorny tree (Zizyphus) 378
Three dimensional environment 387
TI. *see* Tonic immobility (TI)
Tibial dyschondroplasia (TD) 76, 605
Tidal (mammalian) ventilation 476
Tidal volume 476
Tight-fitting shackles 472
Tissue pecking (TP) 511-512
TMS. *see* Transcranial magnetic stimulation (TMS)
Toe pad hyperkeratosis 74
Toe pecking 515
Toe trimming 75
Tonic immobility 424
Tonic immobility (TI) 20
Tonic immobility test 383
Tonic phase 479
- *see also* Neuromuscular spasms

Total welfare index 328
TP. *see* Tissue pecking (TP)
*Trans*-acting eQTL 11
Transcranial magnetic stimulation (TMS) 489
Transitional EEG waveforms 482
Trichotillomania 522
Tryptophan (TRP) 528
Tryptophan-2,3-dioxygenase (TDO). *see* Kynurenine (KYN) pathway enzymes
Twisted legs 604
Tyrosine (TYR) hydroxylase 528

Ultra-wideband (UWB) 159
Unfulfilled motivation
- challenges and limitations 520-522
- to consume feed 518-520
- to explore and search for feed 518

Upright handling method 437
Upright system 474
Usable area 355
USDA-inspected mobile poultry units 439
UV-B illumination 619
UWB. *see* Ultra-wideband (UWB)

Valgus/varus angulation 604
Valgus-varus deformities 604
Vencomatic 'Patio' system 419
Ventilation rate 629
Vent pecking 73, 515
Visible swaying 479
Visual identification 613

Walking 376
Water bath stunning 468
Water consumption 630
Wattles 512
Wearable technologies
- future trends 166-168
- outdoor stocking density, free-range laying hens 164-165
- overview 149-151
- radio-frequency identification technology 151-152
  - and range use of chickens 153-155
  - and resource use in other housing systems 155-157

wearable sensors and accelerometers 157-158
behavioural activities classification 159-160
euthanasia 160-161
monitoring disease 160-161
perching, jumping, falls, and collisions 161-162
physical activity levels 162-164
wearable sensors effects, chicken behaviour 158-159
Welfare assessment improvement for poultry
behavioural indicators 210
poultry vocalisation 211-212
thermal regulation 212
time budget 211
clinical measures 206
bird cleanliness 209-210
body condition score 210
cloacal cannibalism 209
mortality 206-207
pecking damage on comb 208
pecking wounds 209
plumage condition 207-208
future trends 215-216
human-poultry relationship 212
novel object test - fear 212
overview 197-200
thirst in chickens 213
welfare assessment indicators 204-206
welfare assessment protocols 200-202
covering all dimensions of welfare in protocol 202-203
welfare indicators prioritising 213-215
Welfare assessment protocol xix, 202
Welfare impacts 491-492
Welfare improvement of young chickens
hatching practices
food and water provision inside hatchery 274-275
hatching process 274
immune-boosting strategies 277-279
in-ovo sexing 276-277
mode of transport optimization in hatchery 275-276
on-farm hatching 280
short- and long-term consequences, beak treatments 279-280
incubation practices
artificial incubation, commercial hatchery 264-265
embryo to chick development 264
in-ovo chemosensory learning 271-272
in-ovo temperature programming 272-274
light provision during incubation 266-271
natural incubation 264
reducing noise around hatching 265-266
overview 261-262
parental stock welfare and offspring effects 262
nutritional and environmental stressors in breedings 262-263
rearing practices, pullet welfare 280-281
dark brooder use 281-282
early life rearing 281
environmental enrichment and complexity 290
fear and stress reducing, environmental enrichment 292
immune function improvement 291
learning possibilities from older conspecifics 287-289
light period, wavelength and intensity 283-286
olfactory enrichment during rearing 289
reducing risk of injuries 291-292
risk reducing for development, injurious pecking 290
spatial and navigation skills improvement 292-293
stocking density and group size 286-287
Welfare Quality Assessment Protocol 622, 624
Welfare Quality protocol 202
White Leghorn birds 531
Whole-house stunning approach 439
Wing flapping 474
Wood shavings 335
Wood-shavings 628
Working memory 102
World Animal Health Organisation (OIE) 200, 462
Worm run 39

Yolk-free mass (YFM) 271

www.ingramcontent.com/pod-product-compliance
Lightning Source LLC
LaVergne TN
LVHW020629300926
849161LV00004B/648
* 9 7 8 1 7 8 6 7 6 4 2 2 5 *